31st ANNUAL LOSS PREVENTION SYMPOSIUM
American Institute of Chemical Engineers
1997 Spring National Meeting
Houston, Texas
March 9-13, 1997

CONTENTS

LPS 1997
Proceedings of the

31st Annual
Loss Prevention Symposium

Kris Chatrathi
Fike Corporation
Symposium Chair

Gary Page
American Home Products
Symposium Co-Chair

Sponsored by the
American Institute of Chemical Engineers
Safety and Health Division (Group 11)
Area 11a: Loss Prevention

George Brown Convention Center
Houston, Texas
March 9-13, 1997

PROCEEDINGS INTRODUCTION

ACKNOWLEDGMENTS

The Loss Prevention Committee wishes to thank everyone involved with the production of these Proceedings. Special appreciation is given to Ms. Stacey Wilton of the Fike Corporation for her assistance in preparing these Proceedings and to Mr. Chung Lam, of the American Institute of Chemical Engineers, and Ms. Jodi Hauck and the staff of Omni Press for their assistance in printing these Proceedings. We also give special thanks to the authors and session chairs who worked very hard to complete the papers early enough to be included in these Proceedings.

LOSS PREVENTION COMMITTEE
GROUP 11 - AREA 11A

OFFICERS

Everett D. Wixom
Exxon Chemical Company
P.O. Box 45
Linden, NJ 07036
(908) 474-3503
(908) 474-3006 (FAX)
(Chair)

Robert P. Benedetti
National Fire Protection Association
One Batterymarch Park
Quincy, MA 02269-9101
(617) 984-7433
(617) 984-7110 (FAX)
(Vice Chair)

SYMPOSIA CHAIRS

Kris Chatrathi
Fike Corporation
704 S. 10th St.
Blue Springs, MO 64015
(816) 229-3405, ext. 503
(816) 228-9277 (FAX)
kchatrathi@fike.com
(1997 Symposium Chair)

Gary A. Page
American Home Products
One Campus Drive
Parsippany, NJ 07054
(201) 683-2266
(201) 683-4060 (FAX)
(1997 Symposium Co-Chair)
(1998 Symposium Chair)

MEMBERS

J. Steven Arendt
JBF Associates, Inc.
1000 Technology Drive
Knoxville, TN 37932-3341
(423) 671-5812
(423) 966-5287 (FAX)

Henry L. Febo
Factory Mutual Research Corp.
1151 Boston-Providence Tpke.
Norwood, MA 02062
(617) 255-4771
(617) 762-9375 (FAX)

Daniel A. Crowl
Dept. of Chemical Engineering
Michigan Technological Univ.
Houghton, MI 44931
(906) 487-3221
(906) 487-3213 (FAX)

Thomas O. Gibson
Dow Chemical Company
2020 Dow Center
Midland, MI 48674
(517) 636-5535
(517) 636-0350 (FAX)

John A. Davenport
Industrial Risk Insurers
85 Woodland Street
Hartford, CT 06102-5010
(860) 520-7362
(860) 549-5780 (FAX)

Dennis C. Hendershot
Rohm & Haas Company
Route 413 & State Rd.
Bristol, PA 19007
(215) 785-7243
(215) 785-7077 (FAX)
nagdh@rohmhaas.com

David Kirby
Union Carbide Corp.
P.O. Box 8361
South Charleston, WV 25303
(304) 747-3696
(304) 747-5915 (FAX)

Joseph F. Louvar
BASF Corporation
1419 Biddle Avenue
Wyandotte, MI 48192-3836
(313) 246-6354
(313) 246-5031 (FAX)

Richard E. Munson
DuPont Company
1007 Market St, N-2526-2
Wilmington, DE 19898
(302) 774-3646
(302) 774-3410 (FAX)

John Noronha
Eastman Kodak Company
Kodak Park, Bldg. 23, 5th Floor
Rochester, NY 14652-4390
(716) 477-1065
(716) 477-1128 (FAX)

Michael L. Griffin
Procter & Gamble Company
6110 Center Hill Avenue
Cincinnati, OH 45224-1789
(513) 634-4932
(513) 634-7760 (FAX)

Roy E. Sanders
PPG Industries, Inc.
P.O. Box 1000
Lake Charles, LA 70602
(318) 491-4322
(318) 491-4860 (FAX)

Ephraim A. Scheier
Occidental Chemical Company
5005 LBJ Freeway, 15th Floor
Dallas, TX 75244
(972) 404-3686
(972) 404-3201 (FAX)

Richard F. Schwab
AlliedSignal
101 Columbia Rd., Box 1139R
Morristown, NJ 07962-1139
(201) 455-4847
(201) 455-3345 (FAX)

Joseph A. Senecal
Fenwal Safety Systems
700 Nickerson Rd.
Marlborough, MA 01752
(508) 429-3190
(508) 429-2990 (FAX)

Robert W. Ormsby
Air Products & Chemicals, Inc.
7201 Hamilton Blvd.
Allentown, PA 18195-1501
(610) 481-6715
(610) 481-8051 (FAX)

INTERNATIONAL MEMBERS

Malcolm Preston
ICI Engineering Technology
P.O. Box 8
The Heath, Runcorn
Cheshire, England, WA74QD, UK
1928-515695
1928-515660 (FAX)

David R.E. Worthington
DNV Technica-Software Products
Palace House
3 Cathedral Street
London, England SE19DE, UK
44-71-357-6080
44-71-357-6048 (FAX)

EMERITUS MEMBERS

David A. Bosworth
Mobil Research & Development
P.O. Box 1026
Princeton, NJ 08543-1026
(609) 737-4463
(609) 737-5358 (FAX)

William J. Bradford
22 Deer Run Rd.
Brookfield, CT 06804
(203) 775-2076
(203) 775-2076 (FAX)

Stanley S. Grossel
Process Safety & Design, Inc.
41 Sussex Rd.
Clifton, NJ 07012
(201) 779-8579
(201) 779-8579 (FAX)

R. Sethuraman
Nungambakkam, Madras
600-034 INDIA

Trevor A. Kletz
64 Twining Brook Rd.
Cheadle Hulme
Cheadle, Cheshire SK8 5RJ, UK
011-44-161-485-3875
011-44-161-485-3875 (FAX)
T.Kletz@Lboro.ac.uk

Robert C. Mill
873 Village Green
Westfield, NJ 07090
(908) 232-3613

Russell L. Miller
17 Taylor Woods
Kirkwood, MO 63122
(315) 821-6747

Russel Hill
P.O. Box 940
Winterpark, FL 32790
(407) 629-6281
(407) 629-8502 (FAX)

John E. Rogerson
395 East U.S. Rte. 22-3
Maineville, OH 45039
(513) 683-0049

Theodore A. Ventrone
14 Bloomsfield Place
P.O. Box 669
Belle Meade, NJ 08502
(908) 359-3392

PRESENTATION of the WILLIAM H. DOYLE AWARD

The William H. Doyle Award is given to the "best paper" presented at each annual Loss Prevention Symposium. The Loss Prevention Committee awards the author who gives the best paper, considering both technical content and effectiveness of presentation. The criteria include:

1. The paper must be clear and well presented.
2. The paper should have wide applicability to Loss Prevention efforts in the Chemical Industry.
3. The paper should add substantial knowledge to the field of Loss Prevention.
4. The paper should engage and stimulate the intellect of the audience.
5. The paper would have caused Bill Doyle to rise to his feet, ask a penetrating question, make a decisive comment, and lead the applause.

Mr. Francesco Tamanini of Factory Mutual Research Corporation, Norwood, Massachusetts is the William H. Doyle Award winner for 1997 for his paper presented at the 1996 Loss Prevention Symposium. The abstract of Mr. Tamanini's paper appears on the following page.

Previous Winners of the William H. Doyle Award

1987	Dr. Laurence G. Britton	**Thermal Stability for Ethylene at Elevated Pressures**
1987	Dr. Ian Swift	**The Performance of Low Pressure Vents**
1988	Mr. Robert A. Mancini	**The Use (and Misuse) of Bonding for Control of Static Ignition**
1989	Mr. Dean K. Wilson	**Failure Mode Management: A Loss Prevention Philosophy for Programmable Logic Controllers**
1990	Dr. Laurence G. Britton	**Combustion Hazards of Silane and Monosilicon Chlorides**
1991	Dr. Daniel A. Crowl	**Using Thermodynamic Availability to Determine the Energy of Explosion**
1992	A. M. Dowell	**Flashback from Waste Gas Incinerators into Air Supply Piping**
1993	J.E.S. Venart	**To BLEVE or Not to BLEVE: Anatomy of a Boiling Liquid Expansion**
1994	N. E. Scheffler	**Vapor Suppression of Chemicals using Foams**
1995	T. C. Hofelich	**Compatibility Determination Procedures to Comply with legislation HM-183**
1996	R.G. Eierman	**Improving Inherent Safety with Sealless Pumps**
1997	F. Tamanini	**Modeling of Panel Inertia Effects in Vented Dust Explosions**

Recipient of the 1997 William H. Doyle Award

Francesco Tamanini

For his presentation of a paper entitled

Modeling of Panel Inertia Effects in Vented Dust Explosions

by

Francesco Tamanini
Research Division
Factory Mutual Research Corporation
Norwood, Massachusetts 02062

ABSTRACT

Current guidelines for sizing of vents in dust explosions deal with panel inertia effects either by suggesting that the issue be approached experimentally on a case-by-case basis (VDI 3673) or by recommending a maximum mass per unit area (NFPA 68). This empirical approach to the problem is surprising, given that this aspect of explosion venting should be easily amenable to analytical treatment. From this assessment, an analysis of vent panel dynamics was carried out based on a simplified explosion model, which has also been used to develop a generalized vent sizing correlation. The main result of the analysis is the identification of a dimensionless parameter which fully characterizes the effects associated with the inertia of the panel. This inertia parameter includes: the reactivity of the mixture; the volume of the enclosure; the mass of the vent per unit area; the number of equal panels on the vented volume; and a panel shape factor. The analysis has quantified the intuitive expectation that the mass per unit area of the panel is not a property that should be considered in isolation, and that panel inertia effects are more important the more reactive the mixture and the smaller the volume. The predictions from the model have been validated by comparison with available data for both dust and gas explosions. The extensive validation work that has supported their development provides confidence that the design correlations developed from the analysis can be used reliably to account for an effect that is either neglected or referred to testing by existing guidelines.

CALL FOR PAPERS
32ND ANNUAL LOSS PREVENTION SYMPOSIUM
AICHE 1998 SPRING NATIONAL MEETING
NEW ORLEANS, LA MARCH 8-12, 1998

Symposium Chair:
Gary Page
American Home Products
One Campus Drive
Parsippany, NJ 07054
(201)-683-2266/(201)-683-4060 (FAX)

Symposium Co-Chair
Michael Griffin
Procter & Gamble Co.
6110 Center Hill Ave.
Cincinnati, OH 45224-1789
(513) 634-4932/(513) 634-7760 FAX

The 1997 Loss Prevention Symposium is the 32nd in a series of annual meetings sponsored by the AICHE Safety and Health Division's Loss Prevention committee. The purpose of the symposium is to bring together practitioners and other interested persons from industry, academia and government to share experiences, advances in technology, and new ideas on preventing accidents in the chemical process industries. Those interested in presenting a paper in any of the following symposium sessions should contact the appropriate session chair for a description of the session and should submit a 150-200 word abstract by **May 15, 1997.** Original, previously unpublished material is encouraged. Accepted papers will be published in the Symposium Proceedings, available at the meeting. Papers and presentations will be eligible for the Bill Doyle Award, given to the best paper presented at the Symposium.

RISK MANAGEMENT PROGRAMS AND HAZARD ASSESSMENT
Ephraim Scheier
Occidental Chemical Co.
5005 LBJ Freeway, 15th Floor
Dallas, TX 75244
(972) 404-3686
(972) 404-3201 FAX

BUILDING DESIGN FOR EXPLOSION HAZARDS
David Kirby
Union Carbide Corp.
P.O. Box 8361
South Charleston, WV 25303
(304) 747-5915
(304) 747-5915 FAX

INNOVATIVE SOLUTIONS FOR THE STORAGE AND PROCESSING OF HAZARDOUS MATERIALS
John Noronha
Eastman Kodak Co.
Kodak Park, Bldg. 23, 5th Floor
Rochester, NY 14652-4390
(716) 477-1065
(716) 477-1128 FAX

QUANTIFYING CHEMICAL AND REACTIVE HAZARDS AND IDENTIFYING INHERENT SAFETY SOLUTIONS
Michael Griffin
Procter & Gamble Co.
6110 Center Hill Ave.
Cincinnati, OH 45224-1789
(513) 634-4932
(513) 634-7760 FAX

ADVANCES IN FIRE PROTECTION
Joseph A. Senecal
Fenwal Safety Systems
700 Nickerson Rd.
Marlborough,MA 01752
(508) 429-3190
(508) 429-2990 FAX

CASE HISTORIES
Henry L. Febo
Factory Mutual Research Corp.
1151 Boston-Providence Tpke.
Norwood, MA 02062
(617) 255-4771
(617) 762-9375 FAX

- SOME GOOD REASONS -

TO JOIN THE SAFETY and HEALTH DIVISION

- **The Safety, Health, and Environmental fields are in the technological forefront, revolutionizing the way chemical engineering will be practiced in the future.**
- **These fields attract some of the best members of our profession.**
- **Recognizing and explaining risk to the community is essential to the future of the chemical process industry. The community needs your expertise and advocacy on safety and risk issues.**
- **A rapidly developing technology needs your input. YOU need to stay current!**
- **You will associate with others who CARE - care about creating safer working environments and communities and care about their career and about making contributions to our profession.**
- **The Safety and health Division sponsors annual "Best Paper" awards that are presented at the annual Division Dinner at the AIChE Spring National Meeting. You will have the opportunity for such awards and to meet leaders in our field.**

Join now!

You will receive the Division Newsletter and the quarterly Process Safety Progress.

Clip and Send To: **AIChE Safety & Health Division**
c/o W. J. Bradford
22 Deer Run road
Brookfield, CT 06804

NAME:______________________________________

COMPANY:______________________________________

ADDRESS:______________________________________

____

MEMBER of AIChE? ____ YES ____ NO
(Non-Members should apply for membership within 12 months.)

AIChE SAFETY AND HEALTH DIVISION
OFFICERS, DIRECTORS, COMMITTEE CHAIRS

OFFICERS

Joseph F. Louvar
BASF Corporation
1419 Biddle Avenue
Wyandotte, MI 48192-3836
(313) 246-6354
(313) 246-5031 (FAX)
(Chair)

Dennis C. Hendershot
Rohm and Haas Company
P.O. Box 584
Route 413 and State Rd.
Bristol, PA 19007
(215) 785-7243
(215) 785-7077 (FAX)
(1st Vice Chair)
E-mail: NAGDH@RohmHaas.com

Randy Freeman F2E1
Monsanto Company
800 N. Lindbergh Avenue
St. Louis, MO 63167
(314) 694-6068
(314) 694-6138 (FAX)
E-mail: crowl@mtu.edu
(2nd Vice Chair)

Daniel Crowl
Michigan Technological University
Dept. of Chemical Engineering
Houghton, MI 49931
(906) 487-3221
(906) 487-3213 (FAX)
(Past Chair)

William J. Bradford
22 Deer Run Road
Brookfield, CT 06804
203-755-2076
(Secretary-Treasurer)

Marilyn B. Williams
American Institute of Chemical Engineers
345 E. 47 Street
New York, New York 10017
(212) 705-7332
(212) 838-8274
AICHE Representative)

DIRECTORS

Robert P. Benedetti
National Fire Protection Assn.
One Batterymarch Park
Quincy, MA 02269-9101
(617)-984-7433
(617)-984-7056 (FAX)
(1995-1997)

John L. Woodward
Technica, Ltd.
355 E. Campus View Blvd., Suite 110
Columbus, OH 43805
(614)-848-4000
(614)-848-3955 (FAX)
(1995-1997)

DIRECTORS (CON'T)

John E. Murphy
RMT/Jones and Neuse, Inc.
912 Capital of Texas Highway South
Suite 200
Austin, TX 78746-52101
(512) 327-9840
(1996-1998)

Everett D. Wixom
Exxon Chemical Company
P. O. Box 45
Linden, NJ 07036
(908)-474-3503
(908)-474-3006 (FAX)
(1996-1998)

Walter L. Frank
DuPont Engineering
Nemours Building, 5450
1007 Market Street
Wilmington, DE 19898
(302) 774-8040
(302) 774-8038
E-mail: frankw@engg.dnet.dupont.com

Roy E. Sanders
PPG Industries
P.O. Box 1000
Lake Charles, LA 70602
(318) 491-4322
(318) 491-4860

COMMITTEE CHAIRS

Daniel A. Crowl
(Publications)

Joseph F. Louvar
(Continuing Education)

Vasilis (Bill) Fthenakis
Brookhaven National Laboratory
Building 490-D
Upton, NJ 11973
(516)-282-2830
(516)-282-7867 (FAX)
Email: vmf@bnl.gov
(Membership)

John A. Noronha
Eastman Kodak Company
Building 23, 8th Floor
Rochester, NY 14652-4201
(716) 477-1065
(716) 477-4038
(Research-Liaison)

James J. Rooney
JBF Associates, Inc.
1000 Technology Park Center
Knoxville, TN 37932-3341
615-671-5814
615-966-5287 (FAX)
(Programming)

Richard F. Schwab
AlliedSignal, Inc.
P. O. Box 1139
Morristown, NJ 07960-1139
201-455-4847
201-455-3345 (FAX)
(Awards)

EDITORS

31st Annual Loss Prevention Symposium

George Brown Convention Center
Houston, Texas
March 9-13, 1997

American Institute of Chemical Engineers
Safety & Health Division
Area 11a: Loss Prevention Committee

Session 38. **Control Systems for Unit Operation Safety**

Thomas O. Gibson (Chair)
Dow Chemical CO
2020 Dow Center
Midland, MI 48674

David Kirby (Vice-Chair)
Union Carbide Corporation
P.O. Box 8361
South Charleston, WV 25303

Session 39. **Mechanical Integrity and Maintenance of Process Plant Components**

John A. Davenport (Chair)
Industrial Risk Insurers
85 Woodland St.
Hartford, CT 06102-5010

Everett D. Wixom (Vice-Chair)
Exxon Chemical CO
P.O. Box 45
Linden, NJ

Session 40. **Fire and Explosion Prevention and Protection**

Michael L. Griffin (Chair)
Procter and Gamble CO
6110 Center Hill Ave.
Cincinnati, OH 45224-1789

Joseph A. Senecal (Vice-Chair)
Fenwal Safety Systems
7000 Nickerson Rd.
Marlborough, MA 01752

Session 41. **Pressure Relief Systems**

John Noronha (Chair)
Eastman Kodak CO
Bldg. 23, 5th Floor
Kodak Park Site
Rochester, NY 14652-4390

Richard E. Munson (Vice-Chair)
E.I. Du Pont de Nemours and CO
1007 Market St.
Wilmintgton, DE 19898

Session 42. **Use of Consequence Modeling in Risk Reduction**

Ephraim Scheier (Chair)
OxyChem
15th Floor
5005 LBJ Freeway
Dallas, TX 75244

James J. Rooney (Vice Chair)
JBF Associates, Inc.
1000 Technology Dr.
Knoxville, TN 37932-3353

Session 43. **Storage, Processing, and Transportation of Hazardous Chemicals**

Stanley S. Grossel (Chair)
Process Safety and Design, Inc.
41 Sussex Rd.
Clifton, NJ 07012

Joseph F. Louvar (Vice-Chair)
BASF Corporation
1419 Biddle St.
Wyandotte, MI 48192-3836

Session 44. **Case Histories and Miscellaneous Topics**

Robert W. Ormsby (Chair)
Air Products and Chemicals, Inc.
7201 Hamilton Blvd.
Allentown, PA 18198-1501

Roy E. Sanders (Vice-Chair)
PPG Industries, Inc.
P.O. Box 1000
Lake Charles, LA 70602

Index of Papers

USE OF CONSEQUENCE MODELING IN RISK REDUCTION

42A. The Lessons Learned from the Development of a Scenario Based Approach for the Optimization of Passive Fire and Blast Division Design on a New Offshore Platform, B. Poblete, P. Caldwell

42B. Use of Consequence Analysis in Process Hazards Analysis, R. J. Gardner, S. Dharmavaram, J. Klein

42C. Toxic Hazard Reduction with Passive Mitigation Systems, P. Myers, R. Morgan

42D. Protection of Process Buildings and Their Occupants from Toxic and Flammable Releases, C. Fryman

42E. The Explosion Blast Prediction Model in the Revised "Yellow Book", W. Mercx, A. van den Berg

42F. Recent Developments in the Baker Strehlow VCE Analysis Methodology, Q. Baker, C. Doolittle, G. Fitzgerald, M. Jun Tang

STORAGE, PROCESSING, AND TRANSPORTATION OF HAZARDOUS CHEMICALS

43A. A Detailed Reaction Study of Phosphorus Trichloride and Water, G. Melhem, D. Reid

43B. A Fire Risk Analysis Model for Assessing Options for Flammable and Combustible Liquid Products in Storage and Retail Occupancies, J. Hall, Jr.

43C. A Review of Chemical Compatibility Issues, D. Frurip, T. Hofelich, D. Leggett, J. Kurland, J. Neimeier

43D. Decomposition of Thermally Unstable Chemicals in Solvents, S. Chervina, S. Evon, G. Bodman

43E. Developing Chemical Compatibility Charts Using CHEMPAT, D. Leggett

43F. Improving the Effect of Atmospheric Stability Class On Hazard Zone Predictions For Accidental Hazmat Releases, J. Woodward

CASE HISTORIES AND MISCELLANEOUS TOPICS

44A. Carbon Disulfide Incidents During Viscose Rayon Processing, D. Hall, L. Losee

44B. Safety Considerations When Treating VOC Streams with Thermal Oxidizers, F. Self, J. Hill

44C. No Good Deed Goes Unpunished: Case Studies of Incidents and Potential Incidents Caused by Protective Systems, D. Hendershot, A. Dowell

44D. Case Histories of Two Incidents Following Process Safety Reviews, W. Howard

31st LOSS PREVENTION SYMPOSIUM

Session 38. **Control Systems for Unit Operation Safety**

Monday, March 10, 1997
8:00 - 11:05 AM
Grand Ballroom C

Thomas O. Gibson, Chair
Dow Chemical CO
2020 Dow Center
Midland, MI 48674

David Kirby, Vice-Chair
Union Carbide Corporation
P.O. Box 8361
South Charleston, WV 25303

Symposium Introduction - 8:00 AM, Kris Chatrathi, Symposium Chair

Session Introduction - Thomas O. Gibson

38a. **Global Safety Standards for Programmable Electronic Systems in Process Plants,** J. Maggioli, Feltronics Corp. Newark, DE

38b. **Challenges of Applying a Performance based Design Criteria for Safety Instrumented Systems,** C. Thurston, Union Carbide Corp., South Charleston, WV

38c. **Process Control as an Alternative to Pressure Relief**, J. Marshall, The Dow Chemical CO., Freeport, TX; D. Watkins, The Dow Chemical CO, Midland, MI

BREAK

38d. **Out-of-the-Loop Performance Problems and the Use of Intermediate Levels of Automation for Improved Control System Functioning and Safety,** D. Kaber, Mississippi State University, Mississippi State, MS; M. Endsley, Massachusetts Institute of Technology, Cambridge, MA

38e. **Computer Integrated Manufacturing in Batch Chemical Processes - U.K. Perspective,** A. James, Eastman Kodak CO, Merseyside, England

38f. **Software Deviation Analysis: A "Safeware" Technique**, J. Reese, N. Leveson, University of Washington & Software Engineering Corp., Seattle, WA

Concluding Remarks - David Kirby

GLOBAL SAFETY STANDARDS FOR PROGRAMMABLE ELECTRONIC SYSTEMS IN PROCESS PLANTS

V. J. Maggioli
Feltronics Corporation
Newark, DE
Paper 38a

Global Safety Standards for Programmable Electronic Systems in Process Plants

Abstract:

The use of Programmable Electronic Systems (PES) in process plants has gained wide acceptance for basic process control applications. Use of PES in Safety Instrumented Systems lagged behind until a series of guidelines and standards were developed that brought together the producers and users of these products.

This paper provides an overview of today's standards, standards under development, and a sampling of the issues that users of these standards must address. These issues include Safety Integrity Level (SIL), Electrical, Electronic, Programmable Electronic Systems (E/E/PES), and Safety Instrumented Systems (SIS). The intent is to provide a general understanding of the differences found in these standards so that specific problems can be recognized and dealt with effectively.

Introduction:

The potential for harmonization between guidelines and standards from America and comparable international efforts are being addressed with the developing draft International Electrotechnical Commission (dIEC) 1508 and dIEC 1511 standards.

dIEC 1508 is a generic safety standard which embraces a wide range of applications from machine tools to avionics. It combines the work of the International Electrotechnical Committee's (IEC) 65A, Working Groups 9 and 10, which are developing the standard. It will be distributed for final ballot the second quarter of 1997.

dIEC 1511 is a process sector safety standard falling within the dIEC 1508 generic standard. In the USA, ISA standard S84.01 addresses the same issues as dIEC 1511 except from a national perspective. S84.01 has been issued and approved. dIEC 1511 is due for ballot in 1998.

The USA is attempting to harmonize dIEC 1511 and S84.01 while maintaining principles established in dIEC 1508. The USA is not unique in this situation since all countries that have existing safety guidelines and standards are faced with similar situations (Figure 1). All of this is compounded by national, local laws, regulations (Ref. 4), culture differences, different objectives, etc.

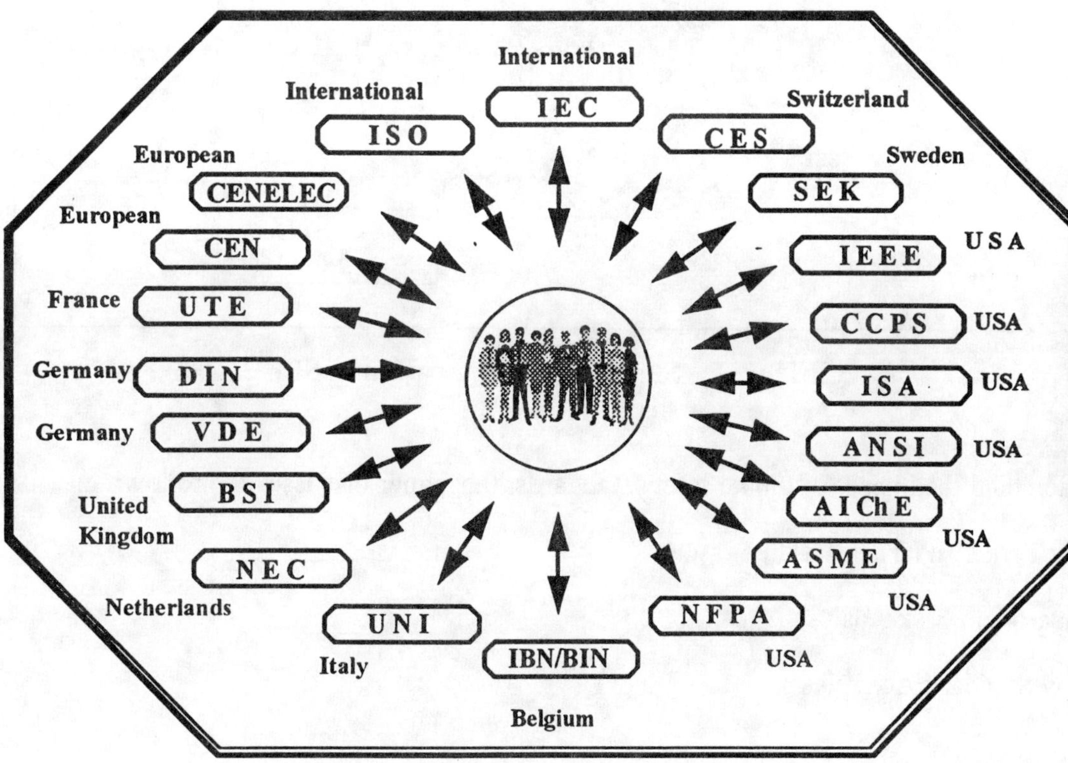

Figure 1 - Standards Globalization

Further, national and regional standards organizations are comprised of academia, users, suppliers, manufacturers, engineers and consultants, each with their understanding of the problem, the solution, and their own specific needs. Combine this with the wide range of applications that require attention and the variable appears limitless. Figure 2 provides an example of how the process sector standard dIEC 1511 utilizes global resources (dIEC 1508, ISA, AIChE, EWICS, etc.) to collect and disseminate information.

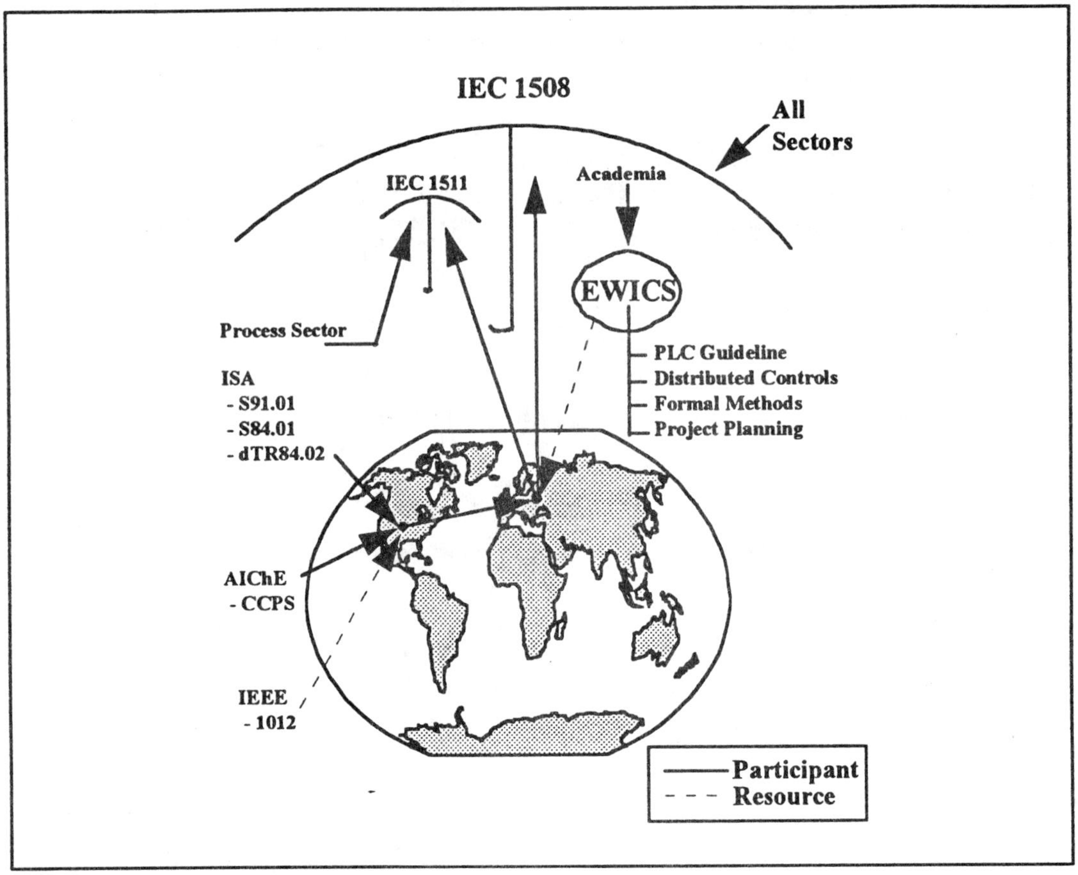

Figure 2 - Standards Participation in SIS

To understand the impact of global safety standards, this paper discusses the following:

- Key Terms and Their Related Issues
- Hazard and Risk Analysis
- New Standard Activities
- Future
- Conclusion
- References

Key Terms & Related Issues:

Following is a list of terms found in the American standard, its comparable global term and the issues arising from the subtle (and non-too-subtle) differences in definition.

Electrical (E)/Electronic (E)/Programmable Electronic Systems (E/E/PES)

USA Definition:

When used in this context, electrical refers to logic functions performed by electromechanical techniques, (e.g., electromechanical relay, motor driven timer, etc.), electronic refers to logic functions performed by electronic techniques, (e.g., solid state logic, solid state relay, etc.), and Programmable Electronic System refers to logic performed by programmable or configurable devices [e.g., Programmable Logic Controller (PLC), Single Loop Digital Controller (SLDC), etc.]. Field devices are not included in E/E/PES.

dIEC 1508 Definition:

A system based on one or more E/E/PE devices, connected to (and including) input devices (e.g., sensors) and/or output devices/final elements (e.g., actuators), for the purpose of control, protection or monitoring.

Associated Issue:

The IEC definition includes sensors and final elements, the USA definition does not. Consequently, mandates may have a very different impact.

Safety Integrity Level (SIL)

USA Definition:

One of three possible discrete integrity levels (SIL 1, SIL 2, SIL 3) of Safety Instrumented Systems. SILs are defined in terms of Probability of Failure on Demand (PFD), see Table 1.

Safety Integrity Level	Probability of Failure on Demand Average Range (PFD avg)
3	10^{-3} - 10^{-4}
2	10^{-2} - 10^{-3}
1	10^{-1} - 10^{-2}

Table 1 - S84.01 SILs

dIEC 1508 Definition

One of 4 possible discrete levels for specifying the safety integrity requirements of the safety functions to be allocated to the safety-related systems. Safety Integrity Level 4 has the highest level of safety integrity; Safety Integrity Level 1 has the lowest (see Table 2).

DEMAND MODE OF OPERATION	
SIL	**Average Probability of Failure on Demand**
4	$>=10^{-5}$ to $<10^{-4}$
3	$>=10^{-4}$ to $<10^{-3}$
2	$>=10^{-3}$ to $<10^{-2}$
1	$>=10^{-2}$ to $<10^{-1}$

DEMAND MODE OF OPERATION	
SIL	**Frequency of Dangerous Failures Per Hour**
4	$>=10^{-9}$ to $<10^{-8}$
3	$>=10^{-8}$ to $<10^{-7}$
2	$>=10^{-7}$ to $<10^{-6}$
1	$>=10^{-6}$ to $<10^{-5}$

Table 2 - dIEC 1508 SILs

Associated Issue:

Note that the IEC definition includes Demand and Continuous Mode while the USA only deals with Demand Mode. In the IEC case, SIL 4 is included while in the USA it is not. The USA S84.01 committee is of the belief that the process sector has very little need for SIL 4 and recognizes it is very difficult to maintain SIL 4 performance with a process Safety Instrumented System (SIS). S84.01 considers continuous mode SISs rare in the process sector, hence the continuous mode SIL table was left out of S84.01. Note that when both tables are used, as in the IEC version, it is not possible to distinguish between a SIL in demand or a SIL in continuous mode without adding a definitive statement (or code) with the SIL.

Safety Instrumented Systems (SIS)

USA Definition:

System composed of sensors, logic solvers, and final control elements for the purpose of taking the process to a safe state when predetermined conditions are violated. Other terms commonly used include Emergency Shutdown System (ESD, ESS), Safety Shutdown System (SSD), and Safety Interlock System.

dIEC 1508 Definition:

Safety Instrumented System is not used in dIEC 1508. Safety-related System is used. A system that implements the required safety functions necessary to achieve a safe state for the Equipment Under Control (EUC) or to maintain a safe state for the EUC; and is intended to achieve, on its own or with other safety-related systems, the necessary level of safety integrity for the implementation of the required safety functions.

Associated Issue:

Safety-related System (SrS) as defined above can be an instrumented or non-instrumented system. To achieve a comparable definition to SIS in dIEC 1508, one would have to use E/E/PE (as applicable) SrS.

Note that the SrS definition implies that non-instrumented safety layers shall have a Safety Integrity Level. S84.01 and dIEC 1511 reserves the use of SIL for instrumented systems only.

Protection Layer

USA Definition:

Engineered safety features or protective systems or layers that typically involve special process designs, process equipment, administrative procedures, the Basic Process Control System (BPCS), and/or planned responses to protect against an imminent hazard. These responses may be either automated or initiated by human actions.

dIEC 1508 Definition:

See Safety-related system definition above.

Associated Issue:

The IEC definition of Safety related System is similar to the USA definition of protection layer. However, implementation of these definitions result in special design control systems (called Safety Instrumented Control Systems) being recognized as layers of protection. This is consistent with the continuous mode SIL presented in dIEC 1508.

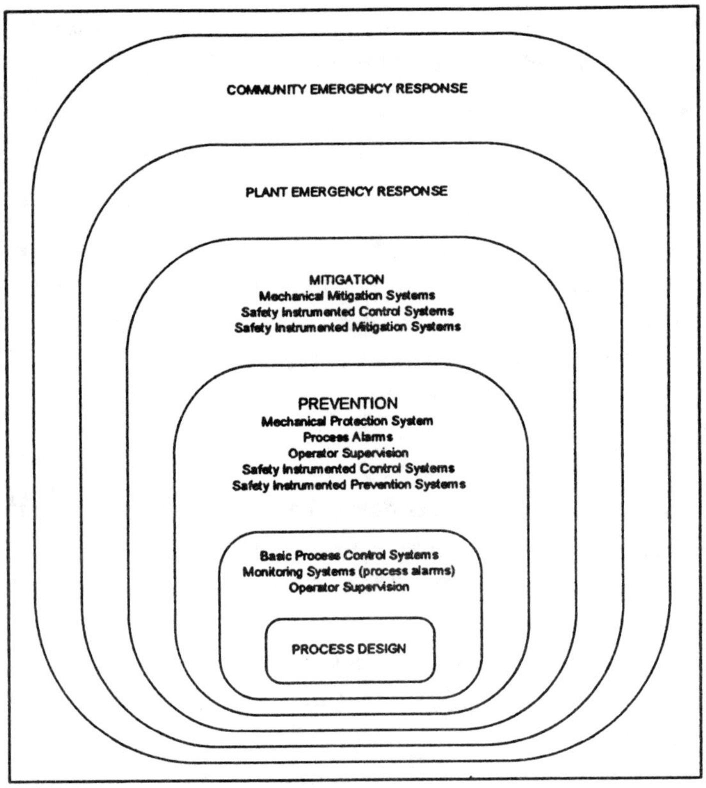

Figure 3 - Layers of Protection

Safety Life Cycle (SLC)

USA Definition:

Sequence of activities involved in the implementation of the Safety Instrumented Systems from conception through decommissioning.

IEC 1508 Definition:

The necessary activities involved in the implementation of safety-related systems, occurring during a period of time that starts at the concept phase of a project and finishes when none of the safety-related systems are any longer available for use.

Associated Issue:

An issue that arose is the limits to which these safety standards will address non E/E/PE issues. S84.01 focuses more on Instrumentation while IEC standards tend to focus more on all layers of protection, PHA, Management, and Assessment issues.

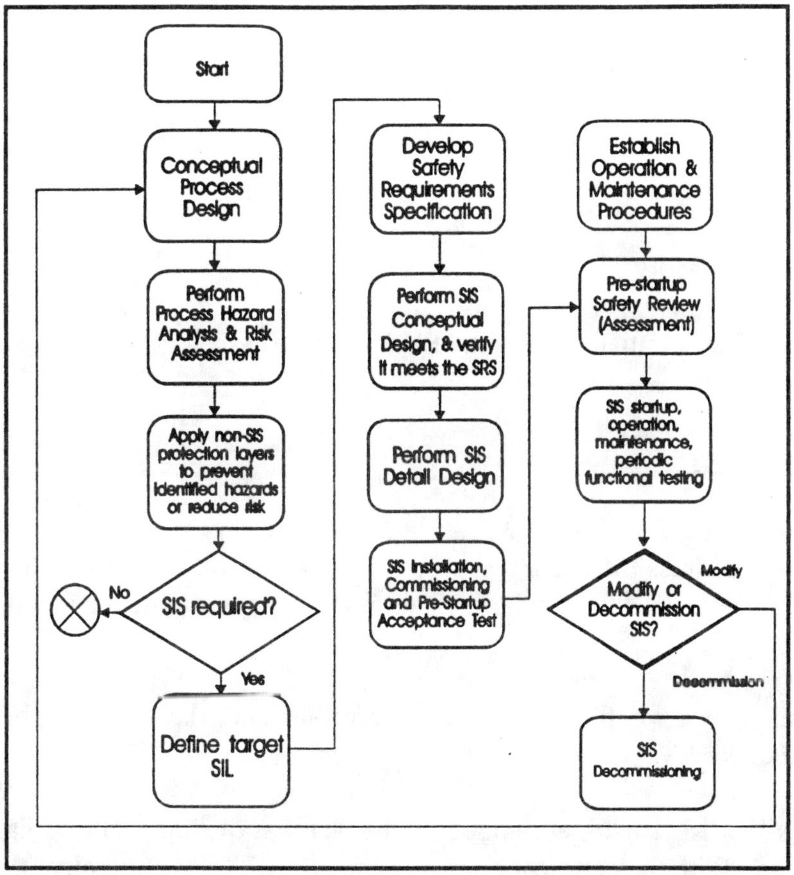

Figure 4 - Safety Life Cycle

Process Hazards Review Team (Hazards & Risk Analysis)

The PHR teams are a group of individuals that bring appropriate expertise to the activity of recognizing/identifying hazards and the design features used to eliminate or mitigate them. Within the process industry, the AIChE has been a leader in developing national guidance in this arena. The impact of global standards on this activity is significant since the PHA defines the SIL for each safety function that will be handled in a SIS.

The standards offer various guidelines (Ref. 1, 2.1, 3.1.3, 3.1.4) to assist the PHA team in that area. In the USA, dTR84.02 (Ref. 2.3) discusses:

- Markov Modeling
- FMEA
- FTA
- Block Diagram

dIEC 1508, Part 5 (Ref. 3.1.1) discusses typical corporate and national approaches such as:

- As Low As is Reasonably Practicable (ALARP)
- Quantitative
- Qualitative
- Hazardous Event Severity Matrix

This area can be confusing initially unless the PHA has a solid background in this area (Ref. 1.1.3).

New Standards Activity

dIEC 1508 is almost ready for USA approval. All countries have worked diligently to define each other's concerns and address them in a constructive manner. The results are encouraging, but lack of approval of sector specific standards may result in approvals with a country specific exclusionary clause.

Control system suppliers are defining their safety controllers per their SIL capability. This improves the confidence level of users in new equipment.

New architectures (1oo1d, 1oo2d, etc.), identify the fact that the logic solvers have high diagnostic coverage. With the certification of systems to a SIL it is imperative that the user understands TUV/FM type certification (e.g., equipment or application) and the effort required to maintain the SIL. SIL is quantitative so an issue becomes how to tie qualitative considerations to it. Sensor(s) and final element(s) performance are key to SIS SIL. To date most of the effort revolves around the logic solver (Ref. 2.4). Combining multiple SIL levels into one logic solver is another area addressed by the new guidelines (Ref 2.2, 3.1.2).

Software certification is leading to more 3rd party certification. Downsizing within much of the process sector is making a market for:

- Hazard and Risk Analysis consultants
- Certification capability
- New equipment usage on SIS since probabilistic analysis not as easily accomplished

SIL classification requires defining whether reference is being made to Demand or Continuous Mode. Further, integration of corporate classification to the standard SIL is required. See Table 3 to understand how one company's SIL(s) can be integrated with the safety standard SIL(s).

Corporate Class	Required PFD Probabilistic	Safety Integrity Level IEC 65A SIL	DIN 19250 AK Class
I	$\geq 10^{-1}$	-	1
II	$\geq 10^{-2}$ - $<10^{-1}$	1	2-3
III	$\geq 10^{-2}$ - $<10^{-2}$	2	4
IV	$\geq 10^{-4}$ - $<10^{-3}$	3	5
V	$\geq 10^{-4}$ - $<10^{-3}$	3	6
VI	$\geq 10^{-5}$ - $<10^{-4}$	4	7-8

Table 3 - Typical Corporate IEC SIL Classification Target

Common cause failures found in today's control systems require that the PHR include analysis of total loss of human machine interface (HMI). The resulting analysis of required action shall be forwarded to the design team. One method used to mitigate this situation is with the use of communication watchdog timer (WDT) which continuously monitors the health of the HMI and communication between the HMI and the SIS.

Future:

Following are the author's future expected changes to process plant instrumentation safety design because of the global standardization:

- Sensors and FCE's will be certified per their SIL capability.
- "Daisey-Chain" type I/O networking will continue to be introduced in applications where they do not contribute to common cause.
- Fieldbus type architecture will gain acceptance in safety applications after it gains acceptance for regulatory control.
- Software tools for operating system certification are being developed in academia as a prelude to industrial usage.
- The BPCS will be upgraded more frequently to serve as a safety layer of protection utilizing continuous mode SILs.
- PC and compact control system logic solver(s) will be included as approved logic solver(s).
- Because global instrumentation suppliers have to make products that meet all standards, their new products and feedback regarding improved methods from other countries will serve as the main catalyst for improved globally consistent standards.

Conclusion:

International safety standards are making excellent progress. Use of these standards requires a thorough understanding of the terms used prior to their implementation. While much headway has been made in the area, there still arises differences that can cause confusion and errors. These are due to different sector needs, different business and legal environments, different cultures, different languages, different operation and plant environments. The positive note is that all participants are trying to improve the safety in their industry. These developing safety standards will help towards that goal as long as a clear understanding of their intent can be realized.

References:

1 USA GUIDELINES

1.1 AMERICAN INSTITUTE OF CHEMICAL ENGINEERS (AIChE)

1.1.1 *Guidelines for Chemical Process Quantitative Risk Analysis, New York, 1989*
1.1.2 *Guidelines for Hazard Evaluation Procedures, New York, 1985*
1.1.3 *Guidelines for Safe Automation of Chemical Processes, New York, 1993*

Available from: **AIChE**
345 East 47th Street
New York, NY 10017
Tel: (212) 705-7657

2 ISA

2.1 ISA-S91.01-1995 Identification of Emergency Shutdown Systems and Controls That are Critical to Maintaining Safety in Process Industries

2.2 ISA-S84.01-1996 Application of Safety Instrumented Systems for the Process Industries

2.3 ISA-dTR84.02-1996 Electrical (E) / Electronics (E) / Programmable Electronic Systems (PES) for Use in Safety Applications - Safety Integrity Evaluation Techniques

NOTE—dTR84.02 is in development; for information, contact ISA.

2.4 Goble, W.M., *Evaluating Control System Reliability Techniques and Applications,* 1992

Available from: **ISA**
P.O. Box 12277
67 Alexander Drive
Research Triangle Park, NC 27709
Tel: (919) 990-9200

3 INTERNATIONAL GUIDELINES AND STANDARDS

3.1 INTERNATIONAL ELECTROTECHNICAL COMMISSION (IEC)

3.1.1 Parts 1-7 IEC draft Publication 1508-1996, Functional safety of electrical/electronic/programmable electronic safety-related systems

3.1.2 IEC draft Publication 1511-1996, Process sector safety standard

NOTE—IEC draft Publications 1508 and 1511 are in development; for more information, contact your national committee.

Available from: **IEC**
P.O. Box 131
3, rue de Varembe
1211 Geneva 20
Switzerland
Tel: 41 22 734 0150

3.1.3 *A Guide to Hazard and Operability Studies,* London, 1977

Available from: **Chemical Industries Association**
King's Buildings
Smith Square
London SW1P 2JJ
England
Tel: 44 171 8343399

3.1.4 *Risk Control and Instrument Protective Systems in the Process Industries,* Warrington, UK, 1980

Available from: **UK Atomic Energy Authority**
Safety and Reliability Directorate
Wigshaw Lane
Culcheth Warrington WA3 4NE
England
Tel: 44 171 925 254486

4 UNITED STATES CODE OF FEDERAL REGULATIONS (CFR)

4.1 29 CFR 1910.119-1992 (Final Rule: February 24, 1992) Process Safety Management of Highly Hazardous Chemicals, Explosives, and Blasting Agents

4.2 40 CFR Part 68 (Proposed rules: October 23, 1993) — Risk Management Programs for Chemical Accidental Release Prevention

Available from: **U. S. Government Printing Office**
Superintendent of Documents
Washington, DC 20402
Tel: (202) 512-1800

Paper 38b

Challenges of Applying a Performance-based Design Criteria for Safety Instrumented Systems

C. W. Thurston, Union Carbide Corporation

Paper 38b was not available for inclusion in the Proceedings.

It will be made available to attendees at the Symposium

Paper Number 38c

PROCESS CONTROL AS AN ALTERNATIVE TO PRESSURE RELIEF

John T. Marshall and Don L. Watkins

Prepared for presentation at the
1997 Spring National Meeting
Technology Transfer for the 21st Century
March 9 - 13, 1997
Houston, Texas

PROCESS CONTROL AS AN ALTERNATIVE TO PRESSURE RELIEF

The design and specification of a pressure relief device can involve a few quick and easy calculations or it can require highly complex and detailed procedures supported by experimental data that sometimes results in pressure relief devices of spectacular proportion. The design procedure begins by developing an awareness of the various codes and regulations pertaining to pressure relief then understanding their applicability to the process design at hand. Next, the chemical process of concern is surveyed to identify the candidates for excessive pressure protection and scenarios developed that can lead to an over pressure condition for each of the candidates. Scenarios will typically consider loss/lack of process control, runaway reactions, fire, thermal relief, or combinations of events from one or more of these categories. In some cases, the next step is to design the pressure relief device, purchase it, install it, register it - and the initial process is done. In other cases, the material released from the over pressured vessel may be toxic or capable of any one of several ignition related events necessitating additional treatment before it can be released to the atmosphere. Scrubbers and flares are typical of these systems and are sometimes dedicated to a single service.

The preferred operating status of a pressure relief device is to never have a demand placed on it. Foam systems, sprinklers, deluge systems, insulation, and drainage are examples of mitigation systems against fire. The best line of defense against runaway reactions is a strong reactive chemical program that identifies, describes, and quantifies potential problem reactions coupled with a design process that utilizes this information to eliminate or minimize the potential for mixing incompatible components and establish operating parameters safely away from initiating conditions when the reacting species must be present. The job of process control is to maintain the process where it was designed to run. This usually includes process start-up, normal operations including correction to deviations from normal set points when they occur, and process shutdown when required. The process control system fails when it loses the capacity to return the process to a normal, safe operating state following a process upset.

The literature warns the engineer against taking credit for process control in upset conditions suggesting that it won't work correctly or will fail entirely under certain conditions[i] A compelling argument in support of this statement is that it is difficult to imagine a process control system that can effectively respond to every over pressure condition - even those not thought of during the design. Yet once the plant is designed and operating, the process control system[ii] is the first defense against potentially dangerous conditions, and, in a rapidly changing process, may be the only defense that can react in time to prevent the demand on a relief device[iii]

As safety and environmental awareness continued to grow and develop, the number and size of relief valve vents requiring treatment grew also. Today it is not uncommon to find that a piece of equipment requires a pressure relief device to protect it from a potential over pressure condition that can exist during a brief time period in its operating history, for example start-up, but does not require protection at other times. If the substance that can be released is either toxic, flammable, or a potential environmental hazard, a mitigation system may be required even when the system is not required by other parts of the process. The process controls are assumed not to work when required, but are designed to ensure the potential event will not occur. As the need for mitigation became more common, more complex, and more expensive, the desire to use process control as an alternative to pressure relief became more intense.

Recent years have seen an increased interest in quantitative risk assessment and process risk management. New concepts have been introduced, many of which are still in debate today. One of the most significant concepts was that of acceptable or tolerable risk. While risk is a part of everyday life, the concept of acceptable risk has sometimes been a tough sell at the corporate level, particularly when the measurement is "the expected number of injuries or fatalities per year". Tolerable risk is a more palatable concept, since it does not require the acceptance of risk, and has made risk management more readily available as a decision making tool.

Decisions involving risk are usually made in one of two ways. One way is to determine the risk associated with an activity and compare that to a standard or criteria for acceptability (or tolerability). Criteria for acceptable risk have been established by several governments, some companies, and multi-company committees operating under the auspices of organizations such as the American Institute of Chemical Engineers (AIChE) or the Major Industrial Accidents Council of Canada (MIACC) creating industry consensus standards for acceptability. The criteria take the form of lines drawn on a FN curve, a statement of acceptability in terms of acceptable risk, and performance requirements for a given risk level. The second way is to determine the risk for each of several alternatives, then use the goal of continued risk reduction as one of the inputs in choosing between them. This alternative does not require criteria of any kind and does not make a statement regarding the tolerability or acceptability of risk

In 1993, the Center for Chemical Process Safety (CCPS) published the book *Guidelines for Safe Automation of Chemical Processes* written by the CCPS Process Control Sub Committee. In the book, concepts such as Basic Process Control System (BPCS), Safety Interlock System (SIS), and SIS Integrity Levels (SIL) were put forward. A particularly significant feature of the book was the stated need for both a BPCS and SIS, where the hardware for each system was separate and independent. Three levels of reliability for safety systems, called SIS integrity levels, are described and differ in requirement by the significance of the risk they manage and by the *availability* (a measure of reliability) of the system to respond to a demand. These are examples of the industry consensus standards for acceptability referred to above. The classification of chemical processes to specific integrity levels is based on general risk criteria with specific requirements left to the discretion of individual companies and organizations. SIS Integrity Level 3 (SIL 3) is the most stringent level identified and calls for an availability greater than 0.999.

While the CCPS Process Control Sub Committee was developing their book, other organizations, such as the Instrument Society of America (ISA) and the International

Electrotechnical Commission (IEC), were developing their own standards of similar content. At the 1994 International Symposium and Workshop on Safe Process Automation held in Houston, Ron Bell of the Health and Safety Executive (HSE) presented a workshop on the underlying concepts of developing an international standard entitled, "Functional Safety: Safety Related Systems". The proposed standard, as envisioned then, was a generic standard focusing specifically at electrical/electronic/ programmable electronic systems and the associated software requirements. The framework for the achievement of Functional Safety was the Overall Safety Lifecycle which takes a potential application from Concept to Decommissioning including definition, specification, installation, commissioning, validation, and operation. The requirements of the safety related systems were determined by first assessing the risk of the equipment under control (EUC) then determining the mitigation impact of the combined safety related systems to arrive at the residual risk of the process being considered. The implication was that the mitigation impact of the safety related systems were jointly and separately adjusted until the residual risk was within a tolerable range. The safety related systems combining to reduce the process risk consisted of 1) external risk reduction facilities, 2) electrical/electronic/programmable electronic systems, and 3) other technology safety-related systems. Figure 1 is an extraction of one of Ron's slides[iv] and presents the idea graphically.

The concept of combining systems to reduce risk was not new, but the realization of the synergy of the three safety systems working together but contributing separately was. The suggestion was, that given a tolerable level of risk, one could enhance the risk mitigating capabilities of one system and accept a lesser capability from another system while still maintaining the overall level of risk reduction from the combined safety systems. Conceptually, it may be possible to eliminate a system in its entirety if one of the other systems can be enhanced to the point of providing risk mitigation at least equal to the two systems working together before change. The safety level of the system is maintained, or improved, by replacing the safety contribution of one system by an improved safety contribution in another system. It is an easy extension of these concepts to provide a process control solution to a scenario normally requiring a pressure relief device for vessel protection.

The Dow Chemical Company uses an in-house developed process control system known as the MOD™5. A MOD™5 unit consists of independent and parallel input/output electronics and processing units that receive signals from one or more field transmitters measuring the process then transmits control information to one or more final control elements regulating the process. The parallel units are referred to as the left and right units and are in constant communication with each other validating input and confirming outputs. Both CPUs read inputs from the field, exercise exactly the same code, and output in parallel to the field. In the event of a CPU failure, watchdog and deadman circuitry ensure that the failed CPU shuts down. The other CPU continues to control the process "bumplessly" without any interruption. The left and right CPUs are directly cabled to dedicated left and right input and output cards. Typically, the left cards provide

Figure 1. Risk and Safety Integrity Concepts

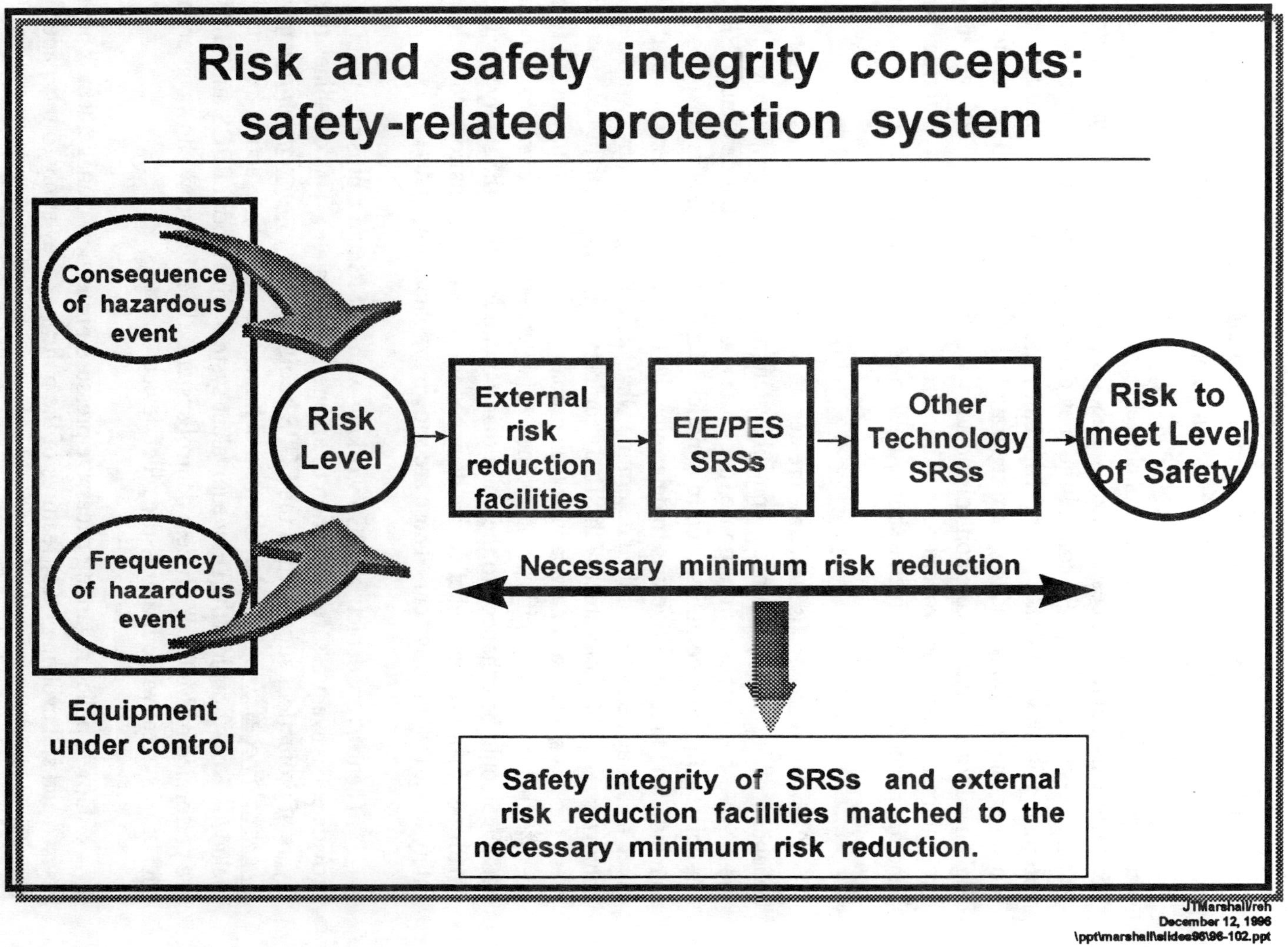

I/O capability for the left computer and the right cards provide I/O capability for the right computer. However, if a left card fails, the right computer provides the needed signal until the left I/O card is repaired, and so forth. The computers are programmed using an in-house developed process control language called DOWTRAN™. DOWTRAN™ allows the use of software aborts which are special program statements that prevent manual override of critical signals such as shutdown commands. In addition, the MOD™5 can also employ hardware aborts which are a set of electronic switches that are designed to interrupt the signal to the final control element putting it in a safe state when a shutdown condition exists in the process. Hardware aborts can accept input from the same field transmitters used for process control and may de-energize the same solenoids on a automated block valve (ABV) that is tripped by the normal process control system. However, the hardware and logic are separate and independent from the hardware and logic used for normal process control and software abort. The MOD™5 system is programmed to provide process control but also functions as a safety interlock system sending independent and diverse shutdown signals to controlled devices in the field on demand.

External consensus standards created an immediate need for us to understand the reliability of our MOD™5 process control system, and, in particular, whether it could meet SIL 3 requirements in response to a sudden and unexpected demand. JBF Associates, Inc. was commissioned to conduct a Fault Tree Analysis with two objectives in mind. The first objective was to determine the reliability characteristics of the MOD™5 given a sudden demand and then determine what modifications, if any, were necessary to achieve a reliability consistent with SIL 3 requirements. The reliability characteristic of merit for this study was unavailability (the probability that the system will fail to perform a specific function at a specific point in time). The second objective was to determine if the MOD™5 could be enhanced to the point that it could take over the safety function performed by a pressure relief device designed for a specific scenario without loss of the safety integrity of the system. These are shown graphically in Figure 2.

The top row depicts a BPCS working with an SIS and a pressure relief device to provide excessive pressure protection to an operating system. This combines the several layers of protections suggested in the CCPS book *Guidelines for Safe Automation of Chemical Processes* except that alarms and human response/interaction have not been included. Row 2 depicts the Dow equivalent system where both BPCS and SIS functions are performed by MOD5™ process control system. The third row shows an enhanced MOD5™ system without a pressure relief device.

Dow's policy regarding excessive pressure protection management is newly revised and still evolving. Key features of the policy, as they apply to this paper, are extracted below:

Figure 2. Expectations of MOD™5 Study

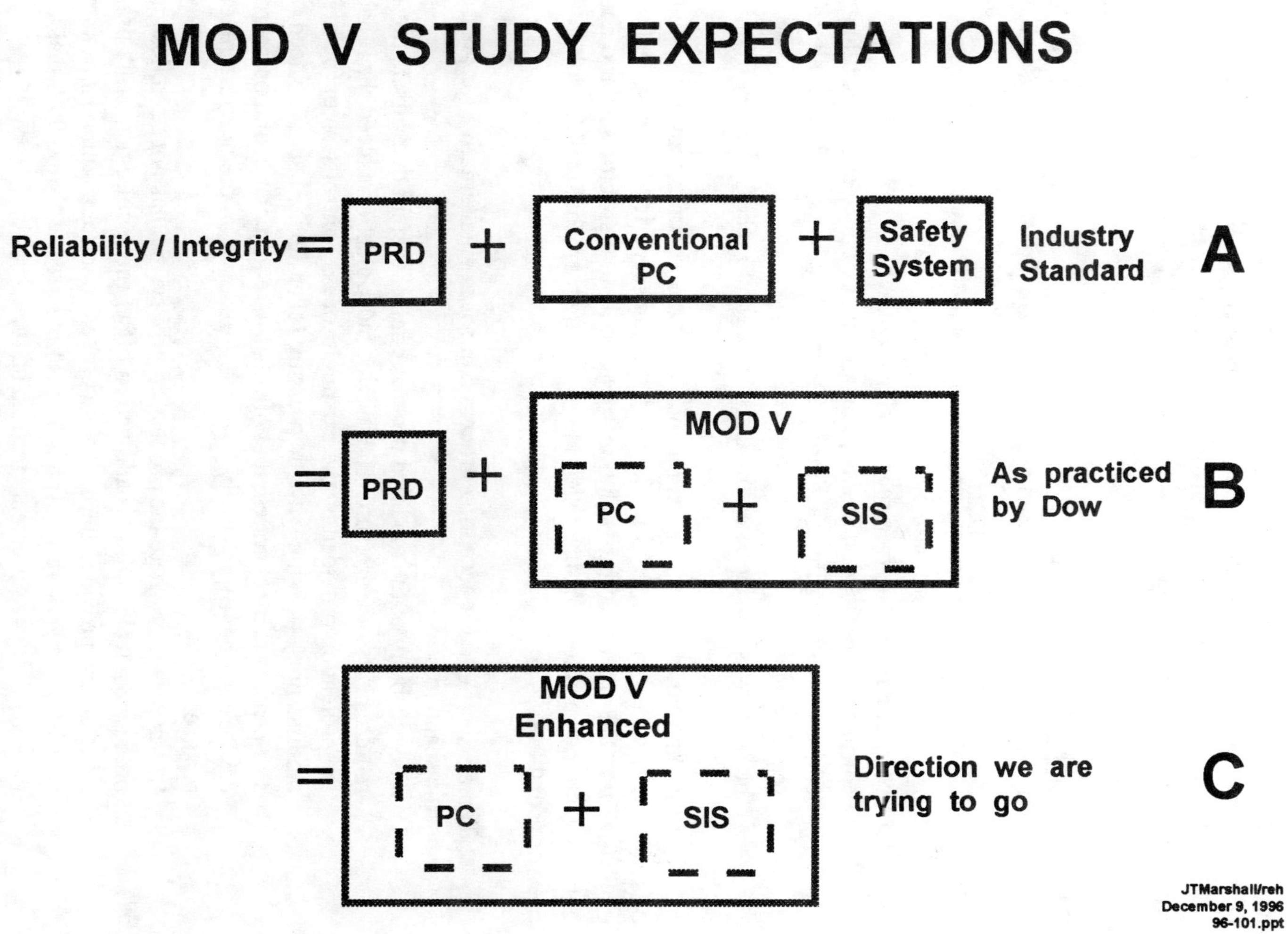

PHILOSOPHY

The Dow Chemical Company will protect its people, the environment and its assets against the consequence of unprotected excessive pressure.

POLICY

Excessive pressure protection systems shall be engineered, installed, registered, inspected, and documented to control hazards due to excess pressure, which have the potential to cause injury to people or damage to property or the environment.

We will meet government regulations applicable for each site or Dow standards, whichever are the most restrictive.

REQUIREMENTS

- Design scenarios shall be based on realistic evaluations of both consequences and likelihood of potential events. Elements of risk, cost, and benefits must be evaluated considering Dow's goal to continually reduce risk.
- Design of excessive pressure protection systems shall:
  - ♦ be approved by qualified relief design engineers.
  - ♦ consider the protection provided by all layers of protection, including: conventional pressure relief devices, control systems, and other alternative relief devices.
  - ♦ be at least as reliable as a conventional pressure relief device system.[v]

Safety Interlock systems, when the interlocks are a process control system, can offset the requirements for pressure relief for a specific scenario. It is essential to emphasize that the process control solution applies only to a specific scenario and does not eliminate the need for pressure relief due to other events that could result in an over pressure condition.

Tentative criteria for accepting process control as an alternative for pressure relief for a specific scenario are:

1) The reliability of the modified process control system that offsets the requirements of a pressure relief device must meet or exceed the reliability of the protection provided by the existing process control system (before modification) working with the pressure relief device being replaced.
2) The process control system that offsets the requirements of a pressure relief device must have a reliability five times greater than the pressure relief device it replaces.
3) In a new design where the process control system has not been designed, a process control solution is allowed to offset the requirements of a pressure relief device provided the complete process control solution [for example field sensor(s) to block valve(s)] has a reliability greater than 0.999, that is, meets or exceeds SIL 3 requirements for reliability.

EXAMPLES

The first example involves a compressor system distributing a potentially hazardous substance through a series of gas distribution lines. The compressor runs continuously and supplies a large number of downstream manufacturing facilities. The system must have a very high on-stream time since none of the downstream users inventory the raw material, but must also have over pressure protection to prevent potential destruction of the compressor and the release of a potentially hazardous substance. The compressor is controlled by our MOD™5 process control system with hardware aborts. During start-up, the compressor can over pressure if the downstream users are slow to open the valves on their end of the transmission line. Once the compressor is up and running, however, there is no realistic scenario for system over pressure since the probability of several users going off stream at the same time is negligibly small. To protect against the over pressure potential during start-up, a pressure relief device (PRD) rated for the full capacity of the compressors is installed on the discharge piping. This is a regulated substance and requires treatment of the relief device discharge. In the usual installation, suitable treatment facilities are within reasonable distance of the compressor. In this installation, however, the nearest treatment facility was a substantial distance away requiring a long and expensive pipeline or the installation of a treatment facility for the sole purpose of treating the vent from the PRDs.

Fault Tree analysis was the method of choice for assessing the reliability of the system with unavailability as the reliability characteristic of merit for the study. Unavailability is the probability that a system will fail to perform a specified function at a specified point in time. The TOP event for the fault tree was:

Loss of Capability to Provide Over Pressure Protection for the Compressor

The study assumed a sudden over pressure condition existed then assessed the capability of the system to return the process to a safe operating state. There were three basic pieces to the fault tree as shown in Figure 3.

The first of these is the pressure relief device. Combinations of human and mechanical failures combine to give an assessment of the capability of the installed device to perform its function on demand. The second piece is the starter relays on the compressor which are opened by the MOD™5. Here again, human and mechanical failures combine to give a probability of failure on demand. The third piece is the process control system, acting as a Safety Interlock System, that sends the trip signal to the starter relays. The system was generally described earlier in this report but a more detailed description follows below.

The compressor discharge pressure is monitored by three different pressure transmitters each on a separate nozzle. The input signal from each transmitter is received by different left and right analog input (AI) cards, one for each of the two computers

Figure 3. Top Event for the Compressor Fault Tree

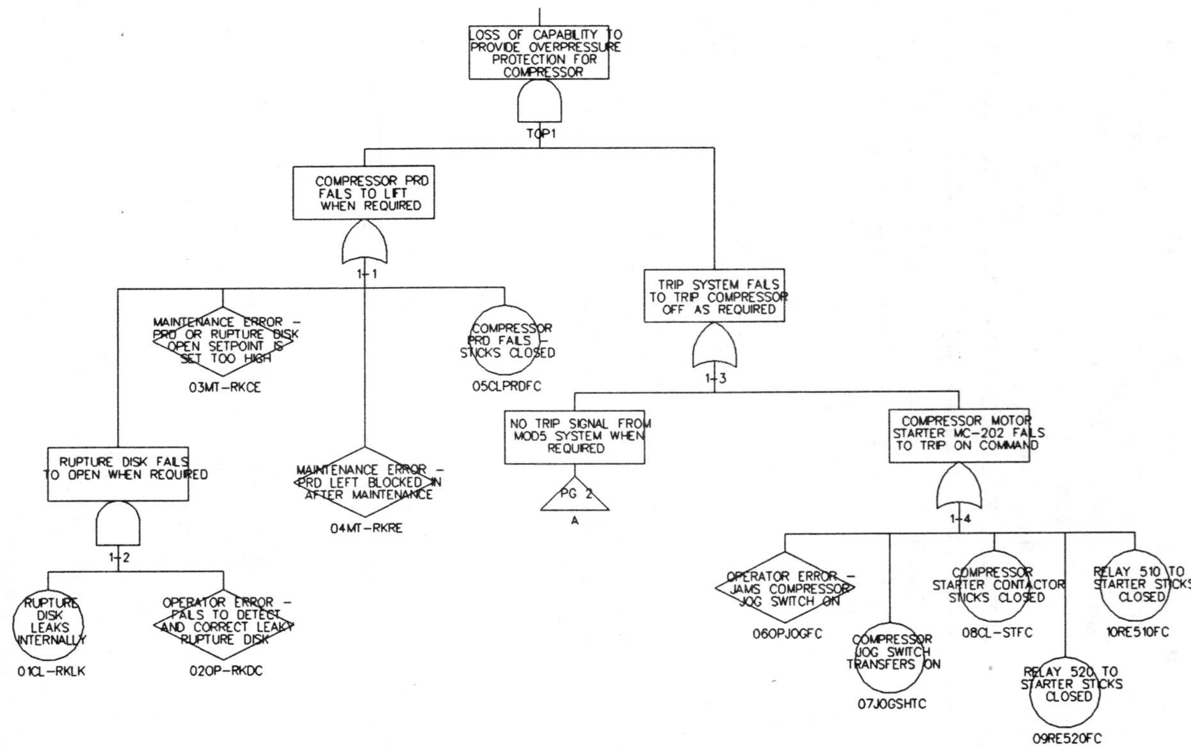

(there are six cards for the three pressure transmitters). Shutdown of the compressor system is initiated when:

1) Two pressure transmitters send a pressure signal that is above the high pressure trip point. A failed transmitter is a vote to shut down.
2) One pressure transmitter sends a pressure signal that is above the high pressure trip point and the other two transmitters send a pressure signal that is within 5 psi of the high pressure trip point.

Conventional analog and digital output signals from the MOD™5 are arbitrated so that the high (energized) signal wins. All outputs are installed such that a de-energized signal from the control system or SIS puts the output in a safe state. In this example, when the output signal to the compressor is de-energized the compressor is off. The output is energized to run the compressor. Because high output wins, this means both computers must send a shutdown signal (de-energized signal) in order for the compressor to shut down. The trip signal for the compressor system is protected by software aborts that prevent manual override of critical signals.

Independent shutdown signals can also be sent by the hardware abort system. The hardware abort card contains two switches in series, one controlled by the left MOD™ 5 computer, the other controlled by the right MOD™ 5 computer. Routing an output signal through an abort card allows either computer to de-energize an output, independent of its neighbor computer.

In this example the hardware aborts use the same pressure transmitters and the same voting logic as the process control section of the MOD™5 system but provide each computer with an independent mechanism for tripping the compressor. The hardware abort system for the compressors consists of five parallel paths that maintain the signal to the final control element. Four of these paths consist of left and right electronic switches in series receiving signals from the left and right computer respectively. The signal to the final control element is normally made, but is interrupted, placing the control element in a safe state, if either the left or right electronic switch opens when commanded by the MOD™5. At least one electronic switch in each of the four hardware abort paths must open in order for the control signal to be interrupted. The fifth path is a key switch that manually bypasses the hardware abort cards creating a hard signal path. Abort keys are under strict administrative controls with their use governed by written procedures.

The MOD™5 system can be defeated by human and mechanical failures, or combinations of both. Many base and undeveloped events were defined in the course of developing the fault tree. Failure rates or probabilities based on Dow internal data for the MOD™5 and public data bases dealing with equipment and human reliability were associated with each of these events. The unavailability of the MOD™5 and the TOP event were determined by solving the fault tree.

The "as-built" solution to the fault tree showed the TOP event had an unavailability of $3x10^{-3}$. The pressure relief piece of the fault tree had an unavailability of

0.05 while the starter relays had an unavailability of 0.012 and the MOD™5 system an unavailability of 0.047. Assessment of the cut set listing and the event importance table showed that the top contributors to the system unavailability were:

1) A maintenance error that results in a too high of a set pressure for the relieving device.
2) An operator error that resets the high pressure trip point to a higher value.
3) Mechanical failure of the starter relays (sticks closed).
4) Mechanical failure of the compressor jog switch (sticks engaged).

Item 1) was not addressed since one of the objectives of the study was to eliminate the relief device. Item 2) was addressed by changing the high pressure trip point from a restricted variable, which requires use of the physical abort key to alter its value, to a literal constant in the machine code which is inaccessible to operations and engineering personnel. Compressor shutdown occurs regardless of set points and control actions in other portions of the code and cannot be overridden by the digital key pad on the MOD™5 panel. This event probability now goes to 0. Items 3) and 4) were addressed by adding series redundant starter relays and jog switches. The relays and jog switches must now fail simultaneously for the system to fail. This squares the probability for the respective events.

With these adjustments made, the TOP event now has an unavailability of 2.75×10^{-5}. The pressure relief piece is unchanged but the unavailability of the starter relay portion now becomes 2.8×10^{-5} and the unavailability of the MOD™5 piece is now 5.17×10^{-4}. The unavailability of the TOP event without the pressure relief portion is 5.45×10^{-4}.

The criteria for acceptance of a process control solution as an alternative to pressure relief requires that:

1) The process control solution must be at least five times more reliable than the pressure relief device it replaces. In this study, the PRD system with an unavailability of .05 is replaced by a process control system with an unavailability of 5.45×10^{-4}, an improvement of almost 100 fold.
2) The reliability of the modified process control system must meet or exceed the reliability of the existing process control system working with the pressure relief system. In this study, a system having an unavailability of 2.97×10^{-3} is replaced by a system having an unavailability of 5.45×10^{-4}. The reliability of the overall system is improved by a factor of 5.
3) The reliability of the modified process control system must meet or exceed SIL 3 requirements for reliability. Clearly, the system does.

The second example considers a mixing vessel where an organic powder is slurried with hot water prior to feeding the mixture to the next step in the process. The hot water for the process is made by blending 235 psi steam and water. The steam and water flow are under MOD™5 control. Each time a new batch is called for, the MOD™5 establishes the water flow then adjusts the steam flow to obtain the needed hot water temperature going to the vessel. There are no automated block valves in either the steam or water lines.

The vessel operates at near atmospheric pressure. A depad valve on the vessel is normally open during the fill step and is capable of venting the vessel at normal fill rates. There are several scenarios involving the inadvertent operation of the depad valve or the steam and water valves that can result in excessive pressure in the vessel. The worst case scenario involves the steam valve failing open while the depad valve is shut. In this case, 53,000 lbs/hr of 235 psi steam can be directly fed to the tank causing a severe over pressure condition. The vessel can be protected by the addition of three 6x8 relief valves mounted on separate nozzles or on a header system capable of handling the flow rate. In the worst case over pressure condition, all three valves must open in order to protect the vessel.

This work extends the original fault tree by adjusting the field instrumentation to reflect the new process while maintaining the internal structure for the MOD™5. Case 1 utilized the standard MOD5 system. The TOP event was labeled:

Loss of capability to control pressure in vessel

Figure 4 is the first page of the fault tree.

The existing fault tree structure began with input from the field transmitters and ends with the output signal going to starter relays as described in the previous example. The original study used triple redundant pressure transmitters with 2-out-of-3 voting as input signals. A similar arrangement was anticipated as part of the final solution but the initial study used existing instrumentation in the assessment of reliability. The system under study uses a pressure transmitter on the vessel and temperature transmitter in the line between the mixer and the vessel to initiate the close sequence for the steam valve. The close sequence for the water valve is initiated by high pressure in the vessel or a water flow to the vessel when the hot water addition step is not current (that is, when water flow is not expected). A single high vote from any one of the devices will cause the steam or water or both valves to block. This is a very reliable (high availability) system but one that can be prone to false trips.

In the first part of this example (Case 1), the protection system for the vessel consists of the control valves on the steam and water line going to the mixer and, independently, three 6 x 8 pressure relief valves mounted on the vessel. The relief valves are sized so that all three must open to protect the vessel in the event the steam valve fails open.

In the second part (Case 2), the relief valves are removed from the system and automated block valves (ABVs) are installed in both the steam and water lines after the control valves. These valves are either open or closed and work with the control valves to provide protection for the vessel. The original plan was to add triple redundant pressure transmitters with 2-out-of-3 voting to control the shutdown sequence. However, the

Figure 4. Top Event Hot Water Drum Fault Tree

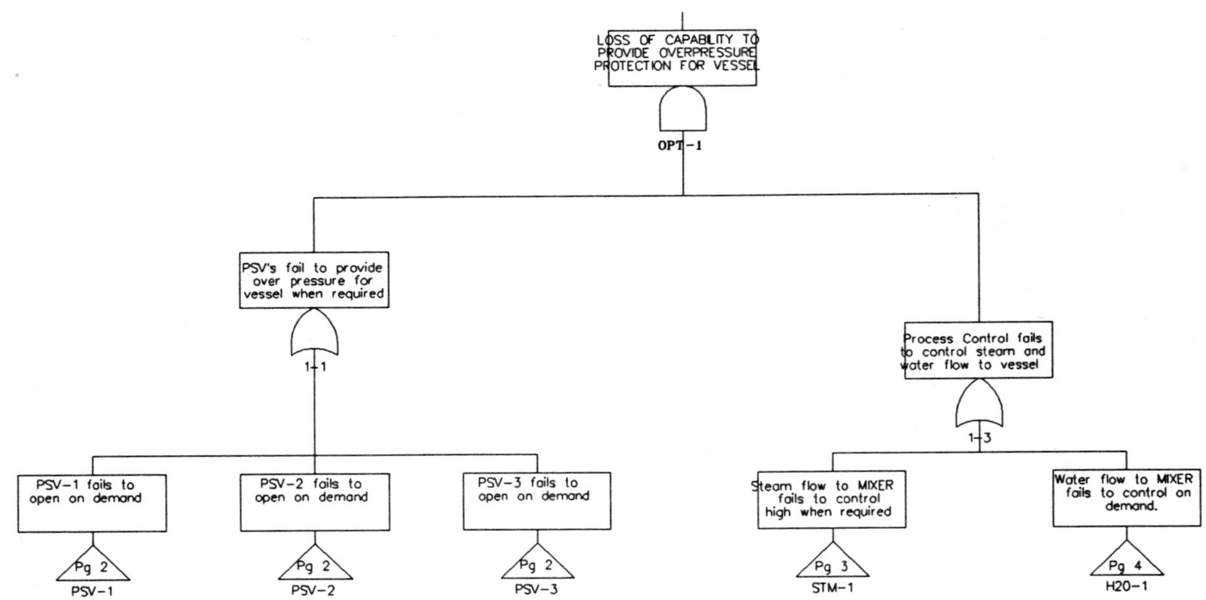

combination of pressure, temperature, and flow measurement were sufficient to achieve the desired reliability. A large number of cut sets were identified in Case 1 including 139 having a probability greater than $1.0x10^{-11}$. However, 99% of the unavailability is accounted for by the top 12 events. These are essentially two event cut sets made up of one event causing a pressure safety valve (PSV) failure and one event causing a process control failure.

The TOP event for Case 1 has an unavailability of 9.64×10^{-3}. The three PSVs have a combined unavailability of 0.084 and the process control side an unavailability of 0.108. On the relief valve side, the dominant event is a maintenance error that results in one or more of the relief valves having an incorrect set pressure. The principal event on the process control side is an operator error where, in response to an over pressure situation, the plant is "saved" by raising the set point through the keypad panel on the front of the MOD™5. The operator error is closely followed by failure of the valves on the steam and water lines to close on demand.

In Case 2, the PSVs are eliminated from the study and an automated block valve (ABV) added in series with the control valve. The unavailability of the TOP event for Case 2 is $5.3x10^{-2}$. This is about twice as good as the process control side of Case 1, but not as good as the combined system. As in Case 1, the main contributor to unavailability is the operator resetting a set point in response to an unexpected pressure excursion. This event can be eliminated by adding a literal constant to the MOD5™ code as described in the first example. Now, this event assumes a probability of zero and the unavailability of the TOP event drops to $2.9x10^{-3}$. This is about three times lower than the base case with pressure relief and process control, but is short of the targeted improvement in unavailability ($.084/5 * .108 = 1.8x10^{-3}$, that is, five times better than the PSVs they replace). Since this is a batch process, the ABVs are exercised with each batch verifying their ability to close on demand. This tests the ABVs and significantly reduces the probability of failure on demand. With this adjustment the unavailability of the TOP event drops to $5.0x10^{-4}$.

The criteria for acceptance of a process control solution as an alternative to pressure relief for the scenario defined in this study were:

1) The process control solution must be at least five times more reliable than the pressure relief device it replaces. In this study, three PSVs with an unavailability of .084 are replaced by a process control system with an unavailability of $5.x10^{-4}$.
2) The reliability of the modified process control system must meet or exceed the reliability of the existing process control system working with the pressure relief system. In this study, a system having an unavailability of $9.64x10^{-3}$ is replaced by a system having an unavailability of $5.x10^{-4}$. The reliability of the overall system is improved by a factor of almost 20.
3) The SIL requirements of this system were not determined but certainly are no higher than SIL 1. However, the reliability of the modified process control system easily exceeded SIL 3 requirements for reliability.

SUMMARY

Process control can be an acceptable alternative to a pressure relief device as protection against excessive pressure in certain instances. It must be emphasized that the alternative solution applies only to a specific scenario and does not eliminate the need for pressure relief due to other events that could result in an over pressure condition. The procedure for qualifying a process control solution as an alternative to pressure relief is intended to be neither quick, easy, or cheap. Instead, it requires a detailed analysis of the entire system to ensure that the final solution moves us toward a safer, more reliable installation. The process does not end with the design of the process control system. Our Excessive Pressure Protection Management System also calls for registering the physical and intellectual aspects of the alternative solution, as if they were a pressure relief device, so that future auditors can ensure that the content of the original package has not changed.

The economic incentives for process control solutions to excessive pressure problems can be considerable, particularly when the solution reduces in size, or eliminates entirely, a scrubber or flare needed to treat the vent. There are also benefits derived from stronger management practices that look after the development and implementation of process control code, and ensure its integrity, by requiring reviews and simulation of all new code and any changes made to it. But perhaps the greatest gain is more reliable protection from excessive pressures in process equipment leading us toward, not away from, safer plants.

---

[i] "Avoid Common Relief Valve Pitfalls", Bravo, Contreras and Jarvis. *Practical Engineering Perspectives Plant Safety,* AIChE, 1996, p. 214.

[ii] This discussion groups the concepts of Basic Process Control Systems and Safety Interlock Systems as defined in the open literature.

[iii] Swain, in his HANDBOOK OF HUMAN RELIABILITY ANALYSIS WITH EMPHASIS ON NUCLEAR POWER PLANT APPLICATIONS (Final Report), Sandia National Laboratories, Albuquerque, NM, Figure 8-1, shows that the probability of error, when control room personnel are required to respond to an abnormal event within minutes of the event's origination, and with normal training, is about 90%. With extensive training the probability of error drops to about 50% but increases to 100% when the training is less than adequate.

[iv] "WORKSHOP A, THE SAFETY LIFE CYCLE Its Role in the Functional Safety of Safety-Related Systems", Ron Bell, *International Symposium and Workshop on Safe Chemical Process Automation,* AIChE, P. 160. Used by permission.

[v] When the excessive pressure protection system is a process control system, the process control system is required to have a reliability five times that of the pressure relief device it replaces.

Out-of-the-Loop Performance Problems and the Use of Intermediate Levels of Automation for Improved Control System Functioning and Safety

by

David B. Kaber[1]

&

Mica R. Endsley[2]

[1]Department of Industrial Engineering, Mississippi State University
Mississippi State, Mississippi

[2]Department of Aeronautics and Astronautics, Massachusetts Institute of Technology
Cambridge, Massachusetts

Prepared for Presentation at the
1997 American Institute of Chemical Engineers' Loss Prevention Symposium
March 1997

Control Systems for Unit Operations Safety Session

Session 38 Paper d

UNPUBLISHED

Out-of-the-Loop Performance Problems and the Use of Intermediate Levels of Automation for Improved Control System Functioning and Safety

David B. Kaber[1] and Mica R. Endsley[2]

[1]Department of Industrial Engineering, Mississippi State University, Mississippi State, MS 39762-9542

[2]Department of Aeronautics and Astronautics, Massachusetts Institute of Technology, 77 Massachusetts Ave., 33-115, Cambridge, MA 02139-4307

Human supervisory control and monitoring of automated systems, as well as, passive system(s) information processing can all be classified as forms of out-of-the-loop (OOTL) performance. Whether the operator's task is to decide if process control intervention is necessary, detect a critical system event, or accept or reject the actions of a computer controller, he or she is removed from direct, real-time control of the system. OOTL performance is a critical issue in overall automated systems functioning because it is associated with numerous negative consequences including: (a) operator failure to observe system parameter changes and intervene when necessary (vigilance decrements); (b) human over-trust in computer controllers (complacency); (c) operator loss of system or situation awareness; and (d) operator direct/manual control skill decay. These consequences have been found to impact human performance under both normal operating conditions and system failure modes, with a greater effect on the later (Endsley & Kaber, in review) leading to serious problems in operator ability to perform their assigned tasks when working with automated systems.

Level of automation (LOA) has been put forth as an approach to ameliorating OOTL performance problems. It is intended to determine the optimal assignment of control between a human operator and computer in order to keep both involved in system operations. LOA considers the capabilities and capacities of both the human and computer controller in determining their optimal coupling. It constitutes a systems approach to resolving OOTL performance problems by minimizing the negative consequences associated with the removal of the operator from active system control, and allows for the strengths of both human decision making and computer processing to be realized. When compared to a technological approach that assesses only the capabilities of the computer in allocating as much responsibility to the machine as possible, and assigning the remaining tasks to the human operator, the advantages can be considerable.

A LOA taxonomy will be presented along with research examining its utility in a dynamic control task. Using LOA to identify optimal combinations of human and computer control was found to produce improvements in system performance under intermediate levels. These levels involve joint human and computer control of various system functions, such as monitoring, planning, and option selection and implementation. Results indicated decreases in the number of system processes/tasks overlooked by operators. These improvements may translate into cost reductions as a result of improved operational safety and are anticipated to be applicable to process control operations.

I. INTRODUCTION

Within the last 20 years, safety in process control and automated systems has become an increasingly important issue due to severe accidents resulting in environmental damage and loss of human life. Incidents motivating safety research in control systems have ranged from nuclear power plant melt-downs, such as that at the Unit 2 Plant of Three Mile Island in 1979 (Ahearne, 1982), to commercial airliner mishaps, including the crash of US Air B-737 on take-off at New York's LaGuardia Airport in 1989 killing two people (National Transportation Safety Board, 1990). These incidents have been attributed to human information processing errors, specifically a lack of operator awareness and understanding of process or automated system states. Such errors are contributed to by problems with: (a) control system interfaces presenting process or task relevant information to operators, and (b) allocation of system functions to human and automated controllers. Failure of interface designs to consider human cognitive and physical capabilities has produced performance pitfalls in process control including poor instrumentation readability and reachability; operator inability to mentally transform information from useless to useful forms; and operator inability to activate controls, interpret codes, locate individual displays and respond to alarms (Woods et al., 1987). Problems in control system function allocation include assigning tasks to human and automated controllers on the basis of computer capabilities and capacities alone and not considering those of the human, as well. This has resulted in, for example, as many tasks as possible being assigned to automated control with a fragmented task set left for operator control (Lockhart et al., 1993). It is often difficult to train operators in this set of tasks. Further the operator, during actual process or system functioning, is usually albeit completely removed from the control loop. Consequently, he or she may be posed with an inappropriate taskload, and is susceptible to boredom, vigilance and complacency.

Both poor system interface design and inappropriate function allocation among human and computer controllers can increase the probability of human information processing errors including misperception of displays, lack of comprehension of the meaning of displayed information and poor decision making in process and systems control. These errors can occur during both normal and abnormal operations, and can lead to degraded performance and accidents.

There exists a need to: (a) develop interfaces that accommodate process control system operator capabilities, and (b) facilitate appropriate system function allocations between human and computer controllers keeping both involved in the control loop. The objective of this paper is to address the latter by: (a) identifying process or automated system roles limiting operator task involvement and promoting the potential for poor human performance and safety; and (b) offering an approach to optimizing human and automation collaboration in process or automated system functioning. The approach was tested in a simulated dynamic control task and results are presented demonstrating improvements in performance and safety under normal operating conditions and failure modes.

II. OUT-OF-THE-LOOP PERFORMANCE

When an operator is removed from a control loop due to allocation of system functions to an automated/computer controller, the level of human system interaction is limited and, consequently, operator awareness of system states may be reduced. This poses a serious problem during normal operations preceding system errors, malfunctions or breakdowns because operators are often slower to respond to such events when removed from a control loop versus actively controlling a system (Wickens, 1992). Further, during failure modes, operators who have been removed from system control may not know what corrective actions need to be taken to stabilize the system and bring it into control. As well, these same operators may have forgotten critical manual control skills for error recovery due to their required absence from the loop. Examples of automated system (e.g., aircraft) mishaps, attributed to human OOTL performance, include the near crash of Air China's Boeing 747 into the Pacific ocean in 1989. The aircraft experienced a gradual engine failure that the human pilot was not aware of because of autopilot compensation (through rudder control) up until the point of failure of the autopilot, itself. Subsequently, the jet stalled and plummeted thousands of feet, being recovered within a few seconds of the ocean surface (Wickens, 1992). Another example is that of the catastrophic crash of Northwest Airlines MD-80 at Detroit Airport in 1987 due to improper configuration of the flaps and slats of the aircraft. All persons were killed but 1 because an automated take-off configuration warning system, which the crew relied on, failed to function (National Transportation Safety Board, 1988). They did not realize the aircraft was improperly configured for take-off and failed to check manually (Endsley, 1996). In both of these cases of automation error, when computer control facilities failed, operators, out of the direct control loop, were unaware of the state of the system, and in at least the one case, were unable to compensate for the failure mode before an accident occurred.

Lack of operator awareness leading to accidents may also arise in process control environments including, for example, petrochemical refining in which a delicate balance of pressure, volume and temperature must be maintained for efficient and safe system functioning. Operators may fail to be vigilant of potential errors (e.g., imbalances) or exhibit overtrust in automated/computer control due to inappropriate process function allocations among themselves and the computer. Inappropriate allocations may constitute assignment of all tasks for which problems have been identified and technological solutions exist to automation, and relegate the human operator from controller to the roles of supervisor, monitor and passive information processor. Unfortunately, these are roles to which humans are ill-suited (Endsley, 1995).

A. Supervisors

As supervisors of automated control, humans do not directly act on the process being controlled, but rather they supervise the behavior of computer controllers between them and the process (Moray, 1986). They observe the actions of a computer and agree or disagree with it. During normal operating conditions, a human supervisory controller monitors displays and looks for signals that intervention may be required to prevent process errors from occurring due to, for example, poor automated decision making. In the event of an error, he or she is required to directly control the process for failure recovery and return to supervisory mode. Human performance under supervisory control (SC) is often poor because of the limited taskload placed on operators under normal conditions reducing control involvement and producing the consequences of loss of process/system awareness and decay of direct control skills, which are critical for error recovery.

B. Monitors

Systems monitor is a role, which is as difficult for humans to fulfill as supervisory controller. It involves waiting to detect a critical process events, such as fluctuation of a pointer on a temperature gauge beyond a certain degree. It requires operators to scan displays without taking any action to change the state of the process unless needed (Moray, 1986). Its sole purpose is to determine whether a process is functioning normally or if human intervention is required. During normal operating conditions, a monitor is required to sample many indicators of process status and evaluate the need for intervention while automated control handles functions that were once carried out by humans (Wickens, 1992). Consequently, the operator has less to do, in terms of direct control, but is often overloaded with process components to monitor. This has the potential of increasing the probability of failing to observe critical events leading to errors. During failure modes, the monitor is often expected to function in a manner identical to a supervisory controller; that is, exercise direct control on the process for recovery.

The requirements and purpose of monitoring are so similar to those of SC that the two are often regarded as being synonymous. Even the OOTL performance problems that monitoring and SC suffer from are similar, yet they do constitute two very different acts. (Supervisory control, unlike monitoring, can involve: (1) choosing appropriate plans for dealing with system errors and failures, (2) executing plans for directing the system to a new goal, and (3) allocating manual control from moment-to-moment between the operator and computer (Moray, 1986).) With respect to OOTL performance problems, humans are poor at monitoring because of complacency and vigilance issues resulting from removal from a control loop and required functioning at a high level of control, like SC. Complacent operators tend to exhibit overtrust in computer controllers to perform process functions flawlessly. This usually occurs in tasks involving highly reliable automation in which it is easy for the operator to instill a great deal of confidence (Parasuraman et al., 1993). Complacency results in reduced process control accuracy or delay in detecting failures (Parasuraman et al., 1993). Human vigilance problems involve failure, on an operator's behalf, to attend to critical process events, such as equipment breakdowns. This usually occurs as a result of prolonged monitoring. There is much evidence that operator ability (speed and reliability) to detect process state changes is dependent upon whether humans are actively involved in direct task control versus simply monitoring or supervising the process (Wickens & Kessel, 1981; Young, 1969).

C. Passive Information processors

The role of passive information processor, much like that of supervisory controller, involves observing the actions of other operators or computer controllers and agreeing or disagreeing with them. The operator's task is to understand the actions of another system controller and thereby accept or reject its actions. The key difference between passive information processing and direct action on the process is that the former involves functions similar to those maintained during process monitoring (e.g., scanning information sources); whereas, the latter involves manual control functions including process planning, decision making, selecting responses and implementing strategies.

During normal operating conditions passive information processors detect the need for manual process intervention and attempt to maintain understanding of the state of the process. In the case that intervention is required, operators must exercise direct control skills for recovery and return to passive functioning. Human performance is often poor during intervention because operators have been relegated to passive information processor under normal conditions. This is due to the lack of control loop involvement inhibiting dynamic process awareness and, consequently, understanding for facilitating process recovery (Endsley & Kiris, 1995).

D. Summary

The lack of operator involvement in process or automated systems control in supervisory modes and passive information processing, as well as complacency and vigilance in process or automated systems monitoring, all contribute to critical human cognitive errors, specifically the loss of operator situation awareness (SA), to which the safety incidents discussed above have been attributed. Further, this lack of control involvement contributes to the loss of human manual control skills for process or automated system error recovery. These pitfalls of OOTL performance have motivated the development of a systems approach to allocating functions, as part of active control, to human and computer controllers intended to maintain involvement of both in ongoing operations. This approach is level of automation (LOA).

III. LEVEL OF AUTOMATION

In an attempt to prevent operators from being reduced to automated control system supervisors, monitors and passive information processors, possibly leading to OOTL performance problems, LOA allocates system functions to human and computer controllers based on consideration of the capabilities and capacities of each under normal operating conditions and failure modes. It requires identification of the functions to be allocated (e.g., systems monitoring, planning, decision making and acting) and the specific tasks within each function that must be performed. (For example, systems monitoring may involve scanning displays for specific information sources, reading instrumentation, transforming system variable samples into useful forms, etc..) These tasks can then be matched to controller sensing and processing abilities. The LOA method has the potential benefits of: (a) maintaining appropriate operator control involvement and taskload to reduce susceptibility to complacency, vigilance and lack of system awareness by taking into account human and computer performance abilities during failures, (b) reducing system errors due to poor automated decision making by relying on human decision making capabilities, and (c) enhancing system performance through computer data processing.

Level of automation differs from alternate approaches to system function allocation such as a technological based method focusing only on computer capabilities and capacities (under normal operating conditions) for assignment formulation. Process and automated system control design using a technological approach has been driven by a desire to reduce costs through reduction of operator workload and, consequently, human staffing requirements (Endsley & Kaber, in review). This usually results in assignment of a majority of tasks to the computer and relegation of the human to one of the identified OOTL performance roles. Unfortunately, the approach also produces the discussed operator performance problems and poor system functioning during failures.

Many different combinations of human and computer for system function performance may result from the LOA approach if controllers are capable of performing multiple functions. Some combinations or levels of automation have been presented in human performance and automated systems literature including those formulated by Ntuen & Park (1988) in the context of a teleoperation system. They developed 5 levels of automation by assigning the functions of "suggesting," "planning," "decision making," "approving," and "implementing" to either a human operator or computer controller. The levels were presented in a taxonomy. Endsley & Kaber (in review) developed a 10-level taxonomy of LOA intended to be applicable to a wide array of dynamic process and automated system control domains, specifically advanced manufacturing, teleoperations, air traffic control and aircraft piloting. These domains have many features in common including: (a) multiple system goals, (b) multiple tasks competing for an operator's attention (each having different relevance to system goals), and (c) high task demands under limited time resources. Further, the domains share common functions for allocation to the human and computer controller comprising: (a) monitoring - scanning displays to perceive system status, (b) generating - formulating options or strategies to achieve system goals, (c) selecting - deciding on a particular option or strategy, and (d) implementing - carrying out the chosen option. Endsley & Kaber (in review) formulated their levels of automation by assigning these functions to human or computer, or a combination of the two, depending on the sensing and processing capabilities of each in the various domains. The resulting levels are presented in Table 1.

Table 1. Endsley & Kaber's (in review) LOA Taxonomy.

	FUNCTIONS			
LEVEL OF AUTOMATION	MONITORING	GENERATING	SELECTING	IMPLEMENTING
1. Manual Control	Human	Human	Human	Human
2. Action Support	Human/Computer	Human	Human	Human/Computer
3. Batch Processing	Human/Computer	Human	Human	Computer
4. Shared Control	Human/Computer	Human/Computer	Human	Human/Computer
5. Decision Support	Human/Computer	Human/Computer	Human	Computer
6. Blended Decision Making	Human/Computer	Human/Computer	Human/Computer	Computer
7. Rigid System	Human/Computer	Computer	Human	Computer
8. Automated Decision Making	Human/Computer	Human/Computer	Computer	Computer
9. Supervisory Control	Human/Computer	Computer	Computer	Computer
10. Full Automation	Computer	Computer	Computer	Computer

IV. EXPERIMENT AND RESULTS

We conducted an experiment to examine the influence of the 10 levels of automation presented in Table 1 on human operator performance and SA in a simulated dynamic control task. The simulation was developed based on a similar task employed by Tulga & Sheridan (1980). It incorporated the common features of the dynamic system control domains studied by Endsley & Kaber (in review) in developing their taxonomy. The simulation required subjects to carry out task processing; select and eliminate targets (boxes of different colors and sizes) moving across a display towards a deadline at its center. All targets moved to the center of the scope with different velocities. Rewards were offered for target eliminations, and penalties were assessed for allowing them to reach the deadline (expire). The operating parameters of the task included: (1) no partial credit was given for incomplete target eliminations, and (2) all targets present on the display, at any given time, could not be eliminated; therefore, operators were required to generate strategies for reducing expirations and collisions.

Thirty university students performed the simulation in four 10-min trials, two of which were used to examine the LOA effect on manual performance during simulated automation failures (dynamic shifts to manual control), and two of which isolated the influence of LOA on operator SA. During the automation failure runs, manual control of the system was allocated to subjects three times (at random intervals) for a fixed period of 1 min. The SA trials involved three simulation freezes occurring at random points in time to administer (Situation Awareness Global Assessment Technique (Endsley, 1988)) queries aimed at assessing operator perception, comprehension and projection of states of the system.

We found that subject performance, measured in terms of the number of tasks addressed, was significantly improved by levels of automation involving computer aiding in the implementation aspect of task functioning. Specifically, under low-intermediate levels ('Action Support' [level 2] and 'Batch Processing' [level 3]) performance was considerably greater than with 'Manual Control' (level 1), and better than with levels of automation that added computer assistance to other task roles, such as strategy generation and selection. Decreases in the number of tasks addressed were observed at levels of automation involving joint human-computer strategy generation ('Shared Control' [level 4], 'Decision Support' [level 5], and Blended [level 6] and 'Automated Decision Making' [level 8]), as compared to levels requiring purely human strategizing (Action Support [level 2] and Batch Processing [level 3]). This finding was attributed to allocation of the generation role to operators possibly causing task overload. Joint human-computer selection of processing plans ('Blended Decision Making' [level 6]) had no significant impact on performance as compared to purely human (Decision Support [level 5]) or computer (Automated Decision Making [level 8]) selection. Also, performance under human selection (e.g., 'Rigid System' [level 7]) did not significantly differ from that under computer selection (e.g., 'Supervisory Control' [level 9]).

Performance measured in terms of the number of target expirations (tasks overlooked) decreased as a function of LOA with the highest number of expirations occurring under Manual Control (level 1). Levels of automation involving purely computer generation of options or joint human-computer generation (Rigid System [level 7] and Automated Decision Making [level 8]), produced the lowest number of target expirations. Although the way in which generation of options was conducted varied considerably between these two levels (in terms of the respective roles of the human and computer in the sharing of the generation option), this did not appear to have a substantial impact on performance.

Operator ability to recover from, and perform during, automation failures was significantly improved with levels of automation requiring human interaction in task implementation (e.g., subject selection of targets one-at-a-time versus queuing them for computer processing or using computer generated strategies). Specifically, the time-to-recovery of task control and the number of tasks addressed under subsequent manual control was worse when operators had been functioning at levels of automation requiring advanced queuing of targets (Batch Processing [level 3] and Automated Decision Making [level 8]). This result was attributed to the ability of operators to focus on future tasks; thus, ignoring the present system state (e.g., computer failure).

Significant differences in operator SA were only observed for system state comprehension queries. It was found that levels of automation not requiring the human to perform strategy selection (Blended [level 6] and Automated Decision Making [level 8], Supervisory Control [level 9], and 'Full Automation' [level 10]) allowed for improved subject task understanding, as compared to all other aided levels and Manual Control (level 1) mode. Other modes of automation corresponded to poorer levels of operator comprehension of task priorities and completion status. This finding was attributed to the added burdens of the selection and monitoring roles being allocated to the human at these levels along with the responsibility of strategy generation, which may have limited time resources available for perceiving targets.

Based on the above results, we concluded the following: (a) the distribution of human versus computer control made a significant performance difference in the implementation and option generation roles of the dynamic control task, but not in the decision making portion (option selection); (b) improvements in system processing were facilitated by computer aiding in the implementation aspect of functioning at low-intermediate levels of automation; (c) reductions in overlooked processes were driven by joint human-computer planning at upper-intermediate levels of automation; (d) levels of automation allowing for advance process planning in the selection role distracted operators from current task events (automation failures) and produced decrements in control response times and task processing ability; and (e) improvements in operator SA, specifically the understanding of task priorities and completion status, were facilitated by reductions in taskload under levels of automation involving human-computer or purely computer strategy selection.

These conclusions demonstrate the named potential benefits of the LOA approach to process and automated system design. Specifically, enhancements in performance due to computer data processing and human cognitive abilities were realized. These occurred at intermediate levels of automation in the form of increased operating efficiency and effectiveness, assessed in terms of completion, and attentiveness to, system processes.

V. CONCLUSIONS

Improvements in process and automated system performance due to the use of intermediate levels of automation, maintaining human and computer involvement in a control loop, may translate into cost reductions as a result of improved operational safety. By limiting operator susceptibility to vigilance, complacency, loss of system or situation awareness and direct/manual control skill decay, due to OOTL performance roles, improved human information processing under normal conditions, and expedient failure mode recovery, may result. This can increase the probability of safe system functioning and reduce safety incident rates. Similarly, such safety enhancements may yield, for example, reductions in machine wear, waste material, labor overtime and environment clean-up, as well as first aid costs, worker compensation claims and law suits for death due to operator negligence.

The named performance and safety benefits of LOA are anticipated to be realizable in process control operations including foundries and smelters, nuclear power plants, and petrochemical refining facilities. Specifically, using LOA to allocate system functions to human and computer controllers on the basis of process control task capabilities (e.g., remote control of robotic arms for carrying crucibles, monitoring and maintaining coolant water levels in reactors or chemical balances between feeder tanks in defractionators, etc.) may increase system error detection and remediation, and, consequently, safe process functioning.

ACKNOWLEDGEMENTS

Parts of this research were completed while the author and co-author held a graduate research assistantship and an associate professorship, respectively, at Texas Tech University, and were presented as a poster at the Second Automation Technology and Human Performance Conference in Cocoa Beach, Florida.

REFERENCES

Ahearne, J. (1982). Keynote address. In: *Conference record for 1981 IEEE standards workshop on human factors and nuclear safety* (R. Hall, J. Fragola & W. Lukas, Eds.). New York: IEEE.

Endsley, M. R. (1988). Design and evaluation for situation awareness enhancement. In: *Proceedings of the Human Factors Society 32nd Annual Meeting*, pp. 97-101. Santa Monica, CA: Human Factors and Ergonomics Society.

Endsley, M. R. (1995). Towards a new paradigm for automation: Designing for situation awareness. In: *Proceedings of the 6th IFAC/IFIP/IFORS/IEA Symposium on Analysis, Design and Evaluation of Man-Machine Systems*, pp. 421-426. Cambridge, MA: MIT.

Endsley, M. R. (1996). Automation and situation awareness. In *Automation and Human Performance: Theory and Applications* (R. Parasuraman & M. Mouloua, Eds.). Hillsdale, NJ: Lawrence Erlbaum.

Endsley, M. R. & Kaber, D. B. (in review). Level of automation effects on performance, situation awareness and workload in a dynamic control task. Submitted to *Ergonomics*.

Endsley, M. R. & Kiris, E. O. (1995). The out-of-the-loop performance problem and level of control in automation. *Human Factors*, 37(2), 381-394.

Lockhart, J. M., Strub, M. H., Hawley, J. K. & Lourdes, A. T. (1993). Automation and supervisory control: A perspective on human performance, training, and performance aiding. In: *Proceedings of the Human Factors and Ergonomics Society 37th Annual Meeting*. Santa Monica, CA: Human Factors and Ergonomics Society.

Moray, N. (1986). Monitoring behavior and supervisory control. In: *Handbook of Perception and Human Performance: Volume II: Cognitive Processes and Performance* (K. R. Boff, L. Kaufmann & J. P. Thomas, Eds.). New York: John Wiley & Sons.

Parasuraman, R. Molloy, R. & Singh, I. L. (1993). Performance consequences of automation induced complacency. *International Journal of Aviation Psychology*, 3(1), 1-23.

National Transportation Safety Board (1988). *Aircraft accident report: Northwest Airlines, Inc., McDonnel Douglas DC-9-82, N312RC, Detroit Metropolitan Wayne County Airport, August 16, 1987* (NTSB/AAR-99-05). Washington, D. C.: Author

National Transportation Safety Board (1990). *Aircraft accident report: US Air, Inc., Boeing 737-400, LaGuardia Airport, Flushing New York, September 20, 1989* (NTSB/AAR-90-03). Washington, D. C.: Author

Ntuen, C. A. & Park, E. H. (1988). Human factors issues in teleoperated systems. In: *Ergonomics of Hybrid Automated Systems I* (W. Karwowski, H. R. Parsaei & M. R. Wilhelm, Eds.), pp. 230-210. Amsterdam: Elsevier.

Tulga, M. K. & Sheridan, T. B. (1980). Dynamic Decisions and Work Load in Multitask Supervisory Control. *IEEE Transactions on Systems, Man, and Cybernetics*, Vol. SMC-10, No. 5.

Wickens, C. D. (1992). *Engineering psychology and human performance* (2nd ed.). New York: Harper Collins.

Wickens, C. D. & Kessel, C. (1981). Failure detection in dynamic systems. In: *Human Detection and Diagnosis of System Failures* (J. Rasmussen & W. B. Rouse, Eds.). New York: Plenum.

Woods, D. D., O'Brien, J. F. & Hanes, L. F. (1987). Human factors challenges in process control: The case of nuclear power plants. In: *Handbook of human factors* (G. Salvendy, Ed.), pp. 1724-1770. New York: Wiley.

Young, L. (1969). On adaptive manual control. *IEEE Transactions on Man-Machine Systems*, MMS-10, 292-331.

BIOGRAPHICAL SKETCH

DAVID KABER is currently an assistant professor of industrial engineering at Mississippi State University. He has been working on issues related to dynamic process and automated systems control for the past 3 years, specifically out-of-the-loop performance and level of automation in the domains of advanced manufacturing, air traffic control, aircraft piloting and teleoperations. Prior to joining Mississippi State in 1996, he held research assistantships sponsored by the Society of Manufacturing Engineers and NASA (Kennedy Space Center), and a fellowship through the U.S. Air Force with areas of study including expert systems in automation, ergonomic equipment design for maintenance tasks and levels of automation in aviation related activities. He received his Ph.D. in industrial engineering from Texas Tech University, with a specialization in human factors and ergonomics, and both a B.S. and M.S. in industrial engineering from the University of Central Florida. He has authored scientific articles and reports on numerous subjects including interface design for teleoperation systems, the role of telepresence in teleoperation performance, the effect of cognitive workload on human performance with complex control systems, and real-time system safety. He is a member of Alpha Pi Mu and an affiliate of Sigma Xi, as well as, a member of several technical societies including the Human Factors and Ergonomics Society, and The Ergonomics Society.

PAPER FOR 31ST ANNUAL LOSS PREVENTION SYMPOSIUM

American Institute of Chemical Engineers
1997 National Spring Meeting
Houston, Texas, USA : March 10-13, 1997

Session on Control Systems for Unit Operation Safety

Title: Computer Integrated Manufacturing in Batch Chemical Processes - UK Perspective

Alun James, Operations Manager
Eastman Kodak Company
Synthetic Chemicals Department, Kodak Limited
Acornfield Road, Knowsley Industrial Park (North)
Merseyside, L33 7UF England

ABSTRACT

Kodak Limited operates a multi-product multi-process chemical batch facility near Liverpool. Computer Integrated Manufacturing (CIM) has been used to link the business level of MRPII through the Manufacturing Execution System (MES) of finite capacity scheduling, Management Information Systems (MIS) and Laboratory Information Management System (LIMS) to the SCADA system for control of the plant, together with acquisition and analysis of the data for process verification.

Mass balances are performed so that predicted overloads of certain waste streams on environment control resources can be properly managed. The plant which consists of a number of glass lined and stainless steel reactors, weigh tanks, hydrogenator, filter dryers, rotary dryers, centrifuges, solvent recovery plant and other ancillary equipment, is controlled by an ABB Taylor computer incorporating a sequence and recipe package within which the phases have been custom designed. A hierarchy of safety controls and alarms have been implemented through the facility. A combination of software checking and physical line testing is used to safely control the transfer of all material from any equipment to any other equipment or for disposal.

Emerging issues on the use of programmable electronic systems (PES) in safety related applications within the UK will be outlined.

INTRODUCTION

Kodak Limited is a wholly owned subsidiary of the Eastman Kodak Company that operates a multi-product multi-process chemical batch plant at Kirkby near Liverpool, England. The facility manufactures, via multi-stage routes, various photographic chemicals which are shipped throughout the world to Kodak locations for eventual coating onto paper or film. The Kirkby site also produces photo-processing chemicals and photocopier toner, and has about 250 employees. The site has achieved ISO 9001, MRPII Class A and IIP certification.

SYNTHETIC CHEMICALS DEPARTMENT

The Synthetic Chemicals Department plant, which produces the photographic chemicals, consists of a number of 2000 gallon glass lined and 3000 gallon stainless steel reactor systems with accompanying weigh tanks for feeding reactive chemicals, a hydrogenator for performing catalytic reductions, automated pressure filter dryers, nutsches and centrifuges for product isolation, rotary vacuum dryers, a solvent tank farm with a solvent recovery unit and various other ancillary equipment. The whole plant is controlled by an ABB Taylor central processing computer incorporating the usual alarm, messaging, trending and graphics functions. A sequence and recipe package has been incorporated within which the various phases have been custom designed. These phases appear to the operator as a mnemonic of up to six alpha characters such as START, SOLVAD or DIST, with up to four parameters for performing functions such as solvent or chemical additions, distillations, hydrogenations, layer separations, drying and so forth. The plant operates a four-shift system 24 hours a day 7 days a week with a Christmas period shutdown. The valves on the plant, including control values, are almost entirely ON/OFF devices without feedback.. The I/O consists of about 2000 digital inputs and outputs with about 200 control loops. The amount of equipment residing in the plant, together with the corresponding I/O, is likely to be increased by 50% over the next three to five years to fill the existing building.

MANAGEMENT SYSTEMS FOR LOSS PREVENTION

Inherent in the running of the operation are several management systems aimed at ensuring that the design of plant equipment, the control systems and manufacturing processes are the best available without incurring excessive cost. All manufacturing processes undergo a critical examination on a three year cycle by the Process Safety Review Committee to ensure that the best current practices are being used. HAZOP studies are done on all newly installed equipment, on all new manufacturing processes and on selected established processes that are deemed to be of a hazardous nature. Risk assessment studies are performed on key tasks to minimise any hazards. All changes to the plant equipment or control software, however large or small, prior to installation are first approved by the Plant Modification Review Committee. Many additional regulatory checks, such as Manual Handling, Pressure Regulations, COSHH and IPC are routinely performed to ensure compliancy. Regular Health, Safety and Environment audits are performed by people ranging from operators to senior management level, all of which set and maintain a high degree of safety culture in the organisation.

COMPUTER INTEGRATED MANUFACTURING

The business processes according to Manufacturing Resource Planning (MRPII) for planning, controlling and costing all resources are managed using the MRPII software product called AMAPS. The production plant is controlled by a Taylor central process control computer and a SCADA system has been installed for the collection of plant data via an Allen Bradley PLC and PC package. This data logging programme, called EDNA (Event Driven Numerical Acquisition) is a very powerful event driven replacement for the traditional analogue strip chart recorder and collects data in digital form as a binary file. A set of software on DOS or Windows platform called KDAT (Kodak Data Analysis Tool) is used to analyse the data collected from one or more sensors during the manufacturing process. Data from several batches can be accessed simultaneously, synchronised with respect to events or overlaid on the same graph with zooming facility. It has a built in set of statistical functions so that rigorous comparisons between batches and between different equipment can readily be made. Calculated results or plant data can be saved in the form of text or graphic files. More advanced tools are also available to progress from univariate (one variable at a time) analysis to multivariate techniques, in order to sift through the huge amount of information available, and to pick out what is important so that it can be translated into useful pictures and graphs.

In order to bridge the rift between the planning and shop floor systems, an intermediate Manufacturing Execution System (MES) has been developed, and this is composed of several facets residing on both VAX VMS and IBM OS/2 platforms. A finite capacity scheduling system (PROVISA) imports the orders and current stock details from the MRPII system and, after scheduling the works orders to the plant work stations over a given horizon using scheduling and user defined rules, the firm planned orders are returned to AMAPS to give a fixed executable plan. When ready to release the works orders for manufacture onto the plant process control computer, the details of the materials and their quantities are down-loaded through the VAX computer to the Taylor. In turn, the status of each work order on the Taylor computer can be relayed to PROVISA to inform production supervision of the relative priorities considering the available resources of equipment and labour.

A wealth of real time information resides within the Taylor computer, and this can be exploited since text messages carrying data is captured in files on either the VAX or PC. This Management Information System (MIS) can be analysed on the VAX computer using Datatrieve, or on PCs using various PC editors. The root cause of any deviations that could result in various types of losses can be speedily identified and permanent fixes implemented.

It is also a valuable tool for improving and optimising plant processes, and for checking on various safety features of the operations. Records of the results of analytical tests for raw materials, in-process samples and final products are stored on the VAX computer using VGLIMS, and these can be used in conjunction with process data for both trouble-shooting and continuous improvement work.

SAFETY AND LOSS PREVENTION

Loss in the context of this paper can refer to physical losses of chemical materials and plant equipment, or to non-physical losses such as safety and time. Numerous design features and control techniques have been engineered onto the plant to ensure the highest standards of safety. Waste stream losses are minimised on a continuous improvement basis and the drive to reduce cycle times and improve quality is never ending. Whenever a deviation is detected on a particular control loop, the first line of actions is a setpoint high or low deviation, or a low level alarm to warn the operator so that he may investigate and possibly take some remedial action. In some cases a hard-wired trip may activate on an annunciator panel requiring the operator to react and acknowledge the alarm before proceeding. The next level of protection involves independent background software programmes that monitor each relevant loop within certain global or process specific limits, and whenever violated, would inform the control room operator and possibly take executive action such as shutting down part of the process, or running the automated emergency stop procedure for that particular vessel. The boiling point of the mixture being processed is continually updated in the computer, so temperature violations during the multi product processing are swiftly and correctly managed. The ultimate safety of the plant relies on the computer independent hard-wired trips and emergency venting systems to catch pots on the roof. Total computer failure, loss of electric power, compressed air or a fire alarm trip causes the entire plant to revert to a pre-determined fail-safe status. After investigation and correction, continuation of the production processes can then be initiated by supervision.

All transfers of materials by pipework around the plant, whatever the vessel type, are controlled by the same general principle. Firstly, a software handshake check is made between the two respective vessels. These checks ensure that the correct works order or batch number resides on each vessel, that the correct material identity is indicated on both vessels and that it ties in with the details on the electronic batch loading record, and that each vessel is at the correct step and stage in their respective sequences. Any violation of these checks results in the appropriate error message and continuation of the process is prevented. Secondly, a physical line pressure test is performed whereby the computer sets up the correct route by opening all but the end valves along the route and closing all the valves that branch off the line. The computer pressurises the complete lines with nitrogen from the source vessel, performs a line pressure and leak test by checking a pressure switch, and finally ensures that the line pressure has been released after opening the valve at the point of destination. In 15 years of operation only one misoperation of this type has been experienced. This was caused by a plant modification which resulted in a nitrogen line being connected at the wrong end of a valve on a process line and part of a batch was lost to drain. The same principle described above applies to the transfer of waste solvents for recovery and to water layer separations from reactors to the drain leading to the waste water treatment plant.

During the separation of two layers in a reactor, such as the discharge to drain of a lower aqueous wash from a hydrocarbon process layer, the computer continually monitors the weight of the reactor and the conductivity of the liquid being separated. Whenever the mass discharged exceeds 15% of the expected weight as given by a function parameter, the computer shuts down the separation, and only allows continuation under manual control. Monitoring is however continued, and in the rare event of 25% of the expected weight being run off, the separation process is again shut down and supervision or technical input is required before continuing the process of separation.

Yet another loss prevention technique revolves around the MRPII AMAPS system to plan and control various plant liquid and gaseous streams to ensure that the caustic scrubbers, solvent recovery and waste water treatment plants will not be over-loaded. Mass balances are performed on all processes and checks are made on waste solvents, waste water, sulphite and sulphate levels as well as chemical oxygen demand (COD). Any foreseen problems of overloading can be readily managed before such an event occurs by simple re-scheduling, slowing down or diluting waste streams.

A plant simulator consisting of a suite of software programmes and a mimic panel representing part of the plant is used to develop software. This can also be used to train control room operators in order to optimise control of the processes.

UK INITIATIVES ON IEC 1508

Significant progress has been made over the past two years in making many large UK companies in the chemical manufacturing sector aware of the impending application and implications of the International Electrotechnical Committee standard IEC 1508. As well as end users, the whole supply chain from the manufacturers of components and control systems to system integrators and implementors have been involved. Programmable Electronic Systems (PES) are being increasingly used in a wide variety of applications in order to exploit their advantages of performance, flexibility and low cost. In particular PESs form part of many systems which involve hazardous processes such as:

Nuclear Power Stations
Chemical and Pharmaceutical Plants
Avionics
Railways

In nearly all sectors, businesses face an unremitting pressure to reduce costs coupled with a need for ever-quicker time to market of shorter life cycle products, and a pressure to maximise the use of capital assets. In addition, there is an increasing regulatory and social requirement for safety and reliability. Businesses are rarely working with perfect data and judgements have to be made. There are penalties for taking too little or too much risk, and the organisations which prosper are those who get their risk-taking "*right*". The use of PESs in safety-related systems needs particular care to ensure that new hazards are not introduced into the overall system. The financial benefits of safe operation are difficult to quantify, but the cost of failure can be catastrophic. However, improved methods of conformity, assessment and certification can reduce development costs while at the same time providing increased system dependability.

A two year project supported by the British Department of Trade and Industry under its Safety-Critical Systems Advanced Technology Programme called FRESCO (Framework for the Evaluation of Safety-Critical Systems) was partnered by Admiral Management Services, ICI Eutech Engineering Solutions, Lloyd's Register, Eversheds Insurance, Civil Aviation Authority and Railtrack, together with many chemical companies and suppliers of electronic components and systems. The project was aimed to establish UK industry requirements for the assessment and certification of safety related software which enables conformance to IEC 1508. This international standard is likely to be published in mid 1997, and covers those aspects that need to be addressed when electrical/electronic/programmable electronic systems (E/E/PES) are used to carry out safety functions or are part of safety-related systems. The standard which has been developed as a generic standard to cover all sectors of industry consists of seven parts:

Part 1	General Requirements
Part 2	Requirements for E/E/PES
Part 3	Software Requirements
Part 4	Definitions
Part 5	Guidelines on the Application of Part 1
Part 6	Guidelines on the Application of Parts 2 and 3
Part 7	Bibliography of Techniques

There are four key concepts within this standard summarised as follows:

The Safety Lifecycle

This structured approach begins with a clear definition of the equipment and processes for which functional safety is being sought. A logical path is provided through hazard identification, risk assessment, risk reduction requirements and the design of protective systems through to commissioning, operation, maintenance and finally decommissioning. Concepts of forward planning, verification and validation are integrated into the Safety Lifecycle.

Safety Management

The standard sets a check-list for those items which need to be in place in order to prepare for and to manage each phase of the Safety Lifecycle, and these are incorporated into a formal Safety Plan. Roles and responsibilities of individuals and organisations, together with their competencies have to be considered.

Design of Safety-related Systems

The standard requires that the safety-related system is designed to meet the specific risk criteria or safety integrity level (SIL). Comprehensive technical guidance is given for design measures and techniques which are appropriate to achieving the specified SIL by the system especially in the application of modern PESs.

Competencies

Whilst the standard does not attempt to detail all the skills and competencies required, it provides guidance on a number of competency factors which need to be addressed, including knowledge of safety engineering principles, regulation and legal issues, qualifications, experience and novelty of design.

Whilst it will not be a mandatory requirement to comply, the fact that it is an international standard will almost certainly mean that it will become the de-facto safety standard applicable to safety related systems. Many suppliers are currently reviewing their practices, procedures and life-cycles to ensure compliance. Several end-users are already specifying to suppliers, design contractors and system integrators the requirement to provide systems that would comply to the standard. In the UK the Health and Safety Executive are beginning to raise comments related to IEC 1508 as they visit chemical companies.

BIOGRAPHICAL SKETCH

Alun James, BSc., PhD (Member of BPICS)

After graduating in Chemistry and obtaining a PhD in Organic Chemistry at Cardiff University, Alun attended Bristol University for a Research Fellowship in Physical Chemistry. He joined Kodak Limited R&D laboratories at Liverpool before moving to production as the Technical Supervisor. He was heavily involved in computerising the batch processes in a new building that was commissioned in 1982, and was the Project Leader for implementing MRPII in 1992. Alongside being the Operations Manager for the production facility, time was found to interface the various computerised systems to give a fully integrated computerised business facility.

Software Deviation Analysis: A "Safeware" Technique*

Jon Damon Reese and Nancy G. Leveson

Dept. of C.S.E.
University of Washington
Box 352350
Seattle, WA 98195, U.S.A.

Safeware Engineering Corp.
7200 Lower Ridge, Unit B
Everett, WA 98203, U.S.A.

{jdreese,leveson}@cs.washington.edu
{jdreese,leveson}@safeware-eng.com

Abstract

Standard safety analysis techniques are often ineffective when computers and digital devices are integrated into plant control. The *Safeware* methodology and its set of supporting safety analysis techniques (and prototype tools) includes modeling and hazard analysis of complex systems where the components may be a mixture of humans, hardware, and software. This paper describes one of the Safeware hazard analysis techniques, Software Deviation Analysis, that incorporates the beneficial features of HAZOP (such as guidewords, deviations, exploratory analysis, and a systems engineering approach) into an automated procedure that is capable of handling the complexity and logical nature of computer software.

*This work was partly funded by NASA/Langley Grant NAG-1-1495, NSF Grant CCR-9396181, and the California PATH Program of the University of California, in cooperation with the California Department of Transportation and the U.S. Department of Transportation.

1 Introduction

The introduction of computer control in plants has created new and unsolved problems in ensuring safety. For the past 16 years, Leveson and students have been studying ways to extend and adapt to software the methods used to control risk in the larger system within which the software is embedded. Although engineers have developed various types of hazard analysis techniques for electromechanical systems, these techniques do not apply when computers are introduced to control dangerous and complex systems. Our goal is to take the basic procedures of system hazard analysis and to translate them into techniques and tools that can be applied to systems containing software and to the software development and validation process.

This work has resulted in an approach, called Safeware, to enhance safety in systems composed of electromechanical, computer, and human components. The basic methodology involves applying software hazard analysis and hazard control procedures throughout software development, based on the identified system hazards [2]. These efforts are closely tied to the system level hazard analysis and control. Early and continuing analysis procedures guide and direct the software as it is developed instead of simply attempting to verify safety after the software is completed. The methodology includes a variety of analysis techniques and tools (see Figure 1). This paper describes one technique, Software Deviation Analysis, that extends basic HAZOP-like analysis to systems containing computer components.

2 Software Deviation Analysis

HAZard and OPerability analysis (HAZOP), a review procedure developed for the British chemical industry in the 1950's, is used widely but it has several limitations when applied to newer, high-technology systems. First, it is time- and labor-intensive [2], in large part due to its reliance on group discussions and manual analysis procedures. Second, HAZOP analyzes causes and effects with respect to deviations from expected behavior, but it does not analyze whether the design, under normal operating conditions, yields expected behavior or if the expected behavior is what is desired.

A third limitation arises from the fact that HAZOP is a flow-based analysis. Deviations from within components or processes are not inspected directly; instead, a deviation within a component (as well as a human error or other environmental disturbance) is assumed to be manifested as a disturbed flow [4]. A purely flow-oriented approach may cause the analyst to neglect process-related malfunctions and hazards in favor of pipe-related causes and effects.

Because HAZOP concentrates on physical properties of the system [4], it is not directly applicable to analyzing computer input and output. Several manual techniques have been suggested to extend HAZOP to incorporate inspection of computer hardware and software. In each of these, the procedure is essentially identical to a standard manual HAZOP except that the guide-words are changed and the model of the system

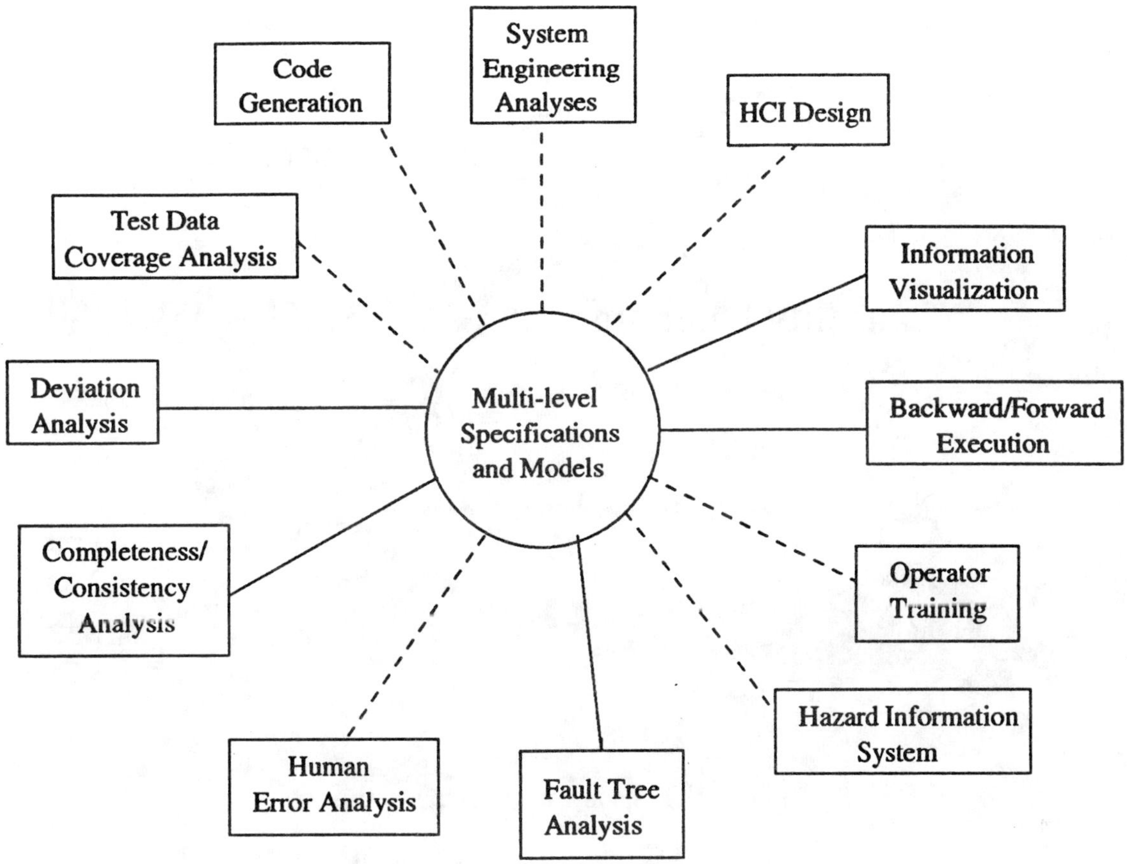

Figure 1: Components and interactions in a CAD environment for safety-critical systems.

may differ from the original pipe-and-process diagram.

These methods suffer from two weaknesses with respect to analyzing software. First, being manual techniques they depend on human understanding of the proposed software, which can be quite limited. Whereas the components of a pipe-and-process diagram usually conform to straightforward and well-understood laws, each instance of a software controller can have a complex and novel behavior: The behavior of pumps and valves is not nearly as complex as software can potentially be.

Second, the manual techniques adhere to the HAZOP principle of identifying deviations in the connections, i.e., the computer inputs and outputs only. Accordingly, they do not provide guidance for following deviations into the control logic. The task of determining how a deviation in a software input is manifested at its outputs is left wholly up to the analyst.

Software Deviation Analysis overcomes some of these deficiencies. Like HAZOP, Software Deviation Analysis is based on the underlying model that accidents are caused by deviations in system parameters. Using a blackbox software or system requirements

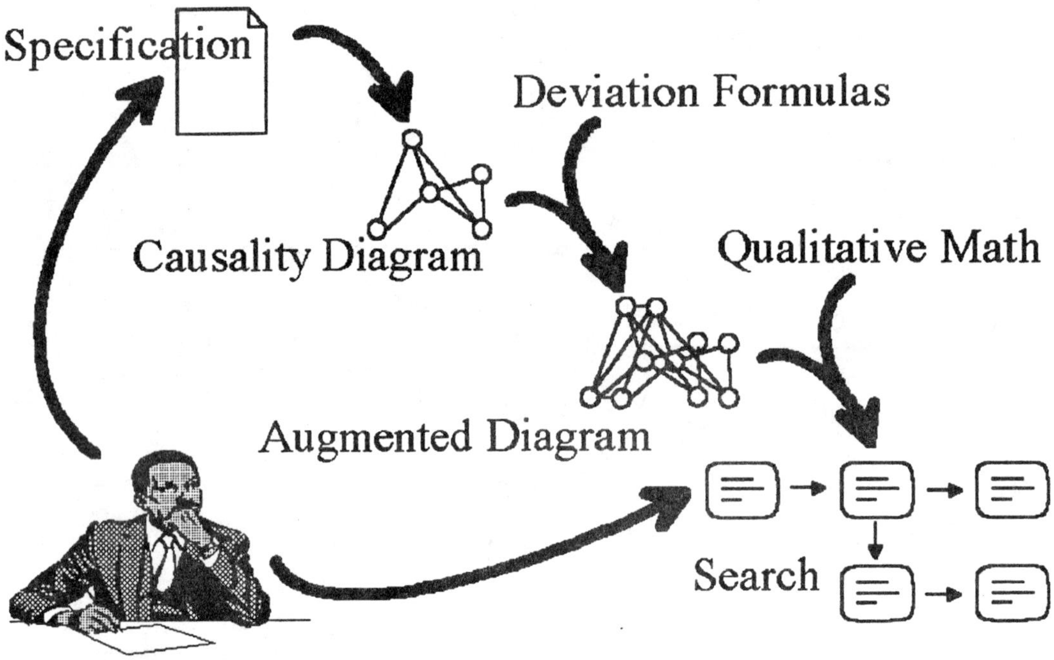

Figure 2: Overview of SDA procedure.

specification, the analyst provides assumptions about particular deviations in software inputs and hazardous states or outputs, and the procedure automatically generates scenarios in which the analyst's assumptions lead to the specified deviations in the outputs.

Figure 2 shows an overview of the SDA procedure. The analyst provides a formal software requirements specification, which the procedure automatically converts into a more basic representation, called a *causality diagram.* The causality diagram is an internal data structure that encodes causal information between system variables, based on the specification and the semantics of the language in which it is written. The simplicity of causality diagrams makes the search algorithm more straightforward and easier to adapt to a new specification or programming language. Causality diagrams may also be helpful to the analyst in understanding how system variables are interrelated.

The automated procedure uses *deviation formulas*, which define how deviations are related. This information is incorporated directly into the causality diagram to create an *augmented* causality diagram. SDA then uses qualitative mathematics on the augmented causality diagram to evaluate deviations. Qualitative mathematics partitions infinite domains into a small set of intervals and provides mathematical operations on these intervals. The use of fixed intervals simplifies the analysis compared to iterations over the entire state space. It also lends itself naturally to the qualitative nature of deviations, such as "slightly too high."

The augmented causality diagram, input deviations, and a list of safety-critical software outputs is passed to the search program, which constructs a tree of states. The state formed by the input deviations is the root of the search tree. Leaves are either deadend searches (in which the state does not contain any deviations) or states containing safety-critical deviations.

The output of the SDA program is a list of *scenarios*. A scenario is a set of deviations in the software inputs plus constraints on the execution states of the software that are sufficient to lead to a deviation in a safety-critical software output. The deviation analysis procedure can optionally add further deviations as it constrains the software state, allowing for the analysis of multiple independent failures (leading to the independent deviations.)

The next sections describe the components of deviation analysis—causality diagrams, deviation formulas, qualitative mathematics, and the analysis procedure. Section 7 provides a simple example.

3 Causality Diagrams

Figure 3 shows an example of a simple feedback system and the corresponding causality diagram. A requirements specification for the software controller contains a black-box model of the relationship (function) between inputs (the measured system variables) and outputs (commands to change the controlled system variables).

The system shown in the figure is a tank equipped with a variable-aperture valve. The system variables are the tank pressure, the flow of material through the tank, and the aperture of the valve. To simplify the example, pressure is computed as the quotient of flow over aperture. The controller increases or decreases the valve opening by ten units if the pressure is above the maximum of 250 units or below the minimum of 100 units, respectively.

The causality diagram in the example contains twelve nodes and sixteen edges. Three of the nodes represent the system variables. *Pressure* is a quotient function, with the numerator edge originating from *Flow* and the denominator edge originating from *Aperture*. *Aperture* is an interesting node in that its current value depends on its previous value, i.e., it has state. The size of the valve aperture is equal to its previous value (indicated by a dashed line) plus one of $\{-10, 0, 10\}$, as provided by the controller.

The behavior of the *Flow* variable is left unspecified. The procedure treats unspecified variables as random inputs to the system and constrains them as needed to propagate deviations.

The remaining nodes comprise the controller (which for process-control software would be a computer, usually performing a much more complex function.) The pres-

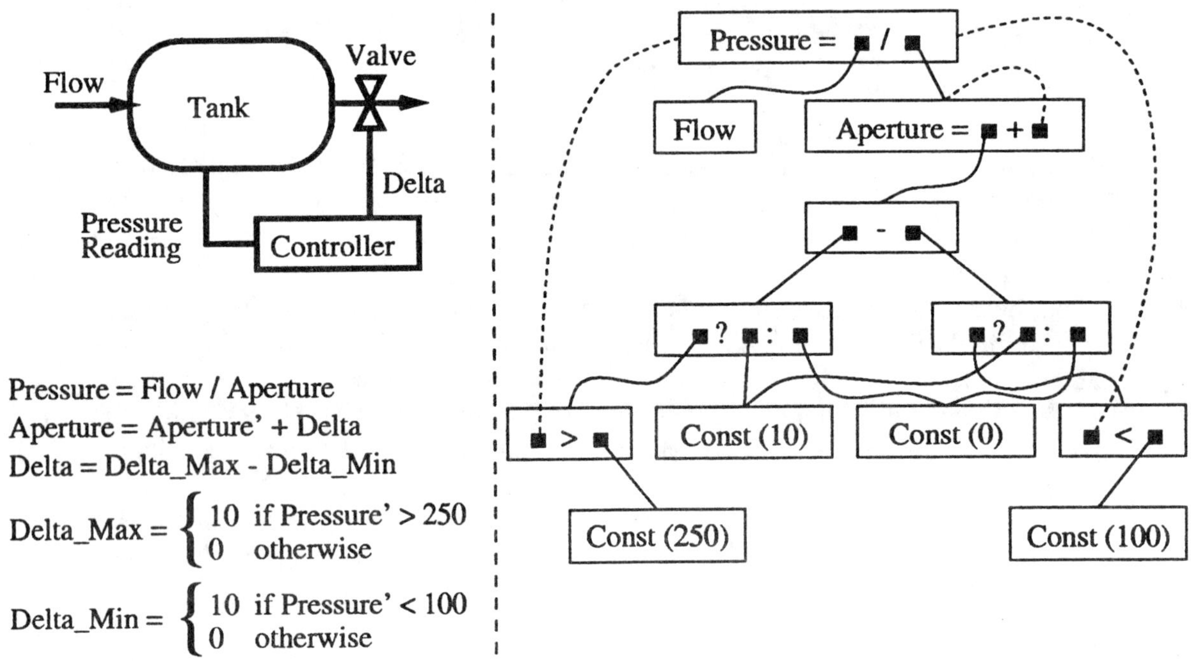

Figure 3: Causality diagram example

sure reading is compared to minimum and maximum values (the "<" and ">" nodes, respectively). Note that each node is a function: For example, the domain of the inequality functions is a pair of numbers and the range is a boolean.

The nodes represented by [□ ? □ : □] are called *selection nodes*. The selection function is defined as follows:

$$b \; ? \; x : y = \begin{cases} x & \text{if } b \\ y & \text{if } \neg b \end{cases}$$

The selection function is an important node in constructing the causal relationships of process-control software, since it maps from a boolean value (e.g., some control decision) to numeric values (e.g., output to an actuator.)

Following the edges from the subtraction node (the output of the controller) backward to the pressure reading, one gets the expression

$$\left\{ \begin{array}{ll} 10 & \text{if } \mathit{Pressure}' > 250 \\ 0 & \text{otherwise} \end{array} \right\} - \left\{ \begin{array}{ll} 10 & \text{if } \mathit{Pressure}' < 100 \\ 0 & \text{otherwise} \end{array} \right\}$$

where *Pressure'* represents the previous value for pressure. This value is output to the valve actuator.

The causality diagram can get complex, but it is generated automatically from a state machine specification. Each node in the specification is linked to its corresponding node in the causality diagram, so that the results of the analysis can be translated from the causality diagram back into the language of the specification, avoiding the need for the analyst to comprehend or even see the causality diagram.

4 Deviation Formulas and Augmented Diagrams

The automated procedure uses *deviation formulas*, which define how deviations are related. This information is incorporated directly into the causality diagram to create an *augmented* causality diagram.

The concept of a deviation needs to be defined both for logical and numeric variables. Booleans are straightforward since they can only take two values; consequently, an actual value is either a deviation from the correct value or it is not. The "exclusive OR" operator satisfies this definition. *Actual* ⊕ *Correct* is *true* (a deviation) when *Actual* is different from *Correct* and *false* when they are the same.

A numeric deviation is defined as the difference between the actual and correct values, i.e., the amount added to or subtracted from the correct value to obtain the actual value:

$$X_d = X_a - X_c,$$

where X is the variable, and the subscripts indicate deviation, actual, and correct values, respectively.[1] For example, if a pressure reading should be 10 p.s.i. but is actually 7 p.s.i., then the deviation is -3 p.s.i.

To relate these values back to the causality diagram, one could assign some correct values and use the relationships expressed in the causality diagram to derive other correct values. Actual values can be derived from other actual values in the same way. For example, $\texttt{Pressure}_a = \texttt{Flow}_a/\texttt{Aperture}_a$.

Deviation values cannot be calculated using the normal relationships between system variables: $\texttt{Pressure}_d$ is not always equal to $\texttt{Flow}_d/\texttt{Aperture}_d$. (In fact, if the actual value of `Aperture` is correct, then $\texttt{Aperture}_d = 0$ and this ratio is undefined.) The causality diagram must be augmented with deviation formulas so that the relationships between deviations are explicitly and properly represented. A deviation formula is the way by which the deviations of a function may be determined from the deviations and actual values of its inputs.

A complete set of deviation formulas and their derivations can be found in [3]. The augmented causality diagram of the tank example is rather larger than the original diagram, and the reader will not be burdened with a complete example, but the fragment representing $\texttt{Pressure}_d$ is shown in Figure 4.

[1]Note that the deviation could be calculated in other ways. For example, X_d could be the ratio $\frac{X_a}{X_c}$. Under this definition, a value of $X_d = -0.5$ would mean that X_a has the opposite sign of and one-half the magnitude of X_c. While this formula is quite useful, X_d does not have a value when $X_c = 0$ and it is virtually meaningless when $X_a = 0$.

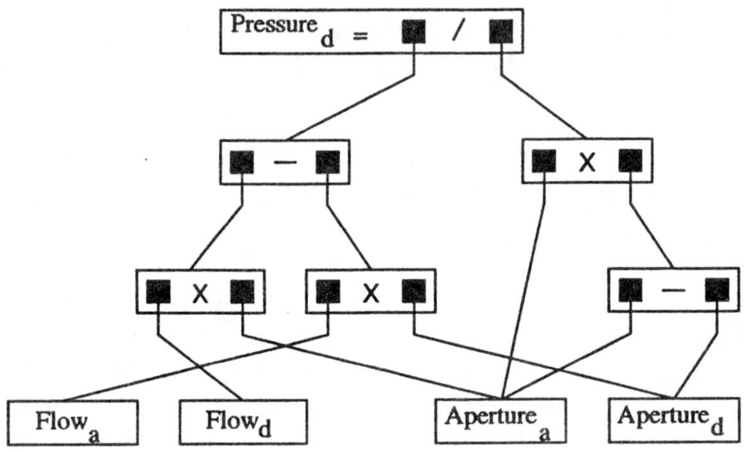

Figure 4: A fragment of the augmented causality diagram for the example, showing the causal relationship between deviation in pressure and the deviations and actual values of flow and valve aperture.

5 Qualitative Mathematics

SDA uses qualitative mathematics on the augmented causality diagram to evaluate deviations. Qualitative mathematics partitions infinite domains into a small set of intervals and provides mathematical operations on these intervals. The use of fixed intervals simplifies the analysis compared to iterations over the entire state space. It also lends itself naturally to the qualitative nature of deviations, such as "slightly too high." The analyst may choose to refer to qualitative set elements, to the intervals they represent, or to symbolic labels such as "a little low," "too high," or "very high."

In a manual HAZOP, analysts need to determine the result of a particular deviation. To do this, they investigate what will definitely occur given the input deviation as well as what else could occur under specific conditions—i.e., they make assumptions about the system state. They must also back-track occasionally to determine whether two separate assumptions are consistent, i.e., whether the scenario they are proposing is realistic. These three types of manual search are also included in the SDA procedure as three functions for each operator appearing on causality diagram nodes: a *forward definite* function, a *backward definite* function, and a *forward assumptive* function.

The forward definite function is simply the operator itself, but applied to sets of intervals. Allowing operations over sets of intervals rather than single intervals greatly reduces the size of the search tree, since a separate branch does not have to be created for each combination of values.

The forward assumptive function is used only on deviation nodes (the nodes created for the augmented diagram). If one of the inputs to a node is a deviation but the value of the node itself is unknown, then the forward assumptive function attempts to find values for the other inputs that will cause the deviation to propagate to the node. For example, suppose that a deviation node is the product of two other nodes, one of which is too high and the other is unknown. In order to propagate the high value, the unknown input is assumed to be negative or positive, but not zero, because a zero will cause the output to be zero, masking the deviation. With the new constraint the output is now either too high or too low.

The backward definite function is essentially the inverse relation of a node's operator. See Reese [3] for definitions of these relations.

The augmented causality diagram, input deviations, and list of safety-critical variables is passed to the search algorithm, which constructs a tree of states. The state formed by the input deviations is the root of the search tree. Leaves are either dead-end searches (in which a state does not contain any deviations) or states containing safety-critical deviations. The output of the SDA procedure is a list of the paths from the root state to all leaves with safety-critical output deviations.

6 Analysis Procedure

The SDA procedure is a forward search procedure—it starts with a deviation in the software's input environment and attempts to find ways in which the deviation can lead to hazardous software outputs. As discussed in the overview, the analysis starts from a system specification, which is converted automatically into an augmented causality diagram. The analyst provides two other pieces of information corresponding to the starting and ending points of the search: (1) an initial system state, including at least one deviation, and (2) the outputs that are safety-critical. The procedure searches forward from the initial state, attempting to find states in which a safety-critical output deviation occurs.

The search procedure is quite complex, and we can only describe it briefly here. The forward and backward definite functions described earlier are used to construct a chain of states representing what will definitely result from the initial state. This chain of states is termed a *scenario* because it describes a sequence of events that the system will follow given the initial state. The chain begins with the initial state and terminates with a state that either contains a safety-critical deviation or no deviation at all. A final state that contains a safety-critical deviation indicates that the analyst's input deviations will always result in a hazardous deviation.

Whether or not the chain leads to a hazardous deviation, the procedure can continue the search by constraining the software state using the forward assumptive function described earlier. Constraints can be added not only to the initial state but to every state in the chain. The forward and backward definite functions are applied to these additional constraints to create another chain of states. The new chain branches from the state to which the constraints were added and ends in either a safety-critical deviation or a dead-end.

Further constraints can be added to the new state chains. The analysis procedure continues in a breadth-first manner, building a tree of state chains, each ending in either a dead-end or hazardous deviation. The depth of each leaf of the tree corresponds to the number of additional assumptions the procedure made to reach that leaf. The procedure finishes when it either runs out of constraints that it can make or the depth reaches some predefined limit set by the analyst. Finally, the procedure provides the analyst with scenarios by tracing each path from the initial state to all ending states that contain hazardous deviations.

SDA has been applied to three real-world examples. It was first applied to the Traffic Alert and Collision Avoidance System II (TCAS II), an avionics system designed to provide pilots with collision avoidance information. The authors found that when an incorrect identifier is received by TCAS (perhaps as a result of a transmission error) there are circumstances in which an evasive maneuver is not displayed to the pilot when it should be. The procedure has also been applied to a developmental aircraft guidance system and a proposed automated highway system with similar results. The time required to perform a search on these examples ranged from about 10 seconds to several minutes.

7 Example of the SDA Procedure

The basic procedure can be illustrated by a simple example from an actual project. First, the analyst provides a formal specification, which in this case is a proposed automated highway system for the California Department of Transportation[2]. The automated highway system (AHS) directs automobiles to form groups within a lane, called *platoons*. Each automobile has a software controller that directs the movement of the car relative to the platoons. Figure 5 shows a page from the AHS specification, written in Requirements State Machine Language (RSML) [1]. The relevant parts will be explained shortly.

Next, the analyst identifies the safety-critical outputs. For simplicity, in this example we will assume that all outputs are critical. The next step is to define the initial input deviations. One of the input variables listed at the top of the AHS specification is `Num_vehicles_in_platoon`, which is the number of automobiles that are in the same platoon as the controller's automobile. Suppose the analyst wishes to find out what happens when this input variable is less than the actual size of the platoon.

For this model, the SDA algorithm identifies two scenarios in which the input deviations lead to a safety-critical output deviation, one of which is presented here.[3]

The initial assumption, as stated above, is that `Num_vehicles_in_platoon` is too

[2]This research was supported by the Department of Transportation, and thus the models we have analyzed so far are not related to the chemical process industry. However, the technique is generally applicable to any control system software.

[3]The search for this example takes approximately 1.2 MB and 25 seconds on an Intel 80486DX2 at 66MHz.

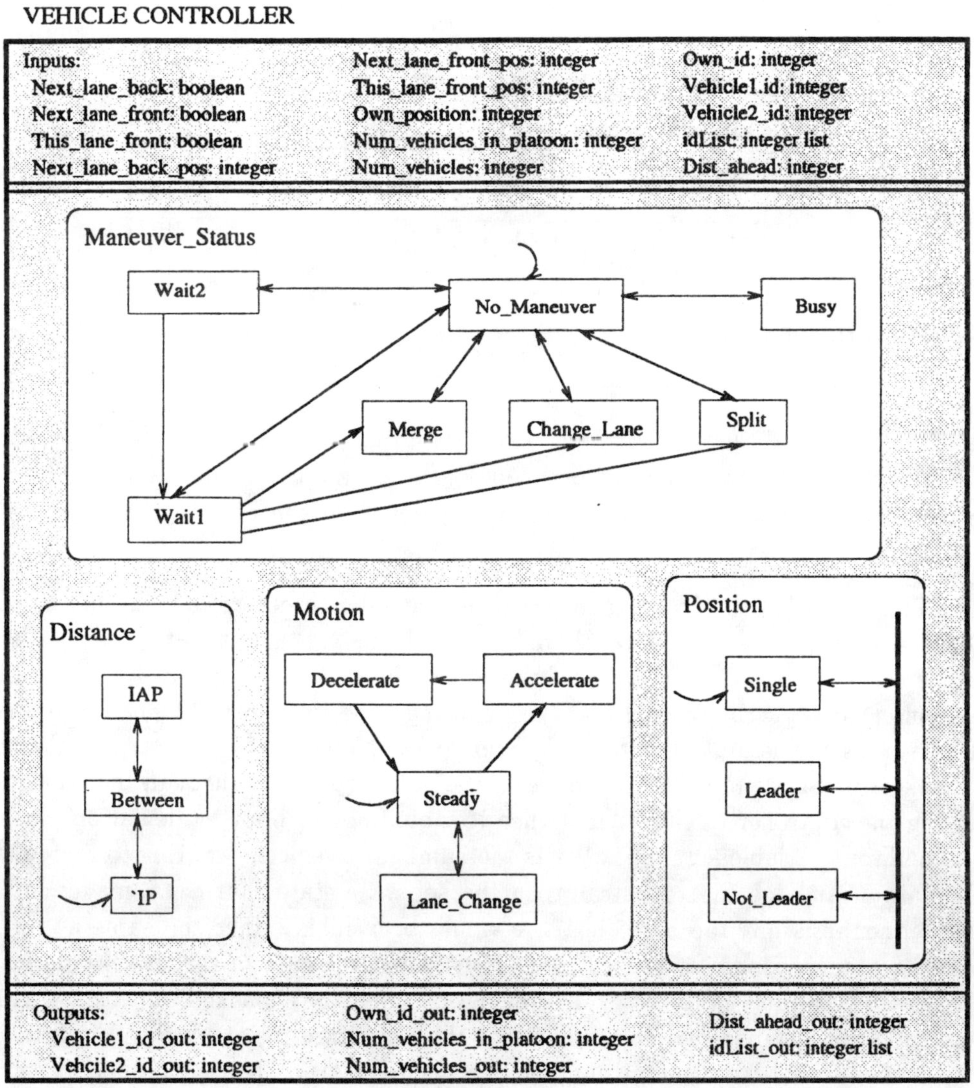

Figure 5: A portion of the Automated Highway System example.

Transition(s): No_Maneuver ⟶ Busy

Location: Maneuver_Status

Trigger Event: Rcvd_req_merge

Condition:

AND	OR	
Position **in state** Single	T	·
Position **in state** Leader	·	T
Distance **in state** IP	T	T
Motion **in state** Steady	T	T
Num_vehicles + Num_vehicles_in_platoon < MAX_VEHICLES_IN_PLATOON	T	T

Output Action: Merge_ok

Figure 6: A transition from the Automated Highway System. The analyst has assumed that the variable `Num_vehicles_in_platoon` is lower than the actual value.

low. This input is used by the automobile's controller to determine whether to commence a maneuver, specified by the transition from **`No_Maneuver`** to **`Busy`** (refer to Figure 6.) This transition is taken if the controller has been requested to merge two platoons (the triggering event) and the guarding condition on the transition permits the platoons to merge.[4] For this scenario, the SDA algorithm constrains the software execution state in order to propagate the initial deviation through the transition, causing the controller to enter **`Busy`** when it should not (a boolean deviation.)

The input variable **`Num_vehicles`** is the number of vehicles wanting to merge with the platoon and the first constraint on the software state that SDA makes is that **`Num_vehicles`** is not too high, i.e., the value received is either the same as or less than the proper value. This constraint ensures that the sum of the two inputs in the fifth row of the table is too low. The deviation could be "masked" if **`Num_vehicles`** were too high. The second constraint that the algorithm makes is that the fifth row is true, namely, the maximum number of vehicles has not been exceeded based on the information provided to the software. The third constraint is that the sum of deviations for **`Num_vehicles_in_platoon`** and **`Num_vehicles`** exceeds the number of empty positions left in the platoon. In other words, the fifth row should be false, which it would be if the deviations were not present. The algorithm presents this logic

[4] The condition is represented by an AND/OR table, which is true if any of its columns is true. A column is true if all of the rows that have a "T" are true and all of the rows with an "F" are false.

in the following way (rearranged slightly for clarity):

```
   MAX_VEHICLES_IN_PLATOON                     -dev(Num_vehicles)
     -value(Num_vehicles)          <     -dev(Num_vehicles_in_platoon)
-value(Num_vehicles_in_platoon)
```

where `value()` is the value read by the controller and `dev()` is the difference between the actual and correct values (negative means too low.) The left-hand side of the inequality is the number of spaces available according to the two inputs read by the controller (`Num_vehicles` and `Num_vehicles_in_platoon`.) The right-hand side of the inequality is the size of the error (the deviation is negative, so the right-hand side is positive.) The inequality therefore shows that the deviation is greater than the number of spaces perceived to be available, and row five would be false if there had been no deviation in the inputs, inhibiting the transition.

The fourth constraint that the algorithm generates is that (1) the controller's automobile is the lead vehicle in the platoon, (2) the distance between the platoons is sufficient, and (3) the automobile is traveling at a constant rate of speed. These are the conditions in rows two through four, respectively, making the second column and the entire condition true.

The final constraint generated by the algorithm is that the triggering event is true, i.e., that a request has been made to merge platoons and the controller is in state `No_Maneuver`. The transition is thus enabled and the output action `Merge_ok` is generated. Both the transition and output action are deviations since they should not have occurred. `Merge_ok` triggers the output that gives permission to merge platoons. The resulting system state is a platoon that exceeds the threshold for safe platoon size. Thus we have identified a scenario, i.e., deviations in one input plus constraints on the software execution state that will lead to a hazardous output.

8 Conclusion

This paper has described a hazard analysis technique, software deviation analysis, that incorporates the beneficial features of HAZOP (e.g., guide words, deviations, exploratory analysis, and a systems engineering strategy) into an automated procedure that is capable of handling the complexity and logical nature of computer software.

We have applied SDA to several realistic systems. Although more extensive experimentation needs to be performed, the procedure appears to provide information that is useful for requirements specification and review. Note that SDA is not intended to replace standard certification methods such as verification and validation. However, as an exploratory procedure, it provides important and timely information to the safety and software analysts.

References

[1] N. Leveson, M. Heimdahl, H. Hildreth, and J. Reese. Requirements specification for process-control systems. *IEEE Transactions on Software Engineering*, SE-20(9), September 1994.

[2] N. G. Leveson. *Safeware: System Safety and Computers.* Addison-Wesley Publishing Co., 1995.

[3] J. D. Reese. *Software Deviation Analysis.* PhD thesis, University of California, Irvine, 1996.

[4] J. Suokas. The role of safety analysis in accident prevention. *Accident Analysis and Prevention*, 20(1):67–85, 1988.

31st LOSS PREVENTION SYMPOSIUM

Session 39. **Mechanical Integrity & Maintenance of Process Plant Components**

Monday, March 10, 1997
1:30 - 5:00 PM
Grand Ballroom C

John A. Davenport, Chair
Industrial Risk Insurers
85 Woodland St.
Hartford, CT 06102-5010

Everett D. Wixom, Vice-Chair
Exxon Chemical CO
P.O. Box 45
Linden, NJ

Session Introduction - John A. Davenport

39a. **A Systematic Approach to Mechanical Integrity Program Development,** C. Freeman, Life Cycle Engineering, N. Charleston, SC

39b. **Application of Safety Interlocks - The Complete Picture,** J. Siebert, E.I. DuPont, Orange, TX

39c. **Criticality-Based Maintenance: Prioritizing Maintenance Activities for Optimum Equipment Reliability,** V.Ciliberti; Berwanger, Inc., Houston, TX

BREAK

39d. **Fully Quantitative Predictive Maintenance/Inspection Planning Optimization for Chemical Process Plant Components,** D. Mauney, Southwest Research Institute, San Antonio, TX; M. Schmidt, Industrial Risk Insurers, Hartford, CT

39e. **Risk Based Prioritization of Maintenance Repair Work,** J. Harnley, Exxon Chemical CO, Baytown, TX

39f. **The Application of Risk Based Inspection Methodology in a Large Chemical Plant,** C. Dykstal, The Dow Chemical CO, Freeport, TX

Concluding Remarks by E. D. Wixom

Paper Number 39a

A Systematic Approach to Mechanical Integrity Program Development

by

Charles J. Freeman

Life Cycle Engineering, Inc.
4360 Corporate Road, Suite 100
North Charleston, South Carolina 29405

Prepared for Presentation at the AIChE 31st Annual Loss Prevention Symposium,
Case Histories and Miscellaneous Topics

March, 1997

Paper Number 39a

A Systematic Approach to Mechanical Integrity Program Development by

Charles J. Freeman

ABSTRACT

Developing a Mechanical Integrity Program (MIP) to meet the requirements of the OSHA Process Safety Management (PSM) standard, 29 CFR 1910.119, requires a systematic approach. The initial stages of the development process are the identification of covered process boundaries, followed by a walk-down of piping and instrumentation diagrams (P&IDs) to verify accuracy and collect equipment information. P&IDs are updated, equipment is numbered and tagged, and an equipment list is developed to identify critical equipment for each covered process. Maintenance, test, and inspection methods are determined for each piece of equipment followed by development of written procedures to cover each maintenance test or inspection identified. A training program is developed and implemented to train personnel maintaining the ongoing integrity of process equipment. In addition, program elements for equipment deficiencies and quality assurance are established through organizational policy and procedures in a Mechanical Integrity Program Manual.

PURPOSE

OSHA created the PSM standard to deal with the risks involved in the storage, handling, and processing of highly hazardous chemicals with emphasis on reducing the number of employee fatalities and injuries associated with catastrophic releases of hazardous substances. The standard consists of fourteen elements. The Mechanical Integrity aspect of the standard is the most inclusive of all the elements and may be the most involved in developing and implementing. Elements of a MIP include the identification and categorization of equipment and instrumentation, inspections and tests, testing and inspection frequencies, development of written maintenance procedures, training of maintenance personnel, the establishment of criteria for acceptable test results, documentation of test and inspection results, and documentation of manufacturer recommendations as to meantime to failure for equipment and instrumentation. The purpose of this paper is to provide guidelines to plant owners and operators for implementing a Mechanical Integrity Program (MIP) that achieves maintenance excellence and the associated cost and production benefits, as well as meets the intentions of OSHA 29 CFR 1910.119, paragraphs (j)(1) through (j)(6), Mechanical Integrity.

RECOGNITION OF NEED

OSHA has determined that 24,939 establishments in 127 industries are affected by the PSM standard. Mechanical Integrity requirements must be defined by each of these establishments individually and a program implemented to a level that satisfies the intent of the standard. In order to define a successful MIP, plant management must first recognize the specific need for compliance, then commit the proper resources to define, develop, and implement an effective program.

INITIAL ASSESSMENT

A performance-based MIP cannot be strictly developed and implemented via cookbook methods. Development of each program should be customized to the particular needs of the processes it will support. The amount of development depends, in large part, on how many of the previously defined MIP elements are already in place. Mechanical Integrity is synonymous with training, inspection, testing, and maintenance. An initial assessment of these areas should be conducted to identify elements in place as well as areas that require attention. This assessment should be objective and encompass all the elements of Mechanical Integrity. Based on the findings of the assessment, areas that require more attention can be prioritized to ensure efficiency in the approach to MIP development.

ALLOCATION OF RESOURCES

Employee participation is an important and sometimes mandatory regulatory element for many facets of PSM. It is important that in-house personnel be involved in all aspects of the MIP development process. This approach will enhance ultimate employee acceptance, identify inherent weak points, resolve related concerns, and better identify potential breaches of integrity. Employee participation will also improve and strengthen an operator's view of maintenance and testing as required by Mechanical Integrity implementation. Again, an important step that plant management can take toward MIP realization is to allocate resources by assigning plant personnel to responsible positions during program development. One such important position may be membership on a Mechanical Integrity steering committee.

A Mechanical Integrity steering committee should be formulated to oversee the progress of MIP development. The purpose of this steering committee would be to administer and direct the various phases of MIP implementation and, following implementation, its purpose shifts to MIP optimization. The committee would primarily function to define, implement, monitor, and improve the program based on feedback during the MIP development process. This committee would have clear ownership of the MIP and would keep the staff and management informed of program status. To facilitate the program on a continuous basis before and after implementation, the committee should consist of personnel from the following groups:

- Process Engineering
- Project Engineering
- Quality Assurance
- Production
- Purchasing
- Electrical & Instrumentation
- Mechanical

It is recommended that program assignments, responsibilities, policies, and procedures be documented in a MIP manual. The manual should be controlled by the Mechanical Integrity steering committee. This manual should provide guidance and direction for individual responsibilities, personnel certifications/qualifications and training, quality assurance, and equipment deficiencies (identification and resolution).

PREREQUISITES FOR MIP DEVELOPMENT

Prerequisites of an effective MIP include development of Process Safety Information, completion of Process Hazard Analyses, and identification of covered system boundaries.

In order to identify and understand the hazards posed by those processes involving highly hazardous chemicals, Process Safety Information (PSI) must first be compiled. This compilation of Information is used to identify, gain understanding, and analyze the covered processes that fall under the constraints of the MIP. The PSI should contain information pertaining to the hazards of the chemicals in the process, the technology of the process, and the equipment used in the process. Resources for obtaining adequate PSI can be found in the form of Material Safety Data Sheets (MSDS), flow diagrams, design specifications, P&IDs, operating procedures, etc. After identifying the covered processes, the next step is to identify the equipment within those processes that are included in the MIP.

Establishing the program boundaries and identifying the equipment included in the MIP are integral parts of conducting a Process Hazard Analyses (PHA). A PHA is required for all processes covered by the PSM standard to identify, evaluate, and control the hazards involved in the process. The analyses may incorporate several methodologies including What-if Scenarios, Checklists, Failure Mode and Effects Analyses (FMEA), Hazard and Operability (HAZOP) Studies, and Fault Tree Analysis. The PHA can be summarized as follows:

- Identify - What can go wrong?
- Analyze - What are the effects?
- Recommend - What should be done about it?

Once the PHA has been completed, the boundaries of the covered process can be clearly established which not only identify the equipment within the process, but ancillary and support systems required to ensure that process integrity is maintained.

PROGRAM DEVELOPMENT

The PSM standard states that Mechanical Integrity applies to the following equipment that is considered common to all processes:

- Pressure Vessels and Storage Tanks
- Piping Systems and Components (such as valves)
- Relief and Vent Systems and Devices

- Emergency Shutdown Systems
- Instrumentation and Controls (such as sensors, alarms, and interlocks)
- Pumps

The first and most critical step in program development is to identify what equipment is in the program. It is the intent of OSHA that at least the equipment previously listed be subject to the requirements contained in paragraph (j). If additional types of equipment are deemed critical to the process (based on the PHA) it is the intent of OSHA that these additional types be considered covered by the standard and treated accordingly. These additional equipment types may include agitators, fire protection systems, ventilation systems, etc. The next stage of MIP development is to verify and collect equipment information to properly identify equipment and components whose integrity must be maintained.

Physically walking down a process and verifying equipment nameplate data is the only way to ensure equipment information found in equipment files accurately reflects the equipment currently installed in the field. A P&ID for the process is a useful tool in trying to identify and locate equipment within the process. For this reason, P&IDs should be up-to-date and accurately portray the process and equipment they represent; this means that they will also need to be an integral part of the plant's Control of Change program . A standard numbering system should be incorporated to number each equipment item that uniquely identifies that piece of equipment. All equipment (including piping and valves) should be identifiable if it is within the covered process and boundaries of the MIP. The assigned number should be permanently fixed to the equipment and reflected on the applicable P&ID. This unique number will be the primary means of locating, identifying, and documenting inspections/tests performed under the MIP. Once all the equipment has been given a unique number identifier, a list can be formulated that encompasses all the equipment that will be covered under the MIP. This list is most commonly referred to as the Critical Equipment List. Now that all the equipment has been identified in the MIP, the next step is to identify all the maintenance requirements necessary to maintain system and component integrity.

INSPECTION AND TESTING

Inspection, test, and maintenance requirements should follow recognized and generally accepted good engineering practices. Some companies have established in-house specifications, company standards, and company procedures that represent good engineering practice and, so long as they meet or exceed more widely published standards, they should be used where applicable. The frequency of inspections, tests, and maintenance tasks should be consistent with applicable manufacturer's recommendations, codes and standards (ASME, API, NFPA, ANSI, etc.), and more frequently if determined to be necessary by prior operating experience.

The PSI identified during Process Hazard Analyses is examined again, but this time for causal criteria rather than the hazardous effect. FMEA results are now analyzed to prevent those failures that produce hazardous effects. Maintenance Engineering specialists utilize their trade to optimize equipment maintenance for maintaining system integrity. The focus shifts to maintenance actions. These can be based on inspection and test results, elapsed time, run time or good engineering practice.

Once maintenance tasks have been identified, they must be documented, together with inspection and test requirements, on a maintenance plan. The equipment maintenance plan identifies the minimum tasks (preventive, predictive, and corrective) required to maintain equipment reliability and function on a specified type of equipment. The equipment maintenance plan should further identify the frequency of inspection and the skill level required to perform the task. The maintenance department should review and approve all specified inspection and test requirements prior to development of written procedures.

Written procedures must be established to yield maximum reliability in maintaining the on-going integrity of process equipment. The planned maintenance procedures developed for MIP inspection, testing, and maintenance requirements should comply with the following guidelines:

- Identify resources (craft or skill, test equipment, and reference documents)
- Identify operating conditions
- Identify safety precautions
- Contain procedural steps to fully accomplish task

- Provide acceptance criteria
- Provide for recording inspection or test results
- Provide for recording the date accomplished and person performing

Procedures should be written in simple words and phrases that will convey the intent of the procedure. Avoid using vague and ambiguous terms that could cause confusion. It is also helpful to provide a section for definitions at the beginning of the procedure when unique or special terms are used within the context of the procedure. The technical validity of the procedure should be reviewed by the maintenance department. A document control system should be put in place for handling any revisions to the written procedures. If changes are required to a procedure, because of equipment change or some other reason, the document control system should ensure that the revised procedure is implemented and used. Any changes to written procedures must be made in accordance with the plant's Control of Change program. The revised procedure should reflect the author's name, authorization date, and revision number. When all the written procedures are in place, the next step is to train personnel on why the procedures exist as well as how to use them.

TRAINING

The development of an effective training program should be an on-going process during all phases of PSM. When developing the framework and format of the training program, initiate a training plan of action that establishes the following:

- Qualification requirements for instructors
- A testing system to evaluate employee job proficiency
- A documentation system to record performance of employee training
- Priorities for training
- A schedule for conducting training
- A training program audit to verify its effectiveness

Each employee involved in maintaining the on-going integrity of process equipment should be trained in an overview of the process and its hazards and in the procedures applicable to their job task to assure that they can perform the job tasks in a safe manner. Appropriate training should be provided to both operations and maintenance personnel to ensure that they understand the planned maintenance program procedures, safe practices, and the proper use and application of special equipment or unique tools that may be required to successfully complete their job tasks. It is the responsibility of management to ensure that employees have received adequate training by maintaining a record which contains the identity of the employee, the date of the training, and the means used to ensure the employee understood the training received. Upon completion of training, the next and perhaps the most difficult step to a MIP is the implementation of the program.

PROGRAM IMPLEMENTATION

A big problem with implementing a MIP is data management. A Computerized Maintenance Management System (CMMS) allows the flexibility of storing, recording, retrieving, and analyzing large amounts of data. A CMMS is a valuable tool for documenting the outcome of inspections and tests. Information stored in the CMMS should include, but not be limited to, original and as-built specifications on covered process equipment, nameplate data, inspection and repair history, correspondence pertaining to the equipment, identified inspection and test requirements and schedules, and written procedures in work order format. The CMMS should provide quick access to relevant information on all covered process equipment, and it must provide cross reference information for logistics, maintenance, operations, engineering, and management personnel to initiate and track elements of the plant's Control of Change program. The computerized system will allow inspections and tests to be properly scheduled and tracked. Along with the implementation of a CMMS, various responsibilities for the management and use of the system need to be addressed:

- Selection of appropriate software
- Assigning responsibility for maintaining and updating system information
- Defining information stored in system
- Training
- Security and accessibility of the data

Once the appropriate data has been entered into the CMMS, inspection, test, and maintenance requirements can be planned and scheduled. Work orders are then generated from the CMMS for the scheduled planned maintenance procedures and assigned to the appropriate craft or skill level for accomplishment. The inspector has been adequately trained to perform the given inspection task and is ready to complete the work order according to the steps specified in the written procedure. The inspector is responsible for collecting the appropriate data to assess the integrity of the equipment. Acceptance criteria specified by the written procedure should aid the inspector in assessing the covered equipment is fit for continued service. After completing the work order, it is signed, dated, and entered into the CMMS as completed. The standard requires that this documentation identify the date of the inspection or test, the name of the person who performed the inspection or test, the serial number or other identifier of the equipment on which the inspection or test was performed, a description of the inspection or test performed, and the results of the inspection or test. The CMMS provides an efficient means for documenting the inspections performed under the MIP.

Policies and procedures should be in place to address equipment deficiencies. The standard requires that equipment deficiencies that are outside acceptable limits be corrected prior to the equipment being placed back into service or that the system be placed in a safe configuration prior to failed equipment use. Equipment deficiencies can be corrected in a safe and timely manner, only if positive action is taken to assure the equipment is safe to operate under its current condition. A system should be incorporated that documents the decision to shutdown or continue operation. This documentation would include an evaluation or analyses done to assure that operation can continue safely. If removal and replacement must be performed, appropriate checks and inspections should also be performed to assure the equipment is installed properly and consistent with design specifications and the manufacturer's instructions and the installation meet Mechanical Integrity requirements. This entails developing policies and procedures to address quality assurance.

At the completion of any inspection or test, the data should be evaluated and analyzed to determine whether or not the covered process equipment is suitable for further operation. During this assessment, it should also be determined when the equipment requires re-inspection. The inspection frequency is determined based on knowledge of the operating process, maintenance priority based on system operating parameters, predicted corrosion rates, design parameters, possible failure modes, historical data, and guidelines from the appropriate inspection codes. In no case shall the interval of inspection be longer than the anticipated remaining life of the equipment.

After implementation, the MIP should be monitored and evaluated on a continuous basis. The Mechanical Integrity steering committee should be assigned this responsibility. The steering committee would also resolve and implement any proposed changes to the MIP. The standard requires that the overall MIP be audited every three years to determine its effectiveness.

CONCLUSION

Elements of a MIP include the identification and categorization of equipment and instrumentation, inspections and tests, testing and inspection frequencies, development of written maintenance procedures, training of maintenance personnel, the establishment of criteria for acceptable test results, documentation of test and inspection results, and documentation of manufacturer recommendations as to meantime to failure for equipment and instrumentation. Implementation of a MIP is an essential activity for continuing production and regulatory compliance. OSHA provides for significant flexibility in program development, however, documentation of program elements is the single most important aspect of the program, once implemented. A well-designed Mechanical Integrity Program can provide the following key advantages:

- Meeting the intent of the law/regulation
- Increased safety within the plant
- Reduced risk and associated costs of releases to the environment affecting the public
- Reduced maintenance costs
- Increased efficiency in performing maintenance
- Improved reliability of production equipment
- Greater productivity and profitability

Paper Number 39b

APPLICATION OF SAFETY INTERLOCKS- THE COMPLETE PICTURE

Joseph F. Siebert
E.I. DUPONT
P.O. BOX 1089
ORANGE, TX. 77630

Prepared for Presentation at the AIChE 31st Annual Loss Prevention Symposium

March, 1997

UNPUBLISHED

APPLICATION OF SAFETY INTERLOCKS-
THE COMPLETE PICTURE

Joseph F. Siebert

ABSTRACT: I will present a review of the newly approved S84 ISA document on the application of safety instrumented systems. S84 presents 15 steps in the life cycle of safety systems. I will present a simplified 3 step approach that helps me better understand S84.

Hello, my name is Joe Siebert and I work for E.I DuPont in Orange, Texas. One of my responsibilities is plant proprietor of our safety interlock standards. I have been asked by DBUG to discuss the newly approved Instrument Society of America's standard on safety interlocks. It is S84 and titled APPLICATION OF SAFETY INSTRUMENTED SYSTEMS FOR THE PROCESS INDUSTRY.

(Refer to slide 1.)

S84 is a very good document that deals with a complex and detailed subject. The standard discusses the 15 life cycle steps of safety interlocks and gives guidance on a few of those steps. ISA's S84 can only give guidance on a few life cycle steps because the topic of safety interlocks is made up of many parts and only a few of those parts apply to Instrument engineering. Guidance for other life cyclce steps are discussed in more detail by the American Institute of Chemical Engineers. The book SAFE AUTOMATION OF CHEMICAL PROCESSES is produced by the AICHE and gives a more in-depth discussion of the first six life cycle steps. The next several steps are properly addressed by the ISA and covered in this document - S84.

I give this as background to give you one user's experience on applying these 15 steps to safety interlocks. Kids - don't do this at home without your mother's permission. This is a complex subject and needs the contributions of a team of diverse, knowledgeable people. Make sure you have qualified people to assist you at this task.

This standard was created by a team of industry experts. It is certainly in alignment with DuPont's corporate standards. It was chaired by an ex-DuPonter. It also has representatives for the European committees on this same subject.

Before we move on to the next slide, *Point one: the main purpose of the activity covered by this standard is to achieve a high-integrity safety interlock design to reduce risk to an acceptable level.* Don't forget this. I have gone down many wrong paths until this became salient in my mind. The end product is an E&I design that is properly maintained and provides a safe plant for the employees and the community.

(Refer to slide 2)

I will not specifically talk about each life cycle step. Instead I will present a simplified paradigm that helps me understand the topics discussed in S84. I find it easier to think of providing appropriate E&I design as a three team activity. I will present these activities as a linear process. In practice, their is a lot of interaction between the teams and several steps require good open communications between the teams. I guess that is the second point I would like you to leave with. *This standard explains what is the product of each work team activity.*

My simplified safety life cycle is:

RECOGNIZE THE HAZARD

E&I DESIGN TO REDUCE THE RISK

MAINTAIN THE DESIGN TO SPECIFICATIONS

Let's look at each work teams tasks in greater detail.

(Refer to slide 3.)

RECOGNIZE THE HAZARD

This main function of this work team is to provide a Process Hazards Analysis. This work team is made up of many individuals - people that experts in Process design, people that are experts in instrument and electrical design, people that are experts in running the process, people that are experts in maintaining the equipment, people that are experts in risk analysis.

(Refer to slide 4)

An important part of risk analysis is an understanding of the concept of layers of protection. The book SAFE AUTOMATION OF CHEMICAL PROCESSES discusses this topic.

The process can be thought of as having many layers of protection. As we add layers of protection, we add more and more protection. These layers increase in safety integrity. This isn't free but it is necessary to reduce the risk to an acceptable level. The inner most layer can be thought of as the basic process controls, operating alarms, qualified people training, non-safety process designs, etc. But these do fail so we need additional layers of protection. That's why we provide critical alarms and manual intervention as another layer of protection. But these also fail and so we need to provide automatic safety interlock systems.

(Sorry, back to slide 3)

The PHA must also provide the appropriate safety integrity level for each safety interlock. We usually have three safety levels. The first level I guess really isn't a level (or level zero depending on your point of view. If you were brought up in the software world of bits you probably start counting with zero. The heat transfer experts also seemed to start with zero - remember the zeroth law of thermodynamics), it is the non-catastrophic level. This protection can be provided in the Basic Process Control system, i.e. one of those inner layers of protection. Level 1 is for catastrophic events and these safety interlocks provide very good protection. Level 3 is also for catastrophic events and provides an order of magnitude better protection. But let's not get down into the weeds too much for this discussion.

Point number three - it is the function of the PHA to provide the E&I design team with the safety integrity level. This is not an E&I function. Let me repeat .... it is a PHA function to provide the safety integrity level to the E&I design team.

(Refer to slide 5)

The E&I design team now really kicks into gear. Since they have the safety integrity level, they can continue on and provide appropriate E&I design to reduce the risk to an acceptable level as determined by the PHA. Things are really starting to work. They have the information they need and they are the experts with the knowledge to do this task.

In their design, they must consider approved hardware for the safety interlock system. They need to provide separation. Let's look at slide 6 quickly.

(Refer to slide 6)

Sometimes a picture is worth a thousand words. Any questions.

(Back to slide 5)

They must consider questions about how to design for redundancy; do we use hardware or software; how do we provide evergreen documentation; don't forget about no automatically resetting interlocks; how frequent do we need to test the safety interlocks; how do we check-out the interlock system; how do we make sure the equipment can be maintained .... which leads us into the last teams activity.

(Refer to slide 7)

The E&I team must provide adequate documentation to install the safety interlock system. It must also provide evergreen documentation to insure the installation maintains its designed safety integrity level. This includes items like other mitigation provided that could back up the safety interlock should it fail, the required periodic test. When you can and can't bypass these interlocks.

The maintenance of the safety interlock system requires appropriate test records. When you do have an overt failure, continuity of repair must be maintained. Also, when you do discover a failure, either overt or covert, you should have an investigation to determine the root cause and correction.

I hope this user's 3 life cyclestep paradigm will help you better understand the complex issue around safety interlock designs. Also remember:

- the main purpose of the activity covered by this standard is to achieve a high-integrity safety interlock design to reduce risk to an acceptable level.
- This standard explains what is the product of each work team activity.
- It is the function of the PHA to provide the E&I design team with the safety integrity level

I know open the floor for any additional questions.

INTRODUCTION

- S84 - Application of Safety Instrumented Systems - is a good document
- One user's experience
- Kids - Don't do this at home without your mother's permission
- Following IEC, Corporate guidelines
- The purpose is to achieve a high-integrity safety interlock design. Don't forget this.

SIMPLIFIED SAFETY LIFE CYCLE

STEPS 1-6

Recognize the Hazard

STEPS 7-12

E&I Design to Reduce the Risk

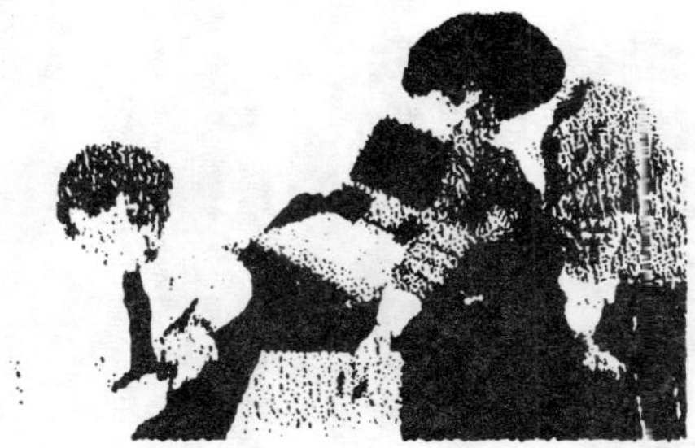

STEPS 14-15

Maintain

Recognize the Hazard

PHA
Layers of Protection
Safety Integrity Level
. . . and many other functions

Typical Layers of Protection Found in Modern Chemical Processing Plants

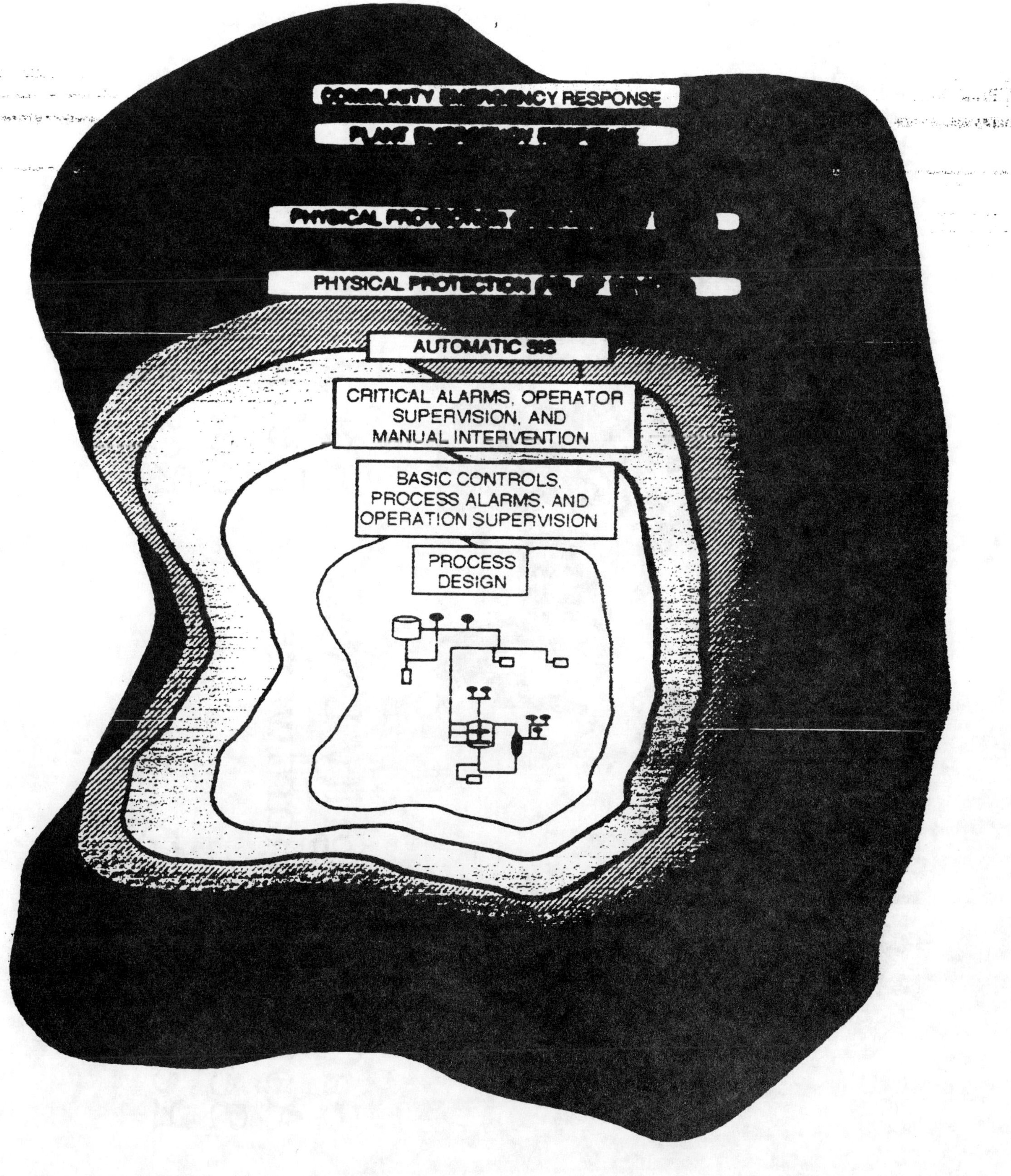

E&I Design to Reduce the Risk

Safety Integrity Level
Approved Hardware
Separation
Redundancy

Hardware vs. Software
Field Devices
Documentation
Reset

SEPARATION

INTERLOCK SYSTEM

PROCESS CONTROL SYSTEMS

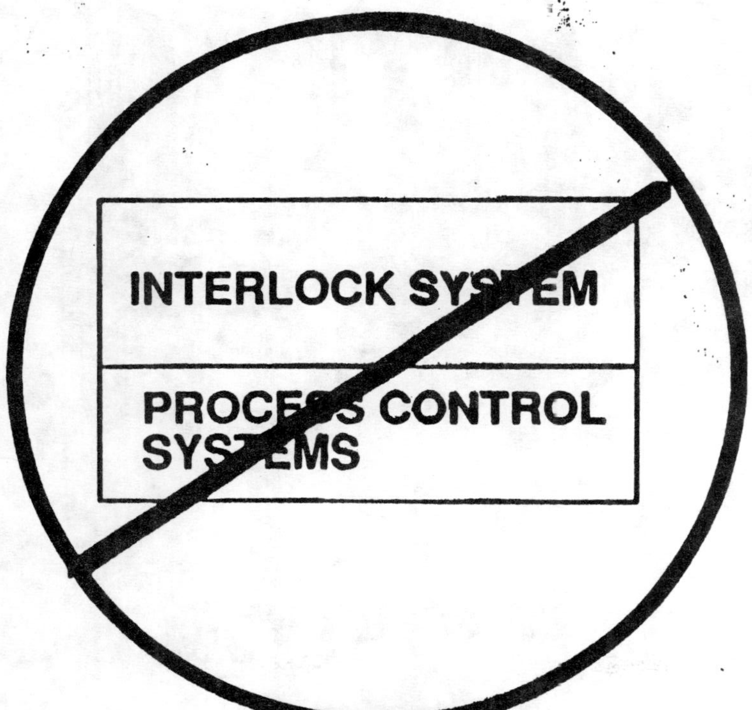

Maintain

Periodic Tests	Test Records
Repair Work	Continuity of Repair
Bypass	High Potential Investigation
Overt Faults	

Paper no. 39c

CRITICALITY-BASED MAINTENANCE: PRIORITIZING MAINTENANCE ACTIVITIES FOR OPTIMUM EQUIPMENT RELIABILITY

V. Anthony Ciliberti
Berwanger, Incorporated
Oil, Gas, and Petrochemical Consulting Division
4615 Southwest Freeway, Suite 925
Houston, Texas 77027
United States of America
Telephone: (713) 775-6456
Facsimile: (713) 850-1025
E-mail: va.ciliberti@berwanger.com

The American Institute of Chemical Engineering 31st Annual Loss Prevention Symposium, Mechanical Integrity and Maintenance of Process Plant Components Session

March, 1997

CRITICALITY-BASED MAINTENANCE: PRIORITIZING MAINTENANCE ACTIVITIES FOR OPTIMUM EQUIPMENT RELIABILITY

V. Anthony Ciliberti
Berwanger, Incorporated
Oil, Gas, and Petrochemical Consulting Division
4615 Southwest Freeway, Suite 925
Houston, Texas 77027
United States of America
Telephone: (713) 775-6456
Facsimile: (713) 850-1025
E-mail: va.ciliberti@berwanger.com

ABSTRACT

Criticality-Based Maintenance (CBM) is a prioritized approach to the maintenance of process equipment in the oil, gas, and petrochemical processing industries. CBM equally weighs process and safety issues in establishing criticality, which gives companies a return on investment for mechanical integrity efforts while establishing compliance with governmental regulations. This approach optimizes the effectiveness of mechanical integrity programs by focusing on the most important, or critical equipment items.

CBM uses a criticality ranking process which is very similar to the methodology used in a process hazard analysis study. All process equipment items are evaluated and then each is given a relative value in an overall criticality hierarchy. With all equipment evaluated for criticality, prioritization of maintenance activities begins.

INTRODUCTION

Maintaining mechanical integrity of process equipment is important in the chemical and hydrocarbon processing industries for continuous output of quality product and for safe, environmentally sound plant operation. Recent trends and developments in industry have increased the difficulty of implementing mechanical integrity programs while on the other hand making their implementation mandatory. Cost reduction measures at corporate and plant levels have reduced the already limited resources used in establishing these programs. Governmental rules and regulations, such as the OSHA Process Safety Management regulation and the EPA Risk Management Plan, now require that companies establish mechanical integrity programs to ensure environmental compliance and the safety and health of plant personnel.

Current maintenance philosophies reflect the need for mechanical integrity, as shown by an emphasis on proactive measures such as predictive and preventive maintenance, failure analysis, corrective maintenance, and reliability analysis. These proactive measures are effective but challenging to implement due to the large number of equipment items that must be addressed with limited resources. To meet this challenge effectively and remain profitable, companies must maximize the utility of their resources. Companies should prioritize mechanical integrity efforts and resources by concentrating on the most important or "critical" equipment items. Compliance programs should not focus strictly on reducing hazards, but should also focus on process-related issues, so as to give a return on investment.

Criticality-Based Maintenance (CBM) is a systematic approach for prioritizing maintenance. All process equipment items are evaluated with equal emphasis on hazard (health, safety, and environmental) and process criteria. Each equipment item receives a composite rating based on the hazard and process inputs. The composite rating is used to establish a Process and Hazard Criticality Ranking (PHCR) for that item. The PHCR value is a relative ranking

Table 1. Hazard Criticality Rating (HCR)

1. Assess the ***Chemical Hazard***. Select the highest value of NFPA fire, health, and reactivity ratings for chemicals handled by said equipment item.
2. Assess ***Other Hazards*** with an arbitrary number (0-4). The following list gives examples but is not all inclusive.
• Highly Hazardous (4): Temperatures > 1000°F or < -50°F, Pressures > 500 PSIG or < 1 ATM where oxygen ingress may cause explosion potential, rotational speeds > 5000 RPM, regulated chemicals, critical safety devices, pollution control systems where failure would result in environmental damage.
• Hazardous (3): Temperatures > 500°F or less than -20°F. Pressures > 250 PSIG, rotational speeds > 3500 RPM, hazardous chemical releases that can be contained, chemicals with moderate vapor-cloud potential and with density greater than air, pollution control systems where failure might result in environmental damage.
• Moderately Hazardous (2): Temperatures > 212°F, pressures > 150 PSIG, rotational speeds > 3500 RPM, chemical releases with potential for long-term health or environmental impact, dilute hazardous chemicals.
• Slightly Hazardous (1): Pressures greater than 15 PSIG. Equipment rotational speeds greater than 200 RPM.
• Not Hazardous (0).
3. ***Hazard Criticality*** (HC) is the greater value of the chemical and other hazards.
4. ***Hazard Risk Reduction Factor (HRR):***
• HRR=0: Potential for an incident is high, or serious health, safety, or environmental consequences might result.
• HRR=1: Potential for an incident is low and minimal health, safety, or environmental consequences would result.
• HRR=4: No possibility exists for an incident, nor for health, safety, or environmental consequences.
5. ***Hazard Spare Equipment Factor (HSE):***
• HSE=1 if an in-line spare or process option exists that mitigates the hazard, otherwise HSE=0.
6. ***Hazard Criticality Ranking (HCR):***
• HCR = HC - HRR- HSE (Round negative numbers to zero).

in an overall criticality hierarchy that is used to determine priorities for maintenance programs, inspections, and repairs.

PROCESS AND HAZARD CRITICALITY RANKING

In a PHCR study, each equipment item is evaluated with "what if it fails" scenarios. This requires personnel with thorough knowledge of the process and equipment under study. Process Hazard Analysis (PHA) teams are an excellent choice. It makes good sense to integrate criticality ranking into a PHA study, with the criticality ranking done at the end of each node, after all key words have been covered and risks discussed.

A PHCR study requires up-to-date equipment records, process and instrument diagrams, and material safety data sheets. Other pre-work includes compiling a study worksheet, training the analysis team on PHCR procedures, and tailoring the ranking tables to fit the facility in which they will be applied. A study worksheet is typically

Table 2. Process Criticality Rating (PCR)

1. Assess the ***Process Criticality (PC).***
• (PC=4) Equipment critical for process capability: failure of equipment will cause complete loss of production capacity or cause unacceptable product quality resulting in total product loss.
• (PC=3) Equipment needed for full process capability: failure of equipment will result in loss of one stream or requires product to have a waiver with delivery.
• (PC=2) Equipment helpful to process capability: Loss of equipment forces recirculation or immediate storage, or causes long-term loss of efficiency and/or reliability.
• (PC=1) Equipment minimally impacts process capability: Loss of product optimization or advanced control.
• (PC=0) Equipment does not affect process capability.
2. ***Process Spare Equipment Factor (PSE):***
• PSE=1 if an in-line spare or process option exists that mitigates the process consequence, otherwise PSE=0.
3. ***Process Criticality Ranking (PCR):***
• PCR = PC - PSE (Round negative numbers to zero).

downloaded from maintenance computer files into a spreadsheet application such as Microsoft Excel. Viewing and entering data on the worksheet is most efficiently done with a computer projection panel. Training the analysis team and tailoring the ranking tables can be done simultaneously and normally takes about one day.

The Ranking Procedure

Assume full rates for the process unit(s) under study. Evaluate each equipment item individually. Following is the procedure for PHCR evaluation:

1. Determine the Hazard Criticality Rating (HCR). Evaluate chemical hazards and physical hazards on a scale from 4-0, four being the worst case. Select the worst case scenario and give credit for factors that mitigate risk. Reference Tables 1 and 3.
2. Determine the Process Criticality Rating (PCR). Evaluate the process impact of equipment failure using the same scale used in step one. Select the worst case and give credit for spare equipment that mitigates risk. Reference Tables 2 and 3.
3. Determine the Process and Hazard Criticality Ranking. Input HCR and PCR values into Table 4 to give a Process and Hazard Criticality Ranking value. The rankings are in alpha-numeric form, with the first character (alpha) identifying the criticality category (A-D, "A" being most critical) and the second and third characters (numeric) identifying the hazard and process ratings respectively. The alpha-numeric designation identifies not only equipment criticality ranking, but also why it is critical with respect to process and hazard criteria.

Average time for evaluation is two minutes per equipment item. When done in conjunction with a PHA study, equipment criticality ranking adds approximately ten percent to the length of the study.

Adjusting Criticality Ranking Based On Product Demand. Process and hazard criticality ranking is evaluated assuming full rates for the process unit under study. However, if product demand dictates lower operating rates, capacity-related process criticality will decrease for all equipment in that unit.

Table 3. Guidelines for Process and Hazard Criticality Ranking.

1. Instrumentation
2. Evaluate criticality of the function which the instrument controls.
   2.1. Shutdowns and alarms have the same criticality as the function criticality.
   2.2. Status/indication instruments with output to the control room have a criticality one (1) less than the function criticality.
   2.3. Status/indication instruments with local output have a criticality two (2) less than the function criticality.
   2.4. Credits for risk reduction and spare equipment still apply where applicable.
3. Two-out-of-three trips receive no spare equipment credit.
4. Mechanical
   4.1. Main/Auxiliary lubrication oil pumps on critical rotating equipment receive no spare equipment credit.
5. General
   5.1. Process chemical hazard
      5.1.1. This is not applicable (N/A) where both of the following apply:
         5.1.1.1. There is no contact with the process chemical.
         5.1.1.2. There is no possibility for release resulting from the equipment function failure.
      5.1.2. For chemicals with no NFPA ratings available, either:
         5.1.2.1. Estimate NFPA ratings and note that this was done in the comments.
         5.1.2.2. Enter "U/A" for unavailable. This information should be input when it becomes available (and ratings changed accordingly).
6. PHC ranking does not include consideration of maintenance work. Maintenance work is covered by safe job procedures.

Table 4. Process And Hazard Criticality Ranking Matrix

	HCR				
PCR	4	3	2	1	0
4	A44	A34	A24	A14	A04
3	A43	B33	B23	B13	B03
2	A42	B32	C22	C12	C02
1	A41	B31	C21	D11	D01
0	A40	B30	C20	D10	D00

The alpha character designates the category, the first numeric value designates the hazard criticality rating, and the second numeric value designates the process criticality ranking. Category "A" is the most critical and category "D" is the least critical.

Table 5. Product Demand Rating (PDR)

(PD=4) Product demand requires process unit to operate from 91 to 100 percent of full capacity.
(PD=3) Product demand requires process unit to operate from 70 to 90 percent of full capacity.
(PD=2) Product demand requires process unit to operate from 40 to 69 percent of full capacity.
(PD=1) Product demand requires process unit to operate at less than 40 percent of full capacity.
(PD=0) Process unit is not in operation.

Table 6. Net Process Criticality Rating (PCR')

	PDR				
PCR	4	3	2	1	0
4	4	3	2	1	0
3	3	2	1	0	0
2	2	1	0	0	0
1	1	0	0	0	0
0	0	0	0	0	0

Quality-related process criticality and hazard criticality change only if the process unit is not in operation.

Incorporation of variable product demand into equipment criticality rankings requires a five-character criticality ranking field. The first three characters indicate the criticality category, hazard criticality rating, and process criticality rating respectively. The fourth character indicates the demand rating, and the fifth character indicates whether or not the equipment item impacts product quality. The combination of process and demand ratings determine a revised alpha value for the criticality category.

Following is the procedure for incorporating product demand into equipment criticality rankings:

1. Using a five-character ranking field, complete the process and hazard criticality ranking as described in Tables 1-4. Flag equipment items that are product quality related with a "Q" in the fifth character of the ranking field (Otherwise leave the fifth character blank). Although demand is variable, assume full unit capacity (product demand rating equal to "4") for the initial evaluation.

2. Update the fourth character of the ranking field as the product demand rating varies. Use Table 5 to determine the product demand rating.

The following steps apply only to equipment items that are not product-quality related:

3. Determine Net Process Criticality Rating (PCR'). Enter the values of PDR and PCR into Table 6 to give a PCR' value.

4. Determine a revised alpha value for the criticality category using PCR' in place of PCR in Table 4. Substitute this alpha value in the first character of the ranking field. Disregard the numeric characters from Table 4 in this step.

Repeat steps 2-4 for all affected equipment each time product demand changes in a particular unit. Use statistical averaging if product demand varies frequently. This process is best done using computerized methods.

CORRELATION OF EQUIPMENT CRITICALITY WITH OTHER MAINTENANCE PARAMETERS

PHCR values are based only on process and hazard criteria. Other parameters such as equipment reliability and repair costs,

Table 7. Criticality and Reliability Index (CRI) values for rotating equipment.

	Mean Time Between Failure (months)			
PHCR	0 - 6	6-12	12-24	> 24
A	a4	a3	a2	a1
B	a3	b3	b2	b1
C	a2	b2	c2	c1
D	a1	b1	c1	d1

Table 8. Rotating equipment inspection intervals based on CRI values.

	Inspection Frequency (days)				
CRI	7	30	90	180	360
a3,a4	V	L			
a1,a2		V	L		
b1-b3			V	L	
c2-d1					V

V = vibration monitoring
L = lubrication sampling/testing

while not a consideration in the ranking study, must be correlated with criticality ranking values when determining certain maintenance priorities. Examples include determination of equipment inspection frequencies and spare parts stocking requirements. Tables 7 and 8 illustrate the procedure for correlating equipment criticality with reliability to determine inspection requirements for rotating equipment.

PHCR values are fixed quantities while the other parameters discussed above are variable and most likely will change with the implementation of mechanical integrity measures. Preventive maintenance inspection, for example, should increase reliability and result in extended inspection windows and a reduction in stock quantities.

APPLICATION OF CBM

Criticality-Based Maintenance applies to any maintenance program or system in which equipment importance can be used as a basis for establishing priorities. One set of criticality ranking tables, if designed properly, will cover rotating equipment, fixed equipment, electrical equipment, instrumentation, piping, and any other types of process equipment. In addition to the examples given above, CBM can be used to prioritize maintenance work orders, equipment-specific training, failure and reliability analyses, and corrective maintenance efforts.

SUMMARY

Criticality-Based Maintenance is a prioritized approach to maintenance of equipment in the chemical and hydrocarbon processing industries. It is a method designed to maximize production output and quality while ensuring safe, environmentally-sound plant operation. CBM is not only a tool for establishing new maintenance programs, but also easily integrates with and improves existing programs. CBM focuses on priority equipment and therefore gives optimum results while minimizing investment.

EXAMPLE

The following illustrates the procedure for criticality ranking and determination of inspection frequencies for a waste water pump; tag number PU-LIFT.

PU-LIFT transports process waste water from a lift station to a waste water storage tank. Failure of this pump to operate can result in process waste overflow into a nearby bayou, which would in turn cause severe environmental consequences. This pump has two in-line spares. Mean time between failure for PU-LIFT is 23 months. NFPA ratings for process waste water are 3,0,0 for health, fire, and reactivity respectively.

1. From Table 1, the Chemical Hazard rating = 3 (Highest of NFPA ratings for health, fire, and reactivity). The Other Hazards rating = 4, due to potential for environmental damage. The Hazard criticality = 4 (greater of the Chemical and Other hazard ratings). The Hazard Risk Reduction factor = 0 (serious consequences would definitely result from a lift station overflow). The Hazard Spare Equipment factor = 1 (in-line spares that mitigate the hazard). The Hazard Criticality rating (HCR) = HC - HRR - HSE = 4 - 0 - 1 = 3.
2. From Table 2, the Process Criticality = 0 (pump not needed in making product). The Process Spare Equipment factor = 1 (in-line spares exist that would mitigate any process consequence). The Process Criticality Rating (PCR) = 0 - 1 = -1, which is rounded up to zero. PCR = 0.
3. From Table 4, the Process and Hazard Criticality Ranking (PHCR) = B30.
4. From Table 7, the Criticality and Reliability Index (CRI) = b2.
5. From Table 8, this pump requires vibration monitoring on 90 day intervals, and lubrication sampling on 180 day intervals.

Table 9 shows a typical study worksheet for equipment criticality ranking and contains data for this example.

Table 9. Process and Hazard Criticality Ranking Worksheet

Equipment Number	Equipment Description	NFPA Ratings H	F	R	O	HC	HRR	HSE	HCR	PC	PSE	PCR	PHCR	Comments
PU-LIFT	Process waste water lift pump	3	0	0	4	4	0	1	3	0	1	0	B30	Failure will cause overflow to bayou: environmental impact
TANK1	Storage tank	4	4	0	0	4	0	0	4	4	0	4	A44	Critical process feed stock.

Fully Quantitative Predictive Maintenance/Inspection Planning Optimization for Chemical Process Plant Components

David A. Mauney

Southwest Research Institute, Materials and Structures Division
P. O. Drawer 28510
San Antonio, Texas 78228-0510

and

Michael E. G. Schmidt, P.E.

Industrial Risk Insurers, Research Dept.
85 Woodland St.
Hartford, Connecticut 06102

Prepared for Presentation at
Mechanical Integrity and Maintenance of Process Plant Components
AIChE Loss Prevention Symposium
March 10, 1997 (session)

Fully Quantitative Predictive Maintenance/Inspection Planning Optimization for Chemical Process Plant Components

ABSTRACT

Predictive maintenance/inspection planning and fitness-for-service analyses are being applied more frequently to address the effect of component aging affecting plant reliability. Component aging is one of the critical issues in an increasingly competitive environment, especially in plant components built and installed during the 50s and 60s. Maintenance optimization is also a critical issue for components, such as those that operate in the creep regime, and therefore age quickly. Maximizing the value received for each predictive maintenance/inspection dollar is the key. This paper introduces the use of decision analysis to unite engineering analysis and economic evaluation for decision making. To illustrate this idea, examples of this approach applied to chemical facility components will be presented. In addition, the concept of priority-ranking multiple component maintenance/inspection projects having maintenance budget and safety constraints in a way that maximizes Net Present Value, will be presented.

Keywords: predictive maintenance; risk based; decision analysis; net present value; aging; maintenance planning; inspection planning; remaining life analysis; probability of failure

1. INTRODUCTION

Increasingly, engineers in the base infrastructure and manufacturing industries in the US are finding it difficult to obtain the resources necessary to maintain large equipment. This is especially critical because much of this equipment is quite old and there are few replacement plans. Modern maintenance management strategies, such as reliability-centered maintenance, and techniques, such as the predictive maintenance technologies, have developed. These strategies and techniques focus on equipment requiring maintenance resources, however, they operate on a strictly engineering level. Therefore, economically unrealistic maintenance activities have sometimes been proposed. Whether proposed activities have been justified or not, engineers have been ill equipped to convincingly "tell their story" to the decision makers. This paper suggests a process that will aid this communication and guide corporate resources into maintenance expenditures that maximize value.

2. COMPETING FOR RESOURCES

Engineers are finding themselves competing against other segments of the corporation for investment resources to support maintenance. The chemical process industry, like other base infrastructure and manufacturing industries, is finding extremely limited resources available for maintenance. Engineers are having a difficult time being heard because they do not normally speak in financial terms and do not usually have the necessary tools to compete in this arena.

Increasingly, the control of financial resources is in the hands of financial decision makers. Competition for corporate resources therefore occurs on the turf of corporate finance, using accepted fully quantitative methods. The methods used in this arena to judge one investment versus another have been financial analysis and decision analysis.

3. EQUIPMENT AGING

Equipment failure rates and, indirectly, maintenance costs, tend to follow a predictable pattern, with a declining initial failure rate ("infant mortality") followed by a long period of relatively constant, lower failure rate, followed by a period if rising failure rate ("wear out"). Much major equipment used in the base infrastructure industry in the United States is in this part of the life cycle. The US went through massive growth in the post World War II era, causing a large growth in base infrastructure industries. This equipment was installed in the 50s, 60s and early 70s. As a result of this construction boom, and because of the way the equipment was designed and operated, this equipment, which includes much of the major equipment in many chemical facilities is in the final part of its life cycle in the 1990s. In fact, a significant amount of chemical industry infrastructure is beyond its originally expected life. The original plan by engineers was to replace equipment at the end of its life cycle, in anticipation of aging. However, competition has prevented the necessary resources being available. This is because the US no longer dominates world markets. As a consequence, corporations are faced with the need to keep aging equipment running with limited resources. Accordingly, there is upward pressure on maintenance costs.

The electric utility industry provides objective data that effectively illustrate this effect. The failure rate (Forced Outage Rate versus calendar time curve for 50 MW to 200 MW coal fired boilers published in 1985) [Heiges, Stoll, 1985] curve, shown in Figure 1, has the classic Weibull life cycle curve shape [Miller, Freund, 1965]. To the left is the decreasing failure rate of the curve, usually called "infant mortality". It is dominated by failures from initial defects, either in manufacture or design. The center section is a period of constant failure rate often called the "honeymoon period." This is the period when minimum prudent maintenance causes the failure rate to remain constant. The right section is the "wearout period," and has an exponentially increasing failure rate because of the dominance of aging failure mechanisms. This life cycle curve is classic and can be used to describe the life of not only engineering components, but almost anything in nature that has a birth, life span, and death.

Industry has been surprised by the large increase in maintenance cost caused by the widespread onset of wearout. During the constant failure rate period, during which most present-day managers gained their experience, it generally seemed that prudent maintenance would hold the failure rate constant. These managers are now faced with serious major component replacement decisions for which their experience provides little guidance.

4. LIMITED MAINTENANCE RESOURCES

Until this point in history, a sound engineering analysis would generally convince management that maintenance expenditures were necessary. However, senior managers and decision makers are less likely to be engineers or well-trained in engineering matters, and for several reasons, corporate resources have tightened. Therefore, engineering analysis alone has lost the appeal necessary to obtain maintenance resources. In short, engineering for engineering's sake is losing its strength.

Any time that resources become more limited, more quantitative methods that use systematic input processes are needed to help make and support to tough decisions. Because choices are not as clear, all possible information must be extracted from inputs, and this information must be processed in a very quantitative manner in an attempt to promote more objectivity and ease of review.

5. COMPETITIVE COMMODITY PRICING

Market globalization, improved transpiration, and production rationalization have changed the product pricing structure and therefore the corporate cash flow. Plants are not competing with facilities that are necessarily similar in any way, rather every facility is or will very soon be competing with the world's lowest-cost producer of its product or of any viable replacement for its product. Therefore, the income side of the ledger is also putting downward pressure on maintenance resource availability.

6. MAINTENANCE RESOURCES CAUGHT BETWEEN AGING EQUIPMENT AND COMPETITIVE PRICING

The limitation on maintenance resources due to lower prices, the less engineering-oriented corporate culture and the increased need for maintenance resources because of aging, has placed maintenance decision making in a dilemma. In the past, when there were more resources, there was more margin to be conservative and to include maintenance actions that may not have been entirely necessary. With current and future limited resources, this degree of conservatism can no longer be tolerated. More fully quantitative techniques are now needed to better discriminate between maintenance actions.

7. DECISION ANALYSIS

Decision analysis, supported by financial analysis and optimization techniques, is already being used for large investment decisions that involve large expenditures. It can be found as part of a course, or as a whole course, in almost every MBA program in the US and in many foreign countries. This decision making methodology was developed almost thirty years ago [Raiffa, 1968] to account for conditions of uncertainty. It is very well suited for maintenance decision making because it addressees the uncertainty involved with maintaining equipment during the wearout part of its life cycle.

Decision analysis employs fully quantitative methods that stand the tests of challenge and review. These methods are classic in that they have been thoroughly tested and are unquestionably accepted by a broad spectrum of practitioners; they are well grounded in the principals of probability theory and of maximizing long term return to stockholders. Their appeal has been the ability to fully quantify and process input information through formal models.

Most companies at the management level are already accustomed to looking at financial decisions on a fully quantitative basis, and accounting for uncertainty in the decision in a fully quantitative manner, using classical methods. In addition, senior management generally expects presentations to be based on these methods. It seems reasonable, therefore, to use the same techniques for maintenance decision making because maintenance is competing for the samc resources.

8. BASIC PREDICTIVE MAINTENANCE DECISION MODEL

8.1 Decision criteria

In decision analysis, the first decision to be made is the criterion to be optimized. Based on the textbooks used most frequently in MBA corporate finance courses, this criterion would be Net Present Value (NPV) [Brealey, Myers, 1991]. This criterion insures that there will be a maximum return for the invested corporate dollar when decisions are being looked at over multiple years. In addition, some of the more recent terms used to emphasize stockholder value of a corporate investment are maximized when NPV is maximized. This makes NPV the most robust of the financial criteria to be used for optimization.

There is a danger, practiced by some engineers, in sometimes using reliability as a criterion for decision making. Reliability is an engineering term, not a financial term. It is not well understood in the battle for corporate resources. Secondly, reliability and NPV analysis can produce the same results when the equipment of concern is in demand a high percentage of the time. However, maintenance decisions for the less-highly utilized equipment are the most difficult. In these cases, NPV will guide the decision toward highest value, rather than highest reliability, because highest reliability is not always needed.

In the case of maintenance, the "Net" of Net Present Value is created by looking at the choice between two maintenance decisions [Mauney, 1993; ASME 1994]. The first possible decision is doing nothing different. That is, run the equipment as it is, performing normal routine maintenance. This we call the "base case." Here, we consider the consequence of shutdown or capacity reduction as a result of keeping the aging equipment operating. The intent of the maintenance action is to avoid this consequence. We call this the benefit of the maintenance action, because we are taking credit for preventing the consequence of shutdown or capacity reduction. The second possible decision is to take a mitigating maintenance action to avoid the potential consequences of failed equipment and shutdown. This we call the "alternative case." It is the cost of taking maintenance action. Here, we look at the cost of the maintenance action, plus the consequence of a shutdown or capacity reduction that might still occur because of the maintenance action not being perfect.

The diagrammatic representation of this is shown in Figure 2 by a decision analysis influence diagram. This diagram is constructed by placing the decision criterion to be optimized on the right. The diagram and its relationships are constructed as we ask the question, "What do we need to know to determine the NPV in this case"? We need to know the benefit and cost side as shown above and below the horizontal line in Figure 2. The consequence of failure without the action is the benefit, and the cost is the cost of the maintenance action plus the consequence of failure with the action. Ultimately, on the far left we have the decision of whether or not to take the action represented in a square. Note that this diagram establishes a relationship between the decision, the specific terms in the maintenance action and its consequences, and the decision criterion of NPV. What you see in this diagram is the cash flow streams created by the relationships in this decision problem.

The "Present Value" part of Net Present Value considers the effect of taxes and the time value of money. Because maintenance decisions on aging equipment have to do with timing, this is an important consideration in any maintenance decision analysis. In taxed corporations, the tax effects on maintenance expenditures, as well as losses declared from shutdowns or capacity reductions are significant. Engineers usually are not concerned with these but, in the financial arena, they need to be accounted for. The same is true for time value of money. This accounts for inflation and the expected return to be passed to the shareholder. The discount rate is usually used, so that the expected return for the invested maintenance dollar has to meet or exceed a minimum desired return to the shareholder over time.

8.2 The link between engineering and finance

Perhaps one of the most important steps in the fully quantitative process is the conversion of engineering analysis results to financial consequences. The barrier to understanding the engineering position on an important maintenance project proposal is that the engineer usually speaks in engineering units, such as time to failure or probability of failure. He talks of the probability of occurrence, not consequence of occurrence. "Consequence of occurrence" is a term that business-oriented managers understand. Engineers need to convert engineering units into financial units.

The term that the decision maker is most concerned with is not whether a component will fail, but what this failure will mean to the health of the company. A more accepted term in engineering and other technical communities with which to describe this is “risk” [ASME, 1994]. Risk is defended as the product of probability of occurrence and consequence of occurrence. In the financial community this same formulation is called the “expected value” of the consequence [Mauney, 1993]. Because of this similarity in definitions, we now have a common link between the engineering world and the financial world. This relationship, defined by risk or expected value of consequence of failure, will be a key in the formulation of the decision model for fully quantitative maintenance decision making.

This key is further represented in the development of the maintenance action decision influence diagram shown in Figure 3. Note the addition of the probability of failure, consequential failure cost, and the conversion of the consequential failure cost node into the expected consequential failure cost nodes. The "with" and "without" maintenance actions cash flow streams create the net value of the decision criterion. The year to perform the maintenance action is added to this figure given that it has been decided to perform the action. The arrows that may seem strange in this figure are the ones pointing from year of performance of the action to the probability of failure without the action, and the one pointing from the probability of failure without the action to the expected consequential cost with action. The reason for these two arrows is to establish the relationship for the expected consequence of failure if the maintenance action is delayed. During the delay to the year in which maintenance action is performed, the expected consequence of failure will be the same as if the project action had not been taken. Note that after the year to perform the maintenance action, the probability of failure with action is used to determine the expected consequential cost of failure with the action.

8.3 Probability of failure with time

The use of "probability of failure" with time is an all inclusive expression of the engineering condition of the equipment. For many years worst case analyses were performed. A probability of failure is set to an absolute certainty — the value of one — with full consequences at that same time for purposes of financial analyses. By intent, this approach is very conservative. Worst case scenario inputs in this type of analysis are meant to account for the uncertainty in the inputs. However, engineering input is limited to the worst side of the distribution rather than using the whole distribution. In addition, engineering, by using single numbers for lives, was, by implication, leading decision makers to believe that the equipment condition was more precisely known than it, in fact, was. The use of full distributions to describe inputs is an all inclusive description of the inputs and their uncertainty, rather than just a restrictive worst value. The advent of probabilistic fracture mechanics started an awareness that a complete expression of the engineering state of the equipment could best be expressed with a probability of failure versus time curve. It is clear from the financial discussions above that an expression of the engineering condition of the equipment in this form directly integrates into already established classical probabilistic, decision, and financial analysis methods.

8.4 Constraints

The further linking of this maintenance decision process to the real world is shown in Figure 4. Note that the maintenance budget limit constraints is added. In some companies there is a corporate policy of a downtime performance limit. This influence diagram shows how this constant is incorporated.

9. OPTIMIZING THE DECISION MODEL

The objective in decision making is to optimize the decision criterion. In this maintenance decision situation we are trying to determine a maintenance action year to maximize NPV, while not exceeding either of the constraints of maintenance budget limit or other limits. The goal is to optimize NPV by maintenance action timing [Mauney, 1993; ASME, 1994].

Operations research methods can perform this function on the decision model described. In fact, the decision model has been constructed in a spreadsheet, and an operations research type optimizer has been written as a spreadsheet macro [Mauney, 1993; ASME, 1994; ASME, 1997].

10. FINANCIAL INPUTS

The financial assumptions used in this type of analysis are usually well established in the financial organization of a corporation. For the engineer, it becomes a matter of seeking them out. These financial assumptions need to be included because they have significant effect on the timing of the maintenance project.

The first set of financial assumptions affect the time value of money. First is the projected inflation rate, to account for the eroding value of the dollar with time. Second is the corporate discount rate used to ensure that any investment is expected to have a return that is at least equal to the planned return for the stock and bond holders.

An assumption seldom thought of is the composite income tax rate and the property tax rate. Of these two, the composite income tax rate is the most significant and has a large effect on the value of a maintenance decision. The effect is that financial loss consequence can be viewed as a business loss, and a significant portion can be written off. The impact of the maintenance cost itself can be greatly reduced for the same reason if the maintenance cost is classified in an expense category. If the maintenance cost is capitalized, it still can be written off but at a much lower rate, over time, through the tax effect of depreciation.

11. ENGINEERING INPUTS

On large maintenance decisions, a lack of credibility is created in the mind of the financial decision maker when excessively large financial consequence estimates, and the worst case probability of failure, are used. However, if a prudent, fully quantitative engineering probabilistic analysis or assessment is performed, and the financial consequences of a failure are appropriately estimated, there is more credibility to the risk or expected value of the consequence provided to the decision maker. As a result, it is appropriate in aging large equipment maintenance decisions that the engineer use all possible technologies to assess the condition of the equipment from the standpoint of failure.

The area of probabilistic fracture mechanics best describes these conditions in most cases. These techniques establish a damage propagation mechanism mathematical model with a failure criterion. This model is then integrated over the range of input distributions with a "Monte Carlo" program or fast probability integration techniques. This integration produces a probability of failure verses time curve [Mauney, 1993; ASME, 1994].

There are less rigorous methods that are fully quantitative for obtaining probability of failure versus time. These methods are good for obtaining preliminary engineering inputs for the maintenance decision analysis. The resulting inputs allow a first level look at the possible allocation of maintenance resources. The first of these methods is the trending of failure history data on a plant component of concern, or on similar components at other plants. A method used for this is standard linear regression analysis using the Weibull distribution function. The reason to use this function is that it was derived to express the reliability of equipment with time. It has been used extensively for this purpose. A second method is a structured interview for future failure projection with plant personnel most familiar with the equipment. This method is called a probabilistic assessment interview, or expert elicitation. It was developed by cognitive psychologists in the decision analysis community over the last fifteen years, and has been extensively used with senior corporate executives on large corporate decision analyses [Mauney, 1993; ASME, 1994; ASME, 1997].

The Bayesian update analysis is a classical method of combining failure history projections with plant personnel structured interviews. The Bayesian method, developed several hundred years ago [Bayes, 1958], focuses on the combination of current probabilistic information of an event with newly found information for the same event, resulting in updated information [Schmitt, 1969]. It is objective and can be used to combine failure history projection data and plant personnel interview projections, using a weighted averaging technique based on the normalized product of the two projections for each year. Using a technique of this type provides better projection of the future failure probability because of the use of both data sources. In addition, using these preliminary techniques allows the decision to be assessed inexpensively as a first iteration in order to determine the need for a higher cost, fully quantitative engineering analysis.

12. EXAMPLE OF SINGLE COMPONENT DECISION

12.1 Primary reformer replacement example

An example of decision analysis supposes a primary reformer tube component in a single path ammonia synthesis unit. The concern is if, and when, to spend \$500,000 for replacement of the entire component. The failure in this example is the burst of any of 200 tube segments, resulting in shutdown of the entire power unit. It is assumed that the unit will be down for six days for repair of each tube burst and the repair cost would be $20,000 and the damage is assumed to be confined to the area of origin. The unit is 1,000 standard tons per day capacity, with an income on an hydrous ammonia per ton of $222.

The probability of failure versus time is based on a remaining life analysis, as shown as the base case in Figure 5(a). This is the expected probability of tube burst versus time curve, given no replacement and given that the unit continues to operate. The alternative case curve assumes that the primary reformer is replaced. Here, the probability of failure curve is lowered due to the replacement action. NPV is generated by decreasing the probability of failurc by replacement of the primary reformer component.

Note from Figure 5(b) that the highest NPV is created by replacement of the component in 1997. Note, also, that the highest NPV is positive, which means that the primary reformer should be replaced, creating the highest value if performed in 1997.

12.2 Boiler feedwater pump example

A second example of applying decision analysis can be shown using a boiler feedwater pump in the same ammonia synthesis unit. If boiler feedwater is not continuously supplied to the boiler, the ammonia synthesis unit would have to be suddenly shut down. The hypothetical decision to be made is if and when to spend $150,000 to replace the boiler feedwater pump. The failure in this example is a failure of the boiler feedwater pump to operate which results in loss of boiler feedwater. It is assumed that the unit will be down approximately 2.5 days for repair and the repair cost would be $20,000 and the damage is assumed to be confined to the areas of origin. The unit is the same capacity as above. The probability of pump failure versus time, based on similar design and service units, is shown as the base case in Figure 6(a). This is the expected probability of failure versus time curve.

The alternative case curve is that the pump is replaced to lower the probability of failure versus time curve. NPV is generated by decreasing the probability of failure by replacement of the boiler feedwater pump.

Note from Figure 6(b) that the highest NPV is created by performing the replacement in 2016. Note, also, that the highest NPV is negative, which means that the replacement should not be performed, creating the highest value if performed during 2016. This conclusion is based on the twenty-year analysis window.

13. CONCEPT OF A MULTIPLE COMPONENT DECISION

In reality, *numerous* maintenance projects are competing for the maintenance budget simultaneously. Maintenance optimization is defined here as the performance of a decision analysis process that maximizes NPV for many maintenance projects simultaneously. In a chemical company, all projects can be annually constrained by the maintenance budget limit, which means that maintenance optimization needs to be conducted with all individual maintenance decision models linked [Mauney, 1993; ASME, 1994; ASME, 1997].

Maintenance optimization has several benefits. The more maintenance projects, and the larger the portion of the maintenance budget that is modeled, the larger the expected value benefit that can be realized. In addition, decision analysis more closely represents the real maintenance budget decisions that are being made, and is performed using a systematic, fully quantitative and reviewable process.

For maintenance optimization, the decision model previously described in Figure 4 is expanded. For each additional project, the associated two probabilities of failure and year of maintenance action nodes will be added to the influence diagram and also to a spreadsheet. The multiple maintenance projects optimization then provides a prioritized list of decision alternatives and years of maintenance action that creates the maximum NPV for the group, while staying within the constraints. This is accomplished in a spreadsheet by inserting different combinations of years of action for each maintenance project seeking that combination for the group that maximizes total NPV without exceeding the constraints. This is accomplished through an operations research algorithm approach.

An example of the benefit of this approach was demonstrated with a PC based spreadsheet model, applied to 140 forced outage critical fossil power plant tube components across a power system [Mauney, 1993]. The estimated resulting NPV of the optimization was in the tens of millions of dollars.

14. CONCLUSIONS

A fully quantitative maintenance decision analysis model can:

1) Be optimized for the timing of the maintenance actions, maximizing the value of the maintenance dollar to the corporation;

2) Create a link between engineering and the financial decision maker as a result of the common definition for risk and expected value of consequence of failure; and

3) Allow engineering to compete for corporate resources during times of tight budgets.

15. ACKNOWLEDGMENTS

The authors would like to acknowledge the support of Southwest Research Institute in the continuing development of this work, the numerous thought provoking discussions with Mr. Dan Ash and, finally, the assistance of Ms. Mary-Sue Searles in the preparation of this manuscript.

16. REFERENCES

ASME Research report, "Fossil Fuel Fired Electric Power Generating Station Applications, Risk-Based Inspection-Development of Guidelines," CRTD, Vol. 20-3, ASME, New York, 1994.

ASME Research Report, "Risk-Based Inspection, Applications Handbook," CRTD, ASME, New York, (to be published 1997).

Bayes, T., "An Essay Towards Solving a Problem in the Doctrine of Chance," Philosophical Transactions of the Royal Society, 53, 1763, pp. 370-418, reproduced with biography of Bayes in G. A. Barnard, "Studies in the History of Probability and Statistics: IX," Biometrika, 45, pp. 293-315, 1958.

Brealey, A. and Myers, S. L., Principles of Corporate Finance, McGraw-Hill, 4th Ed, 1991.

Heiges, H. H. and Stoll, H. G., "Power Plant and Turbine-Generator Upgrading Economics," Proceedings of Fossil Plant Life Extension Conference and Workshop, Electric Power Research Institute, Palo Alto, CA, p. 12-3, August 1985.

Mauney, D. A., "Economic Optimization of Multiple Component Replacement/Inspection in the Power System Environment," ASME Publication, PVP Vol. 251, Reliability and Risk in Pressure Vessels and Piping, Ed. J. H. Phillips, pp. 1-16, July 1993.

Miller, I. and Freund, J. E., Probability and Statistics for Engineers, Prentice-Hall, Englewood Cliffs, NJ, pp. 365-367 and 379-380, 1965.

Raiffa, H., Decision Analysis, Addison-Wesley, Reading, MA, 1968.

Schmitt, S. A., Measuring Uncertainty: An Elementary Introduction to Bayesian Statistics, Addison-Wesley, Reading, MA, 1969.

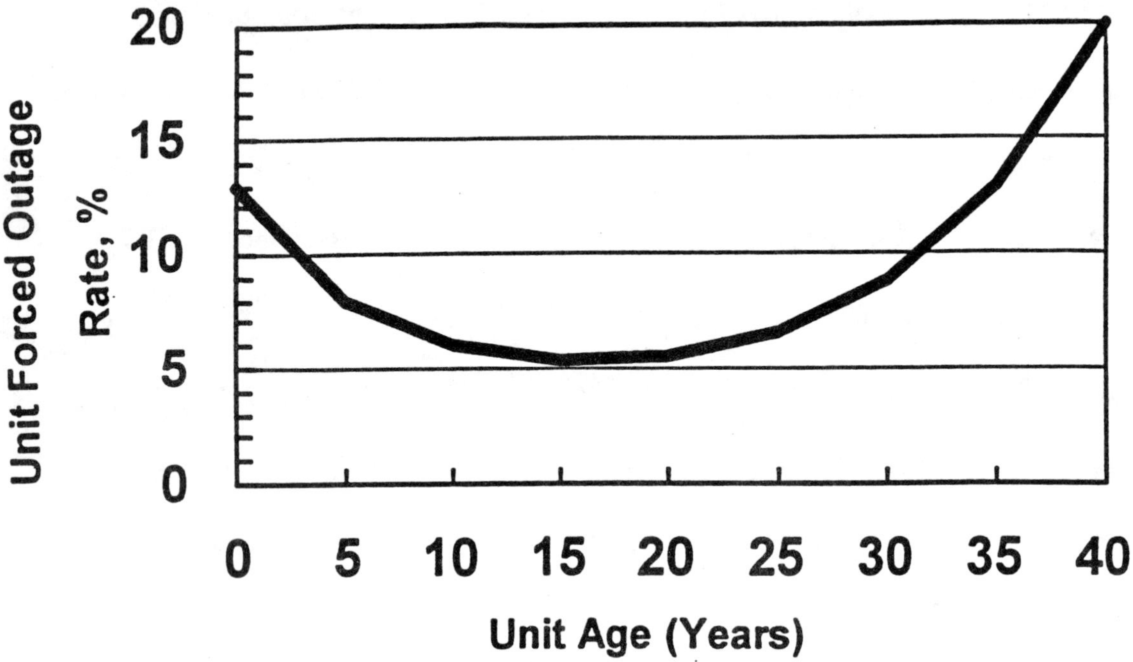

Figure 1. NERC-GADS Data for Forced Outage Rate for Coat Fired Units from 50MW to 200 MW.

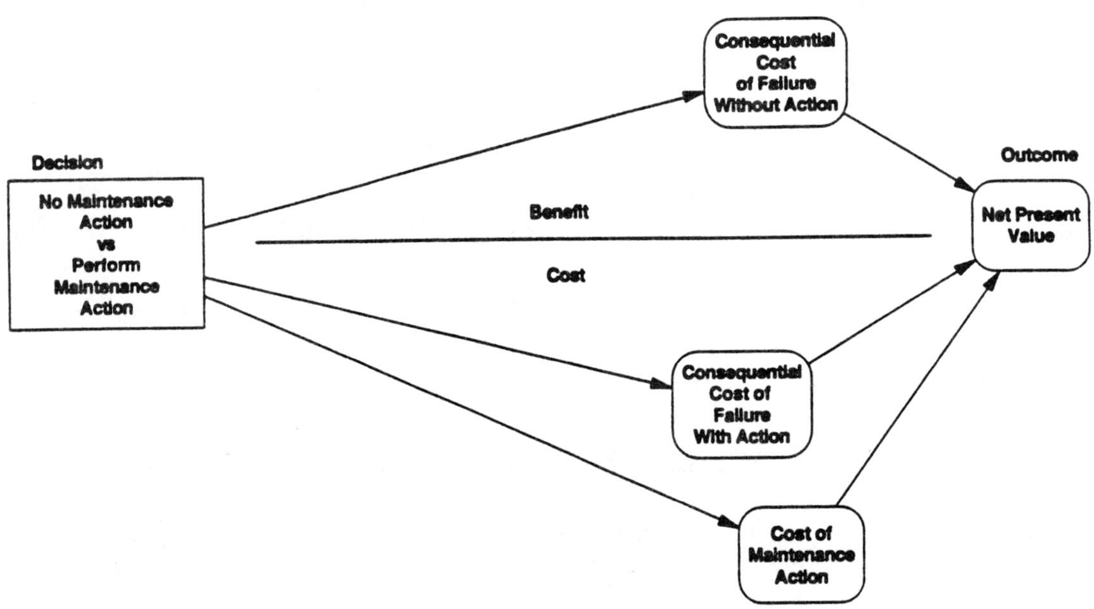

Figure 2. Influence Diagram Emphasizing the Links Between Net Present Value and the Maintenance Action Decision

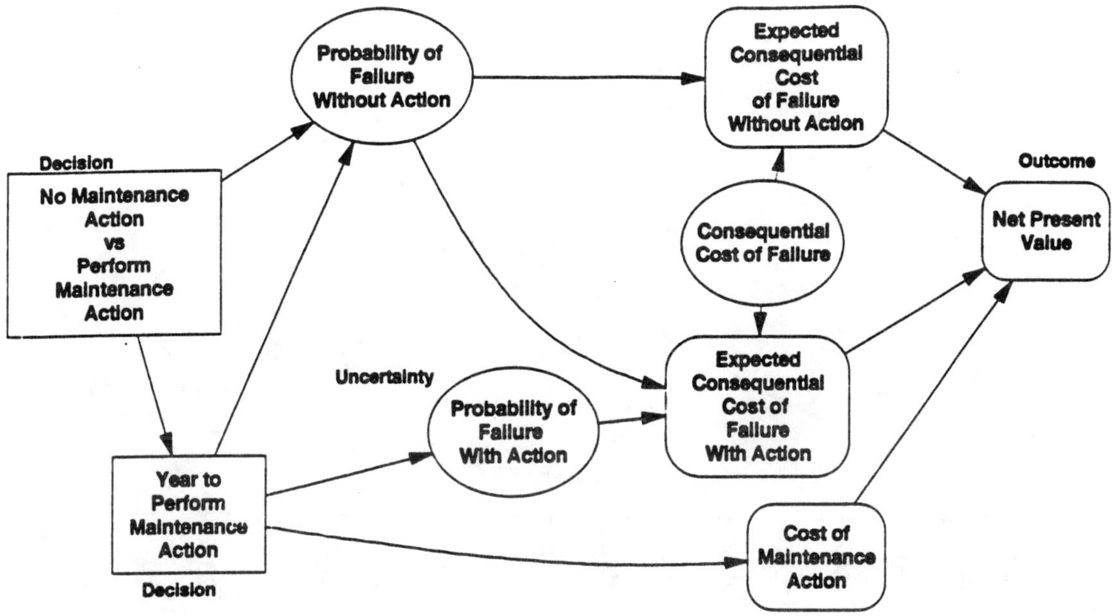

Figure 3. Influence Diagram Emphasizing the Replacement of Consequential Cost with Expected Consequential Cost

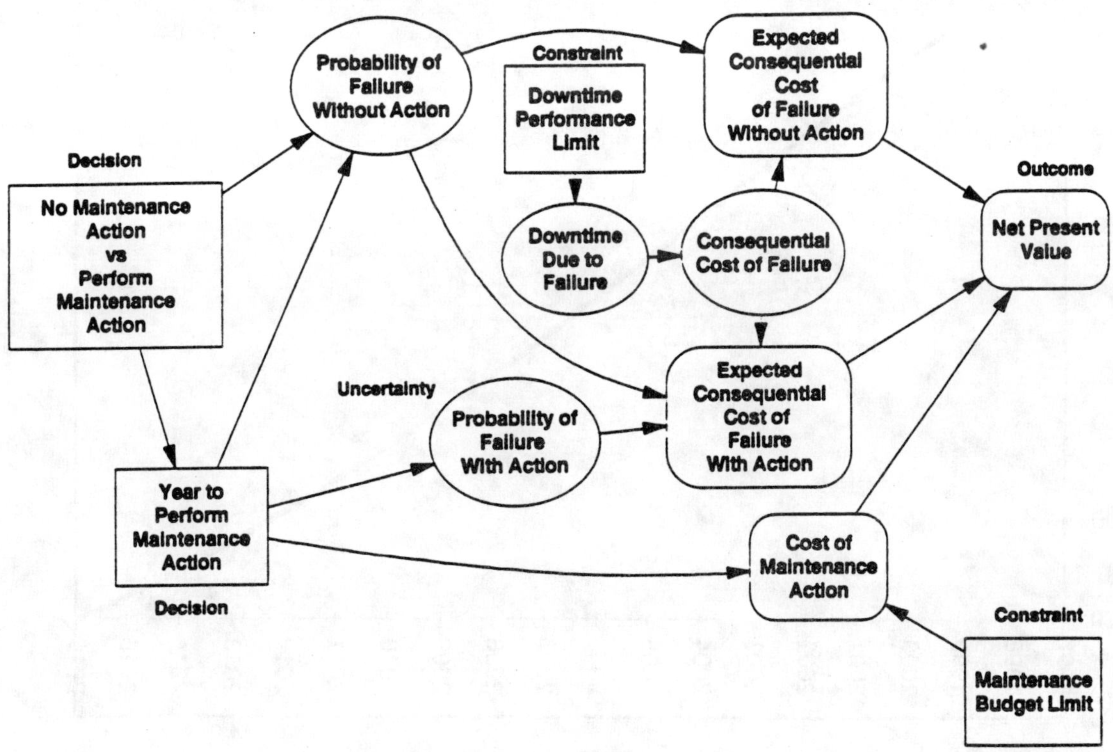

Figure 4. Influence Diagram Emphasizing the Timing of, and the Constraints on, the Maintenance Decision Process.

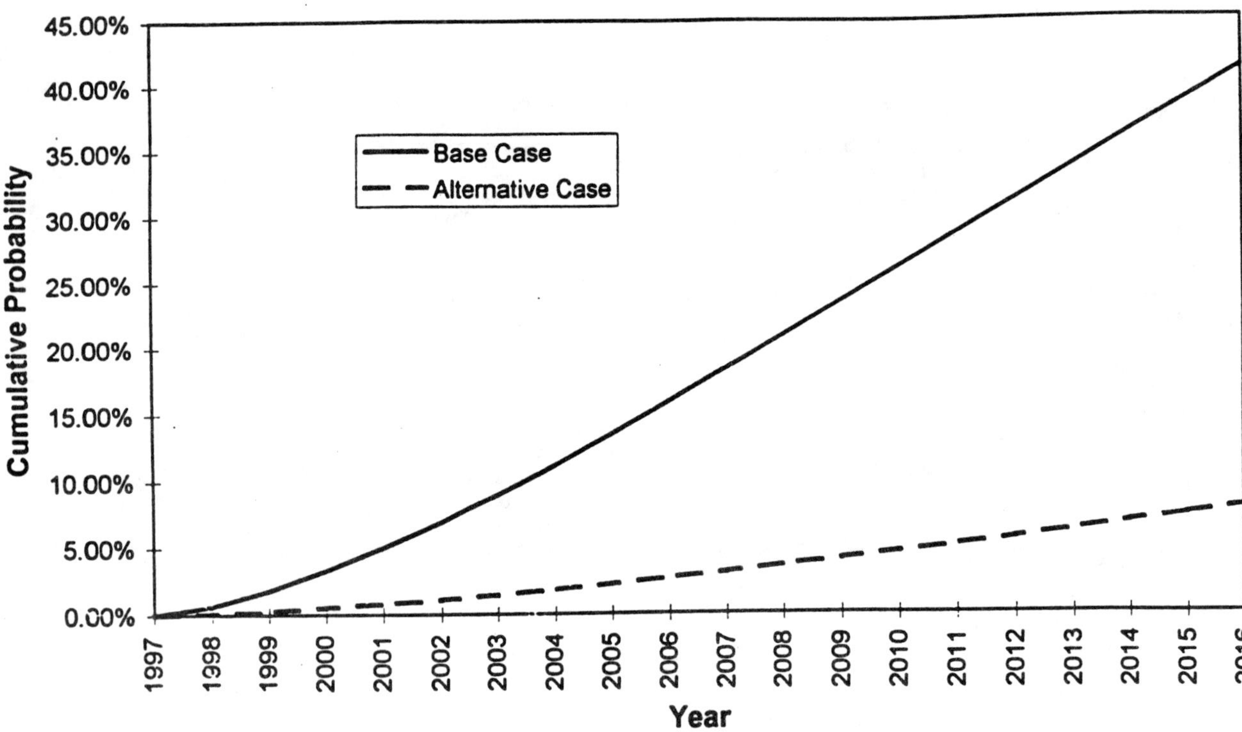

Figure 5(a). Primary Reformer Tube Probability of Failure Vs. Time

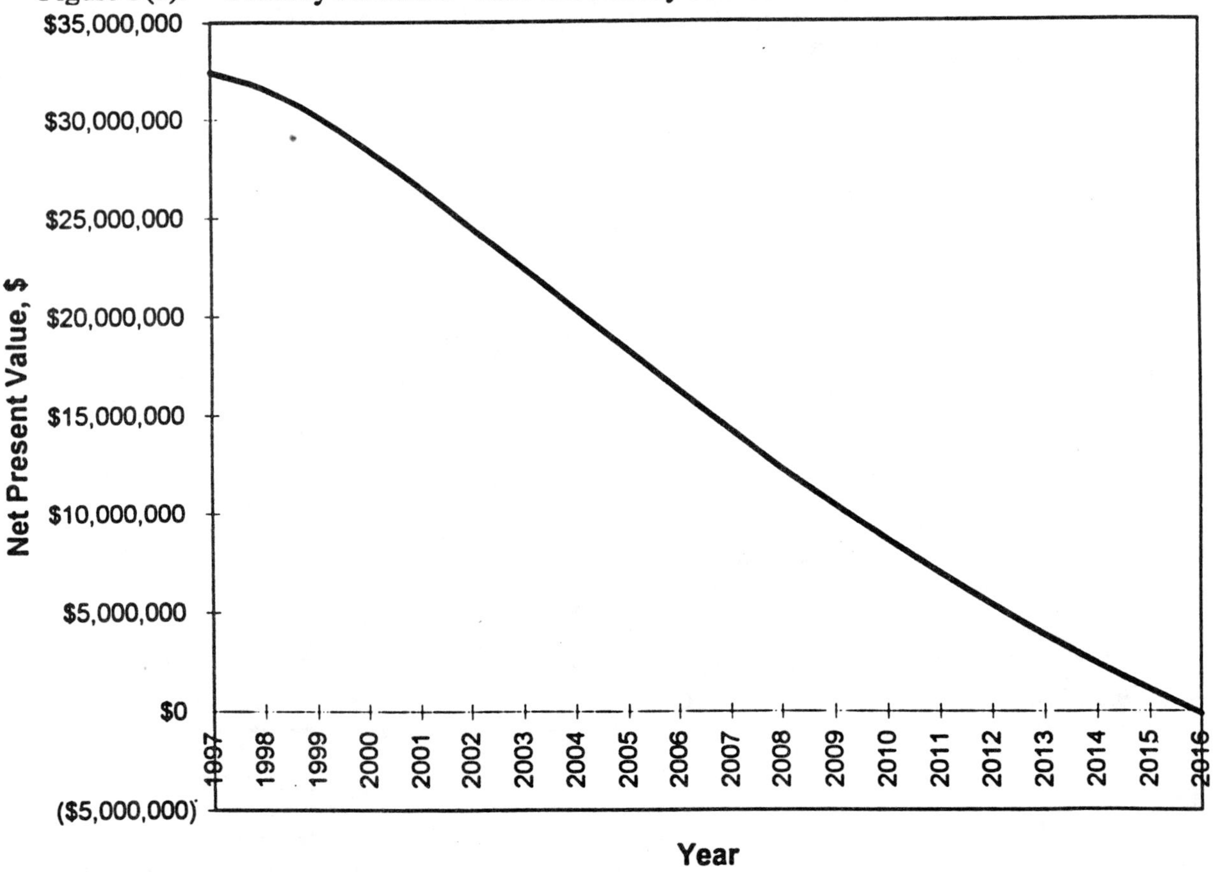

Figure 5(b). Net Present Value Vs. Maintenance Action Year for Primary Reformer Replacement Example.

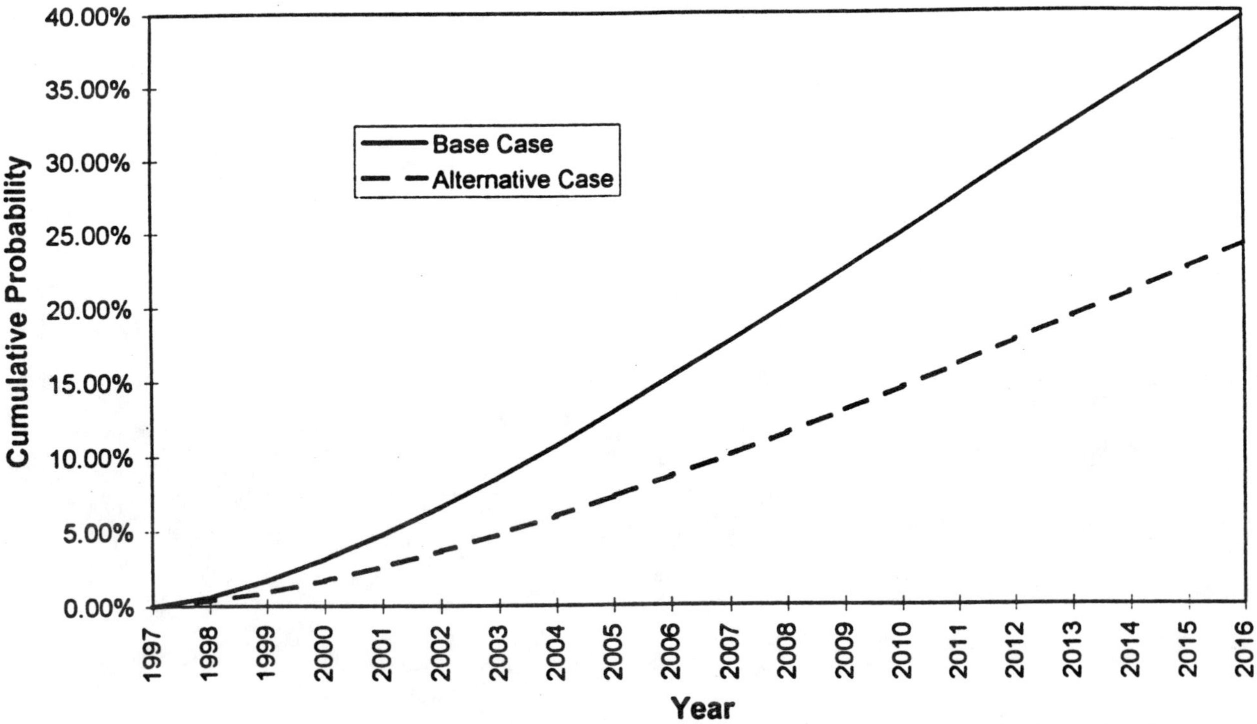

Figure 6(a). Boiler Feedwater Pump Probability of Failure Vs. Time

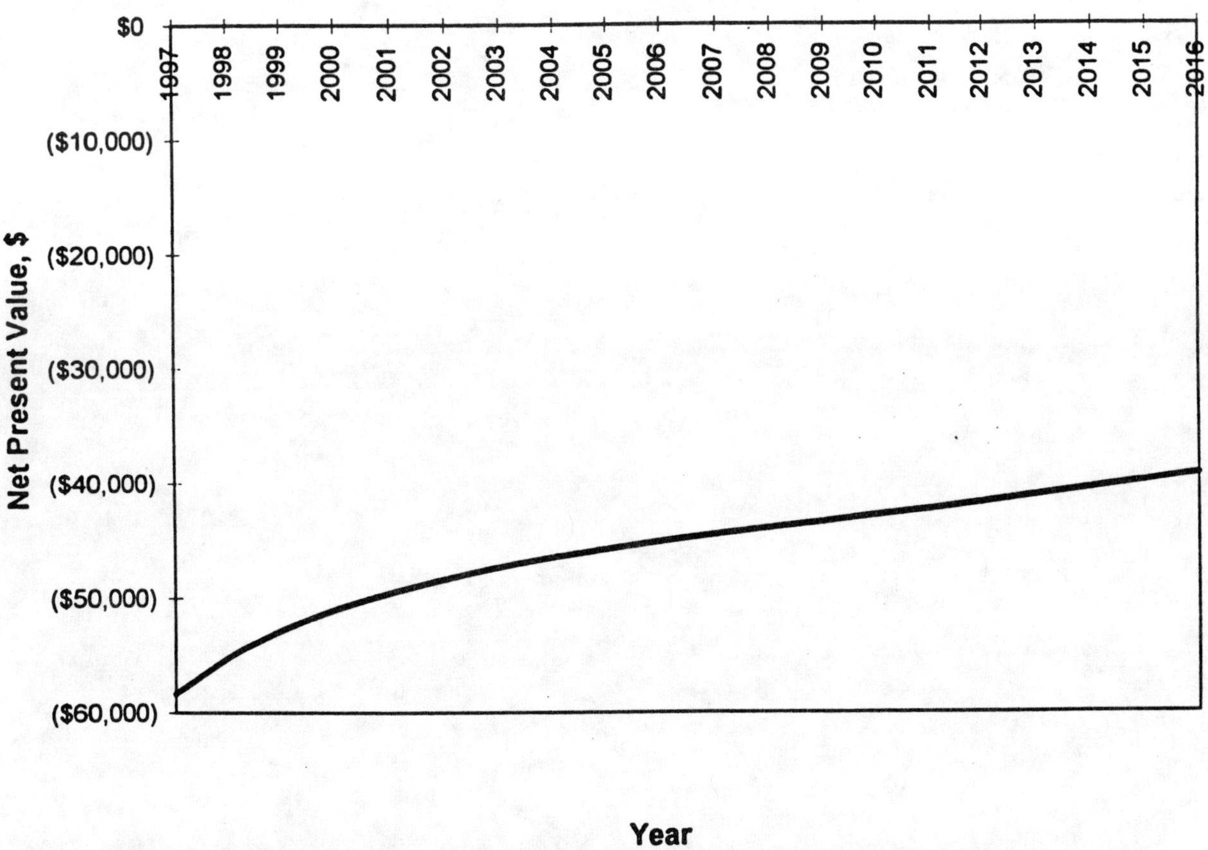

Figure 6(b). Net Present Value Vs. Maintenance Action Year for Boiler Feedwater Pump Replacement

Paper 39e

RISK BASED PRIORITIZATION OF MAINTENANCE REPAIR WORK

John A. Harnly

Exxon Chemical Co.
P.O. Box 4004
Baytown, TX 77522

Prepared for Presentation at the AIChE 31ST Annual Loss Prevention Symposium,
Case Histories and Miscellaneous Topics

March 1997

RISK BASED PRIORITIZATION OF MAINTENANCE REPAIR WORK

by
John A. Harnly
Exxon Chemical Co., Baytown, TX

Abstract

This paper describes the development of a risk ranked Inspection Recommendation procedure that is used by one of Exxon's chemical plants to prioritize repairs that have been identified during equipment inspection.

As part of the Company's Safety Management Practices initiative in the late 1980's, a procedure was put into place to ensure that an Inspector's repair recommendations were properly addressed by the organization. The initial procedures were successful at "systematizing" the documentation and stewardship-to-completion of the Inspector's recommendation, however, there were complications with the original process:

1) The Inspector made a simple High, Medium or Low assessment of the priority/criticality of the recommendation. Frequently, this resulted in disagreements with Operations about the true priority of the recommendation.

2) If there was agreement on the priority of the recommendation, there was still disagreement on the relative rank within the priority - which High priority was the highest priority?

3) With limited funds to spend on repairs, it was (and is) important to make sure that the money was being spent on the highest risk items that had the greatest risk reduction/cost benefit ratio.

To address these concerns, the procedure was modified to incorporate a risk assessment of the recommendation by both the Inspector and Operations. In the new procedure, the Inspector describes the deficiency that he/she finds and assesses the probability of failure within a certain time-frame. Operations must assess the consequences, from an environmental, safety and economics standpoint, were the failure to occur. These assessments are combined in the typical risk equation (risk = probability x consequences) to arrive at a severity index which serves to rank the recommendation relative to the other recommendations. Because Operations participates in the assessment there is very little disagreement about the priority of the recommendation. The severity index puts the recommendations in order so it is quite clear which are the highest priority recommendations. This process has helped to focus the entire organization on those deficiencies that represent the greatest risk with the result that less time and money is spent correcting items that have a low risk/cost benefit ratio, allowing these savings to be used to reduce the higher risks in the plant.

INTRODUCTION

In 1989 the author's company outlined a series of Safety Management Practices that were designed to bring a systematic approach to the safe operation of the company's manufacturing plants throughout the world. These Practices cover virtually every aspect of the operation of the plants from Management of Change to Inspection and Maintenance. The Inspection and Maintenance Practice has several components in it but one specific component requires that Inspection's recommendations for repairs or replacements be documented and effectively communicated to the appropriate management, that accepted recommendations be tracked until completed and that the rationale for rejected recommendations be documented. This paper describes the development and evolution of that Inspection Recommendation (IR) procedure at the author's manufacturing facility.

THE INSPECTION RECOMMENDATION PROCEDURE

In the basic Inspection Recommendation process (Figure 1), the Inspector completes an IR form describing the deficiency that was found and recommending a repair. The IR is forwarded to the Operations Representative who reviews the IR and either approves or rejects it. A rationale is required for any rejection and rejected IRs are reviewed with the Manufacturing Department Manager for confirmation or override of the rejection. Approved IRs are logged into a computer database for future stewardship and the original IR is retained by the Inspection Organization until the deficiency is corrected and the Inspector approves the close out of the IR. Monthly, reports are generated and the status of the IRs are reviewed in a team meeting with Operations and Maintenance where action plans are developed for the next month. Quarterly, reviews are held with upper management to discuss progress on the IR lists and address areas where progress isn't as good as desired.

Initially, the team tried to complete each IR shortly after they were issued. With the increase in inspection activity that was created by the Safety Management Practices, the number of active IRs grew. It was quickly recognized that all the IRs couldn't be completed immediately and many of them didn't need to be completed immediately. Some method was needed to prioritize the IRs. On the original form (Figure 2), a spot was added where the Inspector could make a High, Medium or Low risk assessment of the IR.

This procedure met all the requirements of our Inspection and Maintenance Practice for Inspection Recommendations. From all appearances it seemed that the Inspection Recommendation Procedure was working as planned, however, there were problems:

First, because the Inspector was the only person assessing the IR, it was not uncommon for the other members of the team (operations and maintenance) to challenge the assessment. What the Inspector thought was high risk (based on his/her understanding of the unit operation) the operations people might consider to be a low risk and vice versa. While the Inspector had a good feel for the condition of equipment, he/she didn't necessarily have a good understanding of what could happen if the equipment came out of service unexpectedly. As a result, a lot of time was spent in meetings debating about the correct risk for a particular IR.

Second, when agreement on the risk assessment was finally reached, the team was still confronted with the problem of which IRs to work first. Which of the High IRs was the highest? Working within a budget, decisions have to be made that give you the biggest "bang for your buck". The imprecise High/Medium/Low assessment that was made did not provide a fine enough division to determine which High, Medium or Low IR to work on first.

Finally, there was the intangible issue of IR ownership. Because it was the Inspector who was writing the IR, assessing the IR and closing the IR, the tendency was to view the Inspection Recommendation Procedure as the Inspector's system and, therefore, the Inspector's responsibility. In reality, the IR is a tool that the Inspector uses to inform the equipment owner (Operations) of the condition of his/her equipment. It was important that Operations feel a real ownership for the IR procedure. That was not happening with the current arrangement.

A change was needed, to:

1) remove some of the assessment load from the Inspector;
2) provide a finer division for the priority of the IR; and

3) encourage ownership of the IR by the equipment owner (Operations).

FIRST EVOLUTION - RISK ASSESSMENT

With this first Inspection Recommendation procedure, the entire risk assessment load was on the Inspector. But the traditional risk equation:

Risk = Probability * Consequence

contains two components:

1) the probability that an event will occur, and
2) the consequences if that event does occur.

The Inspector's training makes him/her qualified to assess the probability that a deficiency will lead to an equipment failure, but the Inspector is not necessarily qualified to assess the consequences should the equipment fail. Assessment of consequences is the domain of Operations. They know, better than anyone, the consequences to the unit of a failure of a specific piece of equipment. Sharing the IR assessment responsibility with Operations, by having the Inspector assess the probability of failure and Operations assess the consequences, looked like it could solve the concern for the assessment load on the Inspector and it held promise for drawing Operations into assuming more ownership of the IRs. The challenge was to come up with an assessment process that was quick and simple but provided a meaningful output that could be used to prioritize the work.

Risk assessment can be quantitative or qualitative. The output of a quantitative risk assessment typically will be a number such as damage area (ft^2) or cost impact ($). The number could be used to prioritize a series of items that have been risk assessed, however, true quantitative risk assessment requires a great deal of data both for the assessment of probabilities and assessment of the consequences. Failure frequencies are usually required and tools such as fault trees or decision trees are often used to determine the probability that a certain sequence of events will result in a certain consequence. Quantitative Risk Assessment clearly does not meet the "quick and simple" criteria.

Qualitative risk assessments are less rigorous and the results are often shown in the form of a simple risk matrix where one axis of the matrix represents the probability and the other axis represents consequences. A risk assessment is made using the general definitions provided for the different probability and consequence levels with the risk identified by the intersection of the chosen assessment levels on the matrix. Figure 3 shows a typical 3x3 risk matrix. Example 1 on the matrix is at the intersection of Low Probability and Low Consequence, resulting in a Low-Low risk. Example 2 on the matrix shows a risk of Medium-High. One intuitively recognizes that the Medium-High scenario is a higher risk then the Low-Low but prioritizing these different word combinations can be cumbersome.

If a value is given to each of the probability and consequence levels, a relative value for risk can be calculated. In Example 1, if the low probability is given a value of 1 and the low consequence is given a value of 1, the relative risk for Example 1 would be 1 (1*1). Example 2, with a medium probability (value 2) and a high consequence (value 3) will have a relative risk of 6 (2*3). It is important to recognize that the risk value is a relative number that is meaningless outside the framework of the matrix. Within the framework of the matrix it provides a natural prioritization of items assessed using the matrix. This method meets the "quick and simple" criteria and can provide a value that can be used for prioritization. A variation of this method is what was chosen for assessing IRs.

Figure 4 shows the final format for the probability and consequence levels. For probability of failure (called Failure Potential), the Inspector is required to make an estimate of the time to failure and select from one of four Failure Potential levels. Selecting Failure Potential 1 indicates that the inspector feels that the failure could occur within 5-10 years if it is not corrected. A Failure Potential of 2 means 1-5 years for failure and a Failure Potential of 3 means failure could occur within the year. A Failure Potential of 4 is reserved for those scenarios where the time to failure is not predictable.

The consequence assessment is done in two parts - one directed at the Safety and Environmental consequences and one directed at the Economic impact on the operation (Operating expenses, lost production, etc.). Each of these consequences can have a value of 0 - 3, representing a range from "no concern" to "high concern". The consequences of a failure are determined by a number of considerations some of which include:

- Type of failure (leak, rupture);
- Product characteristics (flammability, toxicity,);
- Disposition of the product (vapor cloud, standing liquid);
- Volume released;
- How quickly the leak can be detected/brought under control;
- Proximity to other units or the community.

The Operations Representative must consider each of these in making the consequence assessment. For example, a light hydrocarbon that is released due to a ruptured pipe that is connected to a large supply drum and is in a congested unit with ignition sources nearby would probably be considered a High Safety/Environmental Consequence. If that same hydrocarbon leaks through a small pinhole on a pipe that is easily isolated and not near any ignition sources, it might be considered a Medium or Low Consequence. Each IR must be evaluated on its own merits and scenario.

The final risk value (called Severity Index) is calculated by adding the two consequence values and multiplying them by the Failure Potential.

$$SI = F * (S + E)$$

Some quick math and you can determine that the Failure Potential ranges from 0 to 24, with 0 representing essentially no risk and 24 representing the highest risk - a failure that could occur at any time and could have high consequences for both Safety/Environmental and Economics.

Notice that the wording for each of these levels uses language such as "Item could fail" and "moderate concern that failure will result in injury" rather than specific comments like "Item will fail" or "Failure will result in injury". It is important that the users of this procedure not be intimidated by it. It was recognized that this is not an exact science and that precise answers were not possible. Moderating the language in this way makes it more palatable to the individuals that are being asked to make these assessments.

After training with the Inspectors and Operations and modifying the Inspection Recommendation form (Figure 5) the new assessment process was implemented. There were several immediate impacts:

1) The change in the monthly meetings was remarkable. Because it was now a shared assessment, there were no more debates on the ranking of the IR. The IR lists that are provided for the meeting are sorted by the Severity Index so the team is able to immediately focus on those IRs that represent the greatest risk to the facility.
2) Communication outside the meeting has also improved. Because the Operations Representative needs information on the type of failure in order to make an accurate consequence assessment, Operations and Inspection typically discuss the IR when it is written. Questions and concerns are usually resolved at that time.
3) The credibility of the IR system received a significant boost. The systematic approach to the assessment gives people the confidence that the assessment is reasonable and that the Inspector is not "crying wolf". In fact, as the system became more entrenched people began to relate more and more to the Severity Index. Now , there is a greater sense of urgency when an IR comes through with a high Severity Index. Resources are quickly assigned to work on it.
4) There is the general sense within the teams that they are truly working those items that represent the greatest risk to the integrity of the plant. The "squeaky wheel" does not drive this work.
5) The Inspectors moved back into their comfort zone. They no longer had the burden of determining the consequences of a failure and being continually on the defensive for their assessment. Overall, the relationship between Inspection, Maintenance and Operation has improved significantly.

SECOND EVOLUTION - ACTION DATES

Implementing the risk assessment process overcame some major stumbling blocks for the IR procedure, however, there was one concern that still needed to be addressed. There is a difference between an item that represents a safety or environmental risk and an item that represents an economic risk. Both of these consequences are important to an operating company but the implications of a safety or environmental incident far outweigh the effects of a loss in production. Just looking at the final value for the Severity Index could cause the team to overlook some IRs that have significant safety consequences.

By way of example:
One IR has a Failure Potential of 3, a Safety/Environmental Consequence of 3 and an Economic Consequence of 0 resulting in a Severity Index of 9. Another IR has a similar Failure Potential of 3 but a Safety/Environmental Consequence of 1 and an Economic Consequence of 3, resulting in a Severity Index of 12. Both of these items could fail within the year but, there is a much greater concern for the item that could cause injury or damage to the environment should it fail. If we strictly worked the high Severity Index items first, we would not work the high safety/environmental consequence IR first. This didn't "feel" right so we looked at ways to resolve this concern.

One approach would be to increase the values for the Safety/Environmental Consequence relative to the values for Economic Consequence. Then, any IR that had a significant safety or environmental consequence would end up with a larger value for Severity Index. This solution, however, doesn't guarantee that the IRs with significant safety/environmental issues will be addressed before they fail. An IR could have a significant safety consequence and still be overlooked due to the quantity of IRs with high Severity Indexes. We wanted to make sure that no IR with a High or Moderate Safety/Environmental Consequence slipped through the cracks.

The solution was to require the assignment of an Action Date for every IR that has a Safety/Environmental Consequence of Moderate or High (2 or 3). The goal is to complete the IR by the Action Date, however, an alternative is to re-inspect the item and extend the Action Date if the inspection supports that option. If the Inspector is not comfortable with extending the date, then Operations must take the actions necessary to minimize or eliminate the risk. This Action Date is a function of the Failure Potential, not the availability of the equipment, therefore, the Inspector sets the date. A key to this process is assigning Action Dates that are consistent with the Failure Potential. If the Inspector assesses the Failure Potential at 1 (5 - 10 years) it makes little sense to set an Action Date for 6 months.

CONCLUSIONS AND LEARNING'S

Some considerations for implementing this assessment method:

1) The assessors must make their assessments independent of one another for the risk assessment to be accurate. If the Inspector assigns a high Failure Potential because he/she thinks that the consequences of the failure are high than the severity index for the IR will be higher than it should be. In contrast, if the Operations Representative lowers his/her consequence assessment because it is not expected to fail for 5-10 years, then its severity index will be lower than it really should be. Both of these tendencies are quite common and should be monitored and corrected in the early going.
2) Multiple recommendations cannot be combined on one IR, unless it is clear that the Failure Potential and Consequence assessments will be identical. Loading up an IR can also be frustrating to the team because some items might be completed fairly quickly and others drag on. As a result, the IR doesn't get closed and the appearances are that there is no progress.
3) When this procedure is initially implemented, it is a good idea to go through some real life examples for the people that will be making the assessments. Not everyone is aware of, or comfortable with, the risk assessment concept so some early "hand holding" will reap significant benefits later on. Occasional "calibration" reviews, where you pick a few IRs and think through the logic that resulted in the assessments with the assessors, can also be beneficial.
4) Action dates must be set consistent with the Failure Potential. The tendency will be to set the dates much shorter than the Failure Potential would indicate is necessary. This will result in maintenance or inspection working something before it is necessary. Not only does it create an additional burden for everyone, the IR process begins to lose credibility because low risk items might be worked ahead of the higher risk items.
5) Participation and ownership by Operations is key. There will probably be an attempt to get the Inspector or the Maintenance Planner to do the consequences assessment. Avoid this. Keep Operations in the loop.
6) The process must have a mechanism to elevate the issue to upper management. Even with a tight procedure, there will be stalemates that require upper management

review and resolution. The procedure should account for this.

7) Regular sit-down reviews (monthly, bi-monthly or, at least quarterly) are key to the success of the procedure. Otherwise it just becomes another list to avoid.
8) Be patient! Each group will have a different perspective of your "angle" for implementing the process. Inspectors will be concerned that needed repairs are being ignored when low risk items migrate to the bottom of the work list. Actually the work that really needs to be done, the high risk items, are worked more quickly and with less distractions with this process. Maintenance will think you are trying to give them more work to do, especially if the list of IRs keeps growing month to month. The reality is that they become more effective with what they work on because the low risk work items fall to the bottom and aren't worked until the high risk items are satisfied. In the past, they were at the mercy of the squeaky wheel. Lastly, Operations will resent the added work load and responsibility that the consequence assessment gives them. You can't argue with them on that point but you can make the point that, by identifying and working the high risk items first, the safety of their work environment is improved.

This method for prioritizing Inspection Recommendations has been extremely effective for the plant but it didn't just happen. It required constant nurturing in the early days and still requires due vigilance to ensure that it continues to operate smoothly. However, considering the reduction in overall risk that is accomplished by working the high risk items first, and the focus that it gives the organization, the effort is worth it.

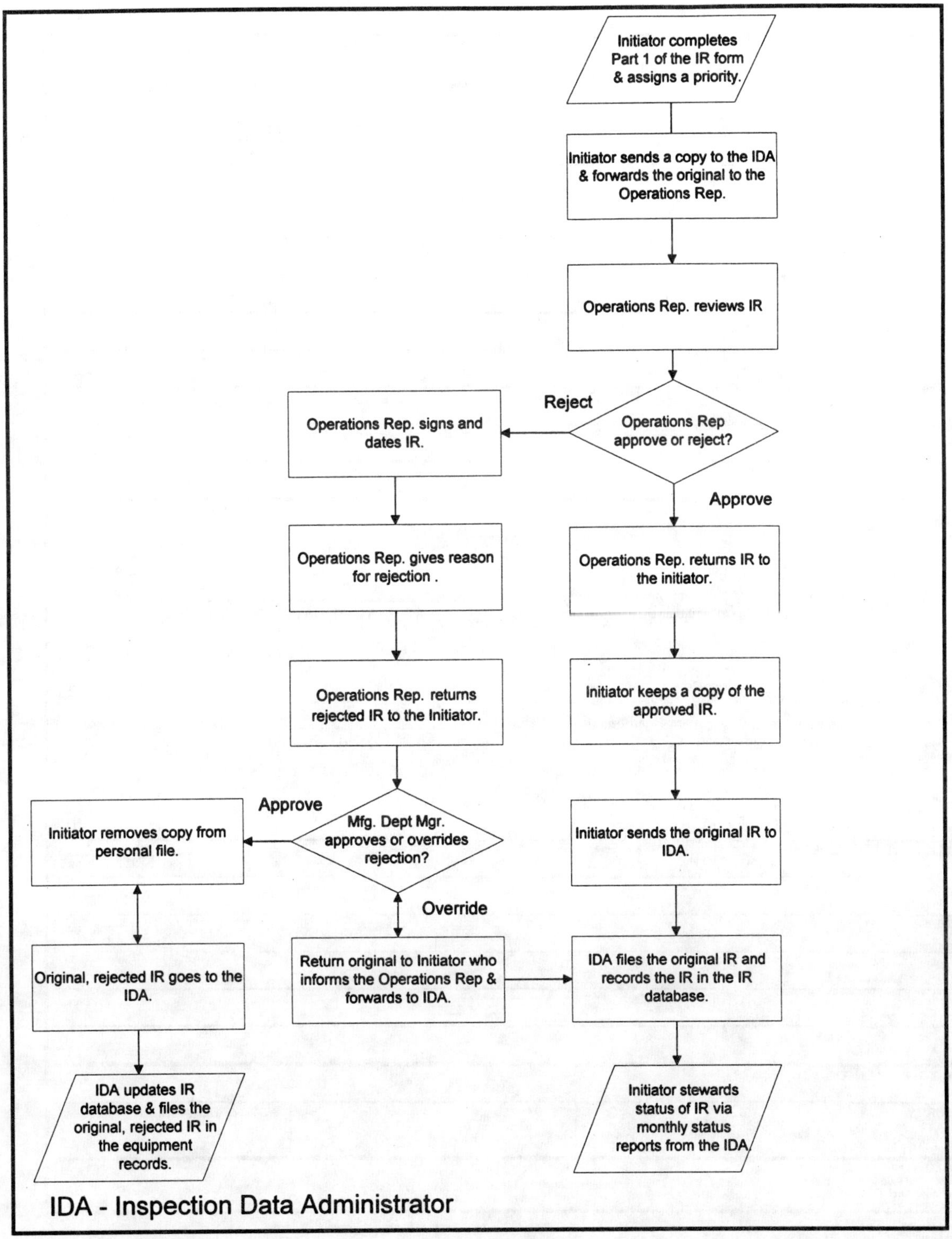

FIGURE 1
INSPECTION RECOMMENDATION PROCESS

INSPECTION RECOMMENDATION

PART 1: BY INSPECTION

DATE:	INSPECTOR:	**NUMBER:**

AREA:	EQUIPMENT NO:

RECOMMENDATIONS:

Risk: High Medium Low

PART 2: ROUTING

INSPECTOR'S SIGNATURE

a) ☐ OPERATION SUPERVISOR
- If approved forward to Maintenance Planner.
- If rejected, provide rationale and return to Inspector.

SIGNATURE/DATE ☐ Approved ☐ Rejected

b) ☐ MANUFACTURING DEPARTMENT MANAGER
- If rejected, review with Manufacturing Department Head
- Manufacturing Department Head has the final approval or rejection authority

SIGNATURE/DATE ☐ Approved ☐ Rejected

RATIONALE FOR REJECTION:

PART 3: BY MAINTENANCE PLANNER

PRIORITY:	ESTIMATED COST:	WORK ORDER NUMBER:

FIGURE 2
The First Inspection Recommendation

CONSEQUENCE \ PROBABILITY	High(3)	Medium(2)	Low(1)
High(3)	(9)	Example 2 (6)	(3)
Medium(2)	(6)	(4)	(2)
Low(1)	(3)	(2)	Example 1 (1)

FIGURE 3
Typical Risk Matrix

FAILURE POTENTIAL (F)	FAILURE CONSEQUENCES SAFETY/ENVIRON (S)	 ECONOMICS (E)	SEVERITY INDEX (SI)
Initiator assigns failure potential as follows: 4 Item could fail in an unpredictable manner. 3 Item has already failed or could fail within one year but not in an unpredictable manner. This includes all temporary repairs. 2 Item could fail in 1 to 5 years. 1 Item could fail with in 5 to 10 years.	**Operations Representative assigns safety/environmental consequences as follows:** 3 High concern that failure will result in injury to personnel or an unacceptable release to the environment. 2 Moderate concern that failure will result in injury to personnel or an unacceptable release to the environment. 1 Low concern that failure will result in injury to personnel or an unacceptable release to the environment. 0 No concern that failure will result injury to personnel or an unacceptable release to environment.	**Operations Representative assigns economic consequences as follows:** 3 High impact on operations if the failure occurs. 2 Moderate impact on operations if the failure occurs. 1 Minimal impact on operations if the failure occurs. 0 No impact on operations if the failure occurs.	**The severity Index is F * (S+E).** The severity index is an assessment of the criticality of the inspection recommendation.It is a relative assessment, not an absolute value, that enables the mechanical organization to prioritize work. By addressing the items with a higher severity index first, the organization corrects the items that have the greatest potential impact on the integrity and operability of the unit.

FIGURE 4
Failure Potential and Consequence Table

Figure 5

INSPECTION RECOMMENDATION

PART 1: BY INITIATOR

DATE:	INITIATOR:	**NUMBER:** (note 1)

AREA:	UNIT:	EQUIPMENT OR PIPING CIRCUIT TAG:

RECOMMENDATION:

PART 2: IR ASSESSMENT (By Initiator and Operations Rep)

Failure Potential (F)	CONSEQUENCES: Safety/Environ (S)	CONSEQUENCES: Economics (E)	Severity Index F* (S + E)	Downtime Required? YES or NO
4 3 2 1	3 2 1 0	3 2 1 0		ACTION DATE: (provide if S = 2 or 3)

PART 3: APPROVALS

a) [ ] OPERATION REPRESENTATIVE
- Assess consequences & calculate severity index.
- If approved, mark if downtime required.
- If rejected, provide reason in comments.
- Sign & date and return to Initiator.

SIGNATURE/DATE [ ] Approved [ ] Rejected

b) [ ] MANUFACTURING DEPARTMENT MANAGER
- If rejected by the Operations Representative, Inspection may review with Manufacturing Department Manager.
- Manufacturing Department Manager has the final approval or rejection authority.

SIGNATURE/DATE [ ] Approved [ ] Rejected

COMMENTS:

PART 4: IR CLOSURE (By Initiator)

Corrective Action Completed and Accepted

SIGNATURE ______________________ DATE ______________________

NOTE 1: The format for this number is MMYY-NNN-XXX, where MMYY in the month & year the IR was written; NNN is the next sequential number for month & initiator and XXX is the initiator's initials.

FIGURE 5

Risked Based Inspection Recommendation

Paper 39f

The Application of Risk Based Inspection Methodology in a Large Chemical Plant

C. Dykstal, The Dow Chemical Company

Paper 39f was not available for inclusion in the Proceedings.

It will be made available to attendees at the Symposium

31st LOSS PREVENTION SYMPOSIUM

Session 40. Fire and Explosion Prevention & Protection

Tuesday, March 11, 1997
8:30 - 11:35 AM
Grand Ballroom C

Michael L. Griffin, Chair
Procter & Gamble CO
6110 Center Hill Ave.
Cincinnati, OH 45224-1789

Joseph A. Senecal, Vice-Chair
Fenwal Safety Systems
7000 Nickerson Rd.
Marlborough, MA 01752

Session Introduction - Michael L. Griffin

40a. **A Detailed Method for the Estimation of Mixture Flammability Limits Using Chemical Equilibrium,** G. Melhem, A.D. Little, Inc., Cambridge, MA

40b. **Vent Access Restriction for Solids Handling Systems,** L. Bernard, F. Brodie, D. Ludwig, A. Ness, K. Weidner; Rohm & HaasEngineering Division, Bristol, PA

40c. **K_G: New Data and Analysis,** J. Senecal, P. Beaulieu, Fenwal Safety Systems, Marlborough, MA

BREAK

40d. **NFPA 30: An Update and a Look into the Future,** R. P. Benedetti, National Fire Protection Association, Quincy, MA

40e. **Determination, Prevention, and Mitigation of Potential Hazards Due to the Handling of Powders during Transportation, Charging, Discharging, and Storage,** N. Jaeger, Ciba - Geigy, Toms River, NJ; R. Siwek, Ciba - Geigy, Ltd., Basle, Switzerland

40f. **Design Practice for Extinguishing Barrier Systems,** R. Siwek, Ciba - Geigy, Ltd., Basle, Switzerland; P. Moore, Kidde Fire Protection, Thame, Oxon. UK

Concluding Remarks by J. A. Senecal

A Detailed Method For The Estimation of Mixture Flammability Limits Using Chemical Equilibrium

G. A. MELHEM
Arthur D. Little, Inc.
20 Acorn Park
Cambridge, Massachusetts 02140

ABSTRACT

This paper presents a general method for the estimation of flammability envelopes for chemical mixtures based on chemical equilibrium. The impact of mixture initial temperature, the presence of diluents and elevated system pressures are implicitly accounted for. The performance of this method is tested against much of the experimental data reported in the literature for systems containing a wide range of chemicals from CHNO compounds to compounds containing sulfur, phosphorus, silicon and halogens.

This method presents a very useful and accurate approach for assessing the flammability envelopes of mixtures where no experimental data is available, and to guide experimental flammability testing work.

FLAMMABILITY LIMITS

Figure 1 summarizes thermochemical estimates of flame temperature for a mixture of methane and oxygen at 1 bar and 25 C. We observe from Figure 1 that both the lower and upper flammability limits occur at a temperature of around 1500 K. We also observe from Figure 1 that the lower flammability limit (LFL) and the upper flammability limit (UFL) do not change significantly over a 500 degrees window. The calculated LFL veries from 3 % at 1000 K to 4.8 % at 1500 K. The calculated UFL varies from 60 % at 1000 K to 68 % at 1500 K.

At the reported literature flammability limits (LFL of 5 % and UFL of 60 %), the methane-oxygen system has a threshold theoretical flame temperature of 1500 K. At this temperature, the combustion reaction is able to generate enough heat to produce a self-sustaining (propagating) reaction.

Many other systems of interest have a similar behavior. Table 1 summarizes estimates of calculated theoretical flame temperatures at both LFL and UFL limits for a variety of chemicals. Most calculated flame temperatures are between the range of 1000 and 1500 K.

The idea of using a threshold theoretical flame temperature to estimate flammability limits is not new (See Stull, [1]). Recently, Melhem and Shanley [2] demonstrated that calculated

Table 1: Calculated theoretical flame temperatures for a veriety of chemical compounds. Combustion occurs in air at an initial temperature of 25 C and an initial pressure of 1 bar.

Compound	LFL(vol %)	UFL (vol %)	T_{LFL} (K)	T_{UFL} (K)
Carbon monoxide	12.50	74.00	1394	1268
Methane	5.00	15.00	1481	1774
Acetylene	2.50	100.00	1268	2831*
Ethylene	2.70	36.00	1370	1216
Ethane	3.00	12.50	1534	1399
Propylene	2.00	11.00	1431	1444
1-Butene	1.60	9.30	1479	1254
1-Hexene	1.20	9.20	1583	1091
CnHn (Mw=200)	0.50	4.70	1543	1000
Hydrogen sulfide	4.30	45.00	1045	1383
Ammonia	15.00	28.00	1632	1825
Ethanol	3.30	19.00	1492	1041
Acetone	2.60	12.80	1541	1229
Dimethyl ether	3.40	27.00	1565	980
Ethyle acetate	2.20	11.40	1571	1047
Acetaldehyde	4.00	36.00	1552	977
Trichlorosilane	7.00	75.00	1260	915
Methyldichlorosilane	3.40	54.00	1543	847
Methyltrichlorosilane	5.10		1602	

* implies self decomposition

adiabatic reaction temperature (CART) is an effective screening tool in evaluating the reactivity hazard potential of many classes of chemical compositions including CH, CHO, nitro, organic nitro, nitrogen compounds (other than nitro and organic nitro compounds), nitrates, redox compositions, etc.

Many combustion reactions leading to the formation of carbon dioxide and water have "freeze-out" temperatures on the order of 1400 K. This "freeze-out" limit is most likely related to the minimum temperature required for carbon monoxide to propagate a self-sustaining flame which is around 1400 K. The concept of a "freeze-out" temperature can be used to estimate flammability limits for multicomponent gas mixtures with diluents and can be used to assess the impact of initial mixture temperature and/or pressure on flammability limits.

Figure 2 illustrates the use of constant theoretical flame temperature to estimate the flammability envelope for a hydrocarbon mixture where water is the diluent. As shown by Figure 2, the selection of a lower temperature limit, such as 1200 K, results in a wider flammability envelope, a conservative estimate for hazard prediction. With few exceptions, the majority of reported experimental flammability limits correspond to adiabtic flame temperatures in the range of 1000 to 1500 K.

Figure 1: Computed theoretical flame temperature for the system methane-oxygen at 1 bar and 298 K using SuperChems Expert version 3.

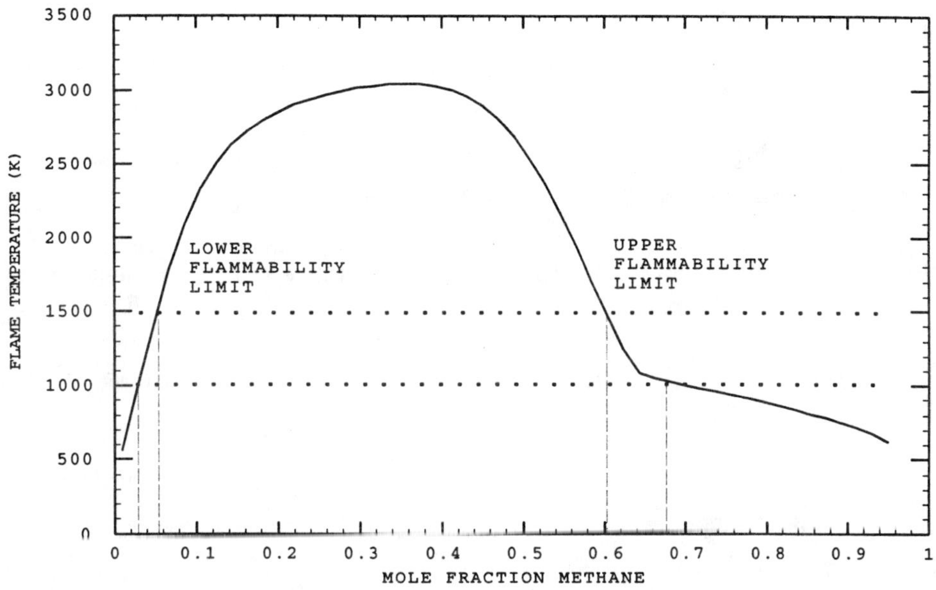

Figure 2: Impact of temperature limit on flammability envelope

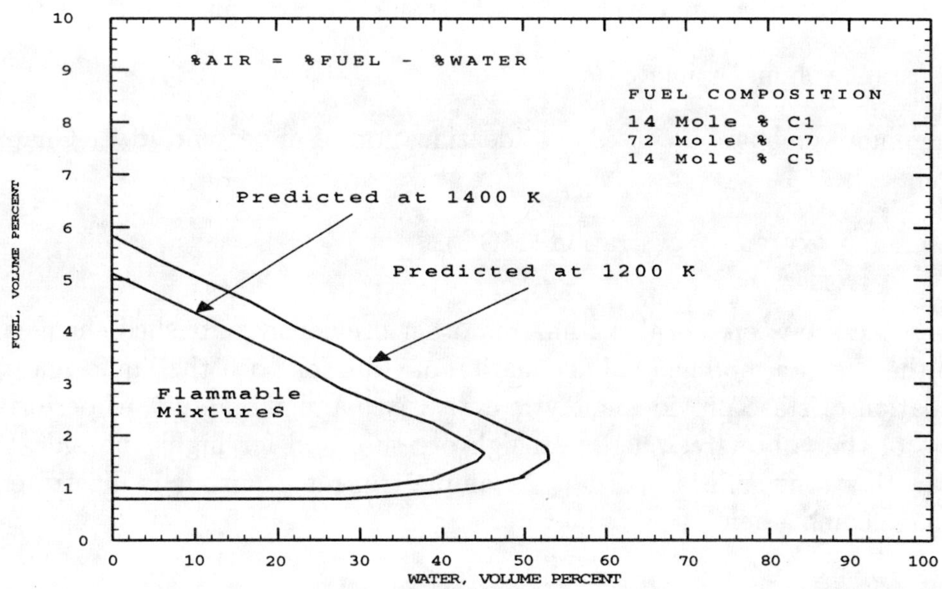

Most compositions yielding adiabatic temperatures lower than these values will not continue to react even if initiated at higher temperatures [3].

The estimation of theoretical flame temperature for vapor and multiphase systems is easily accomplished using direct minimization of the Gibbs free energy. Using a threshold theoretical flame temperature value, of say 1000 K, we can estimate the flammability limits of many vapor and multiphase systems of interest. The 1000 K criterion provides a slightly conservative estimate for the LFL and will provide good guidance for performing empirical flammability testing.

The use of direct minimization of the Gibbs free energy provides many advantages over existing empirical and semi-empirical methods for the:

- estimation of flammability limits where the mixture contains inerts and/or a mixture of fuels,
- estimation of flammability limits where the initial temperature is different from 25 C, and
- estimation of flammability limits where the initial pressure is different from 1 bar.

MULTIPHASE CHEMICAL AND PHYSICAL EQUILIBRIA

An algorithm for direct minimization of the Gibbs free energy is presented in Appenix A. This algorithm was incorporated into SuperChems Expert Version 3 [4] and can be used to perform a wide variety of simultaneous multiphase physical and chemical equilibrium estimates including:

- Constant temperature envelopes, such as the ones presented in this paper.
- Constant volume estimates.
- Hugoniot combustion curves and identification of upper and lower Chapman-Jouguet points.
- Minimum oxygen concentration (MOC).

Key differences between our algorithm and other previously published chemical equilibrium codes is the use of a modified cubic equation of state for both the liquid and vapor phases. This equation of state provides analytic derivatives and the ability to perform direct minimization of the Gibbs free energy at high pressure and for highly non-ideal systems and systems with strong solution effects. A sample truncated output is illustrated in Figure 9 for methane combustion.

Figure 3: Estimation of ethane flammability limits using SuperChems Expert version 3 at a constant theoretical flame temperature of 1400 K

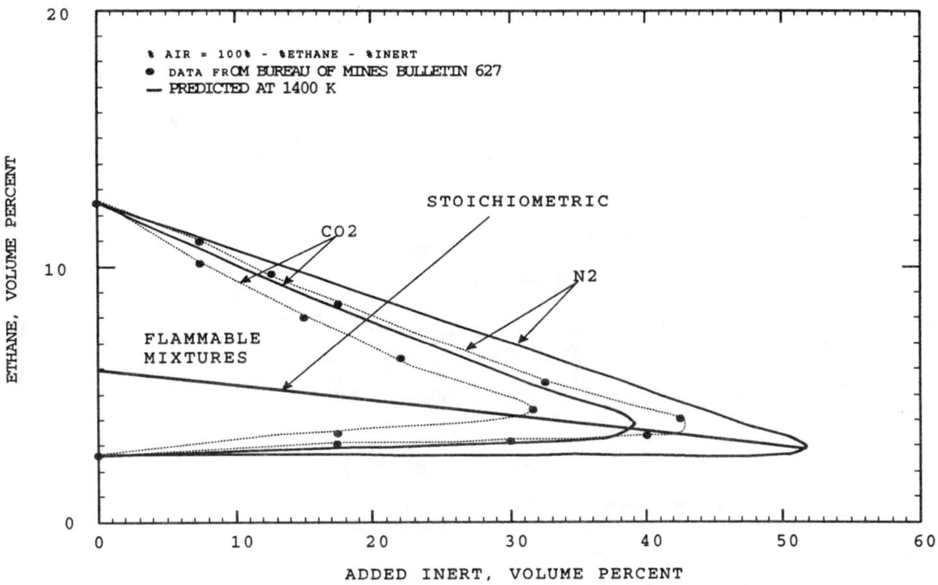

COMPOSITION EFFECTS

Le Chatelier [5] proposed a simple formula for the estimation of mixture flammability limits. This formula applies to both the lower and the upper flammability limits:

$$LFL_{mix} = \frac{100}{\sum_i^C \frac{X_i}{LFL_i}} \quad (1)$$

$$UFL_{mix} = \frac{100}{\sum_i^C \frac{X_i}{UFL_i}} \quad (2)$$

where X_i is the vapor mole fraction, and both flammability limits are specified in volume percent. This simple formula may not provide accurate estimates for mixtures containing inerts or mixtures containing fuels such as acetone and ether leading to "cool flames".

The presence of inerts such as nitrogen, carbon dioxide, etc. prsents a medium which absorbs reaction energy. If enough inerts are present, the reaction cannot provide enough energy to produce a self sustaining propagation.

Figure 3 illustrates the use of constant temperature limit of 1400 K to calculate the flammability envelope for ethane with two different diluents, carbon dioxide and nitrogen. The data compares favorably to values reported by the Zabetakis [6].

Please note that once a flammability envelope is constructed, the minimum oxygen concentration (MOC) is also estimated. The estimated flammability envelope for ethane would have been slightly wider had a lower threshold flame temperature been used.

Figure 4: Impact of initial temperature and water content on flammability limits for gas mixture containing carbon monoxide and hydrocarbons. Envelope is calculated at a threshold flame temperature of 1200 K using SuperChems Expert version 3.

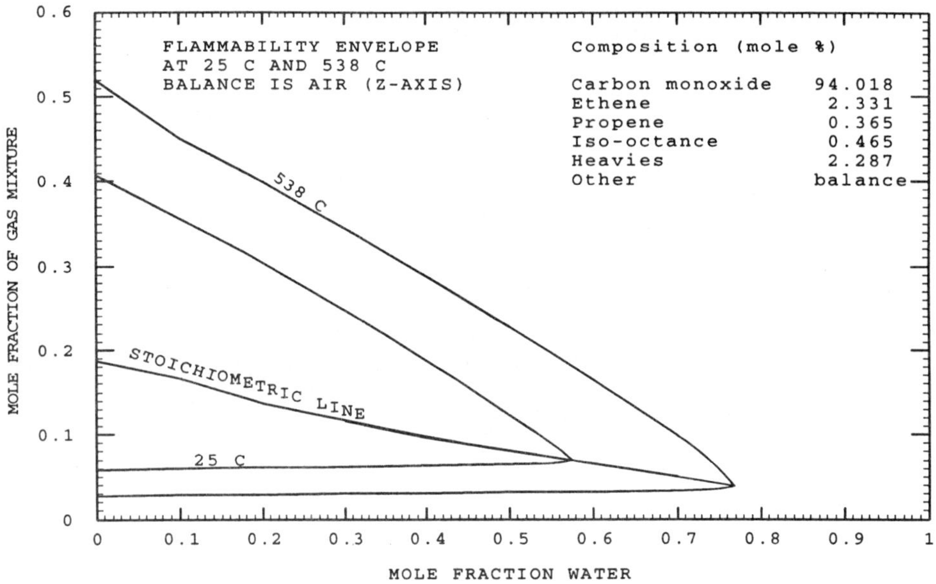

TEMPERATURE EFFECT

An increase in intial temperature of a flammable mixture will enlarge its flammability envelope. Less energy is spent by the combustion reaction to heat the reactants to the threshold temperature required for the reaction to produce a self sustaining flame. Higher initial temperatures will result in wider flammability envelopes. Zabetakis [6] reports the following emiprical relations to assess the imapct of temperature on flammability limits:

$$UFL(T) = UFL(298K) + \frac{0.75}{\Delta H_c}(T - 298) \quad (3)$$

$$LFL(T) = LFL(298K) - \frac{0.75}{\Delta H_c}(T - 298) \quad (4)$$

where T is the initial mixture temperature in Kelvins, LFL and UFL are in volume percent, and ΔH_c is the heat of combustion in kcal/mol. These relationships appear to be valid for straight chain hydrocarbons only.

Figure 4 illustrates the use of the chemical equilibrium algorithm to estimate the impact of initial temperature and water vapor on the flammability of a gas mixture which contains carbon monoxide and other hydrocarbons. The flammability limits for pure carbon monoxide at 25 C and 1 bar are shown in Table 1.

PRESSURE EFFECT

The impact of system pressure on the lower flammability limit is small while its impact on

Figure 5: Effect of pressure on the flammability limits of natural gas at 301 K. Taken from Zabetakis, 1965.

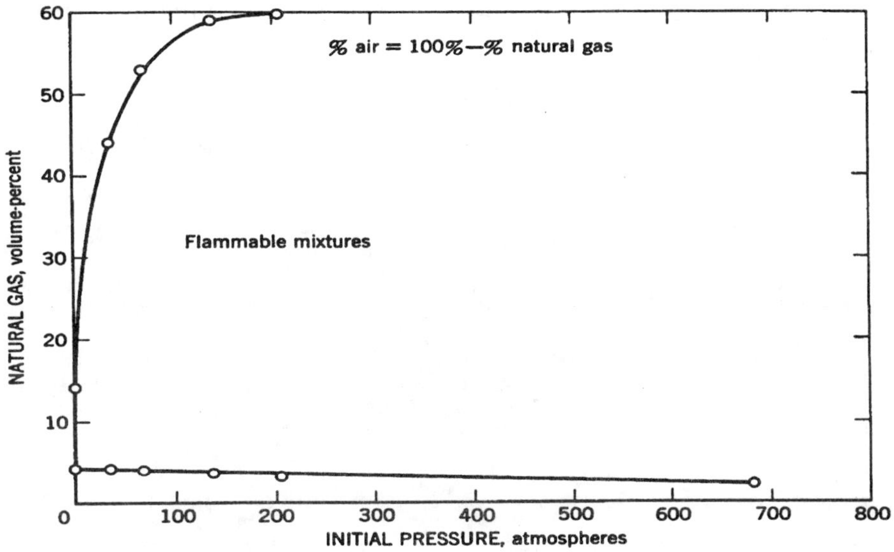

the upper flammability limit is significant. The limits of flammability of natural gas (85 to 95 % methane , 15 to 5 % ethane) are shown in Figure 5. A similar dependence of flammability limits on pressure is expected for other flammable gases. Zabetakis reports the following log-linear fit of the natural gas data:

$$LFL = 4.9 - 0.71 \log P \quad (5)$$

$$UFL = 14.1 + 20.4 \log P \quad (6)$$

where LFL and UFL are both in volume percent and P is the initial system pressure in atmospheres. These relationships may not apply directly to other systems.

Figure 6 illustrates how Gibbs free energy minimization can be used to estimate the impact of pressure on flammability limits. Figure 6 clearly shows that the LFL of natural gas does not vary with pressure while the UFL does. It is also important to note that the equilibrium estimates show a similar trend to the experimentally observed UFL change with pressure. A smaller change is predicted between 100 and 200 bars than from 1 to 100 bars. This trend is clearly shown in the experimental data illustrated in Figure 5.

Figures 7 and 8 also illustrate equilibrium estimates of flammability limits for propane at a flame temperature of 1000 K. Good agreement is shown between the model predictions and the actual data reported in Zabetakis [6]. The equilibrium model predictions are slightly conservative at limting flame temperature values of 1000 K. This should be adequate for most hazard assessment studies. It is interesting to note, that the trend of change of both LFL and UFL are also well predicted by the equilibrium code for propane at 7.91 and 14.8 bars.

Figure 6: Calculated flammability limits at a threshold flame temperature of 1000 K usin SuperChems Expert version 3 with non-ideal gas effects assuming a natural gas composition of 85 % methane and 15 % ethane

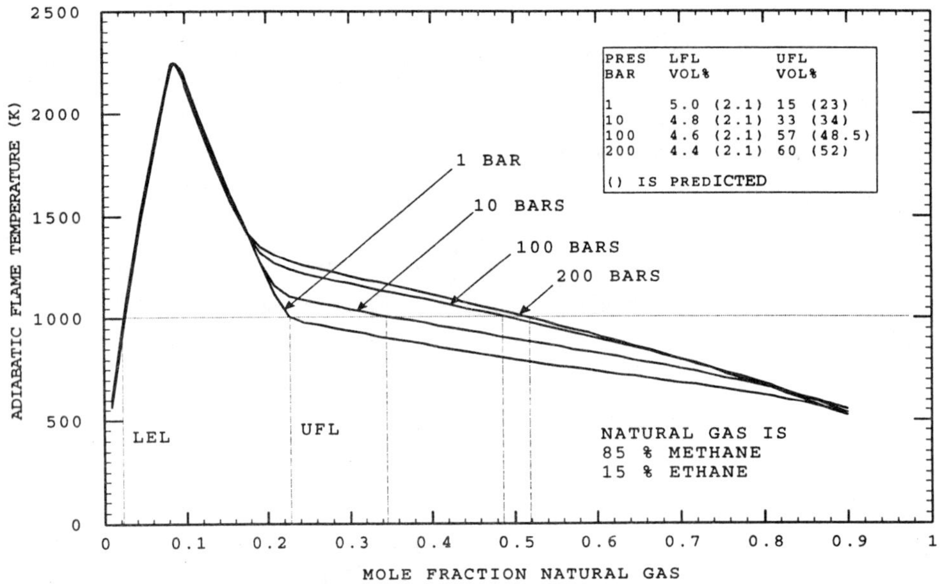

Figure 7: Calculated effects of pressure on the flammability limits of propane at 298 K using SuperChems version 3.

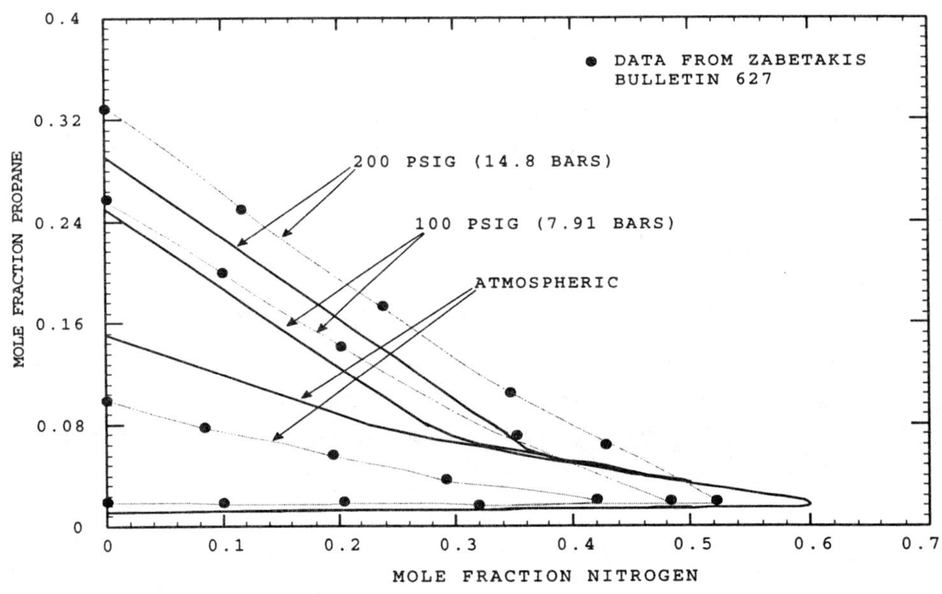

Figure 8: Calculated flammability limits at a threshold temperature of 1000 K usin SuperChems version 3 with non-ideal gas effects for propane

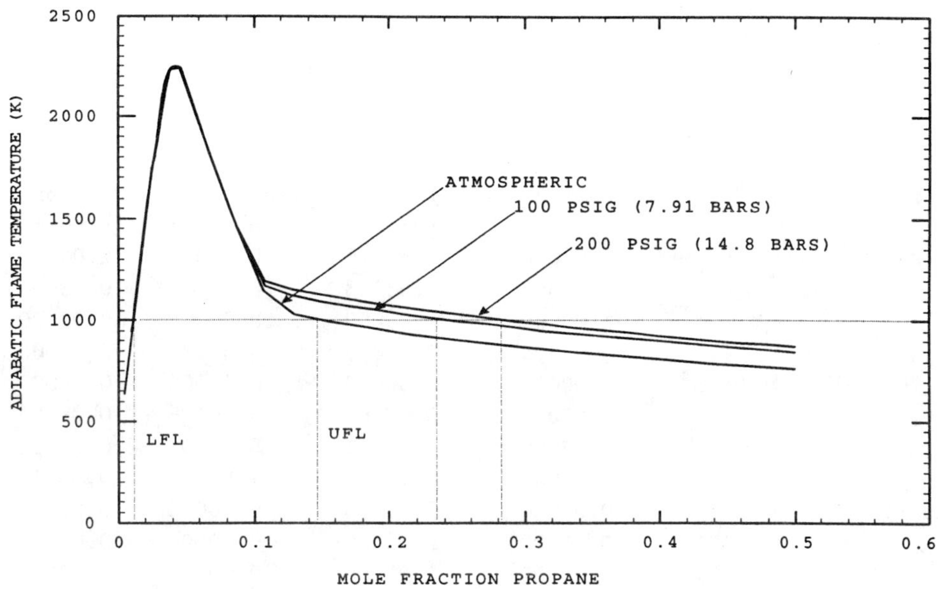

CONCLUSIONS

This paper presents a general method for the estimation of flammability limits for vapor and multiphase systems. The effects of pressure, temperature, inerts, and composition are shown to be well predicted for a veriety of systems. This method should prove to be a very useful tool for fire and explosion hazard assessment studies and for providing guidance to flammability testing.

Figure 9: Typical computer output from SuperChems Version 3 multiphase equilibrium estimator

```
Final system temperature (K) ..........................          298
Final system pressure (Pa) ............................        1E+05
```

Overall material balance:

Compound		Initial	Change	Final	Fraction
CARBON-REF	(S)	0.00000	0.67625	0.67625	1.00000
METHANE	(V)	1.00000	-0.83813	0.16187	0.08094
OXYGEN	(V)	1.00000	-1.00000	0.00000	0.00000
NITROGEN	(V)	0.00000	0.00000	0.00000	0.00000
CARBON MONOXIDE	(V)	0.00000	-0.00000	0.00000	0.00000
CARBON DIOXIDE	(V)	0.00000	0.16188	0.16188	0.08094
HYDROGEN	(V)	0.00000	0.00001	0.00001	0.00001
WATER	(V)	0.00000	1.67624	1.67624	0.83812
NITRIC OXIDE	(V)	0.00000	-0.00000	0.00000	0.00000
NITROGEN DIOXIDE	(V)	0.00000	-0.00000	0.00000	0.00000
NITROUS OXIDE	(V)	0.00000	-0.00000	0.00000	0.00000
HYDROXYL	(V)	0.00000	-0.00000	0.00000	0.00000
MONOATOMIC OXYGEN	(V)	0.00000	-0.00000	0.00000	0.00000
MONOATOMIC HYDROGEN	(V)	0.00000	-0.00000	0.00000	0.00000
METHYL	(V)	0.00000	-0.00000	0.00000	0.00000
HYDROGEN ION +	(V)	0.00000	0.00000	0.00000	0.00000
HYDROGEN ION -	(V)	0.00000	0.00000	0.00000	0.00000
Totals:		2.00000	0.67625	2.67625	

Overall equilibrium balance:

Compound		Fraction	Fugacity	Fugacity Coefficient	
CARBON-REF	(S)	1.00000	1.0000E+00		
METHANE	(V)	0.08094	7.9996E-02	Phi=	1.0015E+00
OXYGEN	(V)	0.00000	4.9580E-11	Phi=	1.0047E+00
NITROGEN	(V)	0.00000	1.3899E-09	Phi=	1.0060E+00
CARBON MONOXIDE	(V)	0.00000	4.9673E-11	Phi=	1.0066E+00
CARBON DIOXIDE	(V)	0.08094	7.9554E-02	Phi=	9.9591E-01
HYDROGEN	(V)	0.00001	6.7612E-06	Phi=	1.0133E+00
WATER	(V)	0.83812	8.1481E-01	Phi=	9.8507E-01
NITRIC OXIDE	(V)	0.00000	4.9654E-11	Phi=	1.0062E+00
NITROGEN DIOXIDE	(V)	0.00000	4.8665E-11	Phi=	9.8621E-01
NITROUS OXIDE	(V)	0.00000	4.9114E-11	Phi=	9.9531E-01
HYDROXYL	(V)	0.00000	5.0032E-11	Phi=	1.0139E+00
MONOATOMIC OXYGEN	(V)	0.00000	5.0032E-11	Phi=	1.0139E+00
MONOATOMIC HYDROGEN	(V)	0.00000	5.0032E-11	Phi=	1.0139E+00
METHYL	(V)	0.00000	5.0032E-11	Phi=	1.0139E+00
HYDROGEN ION +	(V)	0.00000	5.0032E-10	Phi=	1.0139E+00
HYDROGEN ION -	(V)	0.00000	5.0032E-10	Phi=	1.0139E+00

Figure 10: Typical computer output from SuperChems Version 3 multiphase equilibrium estimator (continued)

```
Calculated Reaction Stoichiometry:

Atom Matrix (Input):

  0.0000  0.0000  1.0000  0.0000  0.0000  0.0000
  0.0000  4.0000  1.0000  0.0000  0.0000  0.0000
  2.0000  0.0000  0.0000  0.0000  0.0000  0.0000
  0.0000  0.0000  0.0000  2.0000  0.0000  0.0000
  1.0000  0.0000  1.0000  0.0000  0.0000  0.0000
  2.0000  0.0000  1.0000  0.0000  0.0000  0.0000
  0.0000  2.0000  0.0000  0.0000  0.0000  0.0000
  1.0000  2.0000  0.0000  0.0000  0.0000  0.0000
  1.0000  0.0000  0.0000  1.0000  0.0000  0.0000
  2.0000  0.0000  0.0000  1.0000  0.0000  0.0000
  1.0000  0.0000  0.0000  2.0000  0.0000  0.0000
  1.0000  1.0000  0.0000  0.0000  0.0000  0.0000
  1.0000  0.0000  0.0000  0.0000  0.0000  0.0000
  0.0000  1.0000  0.0000  0.0000  0.0000  0.0000
  0.0000  3.0000  1.0000  0.0000  0.0000  0.0000
  0.0000  1.0000  0.0000  0.0000  1.0000  0.0000
  0.0000  1.0000  0.0000  0.0000  0.0000 -1.0000

Rank of the atom matrix ................................          6
Number of independent reactions ........................         11

  +0.5  CH4 <--->  +0.5  C + H2
  +0.5  CH4 +0.5  O2 <--->  +0.5  C + H2O
  +0.5  O2 +0.5  N2 <--->  + NO
  + O2 +0.5  N2 <--->  + NO2
  +0.5  O2 + N2 <--->  + N2O
  +0.25 CH4 +0.5  O2 <--->  +0.25 C + OH
  +0.5  O2 <--->  + O
  +0.25 CH4 <--->  +0.25 C + H
  +0.25 C +0.75 CH4 <--->  + CH3
  + C +0.5  O2 <--->  + CO
  + C + O2 <--->  + CO2
```

A Gibbs Free Energy Minimization Algorithm

Complex equilibrium analysis has been a widely studied topic in chemical engineering. Its importance derives from a need to determine the equilibrium state of a chemical system, the maximum conversion in a reactor, or in the case of physical equilibrium, the maximum separation. The equilibrium model also serves a useful purpose in determining process conditions.

The computation of the equilibrium state of a system is one of constrained optimization, that is, the minimization of the Gibbs free energy subject to mass or element balance constraints. Recent advances in the field of nonlinear optimization have greatly simplified the solution of this problem. This work investigates the Wilson-Han-Powell's quadratic programming algorithm. Some of the major advantages of this algorithm is its low number of function and gradient evaluations, its ability to handle simple bounds on variables, such as non-negativity constraints and eliminates the need to transform variables in order to avoid singularities.

Equations of state are an attractive choice for representing nonideal systems. Their attractiveness stems from the fact that they can be solved analytically and require no iterative procedures. The Peng-Robinson equation of state and a number of mixing rules have been studied by Melhem et al. [1].

Introduction of a new mixing rule and a new temperature dependency form has made the Peng-Robinson equation of state a reliable, robust and computationally economical method for representing nonideal systems. Stability analysis criterion are used to determine the feasibility of additional phases. This method is direct and simple, and eliminates the need to set up an iterative procedure to determine the number of phases at equilibrium.

A.1 Prior Work

Brinkley [2] in 1947, laid the foundations to a general-purpose algorithm for the computation of chemical equilibria. His algorithm essentially handled only simple ideal systems and was intended for use with hand calculators, although the algorithm does lend itself to implementation on computers. In 1958, White et al [3] working at the Rand Corporation, developed an algorithm that solved the chemical equilibrium problem by the direct minimization of the Gibbs free energy. They were probably the first to use the ideas of nonlinear optimization to solve the problem. They used Newton's method to minimize the Gibbs free energy of an

[1] Melhem, G. A., R. Saini, and B. M. Goodwin, "A modified peng-robinson equation of state,". Fluid Phase Equilibria, 47, pp. 189-237, 1991.

[2] Brinkley, Jr., S. R., "Calculation of equilibrium composition of systems of many constituents," J. Chem. Phys., vol.15, no.2, p.107, 1947. and Brinkley, Jr., S. R., "Note on the conditions of equilibrium for systems of many constituents," J. Chem. Phys., vol.14, no.9, p.563, 1946.

[3] White, W. B., S. M. Johnson, and G. B. Dantzig, "Chemical equilibrium in complex mixtures," J. Chem. Phys., vol.28, no.5, p.751,1958.

ideal system. Boyton [4], Oliver et al. [5], extended the Rand algorithm to handle pure solid species, and Dluzniewski and Adler [6] extended it to handle mixed phases.

Gordan and McBride [7], working at the NASA Lewis Research Center, developed an algorithm that was similar to the Rand algorithm. Again, this algorithm handled only simple ideal systems, as this was justified by the high temperature systems encountered in rocket technology. Probably, the first to implement nonideal models for the solution of complex equilibria were George et al. [8]. They used Powell's method to minimize the Gibbs free energy. In order to avoid singularities they had to transform the variables and eliminate the constraints by introducing allocation functions.

Gautam and Seider [9], working for the U. S. Department of Energy (the ASPEN Project), used the quadratic programming method of Wolfe to extend the Rand algorithm. Their algorithm handled nonideal systems, pure solid species and electrolytic solutions. They also developed a new phase splitting algorithm to determine the number of phases and its distribution at equilibrium. Their algorithm is perhaps the most widely accepted algorithm and is implemented in the ASPEN process simulator.

B Problem Formulation

B.1 Chemical-Physical Equilibrium

Mathematically, the equilibrium problem is to minimize the total Gibbs free energy

$$G^t = \sum_{i=1}^{S} n_i \frac{G_i^\circ}{RT} + \sum_{i=1}^{N-S} \sum_{p=1}^{\pi} n_{ip} \left[\frac{G_i^\circ}{RT} + \ln \left(P \phi_{ip} \frac{n_{ip}}{n_{Tp}} \right) \right] \tag{7}$$

subject to the equality constraints,

$$C_1 \equiv \sum_{i=1}^{S} a_{ki} n_i + \sum_{i=1}^{N-S} \sum_{p=1}^{\pi} a_{ki} n_{ip} = b_k \qquad k = 1, \ldots, R \tag{8}$$

[4] Boynton, F. P., "Chemical equilibrium in multicomponent polyphase systems," J. Chem. Phys., vol.32, p.1880, 1960

[5] Oliver, R. C., S. E. Stephanou, and R. W. Baier, "Calculating free-energy nunmization," Chem. Eng., vol.69, no.4, p.121, 1962.

[6] Dluzniewski, J. H. and S. B. Adler, "Calculation of complex reaction and/op phase equilibria problems," in I. Chem. E. Symposium Series No. 35, (London), Instn. Chem. Eng., 1972.

[7] Gordan, S. and B. J. McBride, "Computer program for calculation of complex chemical equilibrium compositions, rocket performance, incident and reflected shocks, and chapman-jouguet detonations," Tech. Rep., NASA SP-273, 1971.

[8] George, B., L. P. Brown, C. H. Farmer, P. Buthod, and F. S. Manning, "Computation of multicomponent, multiphase equilibrium," Ind. Eng. Chem. Process Des. Dev., vol.15, no.3, 1976.

[9] Gautam, R. and W. D. Seider, "Computation of phase and chemical equilibrium: part 1,2,3," AIChE J., vol.25, no.6, p.991, 1979.

and the inequality constraints,

$$C_2 \equiv n_i \geq 0 \qquad i = 1, \ldots, S, \ldots, N\pi \tag{9}$$

where N the number of chemical species, π is the number of phases, S is the number of condensed solid species, n_{ip} is the equilibrium number of moles of the ith species in phase p, n_{Tp} is the total number of moles in phase p, G_i° is the standard Gibbs free energy of the i species evaluated at the system temperature T, P is the system pressure, a_{ki} is the number of atoms of element k in species i, b_k is the number of gram-atoms of element k, and R is the rank of the atom matrix (usually equal to the number of elements). The element abundance vector $\mathbf{b}$ is calculated as the product of the atom matrix $\mathbf{A}$ and the feed composition $\mathbf{n}^\circ$

$$\mathbf{A}\mathbf{n}^\circ = \mathbf{b} \tag{10}$$

The number of variables NVAR is equal to $((N - S) \times \pi + S)$, the number of equality constraints NMEQ is equal to the rank of the atom matrix R and the total number constraints NCON is equal to (NVAR + NMEQ).

B.1.1 Rank Calculation

The singular value decomposition method (SVD) [10] is used to calculate the the rank of the atom matrix. The atom matrix is decomposed into,

$$\mathbf{A} = \mathbf{U}\mathbf{W}\mathbf{V}^T \tag{11}$$

If the rank of the matrix is less than the number of elements, the dependent rows corresponding to the zero elements of the diagonal vector $\mathbf{W}$ are eliminated. The number of independent chemical reactions is equal to $C = N - R$. The C independent, but non-unique stoichiometric vectors are given by the columns of $\mathbf{V}$ corresponding to the zero elements of the $\mathbf{W}$ vector. This method has a number of advantages. Linear stoichiometric constraints can easily be incorporated into the atom matrix using a procedure similar to that outlined by Smith and Missen [11]. This method also handles the presence of isomers with no modifications to the atom matrix.

Consider the system ethylene, water, ethyl alcohol and dimethyl ether [12], where the alcohol and ether are isomers. With $N = 4$ and $E = 3$, the atom matrix is:

$$\begin{bmatrix} & H_2O & C_2H_4 & C_2H_5OH & C_2H_6O \\ C & 0 & 2 & 2 & 2 \\ H & 2 & 4 & 6 & 6 \\ O & 1 & 0 & 1 & 1 \end{bmatrix} \tag{12}$$

[10] Forsythe, G.E., M. A. Malcolm, and B. B. Moler, Computer Methods for Mathematical Computations. Prentice-Hall, 1977.

[11] Smith, W. R. and R. W. Missen, Complex Reaction Equilibrium Analysis: Theory and Algorithms. John Wiley and Sons, 1982.

[12] Whitwell, J. C. and S. R. Dartt, "Independent reactions in the presence of isomers," AIChE J., vol.19, no.6, p.1114, 1973.

Since the last two columns are the sum of the first two, the rank calculated is 2 and the number of independent reactions is $C = N - R = 2$. The stoichiometry predicted is

$$\nu^T = \begin{bmatrix} 0.632 & 0.632 & -0.316 & -0.316 \\ 0.000 & 0.000 & -0.707 & 0.707 \end{bmatrix} \tag{13}$$

which corresponds to the following two (non-unique) reactions,

$$2C_2H_4 + 2H_2O \rightleftharpoons C_2H_5OH + C_2H_6O \tag{14}$$

$$C_2H_6O \rightleftharpoons C_2H_6O \tag{15}$$

The reaction stoichiometry is only used in calculating the equilibrium K values and individual standard free energy and thus the reactions mechanism is unimportant.

B.1.2 Standard Free Energy Data

The standard free energy data is calculated from the equilibrium constant K which in turn is calculated from heat capacity data and standard heats and free energies of formation data. The equilibrium constant is related to the standard free energy of reaction by:

$$\Delta G^{rxn} = -RT \ln K \tag{16}$$

and ΔG^{rxn} for each of the C independent reactions is related to the individual free energies by:

$$\Delta G_j^{rxn} = \sum_{i=1}^{N} \nu_i G_i^\circ \qquad j = 1, \ldots, C \tag{17}$$

which is a set of C equations in N unknowns ($N > C$). An infinite number of solutions exist for this system and any one of them is a consistent solution. The SVD method is used to solve this underdetermined set of linear equations to give a particular solution to the N pseudo standard free energies.

B.2 Physical Equilibrium

For the case when no reactions occur, the function minimized is

$$\Delta G^t = \frac{G^t - \sum_{i=1}^{N} n_i^\circ G_i^\circ}{RT} = \sum_{i=1}^{N} \sum_{p=1}^{\pi} n_{ip} \left[\ln \left(P\phi_{ip} \frac{n_{ip}}{n_{Tp}} \right) \right] \tag{18}$$

subject to the equality constraints

$$C_1 \equiv n_i^\circ - \sum_{p=1}^{\pi} n_{ip} = 0 \qquad i = 1, \ldots, N \tag{19}$$

and the inequality constraints

$$C_2 \equiv n_i^\circ \geq n_{ip} \geq 0 \tag{20}$$

where n_i° is the number of moles of feed of the ith species.

The number of variables NVAR is equal to $(N \times \pi)$, the number of equality constraints NMEQ is equal to the number of species N and the total number constraints NCON is equal to (NVAR + NMEQ).

C Calculation of the Fugacity

The chemical potential, μ_{ip}, of the ith species in a multicomponent mixture is given by

$$\mu_{ip} = G_i^\circ + RT \ln \frac{\hat{f}_{ip}}{f_i^\circ} \tag{21}$$

The fugacity $\hat{f}_{ip}$ is related to the mole fraction of the ith component in the mixture by

$$\hat{f}_{ip} = \phi_i^p P x_{ip} \tag{22}$$

where ϕ_i^p is the fugacity coefficient and is a function of T, P, and the number of moles of the ith component in the mixture. Thus,

$$\mu_{ip} = G_i^\circ + RT \ln \left(P \phi_i^p \frac{n_{ip}}{n_{Tp}} \right) \tag{23}$$

where $n_{Tp} = \sum n_{ip}$, the total number of moles in the pth phase.

There are two approaches for calculation of fugacity, the activity coefficient approach and the equations of state approach. The activity coefficient model approach is unsuitable for a number of reasons,

1. The activity coefficient model can only approximate nonideality in the liquid phase. An equation of state is used to for vapor phase nonideality.
2. Activity coefficient models give poor results for systems at high pressure and systems containing supercritical compounds.
3. The activity coefficient approach, in general, requires in excess of two binary interaction parameters, a characteristic that leads to errors in extrapolating experimental results.

The equation of state approach does not inherently suffer from these difficulties. The Peng-Robinson equation of state along with a number of mixing rules has been extensively studied by Melhem et al. and has been suitably modified to represent both nonpolar and polar/asymmetric systems. This is the approach used in this work to calculate the fugacity of a component in a mixture.

D The Modified Peng-Robinson Equation of State

The equation of state proposed by Peng and Robinson [13] is:

$$P = \frac{RT}{v-b} - \frac{a\alpha(\omega, T_r)}{v(v+b)+b(v-b)} \tag{24}$$

where b is related to the size of the molecules and a is related to intermolecular attraction. Equation 24 has two adjustable parameters a and b which are obtained as functions of the critical temperature and pressure,

$$a = \frac{0.45724R^2T_c^2}{P_c} \tag{25}$$

$$b = \frac{0.07780RT_c}{P_c} \tag{26}$$

The value of $\alpha(\omega, T_r)$ is determined by regression to fit experimental vapor pressure data. The equation reported for $\alpha(\omega, T_r)$ is equivalent to the one suggested by Soave [14]:

$$\alpha(\omega, T_r) = [1 + m(1 - \sqrt{\frac{T}{T_c}})]^2 \tag{27}$$

where m is a function of the acentric factor, ω:

$$m = 0.37464 + 1.54226\omega - 0.26992\omega^2 \tag{28}$$

To properly correlate phase equilibria, equations of state must be able to reproduce accurately pure component vapor pressure. The original Peng-Robinson predicted accurate vapor pressures for hydrocarbons at reduced temperatures greater than 0.7. To insure accurate vapor pressure predictions for polar substances, Equation 27 is replaced with an equation proposed by Melhem et al.,

$$\ln \alpha_i = m_i \left[1 - \frac{T}{Tc_i}\right] + n_i \left[1 - \sqrt{\frac{T}{T_{ci}}}\right]^2 \tag{29}$$

This equation is capable of reproducing accurately vapor pressure and heat of vaporization data for polar compounds over a wide range of temperature. The m and n factors are empirical constants determined by regression to fit experimental vapor pressure data.

[13] Y., P. D. and D. B. Robinson, "A new two-constant equation of state," Ind. Eng. Chem. Fundam., vol.15, pp.58-64, 1976.

[14] Soave, G., "Improvement of the van der waals equation of state," Chem. Eng. Sci., vol.39, pp.357-369, 1984. and Soave, G., "Rigorous and simplified procedures for determining the pure-component parameters in the redlich-kwong-soave equation of state," Chem. Eng. Sci., vol.35, pp. 1725-1729, 1979.

Equation 24 can easily be extended to represent mixtures,

$$P = \frac{RT}{V - B_m} - \frac{A_m}{V(V + B_m) + B_m(V - B_m)} \tag{30}$$

where A_m and B_m are calculated according to the following mixing and combining rules:

$$B_m = \sum_{i=1}^{N} x_i b_i \tag{31}$$

$$A_m = \sum_{i=1}^{N} \sum_{j=1}^{N} x_i x_j \, a_{ij}^0 (1 - k_{ij}) \tag{32}$$

$$a_{ij}^0 = \sqrt{\alpha_i \alpha_j a_i a_j} \tag{33}$$

The subscript i indicates the ith compound in the mixture. N is the total number of species present and x_i is the mole fraction of compound i. The parameter k_{ij} is a symmetric ($k_{ij} = k_{ji}$) binary interaction parameter obtained from experimental data. In general k_{ij} is a constant. For nonideal systems, however, k_{ij} depends on temperature and small changes in its value can cause large changes in the properties predicted by the equation of state. Moreover, this mixing rule has not been found to do as good a job in predicting binary and ternary vapor-liquid equilibrium than some of the two-parameter mixing rules proposed by Panagiotopolous and Reid [15], Luedecke and Prausnitz [16] and others. The mixing rule proposed by Panagiotopolous and Reid, although emperical in nature, was found to be the most suitable,

$$a_{ij} = a_{ij}^{\circ} \left[1 - k_{ij} + (k_{ij} - k_{ji}) x_i\right] \tag{34}$$

$$A_m = \sum_i \sum_j x_i x_j a_{ij} \tag{35}$$

Here, k_{ij} is not equal to k_{ji}. These binary interaction parameters ar e necessary for accurate modelling of nonlinear systems. This is a limitation, which we believe, greatly diminishes an equation of state's predictive nature as nonlinear regression packages must then be tied in with thermodynamic property generators to obtain these parameters from experimental vapor-liquid equilibrium data.

It has been shown by several authors that it is possible to correlate the binary interaction parameters with pure component properties for the original mixing rule [17]. The mixing rule proposed by Panagiotopolous and Reid is not suitable for predictive methods as the interaction between species i and j is not symmetric. Keeping the compositional dependency

[15] Panagiotopoulos, A. Z. and R. Reid, "Multiphase high pressure equilibria in ternary aqueous systems," Fluid Phase Equil., vol.29, pp.525-534, 1986. and Panagiotopoulos, A. Z. and R. C. Reid, "New mixing rule for cubic equations of state for highly polar, asymmetric systems," in Equations of State: Theories and Applications, pp. 571-582., ACS Symposium Series 300, Chao and Robinson Ed, 1986.

[16] Luedecke, D. and J. M. Prausnitz, "Phase equilibrium for strongly nonideal mixtures from an equation of state with density-dependent mixing rules," Fluid Phase Equil., vol.22, pp. 1-19, 1985.

[17] Reid, R., B. E. Poling, and J. M. Prausnitz, Properties of Gases and Liquids. McGraw-Hill, 1986

of the interaction parameter we propose a new mixing rule, which is similar to the one proposed by Panagiotopolous and Reid, but with symmetrical parameters,

$$A_m = \sum_i \sum_j x_i x_j \left[1 - k_{ij} + l_{ij} x_i\right] a^{\circ}_{ij} \tag{36}$$

where,

$$k_{ji} = k_{ij} \tag{37}$$
$$l_{ji} = -l_{ij} \tag{38}$$

This new mixing rule has been found to do as good a job in predicting binary and ternary VLE as that proposed by Panagiotopolous and Reid. Its suitability in predicting liquid-liquid miscibility limits is illustrated in a following section, and maybe compared to those reported in reference 1 for the other mixing rules.

The fugacity coefficient can be readily calculated using this new mixing rule and is

$$\begin{aligned} \ln \phi_k &= \frac{b_k(Z-1)}{B_m} - \ln(Z-B) \\ &+ \frac{1}{2\sqrt{2}RTB_m}\left[S_1^k - S_2 + S_3^k - \frac{b_k A_m}{B_m}\right]\Psi \end{aligned} \tag{39}$$

where

$$\Psi = \ln\left[\frac{Z + B(1-\sqrt{2})}{Z + B(1+\sqrt{2})}\right] \tag{40}$$

Additional data and all relavent derivatives for the modified Peng-Robinson equation of state is summarized in Section I.

E Phase Splitting

One of the frequently encountered problems in multiphase equilibrium is the determination of the number of phases and their molar distribution at equilibrium. One method is to assume the maximum number of phases at equilibrium, as given by the phase rule. However, due to the fixed nature of the constraints of temperature and pressure, the phase rule does not always give the accurate number of phases at equilibrium. The most popular method is to split a phase into two trial phases and test to see whether a decrease in the Gibbs free energy is observed over a number of iterations. But, this procedure can be rather time-consuming and expensive for multicomponent/multiphase systems.

In order to overcome these problems, the stability criterion outlined by Beegle et al [18] and Heidemann [19] is used to determine, a priori, the feasibility of an additional phase. Extensive

[18] Beegle, B. L., M. Modell, and R. C. Reid, "Leqendre transforms and their application in thermodynamics," AIChE J., vol. 20, no.6, p.1194, 1974.

[19] Heidemann, R. A., "The criteria for thermodynamic stability," AIChE J., vol.21, no.4, p.824, 1975.

use of Legendre transforms is used to represent these criterion and a detailed description of their usage can be found in Beegle et al. The following sections illustrate these criterion, and demonstrate there suitability in predicting miscibility limits in liquid-liquid equilibrium.

E.1 Stability Criterion

The stability criterion can be stated as follows (see Heidmann, 1975):

1. The necessary and sufficient condition for thermodynamic stability in an N component system is,

$$y_{kk}^{(k-1)} > 0 \qquad k = 1, 2, \ldots, (N+1); \tag{41}$$

2. All points on the limit of stability obey,

$$y_{(N+1)(N+1)}^{(N)} = 0 \tag{42}$$

where in thermodynamic notation,

$$y_{11}^{0} = \frac{T}{C_v} > 0 \tag{43}$$

$$y_{22}^{1} = -\left(\frac{\partial P}{\partial V}\right)_{T,n_j} > 0 \tag{44}$$

$$y_{33}^{2} \cdots y_{(N+1)(N+1)}^{(N)} = \begin{vmatrix} \left(\frac{\partial \mu_1}{\partial n_1}\right)_{T,P,n_j} & \cdots & \left(\frac{\partial \mu_1}{\partial n_{(N-1)}}\right)_{T,P,n_j} \\ \vdots & \vdots & \vdots \\ \left(\frac{\partial \mu_{(N-1)}}{\partial n_1}\right)_{T,P,n_j} & \cdots & \left(\frac{\partial \mu_{(N-1)}}{\partial n_{(N-1)}}\right)_{T,P,n_j} \end{vmatrix} > 0 \tag{45}$$

where

$$\left(\frac{\partial \mu_i}{\partial n_j}\right)_{T,P,n_i} = RT\left[\frac{\partial \ln \phi_i}{\partial n_j} + \frac{1}{n_i}\frac{\partial n_i}{\partial n_j} - \frac{1}{n_T}\right] \tag{46}$$

The use of these criterion is simple as analytic derivatives for all terms are available (Section I). Equations 43 and 44 represent the locus of points on the binodal curve and the determinant in equation 45 represents the locus of points on the spinodal curve. If all criterion are positive then the mixture is stable, and if any of the terms are zero then the mixture is metastable or at its limit of stability. If any of the criterion is negative then the mixture is unstable and a presence of an additional phase is indicated. If equations 43 or 44 is negative the mixture is said to be mechanically unstable (usually high temperature systems) and if the determinant is negative the the mixture is said to be diffusionally unstable.

Table 2: Experimental Liquid-Liquid Miscibility Limits

System	ID	T, [K]	P, [atm]	x_1^I	x_1^{II}
Butanol-Water	1	365	1.0	0.260	0.045
Isobutanol-Water	2	364	1.0	0.335	0.055
Methyl ethyl ketone-Water	3	346	1.0	0.640	0.045
Ethyl acetate-Water	4	343	1.0	0.700	0.180
Methanol-Hexane	5	323	1.0	0.885	0.155

Data source:
Ellis, S. A. M. and R. D. Garbett, "A new equilibrium still for the study of partially miscible systems," Ind. Eng. Chem., vol.52, pp.385-388.
Kojima, K. and K. Tochigi, Prediction of Vapor-Liquid Equilibrium by the ASOG Method. Elsevier, 1979

Table 3: Calculated Miscibility Limits Using Stability Analysis Criterion

ID	K_{12}	K_{21}	x_1^I	x_1^{II}
1	-0.08891	-0.11033	0.2871	0.0502
2	-0.10760	-0.12039	0.2577	0.0582
3	-0.14961	-0.18254	0.6510	0.0898
4	-0.16851	-0.22598	0.5216	0.0539
5	0.08755	0.19134	0.7554	0.2748

E.2 Liquid-liquid Equilibrium

The use of the stability criterion outlined above is illustrated in predicting liquid-liquid miscibility limits. Table 2 lists the experimental values for the miscibility limits for five mixtures and table 3 gives the calculated results as well as the binary interaction parameters used for the modified Peng-Robinson equation of state. As shown, the miscibility limits predicted by the stability criterion are accurate.

E.3 Phase Splitting Algorithm

The use of the above outlined stability criterion make the phase splitting algorithm simple. The minimization is carried out for the phases specified by the user. After convergence, the stability criterion is applied to each of the phases. If any of the criterion are violated, an additional phase is added (usually Liquid). However, if the total number of moles of any phase is below a predetermined tolerance, that phase is deleted.

F Successive Quadratic Programming

This work uses the Wilson-Han-Powell Successive Quadratic (SQP) Programming algorithm. SQP algorithms minimize the Langrange function defined in terms of the objective function and equality and inequality constraints. Consider the general nonlinear programming problem:

$$\textit{Minimize } f(n) \qquad n = [n_1, n_2 \ldots n_{\mathrm{NVAR}}]^T \tag{47}$$

subject to

$$\begin{aligned} g_j(n) &= 0 & j &= 1, \ldots, \mathrm{NMEQ} \\ g_j(n) &\geq 0 & j &= \mathrm{NMEQ} + 1, \ldots, \mathrm{NCON} \\ \mathbf{n}_l \leq & \quad \mathbf{n} & &\leq \mathbf{n}_u \end{aligned}$$

where $\mathbf{n}_l$ and $\mathbf{n}_u$ are the lower and upper bounds on the variables, respectively. Writing the Langrange function,

$$\mathcal{L}(n, \lambda) = f(n) + \sum_{j=1}^{\mathrm{NCON}'} \lambda_j g_j(n) \tag{48}$$

where λ are the Langrange multipliers and $\mathrm{NCON}' = \mathrm{NCON} + 2\mathrm{NVAR}$. The bound constraints are formulated as inequality constraints, $g_j(n)$, $j = \mathrm{NCON} + 1, \ldots rmNCON'$.

For the minimization of the Gibbs free energy the Langrange function is given in terms of the objective function (equations 7 or 18), equality constraints (equations 8 or 19) and the inequality constraints (equations 9 or 20). The lower and upper bounds are set to, respectively, 1.0^6 and 1.0^{-8}.

Letting n^* be the optimal solution and λ^* the corresponding Langrange multipliers at the optimal solution, the Kuhn-Tucker conditions for n^* to be a local minimum for the general nonlinear programming problem 47 are [20],

1. $f(n)$, $g_j(n)$ are all twice differentiable at the optimal solution.
2. The constraint sets are linearly independent.
3. The constraints are satisfied at n^*

$$g_j(n^*) = 0 \qquad j = 1, \ldots, \mathrm{NMEQ}$$

[20] Edgar, T. F. and D. M. Himmelblau, Optimization of Chemical Processes. McGraw Hill; 1988

$$g_j(n^*) \geq 0 \qquad j = \text{NMEQ} + 1, \ldots, \text{NCON}$$

$$\mathbf{n}_l \leq \mathbf{n} \leq \mathbf{n}_u \qquad j = \text{NCON} + 1, \ldots, \text{NCON}'$$

4. The Langrange multipliers for the inequality constraints are non-negative

$$\lambda_j \geq 0 \qquad j = \text{NMEQ} + 1, \ldots, \text{NCON}^{\text{ prime}}$$

5. The Langrange multipliers for the active (binding) inequality constraints are zero and greater than zero for the inactive constraints; and the following relation holds,

$$\lambda_j^* g_j(n^*) = 0 \qquad j = \text{NMEQ} + 1, \ldots, \text{NCON}'$$

6. The Langrangian is at a stationary point at the optimal solution

$$\nabla \mathcal{L}_n(n^*, \lambda^*) = 0$$

7. The hessian matrix is positive semidefinite for all active constraints, that is,

$$\lambda^T \nabla^2 \left[\mathcal{L}(n^*, \lambda)\right] \lambda \geq 0$$

The *sufficient conditions* for n^* to be at a local minimum are,

1. The necessary conditions (1) through (6).
2. The hessian matrix of the Langrangian is positive definite, such that

$$\begin{aligned} \lambda^T \nabla g_j(n^*) &= 0 \text{ for active constraints} \\ \lambda^T \nabla g_j(n^*) &\geq 0 \text{ for inactive constraints} \end{aligned}$$

$$\lambda^T \nabla^2 \left[\mathcal{L}(n^*, \lambda^*)\right] \lambda > 0$$

If the function being minimized is a convex function and is being minimized over a convex feasible region, then the optimal solution obtained is a global one.

F.1 The Quadratic Subproblem

The SQP algorithm developed by Wilson, Han and Powell, defines a quadratic subproblem at every iteration to minimize a quadratic approximation of the Langrangian Equation 48. The basic strategy for solution of the general nonlinear programming problem 47 is to minimize a quadratic approximation of the Langrangian in terms of the search direction vector δ, subject to linearized equality and/or inequality constraints as functions of δ. Then a search is carried out in the direction of δ to reduce $f(n)$. It is possible for the solution to the quadratic approximation to wander into infeasible regions even though the solution to the orignal problem 47 is feasible. Moreover, the linearized constraints may be locally redundant, i.e. inactive. To prevent these drawbacks an additional variable, ξ, is introduced and an active set strategy is used to prevent recalculation of gradients of inactive constraints at every iteration. The quadratic function minimized is:

$$Q(n) = \delta^T \nabla f(n^k) + \frac{1}{2}\delta^T \mathcal{H} \delta + \frac{1}{2}\rho\xi^2 \tag{49}$$

subject to

$$\begin{array}{rcll} (1-\xi)g_j(n^k) + \delta^T \nabla g_j(n^k) & \left\{ \begin{array}{c} = \\ \geq \end{array} \right\} & 0 & j = 1, \ldots, \text{NMEQ} \\ g_j(n^k) + \delta^T \nabla g_j(n^k) & \geq & 0 & j = \text{NMEQ}+1, \ldots, \text{NCON} \\ n_l - n^k \leq & \delta & n_u - n^k & j = \text{NCON}+1, \ldots, \text{NCON}' \\ 0 \leq & \xi & \leq 1 & \end{array}$$

where δ is the search direction, $\mathcal{H}$ is a positive definite approximation of the hessian which is updated using the BFGS quasi-Newton formula, and ρ is an additional penality parameter introduced to avoid excessive influence of ξ on the solution of the quadratic subproblem.

If δ^k is the solution of the above quadratic subproblem 49, and u^k the corresponding vector of Langrange multipliers, then a new vector n is claculated from:

$$n^{k+1} = n^k + \alpha^k \delta^k \tag{50}$$

where α is the step length parameter whose value is bounded between 0 and 1. The Langrange multipliers λ are simultaneously updated by:

$$\lambda^{k+1} = \lambda^k + \alpha^k(\lambda^k - u^k) \tag{51}$$

The value of α is that which yields a reduction in the penalty function Φ such that,

$$\Phi(n + \alpha\delta, \psi) < \Phi(n, \psi) \tag{52}$$

where the penalty function Φ is defined as

$$\Phi(n, \psi) = f(n) + \sum_{j=1}^{\text{NMEQ}} \psi_j \left| g_j(n) \right| + \sum_{j=\text{NMEQ}+1}^{\text{NCON}} \psi_j \min(0, g_j(n)) \tag{53}$$

where, ψ is the Kuhn-Tucker multipliers vector which is initially set to the Langrange multipliers, and later calculated as proposed by Powell,

$$\psi_j^{k+1} = \max\left(|\lambda_j|, \frac{1}{2}(\psi_j^k + |\lambda_j|) \right) \tag{54}$$

Table 4: Problem 1: Solution

Species	Feed composition	Solution by White and Seider	Solution by this work
CH_4	15.14	0.875	0.886
H_2O	84.07	61.821	61.860
N_2	0.59	0.590	0.590
H_2	0.20	50.980	50.917
CO	0	6.282	6.298
CO_2	0	7.983	7.956

The line search procedure uses a simple Armijo-type bisection method coupled with quadratic interpolation to determine the optimal step length parameter.

One of the advantages of this algorithm is that it requires only first derivative information. The objective functions, constraints and their derivatives for the chemical-physical equilibrium problem and the physical equilibrium problem are summarized in Section H. Note that the derivatives of the constraints are constant and do not change value from iteration to iteration, and hence they are evaluated only once per simulation.

G Test Examples

The SQP algorithm was tested for a number of problems. To efficiently implement the algorithm, the objective function and constraints were scaled to a common basis. A convergence tolerance of 1.0×10^{-5} was used for all the problems.

1. *Ideal Vapor Phase chemical equilibrium:*
   The SQP algorithm was tested for the steam reforming system described by White and Seider. This gas phase chemical equilibrium problem takes place at $1067°K$ and 1.235 MPa (12.19 atm) with the following reaction:

$$CH_4 + H_2O \longrightarrow CO + 3H_2 \tag{55}$$

$$CO + H_2O \rightleftharpoons CO_2 + H_2 \tag{56}$$

   Nitrogen is also present in the system as an inert. Under these conditions, the gas mixture can be assumed to behave ideally. No interaction parameters were used in the modified Peng-Robinson equation of state. Feed compositions and compositions computed at equilibrium by this work and those reported by White and Seider are tabulated in Table 4. The guessed composition for each species was set to 1.0×10^{-3}. The stability analysis algorithm reported the phase to be stable with the following

Table 5: Literature reported values for Example 2

Species	Feed	Sanderson and Chien Vapor	Sanderson and Chien Liquid	George et al Vapor	George et al Liquid	Gautam and Seider Vapor	Gautam and Seider Liquid
H_2O	0	16.9205	10.2965	42.4642	$< 10^{-8}$	42.4639	0.0
C_2H_6O	50	15.9954	6.7876	7.5353	$< 10^{-8}$	7.5361	0.0
C_4H_8O	0	22.1756	5.0415	42.4642	$< 10^{-8}$	42.4639	0.0
$C_2H_4O_2$	50	7.9204	14.8625	7.5353	$< 10^{-8}$	7.5361	0.0
Total	100	63.01	36.99	100.0	$< 10^{-8}$	100.0	0.0

stability criterion values:

$$-\left(\frac{dP}{dV}\right) = 1.3204 \times 10^{-5}$$

$$\text{Stability Determinant} = 2.5521 \times 10^{11}$$

2. *Nonideal 2-phase chemical equilibrium:*
Ethanol and acetic acid are esterfied to ethyl acetate and water at $358.2°K$ and 1.0 atm. This two phase chemical equilibrium problem was originally solved by Sanderson and Chien, and subsequently by George et al. and Gautam and Seider. Sanderson and Chien report a split of 63.01% vapor and 36.99% liquid. George et al proved that these values obtained by Sanderson and Chein to be inconsistent as they did not satisfy the chemical reaction constraints. The equilibrium values obtained by these authors is tabulated in Table 5. The reaction taking place is:

$$C_2H_6OH + C_2H_4O_2 \rightleftharpoons C_4H_8O_2 + H_2O \tag{57}$$

with an equimolar feed of ethanol and acetic acid. Since this is a fairly nonideal system, binary interaction parameters were regressed from binary VLE data. The following values were used:

$$k_{ij} = \begin{bmatrix} 0.0 & -0.0911 & -0.1690 & -0.1400 \\ -0.0911 & 0.0 & 0.0319 & -0.0315 \\ -0.1690 & 0.0319 & 0.0 & -0.0218 \\ -0.1400 & -0.0315 & -0.0218 & 0.0 \end{bmatrix} \tag{58}$$

$$l_{ij} = \begin{bmatrix} 0.0 & 0.0500 & 0.226 & 0.0604 \\ -0.0500 & 0.0 & 0.0389 & 0.0072 \\ -0.226 & -0.0389 & 0.0 & -0.0155 \\ -0.0604 & -0.0072 & 0.0155 & 0.0 \end{bmatrix} \tag{59}$$

where (i, j) = water, ethanol; ethyl-acetate and acetic acid. This problem was solved using two different guesses. In the first run, a guess of 1.0×10^{-3} was used for all the

Table 6: Solution to Example 2

Species	Solution 1† Vapor	Solution 1† Liquid	Solution 2‡ Vapor	Solution 2‡ Liquid
H_2O	42.1012	0.2975	42.4637	0.00001
C_2H_6O	7.5742	0.0272	7.5363	$< 10^{-6}$
C_4H_8O	42.3349	0.0637	42.4637	$< 10^{-6}$
$C_2H_4O_2$	7.4716	0.1298	7.5363	$< 10^{-6}$
Total	99.4819	0.5181	99.99998	0.000021
$-(dP/DV)$	3.42×10^{-7}	5.38×10^{2}	3.40×10^{-7}	1.01×10^{7}
Stability Det.	2.01×10^{3}	9.12×10^{10}	2.00×10^{3}	2.80×10^{24}
G_{min}/RT	-1.74201		-1.74201	

† Solution with no guesses.

‡ Solution with guesses.

species in all phases. In the second run, a guess of 1, 49, 1, 49 for water, ethanol, acetate and acid respectively, was used for the vapor phase and 1.0×10^{-3} for all species in the liquid phase. The results obtained by both runs is given in Table 6. Also tabulated are the stability analysis data and the minimum obtained at equilibrium for both runs. As expected, convergence was achieved faster when guesses were provided for the components in each phases. However, the minimum obtained for each case was the same.

3. *Nonideal 3-phase physical equilibrium:*
A mixture of 0.3 mol n-butanol and 0.7 mol water is flashed at 355°K and 1.0 atm. This system is known to split into two liquid phases under these conditions. This simple flash problem illustrates the phase splitting algorithm. The following binary interaction parameters were used:

$$k_{ij} = \begin{bmatrix} 0.0 & -0.088799 \\ -0.088799 & 0.0 \end{bmatrix} \tag{60}$$

$$l_{ij} = \begin{bmatrix} 0.0 & -0.11039 \\ 0.11039 & 0.0 \end{bmatrix} \tag{61}$$

where, $(i, j) =$ n-butanol, water. The minimization calculation was started by assuming two phases, a vapor and liquid phase. After 3 SQP iterations, the algorithm converged to an answer the results of which are tabulated in Table 7. At this point, as the stability determinant for the liquid phase was negative, the phase splitting algorithm concluded that the phase was unstable and automatically restarted the calculations by adding an additional liquid phase. The results are given in Table 7. At equilibrium, the vapor phase is non-existent. It was, however, not necessary to eliminate this phase at any stage of the calculations as the algorithm does not have any problems in handling near zero flow rates.

Table 7: Solution to Example 3

Species	2 Phases		3 Phases		
	Vapor	Liquid	Vapor	Liquid 1	Liquid 2
$C_4H_{10}O$	$< 10^{-8}$	0.3	$< 10^{-8}$	0.2931	0.0069
H_2O	$< 10^{-8}$	0.7	$< 10^{-8}$	0.2955	0.4045
Total	$< 10^{-8}$	1.0	$< 10^{-8}$	0.5886	0.4114
$-(dP/dV)$	1.61×10^3	3.47×10^2	1.71×10^3	3.13×10^2	4.29×10^3
Stability det.	3.72×10^{10}	-7.72	3.53×10^{10}	3.28×10^2	5.15×10^4
G_{min}/RT	-1.0111			-1.0243	

4. *Nonideal 2-phase physical equilibrium:*
   A seven component mixture containing propylene, diisopropyl-ether, isopropanol, water, acetone, hexene and n-propanol is flashed at 388.5°K and 40.8 atm. This two phase equilibrium problem was solved by Gautam and Seider. The results obtained by them and by this work is shown in Table 8. The following partial list of binary interaction parameters were used for this system:

$$k_{ij} = \begin{bmatrix} 0.0 & 0.0 & 0.0 & 0.0 & 0.0 & 0.0 & 0.0 \\ 0.0 & 0.0 & 0.04402 & -0.222 & 0.0 & 0.0 & 0.0 \\ 0.0 & 0.04402 & 0.0 & -0.152 & 0.03401 & 0.0 & -0.00675 \\ 0.0 & -0.222 & -0.152 & 0.0 & -0.096 & 0.0 & -0.11374 \\ 0.0 & 0.0 & 0.03401 & -0.096 & 0.0 & 0.0 & 0.0 \\ 0.0 & 0.0 & 0.0 & 0.0 & 0.0 & 0.0 & 0.0 \\ 0.0 & 0.0 & -0.00657 & -0.11347 & 0.0 & 0.0 & 0.0 \end{bmatrix} \tag{62}$$

$$l_{ij} = \begin{bmatrix} 0.0 & 0.0 & 0.0 & 0.0 & 0.0 & 0.0 & 0.0 \\ 0.0 & 0.0 & -0.053 & -0.412 & 0.0 & 0.0 & 0.0 \\ 0.0 & 0.053 & 0.0 & -0.0707 & 0.01832 & 0.0 & -0.0022 \\ 0.0 & 0.412 & 0.0707 & 0.0 & 0.172 & 0.0 & 0.03466 \\ 0.0 & 0.0 & -0.01832 & -0.172 & 0.0 & 0.0 & 0.0 \\ 0.0 & 0.0 & 0.0 & 0.0 & 0.0 & 0.0 & 0.0 \\ 0.0 & 0.0 & 0.0022 & -0.03466 & 0.0 & 0.0 & 0.0 \end{bmatrix} \tag{63}$$

   where, (i, j) = propylene, IPE, IPA, water, acetone, hexene and n-propanol. A partial list was used due to the availability of binary VLE data for only some systems. Inspite of this limitation, the results compare favorably with those reported by Gautam and Seider.

5. *Nonideal 2-phase chemical equilibrium:*
   Benzene and hydrogen are reacted at 500°K and 30 atm to form cyclohexane:

$$C_6H_6 + 3H_2 \rightleftharpoons C_6H_{12} \tag{64}$$

   for a feed containing 1.0, 3.07 and 0.2 moles benzene, hydrogen and cyclohexane respectively. This two phase chemical equilibrium problem is taken from Henley and

Table 8: Solution to Example 4

Species	Feed	Gautam and Seider		This work	
		Vapor	Liquid	Vapor	Liquid
Propylene	78.00	77.9805	0.0195	77.9918	0.0082
IPE	2.14	2.1384	0.0016	2.1141	0.0259
IPA	5.06	4.6382	0.3827	4.7571	0.3029
Water	11.90	4.0035	7.8965	5.7613	6.1387
Acetone	0.071	0.0671	0.0039	0.0703	0.0007
Hexene	2.80	2.7999	0.00001	2.7999	0.00002
n-Propanol	0.03	0.0284	0.0016	0.0262	0.0038

Table 9: Solution to Example 5

Species	Feed Mol	Vapor Mol	Liquid Mol
C_6H_6	1.00	0.000002	0.000007
H_2	3.07	0.053742	0.016287
C_6H_{12}	0.20	0.172855	1.027136
Total	4.27	0.226599	1.043431

Rosen. The binary interaction parameters used were:

$$k_{ij} = \begin{bmatrix} 0.0 & -2.62717 & 0.02413 \\ -2.62717 & 0.0 & -2.95736 \\ 0.02413 & -2.95736 & 0.0 \end{bmatrix} \tag{65}$$

$$l_{ij} = \begin{bmatrix} 0.0 & -0.55343 & -0.00211 \\ 0.55343 & 0.0 & 0.66357 \\ 0.00211 & -0.66357 & 0.0 \end{bmatrix} \tag{66}$$

A guess of 1.0×10^{-3} was used for all components in all phases. The results obtained are tabulated in Table 9.

H Objective Functions, Constraints and Derivatives

H.1 Chemical-Physical Equilibrium

The number of variables, equality constraints and total number of constraints are,

$$\begin{aligned} \text{NVAR} &= S + (N - S)\pi \\ \text{NMEQ} &= R \\ \text{NCON} &= \text{NVAR} + \text{NMEQ} \end{aligned}$$

The objective function is,

$$G^t = \sum_{i=1}^{S} n_i \frac{G_i^\circ}{RT} + \sum_{i=1}^{N-S} \sum_{p=1}^{\pi} n_{ip} \left[\frac{G_i^\circ}{RT} + \ln \left(P\phi_{ip} \frac{n_{ip}}{n_{Tp}} \right) \right]$$

and the constraints are,

$$\begin{aligned} C_1 &\equiv \sum_{i=1}^{S} a_{ki} n_i + \sum_{i=1}^{N-S} \sum_{p=1}^{\pi} a_{ki} n_{ip} = b_k \qquad k = 1, \ldots, \mathrm{NMEQ} \\ C_2 &\equiv n_i \geq 0 \qquad i = 1, \ldots, \mathrm{NVAR} \end{aligned}$$

The gradient of the objective function and the constraints are,

1. For $0 < i \leq S$,

$$\frac{\partial G^t}{\partial n_i} = \frac{G_i^\circ}{RT}$$

and

$$\begin{aligned} \frac{\partial C_1}{\partial n_i} &\equiv a_{ki} \qquad k = 1, \ldots, \mathrm{NMEQ} \\ \frac{\partial C_2}{\partial n_i} &\equiv 1 \end{aligned}$$

2. For $S < ip \leq \pi(N - S)$, and $p = 1, \ldots, \pi$

$$\frac{\partial G^t}{\partial n_{ip}} = \frac{G_i^\circ}{RT} + \ln \left(P\phi_i^p \frac{n_{ip}}{n_{Tp}} \right) + \sum_{j=1}^{N-S} n_{jp} \left[\frac{\partial \ln \phi_j^p}{\partial n_{ip}} + \frac{\partial n_{jp}}{\partial n_{ip}} \frac{1}{n_{jp}} - \frac{1}{n_{Tp}} \right]$$

and

$$\begin{aligned} \frac{\partial C_1}{\partial n_{ip}} &\equiv a_{ki} \qquad k = 1, \ldots, \mathrm{NMEQ} \\ \frac{\partial C_2}{\partial n_{ip}} &\equiv 1 \end{aligned}$$

H.2 Physical Equilibrium

The number of variables, equality constraints and total number of constraints are,

$$\begin{aligned} \mathrm{NVAR} &= N\pi \\ \mathrm{NMEQ} &= N \\ \mathrm{NCON} &= \mathrm{NVAR} + \mathrm{NMEQ} \end{aligned}$$

The objective function is,

$$\Delta G^t = \frac{G^t - \sum_{i=1}^{N} n_i^\circ G_i^\circ}{RT} = \sum_{i=1}^{N}\sum_{p=1}^{\pi} n_{ip}\left[\ln\left(P\phi_{ip}\frac{n_{ip}}{n_{Tp}}\right)\right]$$

and the constraints are,

$$\begin{aligned} C_1 &\equiv n_i^\circ - \sum_{p=1}^{\pi} n_{ip} = 0 \qquad i = 1,\ldots,N \\ C_2 &\equiv n_i^\circ \geq n_{ip} \geq 0 \end{aligned}$$

The gradient of the objective function and the constraints with respect to the mole number of the ith species in the pth phase n_{ip} are,

$$\frac{\partial \Delta G^t}{\partial n_{ip}} = \ln\left(P\phi_i^p \frac{n_{ip}}{n_{Tp}}\right) + \sum_{j=1}^{N} n_{jp}\left[\frac{\partial \ln \phi_j^p}{\partial n_{ip}} + \frac{\partial n_{jp}}{\partial n_{ip}}\frac{1}{n_{jp}} - \frac{1}{n_{Tp}}\right]$$

and

$$\frac{\partial C_1}{\partial n_{ip}} \equiv \frac{\partial C_2}{\partial n_{ip}} \equiv 1$$

Notice, that since the derivatives of the constraints for both the chemical and physical equilibrium problems are constant, they are evaluated only once.

I Analytic Derivatives for the Modified Peng-Robinson Equation of State

I.1 The Modified Peng-Robinson Equation of State

$$P = \frac{NRT}{V - NB_m} - \frac{N^2 A_m}{V^2 + 2NB_m V - N^2 B_m^2}$$

$$\begin{aligned} B_m &= \sum_i x_i b_i \\ A_m &= \sum_i \sum_j x_i x_j a_{ij} \\ a_{ij} &= a_{ij}^\circ \left[1 - k_{ij} + l_{ij} x_i\right] \\ a_{ij}^\circ &= \sqrt{a_i a_j \alpha_i \alpha_j} \\ a_i &= \frac{0.45724 R^2 T_c^2}{P_c} \end{aligned}$$

$$
\begin{aligned}
b_i &= \frac{0.07780RT_c}{P_c} \\
\ln \alpha_i &= m\left[1-\frac{T}{T_c}\right] + n\left[1-\sqrt{\frac{T}{T_c}}\right]^2
\end{aligned}
$$

$$Z^3 + (B-1)Z^2 + (A-3B^2-2B)Z + (B^3+B^2-AB) = 0$$

$$
\begin{aligned}
A &= \frac{A_m P}{R^2T^2} \\
B &= \frac{B_m P}{RT}
\end{aligned}
$$

I.2 Analytic Derivatives

$$
\begin{aligned}
\frac{\partial B}{\partial T} &= \frac{-B}{T} \\
\frac{\partial B}{\partial P} &= \frac{B}{P} \\
\frac{\partial B}{\partial n_j} &= \frac{P}{RT}\frac{\partial B_m}{\partial n_j} \\
\frac{\partial B_m}{\partial n_j} &= \frac{1}{N}\left[b_j - B_m\right] \\
\frac{\partial N B_m}{\partial n_j} &= b_j \\
\frac{\partial A}{\partial T} &= A\left[\frac{1}{A_m}\frac{\partial A_m}{\partial T} - \frac{2}{T}\right] \\
\frac{\partial A}{\partial P} &= \frac{A}{P} \\
\frac{\partial A}{\partial n_j} &= \frac{P}{R^2T^2}\frac{\partial A_m}{\partial n_j} \\
\frac{\partial A_m}{\partial n_j} &= \frac{1}{N}\left[S_1^j - S_2 + S_3^j - 2A_m\right] \\
\frac{\partial N A_m}{\partial n_j} &= S_1^j - S_2 + S_3^j - A_m \\
S_1^j &= 2\sum_i x_i a_{ji} \\
S_2 &= \sum_i \sum_j x_i^2 x_j l_{ij} a_{ij}^\circ \\
S_3^j &= \sum_i x_i^2 l_{ij} a_{ij}^\circ
\end{aligned}
$$

The partial derivative of Z with respect to the flowsheet variable θ is given by:

$$\frac{\partial Z}{\partial \theta} = -\frac{\frac{\partial B}{\partial \theta}\left[Z^2 - (6B+2)Z + 3B^2 + 2B - A\right] + \frac{\partial A}{\partial \theta}\left[Z - B\right]}{3Z^2 + 2(B-1)Z + A - 3B^2 - 2B}$$

If θ is considered to be the n_j, then the partial molar volume may be calculated as:

$$\begin{aligned} \overline{V}_j &= \frac{RT\overline{Z}_j}{P} \\ \overline{Z}_j &= N\frac{\partial Z}{\partial n_j} + Z \end{aligned}$$

The derivative of the fugacity coefficient is calculated directly from the following relation:

$$\frac{\partial \ln \phi_k}{\partial P} = \frac{\overline{Z}_k - 1}{P} = \frac{\overline{V}_k}{RT} - \frac{1}{P}$$

The partial molar volume can also be calculated from:

$$\begin{aligned} \overline{V}_k &= \left[\frac{\partial V}{\partial n_k}\right]_{T,P,n_j} \\ &= -\frac{\left[\frac{\partial P}{\partial n_k}\right]_{T,V,n_j}}{\left[\frac{\partial P}{\partial V}\right]_{T,N}} \end{aligned}$$

The pressure derivatives are readily calculated from the Peng-Robinson equation of state:

$$\begin{aligned} \frac{\partial P}{\partial V} &= -\frac{NRT}{(V - NB_m)^2} + \frac{2N^2 A_m (V + NB_m)}{[V^2 + 2NB_m V - N^2 B_m^2]^2} \\ \frac{\partial P}{\partial n_k} &= \frac{NRTb_k}{(V - NB_m)^2} + \frac{RT}{(V - NB_m)} \\ &- \frac{N(S_1^k - S_2 + S_3^k)}{V^2 + 2NB_m V - N^2 B_m^2} \\ &+ \frac{2N^2 A_m b_k (V - NB_m)}{[V^2 + 2NB_m V - N^2 B_m^2]^2} \end{aligned}$$

I.3 Analytic Derivatives for the Fugacity Coefficient

The fugacity coefficient for the Peng-Robinson equation of State is,

$$\begin{aligned} \ln \phi_k &= \frac{b_k(Z-1)}{B_m} - \ln(Z - B) \\ &+ \frac{1}{2\sqrt{2}RTB_m}\left[S_1^k - S_2 + S_3^k - \frac{b_k A_m}{B_m}\right]\Psi \\ \Psi &= \ln\left[\frac{Z + B(1-\sqrt{2})}{Z + B(1+\sqrt{2})}\right] \end{aligned}$$

The various derivatives are:

$$\frac{\partial \Psi}{\partial \theta} = \frac{2\sqrt{2}}{Z^2+2ZB-B^2}\left[B\frac{\partial Z}{\partial \theta} - Z\frac{\partial B}{\partial \theta}\right]$$

$$\frac{\partial \ln(Z-B)}{\partial \theta} = \frac{1}{Z-B}\left[\frac{\partial Z}{\partial \theta} - \frac{\partial B}{\partial \theta}\right]$$

$$\frac{\partial\left[\frac{b_k(Z-1)}{B_m}\right]}{\partial \theta} = \frac{b_k}{B_m}\frac{\partial Z}{\partial \theta} - \frac{b_k(Z-1)}{B_m^2}\frac{\partial B_m}{\partial \theta}$$

The derivatives of the summation with respect to the mole fractions are as follow:

$$\frac{\partial S_1^k}{\partial n_j} = \frac{2}{N}\left[a_{kj} + \left(\frac{\partial n_k}{\partial n_j} - x_k\right)\sum_i x_i a_{ki}^{\circ} l_{ki} - \frac{S_1^k}{2}\right]$$

$$\frac{\partial S_2}{\partial n_j} = \frac{1}{N}\left[\sum_k (x_k^2 - 2x_k x_j) l_{kj} a_{kj}^{\circ} - 3S_2\right]$$

$$\frac{\partial S_3^k}{\partial n_j} = \frac{2}{N}\left[x_j l_{jk} a_{jk}^{\circ} - S_3^k\right]$$

The derivative of the fugacity coefficient with respect to n_j is calculated as follows:

$$\begin{aligned}
\frac{\partial \ln \phi_k}{\partial n_j} &= \frac{b_k}{B_m}\frac{\partial Z}{\partial n_j} - \frac{b_k(Z-1)}{B_m^2}\frac{\partial B_m}{\partial n_j} + \frac{1}{Z-B}\left[\frac{Pb_j}{NRT} - \frac{B}{N} - \frac{\partial Z}{\partial n_j}\right] \\
&+ \frac{1}{RTB_m(Z^2+2ZB-B^2)}\left[B\frac{\partial Z}{\partial n_j} - \frac{ZP}{RT}\frac{\partial B_m}{\partial n_j}\right]\left[S_1^k - S_2 + S_3^k - \frac{b_k A_m}{B_m}\right] \\
&+ \frac{\sqrt{2}\Psi}{4RTB_m} \\
&\times \left[\frac{\partial S_1^k}{\partial n_j} - \frac{\partial S_2}{\partial n_j} + \frac{\partial S_3^k}{\partial n_j} - \frac{S_1^k - S_2 + S_3^k}{B_m}\frac{\partial B_m}{\partial n_j} - \frac{b_k}{B_m}\frac{\partial A_m}{\partial n_j} + \frac{2b_k A_m}{B_m^2}\frac{\partial B_m}{\partial n_j}\right]
\end{aligned}$$

It is interesting to note that

$$\frac{\partial \ln \phi_k}{\partial n_j} = \frac{\partial \ln \phi_j}{\partial n_k}$$

References

[1] D. R. Stull. Fundamentals of fire and explosion. AIChE Monograph series 10, American Institute of Chemical Engineers, 1977.

[2] G. A. Melhem and E. S. Shanley. On the estimation of hazard potential for chemical substances. *Process Safety Progress*, 15(3):168–172, 1996.

[3] S. R. Brinkley Jr. and W. E. Gordon. Engineering design handbook - principles of explosives behavior. AMC Pamphlet AMCP 706-180, U.S. ARMY, 1972.

[4] G. A. Melhem. Superchems expert version 3: An advanced tool for consequence analysis, thermal hazards assessment and pressure relief design, 1996.

[5] H. Le Chatelier. Estimation of firedamp by flammability limits. Technical Report 8, Ann. mines, 1891.

[6] M. G. Zabetakis. *Flammability characteristics of combustible gases and vapors.* Bulletin 627, Bureau of Mines, 1965.

Vent Access Restriction for Solids Handling Systems

Lionel Bernard, Frank Brodie, Dennis Ludwig, Albert Ness, Karl Weidner

Rohm and Haas Engineering Division
Rt. 413 and State Rd, Bristol, Pa. 19007

Prepared for Presentation at

American Institute of Chemical Engineers
31st Loss Prevention Symposium
Houston, TX

March 11, 1997

Date November 1996

UNPUBLISHED

Vent Access Restriction for Solids Handling Systems

LIONEL BERNARD, FRANK BRODIE, DENNIS LUDWIG, ALBERT NESS, KARL WEIDNER
Rohm and Haas Engineering Division
Rt. 413 and State Rd, Bristol, Pa. 19007

ABSTRACT

A multi-disciplinary team developed a guideline for determining access restriction zones around vented solids handling equipment. The guideline provides a method for ensuring that the discharge from a vented explosion will not cause injury to personnel. The steps in this method include: calculating the extent of external hazards from vented explosions; identifying potential areas where personnel could be exposed to a hazard; identifying ways to eliminate or reduce the hazard area; and establishing and documenting any access restrictions needed. Hazard zone calculations use the latest knowledge from research into fireball size, and external flame and pressure equations in VDI 3673. The Guideline provides guidance for using this information. Options for mitigating or reducing external hazards from vented explosions are also described. As part of the project, the team audited several solids handling systems to look for potential oversights in existing restricted access areas. Some learnings from these audits are reviewed.

INTRODUCTION

After an explosion in a dryer train, a secondary physical explosion in a grinder caused the grinder doors to blow off. The secondary explosion occurred because a suppression system fired when the ductwork and grinder were filled with deluge water. There were no injuries because no one was in the grinder area at the time. Access to the area was not restricted because the grinder and duct work were protected by a suppression system, and so considered safe. The potential for injury existed, however, because the possible consequences of the suppression system activating when the equipment was hydraulically full had not been considered. Some corrective actions taken were: writing a procedure to disarm the suppression system if the deluge water system is activated, and installation of a barrier in front of the grinder doors to protect personnel.

This incident left business and plant managers wondering if there were other potentially unidentified hazards posed to personnel due to explosion venting at plants with solids handling systems. The authors of this paper became members of a team to address that concern. Our objectives were to identify hazards in the solids handling systems of each plant that might require restricted access, and establish principles and policies to protect personnel from these hazards. We audited all solids handling systems in that business for potential hazards to personnel. We also reviewed literature and consulted with several experts in the explosion field on accepted practices for defining hazard zones. From this we developed a methodology to define restricted access areas.

HAZARDS FROM VENTED EXPLOSIONS

Potential hazards to personnel in the vicinity of an explosion vent include:

- exposure to a flame or fireball
- exposure to external overpressure
- exposure to flying debris
- exposure to toxic dusts
- fright (causing people to do something dangerous to escape)

The fireball and external overpressures are the main subject of this procedure.

Housekeeping is the main means of protecting personnel from flying debris. Combustible dusts in front of a vent will disperse and ignite if an explosion is vented, creating a larger fireball and increasing external overpressure. Loose equipment will become flying debris and can cause severe injuries. It is impossible to predict the hazard zone for these hazards. We recommended written procedures for housekeeping practices be put in place and operators trained on them.

In practice, the overpressure from most vessels will be below 2 or 3 psig (0.14 - 0.21 bar-g) and will be very short. Special concerns for external pressure are the presence of buildings with glass windows inside the hazard zone or an elevated rack or walkway where a person can be knocked off by the overpressure.

Exposure to toxic dusts or combustion by-products is not addressed. This must be handled on a case by case basis.

ACCESS RESTRICTION PROCESS

There are six steps to defining and establishing restricted access areas:

- Estimate the Hazard Area
- Eliminate the Need for Access
- Mitigate the Hazard
- Find Alternatives to Venting
- Establish a Restriction Zone
- Manage the Risk

Figure 1 shows the process. It calls for you to identify hazard areas, apply hazard reduction techniques, and restrict access if the hazard cannot be mitigated.

Step 1: Estimate the Hazard Area

Calculating Hazard Zones

Over the last few years, empirical equations for external flames and overpressures have been developed. These come from the experimentation that led to the revised VDI 3673 standard[1] and from the CREDIT Program sponsored by the European Union Environmental Program[2]. Hazard zone calculations are based on the equations from this work.

Flame Length and Fireball Volume Equations

$$L_F = 8\,(V^{1/3}) \quad \text{for vertically discharging vents} \tag{1}$$

$$L_F = 10\,(V^{1/3}) \quad \text{for horizontally discharging vents} \tag{2}$$

$$L_F = 6\,(V^{1/3}) \quad \text{for 2 well separated horizontally discharging vents} \tag{3}$$

$$V_F = 10V \tag{4}$$

$$W_F = (V_F^{1/3}) \quad \text{with no external obstacles} \tag{5}$$

Where:

L_F	Length of flame outside of vessel, meters
V	Vessel Volume, cubic meters
V_F	Volume of fireball outside of vessel, cubic meters.
W_F	Width of fireball outside of vessel, meters

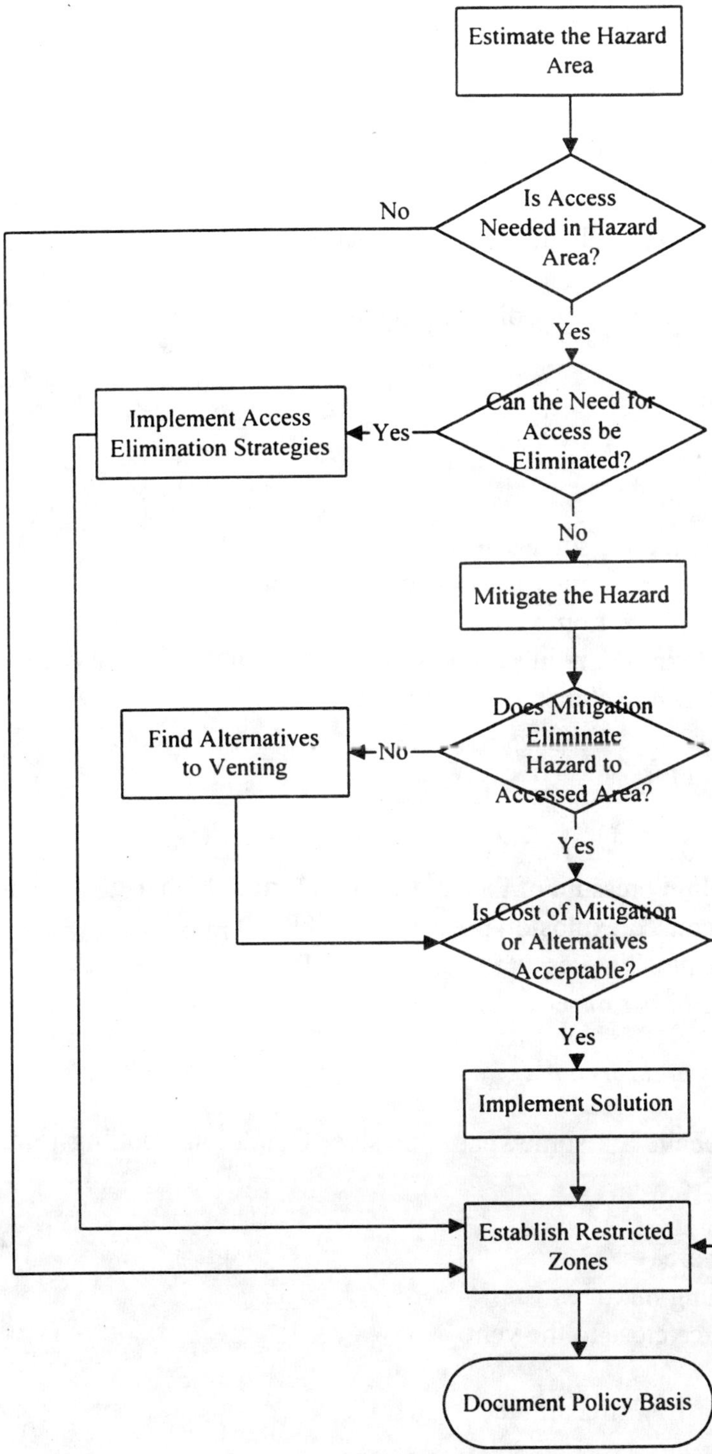

Figure 1. Access Restriction Decision Flowchart

External Pressure Equations

$$p_{max,a} = 0.2\ P_{red,max}\ A^{0.1}\ V^{0.18} \quad (6)$$

$$R_S = 0.25\ L_F \quad \text{for vertically discharging vents} \quad (7)$$

$$R_S = 0.2\ L_F \quad \text{for horizontally discharging vents} \quad (8)$$

$$P_{r,max} = (R_S/r)\ p_{max,a} \quad (9)$$

Where:

A	Vent area, square meters
$p_{max,a}$	Maximum pressure outside vessel, bar-g
R_S	Distance from vent to $p_{max,a}$, meters
$P_{red,max}$	Maximum reduced pressure in vessel, bar-g
$P_{r,max}$	Pressure at distance r (m), bar-g

These equations are valid for:

- Vessel sizes $0.3\ m^3 \leq V \leq 1000\ m^3$
- Static activation pressure of the vent, $0.1 \leq P_{stat} \leq 0.2$ bar-g
- Maximum reduced explosion pressure, $0.1 \leq P_{red,max} \leq 2$ bar-g
- Maximum explosion pressure, $5 \leq P_{max} \leq 9$ bar-g
- $10 \leq K_{st} \leq 200$ bar m/sec

Guidance on Flame Jets and Fireballs

The flame jets and fireballs are somewhat exclusive. Conditions leading to a flame jet are:

- smaller vent area
- rapidly burning mixtures
- ignition source close to the vent

Conditions that lead to a 10V fireball are:

- large vent area
- low vent opening pressure
- slowly emerging flame front
- ignition source remote from vent

In general, the more dust consumed by the explosion inside the vessel, the more likely you are to have a jet flame. Jet flames will be of shorter duration, about 1 or 2 seconds. Therefore, ST2 dusts and higher relief pressures lead to more directional and cleaner jet flames, and ST1 dusts and low relief pressures lead to more spherical and dirtier fireballs.

Figure 2 shows the fireball hazard zone for the simple case of one vent on the side of a vessel. Equation 2 is used to estimate the jet flame hazard zone. Equation 4 is used to estimate the fireball hazard zone.

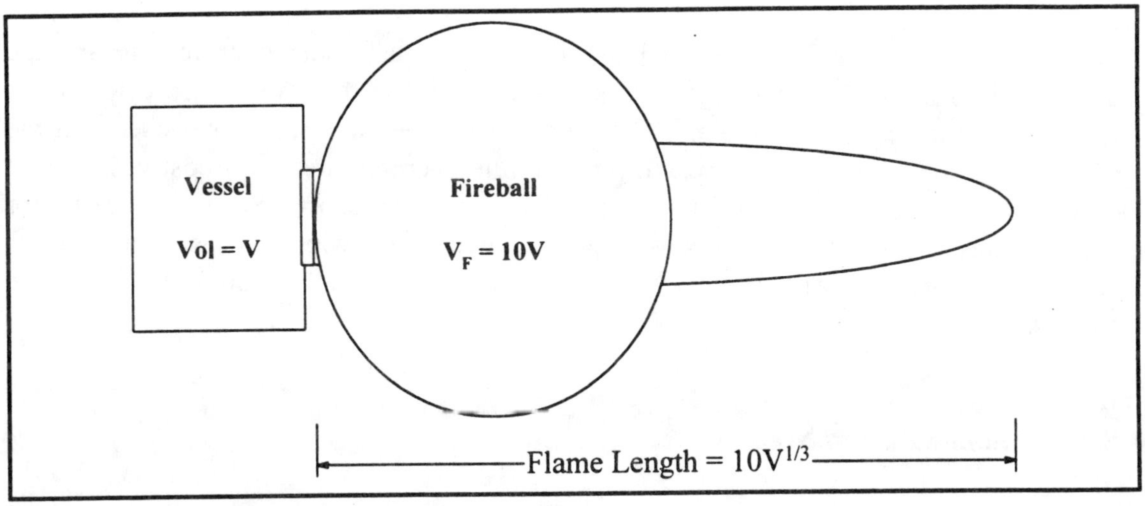

Figure 2 Fireball Hazard Zone

Most large solids handling equipment will be low strength and therefore have a low vent opening pressure. Therefore, fireballs will be the hazard of most concern.

If you have a cubical vessel (L/D ≤ 2) and sized the vents using the NFPA 68 (1994)[3] Nomographs (equivalent to the pre-1995 VDI guideline), you have a conservatively sized vent and can expect the fireball. Equations 4 and 5 correctly estimate the fireball's hazard zone.

If you have a cubical vessel and sized the vent using the 1995 VDI equations, the vents are less conservatively designed. You can expect less fireball and a longer flame jet. Equations 4 and 5 should overstate and therefore give a conservative fireball hazard zone.

As L/D goes above 2, the NFPA 68 (1994) Nomographs are less conservative. A pure flame jet may still not occur because the ignition source may be remote from the vent. Equations 4 and 5 should give a conservative fireball hazard zone.

For L/D > 2, the 1995 VDI equations usually give larger vents than NFPA 68. Use of equations 4 and 5 correctly estimates the hazard zone.

Fireball Hazard Zone - Effect of Multiple Vents:

Equation 3 shows that well separated vents can reduce the flame jet length.

There are no equations for the effect of multiple vents on the fireball size. Based on photographs we recommend assuming the total fireball volume, 10V, is divided equally between each vent.

External Over-Pressure Hazard Zone:

Equations 6 and 7 or 8 can be used to calculate the maximum overpressure and the distance from the vent at which it occurs. Comparing the results of these calculations with existing tables for damage as a function of overpressure[4] provides a check on the potential overpressure hazard zone. In practice, the overpressure from most vessels will be below 2 or 3 psig (0.14 - 0.21 bar-g) and will be very short. Special concerns for external pressure are; the presence of buildings with glass windows inside the hazard zone or an elevated rack or walkway where a person can be knocked off by the overpressure.

Step 2: Elimination of Need for Access:

Elimination of a hazard is always the first choice of options. Examples of ways to eliminate the need for access include:

- Reposition instruments & controls from inside a hazard zone to outside the zone.
- Use remote monitoring to eliminate need for any visual checks operators might need to make
- Use equipment with higher reliability.
- Relocate equipment

When designing a new facility the need for access should be considered in the initial placement of equipment and vents.

Step 3: Mitigate the Hazard

When a system is added or changed to mitigate fireball or overpressure, a management of change review should be held to assess the safety of the equipment change or addition as well as any changes to existing access restrictions. The design basis of any mitigating feature must be documented. Potential methods to reduce or eliminate a hazard zone include:

- *Close unneeded vents*

  Older equipment may have oversized vents. Check the vent size calculation using the latest versions of NFPA 68 or VDI 3673. Increasing the strength of a vessel allows you to reduce the vent area needed. It may be possible to close vents pointed to access areas. Recoil forces must be considered when vents are closed.

- *Barriers or Deflectors*

  It is possible to build barriers or deflectors to block or redirect the flame from areas where access is needed. They must be strong enough to withstand the overpressure, (see equations 6 and 9) and far enough from the vent to avoid impeding venting. Deflectors merely change flame path, therefore it must direct the flame to a safe place or the area must be restricted. Assume the flame is deflected at the angle of the deflector. Dunn[5] gives guidance on design and placement of deflectors.

- *Flameless Venting*

  Vent devices exist that quench flames, but allow pressure out. Examples are the Q-Rohr® (Rembe) and the Flame Quench® (Fike). These devices are more expensive than ordinary vents, but compare their cost to the cost of adding a vent duct or other flame path mitigation techniques.

- *Change equipment layout*

  This is easier during design of new installations than in existing ones, but it is still possible for existing facilities. Hazardous equipment, such as mills, can be located near exterior walls to allow vents to be ducted outside the room without creating vent ducts that are very long (> 3 meters) or with many bends. Locating equipment on a roof and venting upwards is another example. When venting upwards, consider the effects of weather, e.g., water or ice, on the vent. If vent ducting is added to a vessel the vent size will need to be re-calculated to account for the duct.

- *Flame suppression at vent opening*

  Tests have shown that such a suppression or water spray system at the vent opening can reduce the external flame. This will only extinguish the flames, not suppress the explosion overpressure.

- *Containment room*

  Construct a containment room around vented equipment or the vent. The room should be strong enough to contain the pressure or should itself be vented to a safer location.

Step 4: Find Alternatives to Venting:

There may be times when it is not possible to vent to a safe place. In those cases consider replacing vents with suppression, inertion or containment. Over half of all suppression systems installed are on equipment that cannot be vented to a safe location.

When explosion protection is changed from venting to suppression or inertion, a management of change review must be conducted. A formal Process Hazard Analysis (PHA) technique, such as HAZOP or Failure Mode and Effects Analysis (FM&EA) is recommended. Access to areas around equipment protected by suppression, inertion or containment is considered safe.

Step 5: Establish Restricted Zone:

When restricted areas have been identified they need to be clearly marked. Usually a "Restricted Area" sign is posted. Operators must be informed about access restrictions. Painting restricted areas a different color should be considered.

When access is forbidden in certain operating modes, e.g., when on feed, non-operating personnel must check in at the control room before entering the unit. This means access to and from the control room must not be through a hazardous area. An example sign is:

WARNING Explosion Vent Restricted Area Check with Control Operator Before Entering.

Inspections of the unit must check for the presence and condition of the warning signs. A list of warning locations should be developed to aid in this task.

Access Restriction vs. Operating Mode:

Access restriction to a hazard zone can depend on the mode of operation the plant is in. When restrictions are established different modes of operation should be explicitly considered. Points to consider are:

- Is powder moving, e.g. the fans are on?
- Does the powder need time to settle?
- Do vessels have hold up that will create a dust cloud ?

Step 6: Risk Management:

If a vessel cannot be vented to a safe place, and all obvious attempts at mitigation are either unworkable or cost prohibitive, then the risk must recognized and managed.

A formal PHA and even a Quantitative Risk Assessment (QRA) is recommended. If, after a formal PHA or QRA, the risk to personnel exceeds what the plant thinks prudent, business, plant and corporate safety representatives should be brought in to decide if continuing to run the system without further mitigation is acceptable.

AUDIT RESULTS

During the plant audits some common oversights were found. Examples are:

The roofs of nearby buildings were sometimes overlooked. Mechanics occasionally need access for maintenance of equipment such as fans.

Areas that were designated restricted but were infrequently accessed sometimes no longer had restriction signs or the signs were damaged.

There were equipment items that had flexible or temporary connections. The strength of these connections is unknown. They could become unintended vents in areas that are not access restricted.

SUMMARY

As a result of the team's work, a guideline for calculating and establishing hazard zones from explosion vents on solids handling equipment was written. This gives plant and process engineers guidance on how to reduce the risk of injuries to personnel.

References:

[1] VDI 3673 Part 1, (July, 1995) *Pressure Venting of Dust Explosions*, (German/English Issue), Verein Deutscher Ingenieure, Dusseldorf, Germany.

[2] Crowhurst, D., Colwell, S., Hoare, D., Benselium, M., Schramm, S., Friehmelt, V., *Guidelines for Industry: Dust Explosions - Flame and Pressure Effects Outside Vents*, Dust Explosions, London, IBC, October 1995.

[3] NFPA 68, *Guide for Venting of Deflagrations*, 1994 Edition, National Fire Prevention Association, 1 Battery Park, P.O. Box 9101, Quincy, MA. 02269-9109.

[4] CCPS, 1994. *Guidelines for Evaluating the Characteristics of Vapor Cloud Explosions, Flash Fires and BLEVE's*, Center for Chemical Process Safety of the American Institute of Chemical Engineers, New York, NY 10017, p. 348.

[5] Dunn, Geoff, 1992, *Dust Explosion Prevention and Protection, Part 1 - Venting, 2nd Edition,* Institution of Chemical Engineers, Rugby, UK, ISBN 0 85295 293 7, p. 186.

K_G: New Data and Analysis

by

Joseph A. Senecal and Patricia A. Beaulieu
FENWAL Safety Systems
A Division of Kidde Technologies, Inc.
700 Nickerson Road
Marlborough, MA 01752

Prepared for Presentation at

American Institute of Chemical Engineers
31st Loss Prevention Symposium
Houston, TX

March 11, 1997

Abstract

The design of deflagration pressure relief vents is based on correlations developed for various types of combustible materials and for enclosures of different strengths. The primary guideline for deflagration vent design in the US is *NFPA 68 Guide for Venting of Deflagrations*. That document gives guidance for the design of vents for enclosures containing flammable gases, specifically hydrogen, coke oven gas, propane, and methane. Application of the guide to other gases is achieved using the K_G value. Values of K_G are published for a relatively small number of gases, as seen in Table D-1 of NFPA 68. This work presents K_G data on several additional gases obtained in a laboratory scale test vessel along with analysis of the results with respect to published values of fundamental burning velocity.

Introduction

Central to the design and analysis of explosion protection systems is some quantification of the combustion characteristics of the fuel-air mixture. Design guides for venting dust deflagrations (NFPA 68, VDI 3673) use nomographs or equations where K_{St} serves this purpose. Vent design guidance in NFPA 68 for gaseous fuel systems uses nomographs for a few select fuels with extension to other fuels through consideration of differences in $(dP/dt)_{MAX}$ determined by test. Bartknecht (1993) recommends the following equation for vent sizing in gaseous fuel applications:

$$A = ((0.1265 \log K_G - 0.0567) P_{RED}^{-0.5817}) + 0.1754 P_{RED}^{-0.5722} (P_{STAT} - 0.1)) V^{2/3} \qquad (1)$$

where K_G (bar-m/s) is the fuel-air volume normalized rate of pressure rise, P_{RED} (bar g) is the maximum reduced pressure during the vented explosion, P_{STAT} (bar g) is the vent closure release pressure, V is the enclosure volume (m^3) and A is the vent area (m^2). Analysis of vent closure inertia effects (Tamanini) also uses K_{St} for dusts and presumably a similar analysis for gases would use either K_G or S, the burning velocity. The use of S as the fuel characteristic is preferred as it is independent of equipment geometry (see below) but is more difficult to measure than K_G. Values of K_G can be used to estimate S (see below and Appendix A) but results tend to be high. Further, relatively little K_G data has been published. While the use of gas combustibility data is still evolving, it is worthwhile to develop a greater understanding of K_G, including data on new gases and consideration of K_G in relation to vessel geometry and burning velocity.

Objectives of this work were to: (a) develop K_G data on new species likely to be of commercial interest; (b) examine the sensitivity of K_G to variations in stoichiometry; (c) correlate K_G with fundamental burning velocity, S, where values are available; and (d) assess the possibility of using dP/dt data to estimate S.

Apparatus

The tests were carried out in a 22 liter closed, stainless steel combustion vessel with an internal diameter of 304 mm and height of 350 mm giving a 1.15 aspect ratio. The vessel had a water jacket for temperature control, a viewing port and several threaded penetrations located on the corners of the vessel shell, as seen in Figure 1.

The vessel was evacuated using a mechanical vacuum pump through a manually operated ball valve and 0.25 inch, flexible air line tubing. The vacuum was observed using a 0.1 % Heise precision vacuum gauge, with a scale of 0-30 inHg.

Fuels in gas form at ambient conditions were inserted into the vessel through 0.25 inch OD flexible tubing and a 0.25 inch manually operated ball valve.

For fuels that were a liquid at ambient conditions, a method was required to vaporize them before admitting to the vessel. A small, double valved section of 0.5 inch OD stainless steel tubing was used as the sample reservoir. A 0.5 inch manually operated brass ball valve connected this reservoir through 0.5 inch OD stainless steel tubing, 5 inch long, to a 18 inch long section of 0.75 inch OD tubing placed within a Tecam SBS-4 sand bath set for 250 °F. This tubing was connected to the vessel through a 0.5 inch steel, high temperature, manually operated ball valve and 0.5 inch OD tubing. A Thermolyne heating tape, 0.5 inch wide and 2 ft long, was used to heat this tubing to 200 °F. It was powered by a 0-120 V DC Variac variable power supply. Fiberfrax Duraback insulation was used over the heating tape to maintain uniformity along the length of tubing.

The sand bath was monitored using a type K sheathed thermocouple, 0.063 inch diameter, with a Fluke 27 Multimeter and 80TK Thermocouple Module. The 0.5 inch OD tubing temperature was monitored using a bare bead type J thermocouple with an Omega 660 Indicator.

The igniter used in this work was of the corona discharge spark type. It consisted of a group of four 1.9 mm OD graphite rods (No. 4H pencil leads) held together at each end with three or four wraps of 23 gauge copper wire with a 6 mm gap between the wire windings. The wire leads from the igniter were soldered to the power supply lead terminations in the vessel and the igniter was positioned in the center of the vessel. The electrical resistance of the igniter assembly was 1.2 ohms.

The igniter was placed in a circuit which included two series-wired 570 μf capacitors which were charged to a potential of 250 volts. On closure of the ignition relay, which was controlled by an Omron model H5BR electronic timer, the capacitor charge was passed through the igniter assembly, causing a corona discharge to develop over the surface of the graphite rods. The stored capacitive energy was 9 J. The spark duration was measured using an oscilloscope and found to be approximately 80 μs. The I^2R loss in the igniter rods was estimated to be at most 4 J. The energy dissipated in the spark was, therefore, estimated to be 5 J.

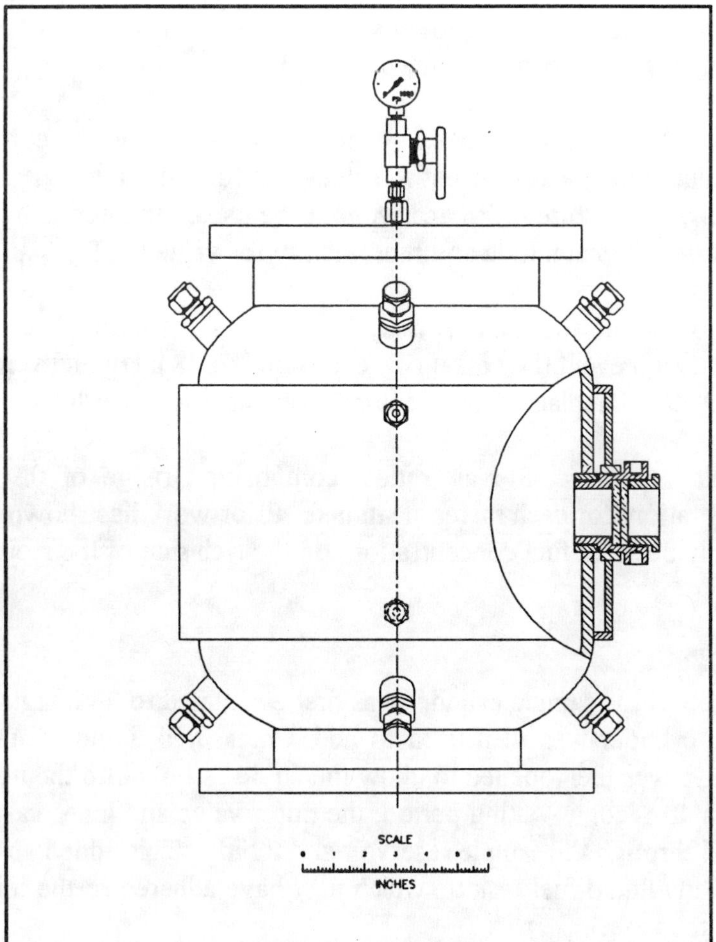

Figure 1. 22 liter stainless steel combustion vessel used for tests.

The vessel pressure was measured using a 0-2000 psig Data Instruments model AB/HP bonded semiconductor strain gauge type pressure transducer with a 22 kHz frequency response. This transducer was used with a Measurements Group 2310 Signal Conditioning Amplifier. Data was collected at a rate of 2 kHz (12 kHz for hydrogen tests) through a Data Translation DT2801A analog to digital data acquisition board using Labtech Notebook software and analyzed on Lotus 123. The vessel temperature was monitored using a type K sheathed thermocouple, 0.063 inch diameter, placed slightly below the center of the vessel, with a Fluke 27 Multimeter and 80TK Thermocouple Module.

Procedure

Gases

The vessel was first evacuated to 28 inHg. The flammable gas was admitted using the Heise vacuum gauge to determine the partial pressure of fuel added. The vacuum ball valve was then fully opened to allow the vessel pressure to rapidly equilibrate to one atmosphere. This step resulted in compressive heating of the gas mixture requiring several minutes of water cooling of the vessel for the gas mixture to cool to a temperature range of 25 to 30 °C.

It was determined that the turbulence induced by the air admission and the convective circulation during temperature equilibration was sufficient to achieve uniform gas mixing. Tests conducted on propane-air and methane-air mixtures with and without the use of an internal mixing fan produced results that were equivalent in magnitude and repeatability for values of P_{MAX} and K_G.

The timer was used to actuate the ignition relay 20 ms after the start of data acquisition. On completion of the combustion event the vessel was evacuated to 28 inHg and vented to atmosphere three times achieving 99.97 % replacement of combustion gases with fresh air.

Tests were carried out at several fuel-air ratios, composition range of 0.9 to 1.3 times the stoichiometric concentration, for each material studied. Prior work has shown that the maximum values of dP/dt normally occur at fuel concentrations on the rich side of the stoichiometric value.

Liquids

The vessel and tubing up to the sample cylinder was first evacuated to 28 inHg and the outer valve closed. A graduated cylinder was then used to add a measured liquid quantity to the sample reservoir. The inner valve was then opened to allow the liquid to flow into the heated tubing within the sand bath. After a 40 second waiting period, the outer valve and inner valve were opened to allow air into the vessel through the sample reservoir and tubing. This added air sweeping through the tubing scavenged any liquid fuel residue which may have adhered to the tube wall, as well as forcing the vapor into the vessel.

At this point, the same procedure as for gases was conducted, except for vessel cooling. The vessel was kept at a minimum temperature that was above the dew point for the liquid being tested.

Table 1 Compilation of K_G and Burning Velocity Data								
	NFPA 68			Bartknecht (1)		**This work**		
Chemical	S (cm/s)	P_{MAX} (bar g)	K_G (bar-m/s)	P_{MAX} (bar g)	K_G (bar-m/s)	P_{MAX} (bar g)	K_G (bar-m/s)	Notes
Acetone						7.3	65	
Acetophenone		6.9	109	7.6	109			2
Acetylene	166			10.6	1415			
Ammonia				5.4	10			3
B-Naphol				4.4	36			4
Butane	45	8.0	92	8.0	92			
isoButane	41					7.4	67	
Carbon Disulfide	58			6.4	105			
1,1-Dichloro ethane						7.7	58	
Diethyl ether	47			8.1	115			
Dimethyl ether						7.9	108	
Dimethyl Formamide				8.4	78			2
2,2-dimethylpropane	39			7.8	60			
Dimethyl Sulfoxide				7.3	112			2
Ethane	47	7.8	106	7.8	106	7.4	78	2
Ethyl Alcohol						7.0	78	
Ethyl benzene		6.6	94	7.4	96			2
Ethylene	80					8.0	171	
Hydrogen	312	6.9	659	6.8	550	6.5	638	5
Hydrogen Disulfide				7.4	45			
Isopropanol	41			7.8	83			2
Methane	40, 37.3	7.05	64	7.1	55	6.7	46	8
Methyl Alcohol	56			7.5	75	7.2	94	2
Methyl Nitrite				11.4	111			
Methylene Chloride				5.0	5			3
Mobiltherm Light Oil				8.0	100			
Octanol				6.7	95			2
Octyl Chloride				8.0	116			2
Pentane	46	7.65	104	7.8	104			2
Propane	46, 41	7.9	96	7.9	100	7.2	76	6, 7, 8
S. African Crude Oil				7.2	49			2
Toluene	41			7.8	94			

Notes: (1) Bartkencht data from 5 L sphere using 10 J ignition energy except as noted. Taken from Bartknecht, 1993. p. 52. (2) Bartknecht value obtained by extrapolation. (3) Bartknecht value ignition energy 100 - 200 J. (4) Bartknecht value at 200 °C. (5) This work: values just before detonation transition. (6) K_G of 75 reported by Bartknecht, 1981. (7) Experimental S value of 41 cm/s for propane from Metghalchi and Keck. (8) Experimental values of S of 37.3 and 41 cm/s for methane and propane, respectively. See Lewis and von Elbe, p. 403.

Data Analysis

Pressure-time data from each test was inspected to determine the maximum pressure value, or P_{MAX}, attained. Rates of pressure rise were calculated by applying a seven point quadratic smoothing technique to the raw pressure-time data. The dP/dt values so obtained were smoothed if necessary and the maximum value determined, $(dP/dt)_{MAX}$. Values of K_G, the volume normalized rate of pressure rise, were calculated as $K_G = (dP/dt)_{MAX} \bullet V^{1/3}$ where $V = 0.022\ m^3$.

Tests were carried out on eleven gas or vapor species. The primary P_{MAX} and K_G results obtained in this work are reported in Table 1 along with data on other species which have previously appeared in the literature.

Comparison With Other Results

The values of K_G reported in this work tend to be lower than those values reported by Bartknecht (1993) or in NFPA 68. There appears to be some degree of inconsistency in the former. Variations in K_G should follow burning velocity. A comparison of results for propane (S = 46 cm/s) and methyl alcohol (S = 56 cm/s) show that the K_G values reported by Bartknecht are inverted in order (100 and 75) while those of the present work (76 and 94) are in proportion with S for these species. The general paucity of published data of this type points to an opportunity for the testing and research community to: (a) review and standardize the method for obtaining and reporting K_G data, and (b) to develop data on a wide range of industrially important gases and vapors.

Comment on Hydrogen Results

The K_G results reported for hydrogen must be viewed as a special case as its combustion behavior is distinctly different from that of other fuels evaluated in this work. A plot of dP/dt vs. P yields, for most fuels, an approximately linear correlation up to the point of $(dP/dt)_{MAX}$ after which point dP/dt declines as wall effects begin to disturb and quench the flame sheet. It is common for the maximum in dP/dt to occur at $(P - P_o)/(P_{MAX} - P_o) \sim 0.6$. The hydrogen results (data acquisition rate = 12.5 kHz) showed a smooth pressure rise up to a pressure of approximately 3.7 bar g after which the remaining unburned mixture appeared to detonate. Ultimate P_{MAX} values were in the range of 25 to 30+ bar g. Shown in Figure 2 are the pressures recorded just before the deflagration to detonation transition (DDT) and the maximum pressures recorded.

A typical data set of P vs. t for hydrogen is shown in Figure 3. The peak rate of pressure rise recorded was not accurately determinable at the data acquisition rate of 12.5 kHz used. Values exceeding 1×10^6 bar/s were recorded. The K_G values reported here were computed from the highest dP/dt value occurring just before DDT. The relevance of K_G values so determined in the design or analysis of safety systems is unclear as the onset of DDT would render the results of such effort useless.

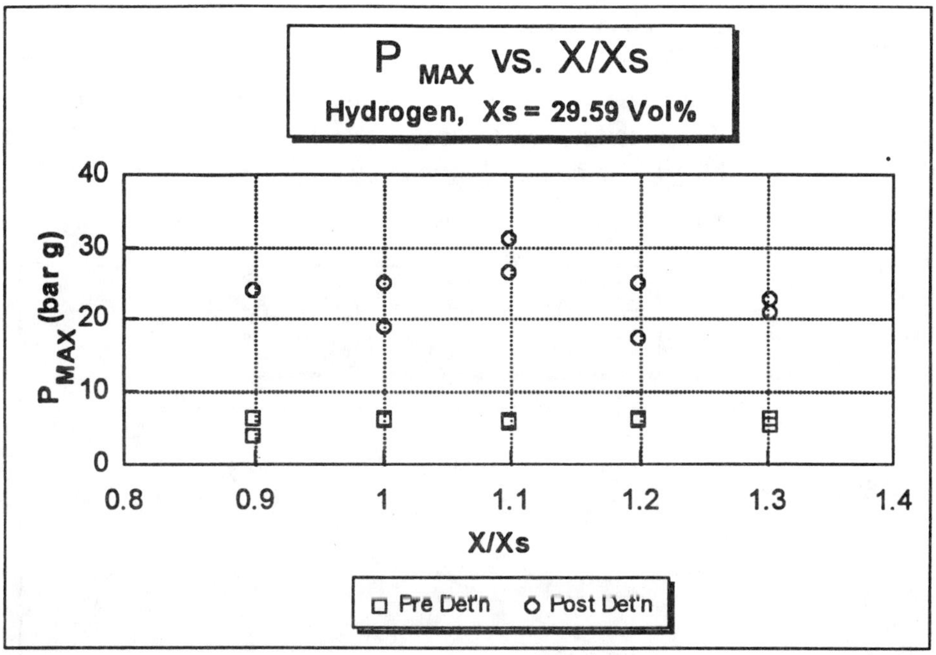

Figure 2. Pressures recorded just before and after H_2 - air DDT.

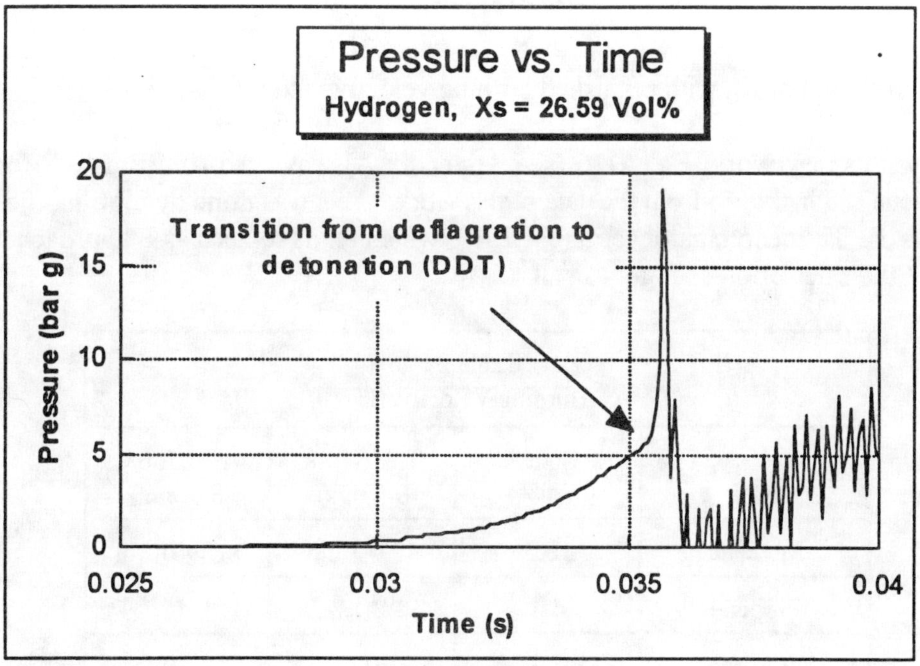

Figure 3. Typical pressure vs. time behavior of hydrogen events.

Burning Velocity, S

Correlation of K_G results obtained in this work with published values of fundamental burning velocity, S, is shown in the Figure 4. Inference of S from K_G from these data would result in an overestimate of S since the plot intercept is at negative K_G. That is, values of S determined from closed vessel deflagrations tend to be high. Nonetheless, use of closed-vessel data may be useful in developing a consistent set of *estimates* of S for which could be refined by application of a correction rule, if one can be discovered.

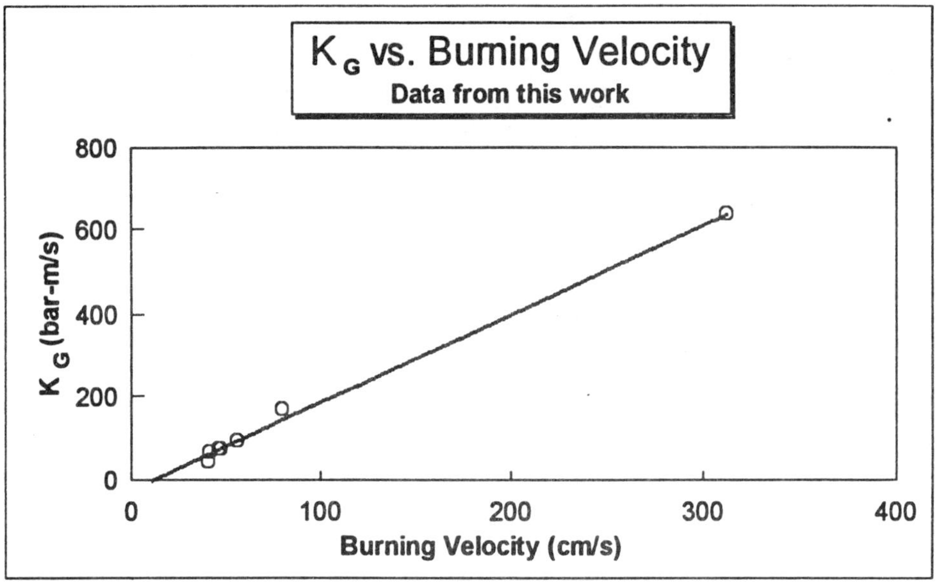

Figure 4. Correlation of K_G with published burning velocity values.

To show how this may work, Eq. A-16 (see Appendix A) was used to estimate S for methane, propane, ethylene and hydrogen from the data of this work. This was done by plotting the right hand side of A-16 vs. P. The denominator term $(1 - P_o/P)$, which tends to zero as P approaches P_o, limits the stability of the calculation as P gets small.

Table 2
Burning Velocity, S

Fuel	Conc. (Vol%)	This Work S (cm/s)	NFPA 68 S (cm/s)
Methane	10.5	43.9 ± 2.5	40
Propane	4.4	52.2 ± 2.3	46
Ethylene	7.2	91.0 ±1.8	80
Hydrogen	35.5	314 ± 55	312

A least squares linear fit was applied to the S vs. P results for $0.3 < P < 2$ (or 3 depending on length of the linear section) bar g. The resulting line was extrapolated to $P = 0$ to obtain an estimate of S. An example of this plot is shown in Figure 5 for the case of propane at 4.4 vol%. The results of the regression analysis, including the standard error of the estimate are given in Table 2.

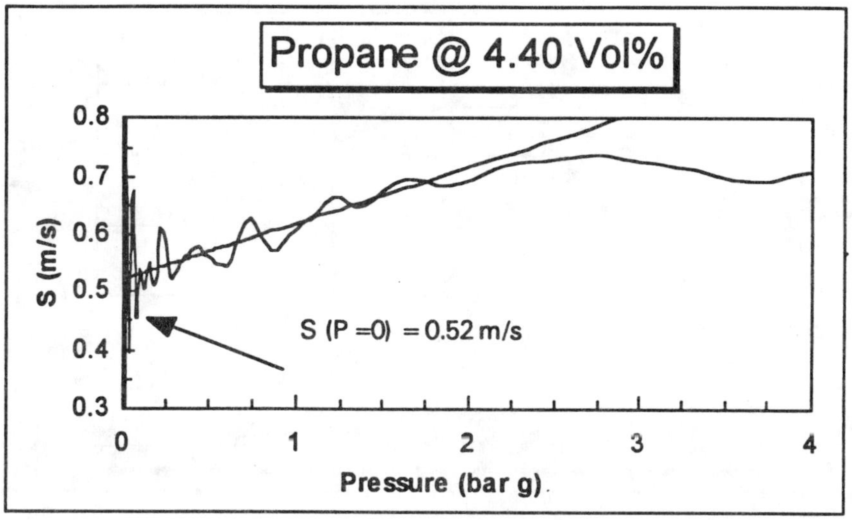

Figure 5. Linear regression of the Equation 1 calculation vs. pressure plot from $0.25 < P < 1.75$ bar was extrapolated to $P = 0$ to get an estimate of S.

Dependence of P_{MAX} and K_G on Stoichiometry

While it is important to know the worst case behaviors of a flammable mixture, it is also useful, especially in risk analysis, to understand the sensitivity of changes in the hazard due to changes in the fuel-air mixture ratio. Shou and Garner show that the maximum burning velocity, S, as reported by several investigators, occurs for propane-air mixtures at an equivalence ratio of 1.1, or at $X/X_S = 1.092$. Here X is the fuel mole fraction and X_S is its value at the stoichiometric composition, 4.03 volume per cent in the case of propane in air. The maximum in K_G value is observed in this work to occur at this mixture ratio for most species. The dependency of K_G on mixture ratio is not the same for all of the species tested. Data presented in Figures 6 through 9 illustrate this point for methane, ethane, propane and isobutane. Figures 6 and 8 also contain burning velocities, S, for methane and propane (Lewis and von Elbe, p. 403) for comparison.

Data development in this work was generally limited to $0.9 < X/X_S < 1.3$. Data over the whole flammable range would be useful in completing a quantitative risk analysis which considers deflagration event severity.

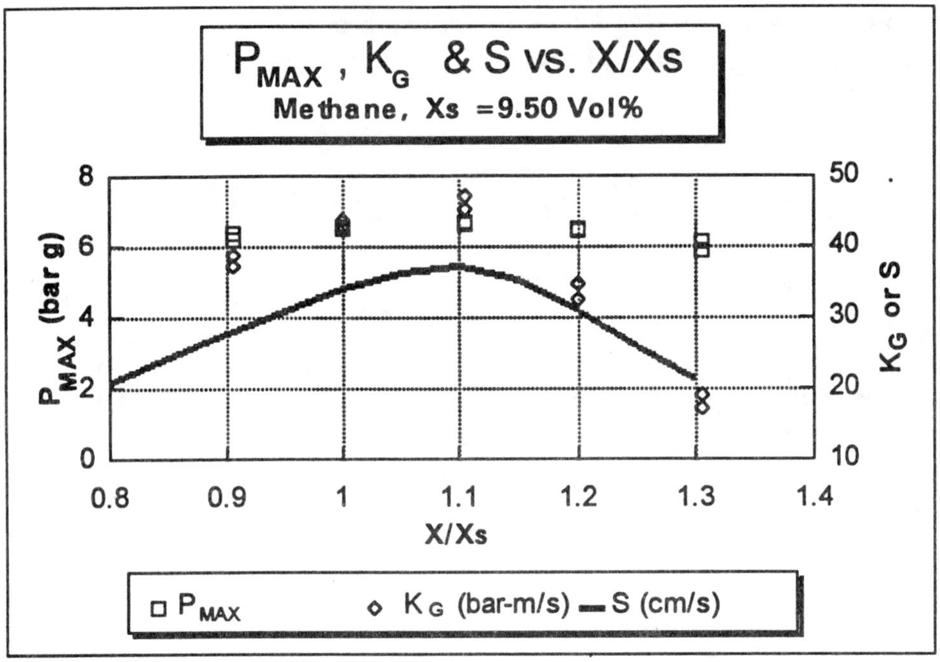

Figure 6. The dependence of P_{MAX}, K_G and S on mixture ratio for methane.

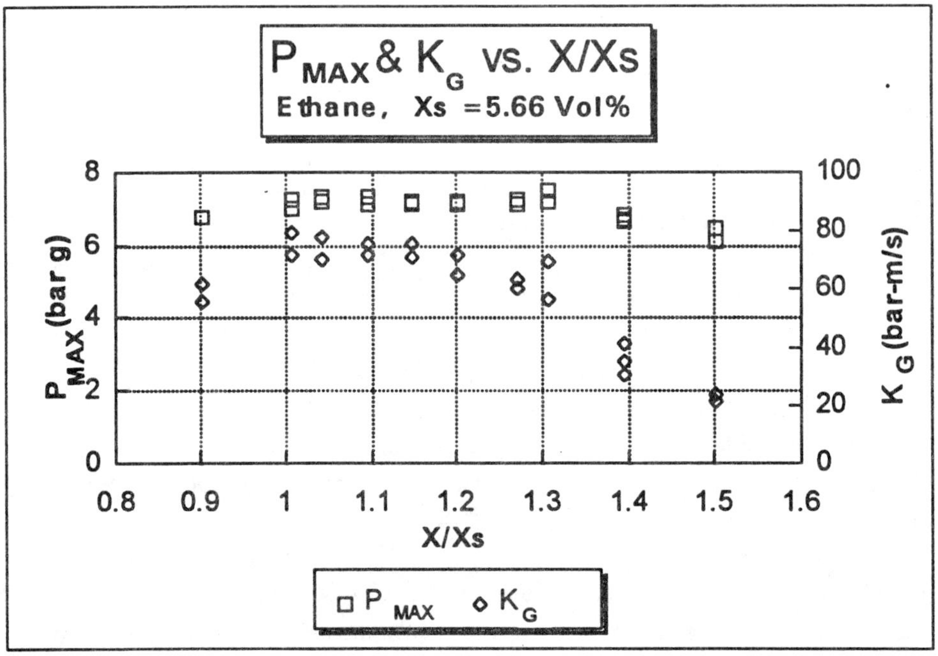

Figure 7. The dependence of P_{MAX} and K_G on mixture ratio for ethane.

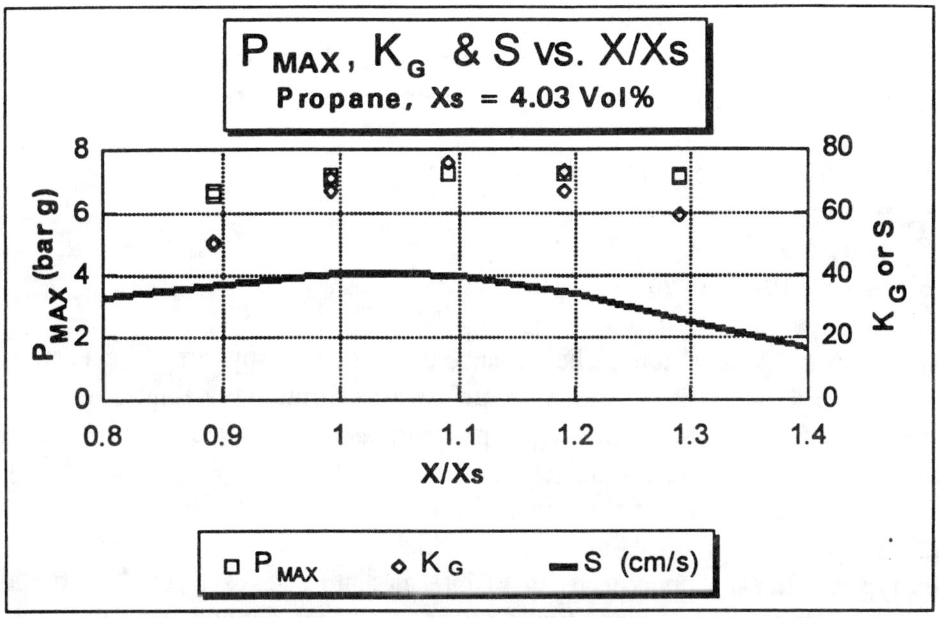

Figure 8. The dependence of P_{MAX}, K_G and S on mixture ratio for propane.

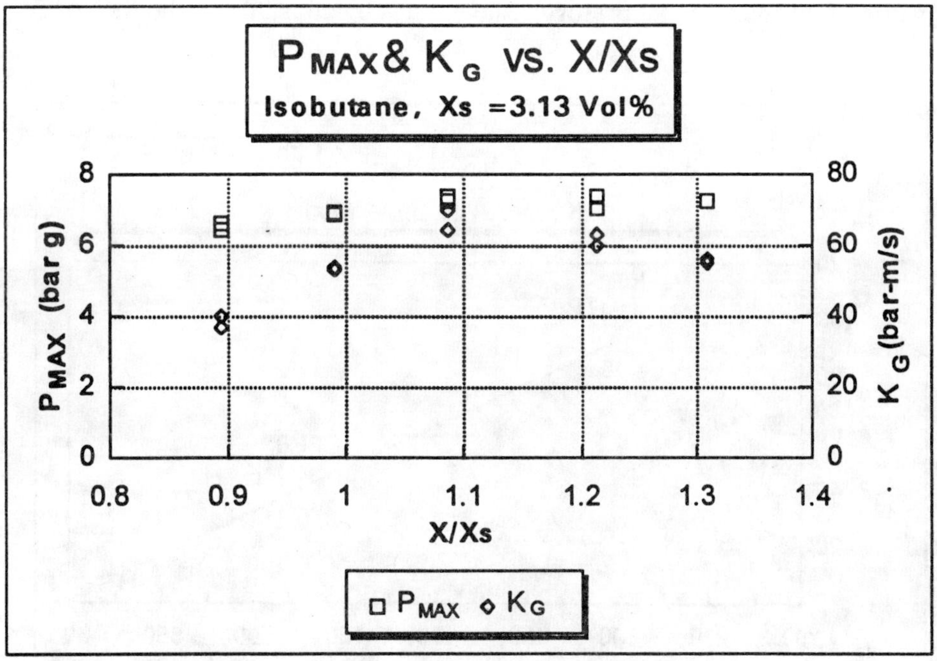

Figure 9. The dependence of P_{MAX} and K_G on mixture ratio for isobutane.

K_G and Test Vessel Size

Data on K_G determinations for four gases obtained in vessels ranging in size from 5 to 250,000 liters is given in Figure B-1 of NFPA 68 in the form of a plot of K_G vs. the log of the vessel volume. The linear regression fits of the form $K_G = A + B \log(V)$ were made of these data to yield the following:

Hydrogen	$K_G = 570 + 43.4 \log(V)$
Methane	$K_G = 53 + 10 \log(V)$
Propane	$K_G = 40 + 54.2 \log(V)$
Pentane	$K_G = 40 + 102 \log(V)$

where K_G is in bar-m/s and V is in liters. The variance of K_G with V appears to be a reflection of the tendency of the flame ball to increase its effective surface area through instabilities in the flame sheet structure. That is, the flame sheet geometry departs from an initially smooth spherical form to one of greater effective surface area across which combustion continues at the burning velocity, S, which prevails at the local conditions of temperature and pressure. Flame sheet instability is likely related to fundamental properties of the fuel-air mixture which bear on the thermal stability of the fuel molecule, the type, diffusivity and reactivity of intermediate species formed. The autoignition temperature is an empirically determined fuel-air property which depends on various fundamental mass transport and chemical kinetic mechanisms. While the data is limited there appears to be a fair correlation of the coefficient "B" of the "log (V)" term of the above K_G relations with the fuel AIT as shown in Figure 10. This simple analysis suggests that it may be possible to estimate the volume effect on K_G values determined in laboratory scale vessels by considering the fuel AIT.

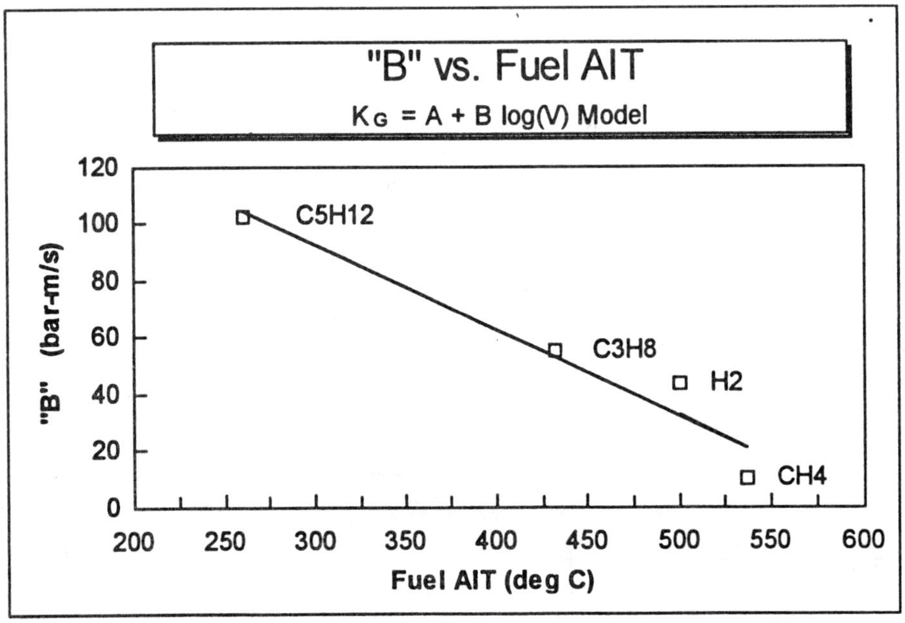

Figure 10. Correlation of "B" with fuel autoignition temperature.

Conclusions

A body of K_G data was developed using a low aspect ratio 22 liter vessel. The K_G values obtained are self consistent and correlate well with burning velocity, S. The sensitivity of P_{MAX} and K_G with fuel concentration was shown to be species dependent to some degree. In the case of hydrogen K_G must be viewed as a pre-DDT limiting property occurring at a pressure rise of about 3.6 bar over the composition range $0.9 < X / Xs < 1.3$ in the apparatus used. The well known relationship between K_G and vessel volume was shown to have a possible relationship to fuel autoignition temperature, a measure of fuel stability, which if validated could lead to a scaling rule for K_G. The closed-vessel pressure-time data obtained was shown to be useful in estimating values of burning velocity, S, with the finding that determinations by this method were 10 to 15 % higher than more rigorously determined values.

References

Bartknecht, W. , *Explosions,* Springer-Verlag, New York, 1981.

Bartknecht, Wolfgang, *Explosionsshutz*, Springer-Verlag, New York, 1993.

Lewis, Bernard and von Elbe, Guenther, "Combustion, Flames and Explosions of Gases," 3rd ed., Academic Press Inc., New York, 1987.

Metghalchi, M and J.C. Keck, "Laminar Burning Velocity of Propane-Air Mixtures at High Temperature and Pressure," *Combustion and Flame*, 38, 2, pp. 143-154, June 1980.

NFPA 68 Guide for Venting of Deflagrations, National Fire Protection Association, Quincy, MA, 1994.

Shou, M. and C.P. Garner, "Direct Measurements of Burning Velocity of Propane-Air Using Particle Image Velocimetry," *Combustion and Flame*, 106, 3, pp. 363-367, August 1996.

Tamanini, Franco, "Modeling of Panel Inertia Effects in Vented Dust Explosions," *Process Safety Progress*, 15, 4, pp.247-257, Winter 1996.

Appendix A

Derivation of a relationship between fundamental burning velocity, S, and closed-vessel deflagration parameters.

In this analysis is it assumed that:

1. the control volume in which the gas-phase deflagration occurs is spherical
2. the control volume is adiabatic
3. the pressure in the spherical test vessel is everywhere uniform
4. the temperature of the burned and unburned gas volumes are uniform at their respective conditions
5. the number of moles of burned gas is not significantly different from the number of moles of unburned gas

In a closed system the fraction of gas burned is equal to the volume the burned fraction in its unburned state divided by the total volume of gas, i.e.

$$X = V_{b,o}/V \tag{A-1}$$

Lewis and von Elbe have shown that the fraction of burned gas in terms of the pressure rise is very nearly given as

$$X = \frac{(P - P_o)}{(P_e - P_o)} \tag{A-2}$$

which states that the fraction of gas burned is equal to the fraction of the total pressure rise achieved. Since the volume of a sphere of radius R is given by

$$V = 4/3\ \pi\ R^3 \tag{A-3}$$

we have

$$r_{b,o}/R = \{(P - P_o)/(P_e - P_o)\}^{1/3} \tag{A-4}$$

The volume of a shell of unburned gas which is about to be burned is (see Fig.A-1)

$$dV_u = 4\ \pi\ r_b^2\ dr_u \tag{A-5}$$

However, the differential radius dr_u about to be traversed by the flame front moving at a velocity S_u into the unburned gas is given by

$$dr_u = S_u\ dt \tag{A-6}$$

In terms of the gas referenced to its unburned state the volume about to be burned is

$$dV_{b,o} = 4\,\pi\, r_{b,o}^{\,2}\, dr_{b,o} \qquad \text{(A-7)}$$

This volume can be related to that volume it occupies at the conditions P and T_u as

$$dV_u = dV_{b,o}\,(T_u/T_o)\,(P_o/P) \qquad \text{(A-8)}$$

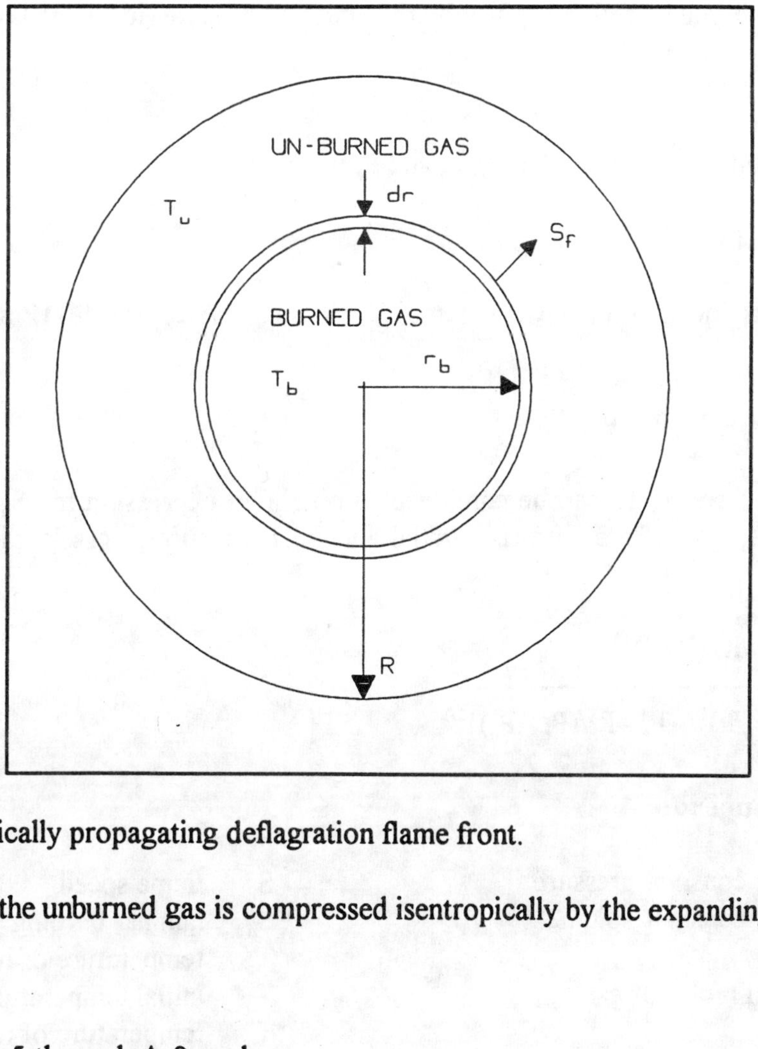

Figure A1. Spherically propagating deflagration flame front.

It is presumed that the unburned gas is compressed isentropically by the expanding burned gas, thus

$$T_u = T_o\,(P/P_o)^{(k-1)/k} \qquad \text{(A-9)}$$

Combining Eqs. A-5 through A-9 we have

$$S_u = (r_{b,o}/r_b)^2\,(P_o/P)^{1/k}\, dr_{b,o}/dt \qquad \text{(A-10)}$$

What is desired, ultimately, is an expression for S_u in terms of known initial conditions and the function P(t) which is determined experimentally. The value of $r_{b,o}$ is given in these terms in Eq. A-4. Differentiating Eq. A-4 we obtain

$$dr_{b,o}/dt = \frac{R\ dP/dt}{3\ (P_e - P_o)^{1/3}(P - P_o)^{2/3}} \qquad \text{(A-11)}$$

An expression is needed, in the appropriate terms for r_b. The total gas volume is made up of the burned and unburned volumes,

$$V = V_b + V_u \qquad \text{(A-12)}$$

The volume of the unburned fraction is given by the equation of state (ideal gas law) as

$$V_u = (1\text{-}X)n_oR_gT_u/P \qquad \text{(A-13)}$$

When Eq. A-13 is multiplied and divided by T_o and P_o we have

$$V_u = (1\text{-}X)\{n_oR_gT_o/P_o\}\ (T_u/T_o)\ (P_o/P) \qquad \text{(A-14)}$$

The term in braces is equal to the initial volume, V. Eqs. A-2, A-3, A-9, A-12 and A-14 can be combined to obtain

$$r_b = R\ \{1 - (P_o/P)^{1/k}(P_e - P)/(P_e - P_o)\}^{1/3} \qquad \text{(A-15)}$$

Eqs. A-4, A-10, A-11 and A-15 can be combined to obtain an expression for S_u in terms of the deflagration pressure, rate of pressure rise, initial and final pressures, vessel radius and ratio of specific heats as

$$S_u = \frac{R\ (dP/dt)\ (P_o/P)^{1/k}}{3\ (P_e\text{-}P_o)\{1 - (P_o/P)^{1/k}\ (P_e - P)/(P_e - P_o)\}^{2/3}} \qquad \text{(A-16)}$$

Nomenclature for Appendix A

C_p	heat capacity at constant pressure
C_v	heat capacity at constant volume
k	C_p/C_v
n_o	initial number of moles of gas
P	pressure
P_o	initial pressure
r	radial position
r_b	radius of burned gas volume at T_b, P
$r_{b,o}$	radius of burned gas volume in unburned state at T_0, P_0
R	vessel radius
R_g	gas-law constant
S_f	flame speed
S_u	laminar burning velocity
T_b	temperature of burned gas
T_o	initial temperature
T_u	temperature of unburned gas
V	vessel volume
$V_{b,o}$	volume of burned gas in unburned state at T_0, P_0
V_b	volume of burned gas at T_b, P
V_u	volume of unburned gas at T_u, P
X	fraction burned

NFPA 30: An Update and a Look into the Future

Robert P. Benedetti
Senior Flammable Liquids Engineer
National Fire Protection Association
One Batterymarch Park
Quincy, MA

Prepared for Presentation at the

American Institute of Chemical Engineers
31st Annual Loss Prevention Symposium
Houston, TX
March 11, 1997

NFPA 30: An Update and a Look into the Future

Robert P. Benedetti
Senior Flammable Liquids Engineer
National Fire Protection Association
One Batterymarch Park
Quincy, MA

ABSTRACT

This paper will describe the major changes to NFPA 30, Flammable and Combustible Liquids Code, that were incorporated into the 1996 edition. The emphasis will be on the changes to Chapter 4, Container and Portable Tank Storage, especially the mandatory fire protection design criteria for storage areas and warehouses that store such liquids.

This paper will also discuss possible future amendments that might be incorporated into the 1999 edition of NFPA 30.

Introduction.

In May, 1996, the Flammable and Combustible Liquids Code Committee of the National Fire Protection Association (NFPA) proposed for adoption by the Association a new edition of NFPA 30, Flammable and Combustible Liquids Code . This new edition was the culmination of two and one-half years' work by the Committee and included one of the most significant changes to that document in some twenty years: the incorporation of mandatory fire protection criteria for warehouses and other inside areas that store flammable and combustible liquids in containers and portable tanks.

The Flammable and Combustible Liquids Code Committee actually consists of four separate Technical Committees that are responsible for developing the technical requirements in the Code and an oversight Technical Correlating Committee, which is responsible for the entire project. The four Technical Committees and their responsibilities are:

- the Technical Committee on Fundamentals, responsible for scope issues, definitions, and the classification of liquids;

- the Technical Committee on Operations, responsible for handling and use of liquids, processing operations, and liquid transfer operations;

- the Technical Committee on Storage and Warehousing of Containers and Portable Tanks, responsible for designating acceptable container types, indoor liquid storage areas, such as warehouses, outdoor storage, storage in other occupancies, such as businesses and retail establishments, and establishing specific fire protection design criteria; and

- the Technical Committee on Tank Storage and Piping Systems, responsible for bulk storage in tanks and interconnecting piping systems.

The Technical Correlating Committee on Flammable and Combustible Liquids serves to ensure that the requirements developed by these four Technical Committees do not conflict with each other and to ensure that the various parts of NFPA 30 are consistent with each other.

NFPA 30, Flammable and Combustible Liquids Code, consists of the following:

- Chapter 1: Scope, Definitions, and Classification of Liquids.

- Chapter 2: Installation of Aboveground and Underground Storage Tanks; Installation of Tanks Inside Buildings.

- Chapter 3: Installation of Piping.

- Chapter 4: Storage of Containers and Portable Tanks.

- Chapter 5: Operations and Facility Design; Incidental Use of Liquids; Loading and Unloading Operations; Marine Transfer; Vapor Recovery Systems; and Management of Fire Hazards.

- Appendices that cover emergency relief venting of tanks; removal from service of underground storage tanks; development of fire protection criteria for inside storage areas and suggested test protocols for further development; and methods for calculating fugitive emissions.

Summary of Major Changes - Chapter 1, General

The first major change involves incorporation into NFPA 30 of the complete text of NFPA 321, Standard on Basic Classification of Flammable and Combustible Liquids[1]. Until now, NFPA 321 has been responsible for defining the several classes of flammable and combustible liquids. Because NFPA 30 is the most widely used document governing storage and handling of flammable and combustible liquids, the Technical Committee on Fundamentals felt that the definitions and classification of liquids should be included in NFPA 30 for the benefit of the user and so they incorporated NFPA 321 into a new Section 1-7, Definition and Classification of Liquids.

With this consolidation, NFPA 30 now becomes the "lead" document for purposes of liquid classification, as well as the areas for which it has traditionally been responsible. All other NFPA documents that address flammable liquids and combustible liquids in any fashion can easily extract the definitions for their own use and the definitions will be uniform throughout NFPA's system.

Another impetus for this change: in December, 1991, the U. S. Department of Transportation (DOT) made some very broad changes to its Hazardous Materials Transportation Regulations[2]. The objective of these changes was to bring U. S. rules in line with those of the United Nations (UN)[3]. While most of the proposed changes dealt with replacing DOT's specification-oriented packaging requirements with more performance-oriented ones, one important change dealt with the definition of "flammable liquid". DOT decided to adopt the UN definition: any liquid having a flash point not greater than 60.5°C (141°F). [Since that time, the UN Committee on the Transportation of Dangerous Goods has agreed to amend this to 60°C (140°F).]

This immediately raised questions of potential conflicts between NFPA 30 and the DOT's Hazardous Materials Transportation Regulations. The Flammable and Combustible Liquids Code Committee had already, in the 1993 edition, taken great care to not use the phrases "flammable liquid" and "combustible liquid" in other than a generic and non-specific sense and to establish rules that applied to specific classes of liquids. Nevertheless, the Technical Committee on Fundamentals determined that it was appropriate to codify the NFPA classification scheme in a more widely used document. Note that there will still be some confusion in the field because containers of Class II liquids that were previously not required to be "red-labeled", will now sometimes carry a

"Flammable Liquid" label. Field inspectors especially will not always be able to rely on container labeling to distinguish between Class I liquids and Class II liquids.

Figure 1 compares NFPA 30's classification scheme with the classification systems used by several other organizations and should be reviewed, as it will be helpful in understanding some of the discussion that follows.

Summary of Major Changes - Chapter 2, Tank Storage

Prior to the 1996 edition of NFPA 30, the maximum allowable internal operating pressure for the typical atmospheric storage tank was 0.5 psig. This operating limit dates from the time when the typical tank had a much flatter roof and the roof supports could not withstand more than this slight overpressure. Tank design standards have improved over the years. The typical tank in use today is stronger and can, in fact, operate above 0.5 psig. But, until recently, there has been no need to do so. Now, stricter environmental regulations are mandating the use of sophisticated vapor collection systems to control emissions from the larger bulk terminal tanks. These systems collect vapors from tank vents and transport them by piping to recovery or flare systems. The vapor collection system, by its nature, imposes a slight backpressure on the tanks, resulting in slightly higher operating pressures than has been the custom.

To address this, the Technical Committee on Tank Storage and Piping Systems has, in Paragraph 2-2.3.2, increased the maximum permissible operating pressure to 1.0 psig, based on the ability of tanks constructed in accordance with Appendix F of API 650, Welded Steel Tanks for Oil Storage[4], to operate safely and satisfactorily up to this pressure. For tanks that are not constructed to Appendix F's specifications, but which operate above 0.5 psig, language was added to require an engineering analysis to verify that they can operate safely at the increased pressure.

Another significant change has to do with new requirements for emergency relief vents for so-called secondary containment-type aboveground tanks, as set forth in Paragraph 2-3.6.1. These tanks are based on similar and highly successful double-wall designs used for underground storage tanks. Double-wall underground storage tanks have proven their worth in preventing underground releases of hazardous materials and petroleum products and have just about become the universal standard for fuel storage at motor vehicle refueling stations. The double-wall feature is now being used for small aboveground tanks, although its value is far less apparent. Aboveground tanks simply do not release their contents as a result of "pin-holing" of their shell. It is far more typical for a release to occur because of an overfill or because of failure or damage to piping connected to the tank.

The secondary containment feature comes in several varieties, from a double-shell design with an annular space to a single shell design mounted in a spill pan and having a shroud that effectively seals the secondary containment volume to the tank shell. The annular space or the secondary containment volume is often used as a receptacle for any overfill. It was this feature that convinced the Technical Committee on Tank Storage and Piping Systems to require an emergency relief vent for the annular space or any other enclosed secondary containment volume, just as is required for the primary vessel. The Technical Committee anticipated that these enclosed volumes might be subject to failure due to the impact of an exposure fire, especially if they contained some liquid because of prior overfill incidents.

The Technical Committee on Tank Storage and Piping Systems completely rewrote and expanded the requirements for temporary and permanent closure of underground storage tanks (Subsection 2-4.4). The requirements generally parallel those of the various environmental regulatory agencies, although the requirements in NFPA 30 are directed at preventing fires and explosions during tank closure or removal. The Technical Committee allowed up to a twelve month temporary closure period, but, after that, the tank must be either returned to full service or permanently closed in place or removed.

The Technical Committee on Tank Storage and Piping Systems added a new requirement (Paragraph 2-5.6.7) for storage tanks in buildings: the connection point between the tank's fill pipe and the hose from a delivery vehicle must be situated outside the building. This applies to any vapor return piping back to the tank vehicle as well. Obviously, this prohibits the practice of running the hose from the delivery vehicle into the building, to be connected to the tank.

Finally, the Technical Committee rewrote, in some detail, the requirements in Section 2-8 for tightness testing of tanks to provide acceptable test methods for single-wall tanks and for the primary vessel and annular space of secondary containment-type designs. In addition, a tightness test is now required to be performed on any tank that has been repaired, has been structurally damaged, or is suspected of leaking.

Summary of Major Changes - Chapter 3, Piping Systems

The first of the three major changes to Chapter 3 involves recognition, in Subsection 3-3.6, for the use of nonmetallic piping systems, including those of secondary containment-type. In general, piping meeting the listing requirements of UL 971, Standard for Nonmetallic Underground Piping for Flammable Liquids[5], and installed in accordance with the listing will be acceptable. Note, however, that there are limitations on where such piping is allowed to be used. These limitations are given in Subsection 3-3.4.

A new set of requirements, in Subsection 3-5.1, governs protection for load-bearing piping supports that are situated where they can be exposed to a severe fire. Some of the allowable means by which such piping supports can be protected include: fire resistive construction; fire resistive coatings; water spray; sloping the ground so liquid spills drain away from the supports; any other means that is acceptable to the authority having jurisdiction.

New Section 3-7 establishes requirements for burial depth and separation of underground piping, including parallel runs of piping. Parallel runs of piping in a common trench must be separated horizontally by a minimum of two pipe diameters, up to a 9 inch maximum, and vertically by 6 inches.

Summary of Major Changes - Chapter 4, Container and Portable Tank Storage

In 1993, Chapter 4 of NFPA 30 was completely reorganized to make it more understandable. With the 1996 edition, additional major improvements have been made. First, Chapter 4 of NFPA 30 now recognizes metal intermediate bulk containers and three types of plastic intermediate bulk containers. Second, more definitive information on spill containment in and drainage from container storage areas is now provided. And, third, but certainly most significant, the fire protection design criteria previously included as Appendix D, Suggested Fire Protection for Containers of Flammable and Combustible Liquids, have been replaced with more extensive and more definitive criteria and these criteria have been added to Chapter 4 as mandatory requirements for new construction of inside liquid storage areas, with an effective date of January 1, 1997.

Over the past several years, intermediate bulk containers [IBCs] have become more popular for shipping liquids. IBCs have some advantages over the typical drum. They have their own drain valve. And, they hold more material in a smaller space: a typical 330 gallon IBC takes up about the same space as a pallet and four 55-gallon drums. However, their acceptance as a safe means of transportation does not necessarily mean they don't present a fire problem in a warehouse. Some IBCs that are allowed for shipping some flammable liquids consist of little more than a thin, flexible bladder, which holds the liquid, surrounded by a metal wire "cage" or heavy corrugated "carton", which gives the assembly stability. The Technical Committee concluded that such containers were just not suitable for safe storage of flammable or combustible liquids. So, until now, NFPA 30 recognized only metal portable tanks up to 660 gallons capacity.

After reviewing information available to it and recognizing that a shift to IBCs is inevitable, the NFPA 30 Technical Committee on Storage and Warehousing of Containers and Portable Tanks decided that metal IBCs could be considered equivalent to metal portable tanks that meet NFPA 386, Standard for Portable Shipping Tanks for Flammable and Combustible Liquids,[6] and could be used for storage of liquids of all classes. The Technical Committee also deliberated granting limited recognition to certain plastic IBCs. Plastic IBCs have gained wide acceptance in industry because of their durability and particularly for their ability to hold corrosive materials without fear of leakage. And, indeed, three types of plastic IBCs are now recognized in this edition of NFPA 30: rigid plastic IBCs of types 31H1 and 31H2 and composite plastic IBCs of type 31HZ1. These IBCs offer some limited degree of fire protection for the liquid inside and are recognized up to 660 gallons capacity, but with the following provisos:

- storage is only permitted for Class II and Class III liquids;
- storage is considered unprotected, regardless of any fire protection system present in the building.

The information on these IBCs appears in Paragraphs 4-2.1(a) and 4-2.1(e) and Table 4-2.3 of NFPA 30, reproduced here in Figure 2. Also, maximum allowable storage heights and storage quantities for these IBCs have been added to Table 4-4.4.1, for indoor unprotected storage, shown here in Figure 3, and Table 4-7, for outdoor storage, reproduced in Figure 4. The previously-mentioned sprinkler design criteria specifically do not address IBCs and will not until full-scale fire test data is available and has been evaluated by the Technical Committee. Only then will guidance be offered for protected storage of plastic intermediate bulk containers.

The next improvement to Chapter 4 of NFPA 30 deals with the requirements for containment and drainage. NFPA 30 has always required some degree of containment and drainage, not only for a spill itself, but also for sprinkler discharge during a fire. Design details were never really given, however. Paragraph 4-4.2.6 of NFPA 30 now requires containment and drainage in certain circumstances for any facility constructed after January 1, 1997. The "certain circumstances" are spelled out in Figure 4-4.2.6, which is reproduced here as Figure 5.

These requirements were developed by a Task Group of the Technical Committee specifically as a complement to the new fire protection design criteria in Section 4-8. Figure 5 illustrates the "decision tree" used to determine when containment or drainage is required and to what parameters they must be designed. The decision tree also clarifies how the characteristics of the liquid and the quantity stored affect the need for containment or drainage.

Note that the primary concern of these requirements is mitigation of the fire hazard resulting from a spill, not protection of the environment. The intent is to prevent a spill from migrating to other parts of the facility, where it might be ignited. If the spill has already been ignited, the objective is to prevent burning liquid from being floated to other areas by sprinkler discharge. These requirements, therefore, do not claim to provide environmental design information. For example, the standard seals used for joints in concrete floors are adequate for the purposes of these requirements, but would likely not meet permeability requirements related to environmental control.

Reviewing the decision tree stepwise shows the following:

For new inside rooms, the decision tree calls for containment or drainage sized for the "largest credible spill". In most cases, this will be one or two of the largest containers present, typically drums. This can easily be accomplished using a 4" sill at the doorway.

Next, the tree shows that containment or drainage is not required if the quantities do not exceed those shown. These limits are the same as the maximum quantities allowed by Subsection 4-5.2.3 of NFPA 30 for a general purpose warehouse and were deliberately chosen so that the requirements would not apply to a facility that isn't designed for flammable/combustible liquids storage anyway. The Task Group felt they could not justify imposing such costly requirements on facilities that store such minimum quantities, and then not at all times. The maximum quantities allowed for general purpose warehouses are a reasonable "floor" above which containment and drainage must be considered.

The decision tree next exempts certain liquids from its requirements:

- Water miscible liquids having less than 50% flammable ingredients. These are exempt because of their reduced risk of ignition and because they can be easily diluted to decrease their flammability.

- Liquids denser than water. Sprinkler discharge or hose streams can blanket the surface of the spill, either to prevent ignition or to extinguish any fire.

- Class IIIB liquids. These are not likely to be ignited if spilled.

- High viscosity liquids. These will not flow beyond the area of the spill and the spill can be more easily controlled.

Assuming a situation justifies containment or drainage, the tree provides guidance for designing the system to handle anticipated flow. The 20-minute duration is based on typical fire department response and assumes the storage area is protected by an automatic sprinkler system that meets the criteria of Section 4-8. Note that the containment/drainage capacity must be increased by 50 percent if the sprinkler system does <u>not</u> meet these criteria because it is expected that it will take longer and more effort to bring the fire under control.

Finally, for the most severe conditions, the system is sized for the worst case condition.

Now, to discuss the most significant improvement to the 1996 edition of NFPA 30, new Section 4-8, Fire Protection and Control.

Previous editions of NFPA 30 did not provide specific design criteria for sprinkler systems to protect flammable and combustible liquid storage. A designer could refer to Appendix D, Suggested Fire Protection for Containers of Flammable and Combustible Liquids, where there was come limited information on sprinkler systems and closed-head AFFF (aqueous film-forming foam) systems. However, as with all appendices in NFPA technical documents, this one included the standard caveat that the information contained was not mandatory and was presented for information purposes only. Besides, little of the information that did appear was based on actual full-scale fire tests.

After the 1993 edition of NFPA 30 was published, the Technical Committee on Storage and Warehousing of Containers and Portable Tanks formed a Task Group to review available information on the behavior of containers of flammable and combustible liquids when subjected to fire and to develop protection criteria based on this information, with objective of incorporating the criteria as mandatory protection requirements.

The Task Group reviewed information on 85 full-scale fire tests summarized in the "Directory of Fire Tests Involving Storage of Flammable and Combustible Liquids In Small Containers", by Nugent[7]. These tests were carried out by Factory Mutual Research Corporation, Underwriters Laboratories, and Southwest Research Institute and include tests conducted using small, intermediate, and large scale configurations that simulated bulk/palletized, rack, and shelf storage typical of those found in warehouse and retail facilities throughout the country. Test fuels included Class IB through IIIB liquids. Container types included relieving-style and non-relieving-style steel, plastic, and glass containers. All of the tests used operating sprinkler systems discharging either water or foam-water. Much of the data is recent and some has been the product of research conducted under the auspices of the National Fire Protection Research Foundation[8,9].

The Task Group identified several variables that were significant determinants of the severity of an ensuing fire and its rate of growth, as well as the ability of the fire protection system to control or suppress the fire. These include:

- the physical properties and the fire hazard properties of the liquid;
- the style of container and its size;
- presence or absence of packaging material (e.g., corrugated cartons);
- the ignition scenario;
- the storage arrangement;
- the design parameters of the fire protection system.

The effectiveness of fire protection systems and container design was judged by:

- the number of sprinkler heads that operated and time of operation;
- the temperature of fire gases at the ceiling and, where available, the temperature of ceiling structural elements;
- the extent of fire travel through the test array;
- whether the fire was able to "jump" the aisle separating the test array from an adjacent target array;
- an assessment of the damage in both the test and target storage arrays.

More detailed discussion of the tests and the interrelationship of these factors can be found in "Fire Tests Involving Storage of Flammable and Combustible Liquids In Small Containers", also by Nugent[10].

Using this data, the Task Group was able to develop fire protection design criteria for a wide range of commodities, container types and sizes, storage arrangements, and automatic sprinkler system design parameters and was able to incorporate the information in Section 4-8 of NFPA 30 as mandatory requirements. The information is presented in a series of tables that contain the mandatory design criteria and several flowcharts that assist in determining when these criteria must be used and when alternative methods can be substituted.

The tables, reproduced here as Figures 6(a) through 14, show the six sets of sprinkler design criteria and the three flowcharts that comprise the core of Section 4-8. The NFPA 30 designations for these tables and charts have been deliberately repeated here to assist the reader. The tables address liquids from Class IB through Class IIIB and both relieving and non-relieving containers. They include palletized, rack, and shelf storage and both water and foam-water sprinkler systems.

Since there was no full-scale test data for Class IA liquids, the earlier fire protection criteria for Class IA liquids in metal containers was retained in the Appendix.

Recognizing that these criteria would impose a tremendous burden on existing facilities if they were forced to retrofit, the Technical Committee decided not to apply the mandatory criteria retroactively, but to make them apply only to facilities constructed after January 1, 1997. In addition, to make the criteria even less burdensome and to provide as much flexibility as possible, the three flowcharts, for metal, plastic, and glass, offer less drastic protection schemes, if the liquids meet certain conditions.

A unique feature of the tables is the "Fire Test Reference". Each entry in these tables has a reference number that appears in the right-most column. This number is keyed to information in Appendix D, Development of Fire Protection Criteria in Section 4-8, which, in turn, identifies the specific full-scale test in Nugent's Directory on which that entry is based. This allows the user to investigate the details of the tests by which these criteria were developed.

One very important consideration: any existing storage facility that has a fire protection system that met previous editions of NFPA 30 is still considered to be adequately protected if all of the following remain unchanged:

- the class of liquid for which the system was designed;
- the maximum quantity of liquid stored;
- the fire protection system design
- the building configuration.

An example case study of how the tables are used is given in this paper's Appendix.

Summary of Major Changes - Chapter 5, Operations

Most of the amendments to Chapter 5 of NFPA 30 were editorial in nature. However, one addition is worthy of note.

A new requirement was added in Subsection 5-3.3.2 that parallels a previously discussed addition to Chapter 3, protection for load-bearing supports. In any facility where liquids are processed, load-bearing supports can be exposed to severe damage should a fire occur and failure can lead to rapid and extensive spread of the original fire. Subsection 5-3.3.2 requires protection for load-bearing supports for the building structure and for process vessels using one of four acceptable options: fire resistive construction; fire resistive coatings; water spray; sloping the ground so liquid spills drain away from the supports. Other means acceptable to the authority having jurisdiction are also allowed.

A Look to the Future

Like any other technical code, NFPA 30 evolves over time. Immediately after a new edition is published the Technical Committees responsible for NFPA 30 begin their work on the next edition. Here follow some of the issues that will be addressed in the next edition of NFPA 30, scheduled for adoption at the 1999 NFPA Fall Meeting.

- Adoption of UN Definition of "Flammable Liquid" . As NFPA 30 becomes more accepted in the international arena, it will have to recognize that the United Nations definitions for the various categories of dangerous materials are already accepted world-wide, at least in the transportation arena. Since NFPA 30 regulates liquids strictly by their class designation, this is not a big issue. Basically, to recognize the UN definitions means only that Class II liquids (flash point: 100-140°F) move from the broad category of combustible liquid to that of flammable liquid.

- New Technology Storage Tanks. NFPA 30 lacks specific installation requirements for secondary containment-type aboveground tanks, fire resistive tanks, and tanks installed in vaults. These new technology storage systems are gaining more market share, as users seek to abandon underground storage for fire- and environmentally-safe aboveground storage. Chapter 2 of NFPA 30 needs to be augmented with requirements for these types of tanks.

- Chapter 2 Editorial Work. Chapter 2 of NFPA 30 needs to be revised in such a manner that each family of storage tank is covered in one self-contained section or chapter. This will eliminate problems involving the lack of necessary cross references to other sections of the chapter. Will this result in much repetition of common installation and design requirements? Obviously, but repetition is a small price to pay to minimize the frustration of casual users and the many enforcement officials who struggle to determine which requirements apply in any specific case. This is a real problem that is documented by the numbers of advisory service calls received on Chapter 2.

- More Comprehensive Identification of Allowable Container Types. Chapter 4 needs to more clearly identify which containers, intermediate bulk containers, and portable tanks are acceptable for indoor storage and which are not. This means that there will need to be a list of those U. S. Department of Transportation-approved container styles that can safely be used to warehouse liquids. It should be recognized that there will be some containers, etc. that are acceptable for transportation, but just not safe enough to be housed in a storage facility.

- More Comprehensive Fire Protection Design Criteria. As stated above, non-metallic intermediate bulk containers are allowed by NFPA 30, at least certain types, but they always considered as unprotected storage, because there is no full-scale fire test data on which to base fire protection design criteria. This should change, as the National Fire Protection Research Foundation is currently engaged in a full-scale fire test program to develop such information. Also, the Technical Committee on Storage and Warehousing of Containers and Portable Tanks will continue to seek out any relevant full-scale test data on which to expand the current information.

- Zone 0. NFPA 70, the National Electrical Code, has now recognized the European concept of Zones 0, 1, and 2 for use in designating hazardous (classified) locations for purposes of electrical equipment installation. The Technical Committee on Operations will be integrating the Zone concept into Subsection 5-9.5, Electrical Installations, of NFPA 30.

REFERENCES CITED

1. NFPA 321, Standard on Basic Classification of Flammable and Combustible Liquids, National Fire Protection Association, Quincy, MA, 1991.

2. Code of Federal Regulations, Title 49, Transportation, Parts 100 - 199, U. S. Government Printing Office, Washington, DC, 1995.

3. Recommendations on the Transport of Dangerous Goods, 9th Revised Ed., United Nations, New York, NY, 1995.

4. API Standard 650, Welded Steel Tanks for Oil Storage, 9th Ed., American Petroleum Institute, Washington, DC, 1996.

5. UL 971, Standard for Nonmetallic Underground Piping for Flammable Liquids, Underwriter's Laboratories, Inc., Northbrook, IL, 1992.

6. NFPA 386, Standard for Portable Shipping Tanks for Flammable and Combustible Liquids, National Fire Protection Association, Quincy, MA, 1990.

7. Nugent, D. P., Directory of Fire Tests Involving Storage of Flammable and Combustible Liquids in Small Containers, Schirmer Eng'g Corp. / Society of Fire Protection Engineers, Boston, MA 1994.

8. National Flammable Liquid Container storage Research Project, National Fire Protection Research Foundation, Quincy, MA, 1990/1992.

9. National Wholesale/Retail Occupancy Fire Research Project, National Fire Protection Research Foundation, Quincy, MA, 1993/1994.

10. Nugent, D. P., "Fire Tests Involving Storage of Flammable and Combustible Liquids in Small Containers", Journal of Fire Protection Engineering, Vol. 6, No. 1, pp 1-10, Society of Fire Protection Engineers, Boston, MA, 1994.

APPENDIX - EXAMPLE OF USE OF TABLE 4-8.2(C) [Figure 8] [See Arrow in Table.]

Basis: Storage of 55-gallon drums of Class IB and IC liquids in double-row racks. Containers are metal, non-relieving style. Liquid is not water-miscible.

What does the table require?

Maximum Building Ceiling Height:	30 ft
Maximum Storage Height:	25 ft
Allowable Sprinkler Type:	Either standard or large-orifice sprinklerheads can be used, but a standard response link must be used.
Ceiling AS Design:	0.30 gpm/ft^2 over 3,000 ft^2
In-Rack AS Design:	One line of sprinklers at every rack level, beginning above the first level of storage.
NOTES:	• Large-orifice heads preferred. • 286°F ceiling sprinkler heads.
Fire Test Reference No.:	2

FROM APPENDIX D:

These criteria are based on full-scale fire tests Nos. S-45 and S-46, appearing in the Directory of Fire Tests Involving Storage of Flammable and Combustible Liquids in Small Containers, D. P. Nugent.[7]

Table A-1-7.3 Comparative Classification of Liquids

Agency	Agency Classification	Agency Flash Point		NFPA Definition	NFPA Classification	NFPA Flash Point	
ANSI/CMA Z129.1-1994	Flammable	< 141°F	< 60.5°C	Flammable Combustible	Class I Class II Class IIIA	< 100°F ≥ 100°F to < 140°F ≥ 140°F to < 200°F	< 37.8°C ≥ 37.8°C to < 60°C ≥ 60°C to < 93°C
	Combustible	≥ 141°F to < 200°F	≥ 60.5°C to < 93°C	Combustible	Class IIIA	≥ 140°F to < 200°F	≥ 60°C to < 93°C
DOT	Flammable	< 141°F	< 60.5°C	Flammable Combustible	Class I Class II Class IIIA	< 100°F ≥ 100°F to < 140°F ≥ 140°F to < 200°F	(37.8°C) ≥ 37.8°C to < 60°C ≥ 60°C to < 93°C
	Combustible	≥ 141°F to < 200°F	≥ 60.5°C to < 93°C	Combustible	Class IIIA	≥ 140°F to < 200°F	≥ 60°C to < 93°C
DOT HM-181 Domestic Exemption[1]	Flammable	< 100°F	(37.8°C)	Flammable	Class I	< 100°F	(37.8°C)
	Combustible	≥ 100°F to < 200°F	≥ 37.8°C to < 93°C	Combustible	Class II Class IIIA	≥ 100°F to < 140°F ≥ 140°F to < 200°F	≥ 37.8°C to < 60°C ≥ 60°C to < 93°C
UN	Flammable	< 141°F	< 60.5°C	Flammable Combustible	Class I Class II Class IIIA	< 100°F ≥ 100°F to < 140°F ≥ 140°F to < 200°F	(37.8°C) ≥ 37.8°C to < 60°C ≥ 60°C to < 93°C
	Combustible	≥ 141°F to < 200°F	≥ 60.5°C to < 93°C	Combustible	Class II Class IIIA	≥ 100°F to < 140°F ≥ 140°F to < 200°F	≥ 37.8°C to < 60°C ≥ 60°C to < 93°C
OSHA	Flammable	< 100°F	(37.8°C)	Flammable	Class I	< 100°F	(37.8°C)
	Combustible[2]	≥ 100°F	≥ 37.8°C	Combustible	Class II Class IIIA Class IIIB[2]	≥ 100°F to < 140°F ≥ 140°F to < 200°F ≥ 200°F	≥ 37.8°C to < 60°C ≥ 60°C to < 93°C (≥ 93°C)

[1] See "NOTE" in A-1-7.3. [2] See *Code of Federal Regulations*, Title 29, 1910.106 for Class IIIB liquid exemptions.

Figure 1

Table 4-2.3 Maximum Allowable Size of Containers and Portable Tanks

Liquids Container Type	Flammable Liquids			Combustible	
	Class IA	Class IB	Class IC	Class II	Class III
Glass	1 pt	1 qt	1 gal	1 gal	5 gal
Metal (other than DOT drums) or Approved Plastic	1 gal	5 gal	5 gal	5 gal	5 gal
Safety Cans	2 gal	5 gal	5 gal	5 gal	5 gal
Metal Drum (DOT Specification)	60 gal	60 gal	60 gal	60 gal	60 gal
Approved Metal Portable Tanks and IBCs	660 gal	660 gal	660 gal	660 gal	660 gal
Rigid Plastic IBCs (UN 31H1 or 31H2) and Composite IBCs (UN 31HZ1)	NP	NP	NP	660 gal	660 gal
Polyethylene DOT Specification 34, UN 1H1, or as authorized by DOT Exemption	1 gal	5 gal	5 gal	60 gal	60 gal
Fibre Drum NMFC or UFC Type 2A; Types 3A, 3B-H, or 3B-L; or Type 4A	NP[1]	NP[1]	NP[1]	60 gal	60 gal

SI units: 1 pt = 0.473 L; 1 qt = 0.95 L; 1 gal = 3.8 L.

[1]Not permitted.

Figure 2

Table 4-4.4.1 Indoor Unprotected Storage of Liquids in Containers, Portable Tanks, and IBCs

	Container Storage			Portable Tank/Metallic IBC Storage			Rigid Plastic and Composite Intermediate Bulk Containers		
Class	Maximum Pile Height (ft)	Maximum Quantity per Pile (gal)	Maximum Total Quantity (gal)[1]	Maximum Pile Height (ft)	Maximum Quantity per Pile (gal)	Maximum Total Quantity (gal)[1]	Maximum Pile Height (ft)	Maximum Quantity per Pile (gal)	Maximum Total Quantity (gal)[1]
IA	5	660	660	—	Not permitted	—	—	—	—
IB	5	1375	1375	7	2000	2000	—	—	—
IC	5	2750	2750	7	4000	4000	—	—	—
II	10	4125	8250	7	5500	11,000	7	4125	8250
IIIA	15	13,750	27,500	7	22,000	44,000	7	13,750	27,500
IIIB	15	13,750	55,000	7	22,000	88,000	7	13,750	55,000

SI Units: 1 ft = 0.3 m; 1 gal = 3.8 L.
[1]Applies only to cut-off rooms and attached buildings.

Figure 3

Table 4-7 Outdoor Liquid Storage in Containers and Portable Tanks

Class	Containers Maximum per Pile (gal)[1,3,4]	Containers Maximum per Pile Height (ft)	Rigid Plastic and Composite IBCs Maximum per Pile gal[5] (gal)	Rigid Plastic and Composite IBCs Maximum per Pile gal[5] Height (ft)	Portable Tanks and Metal IBCs Maximum per Pile gal[1] (gal)[1,4]	Portable Tanks and Metal IBCs Maximum per Pile gal[1] Height (ft)	Distance between Piles or Racks (ft)	Distance to Property Line that Is or Can Be Built Upon (ft)[2,3]	Distance to Street, Alley, or a Public Way (ft)[3]
IA	1100	10	—	—	2200	7	5	50	10
IB	2200	12	—	—	4400	14	5	50	10
IC	4400	12	—	—	8800	14	5	50	10
II	8800	12	8800	12	17,600	14	5	25	5
III	22,000	18	22,000	18	44,000	14	5	10	5

SI units: 1 ft = 0.3 m; 1 gal = 3.8 L.

[1]See 4-7.1.1 regarding mixed-class storage.
[2]See 4-7.1.3 regarding protection for exposures.
[3]See 4-7.1.4 for smaller pile sizes.
[4]For storage in racks, the quantity limits per pile do not apply, but the rack arrangements shall be limited to a maximum of 50 ft (15 m) in length and two rows or 9 ft (2.7 m) in depth.
[5]Storage of Class I liquids in rigid plastic and composite IBCs not permitted.

Figure 4

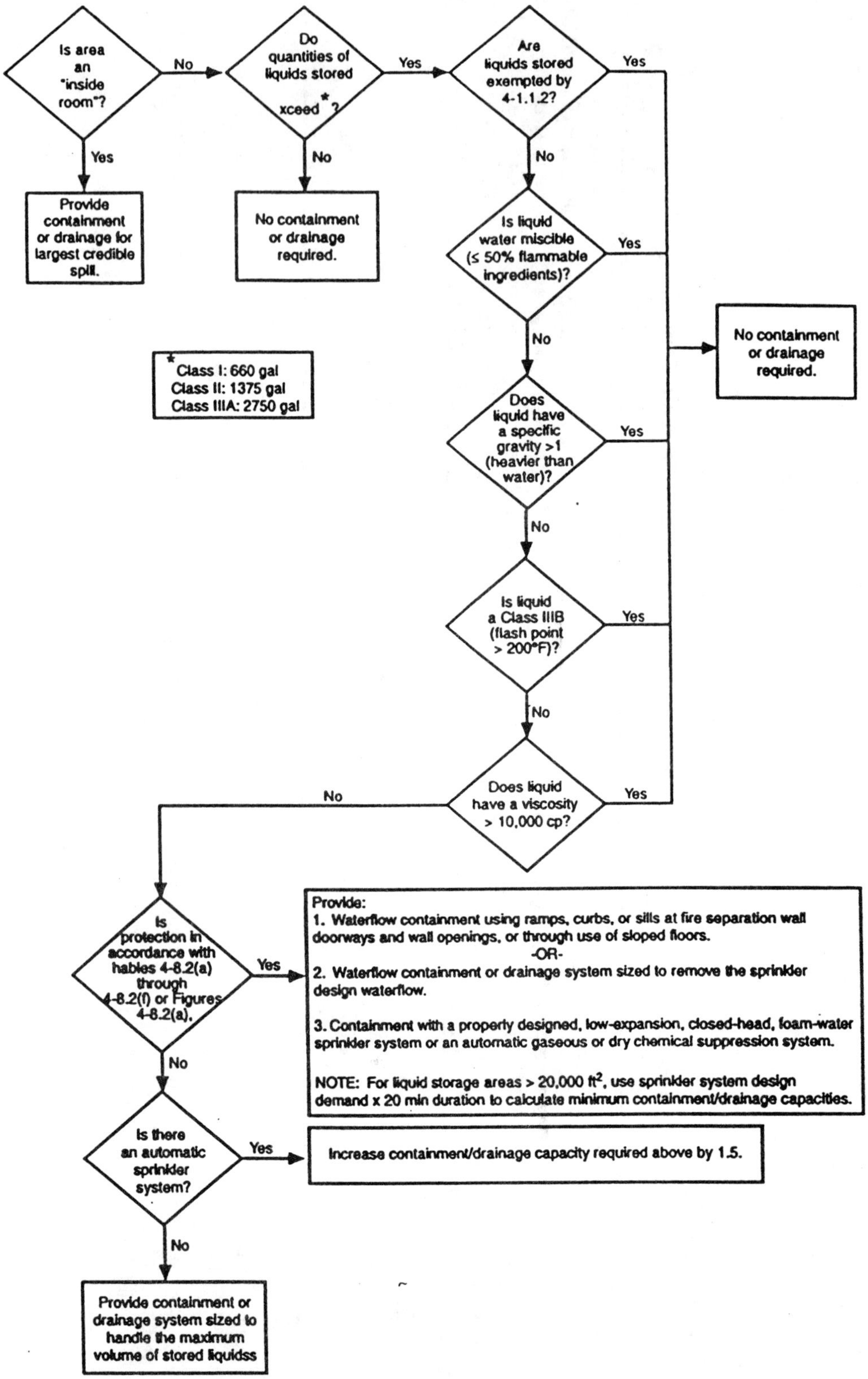

Figure 4-4.2.6 Containment/drainage requirements for inside liquid storage areas.

Figure 5

Table 4-8.2(a) Water Sprinkler Protection Single or Double Row Rack Container Construction—Metal (For Nonmiscible Liquids or Miscible > 50%)

Liquid Class	Container Size and Arrangement	Maximum Storage Height (ft)	Maximum Ceiling Height (ft)	Ceiling: Sprinkler Type: Orifice[1]	Ceiling: Sprinkler Type: Response[2]	Ceiling: Density (gpm/ft²)	Ceiling: Design Area (ft²)[3]	In-Rack Sprinkler Protection	Notes	Fire Test Ref.[4]
Non-Relieving-Style Containers										
IB, IC, II, III	≤ 1 gal	16	30	ELO	QR	0.60	2000	One line 8 ft above floor	(2),(5)	1
		20	30	LO or ELO	SR or QR	0.60	2000	One line 6 ft above floor; one line 12 ft above floor	(2),(5)	2
IB, IC, II, IIIA	≤ 5 gal	25	30	STD or LO	SR or QR	0.30	3000	Every level	(2)	3
IIIB	≤ 5 gal	40	50	STD or LO	SR or QR	0.30	2000	One line every other level, beginning above first storage level	(2),(6)	4
IB, IC, II, IIIA	> 5 gal ≤ 60 gal	25	30	LO or ELO	SR	0.40	3000	Every level	(4)	5
IIIB	> 5 gal ≤ 60 gal	40	50	STD or LO	SR	0.30	3000	One line every other level, beginning above first storage level	(2),(6)	6
Relieving-Style Containers										
IB, IC, II, IIIA	≤ 5 gal	14	18	ELO	QR	0.65	2000	None	(1),(3)	7
		25	30	STD or LO	SR or QR	0.30	3000	One line every other level, beginning above first storage level	(2),(7)	8
IIIB	≤ 5 gal	40	50	STD or LO	SR or QR	0.30	2000	One line every other level, beginning above first storage level	(2),(6)	9
IB, IC, II, IIIA	> 5 gal ≤ 60 gal	25	30	LO or ELO	SR	0.60	3000	One line every other level, beginning above first storage level	(2)	10

Figure 6(a)

Table 4-8.2(a), Continued.

Liquid Class	Container Size and Arrangement	Maximum Storage Height (ft)	Maximum Ceiling Height (ft)	Ceiling: Sprinkler Type: Orifice[1]	Ceiling: Sprinkler Type: Response[2]	Ceiling: Density (gpm/ft²)	Ceiling: Design Area (ft²)[3]	In-Rack Sprinkler Protection	Notes	Fire Test Ref.[4]
Non-Relieving-Style Containers										
IIIB	> 5 gal ≤ 60 gal	40	50	STD or LO	SR	0.30	3000	One line every other level, beginning above first storage level	(2),(6)	11
IB, IC, II, IIIA	Portable tanks	25	30	LO or ELO	SR	0.60	3000	Every level	(4)	12
IIIB	Portable tanks	40	50	LO	SR	0.30	3000	One line every other level, beginning above first storage level	(4)	13

SI units: 1 gal = 3.8 L; 1 ft = 0.3 m; 1 ft^2 = 0.093 m^2; 1 gpm/ft^2 = 40.7 Lpm/m^2.
[1]Larger orifice sprinklers are preferred when installed in accordance with NFPA 13, *Standard for the Installation of Sprinkler Systems*. STD = Standard Orifice, LO = Large Orifice, ELO = Extra-Large Orifice.
[2]SR = Standard Response, QR = Quick Response. Where both are listed, QR is preferred.
[3]Ceiling sprinklers high temperature.
[4]See Appendix D-2 for references to fire tests on which the protection criteria given in this table is based.

NOTES:
1. Double row racks 6 ft wide maximum.
2. Space in-rack sprinklers on maximum 9 ft centers, staggered vertically. Base design on 30 gpm per head, with six hydraulically most remote heads operating in each of upper three levels, or eight hydraulically most remote heads if only one level. Sprinklers are STD or LO, QR, 165°F, operating temperature, with shields.
3. Use pendent-style ELO ceiling sprinklers.
4. Space in-rack sprinklers on maximum 9 ft centers staggered vertically, 30 gpm per head, STD or LO, QR or SR, with shield, 165°F, six hydraulically most remote sprinklers each level (upper three levels) operating. Eight sprinklers operating, if only one level.
5. Protection for uncartoned or case-cut nonsolid shelf display up to $6^1/_2$ ft and storage above in pallets on racking, shelf material, open wire mesh, or 2 in. × 6 in. wooden slats, spaced a minimum of 2 in. apart.
6. A 0.60 density shall be used if more than one level of storage exists above the top level of in-rack sprinklers (LO or ELO orifice for ceiling sprinklers).
7. A 0.60 density/2000 ft2 shall be used if more than one level of storage exists above the top level of in-rack sprinklers (LO or ELO orifice for ceiling sprinklers).

Figure 6(b)

Table 4-8.2(b) Water Sprinkler Protection Bulk or Palletized Storage Container Construction—Metal (For Nonmiscible Liquids or Miscible > 50%)

Liquid Class	Container Size and Arrangement	Maximum Storage Height (ft)	Maximum Ceiling Height (ft)	Ceiling: Sprinkler Type: Orifice[1]	Ceiling: Sprinkler Type: Response[2]	Ceiling: Density (gpm/ft²)	Ceiling: Design Area (ft²)[3]	Notes	Fire Test Ref.[4]
Non-Relieving-Style Containers									
IB,	≤ 5 gal	4	18	STD or LO	SR or QR	0.21	1500	(1),(2)	1
IC,	≤ 5 gal	5	18	STD or LO	SR or QR	0.30	3000		2
II,	≤ 5 gal	$6^1/_2$	30	LO or ELO	QR	0.45	3000		3
IIIA	> 5 gal ≤ 60 gal	5 (1-high)	18	LO or ELO	SR	0.40	3000		4
	≤ 5 gal	18	30	STD or LO	SR or QR	0.25	3000		5
IIIB	> 5 gal ≤ 60 gal	10 (3-high)	20	STD or LO	SR	0.25	3000		6
		18	30	STD or LO	SR	0.35	3000		7
Relieving-Style Containers									
IB, IC,	≤ 5 gal	12	30	ELO	QR	0.60	3000	(3),(4)	8
II,	> 5 gal ≤ 60 gal	5 (1-high)	30	LO or ELO	SR	0.40	3000		9
IIIA	> 5 gal ≤ 60 gal	$6^1/_2$ (2-high)	30	LO or ELO	SR	0.60	3000	(5)	10
	≤ 5 gal	18	30	STD or LO	SR or QR	0.25	3000		11
IIIB	> 5 gal ≤ 60 gal	10 (3-high)	20	STD or LO	SR	0.25	3000		12
		18	30	STD or LO	SR	0.35	3000		13
IB, IC,	Portable tanks	1-high	30	STD or LO	SR	0.30	3000		14
II, IIIA		2-high	30	LO or ELO	SR	0.60	5000		15
IIIB	Portable tanks	1-high	30	STD or LO	SR	0.25	3000		16
		2-high	30	LO or ELO	SR	0.50	3000		17

SI units: 1 gal = 3.8 L; 1 ft = 0.3 m; 1 ft² = 0.093 m²; 1 gpm/ft² = 40.7 Lpm/m².

[1]Larger orifice sprinklers are preferred when installed in accordance with NFPA 13, *Standard for the Installation of Sprinkler Systems*. STD = Standard Orifice, LO = Large Orifice, ELO = Extra-Large Orifice.

[2]SR = Standard Response, QR = Quick Response. Where both are listed, QR is preferred.

[3]Ceiling sprinklers high temperature.

[4]See Appendix D-2 for references to fire tests on which the protection criteria given in this table is based.

NOTES:

1. Using 5-gal cylindrical metal containers with plastic spouts or solid plastic caps and 1-gal F-style (oblong) metal containers.
2. Minimum hose stream demand 250 gpm for 2 hour.
3. Sprinklers shall also be hydraulically calculated to provide a density of 0.80 gpm/ft² over 1000 ft².
4. Use pendent-style ELO ceiling sprinklers.
5. Drums placed on open slotted pallet, not nested, to allow pressure relief from drums on lower levels.

Figure 7

Table 4-8.2(c) Foam-Water Sprinkler Protection Single or Double Row Racks Container Construction—Metal (For Nonmiscible Liquids or Miscible > 50%)

Liquid Class	Container Size and Arrangement	Maximum Storage Height (ft)	Maximum Ceiling Height (ft)	Ceiling: Sprinkler Type: Orifice[1]	Ceiling: Sprinkler Type: Response[2]	Ceiling: Density (gpm/ft²)	Ceiling: Design Area (ft²)[3]	In-Rack Sprinkler Protection	Notes	Fire Test Ref.[4]
Non-relieving-Style Containers										
IB, IC, II, IIIA	≤ 5 gal	25	30	STD or LO	SR or QR	0.30	2000	Every level	(1),(2)	1
IB, IC, II, IIIA	> 5 gal ≤ 60 gal	25	30	STD or LO	SR	0.30	3000	Every level	(1),(3)	2
IIIB	≤ 60 gal	40	50	STD or LO	SR	0.30	2000	One line every other level, beginning above first storage level	(1)	3
Relieving-Style Containers										
IB, IC, II, IIIA	≤ 5 gal	25	30	STD or LO	SR or QR	0.30	2000	One line every other level, beginning above first storage level.	(1),(2)	4
IB, IC, II, IIIA	> 5 gal ≤ 60 gal and portable tanks	25	30	STD or LO	SR	0.30	3000	One line every other level, beginning above first storage level	(1),(3)	5
IIIB	≤ 60 gal	40	50	STD or LO	SR	0.30	2000	One line every other level, beginning above first storage level	(1)	6

SI units: 1 gal = 3.8 L; 1 ft = 0.3 m; 1 ft² = 0.093 m²; 1 gpm/ft² = 40.7 Lpm/m².

[1]Larger orifice sprinklers are preferred when installed in accordance with NFPA 13, *Standard for the Installation of Sprinkler Systems*. STD = Standard Orifice, LO = Large Orifice.

[2]SR = Standard Response, QR = Quick Response. Where both are listed, QR is preferred.

[3]Ceiling sprinklers high temperature.

[4]See Appendix D-2 for references to fire tests on which the protection criteria given in this table is based.

NOTES:

1. Space in-rack sprinklers on maximum 9 ft centers, staggered vertically. Base design in 30 gpm per head, with six hydraulically most remote heads operating in each of upper three levels. Sprinklers are STD or LO, SR or QR, 165°F operating temperature, with shields. Hydraulic design can be reduced to three heads operating per level—three levels operating simultaneously when using a pre-primed foam-water system installed in accordance with NFPA 16A, *Standard for the Closed-Head Foam-Water Sprinkler Systems*, and maintained according to NFPA 25, *Standard for the Inspection, Testing, and Maintenance of Water-Based Fire Protection Systems*.
2. Design area can be reduced to 1500 ft² when using a pre-primed foam-water system installed in accordance with NFPA 16A and maintained according to NFPA 25.
3. Design area can be reduced to 2000 ft² when using a pre-primed foam-water system installed in accordance with NFPA 16A and maintained according to NFPA 25.

Figure 8

Table 4-8.2(d) Foam-Water Sprinkler Protection Bulk or Palletized Storage Container Construction—Metal (For Non-miscible Liquids or Miscible > 50%)

Liquid Class	Container Size and Arrangement	Maximum Storage Height (ft)	Maximum Ceiling Height (ft)	Ceiling: Sprinkler Type: Orifice[1]	Ceiling: Sprinkler Type: Response[2]	Ceiling: Density (gpm/ft²)	Ceiling: Design Area (ft²)[3]	Notes	Fire Test Ref.[4]
Non-Relieving-Style Containers									
IB,	≤ 5 gal cartoned	11	30	LO or ELO	SR or QR	0.40	3000	(2)	1
IC,	≤ 5 gal uncartoned	12	30	STD or LO	SR or QR	0.30	3000	(2)	2
II, IIIA	> 5 gal ≤ 60 gal	5 (1-high)	30	STD or LO	SR	0.30	3000		3
Relieving-Style Containers									
IB, IC, II, IIIA	> 5 gal ≤ 60 gal	6½ (2-high)	30	STD or LO	SR	0.30	3000	(1)	4
IB, IC, II, IIIA	Portable tanks	2-high maximum	30	STD or LO	SR	0.30	3000		5

SI units: 1 gal = 3.8 L; 1 ft = 0.3 m; 1 ft² = 0.093 m²; 1 gpm/ft² = 40.7 Lpm/m².

[1]Larger orifice sprinklers are preferred when installed in accordance with NFPA 13, *Standard for the Installation of Sprinkler Systems*. STD = Standard Orifice, LO = Large Orifice, ELO = Extra-Large Orifice.

[2]SR = Standard Response, QR = Quick Response. Where both are listed, QR is preferred.

[3]Ceiling sprinklers high temperature.

[4]See Appendix D-2 for references to fire tests on which the protection criteria given in this table is based.

NOTES:

1. Drums placed on open slotted pallet, not nested, to allow pressure relief from drums on lower levels.
2. Design area can be reduced to 2000 ft² when using a pre-primed foam-water system installed in accordance with NFPA 16A, *Standard for the Installation of Closed-Head Foam-Water Sprinkler Systems*, and maintained according to NFPA 25, *Standard for the Inspection, Testing, and Maintenance of Water-Based Fire Protection Systems*.

Figure 9

Table 4-8.2(e) Water Sprinkler Protection Single, Double, or Multirow Open Frame Racks Container Construction—Plastic in Corrugated Fibreboard Boxes (For Nonmiscible Liquids on Miscible > 50%)

Liquid Class	Container Size and Arrangement	Maximum Storage Height	Maximum Ceiling Height	Ceiling: Sprinkler Type[1]	Ceiling: Density (gpm/ft²)	Ceiling: Design Area (ft²)[3]	In-Rack Sprinkler Protection	Notes	Fire Test Ref.[1]
IIIB	≤ 5 gal	Unlimited	Unlimited	STD, LO, ELO, large drop, or ESFR	0.20	3000	**Double Row Racks:** Three lines per level with deflectors no more than 9 in. below each barrier and no less than 6 in. above the top of the storage. One line of in-rack sprinklers shall be located in the longitudinal flue and one line shall be located at each face of the racks.	(1), (2), (3), (4), (5)	1
							Multiple Row Racks: Protect as described for double row racks, except that there shall be an in-rack sprinkler located at each intersection of transverse and longitudinal flue, as well as the rack faces.		
							Single Row Racks (up to 5 ft wide): Protect using a barrier above the second tier of storage with a single line of in-rack sprinklers down the center of the rack. Sprinklers shall be spaced 4 to 5 ft apart, at each transverse flue. As an alternate method, for single row racks that will be loaded from one side, in-rack sprinklers can be installed beneath the barrier at each rack upright at the face of the rack on the side from which the rack is loaded. On the opposite face, in-rack sprinklers shall be installed at the transverse flue centered between the rack uprights.		

SI units: 1 gal = 3.8 L; 1 ft = 0.3 m; 1 ft² = 0.093 m²; 1 gpm/ft² = 40.7 Lpm/m²; 1 in. = 2.5 cm.
[1]STD = Standard Orifice, LO = Large Orifice, ELO = Extra-Large Orifice, ESFR = Early Suppression-Fast Response.
[2]See Appendix D-2 for references to fire tests on which the protection criteria given in this table is based.

NOTES:

1. Ceiling sprinkler system design shall be based on demand of surrounding occupancy or minimum design as stated, whichever is greater. If Class IIIB liquid storage does not extend to full height of rack, protection for other commodity stored above barrier shall be in accordance with appropriate standard and based on height of storage of other commodity.
2. Provide plywood (minimum $^3/_8$ in.) or sheet metal (minimum 22 gal) barriers within rack at approximately 10 ft vertical intervals (maximum 12 ft, including over top level of Class IIIB liquid storage). Barrier shall be solid and continuous except at rack uprights where a maximum 12 in. gap is allowed, assuming minimum 8 ft spacing between rack uprights. For Class IIIB liquids having a closed cup flashpoint ≥ 450°F (232°C), a horizontal barrier is not necessary.
3. Longitudinal flue sprinklers shall be located at each transverse flue between pallet loads (approximately 4 ft to 5 ft on centers). Face sprinklers shall be located in the transverse flue at rack uprights on maximum 10 ft centers. If rack uprights are more than 10 ft on center, locate face sprinklers centered between rack uprights and at uprights. In-rack sprinklers shall be quick-response, large orifice with shields, 165°F (74°C), 57 gpm per head, most remote six heads (three on two lines), if one horizontal barrier level; or most remote eight heads (four on two lines), if two or more horizontal barrier levels.
4. Ceiling sprinkler demand shall not be included in hydraulic calculations for in-rack sprinklers. Water demand at point of supply shall be calculated separately for in-rack and ceiling sprinklers and based on the greater demand.
5. If there are adjacent bays of in-rack arrays not dedicated to Class IIIB liquid storage, the barrier and in-rack sprinkler protection shall be extended at least one rack bay (approximately 8 ft) beyond the Class IIIB liquid storage. Additionally, a vertical plywood barrier or similar insulating material shall be installed at both ends of the Class IIIB liquid storage area. The vertical barrier shall extend the full rack width and shall fit flush against any horizontal barriers. Clearance at cutouts for both in-rack sprinkler piping and horizontal rack members shall not exceed 3 in.

Figure 10

Table 4-8.2(f) Water Sprinkler Protection Shelf Storage Container Construction—Metal (For Nonmiscible Liquids or Miscible > 50%)

Liquid Class	Container Size and Arrangement	Maximum Storage Height (ft)	Maximum Ceiling Height (ft)	Ceiling				Notes	Fire Test Ref.[4]
				Sprinkler Type		Density (gpm/ft²)	Design Area (ft²)[3]		
				Orifice[1]	Response[2]				
IB, IC, II, III	≤ 1 gal	6	18	STD or LO	SR or QR	0.19	1500	(1),(2)	1

SI units: 1 gal = 3.8 L; 1 ft = 0.3 m; 1 ft^2 = 0.093 m^2; 1 gpm/ft^2 = 40.7 Lpm/m^2.
[1]Larger orifice sprinklers are preferred when installed in accordance with NFPA 13, *Standard for the Installation of Sprinkler Systems.*
STD = Standard Orifice, LO = Large Orifice.
[2]SR = Standard Response, QR = Quick Response. Where both are listed, QR is preferred.
[3]Ceiling sprinklers high temperature.
[4]See Appendix D-2 for references to fire tests on which the protection criteria given in this table is based.
NOTES:
1. Protection for "gondola"-type shelving 2 ft or less in depth per side.
2. Minimum hose stream demand 250 gpm for 2 hours.

Figure 11

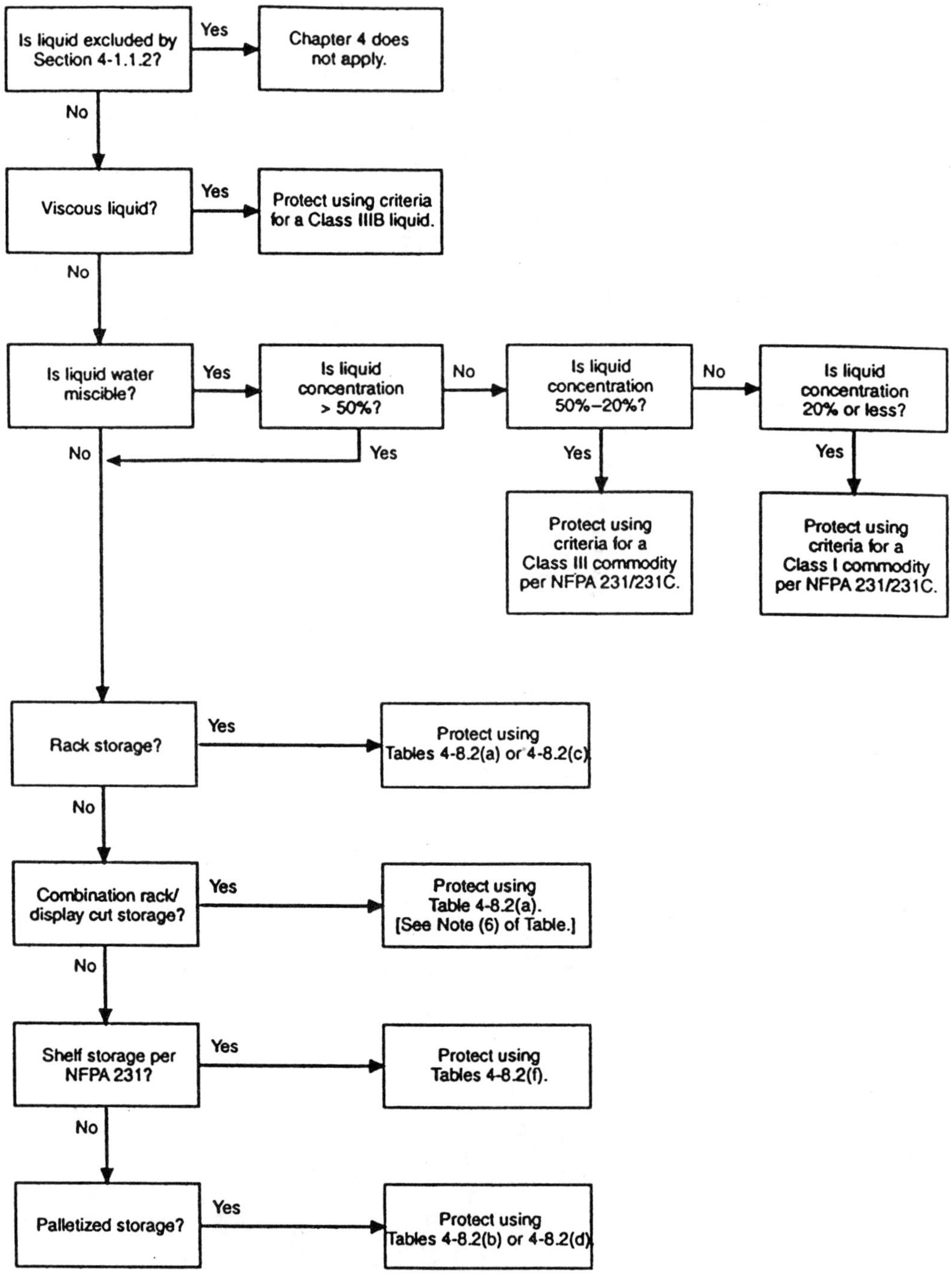

Figure 4-8.2(a) *Fire protection criteria decision tree for metal containers.*

Figure 12

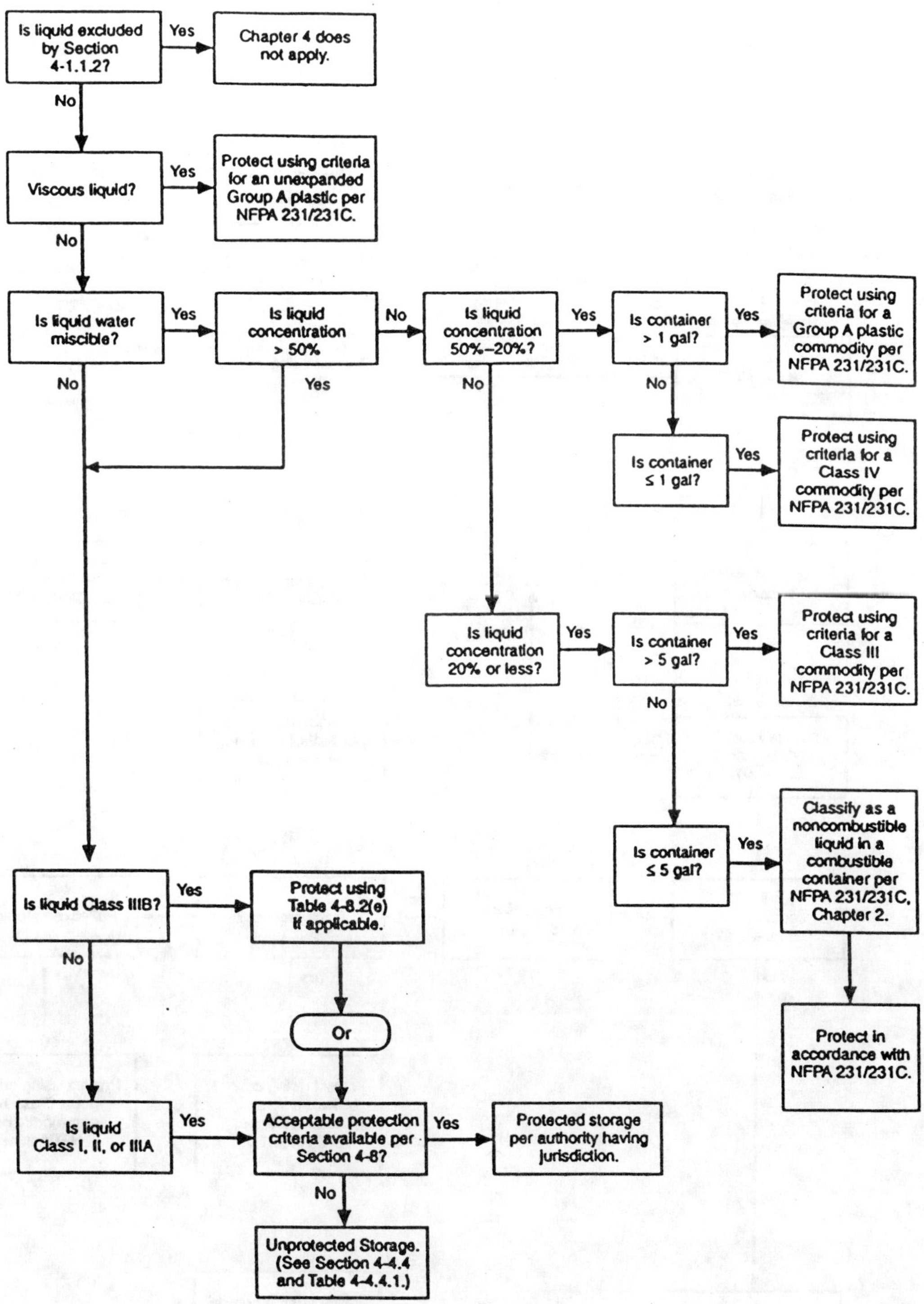

***Figure 4-8.2(b)** Fire protection criteria decision tree for plastic/fiberboard containers.*

Figure 13

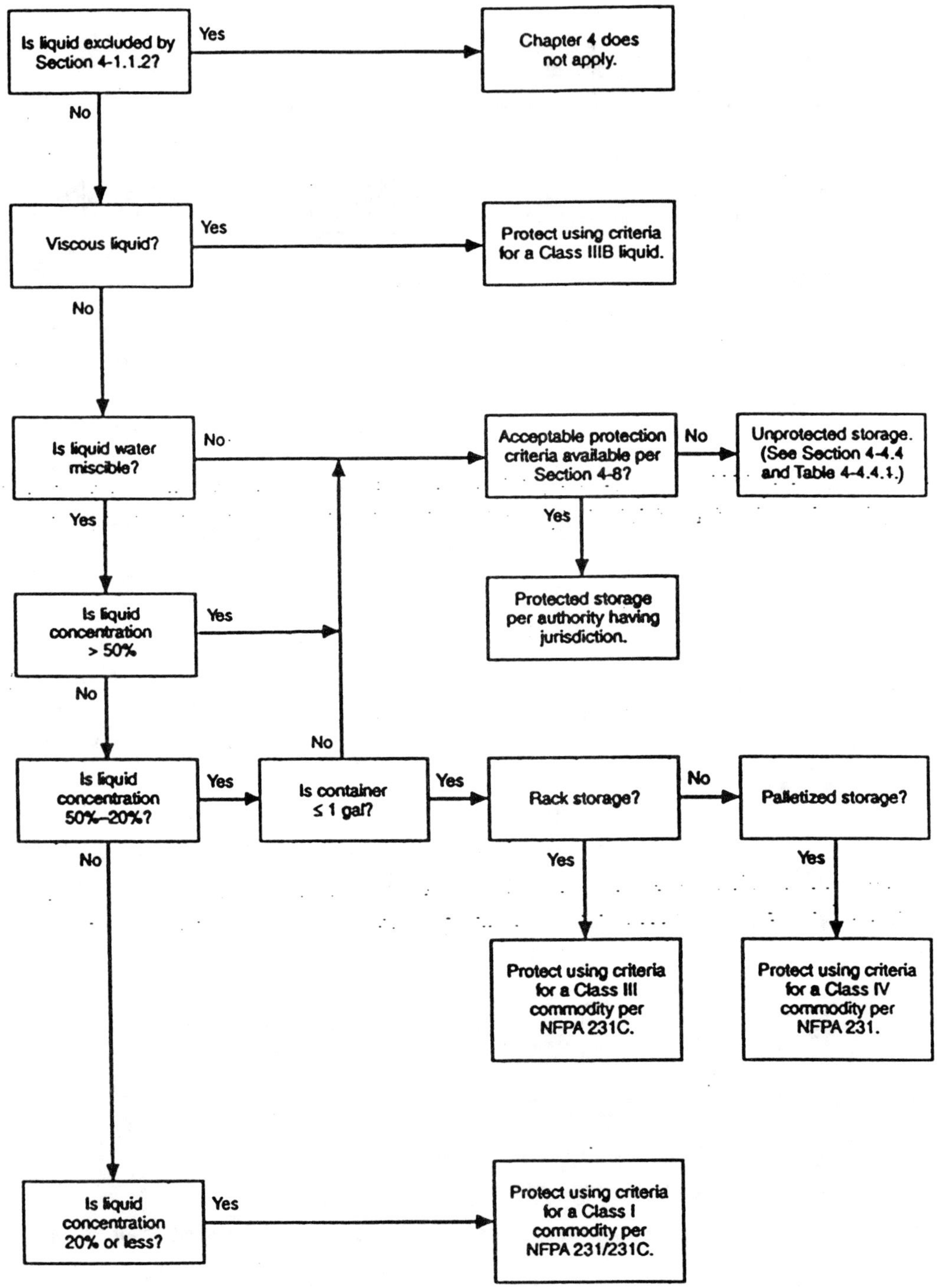

Figure 4-8.2(c) Fire protection criteria decision tree for glass containers.

Figure 14

Determination, Prevention and Mitigation of Potential Hazards due to the Handling of Powders during Transportation, Charging, Discharging and Storage

Norbert Jaeger, Safety Testing Laboratory, Ciba-Geigy Ltd.,
Toms River, New Jersey USA

and

Richard Siwek, Corporate Unit Safety and Environment, Ciba-Geigy Ltd.,
Basle, Switzerland

Prepared for Presentation at

American Institute of Chemical Engineers
31st Loss Prevention Symposium
Houston, TX

March 11, 1997

Determination, Prevention and Mitigation of Potential Hazards due to the Handling of Powders during Transportation, Charging, Discharging and Storage

Norbert Jaeger, Safety Testing Laboratory, Ciba-Geigy Ltd., Toms River, New Jersey USA
and
Richard Siwek, Corporate Unit Safety and Environment, Ciba-Geigy Ltd., Basle, Switzerland

Abstract

The knowledge of the ignition behavior of dust-air mixtures due to electrical sparks (MIE, Minimum Ignition Energy) and hot surfaces (MIT, Minimum Ignition Temperature) is important for risk assessments in chemical production plants. The ignition behavior determines the extent and hence the cost of preventive protection measures.

This paper describes the use of the minimum ignition energy and minimum ignition temperature as very important safety indexes in practice.

Based on the latest results from large scale experiments on pneumatic filling of silos with polymeric materials and new results of full scale filling tests using Flexible Intermediate Bulk Containers (FIBC) manufactured from a variety of materials, guidance can be given to ensure safe operation in different situations such as filling, emptying operations, type of powder handled.

The aim of this paper is to assist people dealing with product. It reflects the present state of the art and current knowledge of the assessment and measures associated with powder handling.

1 Introduction

Thorough knowledge of the ignition behavior of dust-air-mixtures towards electrical sparks (minimum ignition energy, MIE) and hot surfaces (minimum ignition temperature, MIT) is extremely important for assessing the hazards in dust-carrying plants. The ignition behavior determines essentially the extent, and hence the cost of the protective measures, to be used. This is especially true for the use of the protective measure "prevention of ignition sources" and also for the understanding of the ignition phenomena regarding static electricity e.g., brush discharges, bulk surface discharges, spark discharges, propagating brush discharges.

Knowing the minimum ignition energy, the particle size of the product and the product resistivity measures for different operations are recommended depending on the magnitude of the minimum ignition energy, median value, product resistivity and vessel size.

The aim of this paper is to assist people dealing with easily ignitable dust. It reflects the present state of the art and current knowledge of the assessment and measures associated with powder handling.

2 Basic Information

2.1 Minimum Ignition Energy

The minimum ignition energy, MIE, of a combustible substance is the lowest value of the electrical energy stored in a capacitor, which on discharge just suffices to ignite the most readily ignitable fuel-air-mixture at atmospheric pressure and room temperature. To assure a standardized test procedure, a test apparatus known as MIKE 3 (Fig. 1) of the third generation has been specially developed by Kühner AG, Switzerland.

The MIE is usually quoted as a range: The lower value represents the highest energy at which no ignition is found in at least 10 experiments. The higher value, on the other hand, is the lowest energy at which the dust-air mixture is just ignited:

no ignition < MIE < ignition

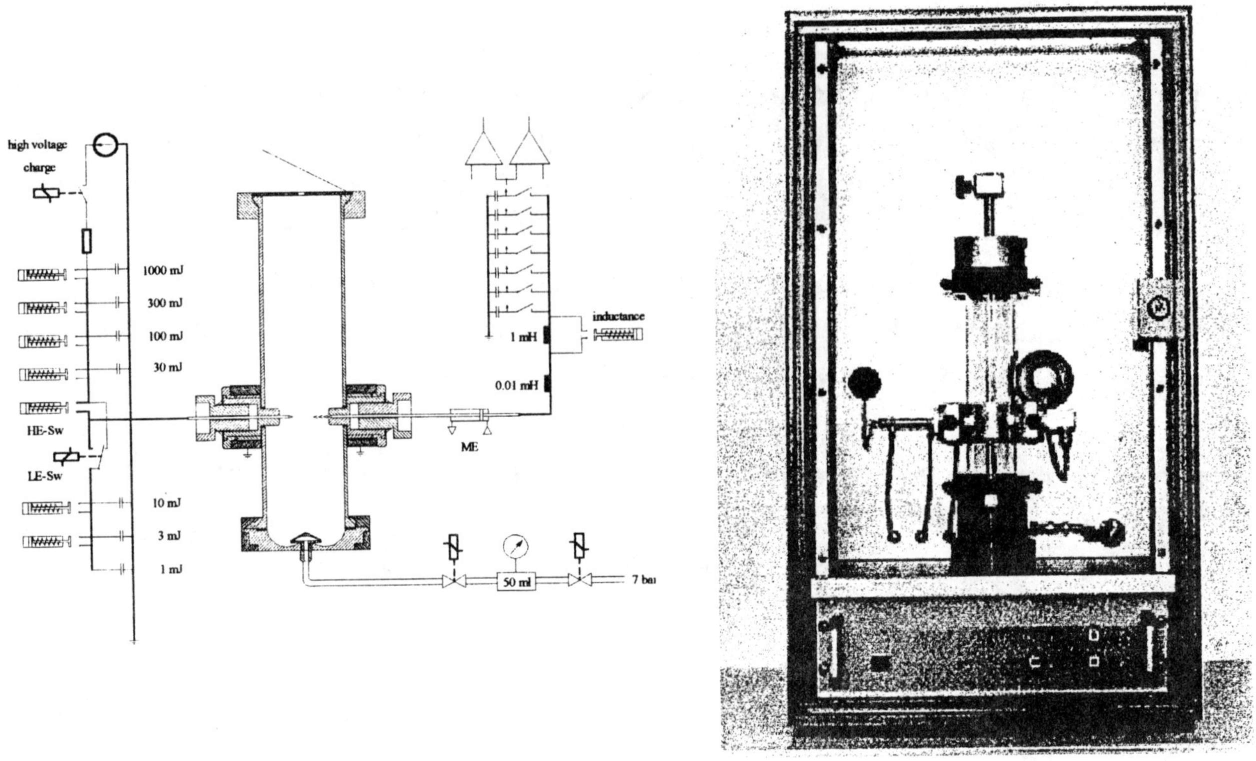

Figure 1 : MIKE 3 apparatus for determination of the minimum ignition energy MIE of dusts

The method of determination of MIE is described in the international standard of IEC and in the nearest future available as an ASTM standard. The MIE is generally obtained with an inductance in the discharge circuit. However, in order to assess the incendivity of electrostatic discharges towards dust-air-mixtures, the MIE must also be determined without an inductance in the discharge circuit. With flammable gases, the influence of the inductance is generally not detectable. Dusts, particularly very easily ignitable dusts, also exist where the inductance has also no influence.

2.2 Minimum Ignition Temperature, MIT

The minimum ignition temperature, MIT, is defined as the lowest temperature of a heated surface at which the most readily ignitable mixture of a dust with air just ignites. It provides information on the ignition behavior of a dust suspension when quickly passing over a hot surface. The MIT can be determined either in the Godbert-Greenwald (GG) furnace (see Figure 2) or in the BAM furnace (German Federal Institute for the Testing of Materials) (see Figure 3)

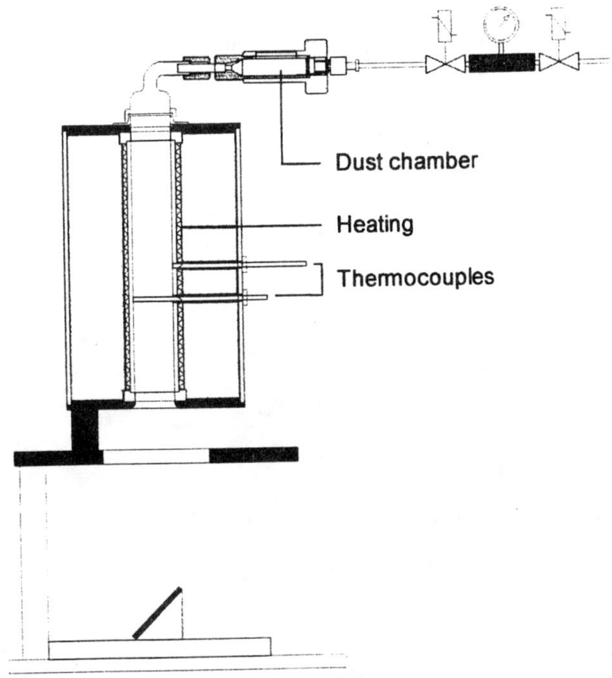

Figure 2: Godbert-Greenwald furnace

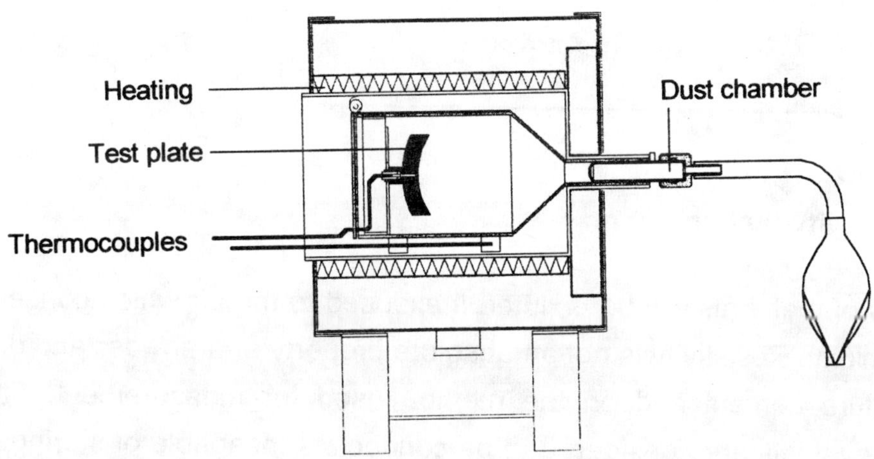

Figure 3: BAM furnace

The BAM furnace is enjoying increasing popularity. The MIT determined in the GG furnace can be converted to BAM-values as follows: MIT(BAM) = 10°C + 0.9 · MIT (GG)

2.3 Powder Volume Resistivity

To characterize a material in view of static dissipative properties **its powder volume resistivity, σ_R,** has to be determined (see Figure 4).

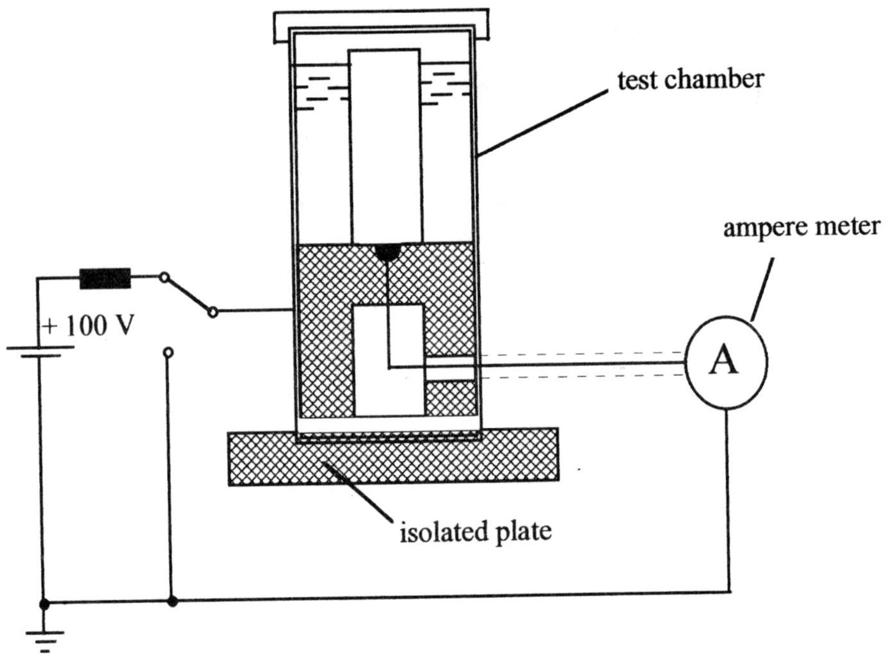

Figure 4: Powder resistivity test chamber

It is a fact that surfaces of materials will be polluted if exposed to the air which consequently includes adsorbing humidity. Resistivity is not an absolute property of a powder and depends very strongly on moisture content and on the method used for measurement. From an electrostatic point of view dusts are considered to be conductors incapable of storing charge until the resistivity exceeds 10^9 Ω·m.

3 Protective Measures

3.1 Principles

Using preventive measures against explosions requires the reliable exclusion of one of the requirements necessary to generate an explosion. Therefore at least one of the sides of the hazard triangle shown in Figure 5 must be eliminated.

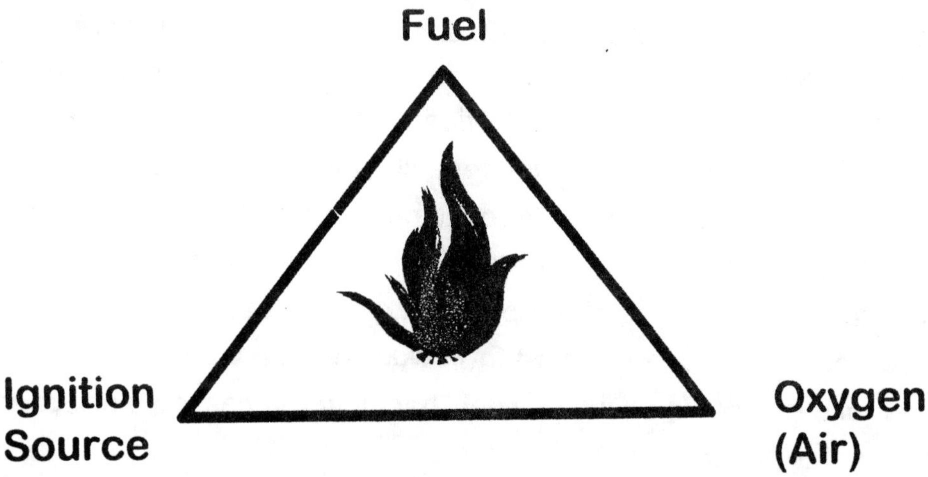

Figure 5: Hazard triangle - Principle of preventive explosion protection

An explosion can thus be excluded with certainty by either

- **avoiding the development of explosible mixtures** (combustible dusts, flammable gases),

or

- replacing the **atmospheric oxygen** by **inert gas**, working in a **vacuum** or using **inert dust**, or
- preventing the **occurrence of effective ignition sources**.

All three measures are summarized under the heading "preventive explosion protection". It should be ensured that at least one of the three conditions is eliminated or so strongly reduced, that an explosion is no longer possible. Their appropriate application prevents an explosion from starting.

When combustible dusts are handled, avoiding an explosive atmosphere by keeping the dust concentration outside the explosive range, is rarely possible due to sedimentation or whirling up of the material being handled. Thus, as a matter of principle, an explosive atmosphere can only be avoided with certainty by reducing the oxygen concentration, e.g., inerting. In practice, however, the possibilities to apply inerting is also limited. For such a situation, avoidance of effective ignition sources and /or explosionproof design are the only measures available.

The use of avoidance of effective ignition sources as the sole protective measure is acceptable, if it has been established, that the occurrence of effective ignition sources is excluded with certainty under all normal operating conditions, as well as possible deviations. Trivial ignition sources (welding, smoking etc.) must be excluded by organizational means. Ignition sources which could result from processes introducing energy into the product being handled are not the topic of this paper. This holds also for products that have a tendency to form glowing particles and, if in the course of the process, glowing particles could form or entrain. An ignition source not to be underestimated in powder operations is electrostatic charging. This is a normal occurrence in most powder handling processes and operations and therefore the topic of this paper.

Below, protective measures are indicated for typical powder handling operations and processes. The following product and plant properties are important for an accurate hazard assessment:

- Minimum ignition energy, MIE, of the bulk material (measured without inductance in the discharge circuit),
- Minimum ignition temperature, MIT, of the bulk material,
- Volume resistivity σ_R of the powder,
- Particle size distribution of the bulk material and mean value, M,
- Volume and shape of the silo or container (volume and shape of the product heap and of the dust cloud).

Unless otherwise stated, the following sections are based on the assumption that the bulk materials are handled without flammable gases or vapors being present.

3.2 Filling and Emptying Operations

In filling and emptying, ignition hazards are of prime importance owing to electrostatic charging. The hazard comprises possible charge accumulation on not only plant units incl. drum and container, but also - in the case of insulating bulk material - on the bulk material itself which is shown in Figure 6.

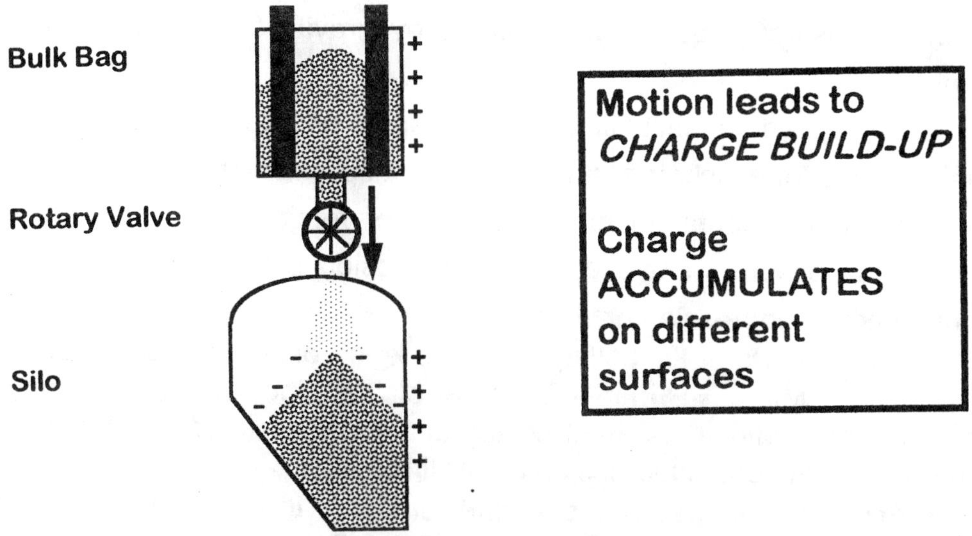

Figure 6: Filling (charging) and Emptying (discharging) Operations

With regard to the ignition hazards, assuming the insulating bulk material carries a charge, filling represents the most hazardous operation for the following reason. In each filling operation, the bulk material undergoes dispersion (e.g. gravity feed, pneumatic transport, etc.) and can therefore be charged in the separation processes occurring in transport. The bulk material and hence its associated charge is then "packed" in a small space in which the charge is not able to flow to ground sufficiently quick, even with a conductive and grounded receiver. This generates a high space charge density and hence a high electric field. Consideration must also be given to the problems associated with heat accumulation and the possibility of entrainment of smoldering lumps.

Based on the product and plant properties mentioned above, the decision tree shown on the next pages leads to information on the required safety measures.

① Use only conductive components in plant and earth.

② Use only conductive containers or those capable of discharge (The leakage resistance to earth point from any point on the surface of the container must be less than 100 Mohm) and earth.

③ Use only conductive containers or those capable of discharge or which cannot be charged (containers having at least on one side a surface resistance measured according to DIN 53482 of less than 100 GOhm at 30% rel. humidity and more than 100 Mohm at 65% rel. humidity) and earth.

④ Use additional explosion protection measures (e.g., inerting, containment, suppression or venting) or seek advice from experts.

⑤ No additional measures required when filling by gravity from conductive, grounded installation.

⑥ Seek advice from experts or limit filling rate to 5 $kg \cdot s^{-1}$ and use conductive containers or containers capable of discharge and earth or use additional explosion protection measures (e.g. inerting, containment, suppression or venting).

⑦ Seek advice from experts or use additional explosion protection measures (e.g., inerting, containment, suppression or venting).

⑧ Seek advice from experts or limit filling rate to 5 $kg \cdot s^{-1}$ or restrict container diameter to $D < 0.612 \times M^{-0.435} \times MIE^{0.298}$ (D in m = Diameter of Container: range of validity: 0.5-6 m; M in mm: range of validity: 0.1-5 mm) or use additional explosion protection measures (e.g., inerting, containment, suppression or venting).

⑨ Seek advice from experts or restrict container diameter to $D < 0.612 \times M^{-0.435} \times MIE^{0.298}$ (D in m = Diameter of Container: range of validity: 0.5-6 m; M in mm: range of validity: 0.1-5 mm) or use additional explosion protection measures (e.g., inerting, containment, suppression or venting).

⑩ No additional safety measures required.

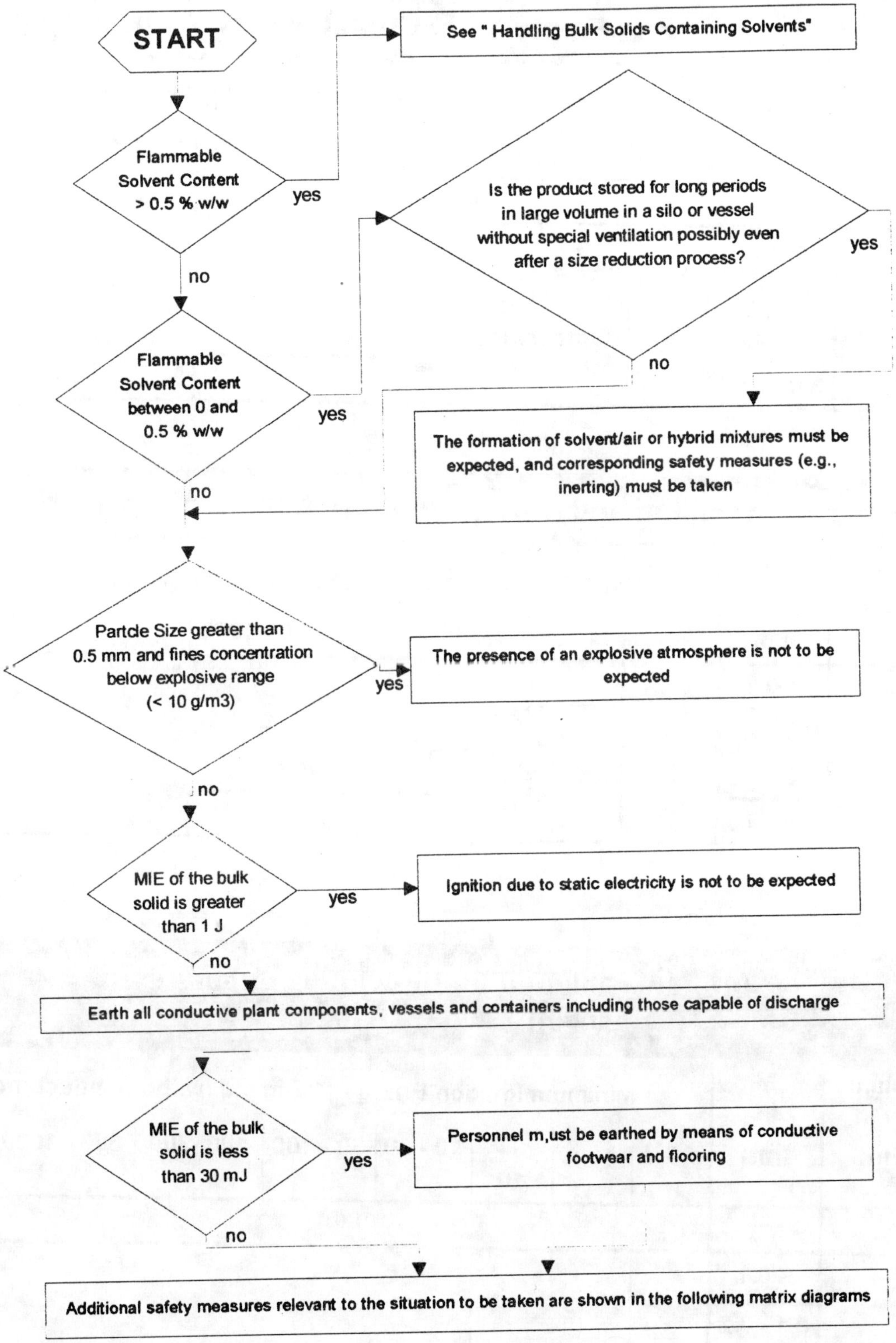

The matrix mentioned in the decision tree are shown in the following Tables for different container volumes.

Measures for Filling and Emptying of Containers of $V < 0.2\ m^3$

Powder volume resistivity in MOhm·m	M in mm	Minimum Ignition Energy MIE in mJ (without inductance) < 1*)	1 - 3	3 - 10	10 - 30	30 - 1000	> 1000
< 1	any	① ②					
1 - 10'000	< 0.5						
	0.5 - 1.5	①					
	1.5 - 3.0	③	⑤otherwise⑥		⑩		
	3.0 - 6.0	④					

Measures for Filling and Emptying of Containers of $V < 0.2\ m^3$

Powder volume resistivity in MOhm·m	M in mm	Minimum Ignition Energy MIE in mJ (without inductance) < 1*)	1 - 3	3 - 10	10 - 30	30 - 1000	> 1000
> 10'000	< 0.5						
	0.5 - 1.5	①					
	1.5 - 3.0	③	⑦		⑩		
	3.0 - 6.0	④					

Measures for Filling and Emptying Containers of $0.2\ m^3 < V < 2.0\ m^3$

Powder volume resistivity in MOhm·m	M in mm	Minimum Ignition Energy MIE in mJ (without inductance) < 1*)	1 - 3	3 - 10	10 - 30	30 - 100	100 -300	300 - 1000	> 1000
< 1	any	① ②							
1 - 10'000	< 0.1								
	0.1 - 0.2								
	0.2 - 0.5	①				⑩			
	0.5 - 1.0	②		⑤otherwise⑥					
	1.0 - 2.5	④							
	2.5 - 5.0								

Measures for Filling and Emptying Containers of 0.2 m³ < V < 2.0 m³									
Powder volume resistivity in MOhm·m	M in mm	Minimum Ignition Energy MIE in mJ (without inductance)							
		< 1*)	1 - 3	3 - 10	10 - 30	30 - 100	100 -300	300 - 1000	> 1000
>10'000	< 0.1		②						
	0.1 - 0.2			②					
	0.2 - 0.5	①			②			⑩	
	0.5 - 1.0	②	⑤otherwise⑥			②			
	1.0 - 2.5	④					②		
	2.5 - 5.0							②	

*) It is assumed, that these type of dusts are ignitable by brush discharge. This must be check by FETES. If no ignition by brush discharge, use measures according 1 mJ < MIE < 3 mJ.

Measures for Filling and Emptying of Vessel/Silo of V > 2 m³				
Powder volume resistivity in MOhm·m	M in mm	Minimum Ignition Energy MIE in mJ (without inductance)		
		< 10	10 - 1000	> 1000
< 1	any		①, ②	
1..10'000	< 5	①, ②, ④	②, ⑧	⑩
> 10'000	< 5		②, ⑨	

3.3 Loading Bulk Materials into Flammable Solvents

Whenever it is possible, bulk materials should be fed into flammable solvent under closed conditions and in an inert atmosphere (see Figure 7). This can be realized by means of a bucket wheel, a two-valve system, a feed screw or by conveying by inert gas. If such a closed introduction system is not possible, as the next best thing, the solvent should be cooled below its flash point (T ≥ 5K) before loading the bulk material, or the bulk material introduced first and then the flammable solvent.

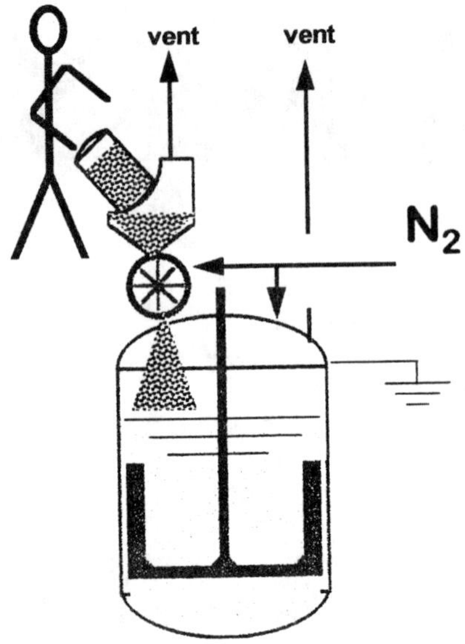

Figure 7: Powder charging into a vessel with a precharged solvent

If the open addition of bulk material into previously loaded flammable solvent at a temperature greater than the flash point less 5°C is unavoidable, then the following points must be considered:

- personnel must be grounded.

- all aids to introduction, such as funnels, charging chutes, etc., must be made in conducting materials and grounded during the transfer.

- containers for solvent and powders must be of conductive material and grounded during the charging process.

- conductive vessels for solvents and bulk material with an insulating internal coating of up to 2 mm maximum may be used, if they are grounded during filling and emptying.

- normal loose plastic sacks or plastic liners should not be used. Plastic liners or plastic sacks with at least one-sided non-chargeable (surface resistance according to ASTM $<10^{11}$ ohms at 30% relative humidity) are allowed (e.g. plastic sacks with an anti-static coating, paper sacks or paper sacks lined one side with a normal plastic coating, if the coating thickness is less than 2 mm).

3.4 Mixing

The occurrence of effective ignition sources in mixing is generally not anticipated, even with dusts exhibiting high ignition sensitivity, provided the following conditions are met:

- In the filling and emptying of the mixer, the considerations and measures that apply are the same as those in the filling and emptying of containers.

- In the filling and emptying of the mixer, the mixing elements must be at a standstill or run at a circumferential speed (relative speed) which does not exceed 1 $m \cdot s^{-1}$. This restriction must be assured by technical safeguards.

- In the closed condition and with a fill level of 70 vol. % or more, the circumferential speed of the mixing elements is no longer restricted (see Figure 8).

- No insulating inner coatings with high electrical breakdown strength (breakdown voltage must be less than 4 kV) may be affixed to the inner wall of the mixer (Product build-up must be checked).

- Circumferential speeds up to 10 $m \cdot s^{-1}$ can be tolerated in filling and emptying and with a mixer not filled to 70 vol. % provided the minimum ignition temperature, MIT, of the processed product lies above the limit values specified in the Table below.

MIE[1)] [mJ]	< 1	1-3	3-10	10-30	30-100	100-300	300-1000	> 1000
MIT [°C]	do not process	530	500	465	430	395	360	325

[1)] The MIE values must be determined with an additional inductance in the discharge circuit.

- Nauta mixers with bottom support of the helical screw can heat up during operation and care must be exercised with substances capable of spontaneous decomposition.

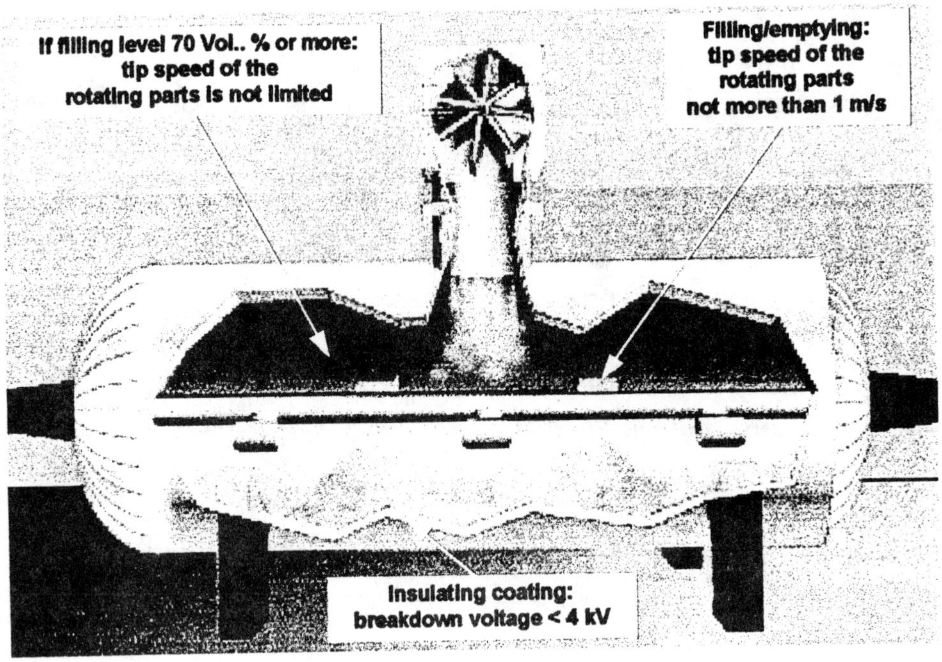

Figure 8: Mixing Operation

3.5 Dust Separation

In the case of dust separators, especially in filters, the dust explosion hazard must not be underestimated. The probability of occurrence of a fine dust atmosphere sensitive to ignition is large. In addition to the entrainment of ignition sources, which must be assessed based on the product properties (e.g. formation of smoldering lumps) and upstream operation, the danger of ignition through electrostatic charging is of prime importance (See Figure 9).

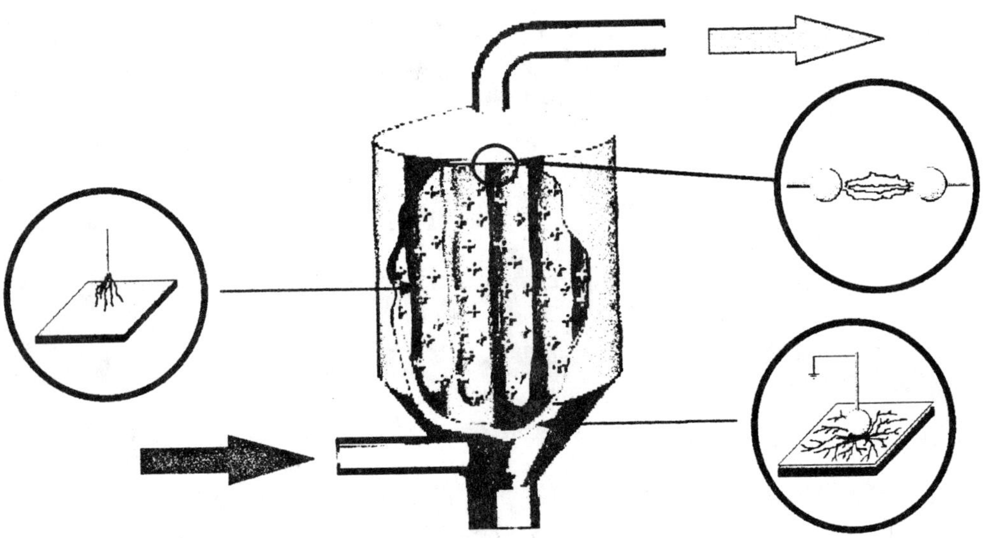

Figure 9: Electrostatic Ignition Sources (Dust Collector)

Electrostatic charging must be inhibited by the following measures:

- Grounding of all conductive apparatus parts. Particular attention must be paid to the grounding of all conductive parts which could possibly be insulated from ground if a filter cloth made of insulating material is used (e.g. filter supports, clamps). This must be specially checked after repair and maintenance work.
- With a MIE < 3 mJ or in the presence of flammable gases or vapors in the air being cleaned, electrically conducting filter materials must be used, except when the protective measure "inerting" is employed. Continuity of the conductivity and safe grounding must be checked. Multiple washings can have an adverse effect on the continuity of the filter material conductivity and thus require repeat checking.
- All inner walls on which dust can impact at high speed must not have any insulating inner coatings with a high electrical breakdown strength (breakdown voltage must be less than 4 kV; periodic check).

In general, with dusts with a MIE < 10 mJ it is advisable to implement explosion protection measures which go beyond the avoidance of effective ignition sources or to consult the responsible specialist departments. It should further be noted that the fan must be installed on the clean air side and that dust deposits must be avoided in the pipe and fan housing (periodic check or install a dust control unit).

3.6 Use of Flexible Intermediate Bulk Container FIBC

Flexible intermediate bulk containers are used on ever increasing scale in the powder handling industry. Depending on their hazard situation at the location where they are used, they must meet different requirements (see following Table) in order to avoid ignition hazards caused by electrostatic charging.

Environment / **Bulk material**	***no explosible atmosphere***	***explosible dust atmosphere***	***flammable gases or vapors***
MIE* > 1 J	**A**	**B**	**C**
3 mJ < MIE* < 1J	**B**	**B**	**C**
MIE* < 3 mJ	**C**	**C**	**C**

MIE*, measured without inductance in the electrical circuit

A: No special requirements.

B: Breakdown voltage of the FIBC wall material must not exceed 4 kV in order to prevent propagating brush discharges

C: The bag material including the slings must be electrostatic conductive. The resistance to the ground measured at any bag location (inside and outside) must be less than 100 Mohm (10^8 Ω). The flexible bulk bag must have a grounding tap. The conductivity and the necessity for grounding must be clearly marked.

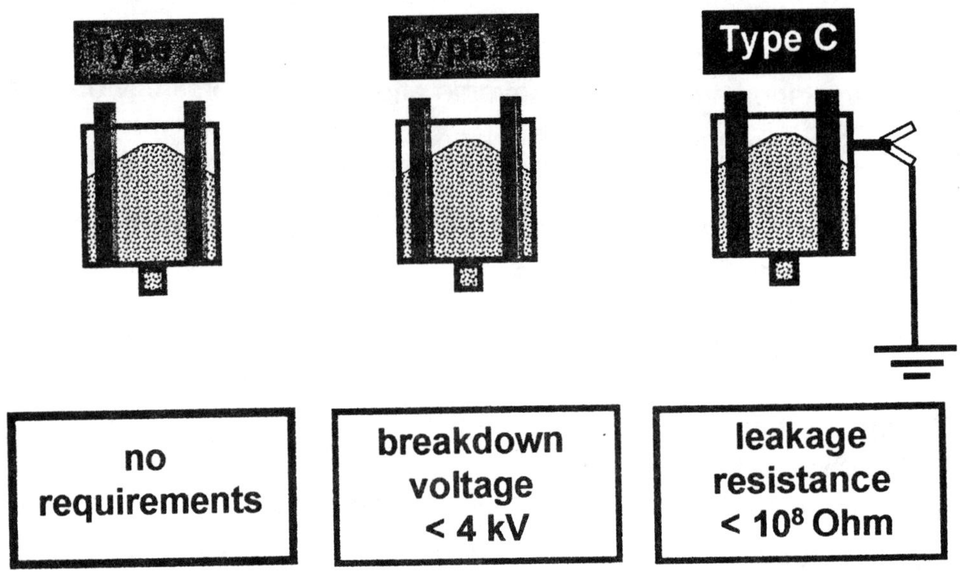

Figure 10: FIBC requirements

Most FIBC's on the market today are made of polypropylene ribbon fabric. To pass the type B classification the following requirements are recommend:

- any inner PE coating/liner present is not thicker than 20-30 μm
- and the inliner is not made of plastic.

FIBC's on the market meeting the requirement of type C are constructed as follows:

- The basic fabric consists of conductive material (e.g., plastic with sufficient admixture of carbon)
- The basic fabric consists of non-conductive material, but web contains interwoven threats of conductive plastic material which are interconnected.
- The basic fabric consists of non-conductive material, but the web contains interwoven metal threats which are interconnected.

- The basic fabric consists of non-conductive material, but the FIBC has an internal conductive coating.

To meet the specification as a type C bag, the following requirements are recommend:

- The FIBC must have an indication of its conductivity and the grounding obligation during charging and discharging.
- The FIBC must have a clearly marked area for the attachment of the grounding clamps.
- The lifting traps must also be made of conductive material and have a leakage resistance of less than $10^8\,\Omega$ to the FIBC body.

Using a type C bag requires a permanent grounding of the bag during the whole period, the bag is filled or discharged (see Figure 11).

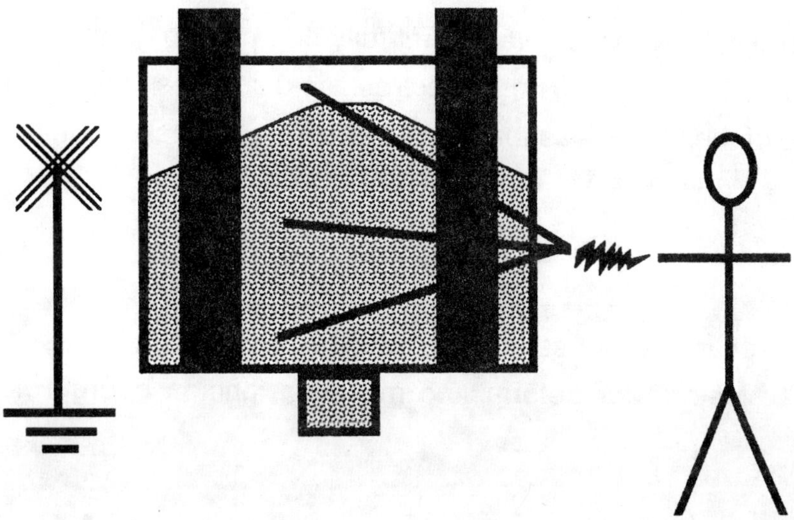

Figure 11: Electrostatic Discharges from an ungrounded type C bag.

The discharge from an ungrounded bag aligns to a single discharge point. Such a discharge is strong enough to ignite dust clouds.

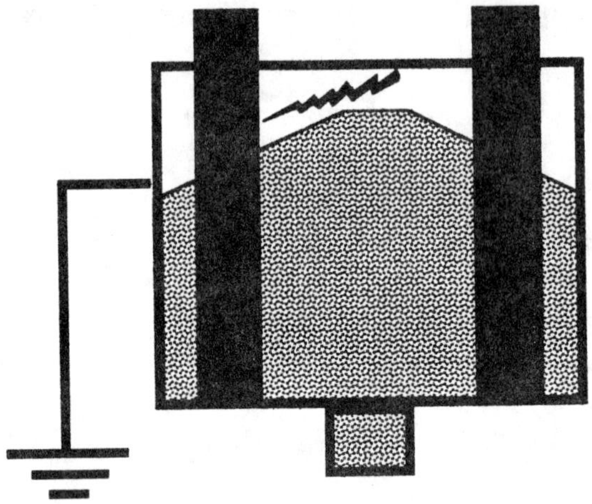

Figure 12: Electrostatic Discharge inside a grounded type C bag

The generated charge in the product pile cannot fully dissipate to ground. Small discharges can occur along the surface of pile. These electrostatic discharges are to weak to ignite dust clouds if the volume of the bag is less than 2 m^3 but strong enough to ignite solvent vapors.

4. References

/1/ IEC: Electrical apparatus for use in the presence of combustible dust, Part 2: Test methods, Section 3: Method for determining minimum ignition energy of dust/air mixtures (draft), 1992

/2/ Siwek, R., Cesana, Ch.: Ignition Behavior of Dusts: Meaning and Interpretation, Process Safety Progress (Vol. 13, No. 5), April, 1995

/3/ N. Jaeger: Explosionen und Emissionen, Handbuch des Umweltschutzes und der Umweltschutztechnik, Band 3 (649-670), Springer Verlag, 1996

/4/ Cesana, Ch., Siwek R.: MIKE 3 minimum ignition apparatus, Kühner AG, CH-4127 Birsfelden, 1992

/5/ IEC 31 H (CO)4: Methods for determining the minimum ignition temperature of dusts. Part 2: Dust cloud in a furnace at a constant temperature, 1984.

/6/ VDI guideline 2263: Dust Fires and Dust Explosions, part 1; Test Methods for the Determination of the Safety Characteristic of Dust. VDI-Verlag, Düsseldorf, 1990

/7/ Lüttgens, G., Glor, M.: Understanding and Controlling Static Electricity, Expert Verlag 1988

/7/ Bartknecht, W.: Explosionen, Grundlagen und Anwendungen, Springer-Verlag, 1993

/9/ Glor, M., Maurer, B.: Ignition tests with discharges from bulked polymeric granules in silos, J. Electrostat. 30 (1993) 123-134

/10/ Glor, M, et al., "Recent Developments in the Assessment of Electrostatic Hazards Associated with Powder Handling, Proceedings of the 8th Int. Symposium "Loss Prevention and Safety Promotion in the Process Industries" Antwerp, Elsevier Science Publishers B.V., Amsterdam, 1995

/11/ Siwek, R.: Dusts: Explosion Protection, 7th edition of Perry's chemical Handbook for Chemical Engineering, USA, 1996

/12/ Siwek, R, Cesana, Ch.: Methods for Determination of the Explosion Characteristics According to International Standards, Proceedings of the "First International Seminar on Fire and Explosion Hazard of Substances, Venting of Deflagrations", July 17-21, Moskau, Russia, 1995

/13/ Siwek, R.: Moderner Explosionsschutz, Brasilien, 1995

/14/ Siwek, R, Cesana, Ch.: Practical Application of the Minimum Ignition Energy and the Minimum Ignition Temperature, Proceedings of the 2nd International Conference and Exhibition on Health, Safety and Loss Prevention in the Chemical and Process Industries, Singapore, 4-8 December 1995

/15/ Siwek, R.: Determination of technical safety indices and factors influencing hazard evaluation of dusts, Journal of Loss Prevention in the Process Industries, Elsevier Science Ltd., 1995

/16/ N. Jaeger, R. Siwek: Safe handling of easily ignitable dusts, 5th World Congress Of Chemical Engineering, San Diego 1996

Design Practice for Extinguishing Barrier Systems

Richard Siwek and Peter E. Moore

Siwek, R.; Corporate Unit Safety and Environment, Explosion Technology, K-32.3.04
Ciba-Geigy Ltd., Basle, Switzerland

Moore, P.E., Kidde International, Industrial Fire Protection Division,
Thame Park Road, Thame, Oxon. OX9 3RT UK

Prepared for Presentation at

American Institute of Chemical Engineers
31st Loss Prevention Symposium
Houston, TX

March 11, 1997

Design Practice for Extinguishing Barrier Systems

Richard Siwek and Peter E. Moore

Summary

An extinguishing barrier comprises an optical flame sensor and a HRD-Suppressor located downstream of the detected flame front. The effectiveness of an extinguishing barrier is based on its ability to detect an explosion in a pipeline by means of an optical flame detector whose tripping signal is amplified and then very quickly actuates the detonator-actuated valves of the pressurized HRD-Suppressors.

Extensive practice-related tests in pipelines, having different cross sections and length, in pipelines connected with different vessels, have shown that extinguishing barriers can be used without reservation to halt or stop an explosion in practice. The amount of suppressant agent required depends on the nature of the combustible dusts, the nominal diameter of the protected pipeline, the explosion velocity and the maximum reduced explosion overpressure in the vessel. Most significantly, the theoretical understanding of explosion propagation and extinguishing has led to computer design guidance which has simplified system design. Explosions can be combated effectively in pipelines up to diameter 2500 mm.

1 Introduction

An extinguishing barrier comprises an optical flame sensor and a HRD-Suppressor located downstream of the detected flame front. The effectiveness of an extinguishing barrier is based on its ability to detect an explosion in a pipeline by means of an optical flame detector whose tripping signal is amplified and then very quickly actuates the detonator-actuated valves of the pressurized HRD-Suppressors (Fig. 1).

In case an equipment is protected by a design measure e.g., containment, explosion suppression or explosion venting /1/, the usual explosion pressure sensor with a corresponding low activation pressure can also be used to initiate the triggering mechanism for the extinguishing barrier. The suppressant agent - preferably suppressant powder - is discharged into the pipeline and forms a thick blanket which extinguishes the incipient flame. This type of barrier does not impede product throughput down the pipeline.

Since, with explosions in pipelines, the detection and quenching of the explosion flames is important, optical IR-sensors which have a relative low sensitivity to daylight are chosen for the activation of the barrier system. However daylight into the pipe in the vicinity of the sensor must be avoided.

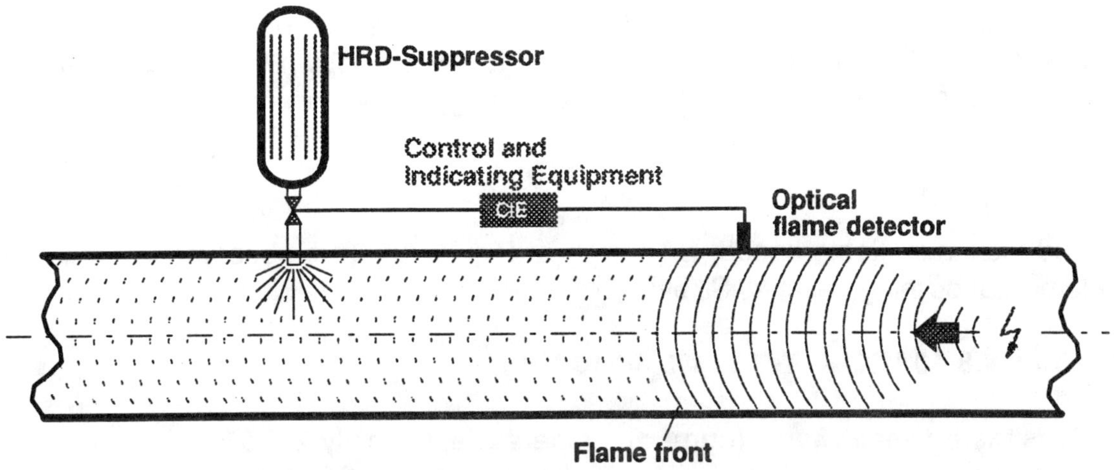

Figure 1. Automatic extinguishing barrier

Pressure detectors or pressure sensors are not suited for the case on hand because there is no distinct separation between the pressure and flame fronts for explosions in pipelines.

There is a definite distance between the installation sites of the optical detector and the extinguishing barrier to ensure that the suppressant agent acts directly on the flame. The amount of suppressant agent required depends on the nature of the combustible fuels, the nominal diameter of the protected pipeline, the flame velocity and the maximum reduced explosion overpressure in the vessel.

2 Design Parameters

2.1 Base Parameters

This ISO test method /2...4/ establishes representative worst case conditions of fuel concentration, homogeneity and turbulence for determination of the explosion indices of the fuel and describes two principle explosion indices:

- P_{max} - Maximum explosion overpressure at optimum fuel concentration.
- K_{max} - Maximum explosion constant defining the maximum rate of pressure rise with time $(dP/dt)_{max}$ of an explosion at optimum fuel concentration in a volume V, according to the Cubic Law equation:

$$K_{max} = V^{1/3} \cdot (dP/dt)_{max} \qquad (1)$$

For compact vessels, P_{max} and K_{max} are essentially independent of vessel volume, V. From their determination it is possible to quantify the maximum rate of pressure rise of a representative worst case explosion of the fuel in any compact volume, V. For most practical applications this determination of the explosion serves as a basis for assessing the requirements of explosion suppression.

The derived maximum explosion constant K_{max} is a direct measure of the explosible gas and dust fundamental adiabatic burning velocity, S_u /5/. It can be shown that for explosibility determinations in accordance with the ISO test methodology /2...4/:

$$S_u \sim 4.8 \cdot 10^{-3} \cdot K_{max}. \qquad (2)$$

The observed maximum flame speed, S_f, occurs at $P = P_0$ /5/. This observed value is related to the fundamental burning velocity, S_u by:

$$S_f = S_u \cdot T_b / T_u = 4.8 \cdot 10^{-3} \cdot K_{max} \cdot T_b / T_u = 3.84 \cdot 10^{-2} \cdot K_{max} \qquad (3)$$

Where: T_b is the adiabatic flame (burned) temperature, typically 2400 K
T_u is the initial (unburned) fuel temperature, typically 300 K.

For the example of a propane, St 1 and St 2 limit dust explosion we can determine the following base parameters of the explosion in a closed vessel (Table 1):

Table 1. Base parameters of fuel explosion in closed vessels

Base Parameters	Propane	St 1	St 2
Maximum explosion overpressure P_{max} [bar]	7.9	9	9
Maximum explosion constant K_{max} [$m \cdot bar \cdot s^{-1}$]	100	200	300
Fundamental adiabatic burning velocity S_u [$m \cdot s^{-1}$] (2)	0.48	0.96	1.44
unburned fuel temperature T_u [K]	300	300	300
adiabatic flame temperature T_b[K]	2400	2400	2400
Maximum flame speed S_f at $P = P_0$ [$m \cdot s^{-1}$] (3)	3.8	7.7	11.5

2.2 Barrier Parameters

2.2.1 Pipe nominal diameter of $150 \leq DN \leq 700$-mm

The basis for the design of extinguishing barriers is established from experimental trials /6/. Figure 2 shows the correlation between the maximum explosion constant, K_{max}, of a dust and the explosion velocity, v_{ex}, at the 10 m mark with the flame detector at the 1-m mark. The values were determined in a 20 m long pipeline having nominal diameter between DN 200 to DN 400. For larger pipe areas, the values for the explosion velocity move downwards.

The relationship between the explosion velocity, v_{ex}, and the maximum explosion constant, K_{max}, can be described according to the following empirical equation (4):

$$v_{ex} = 29 \cdot K_{max}^{0.45} \quad (4)$$

Where: v_{ex} is the explosion velocity in $m \cdot s^{-1}$.

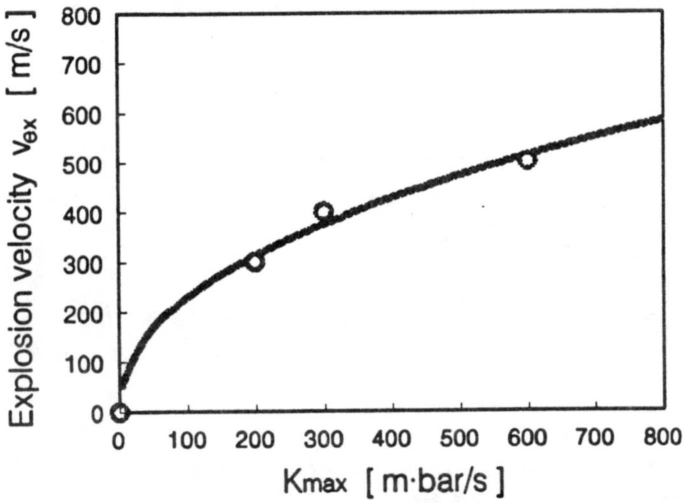

Figure 2. Correlation of maximum explosion constant, K_{max}, with explosion velocity, v_{ex}

Intensive studies /6/ have shown (Fig. 3) that the suppressant requirement at the barrier location must be adjusted linearly with the explosion velocity. Especially at high explosion velocities the HRD-Suppressors with 4 kg content are more effective than the ones with 2 kg. For a given explosion velocity (4 kg up to 150 $m \cdot s^{-1}$ and for 2 kg up to 300 $m \cdot s^{-1}$), the sectional density of suppressant, SDs is constant in order to force the disruption of the explosion, independent of the pipe diameter.

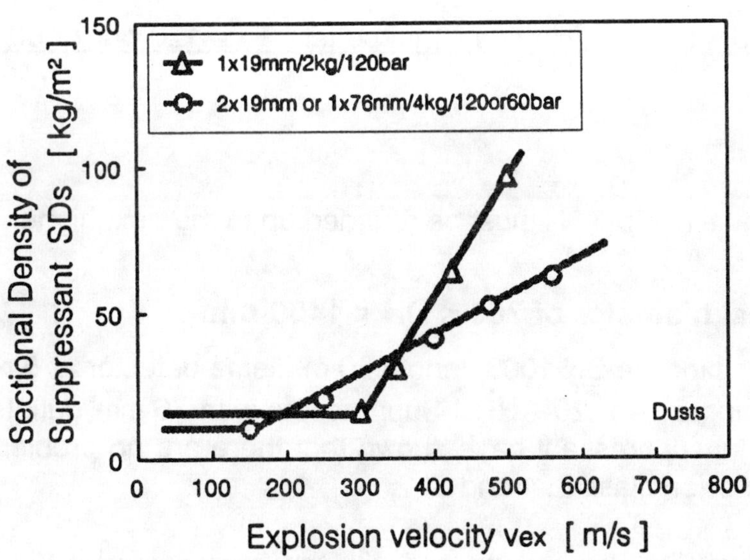

Figure 3. Correlation of explosion velocity, v_{ex}, at location of extinguishing barrier with sectional density of suppressant, SDs

The empirical equations (5 and 6) developed from Figure 3 present the experimentally derived sectional density of suppressant, SDs, needed across the duct as a function of the explosion velocity respectively K_{max}:

<u>5-ℓ-HRD-Suppressors with 1x19 mm outlet, Ps = 120 bar and 2 kg suppressant</u>

$$SDs = (0.4 \cdot v_{ex}) - 104 = (11.6 \cdot K_{max}^{0.45}) - 104 \tag{5}$$

<u>5-ℓ-HRD-Suppressors with 2x19-mm outlet or 1x76-mm outlet, P_s = 120 bar or 60 bar and 4 kg suppressant each</u>

$$SDs = (0.13 \cdot v_{ex}) - 10 = (3.77 \cdot K_{max}^{0.45}) - 10 \tag{6}$$

Thus the minimum suppressant charge, M_s, can be calculated from the empirical equation (7):

$$M_s = SDs \cdot DSA \tag{7}$$

Where: Ms is minimum suppressant charge in kg,
SDs is the sectional density of suppressant in $kg \cdot m^{-2}$,
DSA is the duct sectional area in m^2.

Based on the minimum suppressant charge, the minimum number of HRD-Suppressors, Ns, required to effectively extinguish an explosion of a defined intensity in a given pipe, can be determined from the following equations (8) and (9):

<u>5-ℓ-HRD-Suppressors with 1x19 mm outlet, Ps = 120 bar and 2 kg suppressant</u>

$$Ns = Ms : 2\ kg \tag{8}$$

<u>5-ℓ-HRD-Suppressors with 2x19-mm or 76-mm outlets, P_s = 120 bar or 60 bar and 4 kg suppressant each</u>

$$Ns = Ms : 4\ kg \tag{9}$$

Note:
The calculated minimum number Ns must be rounded up to the next higher integer value.

2.2.1 Pipe nominal diameter of 700 < DN < 1400-mm

New results /7/ from a pipeline DN 1000, length 20 m, flame detector at 5 m mark, flame barrier at 15 m mark, using 5-ℓ- or 20-ℓ-HRD-Suppressors with 76 mm outlets, P_s = 60 bar and 4 kg respectively 16 kg suppressant have shown that there are no problems in fighting dust explosions of the explosion class St 1 and St 2.

The test results shown (Fig. 4) are compared with the ones shown in Fig. 3 using 5-ℓ-HRD-Suppressors with 2x19-mm, P_s = 120 bar and 4 kg suppressant. Figure 4 correlates the explosion velocity with sectional density of suppressant in different pipelines. Again the suppressant requirement at the barrier location must be adjusted linearly with the explosion velocity. For a given explosion velocity $v_{ex} \leq 150\ m \cdot s^{-1}$ for the small pipe and $v_{ex} \leq 200\ m \cdot s^{-1}$ for the larger pipe, the sectional density of suppressant is constant.

The empirical equations (10) developed from Figure 4 present the experimentally derived sectional density of suppressant, SDs, needed across the duct as a function of the explosion velocity respectively K_{max} if 5-ℓ-HRD-Suppressors with 1x76-mm outlet, Ps = 60 bar and 4 kg suppressant or 20-ℓ-HRD-Suppressors with 1x76-mm outlet, Ps = 60 bar and 16 kg suppressant are used:

$$SDs = 0.08 \cdot v_{ex} - 3 = 0.08 \cdot (29 \cdot K_{max}^{0.45}) - 3 \quad (10)$$

The minimum suppressant charge, Ms, can be calculated from the empirical equation (7).

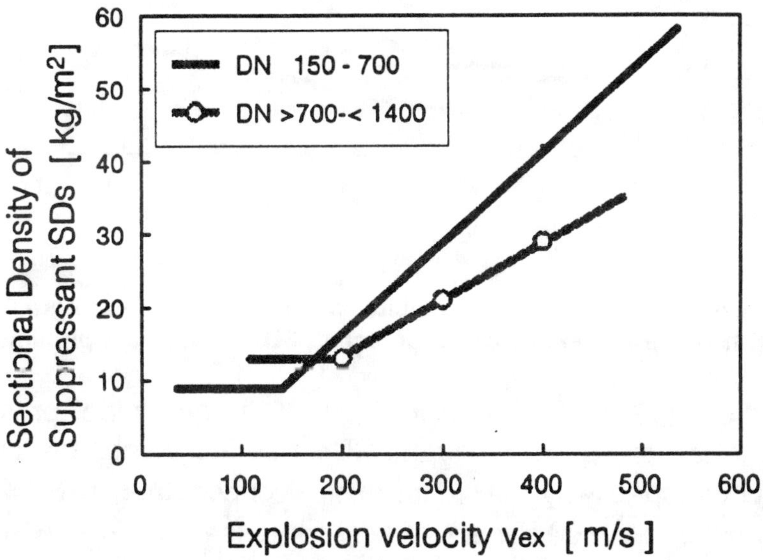

Figure 4. Correlation of explosion velocity, v_{ex}, at location of extinguishing barrier with sectional density of suppressant, SD_s in different pipelines

Based on the minimum suppressant charge, the minimum number of HRD-Suppressors, Ns, required to effectively suppress an explosion of a defined intensity in a given pipe, can be determined from the following equations (9) for 5-ℓ-HRD-Suppressors and (11) for :

<u>20-ℓ-HRD-Suppressors with 1x76-mm outlet, Ps = 60 bar and 16 kg suppressant:</u>

$$Ns = Ms : 16\ kg \quad (11)$$

Note:
The calculated minimum number Ns <u>must</u> be rounded up to the next higher integer value.

2.2.3 Pipe nominal diameter of 1400 ≤ DN ≤ 2500-mm

The basis for the design of extinguishing barriers for large pipelines is also established from experimental tests /6/.

Figure 5 correlates the explosion velocity with sectional density of suppressant in different pipelines. Again the suppressant requirement at the barrier location must be adjusted linearly with the explosion velocity. For given explosion velocities in the range of $v_{ex} \leq 150\ m \cdot s^{-1}$ to

$v_{ex} \leq 200\ m \cdot s^{-1}$ the sectional density of suppressant is constant.

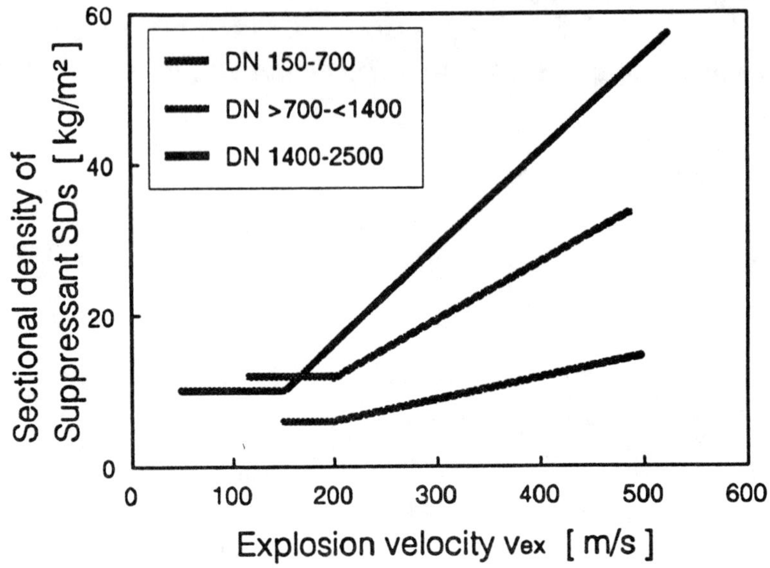

Figure 5. Correlation of explosion velocity, v_{ex}, at location of extinguishing barrier with sectional density of suppressant, SDs, in different pipelines

The empirical equations (12) developed from Figure 5 (largest pipe) present the experimentally derived sectional density of suppressant, SDs, needed across the duct as a function of the explosion velocity respectively maximum explosion constant, K_{max}, if 5-ℓ-HRD-Suppressors with 1x76 mm outlet, Ps = 60 bar and 4 kg suppressant or 20-ℓ-HRD-Suppressors with 1x76 mm outlet, Ps = 60 bar and 16 kg suppressant are used:

$$\mathbf{SDs = 0.03 \cdot v_{ex}} \qquad \textbf{(12)}$$

Again, the minimum suppressant charge, Ms, can be calculated from the empirical equation (7).

Based on the minimum suppressant charge the minimum number of HRD-Suppressors, Ns, required to effectively extinguish an explosion of a defined intensity in a large pipe, can be determined from the following equations (9) and (11).

2.3 Design Parameters

Having determined the minimum suppressant charge, Ms that is to be deployed across the duct section or the minimum number of HRD-Suppressors, Ns, it is necessary to locate the barrier with respect to the expected point of ignition. Extinguishing barrier location is determined from the flame velocity, v_f, and not from the worst case accelerated explosion velocity, v_{ex}.

The flame velocity, v_f, is the sum of the contributing velocity elements S_f (flame speed derived from K_{max}) and v_p (velocity due to pressure piling) and can be calculated according to the following equation (13):

$$v_f = S_f + v_p(t) \tag{13}$$

Using the equation (3) for the flame velocity, then the equation (13) changes to equation (14):

$$v_f = 3.84 \cdot 10^{-2} \cdot K_{max} + v_p(t) \tag{14}$$

Where: v_f is the flame velocity in $m \cdot s^{-1}$,
K_{max} is the maximum explosion constant in $m \cdot bar \cdot s^{-1}$.

Taking into account the forward air velocity, v_{air}, - if existing in the practice - the equations (13) change to equation (15):

$$v_f = S_f + v_p(t) + v_{air} \tag{15}$$

Where: v_{air} is the forward air velocity in $m \cdot s^{-1}$,
v_{air} is 0 $m \cdot s^{-1}$ for an upstream placement of the extinguishing barrier.

For an explosion pressure detector we can describe an effective explosion detection pressure P_a, and corresponding time to detect an explosion, t_a. If a pressure detector is mounted on the vessel and is the selected means of triggering, then for safe system design it should be assumed that the explosion is ignited at worse case location for flame propagation down the pipeline i.e., at the inlet port of the interconnecting duct. At the point of detection, P_a, the flame can be assumed to have propagated down the pipeline a distance d_a where:

$$d_a = v(t_a) \cdot t_a = (S_f + v_{air} + \int_{P_o}^{P_a} v_p \cdot dP \; (P = f(t))) \cdot t_a \tag{16}$$

If detection is by a flame detector mounted on the interconnecting pipeline at the inlet, then for safe system design it should be assumed that the explosion is ignited at the worse case location for explosion detection, i.e., on the opposite wall at a distance d_z from the inlet part. Flame detection will occur, provided that the detector response time is fast relative to the event, only when the flame reaches the inlet port. The corresponding explosion overpressure can be estimated from the projected explosion/time characteristic using an estimate of t_a:

$$t_a \sim d_z / S_f \tag{17}$$

For the estimation of the minimum distance, d_{min}, of the extinguishing barrier from the flame sensor installed in a pipe, the extinguishing barrier establishment time, t_b, has to be known. The time t_b is the time between the activation of the fast acting valve of the HRD-Suppressor and a blanket of suppressant is formed in the pipe. This time includes also the electronical delay time of the used sensor and the control and indication equipment. From a knowledge of this extinguishing barrier establishment time, t_b, the mentioned minimum distance of the extinguishing barrier from the flame sensor installed in a pipe can be determined as:

$$d_{min} = v(P) \cdot t_b = \tag{18}$$

Where: d_{min} is the minimum distance in m,
t_b is the extinguishing barrier establishment time in s.

The extinguishing barrier establishment time is depending upon the type of HRD-Suppressors and lies normally between t_b = 0.025 s and 0.1 s.

If a pressure sensor is installed in a equipment and is the selected means of triggering, then the minimum distance, d_{min} from the extinguishing barrier to the connection point equipment/pipe can be estimated as follows:

$$d_{min} = v(P) \cdot t_b + d_a \tag{19}$$

Where: d_a is the distance the flame has propagated down the pipeline in m.

The determination of v(P) demands a solution to the integral equation:

$$v(P) = S_f + v_{air} + \int_{P_a}^{P_{max}} v_P \cdot dP \; (P = f(t)) \tag{20}$$

Note that where the equipment is fitted with an explosion protection system (suppression or venting) P_{max} becomes $P_{red,max}$.

v(P) is a complex function that becomes sonic velocity limited as flame velocity approaches the sonic velocity limit in the medium. The density of the medium within the pipeline does not remain invariant through the explosion event. Meaningful determination of v(P) has no simple solution. It demands a complex finite element treatment, with the final determination being very specific to a practical circumstance.

In difficult situation, it is recommended to install both sensors - pressure and flame - and they must be switched in a OR-logic to activate the extinguishing barrier (Fig. 6). In this case the minimum distance from the extinguishing barrier to the connection point equipment/pipe must be estimated from the equation (19).

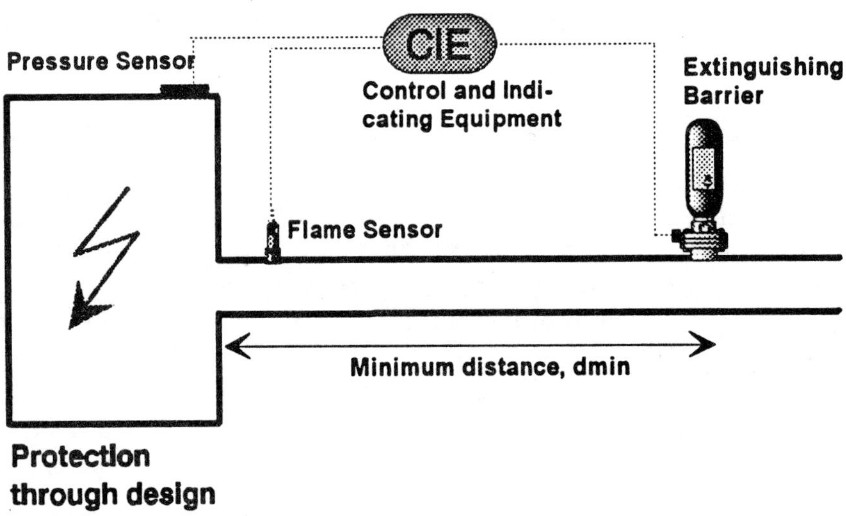

Figure 6. Principle of an extinguishing barrier using both, pressure and flame sensors

The pressure sensor provides the earliest detection of an explosion in the interconnected vessel, whereas the flame sensor provides assured detection, even for lazy flames, as they propagate down the pipeline towards the extinguishing barrier.

3 Practical Experience

The flame velocity, v_f, is a very important value for the design of extinguishing barriers (see section 2.3). The distance from an extinguishing barrier to the connection point equipment/pipe is typically in the range of 5 to 10 meters. Therefore, for safety reasons, the flame velocity at the 10 m mark must be known for the design of the extinguishing barriers.

Examples for the combination vessel-pipeline are:

- Combination of storage, mixing and process vessels with pipelines,
- Silos, milling and drying devices with downstream dust collectors,
- Local and general dust collection.

In the following for different type of test arrangement the measured flame velocity v_f, derived from experiments are reported.

Vogl /8/ investigated the explosion behavior in pipelines of DN 100 and DN 200 using the pneumatic conveying system and different K_{max} values. For the worst case scenario, that is: optimum dust concentration, location of ignition sources at the center of the pipe, conveying speed 30 m·s^{-1} and DN 200, the flame velocity at 10-m mark of the pipe can be calculated as follows:

$$v_f = 7.3 \cdot K_{max}^{2/3} \quad (21)$$

The correlation of maximum explosion constant with flame velocity in a pipeline of DN 200 at the 10 m mark is shown in Figure 7.

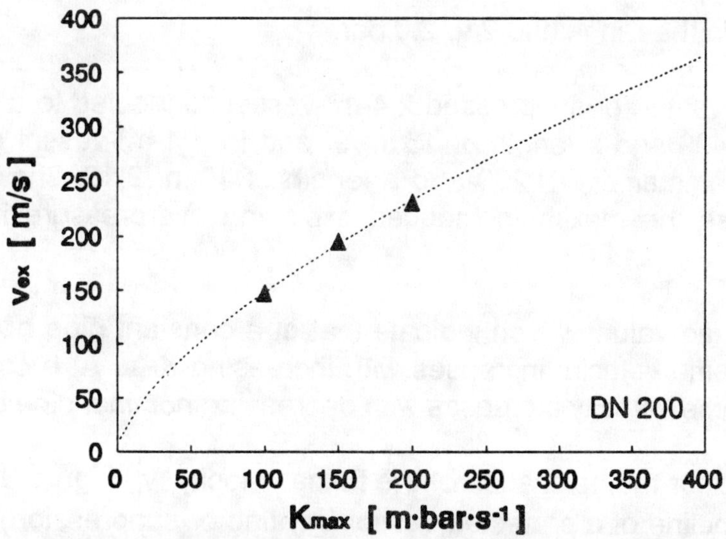

Figure 7. Correlation of maximum explosion constant, K_{max}, with flame velocity, v_f - DN 200, conveying speed maximum 30 m·s^{-1}-

Vogl /8/ investigated also the explosion transfer from a 9.4-m³-vessel into a pipeline of DN 100 and 200 using the pneumatic conveying system. He created different maximum reduced explosion overpressure by using the explosion venting technology with different vent areas. During these experiments the maximum reduced explosion overpressure was in the range of $P_{red,max}$ = 0.2 to 2.0 bar.

The correlation of the maximum reduced explosion overpressure with flame velocity in a pipeline of DN 200 (worst case) again at the 10-m mark using a dust belonging to the dust explosion class St 1 is shown in Figure 8.

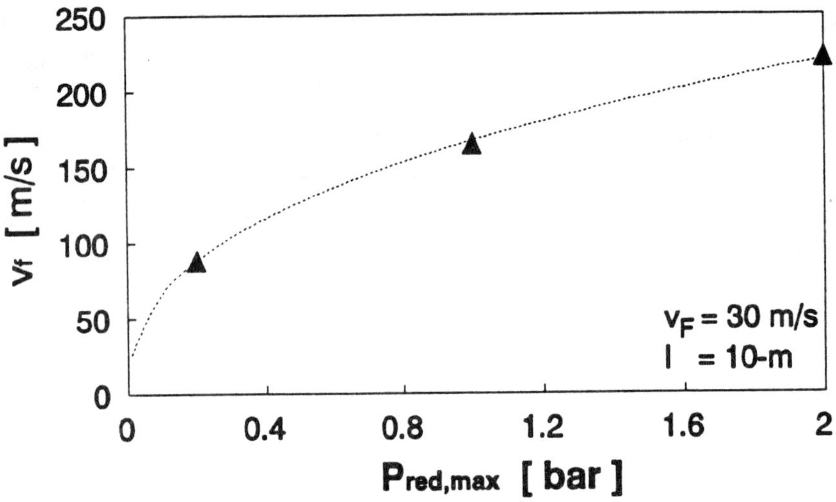

Figure 8. Correlation of maximum explosion overpressure, $P_{red,max}$, with flame velocity, v_f - St 1 Dust, DN 200, maximum conveying speed 30 m·s^{-1} -

In this case the flame velocity at 10 m mark of the pipe for different maximum reduced explosion overpressure can be calculated as follows:

$$v_f = 163 \cdot P_{red,max}^{0.4} \qquad (22)$$

Where: $P_{red,max}$ is in the range of 0.2 to 2.0 bar.

Explosion tests in a vented or suppressed 2.4-m³-vessel connected to a pipe having a nominal diameter of DN 400 and a length of 15 m /9/ and in a 1-m³-vessel connected to a pipe having a nominal diameter of DN 200 and a length of 10 m /9,10/ show the relationship in Figure 9. In this case the maximum reduced explosion overpressure in both vessels was $P_{red,max} \leq 2$ bar.

The data are estimated values which indicate that at a constant pipe nominal diameter, DN, the values for the flame velocity increases with increasing K_{max}. At a constant value of K_{max} the values for the flame velocity increases with decreasing nominal pipe diameter.

As an orientation aid for the estimation of the flame velocity, v_f, Figure 9 can be used. At the 10-m mark of the pipeline of a protected vessel (venting or suppression) with a $P_{red,max}$ equal to or less than 2 bar overpressure, the flame velocity, v_f, can be estimated according to the following equation (23):

$v_f = \{(0.42 - 7.5 \cdot 10^{-4} \cdot DN) \cdot K_{max}\} + 60$ (23)

Where: $P_{red,max}$ in the protected vessel is equal to or less than 2 bar overpressure.

Note:
For DN < 200 the estimated flame velocity for DN 200 and for DN > 400 the one of the DN 400 must be used.

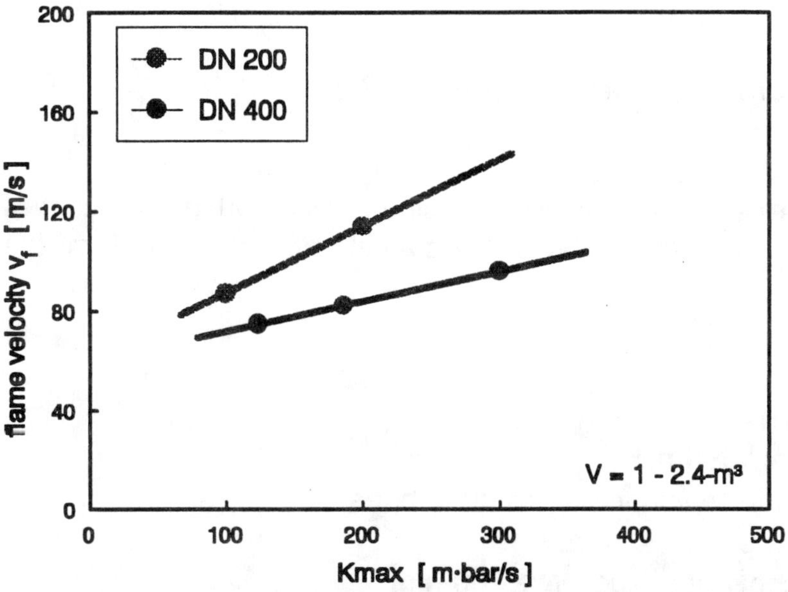

Figure 9. Correlation of nominal diameter, DN, of a pipe and the maximum explosion constant, K_{max}, with flame velocity, v_f,

In this case, the diameter of the pipe must correspond to a vent area which results in a maximum reduced explosion overpressure of $P_{red,max}$ = 2 bar, when a St 2 dust (P_{max} = 9 bar, K_{max} = 300 m·bar·s-1) is vented through this area and the static activation overpressure of the bursting disk is P_{stat} = 0.1 bar. In this case the minimum nominal diameter DN of the pipe can be calculated from the area A as follows:

A = 0.06 · V0.753
(24)

4 Guideline for the Practice

Practical application of explosion protection by an explosion extinguishing barrier remains more dependent on the interpolation and extrapolation of data from test results than it does on the basis analysis of the barrier distance requirement. This fact is a consequence of the complexity in deriving a solution for the equation (20) for v(P). The following equations for the estimation of the flame velocity, v_f, are recommended. In case of difficulties in applying the equations, experts may be contacted for advice.

Vessels, Silos combined with a pneumatic conveying system:
Volume of vessel, silo V < 9-m³:

For pneumatic conveying system with pipelines DN ≥ 100 and a conveying speed up to 30 m·s-1 in vessels, silos having a volume smaller than 9-m³, the flame velocity, v_f can be calculated as:

$$v_f = 7.3 \cdot K_{max}^{2/3} \tag{21}$$

This equation is valid for:
Volume of the vessel V < 9-m³,
- conveying speed up to vF ≤ 30 m·s-1,
- diameter of the conveying pipe DN ≥ 100-mm.

Volume of vessel, silo V ≥ 9-m³:

For pneumatic conveying system with pipelines DN ≥ 100 and a conveying speed up to 30 m·s-1 in vessels, silos having a volume equal to or greater than 9-m³ the flame velocity, v_f can be calculated as:

$$v_f = 163 \cdot P_{red,max}^{0.4} \tag{22}$$

This equation is valid for:
Volume of the vessel V ≥ 9-m³,
- maximum explosion constant K_{max} ≤ 200 m·bar·s-1,
- conveying speed up to vF ≤ 30 m·s-1,
- diameter of the conveying pipe DN ≥ 100-mm,
- maximum reduced explosion overpressure $P_{red,max}$ ≤ 2.0 bar.

Vessels, silos combined with pipelines:

For vessels protected with explosion venting or explosion suppression, connected with a pipe with nominal diameter, DN, corresponds to a minimum area of A = 0.06·V0.753 the flame velocity, v_f, including the forward air velocity, v_{air}, - if existing in the practice - can be calculated as:

$$v_f = \{(0.42 - 7.5 \cdot 10^{-4} \cdot DN) \cdot K_{max}\} + 60 + v_{air} \tag{23}$$

Note:
For DN < 200 the estimated flame velocity for DN 200 and
for DN > 400 the estimated flame velocity for DN 400 must be used.
Never use a negative value for v_{air}.

This equation is valid for:
nominal diameter DN of a pipe that corresponds to a minimum area of A = 0.06·V0.753,
- maximum reduced explosion overpressure $P_{red,max}$ ≤ 2.0 bar.

REFERENCES

/ 1 / Siwek, R.: Dusts: Explosion Protection, 7th Edition of Perry's Chemical Handbook for Chemical Engineering, USA, 1997

/ 2 / ISO-Standard: International Organization for Standardization: Explosion Protection Systems - Part 1: Determination of Explosion Indices of Combustible Dusts in Air, ISO 6184/1, 1985

/ 3 / ISO-Standard: International Organization for Standardization: Explosion Protection Systems - Part 2: Determination of Explosion Indices of Combustible Gases in Air, ISO 6184/2, 1985

/ 4 / ISO-Standard: International Organization for Standardization: Explosion Protection Systems - Part 3: Determination of Explosion Indices of Fuel/Air Mixtures other than Dust/Air and Gas/Air Mixtures, ISO 6184/3, 1985

/ 5 / Moore, P.E., Siwek, R.: Triggering Barrier Explosion Isolation Procedures, VDI-Berichte 1272. VDI-Verlag GmbH, Düsseldorf 1996

/ 6 / Bartknecht, W.: Explosionsschutz - Grundlagen und Anwendung, Springer Verlag Berlin, Heidelberg, New York, 1993.

/ 7 / Siwek, R.: Test with extinguishing barriers in a 1000-mm diameter pipe, Explosion Technology, Ciba-Geigy Ltd., Basle, 1995

/ 8 / Vogl, A.: Explosionsübertragung aus Behältern in Rohrleitungen pneumatischer Anlagen, VDI-Berichte 1272. VDI-Verlag GmbH, Düsseldorf 1996

/ 9 / Siwek, R.: A Review of Explosion Isolating Techniques, Proceedings of the EuropEx 2nd Word Seminar, Gent - Belgium, 1996

/10/ Bartknecht, W.: Prüfung des Ansprech- und Schliessverhaltens eines explosionsdruck- und fremdbetätigten Ventex-ESI-Ventils DN 200, Bericht 2/92, BGN, Mannheim, 1992

31st LOSS PREVENTION SYMPOSIUM

Session 41. Pressure Relief Systems

Tuesday, March 11, 1997
1:30 - 5:15 PM
Grand Ballroom C

John Noronha, Chair
Eastman Kodak CO, Rochester, NY
Bldg. 23, 5th Floor
Kodak Park Site
Rochester, NY 14652-4390

Richard E. Munson, Vice-Chair
E.I. du Pont de Nemours & CO,
N2526-2
1007 Market St.
Wilmington, DE 19898

Session Introduction - J. Noronha

41a. **Batch Runaway Reaction Relief: Re-Evaluation of "Credible" Scenario,** J. Singh; Hazard Evaluation Lab, Inc., Monmouth Junction, NJ

41b. **Why DIERS Should be used in Risk Assessment - Call for BenchMarking,** J. Noronha, Eastman Kodak CO, Rochester NY

41c. **Pressure Relief System Documentation: Equipment Based Relational Database is Key to OSHA 1910.119 Compliance,** P. Berwanger; R. Kreder; Berwanger, Inc., Houston, TX

BREAK

41d. **An Overview of SuperChems for DIERS for Emergency Relief Systems and Vent Containment Design,** G. Melhem, A.D. Little, Inc., Cambridge, MA; H. Fisher, Union Carbide Corp., South Charleston, WV

41e. **Emergency Relief Adequacy for Acrylic Polymerization Processes,** A. Balchan; D. Paquet Jr., DuPont Experimental Station, Wilmington, DE; J. Klein, DuPont Marshall Lab, Philadelphia, PA

41f. **QRA of Chemical Reaction Systems - The State of the Practice,** J. Farquharson; S. McNutt; H. Paula; M. Roberts, JBF Associates, Inc., Knoxville, TN; A. Waller, ARCO Chemical CO, Newton Square, PA

Concluding Remarks by R.E. Munson

BATCH RUNAWAY REACTION RELIEF: RE-EVALUATION OF "CREDIBLE" SCENARIO

Jasbir Singh
Hazard Evaluation Laboratories, Inc.
7 Deer Park Drive, Suite M
Monmouth Junction, NJ 08852-1921

Many chemical reactions in the fine chemical and related industries are performed in a semi-batch manner under isothermal conditions. Relief system design evaluation for runaway reactions, using DIERS methodology for example, is frequently studied adiabatically using data derived from batch mode testing. In many cases, however, evaluation of the process isothermally under the realistic semi-batch mode can be extremely useful, often helping to greatly reduce the vent size as a result of a better understanding of the process. In some cases, it is even possible to eliminate runaway reaction as a viable case for relief sizing. Such evaluations of venting are fully consistent with DIERS methodology.

This paper will present the range of process variables that can influence the vent size and specifically highlight the role that isothermal calorimetry can play in this determination. Methodology for vent sizing using relevant isothermal and adiabatic data will be presented to highlight the main points, using several examples.

1. INTRODUCTION

Relief sizing methodology for batch-type chemical reactors has evolved considerably over the last 10 years, particularly where exothermic chemical reactions are involved, and much of the credit for this goes to the AIChE'S DIERS group (ref. 1). This technology improvement has not entirely been welcomed by the chemical industry, due to the fact that its application often leads to vent sizes that are difficult to accommodate, particularly on existing vessels. The widespread review of relief sizing, promoted by OSHA regulations, has only served to highlight the problems.

At least part of the difficulty is caused by lack of appreciation of the impact on relief sizing of reaction kinetics, and the precise way in which modern plants are operated. Thus, while many processes are run in a semi-batch manner, relief vents continue to be sized for the most pessimistic case of a "full-charge" runaway.

The reality is that a combined knowledge of the chemistry and the process controls/interlocks that are in place can help to justify venting scenarios that are far less onerous than the "full-charge" assumption. The vent size needed for a particular process and reactor vessel is frequently regarded as a fixed value, when in fact, it depends entirely on the definition of 'credible' worst case scenarios. By suitable modification of the chemistry and/or improvements in process control, the incidents that are 'credible' can (within limits) be adjusted to suit venting arrangements that are feasible.

The objective of this paper is to present characteristics of the chemistry that influence the above approach to vent sizing and indicate how aspects of process control can directly (and objectively) help to define an acceptable 'credible' worst case, which is much less severe than the actual worst case.

2. SEMI-BATCH CHEMISTRY

Controlled dosing of a key chemical into a stirred vessel has the well recognized advantage of minimizing the amount of energy available for a potential exothermic runaway. The extent of safety improvement resulting from semi-batch operation depends on the rate of reactant addition on the one hand, and consumption by chemical reaction on the other. It is important to recognize how these two rates compare and, therefore, how much (if any) unreacted dosed material is accumulating in the reactor vessel.

Consider, for example, the esterification reaction between a methanol and acetic anhydride. This has been studied extensively over the years, and is known to occur relatively slowly at modest temperatures, but can accelerate rapidly if the temperature is allowed to increase. At 40°C for example, if anhydride is dosed to methanol, considerable accumulation of unreacted anhydride can occur. The results from an isothermal calorimetry experiment in which acetic anhydride is dosed to methanol over ~100 minutes are shown in Figure 1. The following information is present in Figure 1:

- Feed: this is the accumulated amount of acetic anhydride dosed to methanol.
- Power: this is the instantaneous heat generation rate as the reaction proceeds.
- Total Energy: this is integration of power, and is essentially the total heat released at any time.

The data is derived from a 1 liter stirred reactor vessel, which is held at 40°C throughout the experiment. The heat output increases as the dosing proceeds, and then decays after feed has stopped. The total heat obtained after complete dosing is more than 70% of the total, showing that much of the anhydride in the vessel reacted out slowly over time.

At higher temperatures, the rate of anhydride consumption is much greater, and, therefore, the accumulation much reduced. Another way of reducing the accumulation is to catalyze the synthesis; for this reaction it is easily achieved with a few drops of sulphuric acid. The heat output from a catalyzed synthesis is shown in Figure 2. This is also performed at 40°C, and it is clear that the "accumulation" is very small (~ 30%) compared with the uncatalysed data (over 70%).

Another reaction that is interesting is presented in Figure 3. This is an oligomerization reaction to which caustic solution is added over four hours. In Figure 3, the heat output is plotted as a function of time, while reactor temperature is held constant at 25°C. It is clear that no heat output is observed for around 60 minutes, after which there is a sudden burst of energy. This is then followed by a period of almost constant heat output.

The sudden heat output coincides with an observed dramatic mixing of the reactants which were previously in two separate phases (though vigorously stirred). The mixing is induced by formation of product; until there is enough product, the reaction rate is mass transfer limited and very slow, but after a certain point, it is kinetic-controlled and very fast.

Commercially, this reaction is run by recycling a small amount of product which 'catalyses' the mixing, and, therefore, produces a smooth heat output profile (see Figure 4). There is no accumulation of reactant (i.e., the power falls off rapidly as feed is stopped) and, therefore, no sudden output of heat in the 'catalyzed' operation.

It is not always necessary to use a catalyst in order to obtain the energy output curve in Figure 4. An example of an industrial synthesis reaction where this occurs naturally is shown in Figure 5. The power output at 75 and 90°C is plotted while one of the components is dosed over 160 minutes. The "square-shaped" heat output trace is characteristic of fast kinetic reactions (due to a sudden increase in heat output as feed is commenced, and an equally dramatic fall when feed is stopped) and confirms that accumulation is negligible.

The heat output data presented in Figures 1 to 5 was obtained from the SIMULAR reaction calorimeter. This is essentially a 1 liter bench-scale computer controlled reactor with special facilities to monitor heat output at the same time. The basic instrument layout is shown described in Appendix I (ref. 2).

The heat output rate from semi-batch reactions can conveniently be described as being either kinetic-controlled or dose-controlled. The uncatalyzed esterification reaction is an example of the former: heat output rate is limited by the rate of reaction (i.e., kinetics), and not by rate at which the dosed component is being added. If the same reaction is performed at a higher temperature, the heat output is increased, but if the temperature is held constant and the feed rate significantly altered, the heat output will be only marginally influenced.

The synthesis reaction (Figure 5) and the 'catalyzed' oligomerization (Figure 4) are both dose-controlled; if the feed rate is doubled, the heat output rate will increase in proportion. Changing the operating temperature by moderate amounts (thus altering the kinetics) will have little effect on the heat output rate - this is clear from the synthesis reaction data in Figure 5, where a change of 15°C in operating temperature is shown to have no effect on the power output.

3. DEFINITION OF RELIEF SIZING CASE

The above three examples represent a reasonable cross-section of semi-batch reaction types. Depending on the degree of automation, reliability of equipment, maintenance standards and the expertise of operators, it is possible to define the most "credible" basis for relief sizing, for each reaction type.

For a kinetic-controlled reaction, where accumulation of reactant cannot be avoided, the most reasonable assumption is to size the vent at least to cope with the runaway that would occur from the full amount of accumulated reactant. Thus, for the uncatalyzed esterification example, this would imply that ~ 75% of the batch anhydride is mixed with the full amount of methanol and that this mixture runs away exothermically.

For dose-controlled reactions, a case can be made for total exclusion of a true runaway, by arguing that reactant accumulation is prevented by instrumentation and control. This would leave only the vapor generated by the "steady-state" heat output rate to be accommodated by the vent. This will normally be much less onerous (by an order of magnitude, or more) than even a moderate exothermic runaway.

It should be recognized that the basis for relief sizing described above for dose-controlled reactions does not represent the absolute worst case. Consider for example the following scenarios:

- Agitation is temporarily ineffective, or deliberately turned off, while dosing is maintained resulting in negligible or no reaction. Then, at a later stage, mixing is resumed (or the accumulation of reactant produces sufficient reaction and then mixing) causing a sudden batch-type runaway.

- Dosing is started when the reactor is at a substantially lower temperature than normal (for example, because the temperature sensor is faulty), leading to negligible or no reaction. After some considerable delay, either the reactor is warmed up to the correct temperature, or the reactant concentration builds up to a point where significant reaction occurs even at the low temperature; in either case, accumulation followed by a runaway can result.

The above possibilities (and others) have led to industrial accidents, and are therefore quite credible. However, in a modern well-instrumented plant with proper interlocks and good operator training, these can be rendered "remote," and justifiably eliminated as a basis for vent sizing.

Note that the 'catalyzed' oligomerization and the synthesis reaction are both close to being "dose-controlled," and vent sizing could be based on steady-state heat output alone. It should be noted, however, that this basis for relief sizing is more easily justified for the organic synthesis system than the catalyzed reaction. This is because in the case of the latter, it requires an important additional step in order to realize the dose-controlled chemistry, namely, the addition of a catalyst. If the catalyst is not added and then dosing commenced, it would behave just like a kinetic-controlled reaction with reactant accumulation. Worst still, imagine the consequences if the operator now remembers that the catalyst has been omitted and adds it at this later point!

4. QUANTIFICATION OF DIFFERENT SCENARIOS

When the true worst case runaway scenario presents problems in terms of the engineering or operating feasibility (which happens frequently), then it is necessary to take advantage of the particular characteristics of the process in order to move to a less severe runaway. However, such a move should be carefully considered and properly justified.

Consider as an example the oligomerization process and vent sizing for this chemistry. At least four separate scenarios can be presented:

- **Case 1:** steady-state heat output only, with no accumulation (hence, no runaway)
- **Case 2:** runaway represented by a failure to recycle the product (i.e., add catalyst), and, therefore, accumulation of reactant until sufficient product is formed to induce sudden reaction.
- **Case 3:** full batch accumulation due to a failure to add catalyst, the assumption being that the sudden 'kick' due to some product formation does not occur until virtually all the charge has been added.
- **Case 4:** full batch accumulation, followed by catalyst addition, (i.e., the same situation as above, but the operator adds the recycle product before the reaction self-catalyses).

These four situations are all possible and have different consequences in terms of the rise in pressure and temperature and the resulting vent size. The temperature rise for each case, following a process mishap, is plotted in Figure 6. The following points are noteworthy:

a) the steady-state heat output (Case 1) leads to a relatively slow and linear rise in temperature. The rate is governed by the feed rate. The temperature rise will continue until the vapor pressure reaches the rupture disc opening pressure.

b) the experimental data for the second case is obtained by mixing the appropriate quantities of the reactant in an adiabatic calorimeter, and allowing the reaction to runaway. Only 35% of the dosed component was added, representing the assumed accumulation.

c) Case 3 is the same as the one above, but the full batch amount of chemicals is mixed instead of only 35% of the batch quantity.

d) the worst situation (Case 4) is represented by mixing the components as in Case 3, but a small amount of product (i.e., catalyst) is added at the same time to induce immediate reaction.

The last three scenarios represent exothermic runaways of differing severity, and the information was generated using a low thermal capacity adiabatic calorimeter (PHI-TEC II). This is a device with a reactant test cell of ~110 ml volume and one of the main applications is accurate experimental simulation of a thermal runaway. The low thermal capacity test cell ensures that the rates of pressure and temperature rise measured accurately reflect a large-scale plant; see Appendix II (ref. 3)

The rate of temperature rise gives the most informative indication of the severity of a reaction and the information from Figure 6 is re-plotted in Figure 7 in the form of rate of temperature rise (dT/dt) as a function of temperature. It is clear that, as expected, hazard increases as you move from Case 1 to Case 4, and this will be directly reflected in the vent size needed to protect a reactor vessel.

The vent sizes based on a reactant charge of 10,000 kg in a $12m^3$ vessel, with relief set pressures of 0.7 barg and higher are presented in Figure 8. This compares the vent diameter needed for the different situations discussed above, and shows how the venting requirements change with the heat output rate. (The calculations were performed using the DIERS short-cut equatons for vapor-pressure systems with homogenous two-phase flow, ref. 1.)

5. VENT SIZING FOR KINETIC CONTROLLED REACTIONS

The relationship between the four scenarios discussed above, and the equipment and/or operator errors necessary to realize each case, are summarized in Table 1. A case '0' has been included in Table 1 to represent the situations where no venting is required. As you move down to Case 4, the number of implied equipment failures increases, as does the associated vent size. (The list of failures in Table 1 is not intended to be exhaustive, but simply represents a comparison between the different cases and is included only to illustrate important features.)

As the number of failures necessary to realize a particular scenario increases, the chance of realizing that scenario decreases. Thus, in Table 1, Case 0 is much more likely than Case 4. The precise frequency can be calculated by performing a detailed analysis of the combination of events necessary to achieve each scenario and then mathematically evaluating the probability of this combination. The results depend on the trips, monitors, alarms, etc. which are in place and is, therefore, quite specific to each plant.

For a hypothetical plant, the relationship between vent size and event frequency is plotted in Figure 9. This shows that for low frequency (i.e., 'remote') events, a large vent size is required, while for the more frequent events, only a small vent is needed. Thus, for the hypothetical plant being considered, vent sizes larger than 2 inches are only required for events that have a frequency lower than 0.001 per year.

If the reactor in question already has a vent, one way to proceed would be to use Figure 9 to determine the frequency of failures that are not being accommodated, and decide whether this is acceptable. A more informed method of reaching a decision is to extend the frequency calculations and translate the numbers in Figure 9 into a fatality risk (say, to operators). Thus, a high vent area would imply a low risk of a fatal incident involving workers at the plant. The advantage of doing this is that calculated figures can then be compared with generally accepted values for fatality risk in the chemical industry.

As an example, consider a vessel fitted with a 2 inch rupture disk, corresponding to failure scenario 1. The objective would be to determine the fatality risk when failure scenarios 2, 3 and 4 (plus others) take place, all of which cannot be properly accommodated by the vent on the reactor. The actual relationship between vent size and fatality risk for this hypothetical case is plotted in Figure 10. A risk value that might be accepted is ~$5x10^{-5}$; thus, in this case it appears that the vent size should be increased to ≈33". (Another way to interpret this result is to say that the definition of 'credibile' should be extended to scenario 4, instead of just 1.).

Instead of simply increasing the vent size, it is possible to take two other approaches:

a) accept the small vent size, realizing that you are operating at a risk level somewhat higher than normally expected, or preferably

b) make control/instrumentation modifications which will improve the reliability, and thus reduce the fatality risk. For example, the addition of redundant monitors or duplication of trips might make a sufficient improvement to the risk that the existing (2 inch) vent is acceptable. Thus, with improved instrumentation, scenario 1 would become an acceptable basis for vent sizing and scenarios 2 to 4 would be non-credible.

In Figure 10, therefore, two lines are drawn, one corresponding to "basic" plant where the 2" rupture disk would be unacceptable, and an "enhanced" plant, where the same rupture disk would be justified.

6. CONCLUSIONS

Vent sizing for exothermic runaway reactions involving batch reactors requires the use of DIERS methodology and frequently involves the release of a two-phase mixture of gas (or vapor) plus liquid. The flow rate to be vented (and hence, the necessary vent size) depends on the failure scenario that is assumed to cause the runaway reaction. Over-simplification of the venting scenario frequently leads to impractically large vents, but there is little guidance on how the situation can be improved without unacceptable compromise of safety.

It is suggested in this paper that one approach to this problem is to look more carefully at the normal process operation, both in terms of the chemistry involved and the way the plant is engineered. Since many processes are operated in a semi-batch manner where one of the key reactants is dosed over a period of time, it is possible to justifiably reduce the severity of the exotherm for which the vent is sized. Thus, for a reaction involving fast kinetics, it may be possible to size a vent only for the steady-state heat output, or even no vent at all.

In adopting this approach, however, it is necessary to consider the likelihood of the more severe events that are being ignored (i.e., not regarded as credible). Ideally, this should be done through quantitative assessment of the risk, taking full account of all operator interactions and alarms, trips, etc. in place. This level of detail, however, is not essential, and in many cases a thorough, but methodical review, may be adequate.

REFERENCES

1. H.G. Fisher, et. al., Emergency Relief System Design Using DIERS Technology, AIChE, New York, 1992.

2. J. Singh, Reaction Calorimetry for Process Development: Recent Advances, *Process Safety Progress*, Vol. 16, No. 2, 1997 (to be published).

3. J. Singh, Reliable Scale-up of Thermal Hazards Data Using the PHI-TEC II Calorimeter, *Thermochimica Acta*, 226 (1993) 211-220.

Runreac.wpd - article

TABLE 1

COMPARISON OF VENT SIZING SCENARIOS

RELIEF CASE	COOLING FAILURE	FEED CONTINUES	INITIALLY NO CATALYST	FULL CHARGE BEFORE REACTION	CATALYST ADDED LATE	VENT DIAMETER (NOTE 1)
0	YES	NO	NO	NO	NO	0
1	YES	YES	NO	NO	NO	2
2	YES	YES	YES	NO	NO	7
3	YES	YES	YES	YES	NO	18
4	YES	YES	YES	YES	YES	40

NOTE 1: Vent dia. in inches
Reactor Vol. = 12m^3
Reactor Charge = 10,000 kg
Vent Set Pressure = 2.0 barg
Overpressure above set pressure = 20%

Fig 1 : Uncatalysed Semi-Batch Esterification
Power Output, Total Energy and Feed Vs Time

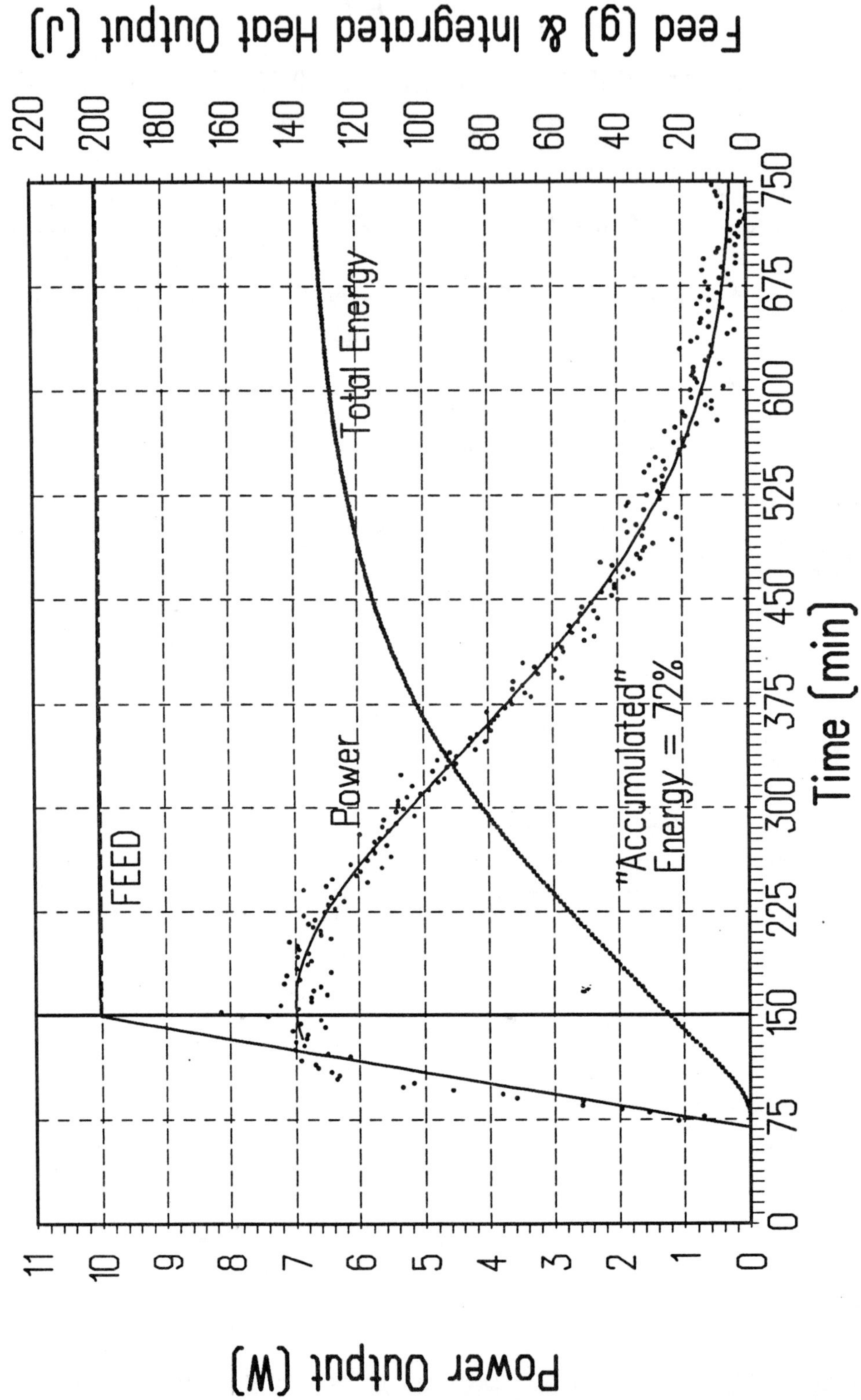

Fig 2: Catalysed Semi-Batch Esterification
Power Output, Total Energy and Feed Vs Time

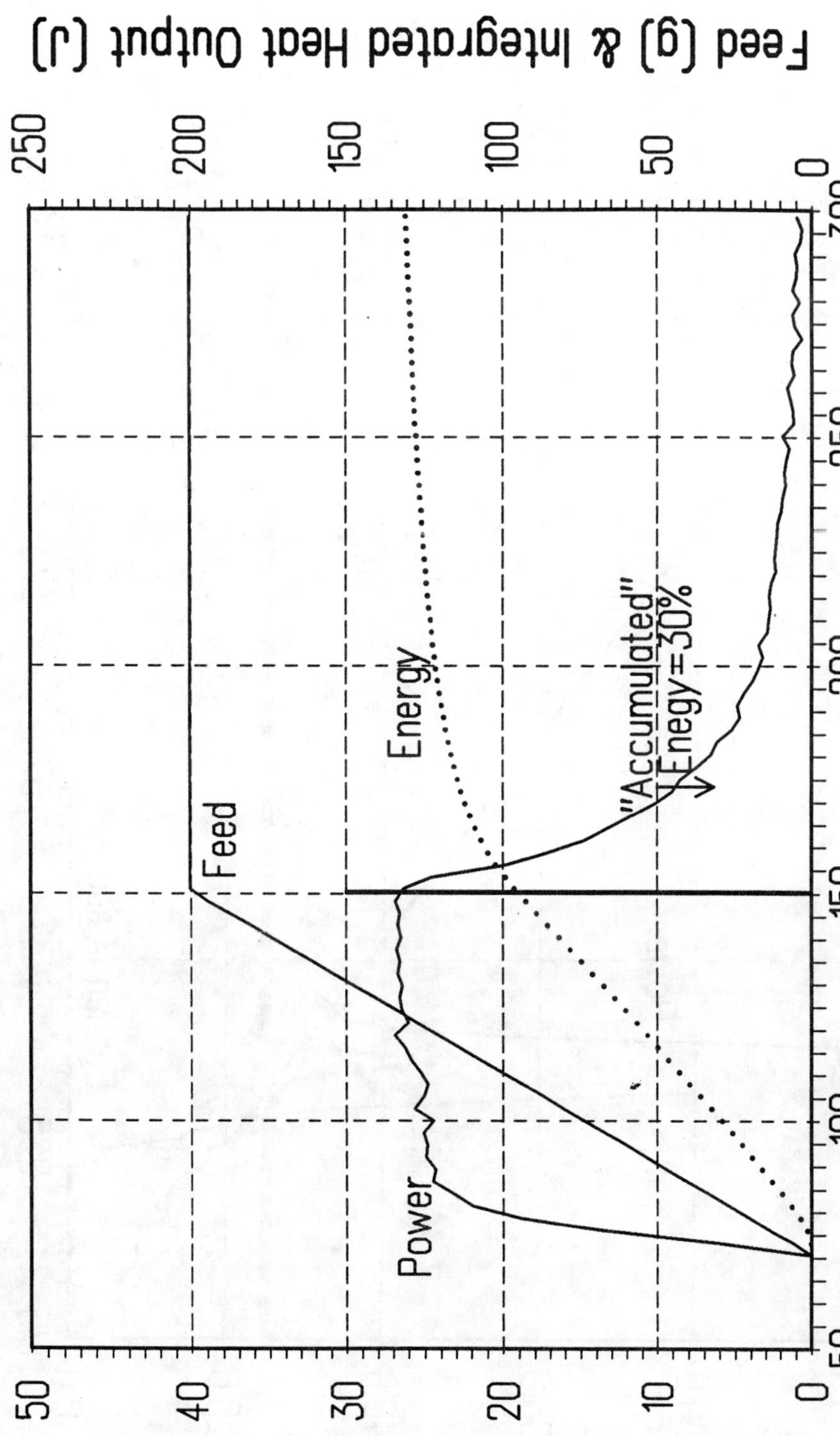

FIG 3: UNCATLYSED ISOTHERMAL SEM-BATCH OPERATION
REACTOR TEMPERATURE, HEAT OUTPUT & FEED RATE

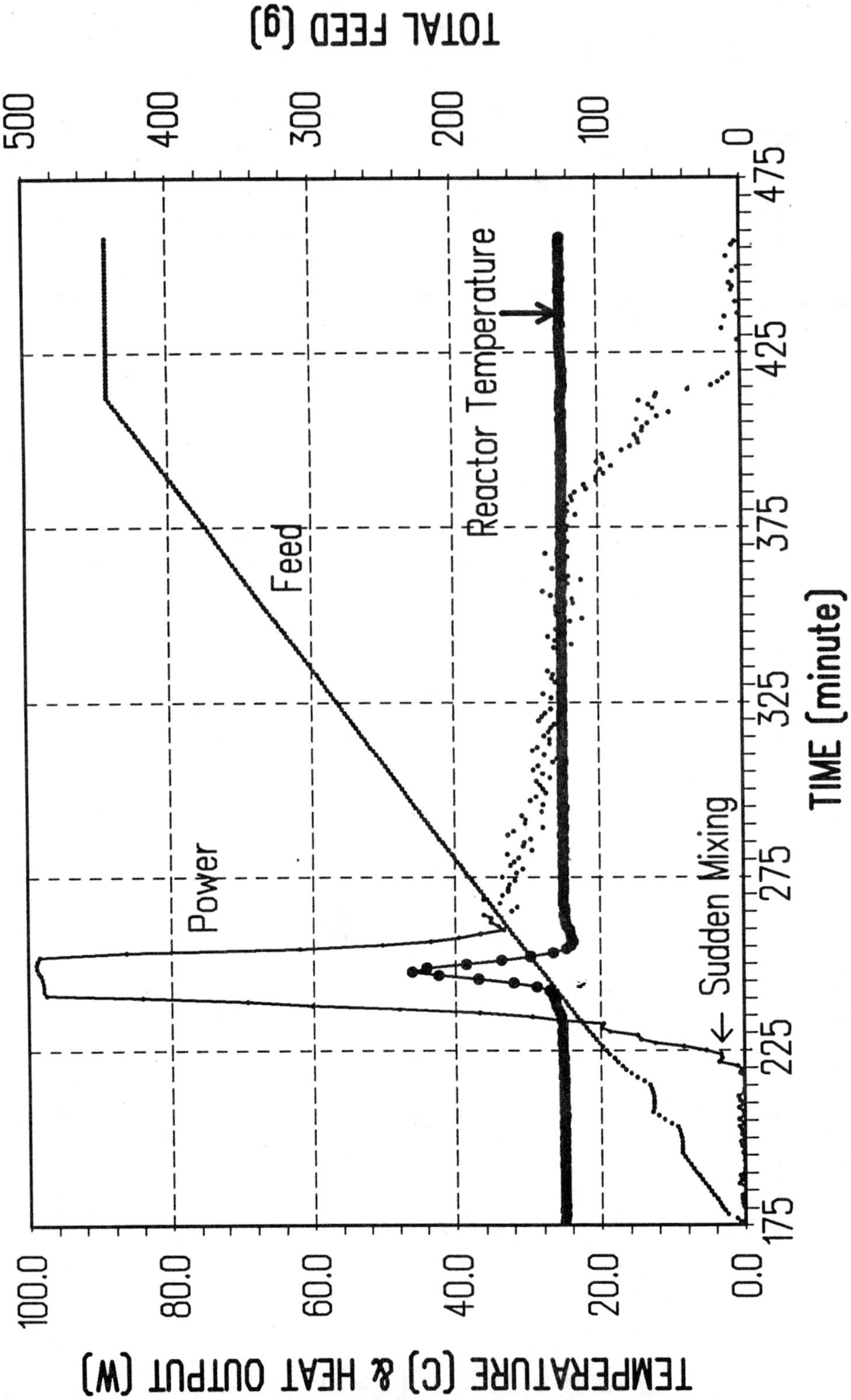

Fig 4: HEAT OUTPUT FOR SEMI-BATCH CATALYSED AND UNCATALYSED REACTIONS

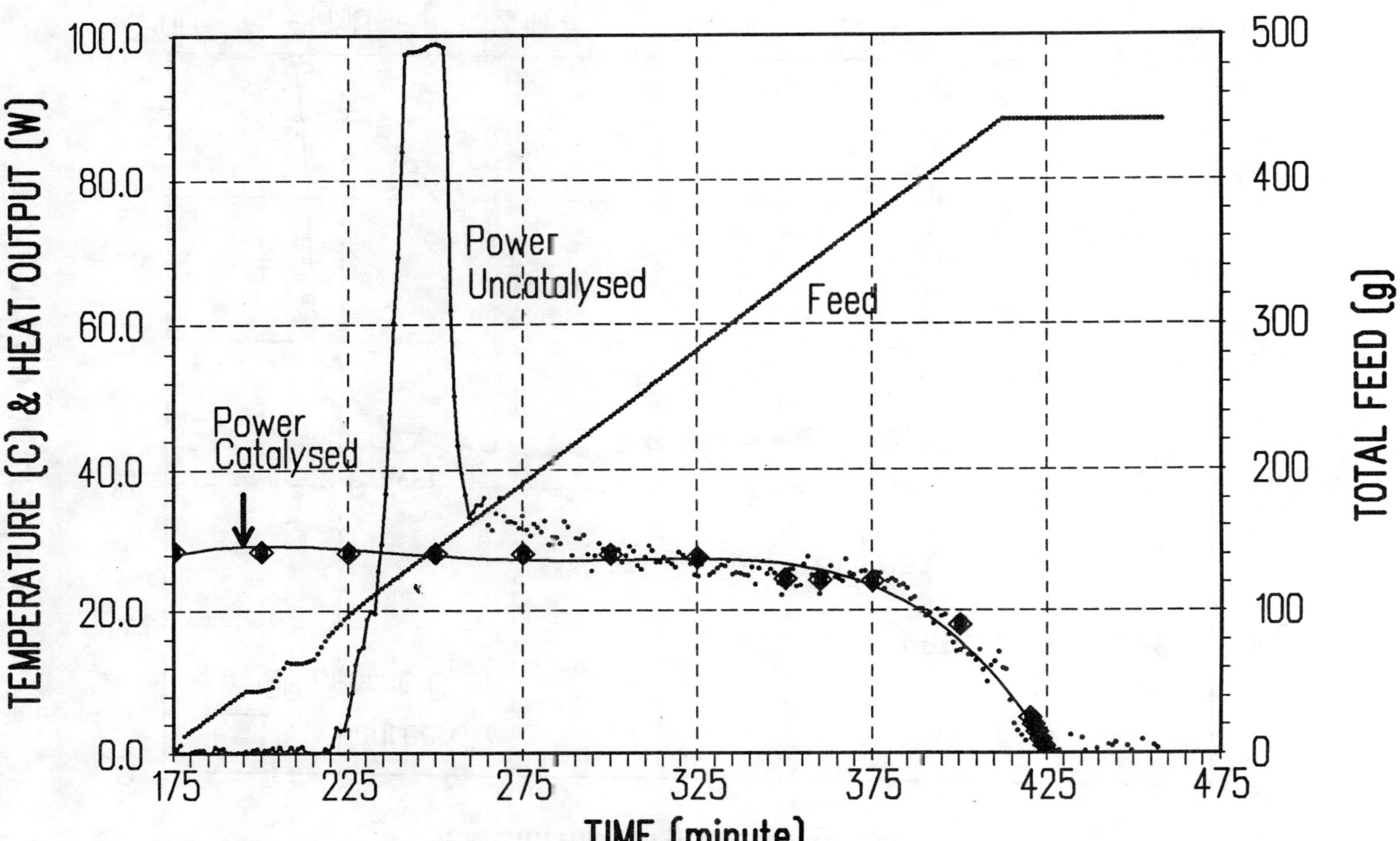

"Catclyst" is Recycled Product

Fig 5: Heat Output at 75 & 90 deg Celcius
Isothermal Organic Synthesis

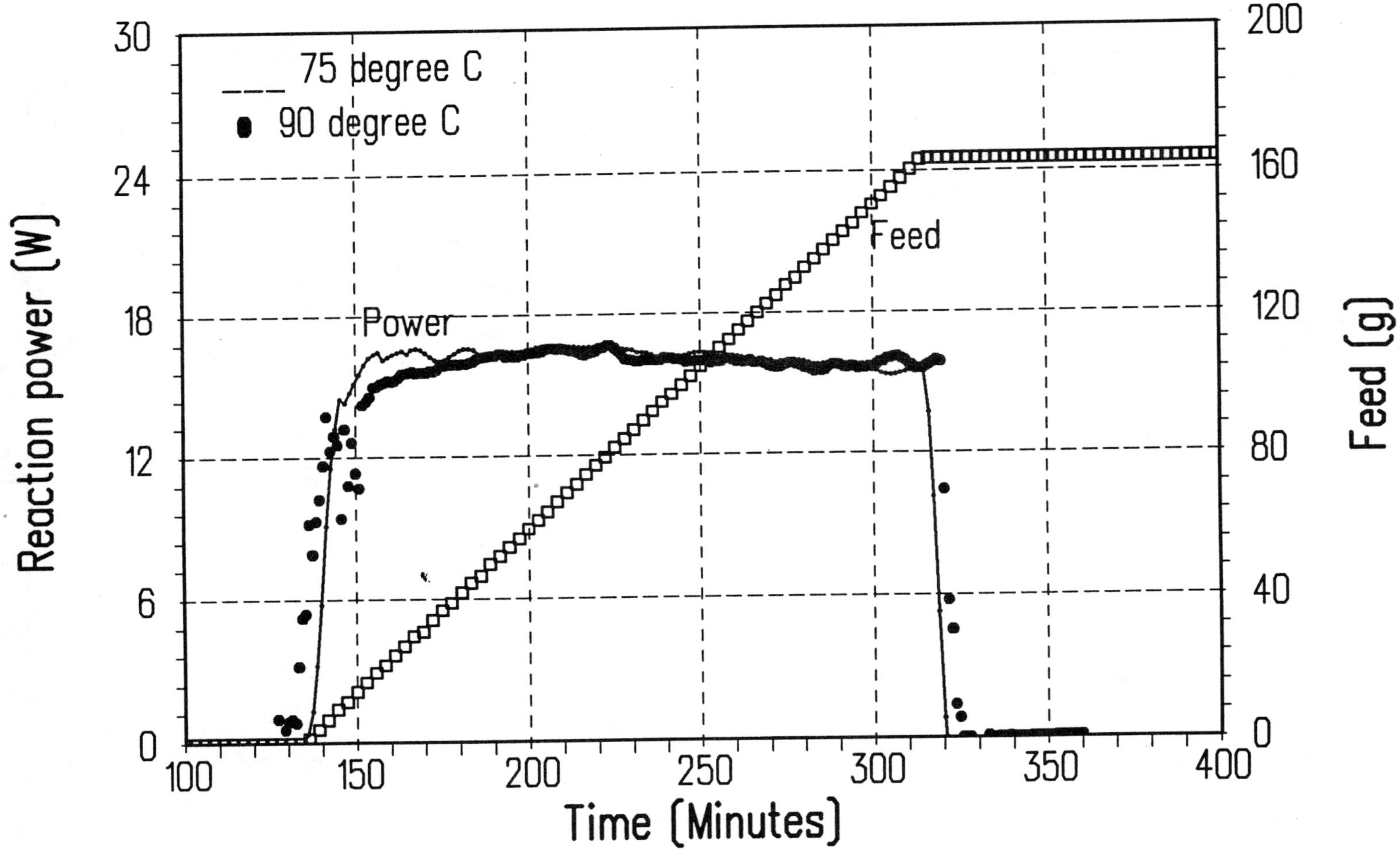

Fig 6 : TEMPERATURE RISE DATA FOR DIFFERENT RUNAWAY REACTION SCENARIOS

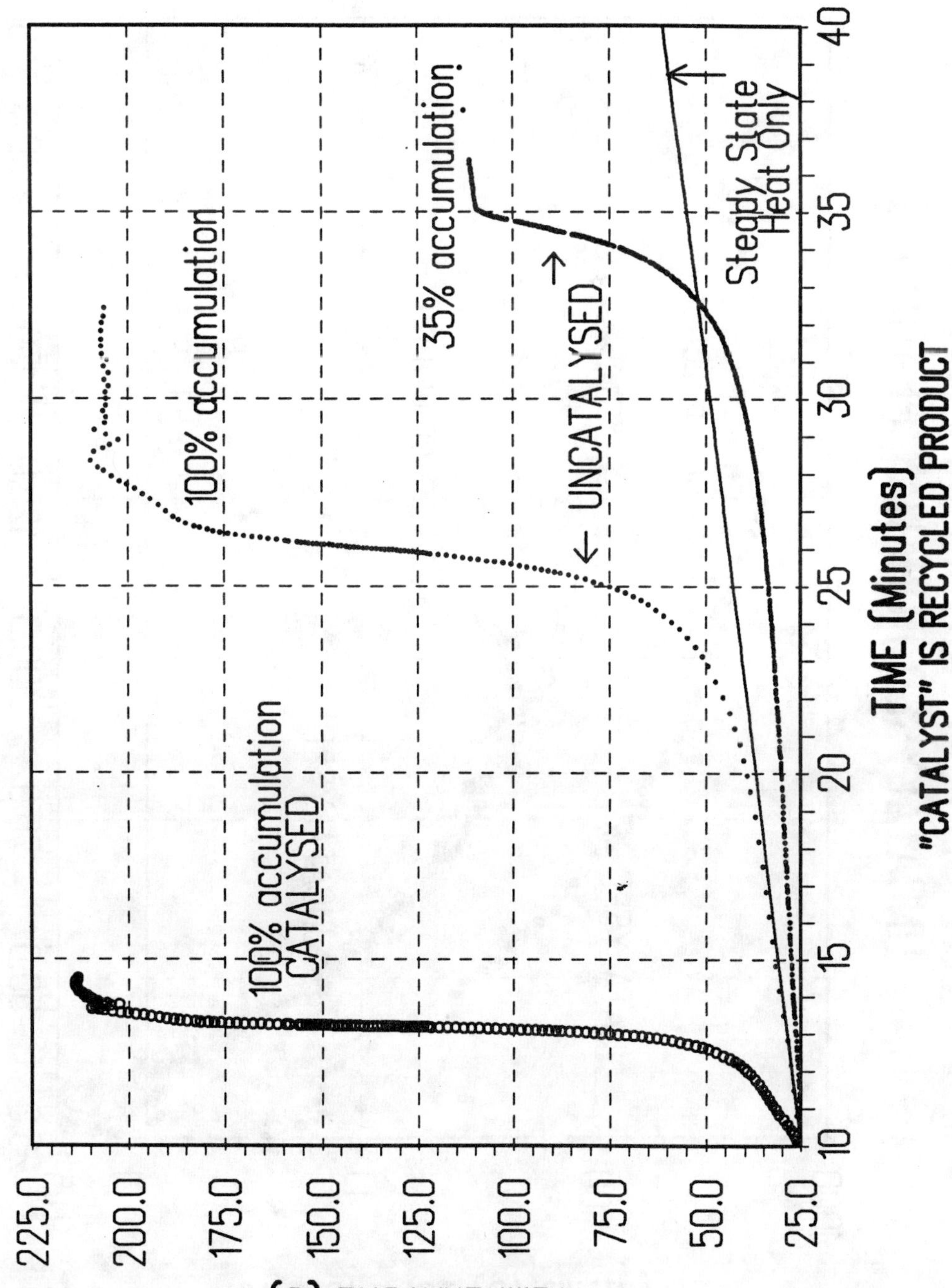

Fig 7 : RATE OF TEMPERATURE RISE FOR CATALYSED & UNCATALYSED RUNAWAY REACTIONS

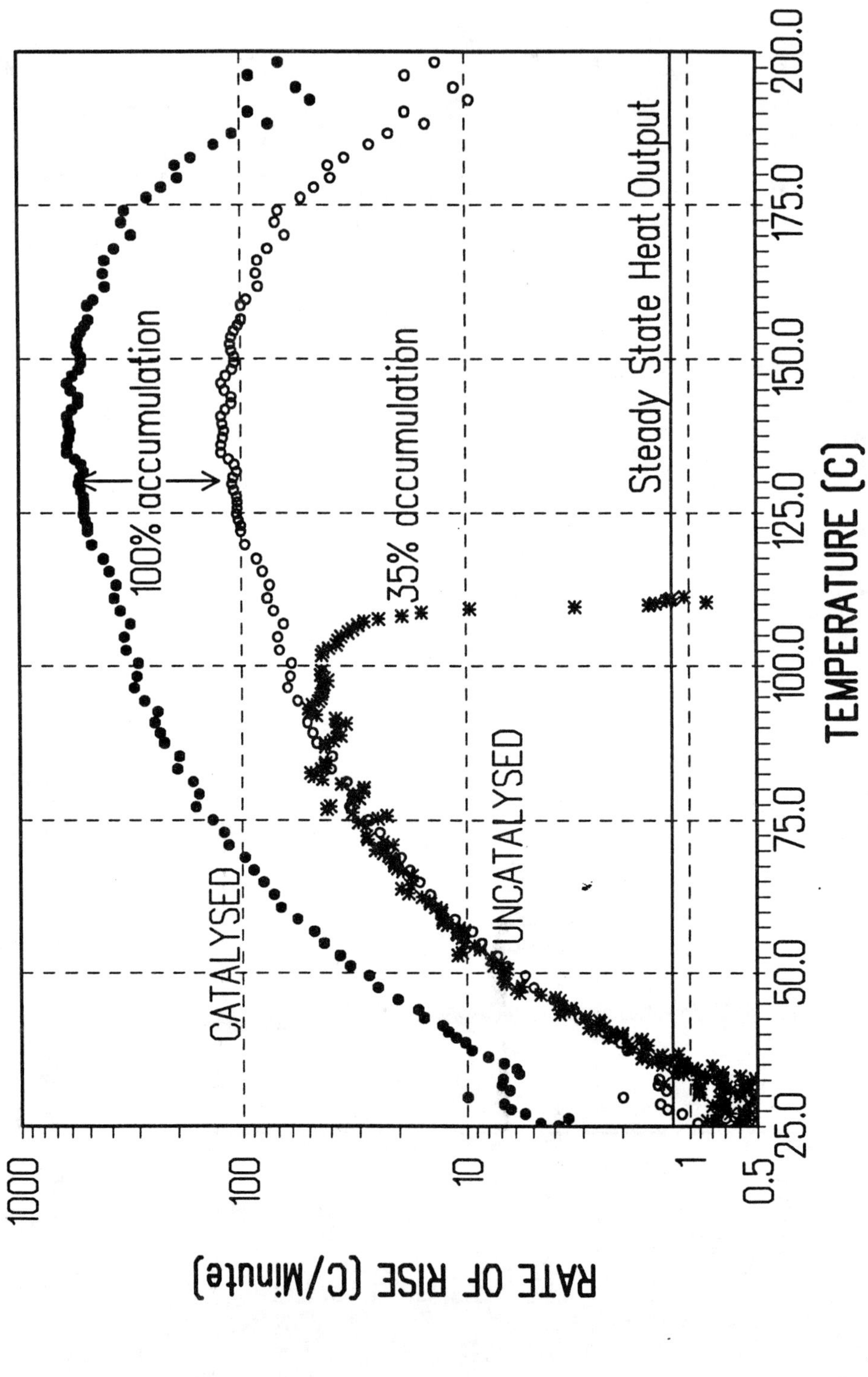

Fig 8: VENT SIZE DEPENDENCE ON DIFFERENT SCENARIOS

Reactor Charge 10,000kg, Volume 12 cu.meter

Overpressure Above Set Pressure = 20%

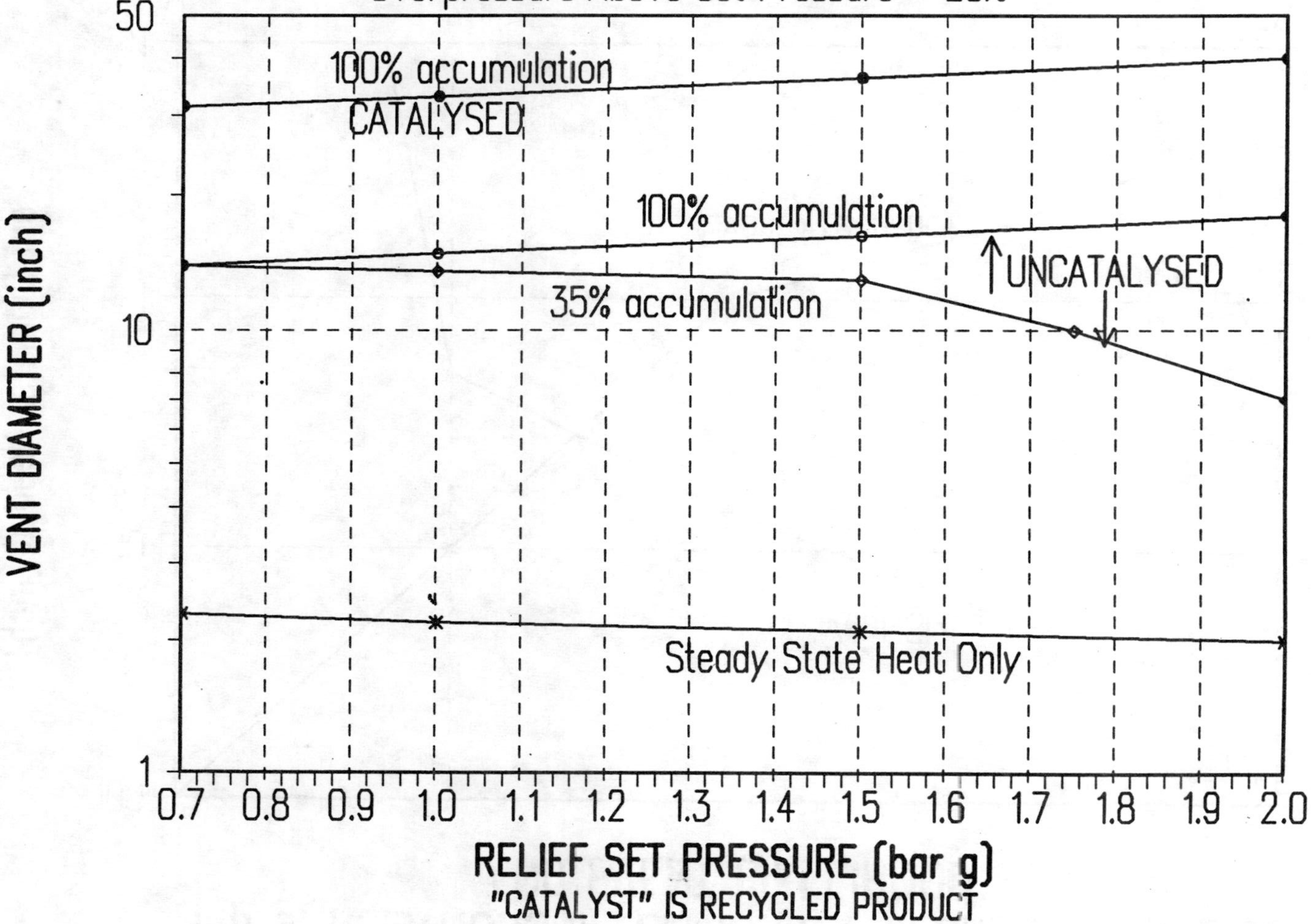

Fig 9: RELATIONSHIP BETWEEN FAILURES, VENT SIZE AND EVENT LIKELIHOOD

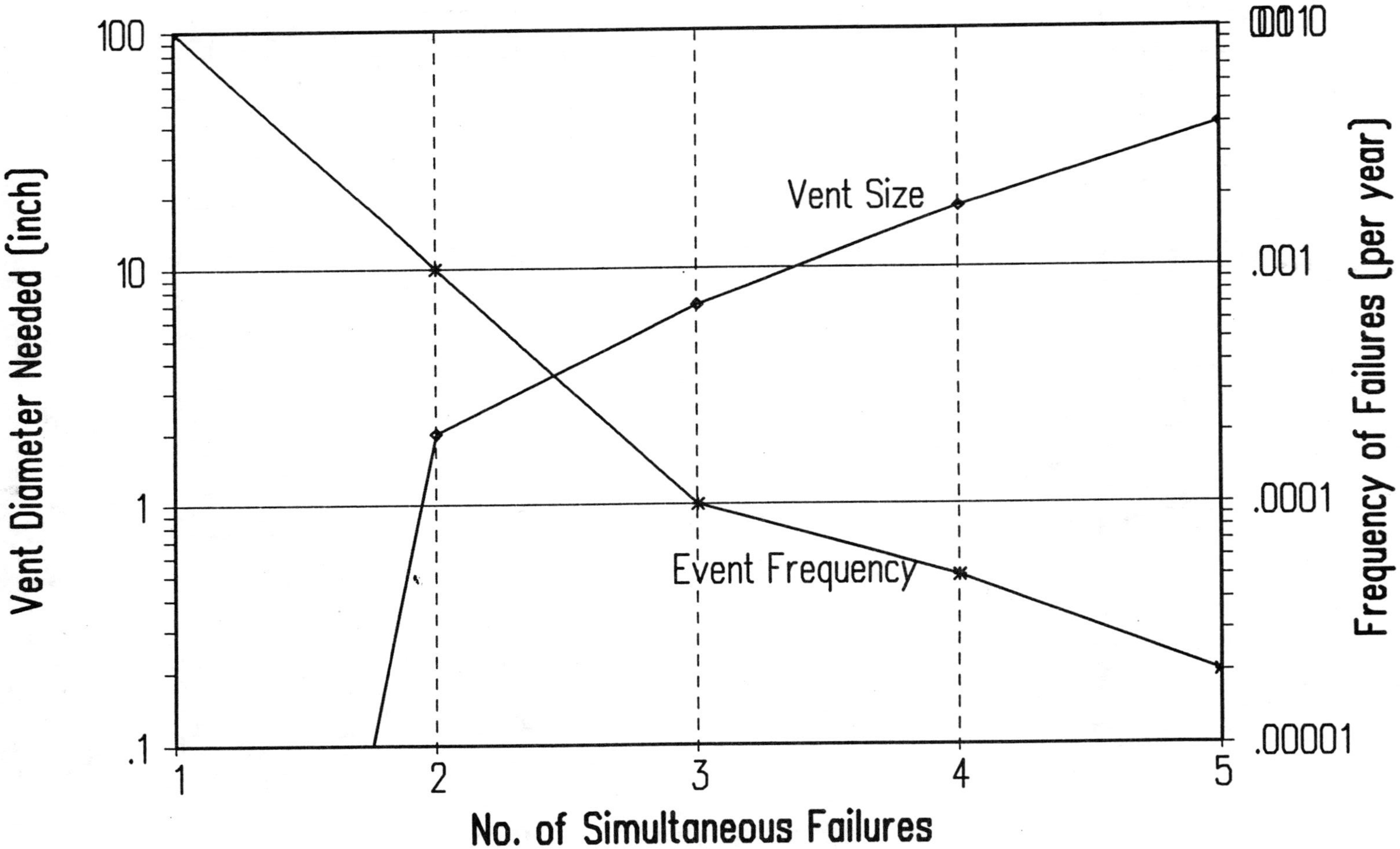

Fig 10: RELATIONSHIP BETWEEN RISK AND VENT SIZE

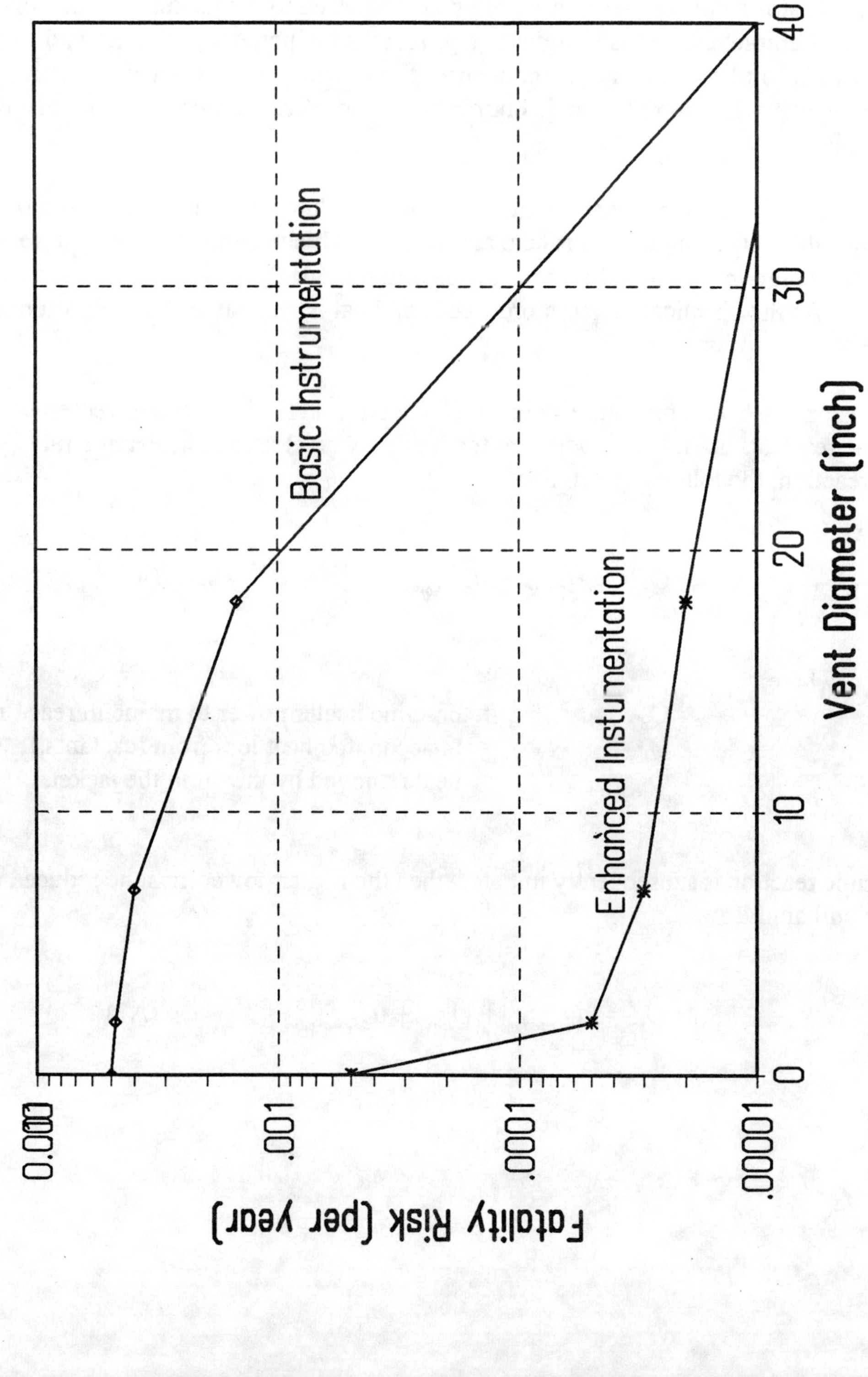

APPENDIX I

SIMULAR REACTION CALORIMETER

The SIMULAR Reaction Calorimeter was employed to evaluate the thermochemical data for semi-batch reactions under isothermal conditions. This equipment can be used as a calorimeter and/or an automated reactor. A schematic detail of the SIMULAR reactor is provided in Figure 1A. The instrument consists of a reaction vessel and surrounding heating/cooling jacket. Silicone oil is circulated through the reaction vessel jacket and a circulator/heater/chiller unit before being returned to the jacket. For this experiment, a reflux condenser is also employed. Sensor signals (e.g. reactor temperature, oil inlet and outlet temperatures, condenser inlet and outlet temperatures, condenser flow, calibration heater power, and stirrer speed) are logged on a computer which monitors and controls the reactor performance in accordance with a pre-defined experimental plan.

Most of the reported experiments were carried out using the "power compensation" method of calorimetry. In this technique the jacket temperature is held at conditions which maintain a temperature difference (~20°C) below the reactor solution. A calibration heater is used to provide the power necessary to maintain the reactor at temperature. As the chemical reaction proceeds and heat is generated the calibration power is varied to maintain constant conditions.

Consider a reactor vessel at temperature Tr in which the oil jacket is fixed at a lower temperature Tj. In order to maintain the reactants at Tr, heat is added to the liquid by an electrical heater at a rate Qp. In the absence of chemical reaction, the following will apply:

$$Q_{p,o} = Q_{loss} + Q_{removed} \qquad \text{(A1)}$$

where:

$Q_{p,o}$ = baseline heater power to maintain reactor,
Q_{loss} = unaccounted heat loss from reactants,
$Q_{removed}$ = heat removed by the oil in the jacket,

If an exothermic reaction is subsequently initiated, then the heater power must be reduced and the following heat balance will apply:

$$Q_p = Q_{reaction} + Q_{loss} + Q_{removed} \qquad \text{(A2)}$$

Appendix I cont'd...

Combining equations (1) and (2):

$$Q_{p,o} - Q_p = Q_{reaction} \qquad \text{(A3)}$$

Thus, the reaction heat output rate $Q_{reaction}$, may be deduced directly from the difference in electrical power before reaction begins and that during reaction.

The heat flow method of calorimetry can also be used with the SIMULAR unit, but is less common.

Fig A1 Reaction Calorimeter Flowsheet

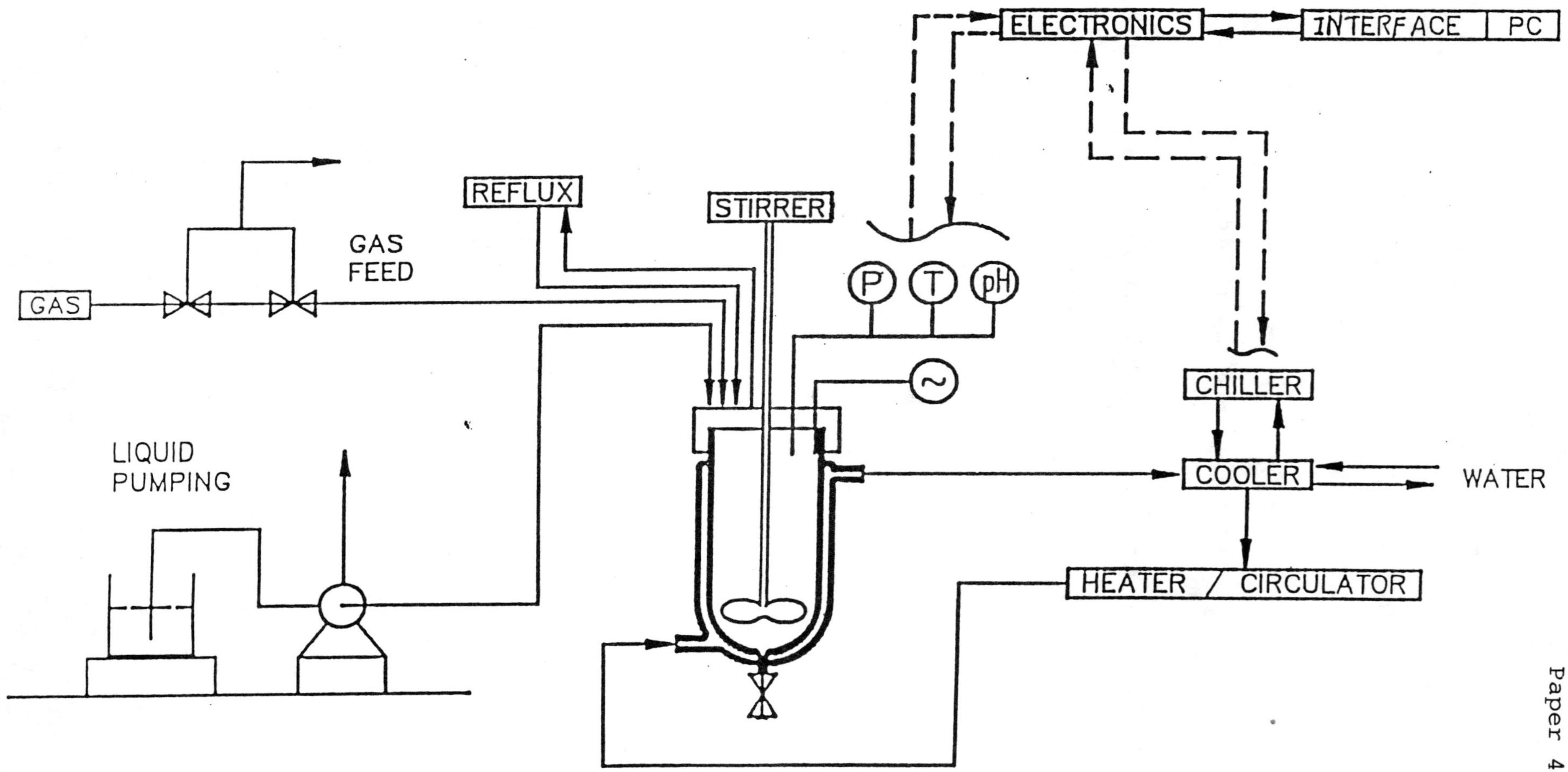

APPENDIX II

RUNAWAY REACTION TESTING USING THE PHI-TEC CALORIMETER

A bench-scale adiabatic calorimeter, PHI-TEC, developed by HEL, was used to perform the runaway reaction experiments (see Figure A2). At the heart of the calorimeter is a sample container, or test cell, which is surrounded by three metal guard heaters : these are top, bottom and circumferential units. When a test sample undergoes exothermic reaction leading to a rise in temperature this is detected by a type K thermocouple in the test cell. This temperature is the set point for the guard heaters which are then independently controlled to match the sample temperature. Heat losses are thereby virtually eliminated and as a consequence the sample is maintained in an adiabatic environment. The guard heaters can track temperatures at up to ~200°C min^{-1}.

Different test cells are available. All of these have relatively thin walls which result in a low thermal mass of the cell relative to that of the test sample. This is normally expressed in terms of the Phi-factor.

$$\textit{Phi factor} = \phi = 1 + \frac{\textit{thermal mass of test cell}}{\textit{thermal mass of sample}} = 1 + \frac{(MC_p)_t}{(MC_p)_s} \tag{A4}$$

where M is the mass and C_p the specific heat, subscript 't' refers to the test cell and subscript 's' to the sample material.

The value of the Phi-factor in large scale plant is close to 1.0: by using thin walled test cells PHI-TEC is able to achieve very similar values. In order to prevent the test cell from rupturing when the sample pressure rises, nitrogen is admitted to the PHI-TEC pressure chamber around the calorimeter assembly, see Figure A2. As the reaction proceeds the rate of this pressurization is controlled so that only a small pressure differential is allowed to develop across the test cell wall. The pressure compensation system will track the sample pressure at up to ~ 100 bar min^{-1}.

FIGURE A2 THE PHI-TEC ADIABATIC CALORIMETER FOR THE ASSESSMENT OF REACTIVE MATERIALS

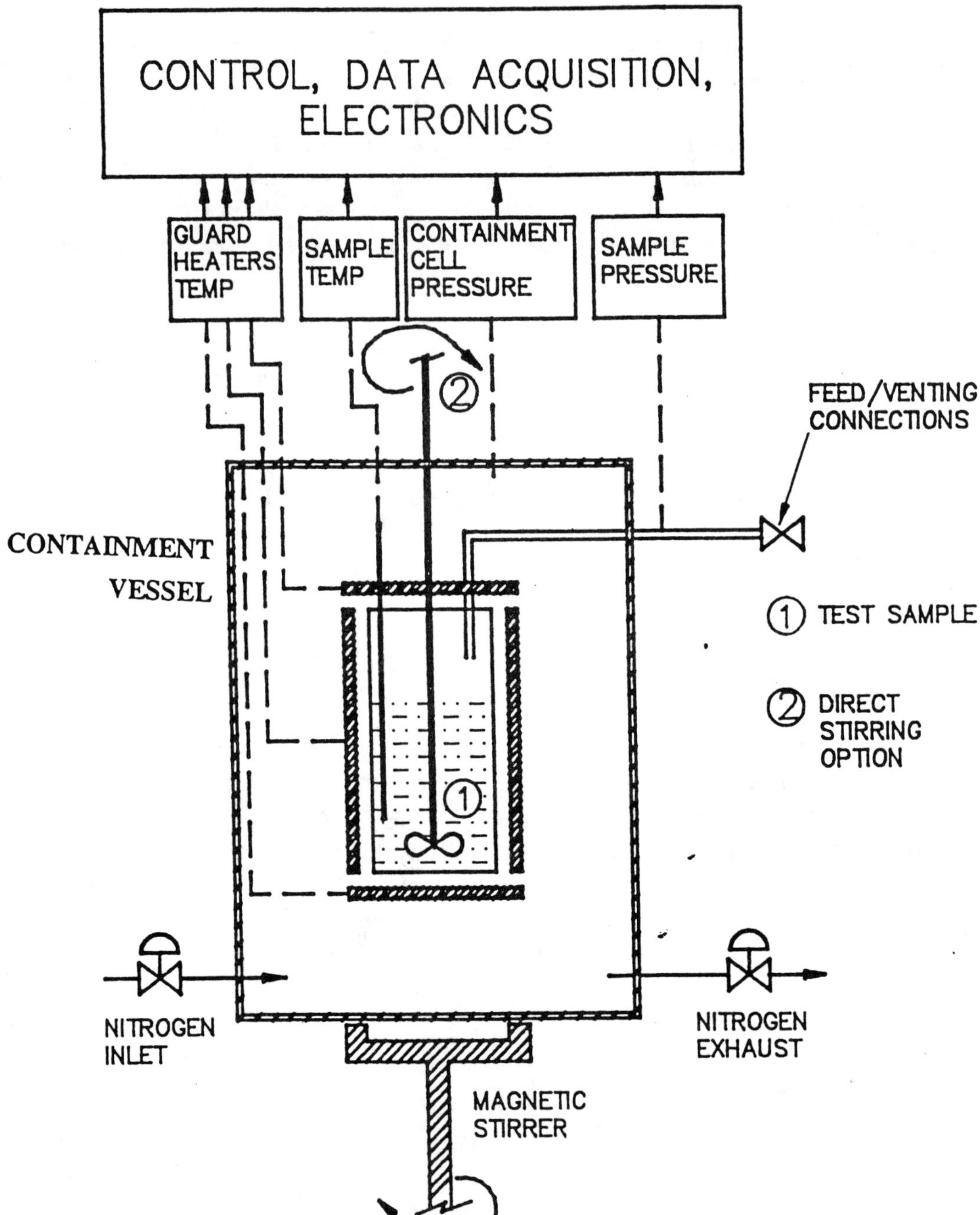

WHY DIERS SHOULD BE USED IN RISK ASSESSMENT - CALL FOR BENCHMARKING

A Presentation for

31st Annual Loss Prevention Symposium
AICHE 1997 Spring National Meeting
Houston, Texas
March 1997

by

John Noronha, Chair, DIERS Risk Subcommittee
c/o Eastman Kodak Company
Engineering & Project Management Division
Rochester, NY 14652-4390
Phone (716) 477-1065

Why DIERS Should Be Used in Risk Assessment - Call for Benchmarking

by

John Noronha, Chair, DIERS Risk Subcommittee
c/o Eastman Kodak Company, Engineering & Project Management Div.
Kodak Park Site, Rochester, NY 14652-4390
Phone (716) 477-1065, Fax: (716) 477-1128

ABSTRACT

The Design Institute for Emergency Relief Systems (DIERS) Users Group has a membership of over 110 companies worldwide. It is an affiliate of the American Institute of Chemical Engineers. Its primary mission is to advance the technology and practice in the design of emergency relief systems for runaway reactions.

This paper is to solicit comments on a ""Proposed DIERS Benchmarking Survey on Runaway Risk Analyses and Reduction Methods (Venting, Prevention, and Protection)." This presentation will outline the need to use DIERS in risk assessment, the apparent reasons why DIERS is not used; and the need for benchmarking. Finally, it will propose an outline of the proposed benchmarking survey (see Section IV).

This proposed survey is aimed at primarily non-DIERS users; so we look forward to your comments before we will finalize the details of the benchmarking questionnaire at the next DIERS Users Group meeting in April '97. The survey may be sent through a mass-mailing to many AIChE members. We would appreciate it if you would respond to this survey. The results of the survey will probably be published.

I. PURPOSE OF THIS PROPOSED SURVEY

Even though the DIERS technology was first published eleven years ago to estimate runaway venting pressures, there appears to be a significant percentage of chemical risk technologists who do not consistently use DIERS or other technologies. DIERS and other related technologies are used to estimate the vent size to limit the venting pressures. Generally, the limit is 1.2 (MAWP), but sometimes above depending on the preventive techniques used and likelihood of a runaway. (see Section IIc)

For those who do use it, many use a variety of venting technologies and design for large vents and stronger vessels. For those who do not use it, many use a variety of prevention and/or protection methods.

There appears to be two reasons why design is often based only on prevention:

1) Lack of confidence and complications in designing relief systems. The original DIERS Methods and even the Simplified DIERS Methods are considered very complicated and time consuming. (This is the main emphasis of our DIERS Users Group and, therefore, of this memo.)

2) The need for emergency effluent handling adds considerable cost and complications.

Section II explains more details of these two reasons of designing based on prevention. Section III discusses the benefits of large vents and high design pressures. Many of the broader issues of risk assessment were addressed in the 1994 DIERS Report, "Risk Consideration for Runaway Reactions," Appendix 1 entitled "Addressing Some Issues Faced in Runaway Risk Assessment." (Ref. 1)

It is probably a good time now to bench mark various risk reduction methods used, from a risk and cost benefit perspective. These benchmarking surveys can be slow. A previous DIERS survey had to be done through AIChE legal staff so that companies could respond confidentially. It took about eight months. However, we feel this is a worthwhile effort as it will promote the consistent use of the DIERS and other venting technologies in risk assessment; or at least bench mark how other companies are basing their runaway risk assessments.

II. MORE BACKGROUND OF WHY DIERS SHOULD BE USED AND WHY IT IS APPARENTLY USED INCONSISTENTLY

a) Fig. 1 is the DIERS Risk Matrix. (Ref. 1) It addresses:
Risk =f(Likelihood, Consequences);
Likelihood =f(Causes, Preventive Measures);
Consequences =f(Runaway Venting Pressures, Protection);
Protection =f(Vent Size, Vessel Design Pressures);
Venting Pressures = f(Chemistry, Batch Size, Vent Size, Vessel Size, etc.).

b) We, at DIERS, believe that the potential consequences are the dominant aspects of catastrophic risk. If a runaway is considered credible, we should design out unacceptable consequences, (e.g. loss of life, loss of business viability).

c) Many companies are not aware that NFPA (Ref. 2,3) recognizes the potential to have venting pressures above 1.2 (MAWP) of a pressure vessel designed to ASME Codes (Ref. 4). The DIERS Risk Guidelines (Ref. 1) recommend appropriate additional preventive and protective measures when the venting pressures are high (Figs. 1,2,&3).

d) Before DIERS was published in 1985, there was no generally accepted guidelines to estimate venting pressures. So most companies had to base their decisions primarily on prevention (see Para f for another reason). Only a few of these users provided vent and pressure vessel protection as recommended by the 1965 Factory Insurance Association (FIA) guidelines (Ref. 5) before DIERS was published. The former FIA curves (Fig. 4) were based on limited industrial loss experience. For example, Fig. 4 specifies 750 gal/8"/100 psig for Type C chemistries. For the most severe Type D chemistries, it specifies 750 gal/18"/100 psig.

e) As indicated above, FIA curves were based on limited industrial loss experience. Note that FIA withdrew their guidelines after DIERS was published in 1985 (Ref. 6). As the DIERS results later indicated, the FIA curves broadly addressed the chemical severity classification, vessel volume, and the vent's size and setting; but they did not specify many significant variables such as batch size, catalysts, diluents, and temperatures. There were several catastrophic incidents even with vessels designed for the FIA based vents.

f) <u>Many companies felt the need to estimate venting pressures to avoid catastrophes</u>. However, the problem was complex as it involved runaway kinetics and two-phase critical flashing flow. Therefore, DIERS was formed in 1979 to do research and develop practical guidelines. We would like to know how many (but not necessarily which) companies are using the DIERS technology are used to estimate the vent size to limit the venting pressures to about 1.2 (MAWP) or, sometimes, to estimate the consequences of a runaway reaction (see Section IIc).

g) <u>However, the original 1985 DIERS Methods and even the Simplified Methods are very complicated and time consuming</u>. <u>So most users continue to base their risk decisions primarily on prevention as they did before DIERS was published</u>; and most of them do not even use vent protection guidelines such as the FIA curves because they do not have confidence in or are not aware of the FIA curves and depend almost entirely on prevention.

h) There is another reason why users base their risk decisions primarily on prevention. <u>By accepting that an event is not prevented, one implies the need for emergency effluent handling</u>. This can add significant cost and complications.

III. WHAT ARE THE BENEFITS* OF LARGE VENTS AND HIGH DESIGN PRESSURES?

a) They allow larger batch sizes for a given risk (Fig. 5).
b) For non-dedicated vessels, they allow flexibility.
c) The reduced consequences lead to the need for fewer better and extraordinary preventive and protective measures, which leads to simpler and more practical measures.
d) They lower inherent risks.

*NOTE: Those, who do not base risk on consequences, see no benefit in these issues. In fact, for many of them, large vents and high vessel design pressures are considered non-standard, expensive and complicated; and are attributed to lead to more expensive emergency effluent handling and related complications.

IV. "PROPOSED BENCHMARKING OF RUNAWAY RISK ANALYSES AND REDUCTION METHODS (VENTING, PREVENTION, AND PROTECTION)."

a. Do companies have internal resources or use outside consultants to apply DIERS or other consequence technologies to estimate the vent size, design pressure, and batch size for a given exothermic risks?

*b. If they use DIERS, do they accept venting pressures above the 1.2 (MAWP) level with correspondingly increased preventive and protection measures (See Section IIc)?

c. If they use DIERS, what methods of analyses do they use and what are their bases (process severity, process complexity, and production limitations computed by a previous level of analyses)? What is the average time and cost of your analysis for reactive and nonreactive systems (see Fig. 6)?

*d. If companies do not use DIERS to estimate consequences, do they base risk decisions on prevention and, if so, what prevention techniques do they use?

e. If they do not use DIERS to estimate consequences, do they use protection designs such as FIA curves' vent size and design pressure (recognizing that they have been withdrawn) (see Fig. 4)?

*f. If they have non-dedicated batch reactors, what are the typical sets of conventional, better, and extraordinary preventive/protective methods used (see Fig. 3)?

*NOTE We recognize that the survey (especially items b,d, and f) would be difficult to assess because risk reduction does involve a wide variety of preventive/protective techniques. Many responses may involve reference to several alternatives; or perhaps or to their internal standards which may not be readily available to other companies.

<u>References</u>:

1. DIERS Guidelines for "Risk Considerations for Runaway Reactions," AIChE, April 11, 1994 (Rev. 2)

2. NFPA 69-1992 "Explosion Prevention Systems," Section 5.3.3 on Deflagration Pressure Containment

3. NFPA-68-1994 "Guidelines for Venting of Deflagrations" Section 3-3-3

4. ASME Codes, Section VIII

5. Factory Insurance Assn., Engr., Bul N-53, April 9, 1965 (withdrawn in 1986)

6 Fisher, H.G., Forrest, H.S., Grossel, S.S., Huff, J.E., Muller, A.R., Noronha, J.A., Shaw, D.A., Tiley, B.J., "Emergency Relief System Design Using DIERS Technology," AIChE/DIERS Project Manual, TP155.5.E.63 (1992)

FIGURE 1 - DIERS RISK MATRIX

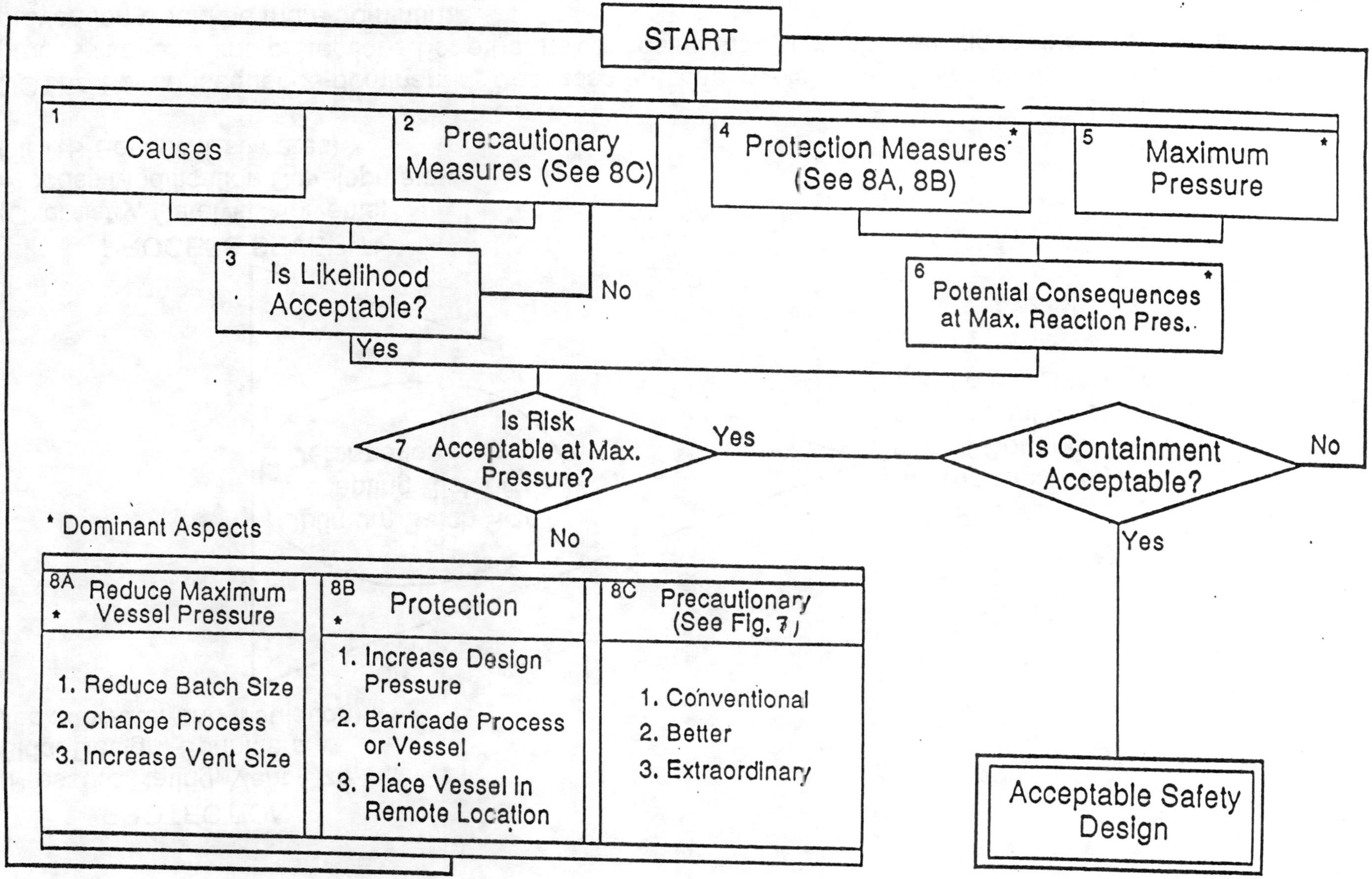

FIGURE 2

TEAM APPROACH - ALTERNATE METHODS FEASIBLE FOR OPTIMUM BATCH SIZE AT ACCEPTABLE RISK

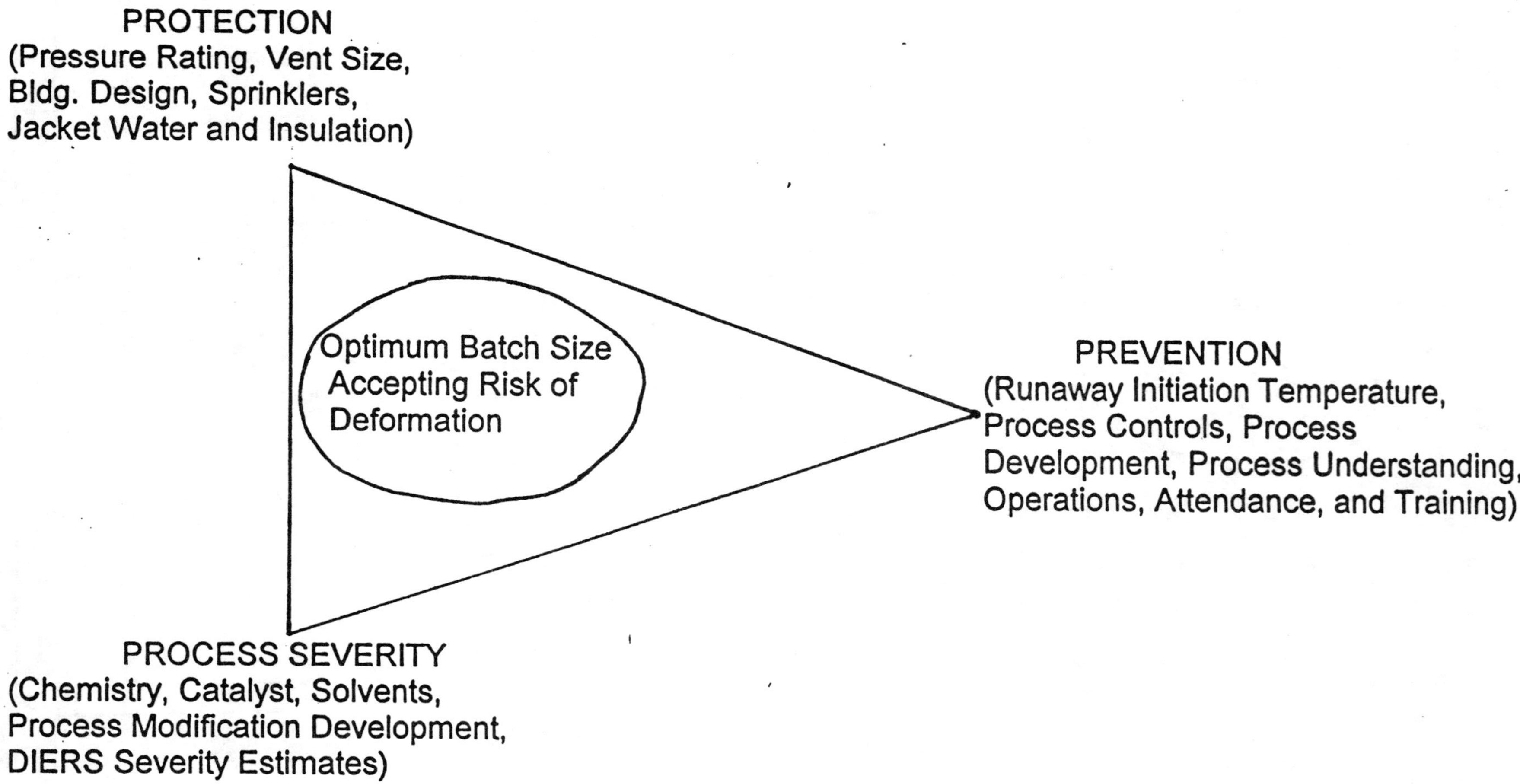

NOTE: For consequence-dominant risk assessments (as in this case), the dominant aspects are typically process severity and protection. For other risk assessments, such as reducing frequency of incidents, prevention would be more dominant.

FIGURE 3

Examples of Relative Types of Preventive and Protective Measures

Conventional	Better	Extraordinary
• Steam Heating	• Tempered Water Heating (Matching Process Requirements)	• Specialized Early Instability Detection (i.e. OLIWA)
• Cooling Systems	• Emergency Cooling	• Quench Reaction with Inhibitor or Water
• Fire Dept. Available	• Active Fire Protection (Sprinklers)	• Passive Fire Protection Such as: - Fire Proof Jacket -High Temp Insulation
• Process Safety Writeup	• Better Understanding of Process	• New Technologies to Evaluate Process
• Operator Training	• Specially Trained Operator	• Continually Supervised Operator

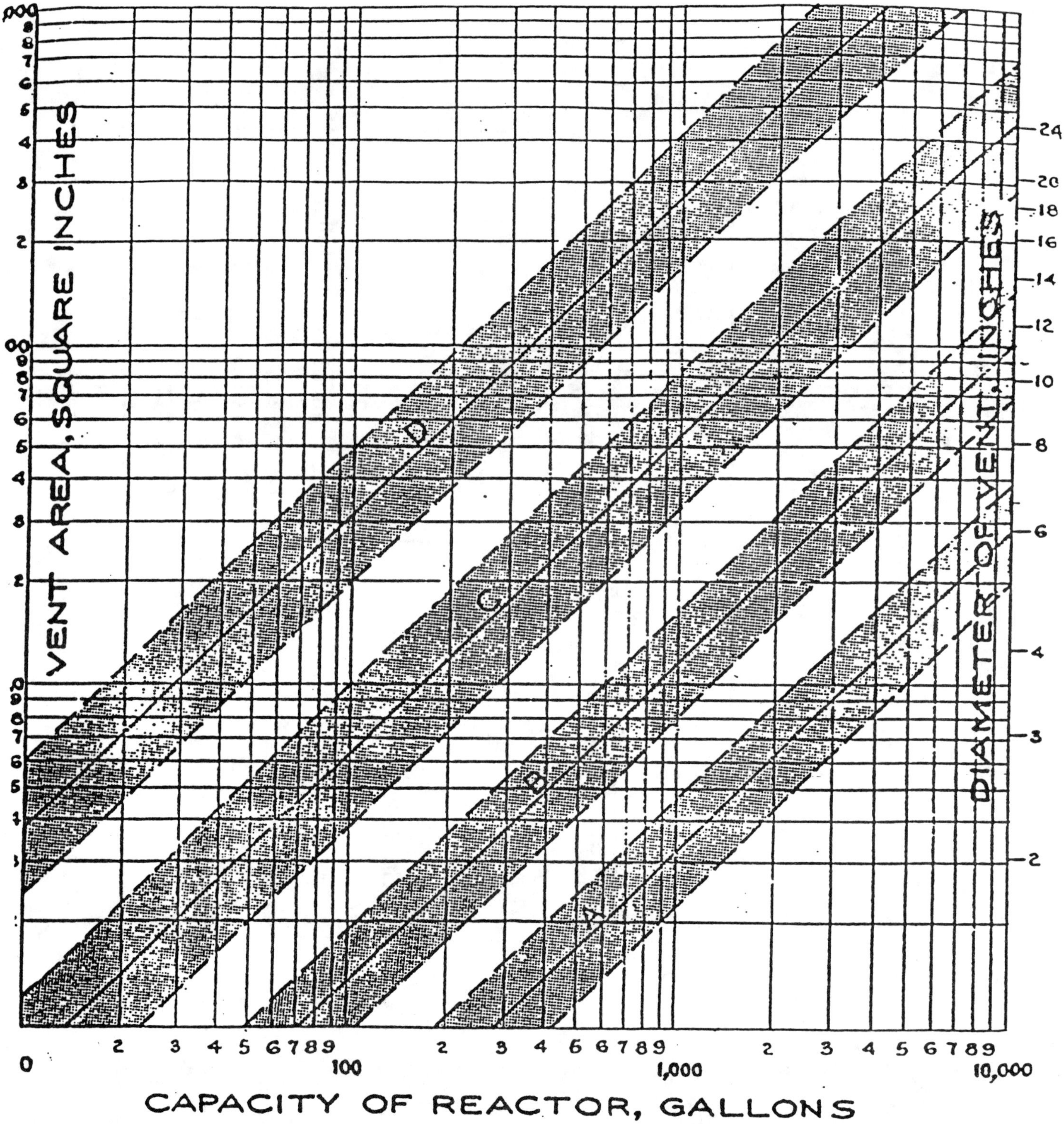

Former FIA - VENT AREA VERSUS REACTOR CAPACITY
Curves FIGURE 4
(ref 5)

$$M_o = \left[V_P \, \Delta P_P \left(\frac{D_P^{\,2} M_S P_P}{2.07 \; T_T \left(\frac{dP}{dt}\right)_T} \right)^2 \right]^{1/3}$$

P - Plant
T - DIERS Test

dP/dt = 500 psi/min $\quad M_T$ = 25 gms $\quad T_T$ = 348K

Vessel Size Gals P_d = 125 psig, Ps = 15 psig	Vent Dia (D_P) Inches	Maximum Allowable Mass (KG) DPC PP=1.2 PS	Maximum Allowable Mass (KG) Non-DPC PP=1.2 Pd
50	2	10	40
100	3	20	80
100	4*	30	120
200	4	40	150
500	6	90	350
750	8	150	600
750	10*	210	1100

(P_d = 150, P_s = 25)

*Special Sizes

: Hypothetical Venting Analysis Gassy Reactions

FIGURE 5

FIGURE 6

Applicability, Timeliness, and Cost of Various DIERS Methods

Description	Cost ($)	Time	Process Complexity	Process Severity	Production Restrictions
By Inspection	x	1 Hour	Simple	Low	--
Screening Methods	10x	Thermal Tests + 6 Hrs. Analysis	Moderate	Moderate	Most
Simplified DIERS Methods	40x	Vent Tests + 20 Hrs. Analysis	Complex	High	Less
Complex DIERS Methods	100x to 200x	Computer and Other Analysis and Vent Tests	Very Complex	Very High	Least

Pressure Relief System Documentation: Equipment Based Relational Database is Key to OSHA 1910.119 Compliance

P.C. (Pat) Berwanger, P.E.
R.A. (Rob) Kreder
Berwanger, Incorporated
Houston, Texas

BIOGRAPHICAL INFORMATION

Pat Berwanger is Principal Engineer of Berwanger, Inc., a Houston based multi-disciplinary engineering firm serving the oil, gas, and petrochemical industries. His firm authored the Pressure Protection Manager©™ (PPM) software which is now widely used in the industry to design, verify, and maintain pressure relief system information for OSHA 1910.119 compliance.

Mr. Berwanger graduated Summa Cum Laude from Vanderbilt University's School of Engineering in 1981 with majors in Chemical Engineering, Chemistry, and Mathematics.

Pressure Relief System Documentation: Equipment Based Relational Database is Key to OSHA 1910.119 Compliance

P.C. (Pat) Berwanger, P.E.
R.A. (Rob) Kreder
Berwanger, Incorporated
Houston, Texas

ABSTRACT

Under OSHA 1910.119, all Process Safety Management (PSM) facilities are required to keep their pressure relief system design information current. This article demonstrates why a pressure relief system design verification effort must be based on an equipment list, rather than a relief device list, in order to ensure that every piece of equipment is adequately protected. The formerly common practice of simply checking the design bases of all existing relief devices is deficient since this technique does not systematically ensure that every piece of equipment is protected.

The "Berwanger Method" is a step by step process for designing or analyzing a pressure relief system to meet OSHA 1910.119 Process Safety Information (PSI) and Process Hazard Analysis (PHA) mandates. The method uses a relational database which tracks the relationships between protected equipment, potential overpressure scenarios, and protective devices.

Pressure Relief System Documentation:
Equipment Based Relational Database is Key to OSHA 1910.119 Compliance

P.C. (Pat) Berwanger, P.E.
R.A. (Rob) Kreder
Berwanger, Incorporated
Houston, Texas

Introduction

Under OSHA 1910.119, all Process Safety Management (PSM) facilities are required to keep their pressure relief system design information current. This article demonstrates why a pressure relief system design verification effort must be based on an equipment list, rather than a relief device list, in order to ensure that every piece of equipment is adequately protected. The formerly common practice of simply checking the design bases of all existing relief devices is deficient since this technique does not systematically ensure that every piece of equipment is protected.

The "Berwanger Method" is a step by step process for designing or analyzing a pressure relief system to meet OSHA 1910.119 Process Safety Information (PSI) and Process Hazard Analysis (PHA) mandates. The method uses a relational database which tracks the relationships between protected equipment, potential overpressure scenarios, and protective devices.

The challenge facing an operating company does not end once the design basis has been "verified" - the design basis information must also be maintained and be readily accessible to avoid costly reinvention of the wheel down the road. The "Berwanger Method" also addresses these maintenance issues.

Pressure Relief Systems and OSHA 1910.119: Three Goals

As process safety managers, we generally have three goals for a pressure relief system as it relates to OSHA 1910.119:

Goal #1: Ensure the plant is adequately protected against all potential overpressure scenarios.

To ensure protection, we must address all the potential overpressure scenarios associated with every piece of equipment.

Goal #2: Maintain the safe design as changes are made to the plant.

To maintain the safe design, we must have a realistic and reliable information management system.

Goal #3: Implement the verification and maintenance effort at low cost.

To minimize costs, we must carefully define tasks and then execute each task with the most efficient people and tools.

Why the Old Pressure Relief System Data Structure Does Not Work

To understand why the old pressure relief system data structure, which is based on a relief device list and a filing cabinet full of relief device data sheets (see Figure 1) does not work, take a look at the typical distillation column example shown in Figure 2.

Five potential overpressure scenarios have been identified:

- inlet control valve failure
- blocked outlet
- loss of overhead coolant
- reboiler tube rupture
- external fire

For these five potential overpressure scenarios, we have a single PSV.

Thus, this example helps illustrate the natural relationships (i.e. data structures) that exist between pieces of equipment, overpressure scenarios, and relief devices. Figure 3 shows that, in general, for each piece of equipment, there can be one or more overpressure scenarios and relief devices.

Now, let us remove the relief device from the distillation column example - as illustrated in Figure 4. And then, ask ourselves a few questions:

- If we are verifying overpressure protection in our plant by working our way down a PSV list, at what point in this "process" would we discover that a piece of equipment without a relief device (such as the distillation column) was unprotected?

  Answer: There is not any point in this "process" for determining that a piece of equipment is without protection.

- Where would we store information about potential overpressure scenarios (that could effect the distillation column) if there was no relief device?

  Answer: We certainly could not store the information with the relief device data sheet. However, we could store the information with the equipment file.

- If we changed a relief device, does this change the potential overpressure scenarios?

  Answer: No, the potential overpressure scenarios would still be there.

- In light of the above, does it make sense to use a relief valve list, rather than an equipment list, as the basic checklist for relief system verification? Does it make sense to store potential overpressure scenarios with the relief device data sheets rather than the equipment?

  Answer: No, it doesn't - but, this is what a lot of folks have been doing.

An Equipment Based Data Structure Solves the Problem

The key to resolving our dilemma is the equipment information form in Figure 5, which lists protective devices as well as overpressure scenarios. If we prepare such a form for each piece of equipment in the plant, we have a means of tracking all the overpressure scenarios as well as the protective devices.

With this in mind, a physical metaphor for our relief system information would consist of the two filing cabinets shown in Figure 6. One for equipment information forms and the other for relief devices.

Of course, it would be even more efficient to store the relief system design information electronically in a relational database with the basic structure shown in Figure 7. In electronic format, we do away with the paper, and the information can be accessed from a PC and/or over a computer network.

A Step by Step Procedure for Analyzing a Relief System

OK, now we know how we can store and access the relief system design information. The next question, how do we create the information in the first place? The four step process described below has proven most efficient for both new designs and verification of existing facilities.

Step 1: For each piece of equipment, an "expert" identifies all potential overpressure scenarios.

This is unquestionably the most critical step in the process. Only after an overpressure scenario has been identified by the "expert" can it be addressed. Table 1 of API 520, "Bases of Relief Capacities Under Selected Conditions" provides a widely recognized checklist for this purpose (see Figure 8).

For the most part, the remaining three steps can be reduced to computerized algorithms. Except for quality control, the "expert's" work is largely done.

Step 2: For each overpressure scenario identified in Step 1, quantify the required relief rate.

The required relief rates for the majority of overpressure scenarios, such as tube rupture, external fire, control valve failure, etc... can be "canned". The use of computerized algorithms, so long as their applicability are subject to the "expert's" approval, not only greatly improves efficiency, but also improves quality and consistency.

Step 3: For each required relief rate in Step 2, verify that there is a relief device of adequate capacity.

Again, relief valve sizing can largely be accomplished with a computer. For years, API 520 has provided the basis of most relief valve sizing equations. More recently, the Design Institute of Emergency Relief Systems (DIERS) has provided more accurate, and complicated, correlations for two phase flow situations. All these correlations are subject to computerization.

Step 4: With due consideration of all the scenarios identified in Step 2, and the actual relief device capacities determined in Step 3, analyze the relief headers.

Having information on all the individual scenarios quantified in Step 2 available in a relational database greatly simplifies the process of combining individual scenarios to arrive at "global" scenarios which may be of most concern to the relief header network.

For example, if one wants to identify all the external fire scenarios for equipment in a specific area of the plant - which have the potential to occur simultaneously - one could query the database for a detailed listing of scenarios for those pieces of equipment as well as the effected relief devices.

Likewise for other "global" scenarios such as cooling water failure, or instrument failure.

Rounding Out a Complete Relief System PHA

Insofar as Process Hazard Analyses (PHA's) are concerned, OSHA 1910.119 does not limit itself to the "engineering" analysis which has been described above. Other elements which must be addressed are:

- hazards of the process (paragraph e.3.i)
- previous incidents (e.3.ii)
- engineering and administrative controls (e.3.iii)
- consequences of failures (e.3.iv)
- facility siting (e.3.v)
- human factors (e.3.vi)
- qualitative evaluation of safety and health effects (e.3.vii)
- team/employee input (e.4)

The author has addressed these other elements by:

- developing a written Process Hazards Analysis procedure, approved by company management, which clearly sets forth the scope, requirements, relevant documents, materials and equipment, and procedure for performing the PHA.
- performing operating employee interviews, and otherwise inviting employees to express their opinions, to ensure that their concerns are addressed
- review of previous incident reports
- review of previous PHA reports
- conducting a plant management review session, during which, for each concern raised, a qualitative risk is assessed, a timely plan for resolution developed, and responsibility for resolution assigned.
- issuing a detailed report which is then used to effectively communicate the substance and findings of the PHA.

Summary

Potential overpressure scenarios exist whether or not relief devices are there to protect the effected equipment. To ensure that every piece of equipment is protected it is essential that the pressure relief system design verification be based primarily on the equipment list. The formerly common method of simply checking existing relief devices is inadequate since this technique does not systematically ensure that every piece of equipment is protected.

It is most efficient to store relief system design information in a relational database where overpressure scenarios and relief devices are associated with the effected equipment. This system can be implemented electronically and made available across a computer network. Most of the necessary calculations can be "canned" which improves efficiency and quality control.

Relief systems can be designed and documented using the four step procedure which forms the heart of the "Berwanger Method".

Other OSHA mandated Process Hazard elements such as facility siting, consequence evaluation, and human factors have been integrated into the "Berwanger Method", so that in its entirety, the method can be used to meet OSHA 1910.119 PHA requirements.

Traditional Relief System Documentation

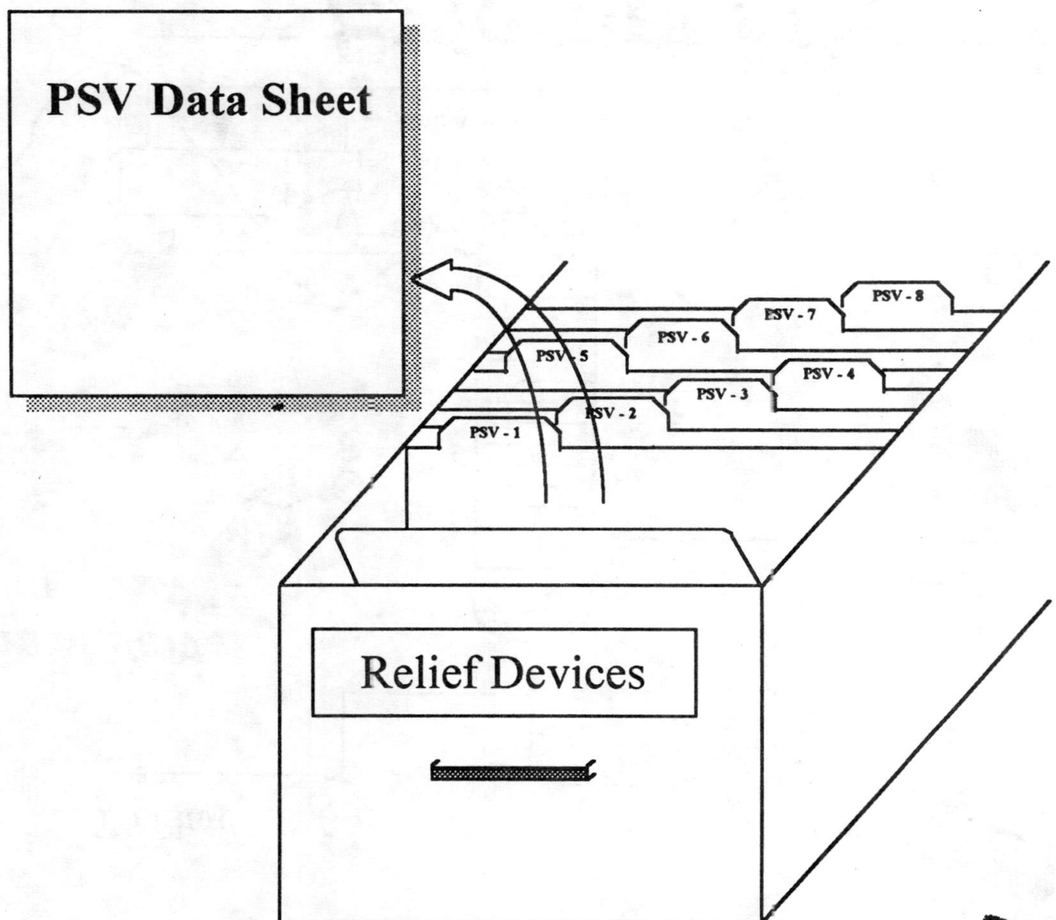

Figure 1

Berwanger, Inc.
Oil, Gas & Petrochemical Consulting

Example Distillation System

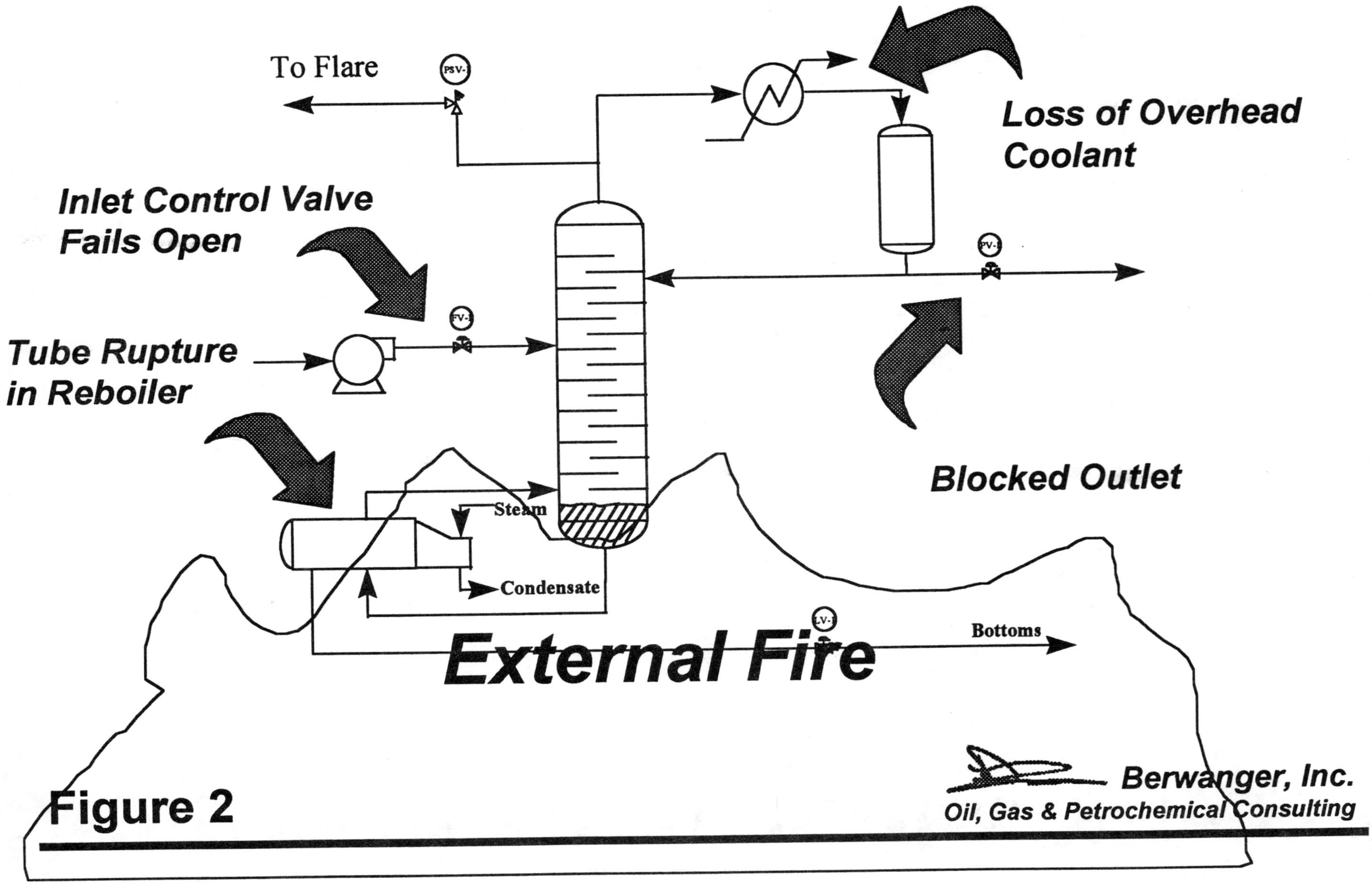

Figure 2

Relationships between Equipment, Overpressure Scenarios, and Relief Devices

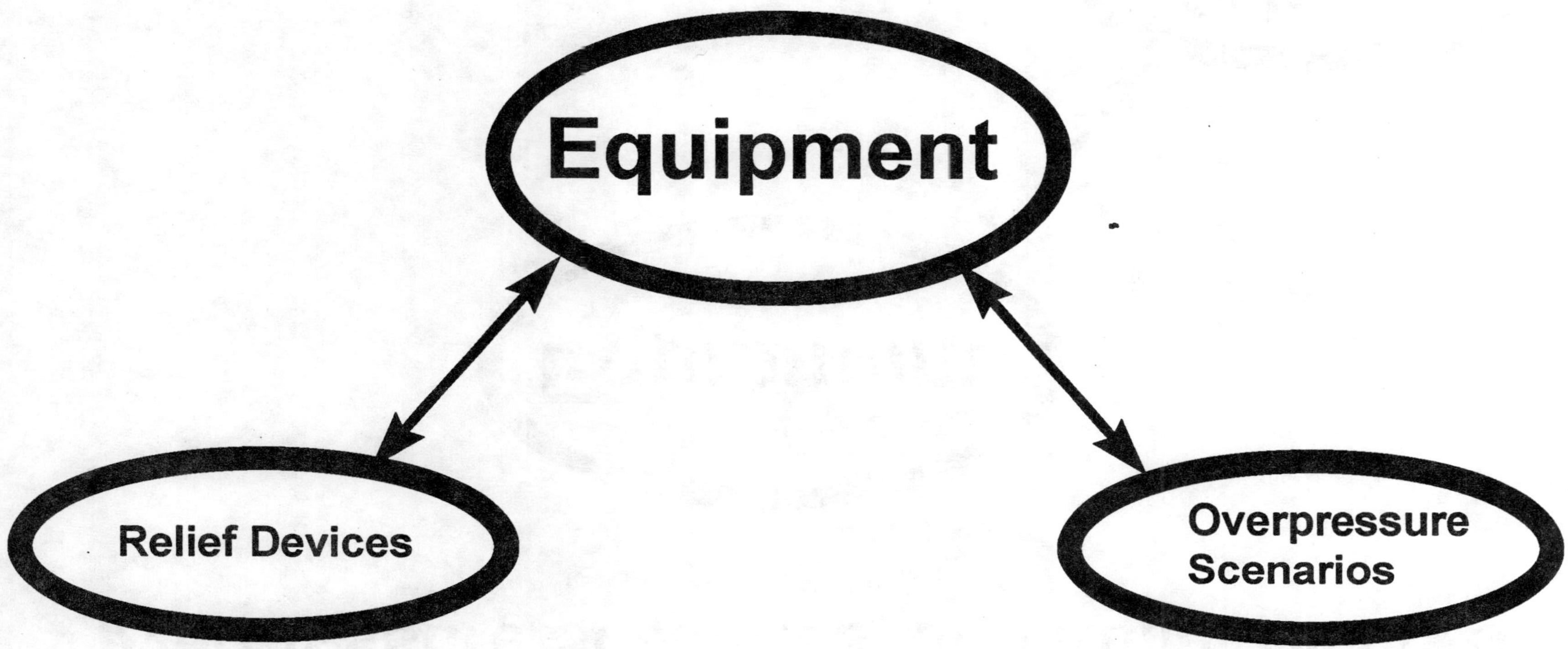

Figure 3

Berwanger, Inc.
Oil, Gas & Petrochemical Consulting

Relationships between Equipment, Overpressure Scenarios, and Relief Devices

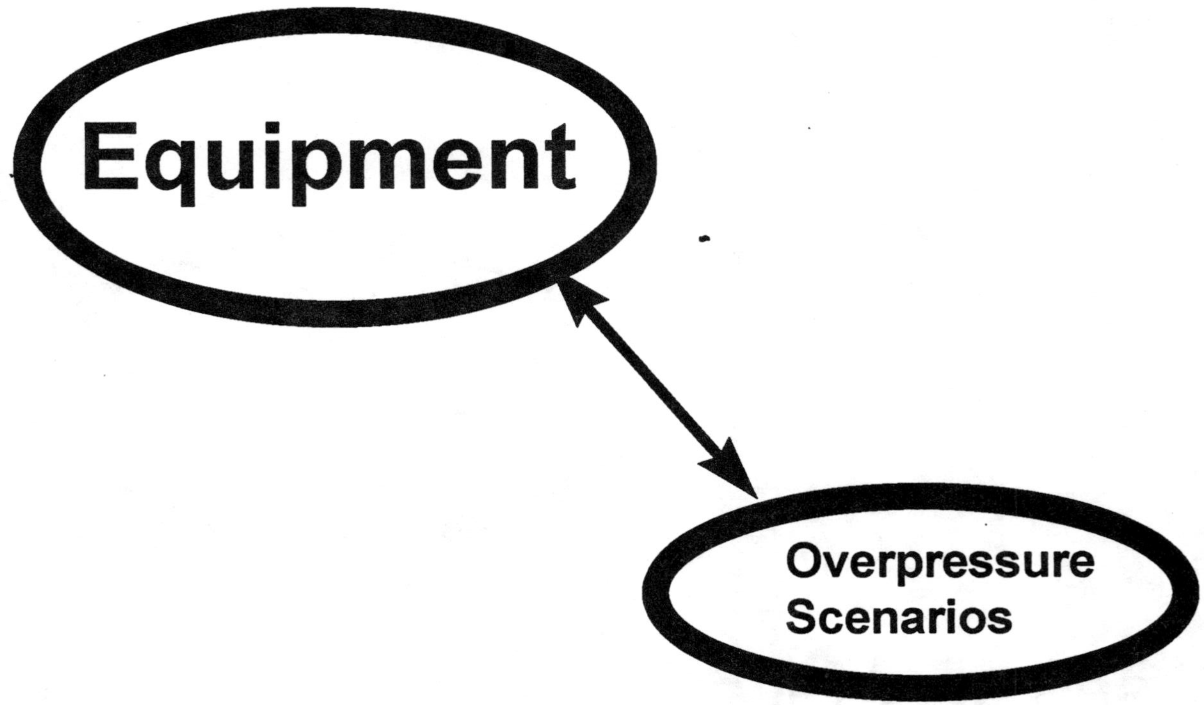

Figure 4

Berwanger, Inc.
Equipment Information Form

Revision Record
Tag No: V-1
No: 01 **Date:** 04/19/95 **Initials:** AAA
Propylene Purification Vessel

Reference Information

Manufacturer:	Vessels, Inc.	**PFD No:**	100-1A
Nat'l Board No:	4235	**P&ID No:**	200-1A
		Iso No:	300-1A

Equipment Description

Type:	Vertical Vessel (Cylinder)	**Height/Length:**	20.0 ft
Material:	Carbon Steel	**Width/ID:**	5.0 ft
Mode:	Continuous	**Equip Volume:**	400.0 ft3
Contents:	Oxo-Purge	**Fluid Volume:**	400.0 ft3

Conditions

Normal Operating Pres:	350 psig	**Normal Operating Temp:**	150 degF
Max. Operating Pres:	360 psig	**Max Operating Temp:**	150 degF
Max. Working Pres:	400 psig	**Max Working Temp:**	150 degF

Other

Is equipment insulated?	Yes	**Insul. Mat'l:**	1" Ca Silicate
Is equipment cooled?	No	**Capacity:**	
Is fluid agitated?	No		

Equipment Notes

The vessel could be overpressured in the event of thermal expansion due to ambient temperature rise while the fluid is blocked in. The relief requirement is nominal compared to the external fire relief case.

In the event of an external fire, the initial relief will be flashing liquid. Upon formationof a vapor space, the relief will be vapor. The relief device should be sized for the case requiring the largest orifice area.

Protective Devices	Causes of Overpressure		
PSV-1	External Fire: V. Vessel w/ Liquid	57469	lb/hr
PSV-2	Failure of Automatic Controls: Inlet	25093	lb/hr
	Thermal Expansion	416.9	lb/hr

Berwanger, Inc.
Oil, Gas & Petrochemical Consulting

Figure 5

New Relief System Files

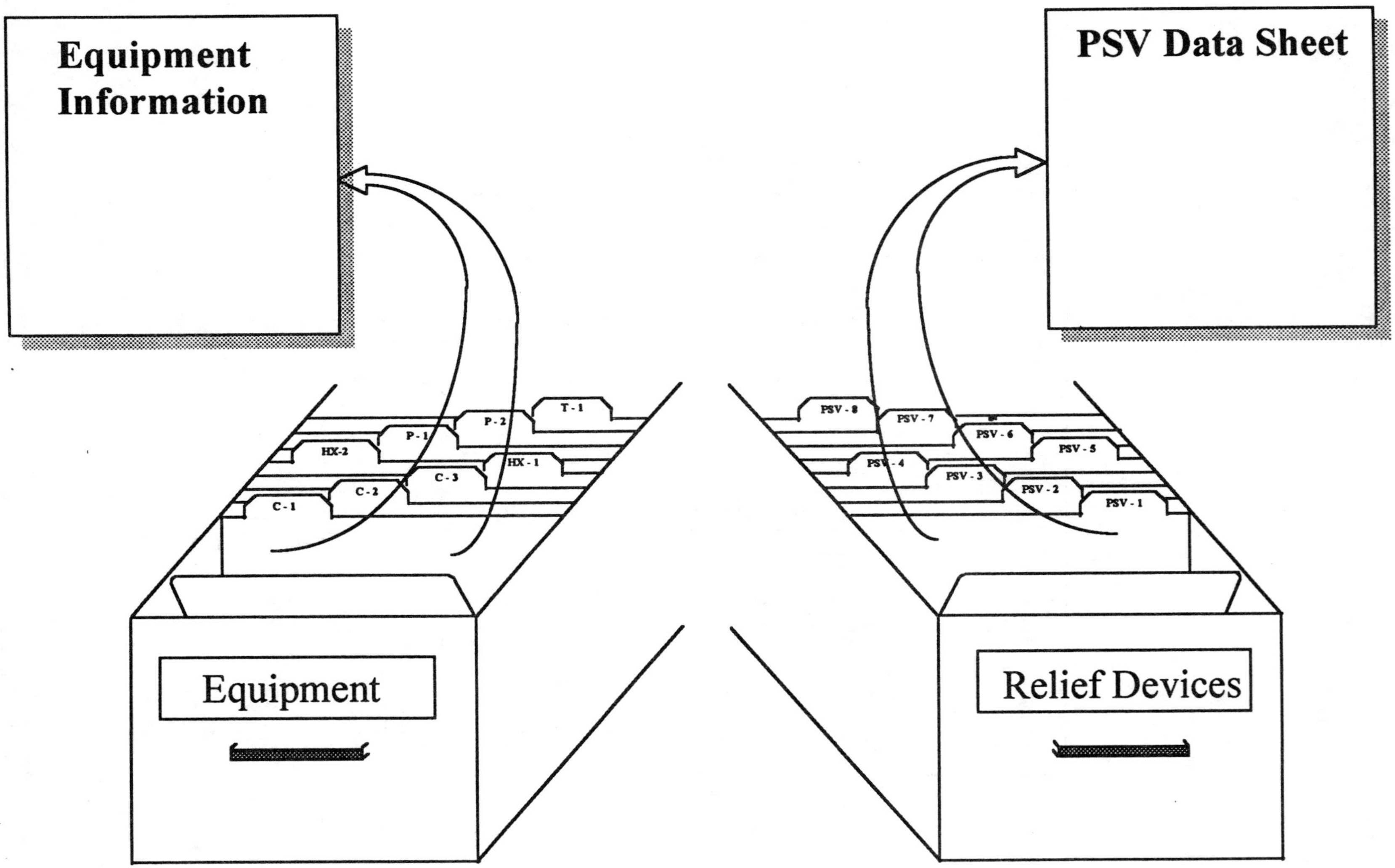

Figure 6

Berwanger, Inc.
Oil, Gas & Petrochemical Consulting

Electronic Relational Database Structure (patent pending)

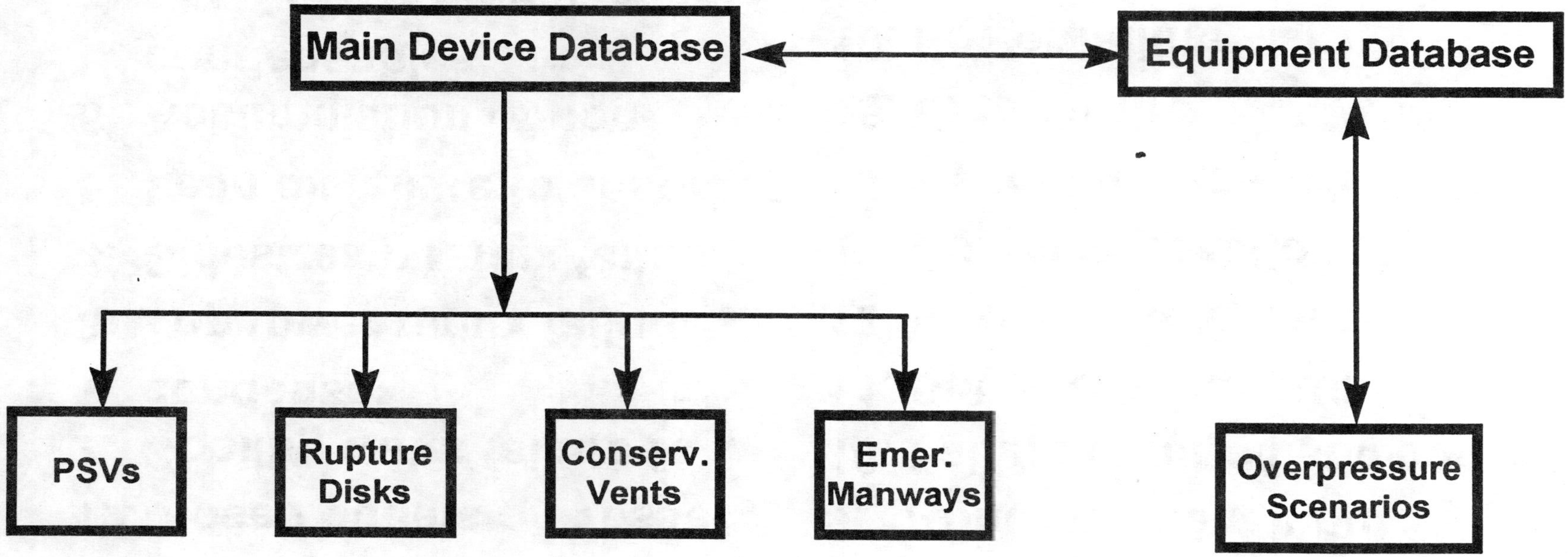

Figure 7

Berwanger, Inc.
Oil, Gas & Petrochemical Consulting

API 520 Table 1

1. Closed outlets on vessel
2. Cooling water failure to condenser
3. Top tower reflux failure
4. Sidestream reflux failure
5. Lean oil failure to absorber
6. Accumulation of non-condensables
7. Entrance of highly volatile material
8. Overfilling storage or surge vessel
9. Abnormal heat input
10. Failure of automatic controls
11. Split exchanger tube
12. Internal explosions
13. Chemical reaction
14. Hydraulic expansion
15. Exterior fire
16. Power failure

Figure 8

Berwanger, Inc.
Oil, Gas & Petrochemical Consulting

Berwanger, Inc. Presents:

The Emerging Industry Standard:

Equipment Based Pressure Relief System sign and Documentation

Presentation Slides

When we are finished, you should understand:

- the “Berwanger Method”, a step by step process to meet relief system PSI and PHA requirements
- your relief system evaluation must be “equipment list” based to ensure that all equipment is protected (the relief device data sheets alone cannot do it)
- this is also the least costly way to ensure safe operations and OSHA compliance

What we will talk about ...

- Three goals and three keys to success
- The new "equipment based" data structure
  (why the old relief device data sheets are inadequate)
- Execution plan
- OSHA mandated PHA elements
- Results

Three Goals ...

Goal #1:

ENSURE plant is adequately protected against all potential overpressure scenarios

Goal #2:

MAINTAIN safe design as changes are made to the plant

Goal #3:

Implement at low COST

... and Three Keys to Success

Goal #1: ENSURE protection

- check each potential overpressure scenario associated with each piece of EQUIPMENT

Goal #2: MAINTAIN safe design

- implement an information management system that is based upon protected EQUIPMENT (not relief devices)

Goal #3: Implement at low COST

- define tasks and then execute each task with most efficient people and tools

Example Distillation System

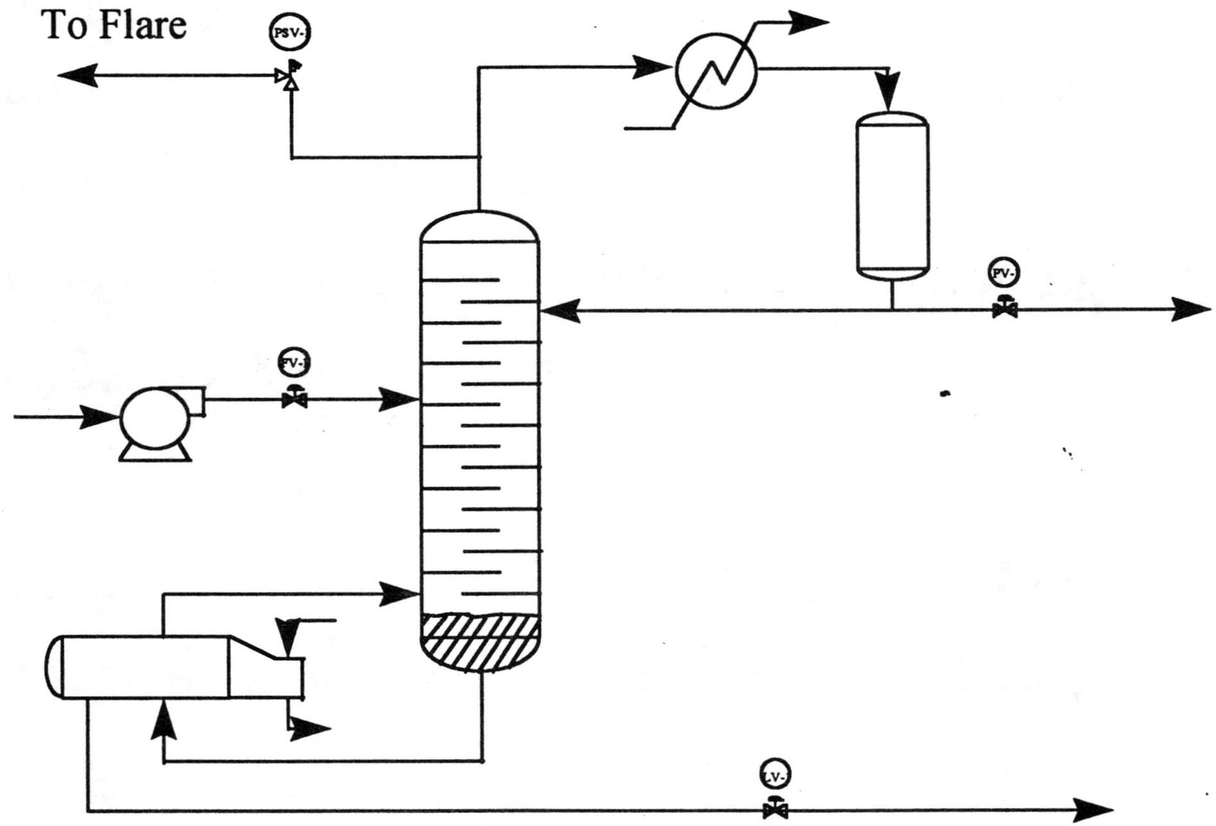

Berwanger, Inc.
Oil, Gas & Petrochemical Consulting

Example Distillation System

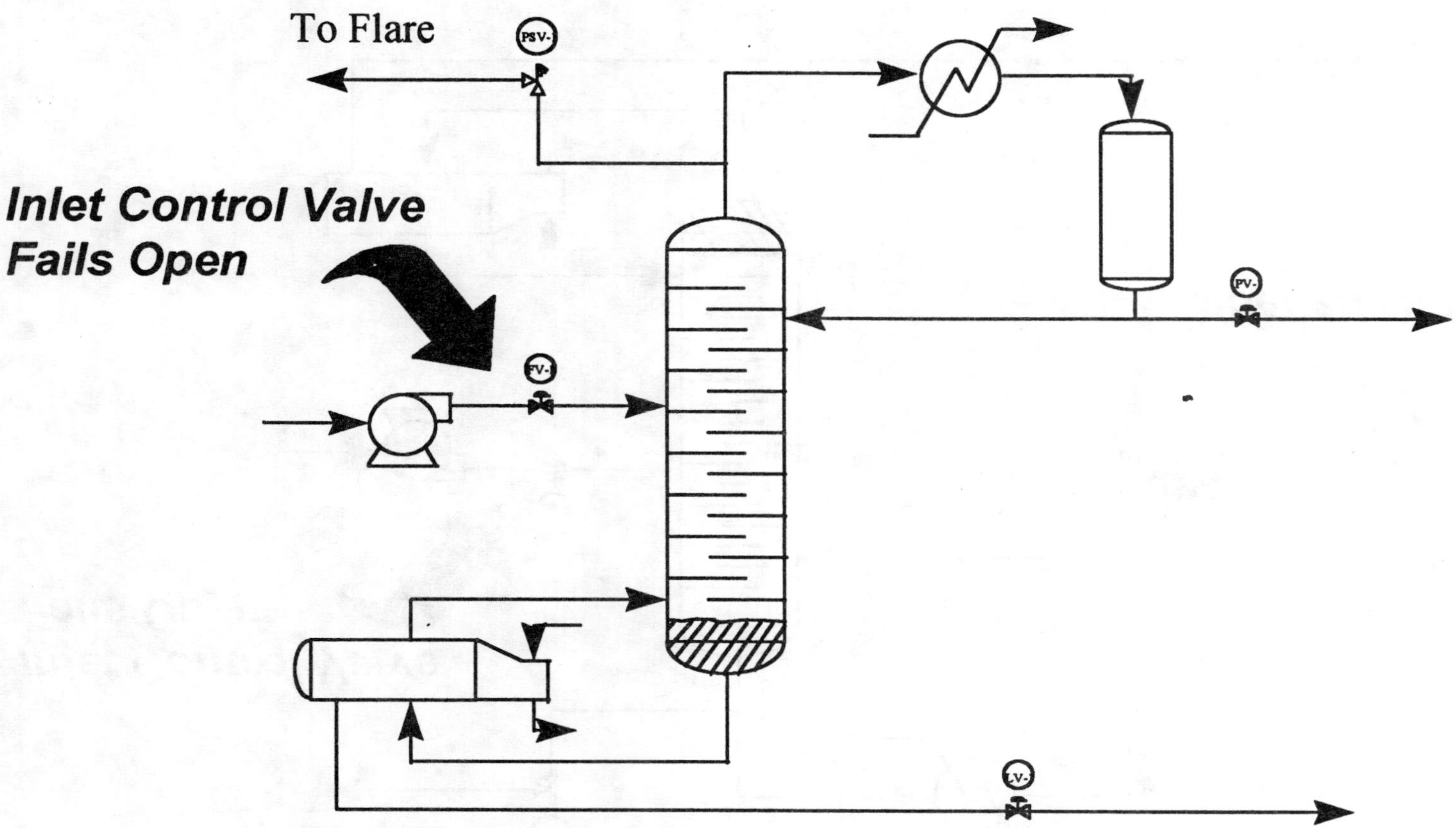

Berwanger, Inc.
Oil, Gas & Petrochemical Consulting

Example Distillation System

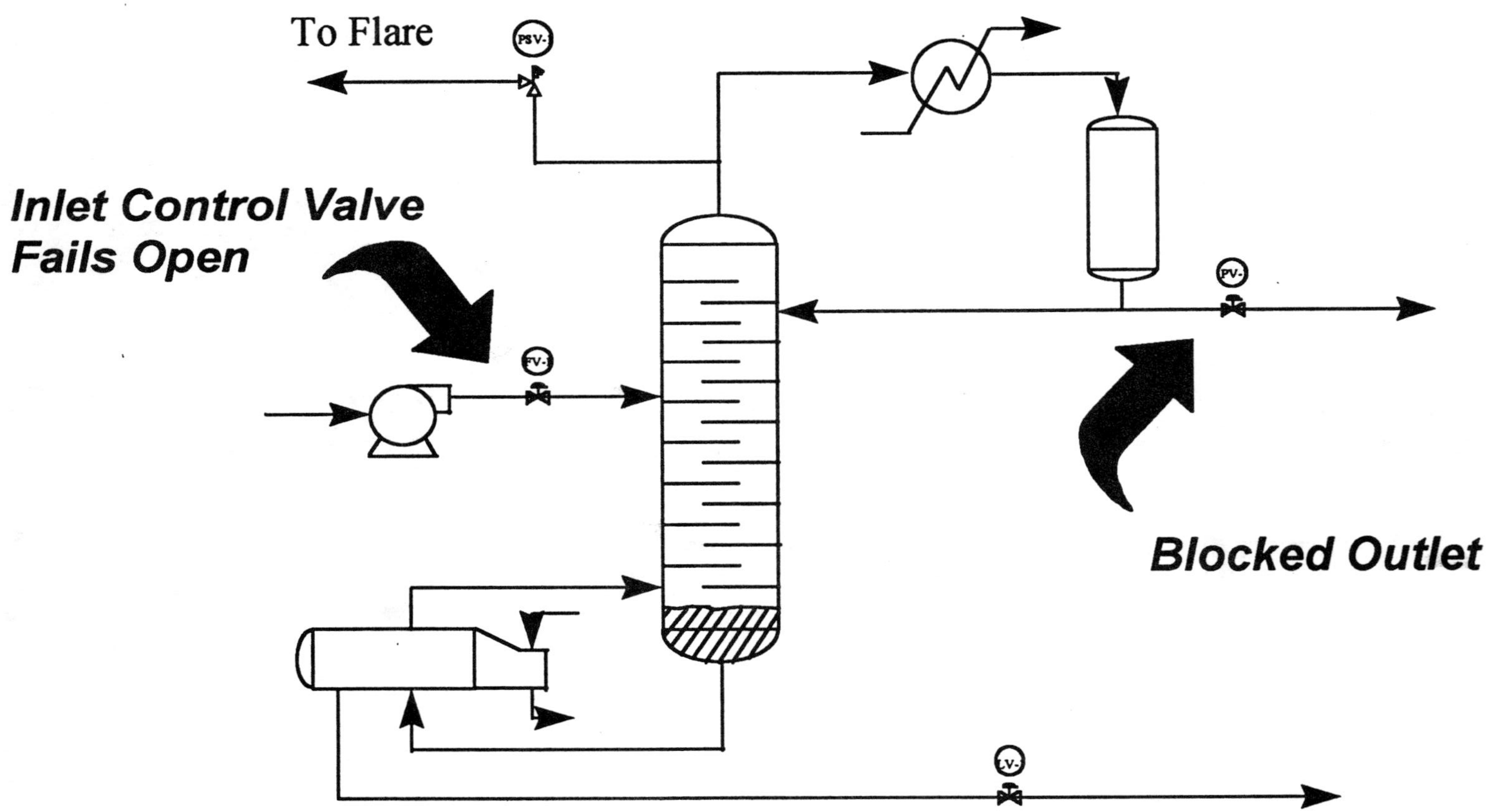

Berwanger, Inc.
Oil, Gas & Petrochemical Consulting

Example Distillation System

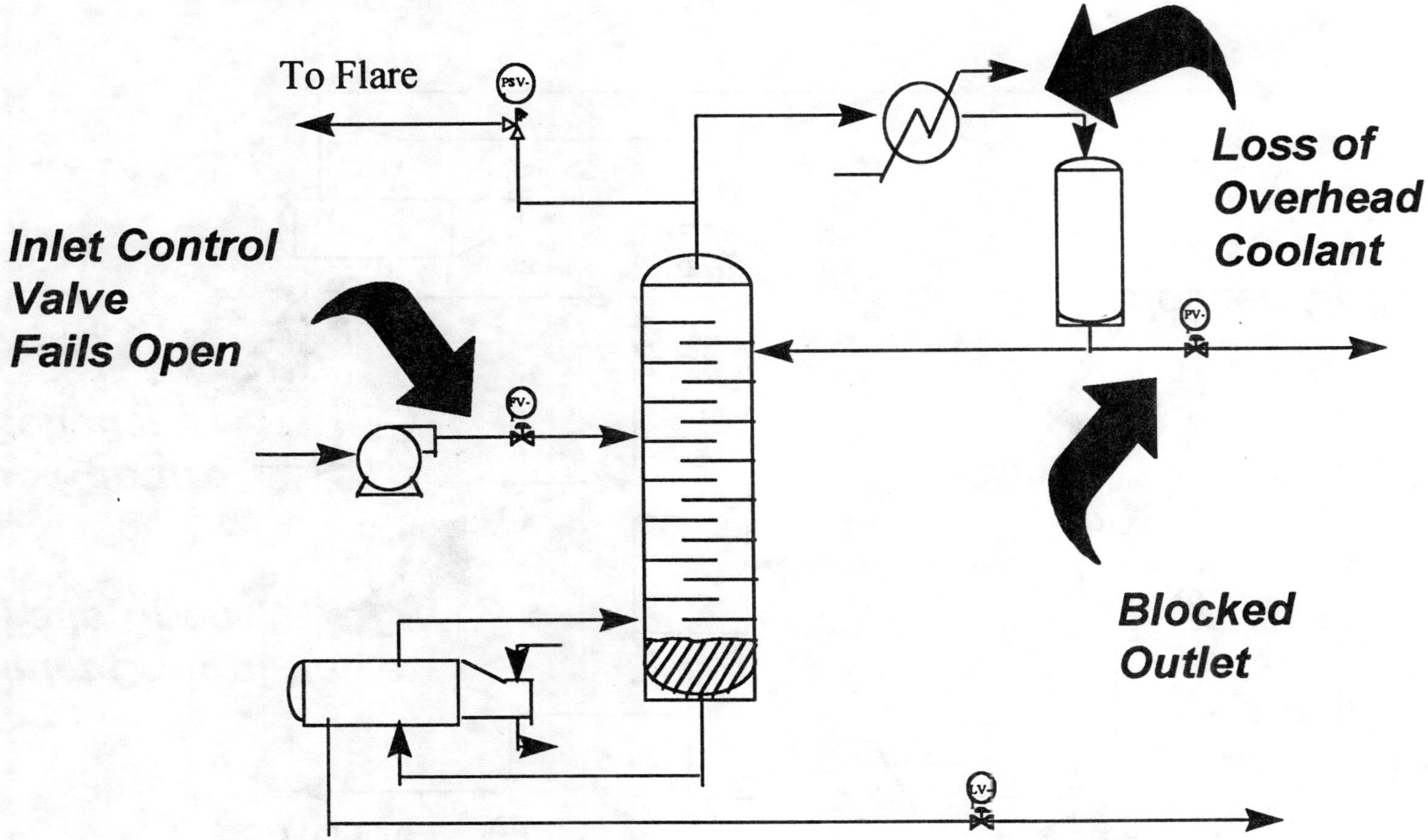

Berwanger, Inc.
Oil, Gas & Petrochemical Consulting

Example Distillation System

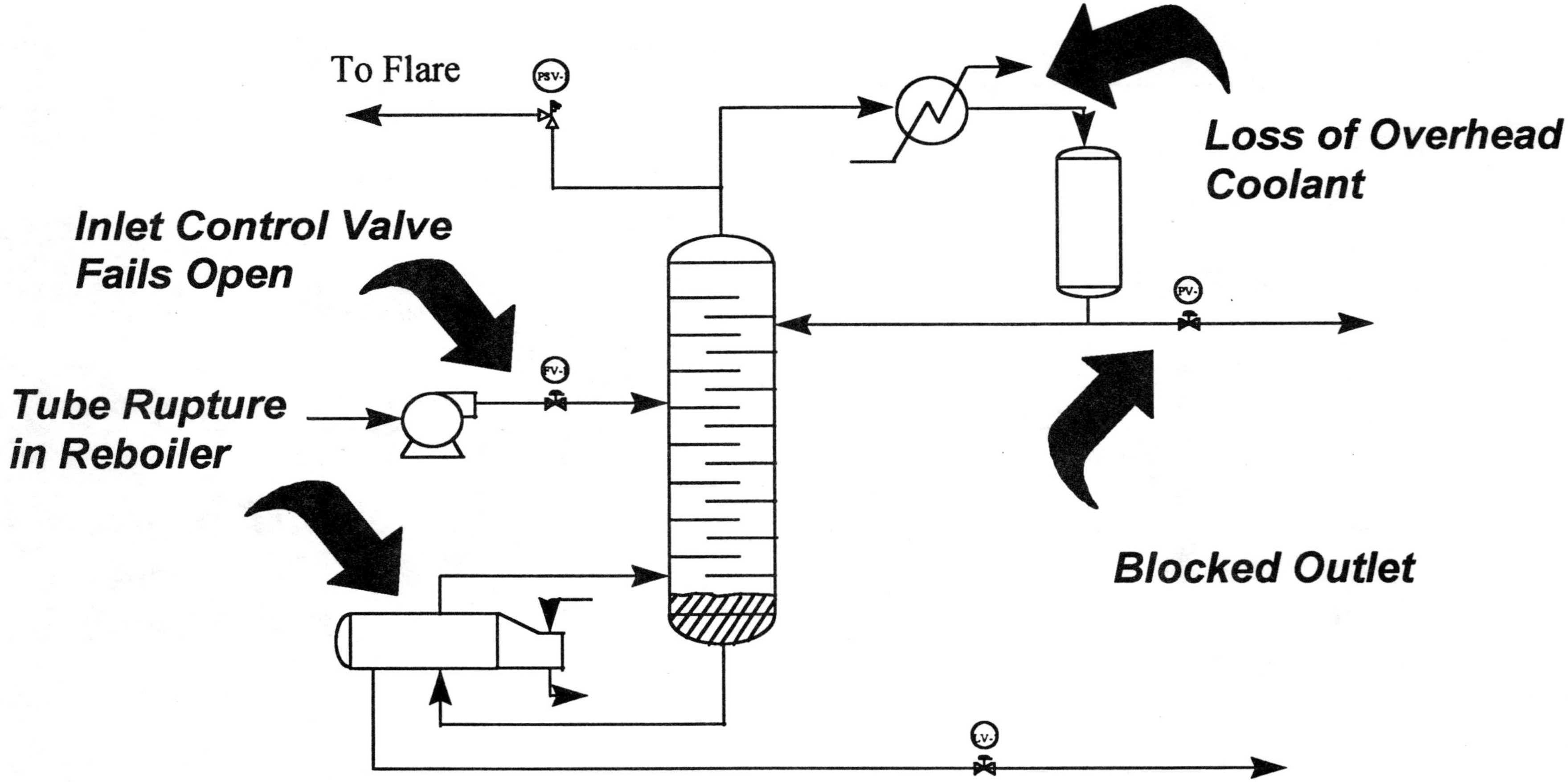

Berwanger, Inc.
Oil, Gas & Petrochemical Consulting

Example Distillation System

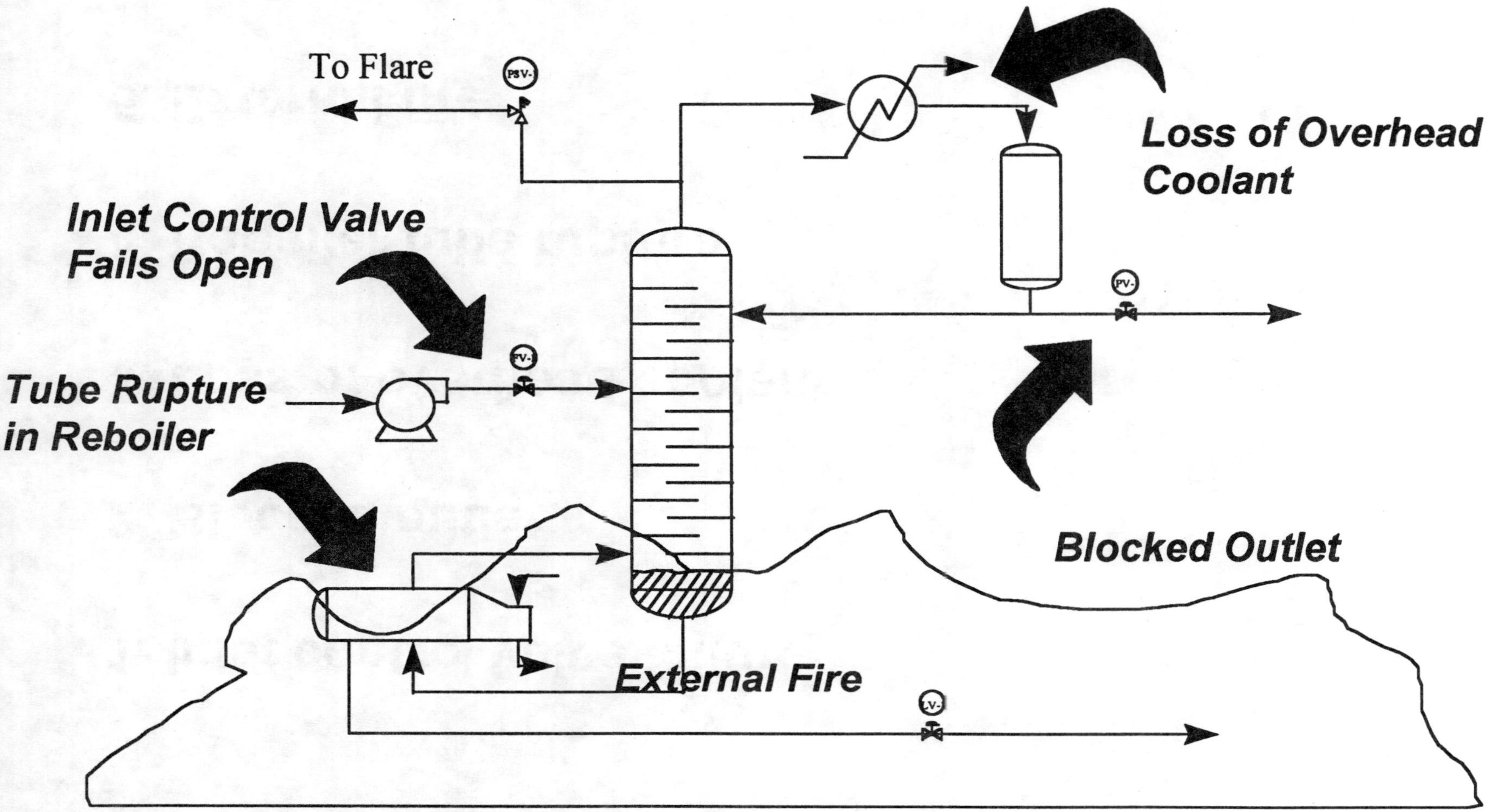

Berwanger, Inc.
Oil, Gas & Petrochemical Consulting

Overpressure Scenarios

1. Inlet control valve failure
2. Blocked outlet
3. Loss of overhead coolant
4. Reboiler tube rupture
5. External fire

Relationships between Equipment, Overpressure Scenarios, and Relief Devices

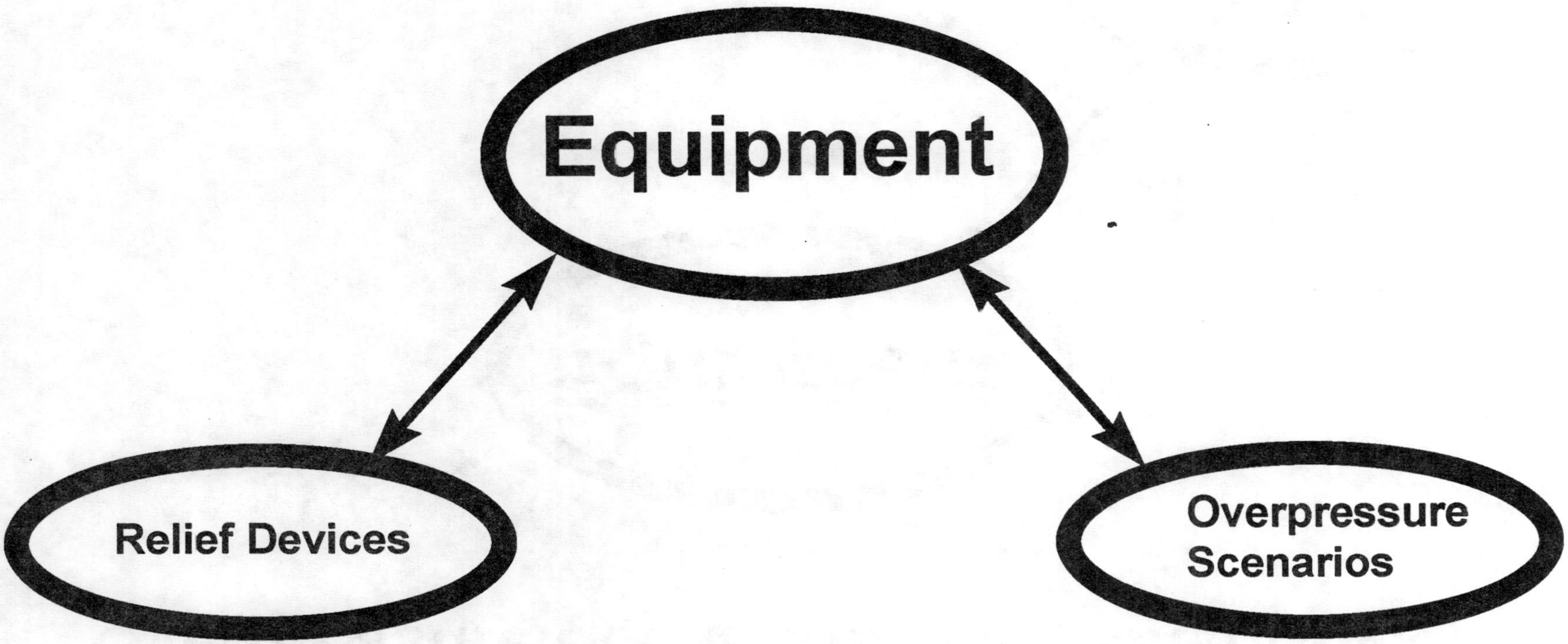

Relationships between Equipment, Overpressure Scenarios, and Relief Devices

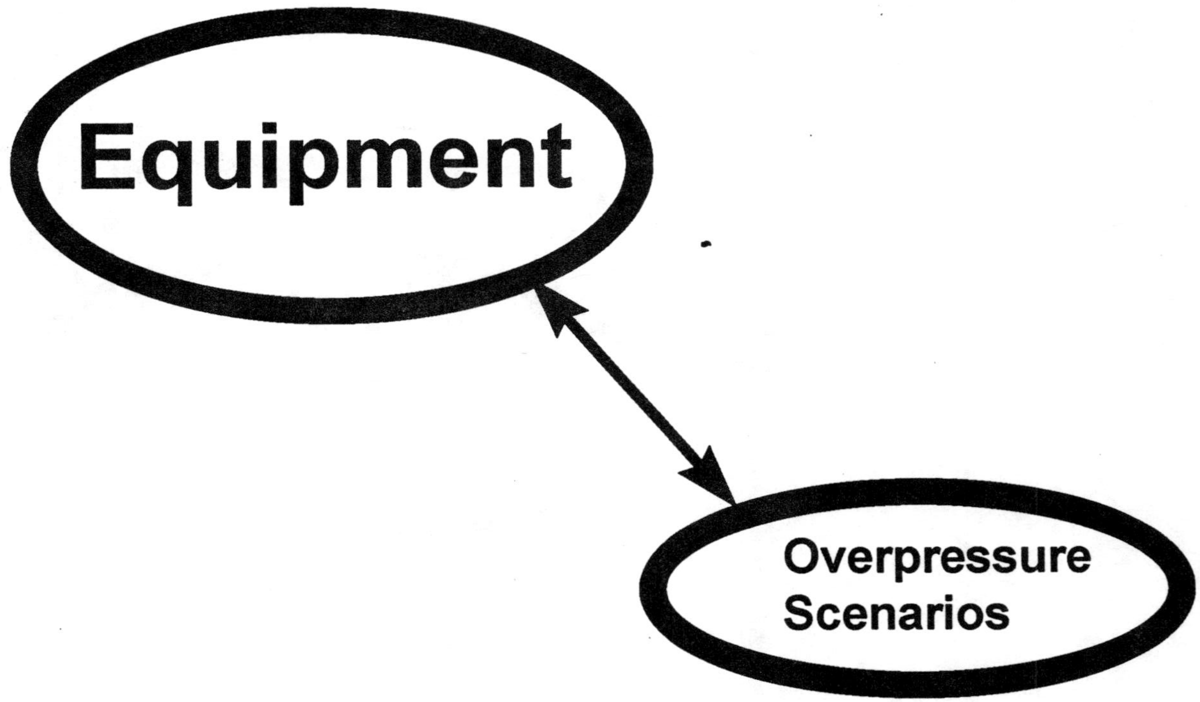

Paradigm Shift from Relief Device to Protected Equipment ...

Premise:

Overpressure scenarios are directly associated with pieces of equipment - and exist even if no relief devices (or their data sheets) are present.

Conclusion:

A relief device list (or the relief device data sheets) cannot serve as the primary checklist for confirming overpressure protection.

Traditional Relief System Documentation

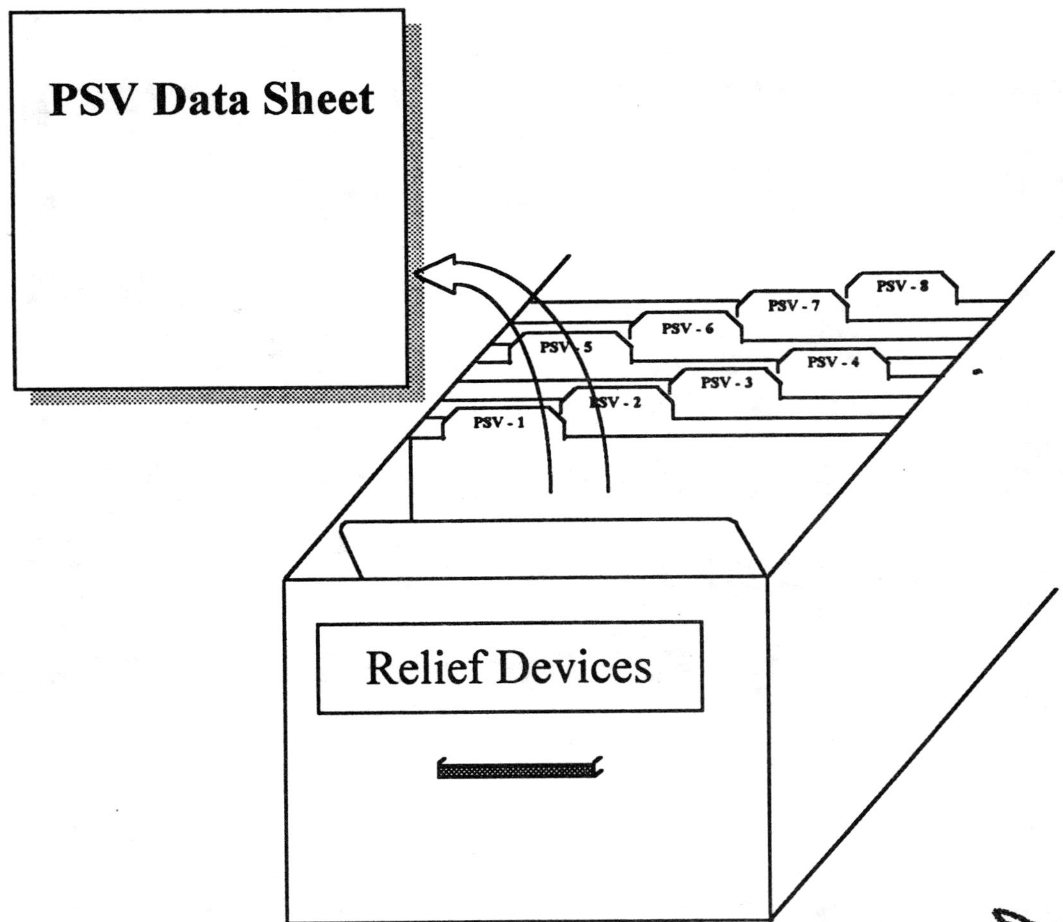

Berwanger, Inc.
Oil, Gas & Petrochemical Consulting

Berwanger, Inc.
Oil, Gas and Petrochemical Consulting

PROJECT: Relief System Presentation
CUSTOMER: Audience
CUSTOMER JOB NO: 10
PLANT: XYZ
LOCATION: Houston

PRESSURE RELIEF VALVES			
NO	BY	DATE	REVISION
0	PCB		

SHEET ____ OF ____
SPEC. NO. REV.
CONTRACT DATE
REQ P.O.
BY CHK'D APPR.

GENERAL	1	TAG NO.		PSV-001	
	2	SERVICE		Natural Gas	
	3	LINE NO./VESSEL NO.		3"-PR-001-A1	
	4	FULL NOZZLE/SEMI NOZZLE		Full	
	5	SAFETY OR RELIEF		Safety	
	6	CONV., BELLOWS, PILOT OP.		Conventional	
	7	BONNET TYPE		Closed	
CONN	8	SIZE: INLET	OUTLET	3	4
	9	FLANGE RATING OR SCREWED			
	10	TYPE OF FACING		Raised Face	
MATERIALS	11	BODY AND BONNET		Carbon Steel	
	12	SEAT AND DISC			
	13	RESILIENT SEAT SEAL			
	14	GUIDE AND RINGS			
	15	SPRING			
	16	BELLOWS			
	17	BLANK 1			
OPTIONS	18	CAP: SCREWED OR BOLTED			
	19	LEVER: PLAIN OR PACKED			
	20	TEST GAG			
	21	BLANK 2			
	22	BLANK 3			
	23	BLANK 4			
BASIS	24	CODE		API 521	
	25	FIRE		No	
	26	BLANK 5			
	27	BLANK 6			
FLUID DATA	28	FLUID		Natural Gas	
		FLUID STATE		Vapor	
	29	REQUIRED CAPACITY		10.7 MSCFD	
	30	MOL. WEIGHT	OPER. SP. GR.	20	-
	31	OPER. PRESS.	SET PRES.	350 psi	500 psi
	32	OPER. TEMP.	REL. TEMP.	98 °F	98 °F
	33	BACK PRESSURE: CONSTANT		3.0 psi	
	34	VARIABLE		11.0 psi	
	35	TOTAL		14.0 psi	
	36	% ALLOWABLE OVERPRESSURE		10	
	37	OVERPRESSURE FACTOR			
	38	COMPRESSIBILITY FACTOR		0.93	
	39	LATENT HEAT OF VAPROIZ.		-	
	40	RATIO OF SPECIFIC HEATS		1.25	
	41	OPER. VISC.		0.05 cP	
	42	BAROMETRIC PRESSURE		14.65 psia	
	43	BLANK 7			
	44	BLANK 8			
	45	CALC. AREA SQ. IN.		0.64 in2	
	46	SELECTED AREA		1.287 in2	
	47	ORIFICE DESIGNATION		J	
	48	MANUFACTURER		Crosby	
	49	MODEL NUMBER			

NOTES

ISA FORM S20 S3

BLANK 4				
CODE		API 521		
FIRE		Yes		
BLANK 5				
BLANK 6				
FLUID		Natural Gas		
FLUID STATE		Vapor		
REQUIRED CAPACITY		10.7 MSCFD		
MOL. WEIGHT	OPER. SP. GR.	20	-	
OPER. PRESS.	SET PRES.	350 psi	500 psi	

OPER. VISC.	0.05 cP	
BAROMETRIC PRESSURE	14.65 psia	
BLANK 7		
BLANK 8		
CALC. AREA SQ. IN.	0.64 in2	
SELECTED AREA	1.287 in2	
ORIFICE DESIGNATION	J	
MANUFACTURER	Crosby	
MODEL NUMBER		

Why a PSV datasheet alone is inadequate...

- Data sheet does not address unprotected equipment
- Data sheet identifies only a single overpressure scenario per device

(data sheet is primarily intended for valve purchase - not design basis documentation)

Solution ...

- Use the equipment list as the primary "checklist"
- Implement a data management system that links multiple overpressure scenarios and relief devices to their protected equipment
- EXAMPLE: Equipment Information Form

(data management system can be paper or electronic - but electronic is more efficient)

New Relief System Files

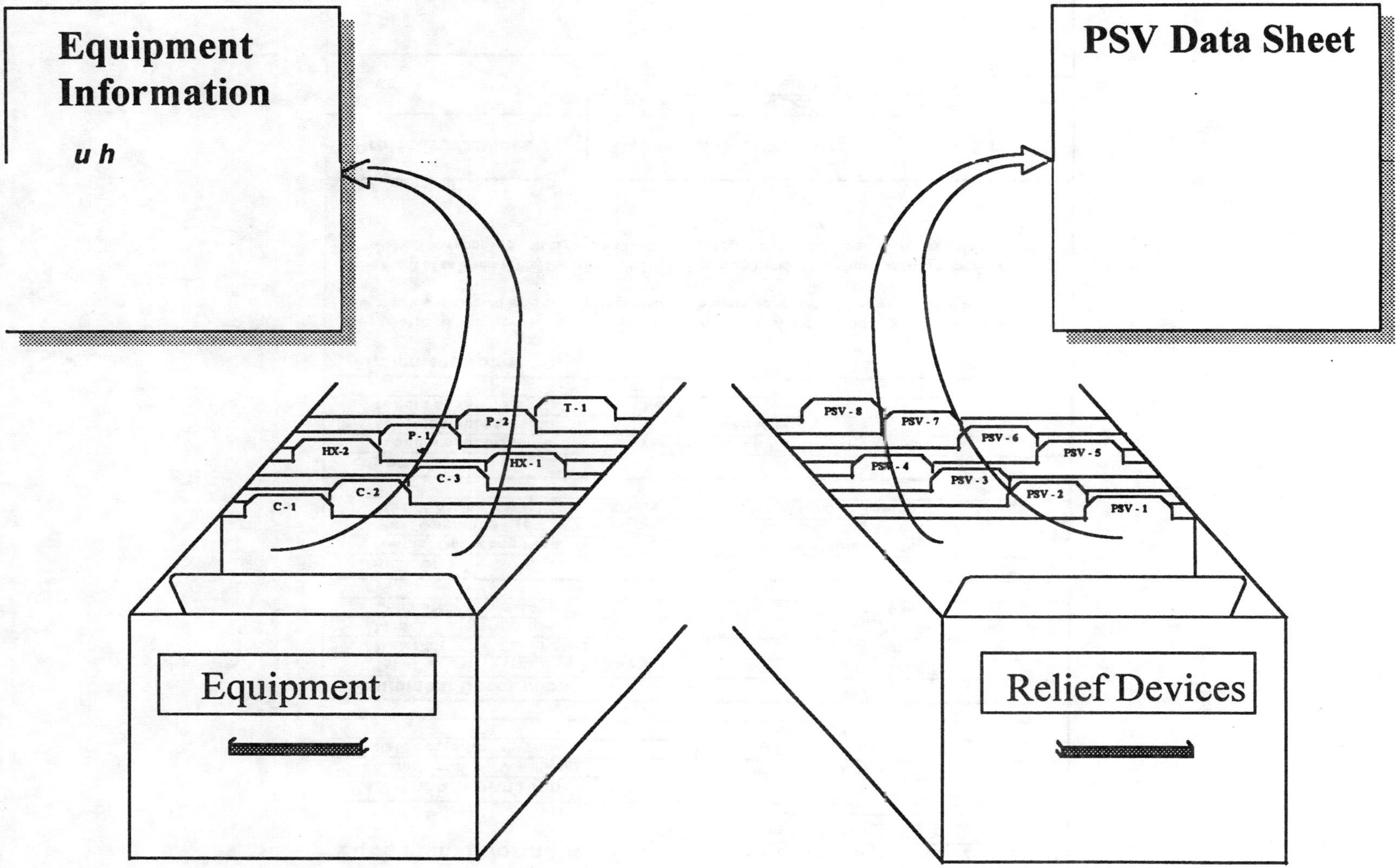

Berwanger, Inc.
Oil, Gas & Petrochemical Consulting

Berwanger, Inc.
Equipment Information Form

Revision Record
Tag No: V-1
No: 01 **Date:** 04/19/95 **Initials:** AAA
Propylene Purification Vessel

Reference Information

Manufacturer:	Vessels, Inc.	**PFD No:**	100-1A
Nat'l Board No:	4235	**P&ID No:**	200-1A
		Iso No:	300-1A

Equipment Description

Type:	Vertical Vessel (Cylinder)	**Height/Length:**	20.0 ft
Material:	Carbon Steel	**Width/ID:**	5.0 ft
Mode:	Continuous	**Equip Volume:**	400.0 ft3
Contents:	Oxo-Purge	**Fluid Volume:**	400.0 ft3

Conditions

Normal Operating Pres:	350 psig	**Normal Operating Temp:**	150 degF
Max. Operating Pres:	360 psig	**Max Operating Temp:**	150 degF
Max. Working Pres:	400 psig	**Max Working Temp:**	150 degF

Other

Is equipment insulated?	Yes	**Insul. Mat'l:**	1" Ca Silicate
Is equipment cooled?	No	**Capacity:**	
Is fluid agitated?	No		

Equipment Notes

The vessel could be overpressured in the event of thermal expansion due to ambient temperature rise while the fluid is blocked in. The relief requirement is nominal compared to the external fire relief case.

In the event of an external fire, the initial relief will be flashing liquid. Upon formationof a vapor space, the relief will be vapor. The relief device should be sized for the case requiring the largest orifice area.

Protective Devices	Causes of Overpressure		
PSV-1	External Fire: V. Vessel w/ Liquid	57469	lb/hr
PSV-2	Failure of Automatic Controls: Inlet	25093	lb/hr
	Thermal Expansion	416.9	lb/hr

while the fluid is blocked in. The relief requirement is nominal compared to the external fire relief case.

In the event of an external fire, the initial relief will be flashing liquid. Upon formation of a vapor space, the relief will be vapor. The relief device should be sized for the case requiring the largest orifice area.

Protective Devices	Causes of Overpressure	
PSV-1 PSV-2	External Fire: V. Vessel w/ Liquid	57469 lb/hr
	Failure of Automatic Controls: Inlet	25093 lb/hr
	Thermal Expansion	416.9 lb/hr

Electronic Relational Database Structure (patent pending)

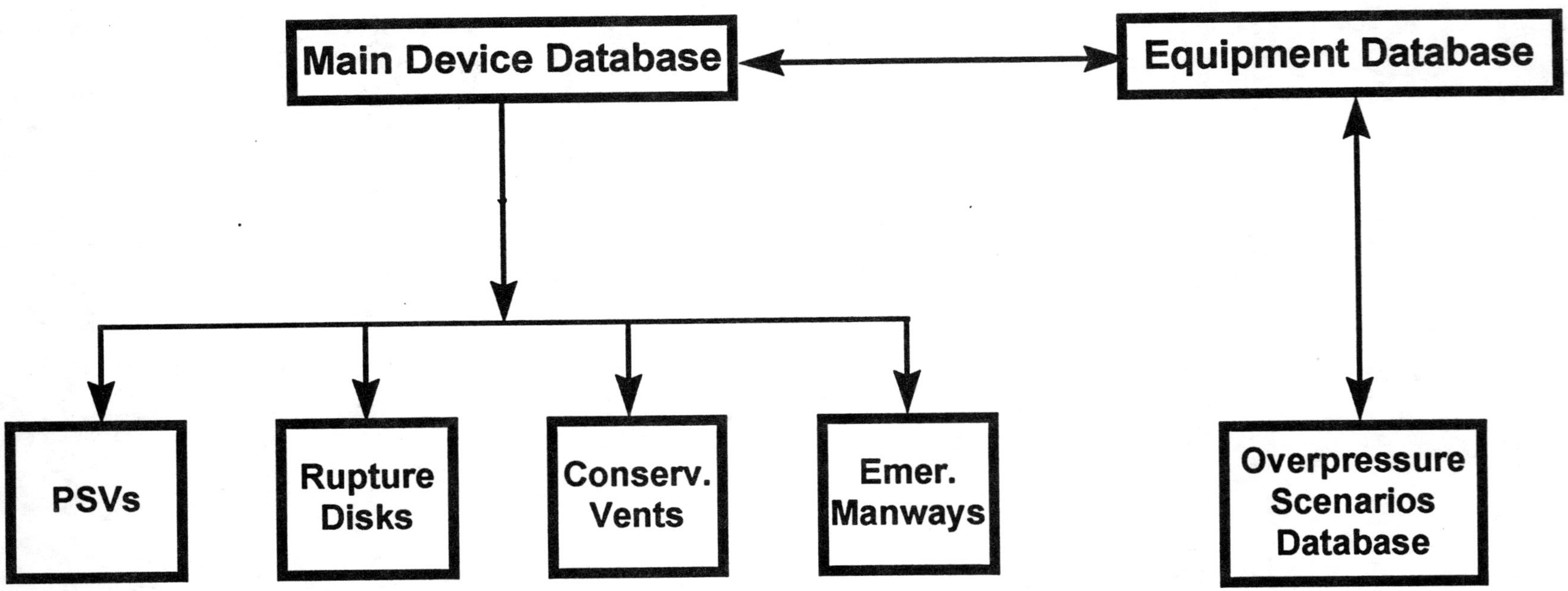

Execution Plan (patent pending)

Step 1: "Expert" identifies overpressure scenarios

Step 2: Quantify required relief rate for each scenario

Step 3: Check relief device capacities

Step 4: Evaluate relief headers

Step 1: Scenario Identification

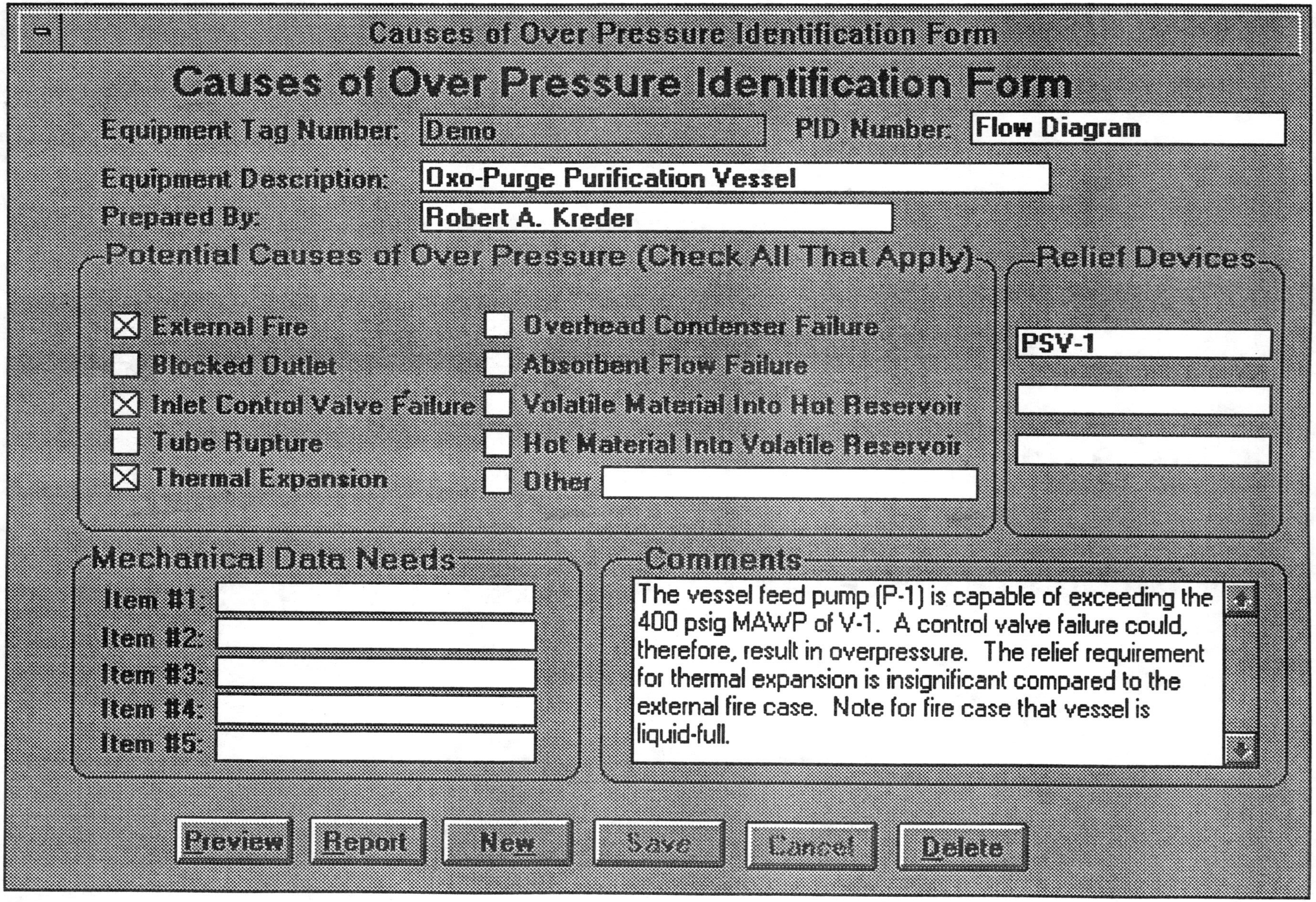

Step 2: Required Relief Rate Calculation

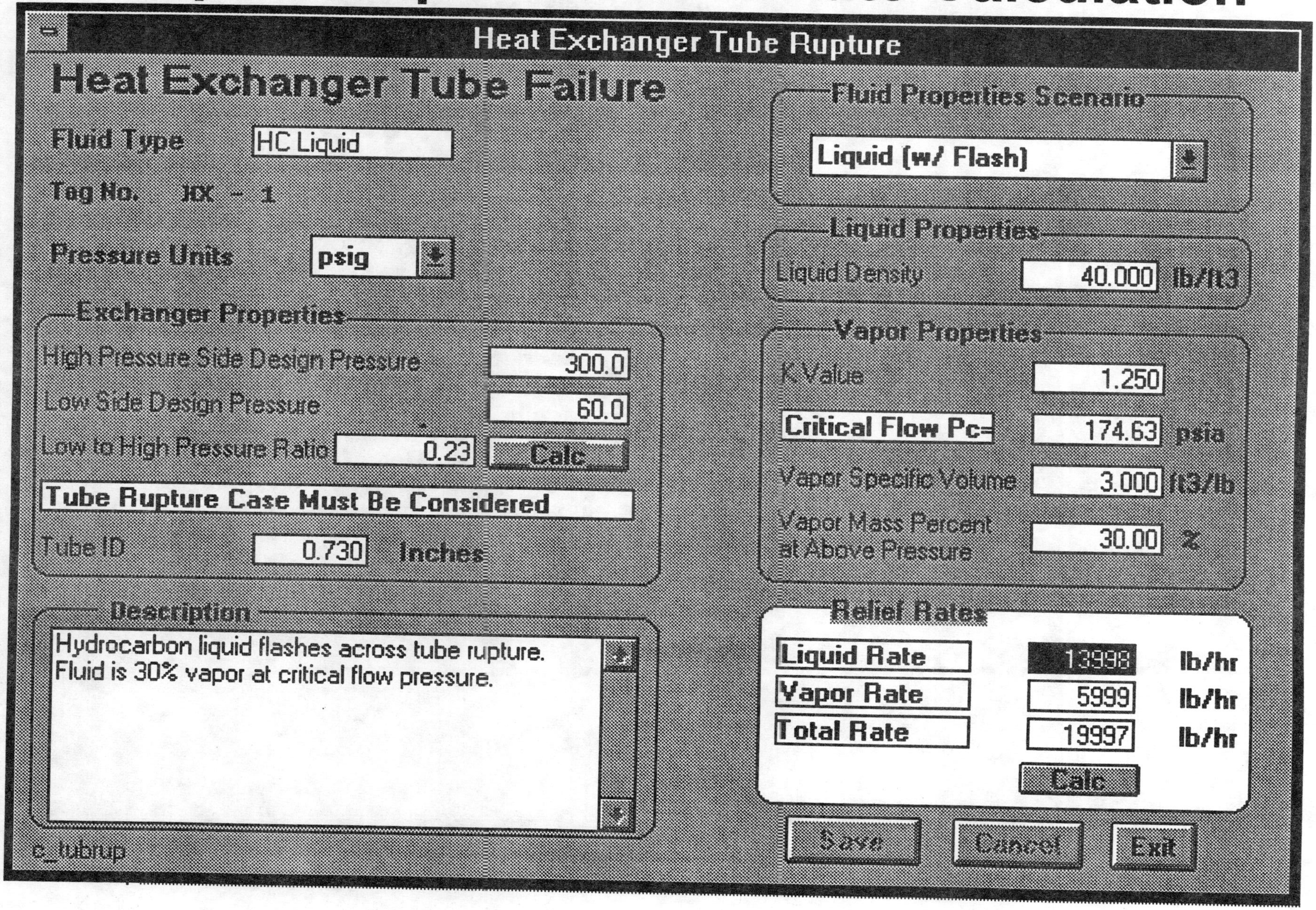

Step 3: Relief Valve Capacity

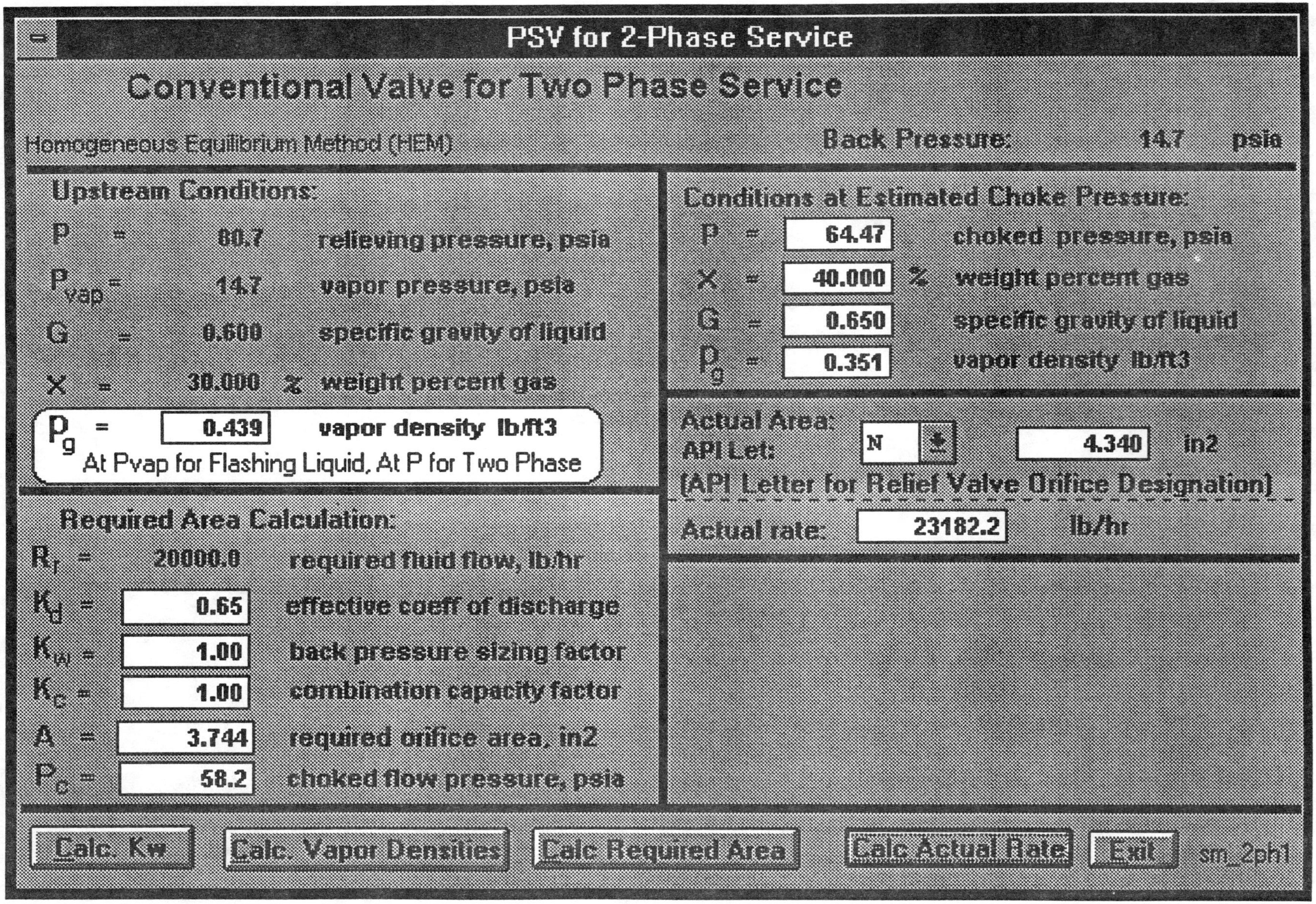

Step 4: Evaluate Relief Headers

- Create hydraulic model of relief headers
- Sum individual scenarios to arrive at global scenarios (example: fire circles)
- Evaluate backpressures for global scenarios
- Evaluate flare scrubber performance
- Evaluate radiation effects from flare

The results of a fundamentally different approach ...

- Most thorough evaluation in industry
- OSHA compliance
- Only all-electronic and maintainable system
- Fewer findings in future PHA's
- Lowest cost in industry (history shows)

OSHA Mandated PHA Elements

- hazards of the process
- previous incidents
- engineering and administrative controls
- facility siting
- human factors
- qualitative consequence analysis
- employee involvement
- follow-up and validation

At Berwanger ...

We look at each and every job as an opportunity to find a better, more efficient way of achieving our client's objective.

Berwanger, Inc.
Oil, Gas & Petrochemical Consulting

An Overview of SuperChems for DIERS for Emergency Relief Systems and Vent Containment Design

G. A. MELHEM
Arthur D. Little, Inc., 20 Acorn Park. Cambridge, Massachusetts 02140

H. G. FISHER
Union Carbide Corporation, South Charleston, West Virginia, 25303

ABSTRACT

SAFIRE is a computerized tool developed by the Design Institute of Emergency Relief Systems (DIERS) of AIChE. This computer program is used for ERS design and has undergone several minor revisions since its release. Numerous copies have been sold and are being used by several companies to design emergency relief systems.

DIERS is currently revising SAFIRE to include several major enhancements to its engine and user interface. This paper will present an overview of the new computer code/technology, validation, and will illustrate its many advantages on many industrial applications.

The new computer code, SuperChems for DIERS, is a dynamic simulator, capable of performing ERS and vent containment design for complex geometries and multiphase reaction systems. In addition, SuperChems for DIERS is an equation-of-state based code which provides several benefits over existing non-equation-of-state based methods in ERS design for systems involving supercritical reactions like polymerizations of butadiene and acrylonitrile, solution effects such as HCl/Water, and apriori determination of phase-splitting.

This new computer code allows the user to dynamically simulate several common configurations for vent containment design. For example, the user is able to simulate a vessel discharging a two-phase mixture into a quench/vent where the catch/vent tank will vent to a stack or a scrubber. Unit operations available include separators (horizontal and vertical), cyclones, etc. Impact of back pressure and continuing reaction in the vent containment system is accounted for in the dynamic simulations.

BACKGROUND

During the 1st quarter of 1996, the Design Institute for Emergency Relief Systems (DIERS) Users Group awarded Arthur D. Little Inc. a contract to provide an improved next generation version of SAFIRE. Highlights of the new computer code as outlined by the request for proposal are summarized below:

1. Provide a general purpose thermodynamic and transport properties generator with implicit corrections for non-ideal behavior in both the liquid and vapor phase. The properties generator should also be able to provide temperature and pressure dependent derivatives for all properties of interest. This generator should be detailed enough so that heat of solution effects and vapor/liquid non-ideal equilibrium are implicit. This generator should be able to provide all properties and all derivatives required by SAFIRE models. The generator should be equation of state (EOS) based.

2. Create/design an interface allowing the properties generators to access a thermophysical properties database, so that properties do not have to be manually added. A databank manager should be provided in order to allow the users to input/modify their own compounds, if needed.

3. Revise all flow models such that the new revision will address the following items:

   (a) inclined flow
   (b) subcooled flow
   (c) sudden expansion/contraction
   (d) piping segments with varying diameter and orientation
   (e) viscous 2-phase flow through SRV and pipes
   (f) continuing chemical reaction in piping
   (g) detailed energy balances for vessels and piping
   (h) detailed momentum balance for piping
   (i) implicit vapor/liquid equilibrium relations
   (j) the flow models and the vessel balances should be equation oriented.

4. Provide a suitable stiff differential/algebraic (DAE) equation-solver to replace existing integrators. Proposed schemes should include Gear's or Michelsen's methods.

5. Provide a user-friendly menu-driven interface including graphics plotting capabilities and report generation.

6. Provide required documentation including an Operations Guide, a Users Guide, and a Reference Manual.

7. Validate all new models using experimental data, where available.

Arthur D. Little Inc. (ADL) had already developed a computer code which contained all the above specifications. This code took about five years to develop and is known as SuperChems Expert Version 3. The ERS design portion of SuperChems was customized for the DIERS Users Group and is currently undergoing final beta testing.

In addition to satisfying the criteria specified above, SuperChems for DIERS is capable of tracking continuing reactions in piping and vent containment systems, providing relief sizing for long pipes, etc. SuperChems is an equation-of-state based code which provides several benefits over existing non-equation-of-state based methods in ERS design for systems involving:

1. Supercritical reactions like polymerizations of butadiene, acrylonitrile, etc.
2. Solution effects such as HCl/Water, etc.
3. Apriori determination of phase-splitting

SuperChems contains a databank of more than 3000 binary systems with equation-of-state binary interaction parameters derived from experimental vapor/liquid and liquid/liquid data. There is also a VLE/VLLE data package. This is a versatile utility which allows the estimation of binary interaction parameters for the equation of state composition dependent mixing rule. Source of data can be one of three data types:

1. Azeotropic data
2. TPXY measured data (SuperChems also include three thermodynamic consistency utilities for X and Y calculations)
3. Mutual solubility data
4. Henry's law constants (mostly used for gas solubility in liquids)
5. Activity coefficients model parameters
6. Infinite dilution activity coefficients

SuperChems has an extensive database with over 1200 chemicals. The database contains 39 thermophysical properties with temperature dependent properties and data quality parameters. The code also has a detailed regression package (linear/non-linear) which allows the reduction of tabular data to equation forms supported by the databanks.

COMPUTER IMPLEMENTATION

ERS design does not stop at the estimation of the vent size. The ERS effluent must be treated if it is toxic and/or flammable or if it presents an environmental impact. While homogeneous flow (no slip) is typically used for sizing the vent, slip flow must be used to establish correct pressure drops and safety/environmental impact. SuperChems for DIERS is a computer code which allows the integral evaluation of relief dynamics and dowstream effects.

For example, using SuperChems for DIERS, we can evaluate the time dependent history of pressure, temperature and composition in a reactor vessel as the relief occurs. Simultaneously the effluent is expanded, and dispersed to established (regulatory or internal) limiting damage criteria. Typically, many options are evaluated before a final ERS design is selected. This

Figure 1: ERS/Mitigation System Effectiveness Evaluation Using SuperChems for DIERS

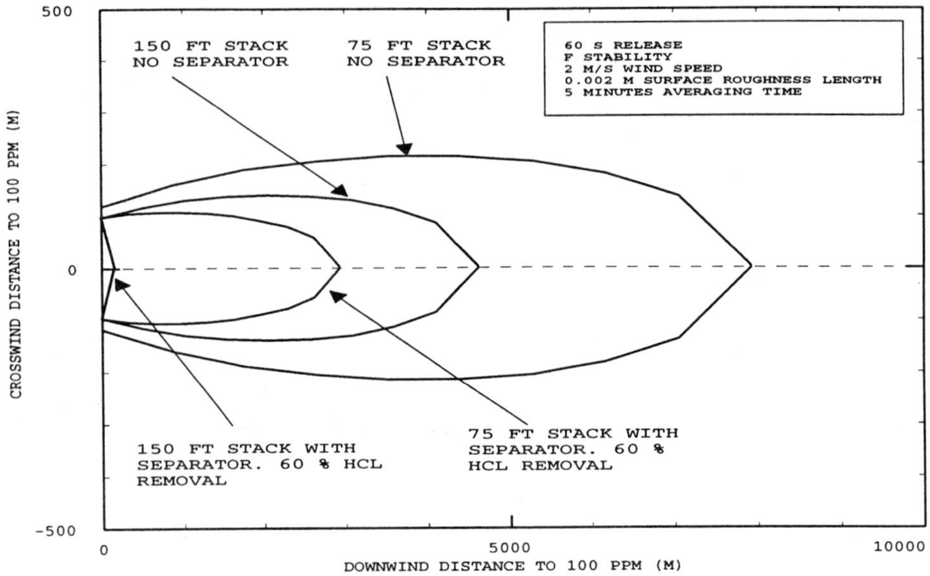

includes separation equipment, flares, stacks, etc. Figure 1 illustrates the use of SuperChems for DIERS to assess various mitigation measures following an alkyl chloride/water reaction which produces hydrogen chloride.

The most useful aspect of SuperChems for DIERS is its scenario/object driven architecture. Once an object (such as a vessel or piping configuration) is defined it can be used by one or more scenarios. Once a scenario is defined it can be duplicated and used to perform what-if or sensitivity analysis.

The detailed algorithms for SuperChems are published in references [1] and [2]. This paper will focus on providing examples and benchmarks for SuperChems for DIERS.

QUENCH TANK DESIGN FOR PCL3/WATER

This example deals with a 5,000 gal reactor where phosphorus trichloride (PCl_3) is used. A scenario was identified where it is possible for a heel of phosphorus trichloride (2,700 kgs) to remain in the reactor undetected (below detection level) at 40 C, and for an operator to attempt to flush the vessel with water. This can lead to the generation of gaseous hydrogen chloride and excessive system pressure.

Water can be introduced into the reactor at the rate of 15 kg/s for 38 seconds. The reactor has a 12 inch rupture disk set at 20 psig. The effluent is discharged into a quench tank. The quench tank has a volume of 10,000 gal and initially contains 24,000 kgs of water at ambient conditions. This is illustrated in Figure 2. The reaction rate and characteristics of PCl_3/H_2O reaction are described by Melhem and Reid in paper 43a which appears elsewhere in these proceedings.

Figure 2: A Simple Reactor/Quench Tank Arrangement

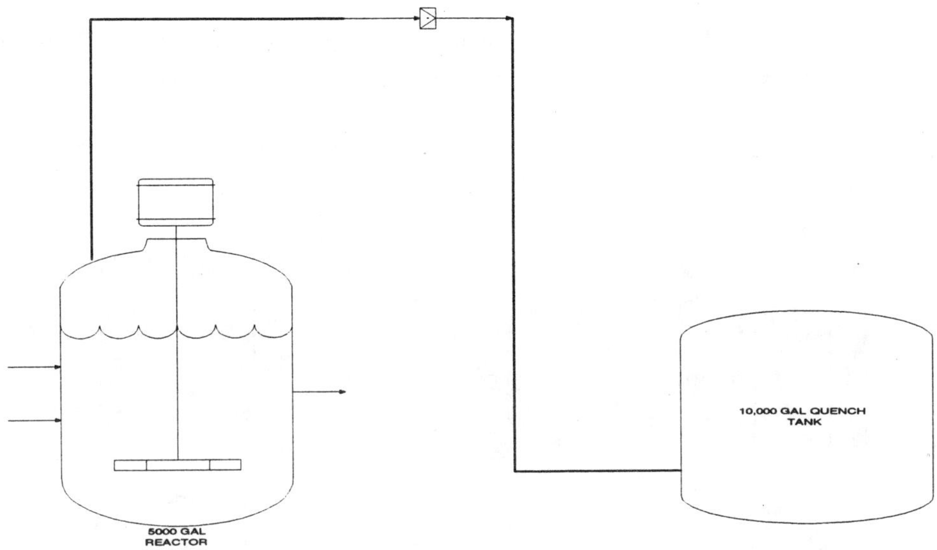

Figure 3 illustrates the calculated time history of pressure, temperature, and inidividual component flow rates for the reactor and the quench tank. Please note that both HCl and PCl_3 are discharged from the reactor and that PCl_3 reacts with the quench tank water to form phosphorous acid and HCl. One should also note that the temperature and pressure rise in the quench tank are caused primarily by the hydrogen chloride heat of solution.

FIRE EXPOSURE WITH MULTICOMPONENE VAPOR FLOW

The following example illustrates the use of SuperChems for DIERS to size a safety valve for a vessel containing a mixture of water, ethanol and acetone. Important vessel and fire exposure data are summarized in Figure 4 and Table 1. The vessel is a vertical cylindrical vessel and vapor flow is expected. This example illustrates the types of data that can be obtained from SuperChems for DIERS.

SuperChems is an equation of state based computer code. The equation of state is used to generate thermodynamic data as well as physical and chemical equilibrium data. Binary interaction parameters (BIPS) for non-ideal systems are either estimated from group contribution methods such as UNIFAC or estimated from experimental vapor/liquid equilibrium data. The binary interaction parameters for the system acetone, water and ethanol were obtained from SuperChems's extensive databank and are based on experimental data. This data is summarized in Table 2. The simulation also involved the use of nitrogen which was present in the vapor space prior to venting. Binary interaction parameters involving nitrogen were set to 0.

Figures 5 through 10 illustrate the results obtained using SuperChems. Two designs using a 2x3J and a 2x3H conventional valves were considered. The simulation results show that

Figure 3: Estimated Transient Profiles for Pressure, Temperature, Mass and Flow Rate Using SuperChems for PCL3/Water Quench Tank Design Problem

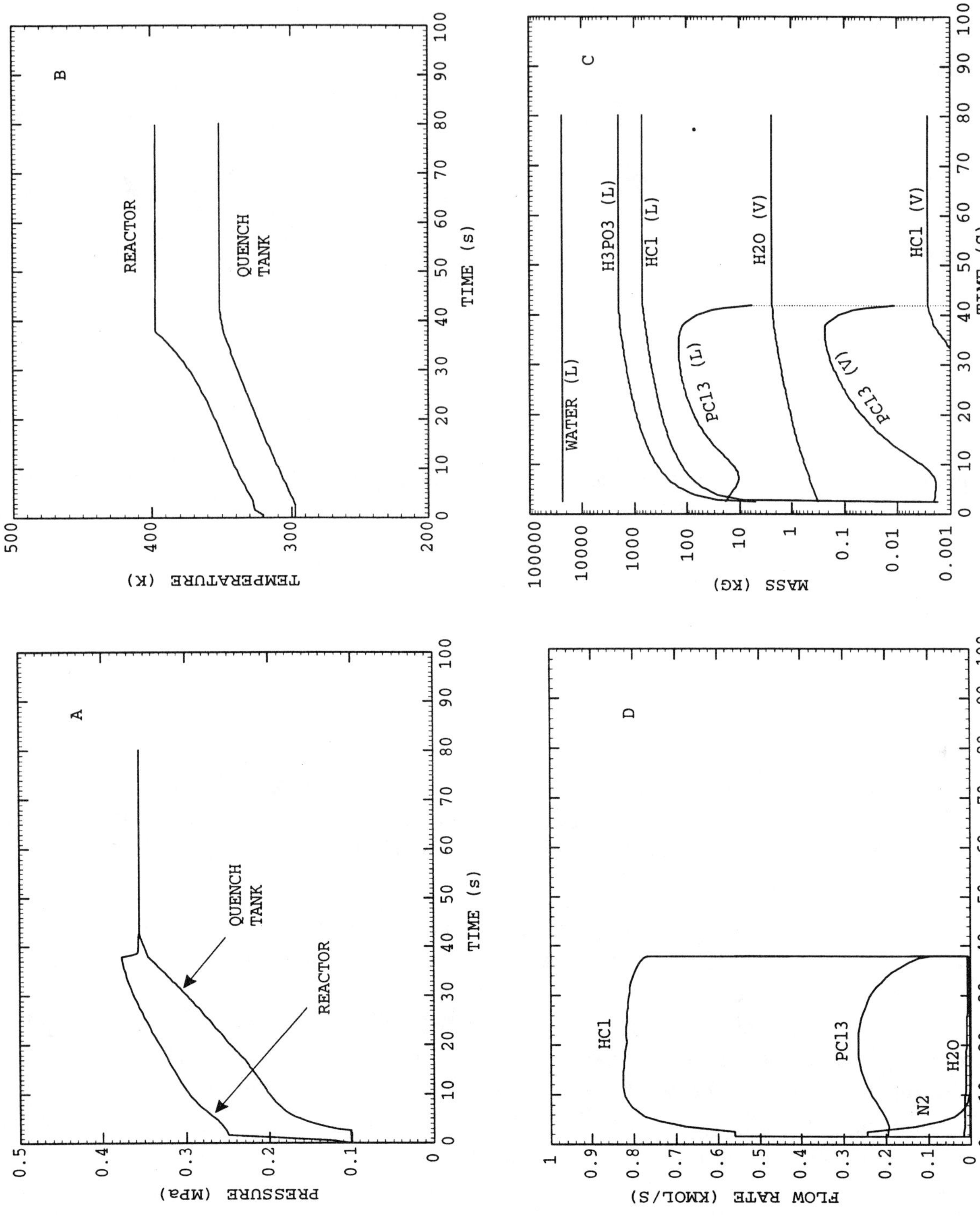

Figure 4: Fire loading on a vessel containing a mixture of acetone, ethanol and water

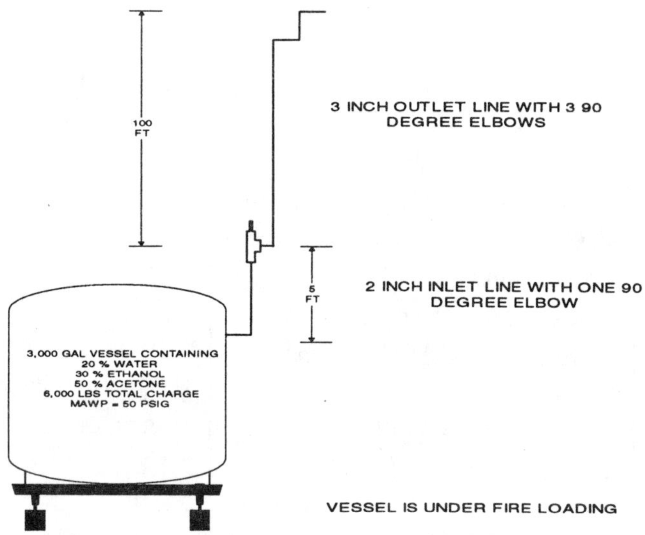

the 2x3J valve is slightly oversized while the 2x3H valve produces a vessel pressure which exceeds 72.5 psia.

Four cases were considered by the computer simulation. These included:

1. 2x3J (1.287 in^2 orifice) valve with a 2 inch inlet line
2. 2x3J valve with a 3 inch inlet line
3. 2x3H (0.785 in^2 orifice) valve with a 2 inch inlet line
4. 2x3H valve with a 3 inch inlet line

Figure 5 illusrates the calculated pressure/time profile calculated by SuperChems. All valves remain closed until 10 % overpressure is accumulated in the vessel at which time the valves fully open. The 2x3J opens and re-seats initially as nitrogen is discharged from the vapor space of the vessel. The 2x3H valve opens and remains fully open as the vessel contents are vaporized and discharged.

Figure 5 indicates that while the 2x3H valve is undersized the 2x3J valve should be adequate. SuperChems also computes the percent inlet pressure drop with respect to the valve set pressure (absolute). This is illustrated in Figure 6. The percent inlet pressure drop calculated using 2x3J valve with a 2 inch inlet line exceeds the 3 % value recommended by ASME/API. Please note that the percent inlet pressure drop reported in Figure 6 includes the accelerational loss as well as the frictional loss. The ASME 3 % rule applies to irrecoverable frictional pressure lossess only.

Based on the simulation results, we conclude that a 2x3J with 3 inch inlet line should be adequate for the selected design basis and will also satisfy the recommended 3 % percent inlet pressure drop criterion.

Table 1: Important vessel and fire exposure data

Vessel total charge (kg)	2,722 (6,000 lbs)
Vessel weight fraction water	20 %
Vessel weight fraction ethanol	30 %
Vessel weight fraction acetone	50 %
Vessel metal weight (kg)	2,264
Vessel MAWP (psig)	50
Vessel 1.21 MAWP (psia)	75.2
Vessel type	Vertical Cylindrical
Vessel volume (m3)	11.355
Vessel Diameter (m)	2.13 m
Vessel total surface area (m2)	28.45
Fire protection	Approved drainage and insulation
Valve type	Conventional, Kd = 0.864
Valve set pressure (psig)	50
Valve piping	2 inch inlet line, 5 ft long with one 90 degree elbow 3 inch outlet line, 100 ft long, vertical, with three 90 degree elbows
Piping roughness (mm)	0.0457
Fire Flux (kW/m2)	NFPA 30 flux (15,700)
Fire duration (hr)	4

Table 2: Binary interaction parameters data for the system water, ethanol and acetone

Binary System	k_{12}	λ_{12}
Water-Ethanol	-0.08038	0.04809
Water-Acetone	-0.20594	0.15286
Ethanol-Acetone	0.02586	0.01334

Figure 5: Calculated vessel pressure as a function of time using a 2x3J and a 2x3H valve

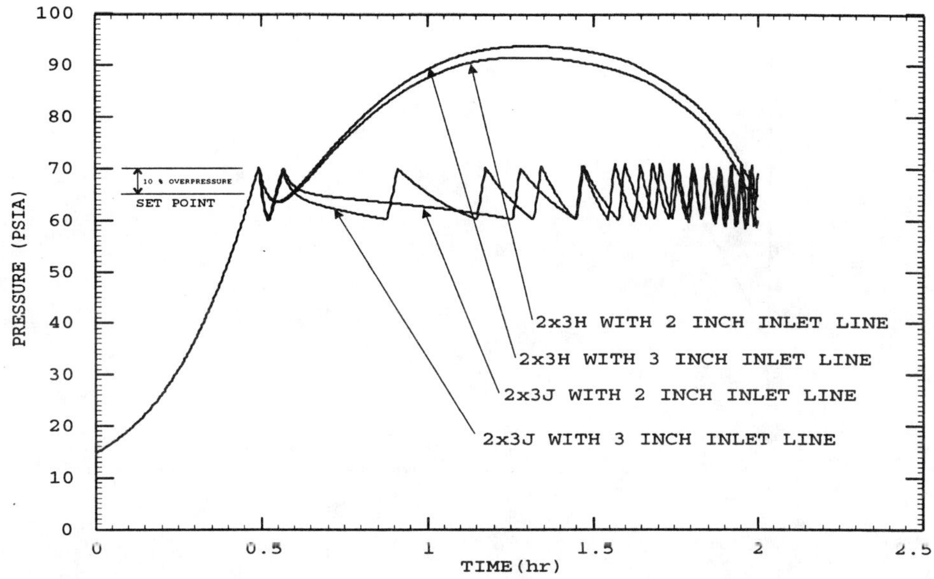

Figure 6: Calculated safety valve inlet line pressure drop as a function of time

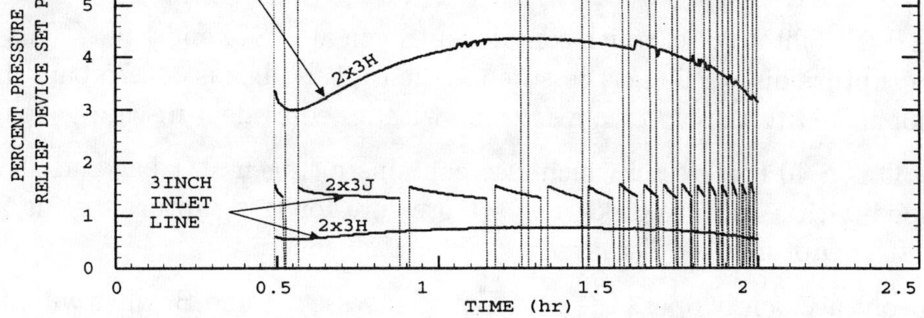

Figure 7: Calculated vessel temperature as a function of time using a 2x3J valve with a 3 inch inlet line

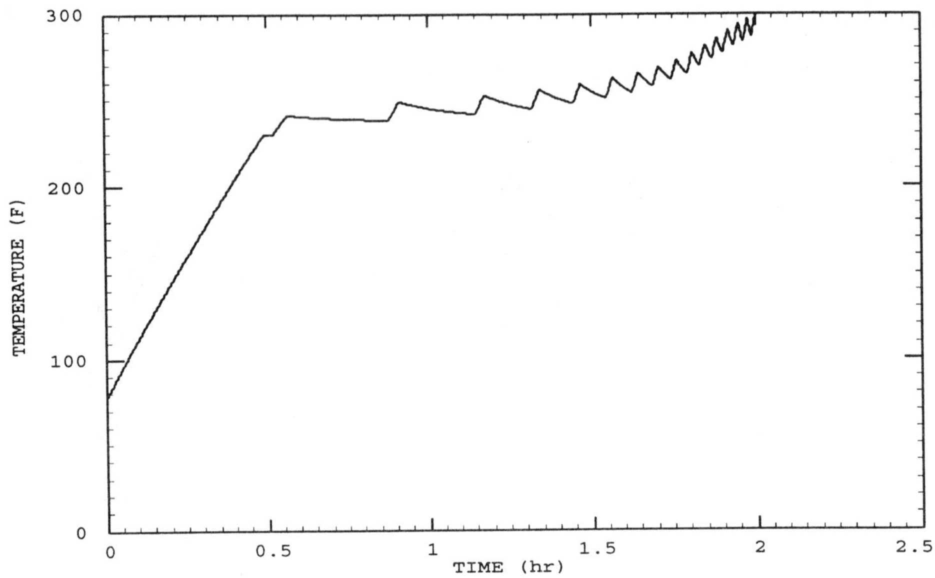

Figures 7, 8, 9 and 10 illustrate the computed profiles of temperature, reaction force, individual component flow rates, and vessel composition as a function of time. The results indicate, as expected, that the vessel mass is preferentially depleted over time, i.e. light components are vaporized first.

DYNAMIC SIMULATION

The following example illustrates the use of SuperChems to model the dynamics of pressure relief when multiple vessels are connected together. Figure 11 shows the connectivity of the vessels considered in this example.

Vessel A is a 40 liter vessel which contains nitrogen at 414.7 bars (6,000 psig) and 298 K. Vessel A is connected to vessel B via a short 1/4 inch line with an equivalent discharge coefficient of 0.6.

Vessel B is a 5 liter vessel which contains nitrogen at 298 K and 1 bar. This vessel is equipped with a rupture disk. The set pressure for the rupture disk is 345.75 bars (5,000 psig). Vessel B is connected to vessel C via a 1/2 line with an equivalent discharge coefficient of 0.5.

Vessel C is a 40 liter vessel which also contains nitrogen at 1 bar and 298 K. This vessel is equipped with a rupture disk. The set pressure for this rupture disk is 276.87 bars (4,000 psig). The rupture disk discharges into the atmosphere.

If someone accidently opens the valve between vessels A and B, what will be the final pressure in the system. Will the rupture disk on vessel C burst ?

Figures 12 and 13 displays the iteration history and how SuperChems achieves convergence.

Figure 8: Calculated impulse at pipe inlet and pipe outlet as a function of time using a 2x3J valve with 3 inch inlet line

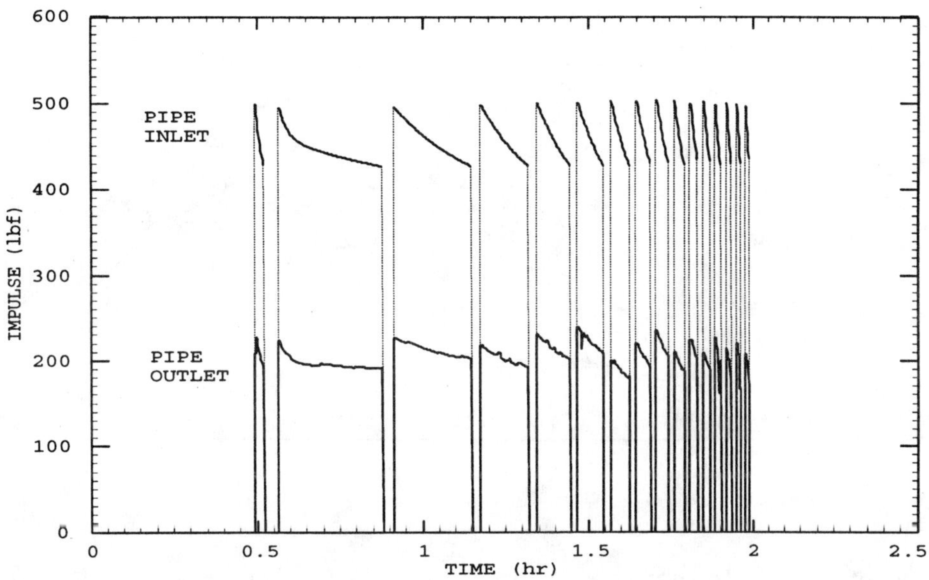

Figure 9: Calculated vapor flow rate of mixture constituents as a function of time using a 2x3J valve with 3 inch inlet line

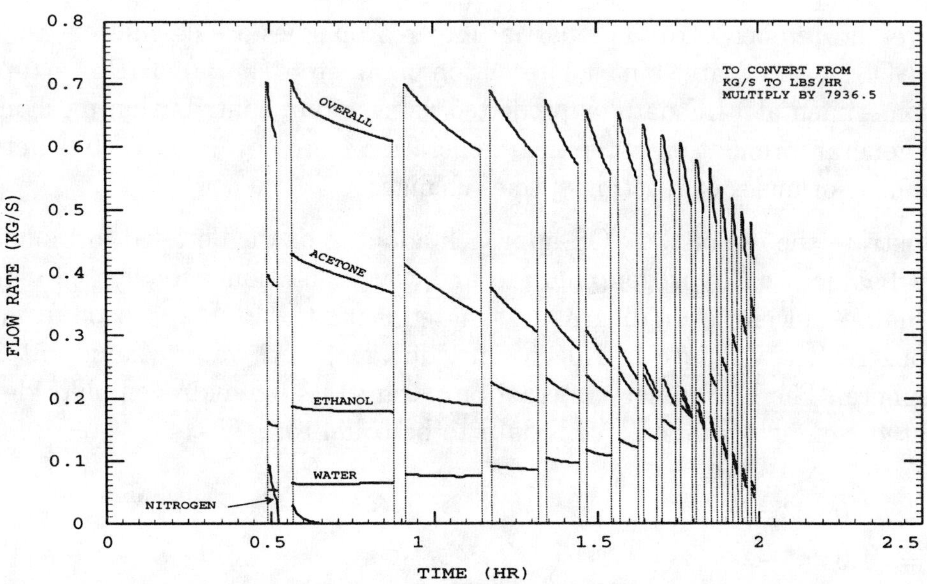

Figure 10: Calculated vessel mass of individual mixture constituents as a function of time using a 2x3J valve with a 3 inch inlet line

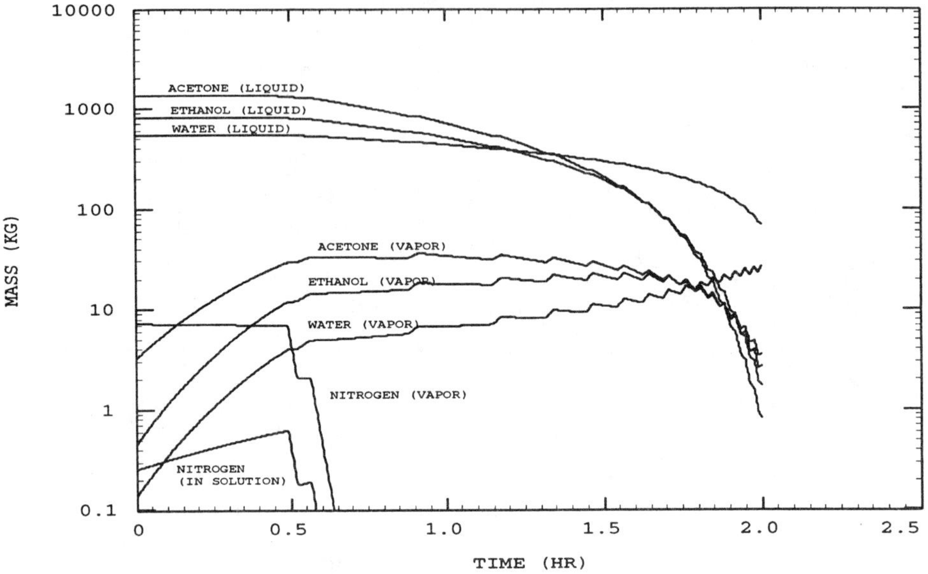

It is interesting to note that the rupture disk on Vessel C does not rupture for the converged solution while it was predicted to rupture in earlier iterations.

NON-IDEAL SOLUTION BEHAVIOR

The constant volume mass and energy balance equations derived in the previous sections are general and implicitly represent fluid non-ideal behavior caused by solution behavior, system pressure and temperature. The partial molar properties are calculated from an equation of state (EOS) with suitable binary interaction parameters (BIPs). BIPs are typically estimated from experimental VLE data or predicted using group contribution methods. Also implicit in the general formulation is the heat generation or removal caused by chemical reaction as the number of moles of reacting species changes.

We illustrate the use of the EOS approach adopted in the detailed formulation of mass and energy balances on hydrogen chloride and water solution behavior. Figure 14 illustrates an enthalpy-concentration diagram estimated using the EOS approach for hydrochloric acid solutions relative to pure hydrogen chloride (gas) and water (liquid) at 273.15 K and 1 atmosphere. The EOS binary interaction parameters for hydrogen chloride and water were estimated from experimental VLE data to be equal to:

$$k_{1,2} = -0.803 = k_{2,1} \tag{1}$$

$$\lambda_{1,2} = 1.05 = -\lambda_{2,1} \tag{2}$$

The predicted data shown in Figure 14 is in excellent agreement with the experimental data

Figure 11: Dynamic simulation example using three connect vessels.

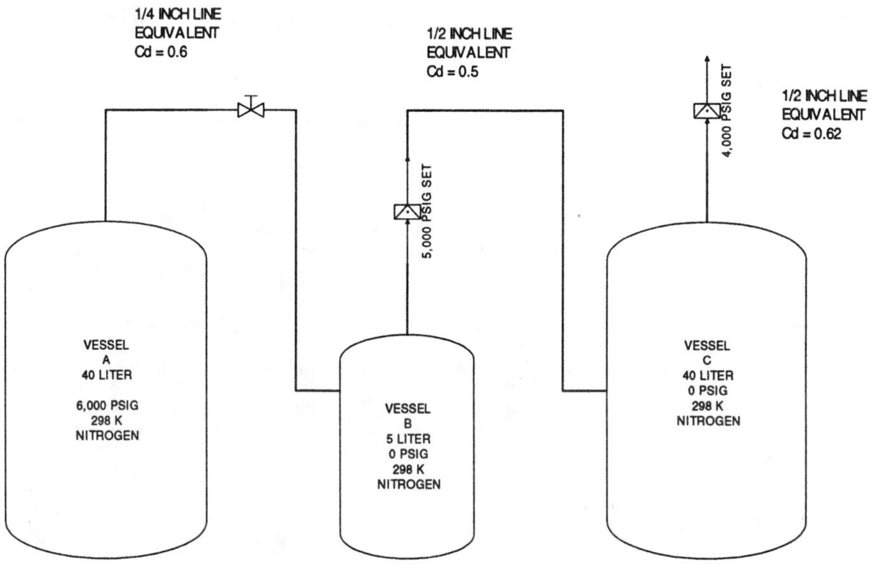

on heat of solution for hydrogen chloride - water solution reported by Hougen and Watson in reference [3].

Figure 14 is useful for the estimation of final mixture temperature when mixing solutions of varying hydrogen chloride fraction. For example, if we were to mix 0.263 kmol of a 10 % by weight (5.26 mol %) mixture of hydrogen chloride and water at 288 K witha mixture containing 30 % by weight (17.6 mol %) hydrogen chloride and water, what would be the final mixture temperature ?.

The enthalpy of each of the two solution mixes is read from Figure 14. The 10 % solution enthalpy is -3.5 MJ/kmol or -0.92 MJ, and the 30 % solution enthalpy is -8.00 MJ/kmol or -3.024 MJ. The mixture enthalpy is calculated to be -3.944 MJ or -6.15 MJ/kmol. Upon mixing, the final mixture composition would be 12.53 mol % hydrogen chloride. Using a mole fraction of 12.53 % and a molar enthalpy of -6.15 MJ/kmol, we can read from Figure 14 a temperature of 305.5 K or 89.9 F. Hougen and Watson [3] report a temperature of 90 F for the same problem.

We consider another example to illustrate how non-ideal solution behavior can alter the pressure-temperature behavior of systems where hydrogen chloride and water are generated by a chemical reaction. Two cases are considered:

(a) A vessel initially contains 1000 kg of water (55.5 kmol) at 300 K and 1 bar. The total vessel volume is 9.2 m^3. 622 kg (5.5 kmol) of liquid chloroacetyl chloride (CAC) is pumped into the liquid space at 0.00925 kmol/s. The CAC entering the vessel is at 300 K and 1 bar and is immeditely mixed with the water due to the action of an agitator. Mixing is sufficient to cause an immediate reaction between CAC and water to generate hydrogen chloride and chloroacetic acid:

$$C_2H_2Cl_2O + H_2O \rightarrow C_2H_3ClO_2 + HCl \tag{3}$$

Figure 12: SuperChems iteration history as the material and energy balance is converged for vessels A, B and C.

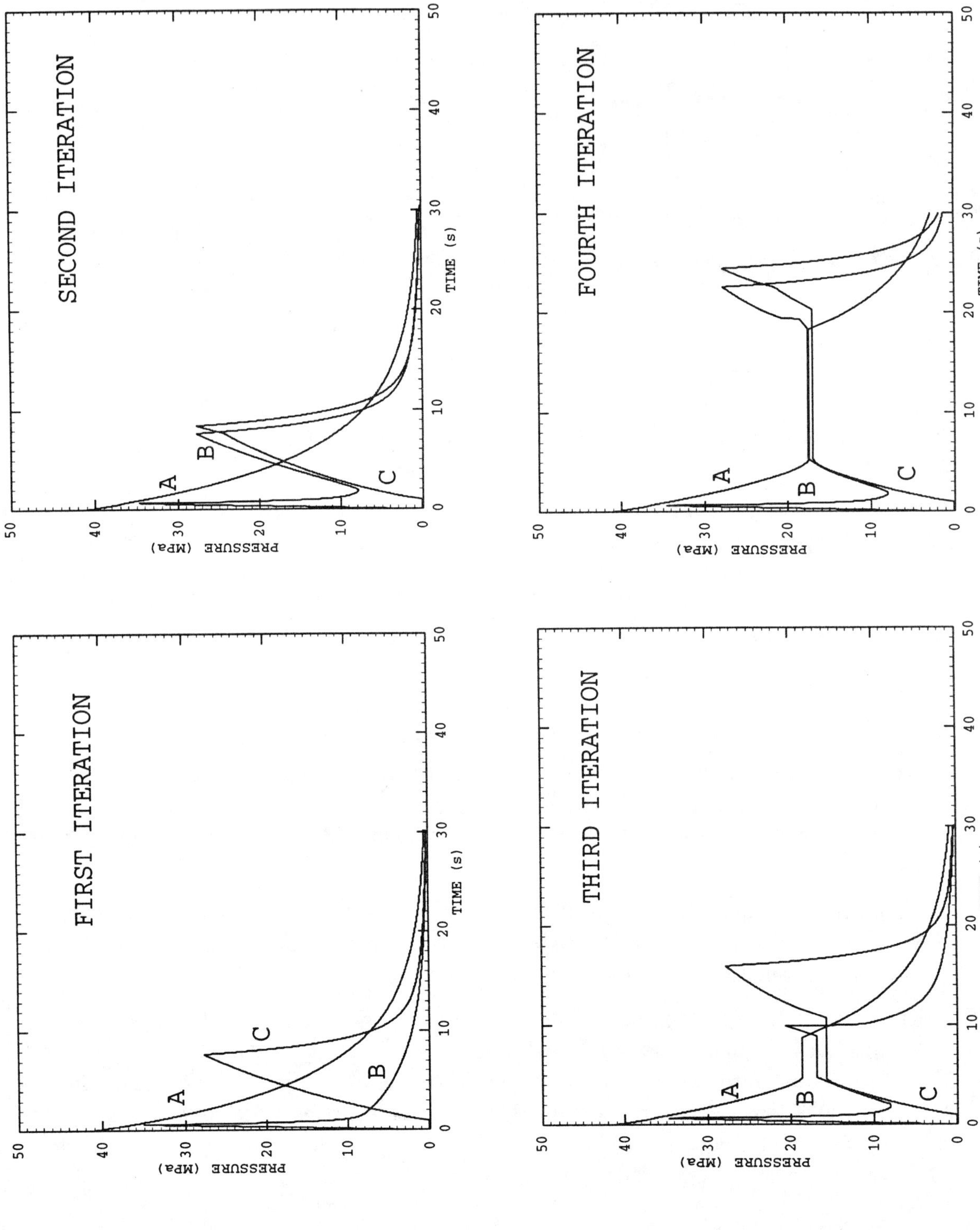

Figure 13: Final pressure profiles in vessel A, B, and C after convergence is achieved.

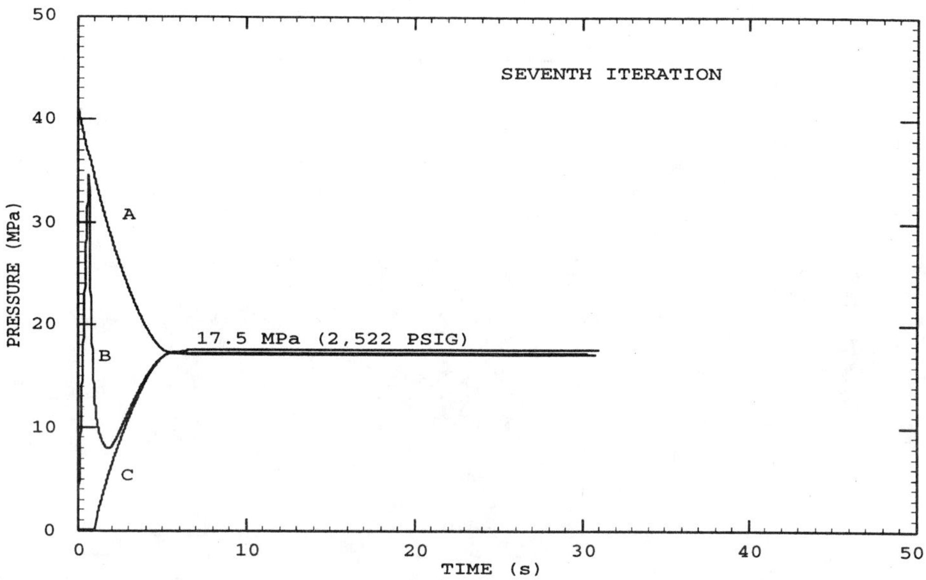

The vessel vapor space contains nitrogen.

(b) The case is similar to case a. The vessel initially contains 622 kg of CAC and water is pumped into the vessel at the rate of 0.00925 kmol/s.

Figure 15 illustrates the behaviour of pressure-temperautre for both cases. For case (a), as CAC is pumped into the vessel, it mixes with water and immediately reacts to generate hydrogen chloride which goes into solution as the ratio of water to CAC is 10 to 1. A peak pressure of 2.2 bar is reached at 100 C. For case (b), water is pumped into the vessel, mixes with CAC and generates hydrogen chloride which has a limited solubility in chloroacetyl chrloride and CAA. As a result, the pressure and temperature increase steadily. As the CAC is depleted and more water is added, the hydrogen chloride goes into solution. The final vessel conditions are the same as in case (a) with a peak pressure of 20.2 bar. Figure 16 shows a similar behavior for the same system at 1 bar. Figure 16-a illustrates the effect of water to CAC molar ratio on the combined net heat of solution and reaction. At a molar ratio of 1, the heat of reaction is -25 MJ/kmol. Figure 16-b indicates that all evolved hydrogen chloride is gaseous when the molar ratio of water to CAC is less or equal to 1, i.e. CAC is in excess or stoichiometric. As excess water is added, the hydrogen chloride goes into solution. At a molar ratio of water to CAC greater or equal to 4, all evolved hydrogen chloride is retained in the liquid phase. The heat of solution effects are shown in Figure 16.

The pressure-temperature behavior is a key relation for pressure relief design. Non-ideal solution behavior is plays a key role in the interpretation of experimental data and scale-up. This is similar to concepts taught in general chemistry about adding acid to water !.

Figure 14: Enthalpy-concentration chart of hydrogen chloride - water solutions relative to pure hydrogen chloride gas and water liquid at 273 K and 1 atmosphere.

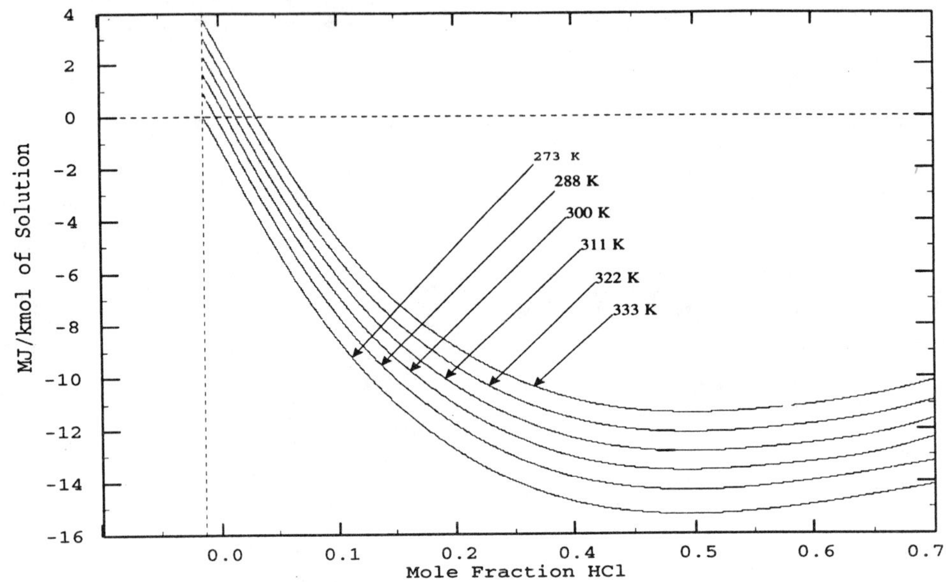

PEAK LIQUID AND VAPOR FLOW DURING RELIEF

The design of separation equipment for two-phase flow requires the specification of a peak vapor flow rate [4]. For many reactive systems, peak vapor/liquid flow may be reached later during relief. Figure 17 illustrates this point by using SuperChems to simulate the relief transient of the decomposition of $C_9H_{13}N$. As shown by Figure 17 at the start of relief, the cumulitive (all mixture components are included) liquid flow rate is around 250 kg/s and the cumulitive vapor flow rate is around 1 kg/s. The peak vapor flow takes place at 240 s during relief and is around 5 kg/s with a correponding liquid rate of 100 kg/s. It is the peak vapor flow value that should be used as input for design and not the initial value. Most simplified/shortcut methods are not capable of providing such data.

IMPACT OF PIPING CONFIGURATION ON FLOW

In order to illustrate how piping configuration might impact the flow capacity of a relief line, we consider a simple example involving saturated water. Two-phase homogeneous flow occurs in a piping configuration consisting of two equal length segments having the same diameter (2 inch). The pipe inlet conditions are as follows:

- Temperature = 151.9 C
- Pressure (saturation) = 500000 Pa
- Vapor quality = 0.001

Figure 15: Non-ideal solution behavior impact on closed vessel pressure-temperature behavior for chloroacetyl chloride / water system

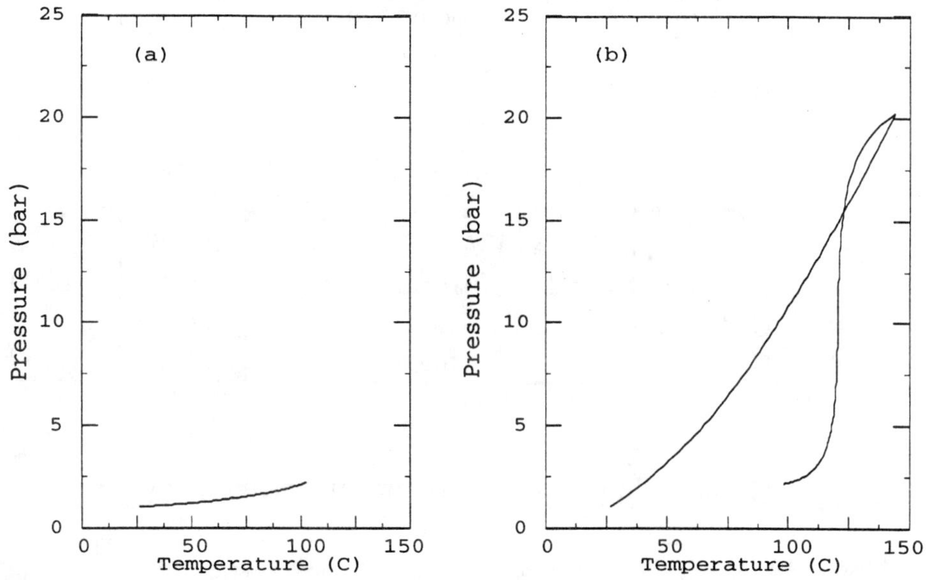

- Pipe roughness = 4.57×10^{-5} m

Two cases are considered:

A Pipe consists of a horizontal segment followed by a vertical segment

B Pipe consists of a vertical segment followed by a horizontal segment

Table 4 summarizes the computed flow rates by SuperChems for both cases A and B for different pipe lengths. More flow is predicted for case A than for case B. A flow method that does not account for the gravity term in the momentum equation would predict the same flow rate for both cases.

Case A arrangement should lead to more flow because less pressure drop is exhibited in the horizontal segment without much change in the vapor quality. For case B, however, the vapor quality increases faster in the first segement (90 deg orientation) due to more pressure drop and thus leads to less flow. Figures 18 and 19 illustrate the pressure and vapor fraction profiles for the 244 m long pipe.

SUBCOOLED LPG FLOW

Figure 20 illustrates an experimental setup used by Shell Research [5] for the estimation of subcooled LPG flow. Pressure in the vapor space is maintained at 880000 Pa by using a nitrogen pad. Vessel and piping dimensions are also shown in Figure 20. SuperChems was used to calculate the flow rate and pressure/void fraction profile in the piping. Table 5

Table 3: Equation of state binary interaction parameters used to describe the system chloroacetyl chloride, water, chloroacetic acid, hydrogen chloride and nitrogen

Component 1	Component 2	$k_{1,2}$	$\lambda_{1,2}$
Water	Hydrogen chloride	-0.8030	-1.0500
Water	CAC	-0.0513	-0.1950
Water	CAA	-0.1110	0.0417
Water	Nitrogen	0.0000	0.0000
Hydrogen chloride	CAC	0.0021	-0.0081
Hydrogen chloride	CAA	-0.0267	0.0023
Hydrogen chloride	Nitrogen	0.0000	0.0000
CAC	CAA	-0.0135	0.0054
CAC	Nitrogen	0.0000	0.0000

Table 4: Impact of piping configuration on flow capacity

Seg. 1 Angle (deg)	Seg. 2 Angle (deg)	Seg. 1 Length (m)	Seg. 2 Length (m)	Flow Rate (kg/s)
0	90	2.5	2.5	5.547
90	0	2.5	2.5	5.410
0	90	24.45	24.45	2.973
90	0	24.45	24.45	2.529
0	90	244.0	244.0	0.864
90	0	244.0	244.0	0.540

illustrates the calculated flows using homogeneous equilibrium and Moody slip. The reported experimental flow value is closer to the that predicted using slip equilibrium flow. The same behavior is observed for subcooled water flow illustrated in the next example based on the experiments of Nielsen [6].

Figures 21 and 22 illustrate pressure and void fraction profiles during vessel blowdown. It is apparent from these graphs that large changes in pressure and void fraction occur at the point at which the pipe diameter reduces from 150 mm to 52 mm.

LPG CYLINDER BLOWDOWN

This benchmark deals with the blowdown of a cylinder containing propane. The experiment was conducted by Melhem et al. in 1991 [7]. The pressure profile in the cylinder was recorded

Figure 16: Non-ideal solution thermodynamics for the system chloroacetyl chloride, water, chloroacetic acid and hydrogen chloride. Basis = 1 kmol of chloroacetyl chloride, 298 K, 1 bar, open volume.

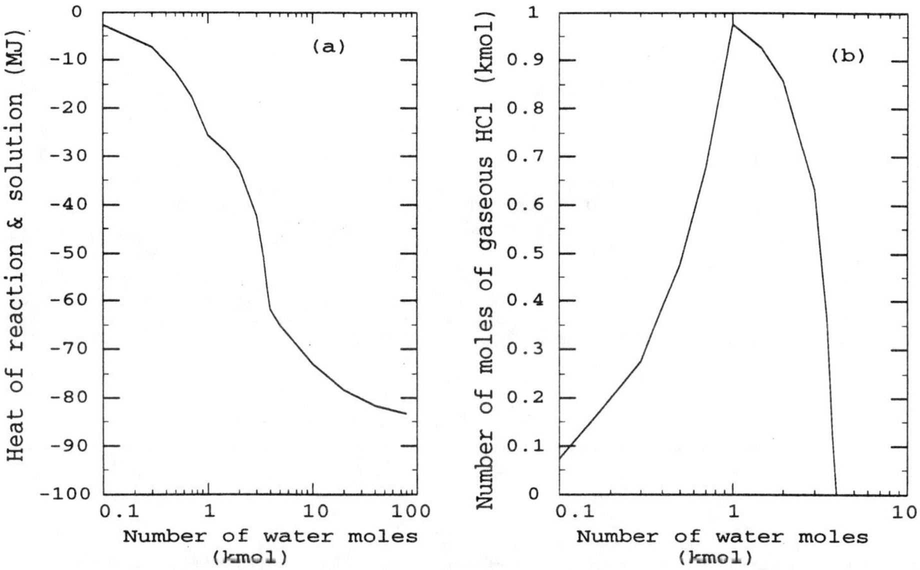

Table 5: Comparison between calculated subcooled LPG flows by SuperChems and experimental data

Flow Method	Flow Rate (kg/s) $\epsilon = 1.5 \times 10^{-6}$ m	Flow Rate (kg/s) $\epsilon = 2.5 \times 10^{-5}$ m
Experimental	21.4	
Homogeneous	14.5	13.8
Moody slip	19.2	18.0
Theoretical	28.2	28.2

as a function of time and the total mass of propane remaining in the vessel was measured at 180 s. The experiment was conducted on July 15, 1991 at 2:00 pm at NASA's Stennis test site in Mississipi. Table 6 summarizes data pertinent to the test.

Model results are summarized in Figure 23 and 24. The data shows excellent agreement between measured and predicted data.

NIELSEN'S SUBCOOLED/FLASHING WATER FLOW

SuperChems was used to predicted subcooled/water flow data experimentally measured by Nielsen [6]. A 1.98 cubic meters, vertical cylindrical vessel having a diameter of 1.078 m and two ASME 2:1 elliptical heads contains water at a temperature of 176.1 C. The total (water vapor plus hydrostatic) pressure in the vessel is (9.3 bara) 930000 Pa. Horizontal

Figure 17: A typical vapor/liquid discharge profile during a process induced runaway reaction

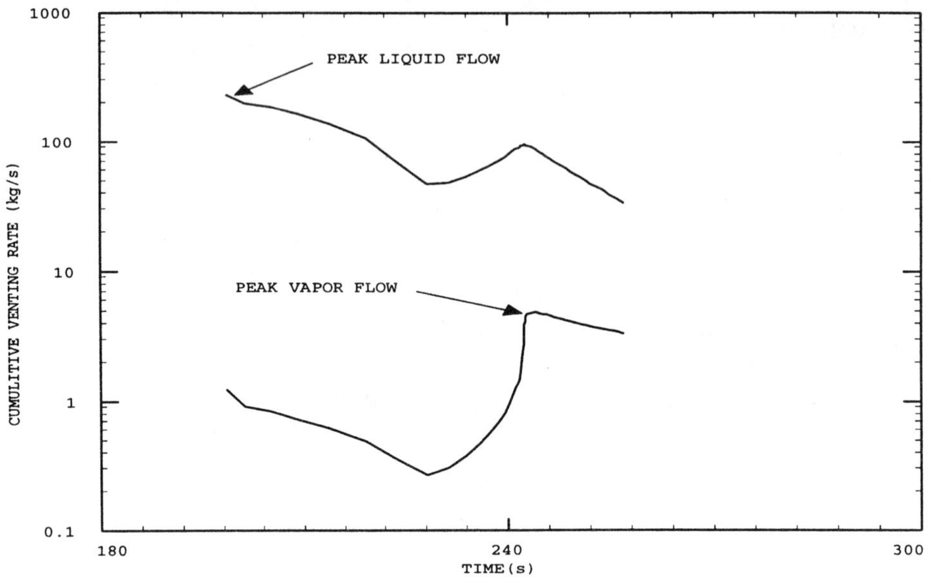

pipes having square-cut entrances and the internal diameters, lengths and pipe roughness values shown in Table 7 drain the vessel.

These drain pipes are side-mounted at 0.1 m above the tangent line of the bottom head. The 1145.4 kgs of water above the drain pipe results in the initial hydrostatic head of 0.131 bara. Assuming both homogeneous-equilibrium and slip-flow, the initial flow rates were calculated by SuperChems and are tabulated in Table 8. Again, this data indicates that some degree of slip is present during discharge.

INCLINED PIPE FLOW

We consider the following problem involving an inclined pipe with all liquid flow. The following assumptions are made:

- Both temperature and density are constant.
- Flow reaches steady state, i.e. initial transients following flow actuation are not accounted for.
- Pipe area is constant.
- Pipe discharges to atmosphere with no exit losses.

Starting with a differential momentum balance for a single phase flow, we can show that pressure drop in the pipe has three components: accelerational, frictional and gravitational:

$$\frac{1}{\rho}\frac{dP}{dz} = -u\frac{du}{dz} - \frac{2u^2 f}{D} - g\sin\theta \tag{4}$$

Figure 18: Impact of piping configuration on pressure

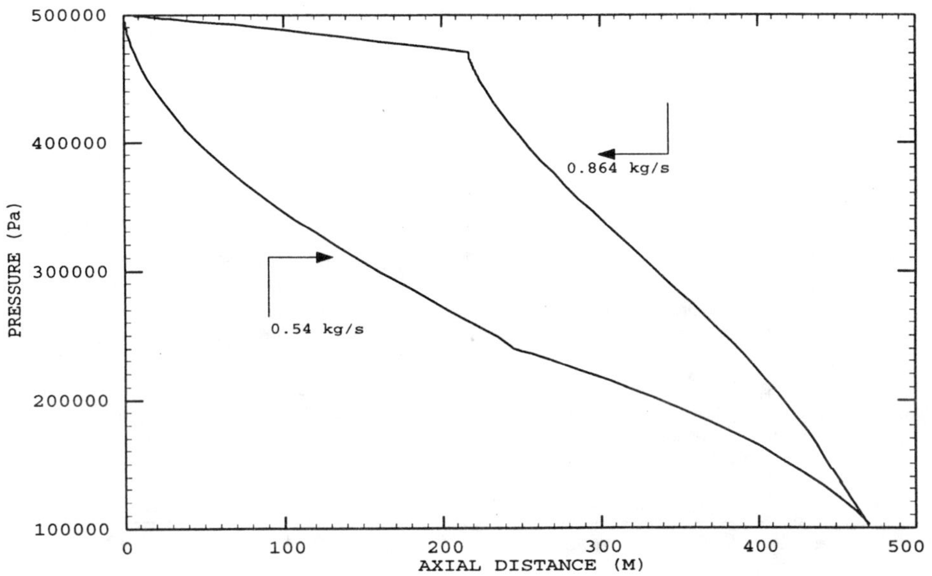

For most liquid flows, the term du/dz is small and can be ignored for illustration purposes. The term dP/dz must be less or equal to zero in order for flow to occur. At the limit:

$$\frac{dP}{dz} = 0 = -\frac{2u^2 f}{D} - g\sin\theta \tag{5}$$

where θ is the pipe angle with respect to the horizontal. For small diameters, the first term dominates and dP/dz is always negative. For large diameters, dP/dz can reach zero at a mass flux of:

$$G = \rho\sqrt{\frac{-gD\sin\theta}{2f}} \tag{6}$$

Integrating the mechanical energy balance without constraining the sign of dP/dz will lead to an integral balance that can be used for inclined flow with diameters large enough to cause a positive dP/dz term. The integral balance will lead to an overall pressure drop but will fail to illucidate the pressure behavior between the inlet and the outlet points. The integral balance will only indicate that flow will increase as pipe length increases.

The system will maximize mass flow such that the Gibbs free energy is minimized between the inlet and the outlet point. This will require that the overall pressure drop to be negative and will allow pressure to increase in order to satisy the overall pressure drop constraint.

We will illustrate this using an actual example. Consider water flow at 298 K and 2 bars in an inclined pipe. Several cases are considered with pipe diameters ranging from 2 to 8 inches and pipe inclination ranging from -45 to -90 degrees (vertical downward). An entrance loss equivalent to 1.5 velocity heads is assumed.

Figure 19: Impact of piping configuration on vapor quality

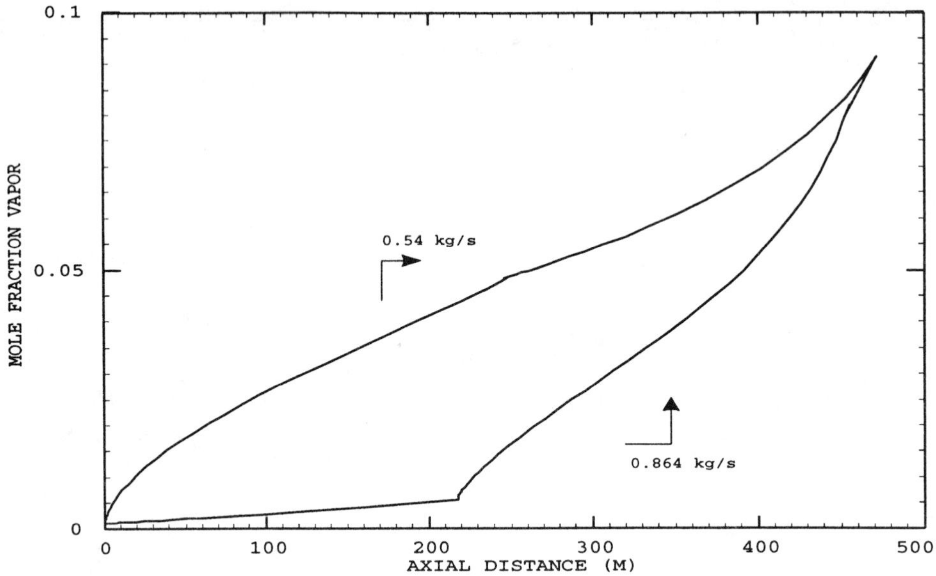

Table 9 illustrates maximum flow capacity of the pipe for various conditions. Figure 25 illustrates the pressure profile along the pipe axis and shows that while the overall pressure drop is negative, the pressure across the pipe entrance is allowed to decrease and than increase to ambient pressure at the outlet. Clearly, this indicates the the pressure will fall below the bubble point and will lead to the formation of an unstable flow regime with continuing bubble formation and collapse as flow occurs. This might present several implications for bottom venting with reactive chemicals such as peroxides unless sufficient air ingestion is allowed to take place during flow.

SENSITIVITY TO FIRE LOADING

The following example illustrates the use of SuperChems to establish the sensitivity of both reactive and non-reactive systems to fire exposure.

To illustrate the impact of fire loading on ERS design for a non-reactive system, we consider a propane storage vessel which initially contains 600 kgs of propane at 292 K and 7.985 bars. The vessel characteristics are shown in Table 10. The pressure history is estimated for three external heat flux values, 5, 25, and 100 kW/m2. Flow from the relief valve is assumed to be homogeneous two-phase without slip.

As shown by Figure 26, the relief valve is oversized for a 5 kW/m2 heat flux value. This is apparent due to valve cycling. Under a 100 kW/m2 heat flux the pressure level in the tank reaches two times MAWP. Valve cycling occurs when the vessel contents are completely vaporized. For the high fire flux case, the relief valve is undersized, however, vessel integrity is not seriously compromised.

For vessels containing reactive chemicals the rate of energy/pressure accumulation is signifi-

Figure 20: Subcooled LPG flow experimental setup

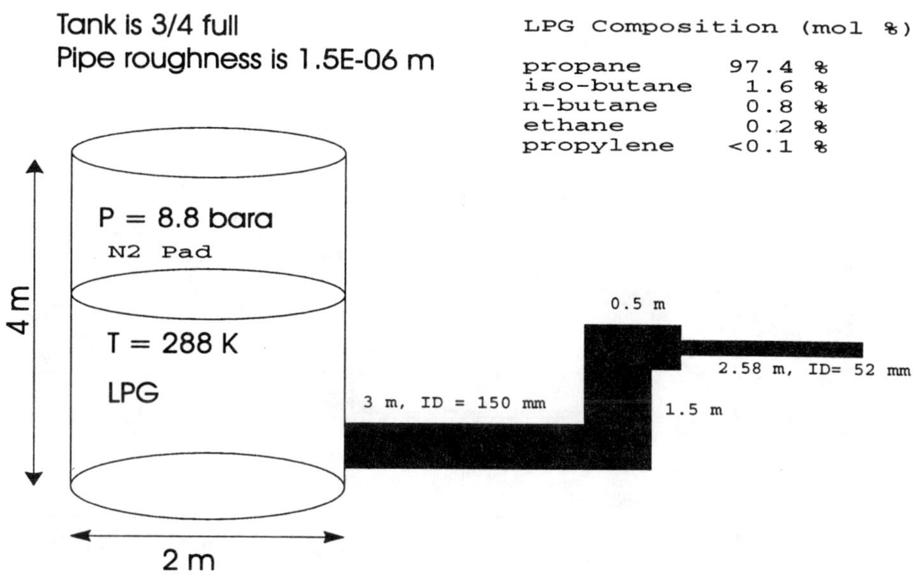

cantly increased because the reaction temperature reaches onset with less reactant consumption. The effect of fire on reaction rates is highly nonlinear.

To illustrate this highly nonlinear effect of fire loading on vessel temperature and pressure behaviors we consider a steel storage vessel containing a hydrogen peroxide / water solution that is exposed to an external heat flux varying between 5 and 100 kW/m^2. The vessel is equipped with a rupture disk (3 inches in diameter) at a set pressure of 3 bars. We assume that no piping is attached to the rupture disk. The vessel volume is 1 m^3 and has an MAWP of 10.34 bars.

The peroxide/water solution consists of 159 kg of hydrogen peroxide and 716 kgs of water. The vessel initial temperature and pressure are 10 C and 1 bar respectively. Hydrogen peroxide decomposes in the liquid phase and forms oxygen and water according to the following stoichiometry:

$$H_2O_2 \rightarrow \frac{1}{2}O_2 + H_2O \tag{7}$$

The reaction is first order and has a temperature dependent rate described by the following expression:

$$K = 1.5 \times 10^{12} \exp\left(-\frac{12,832}{T}\right) \tag{8}$$

where K is in s^{-1} and T is in Kelvins.

The pressure history is shown in Figure 27. Homogeneous equilibrium two-phase flow is used to estimate the flow from the rupture disk upon actuation. At an external heat flux value of 5 kW/m2 (simulating a process upset) the 3 inch rupture disk is properly sized. As

Figure 21: Calculated pressure profile in piping during vessel blowdown

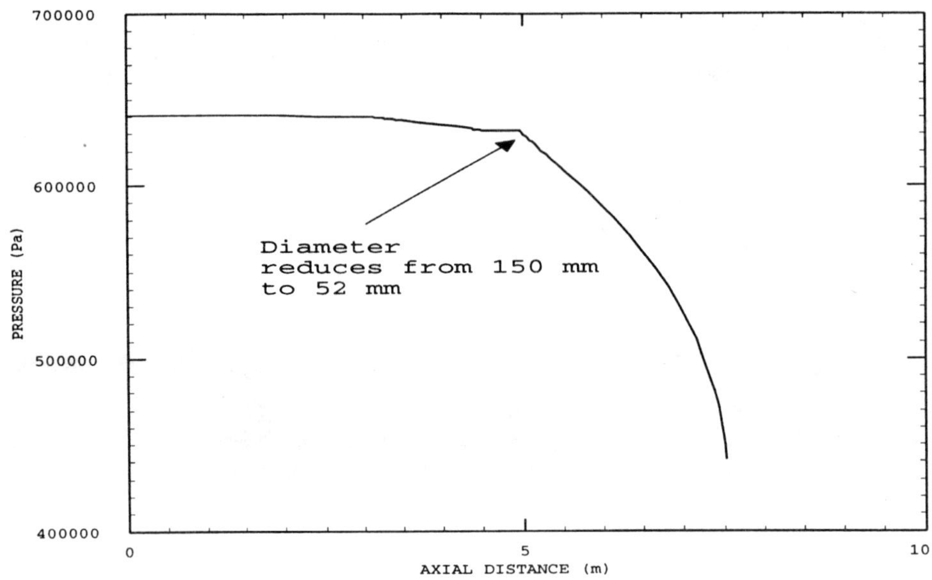

shown in Figure 27, higher external heat flux values (simulating a fire) lead to much faster reaction rates and higher pressure levels in the vessel. With the existing rupture disk the vessel pressure reaches 3 times MAWP with an external heat flux of 25 kW/m2.

Figure 28 illustrates the maximum pressure level reached in the vessel during runaway reaction conditions caused by a fire exposure with a a flux of 100 kW/m2. Rupture disk sizes of 4, 6, and 8 inches are shown. The effect of rupture disk size on the final pressure level reached in the vessel can be more pronounced with piping attached to the rupture disk. This is attributed to mass flow reduction effects resulting from continuing reaction in the piping [2].

CONCLUSION

This paper presents an advanced modeling approach and computer software which significantly improves predictions of reaction rates and critical data that engineers need to design effective pressure relief systems.

The comprehensive approach and user friendly computer code developed in this paper provide a reliable design basis for difficult systems, including highly energetic and nonideal reactions, systems with continuing reactions in piping and containment vessels, etc.

Figure 22: Calculated void fraction profile in piping during blowdown

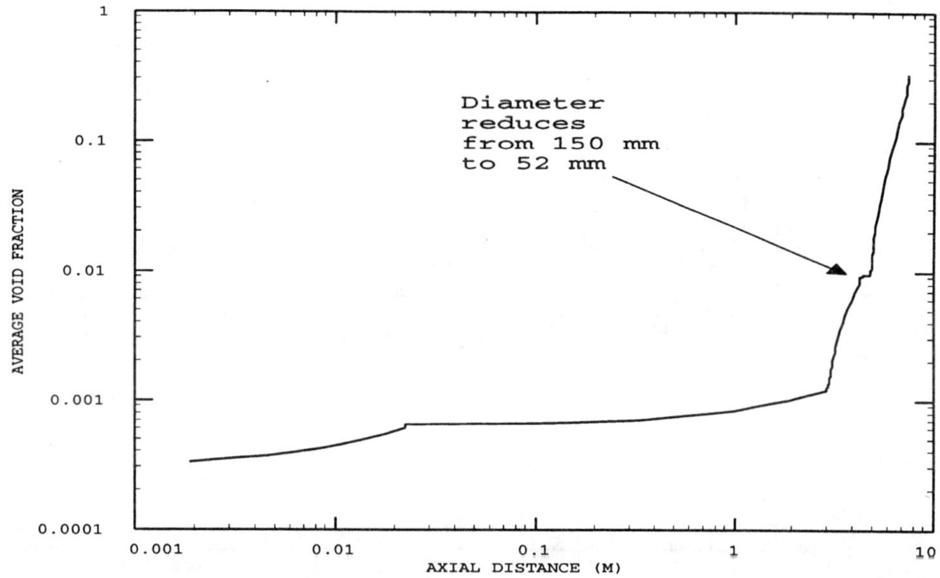

Table 6: Propane cylinder experimental data summary

Variable	Value
Cylinder weight (empty)	30.4 kg
Cylinder volume	0.102 m^3
Cylinder charge	43.7 kg
Initial temperature	308 K
Initial pressure	16 bars
Propane mole fraction	97.3
Butane mole fraction	1.7
Other	1.0
Release orifice size	3/8 inch ID
Release piping	50 ft copper tubing
Copper roughness	0.0000457 m (assumed)
Ambient temperature	103 F
Relative humidity	95 percent
Date	July 15, 1991 at 2:00 pm
Flow phase	Gas flow only
Total mass vented	8.0 kg in 180 s

Figure 23: Propane cylinder blowdown: comparison of experimental data and model predictions for pressure profile and total mass vented

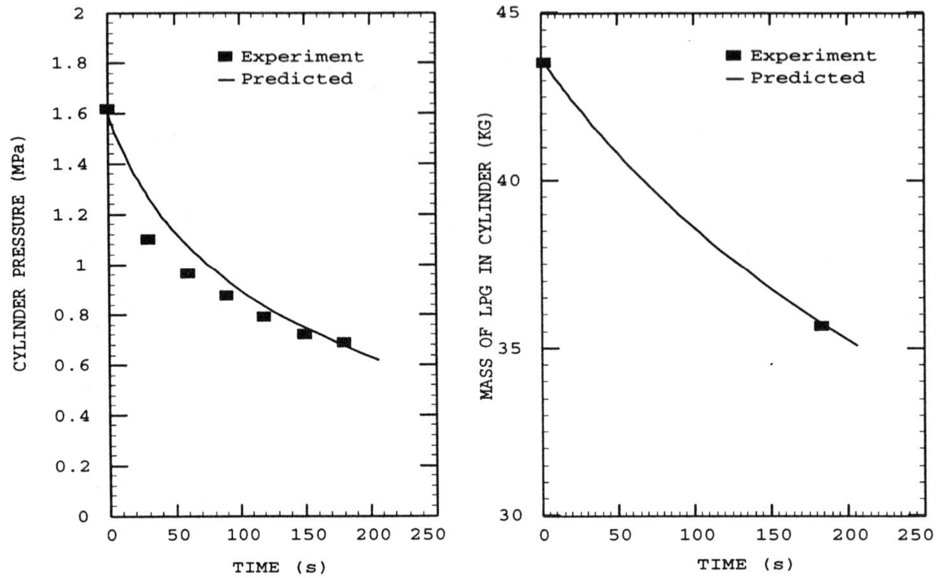

Figure 24: Propane cylinder blowdown: model predictions of venting rate and flow reduction due to the presence of copper tubing

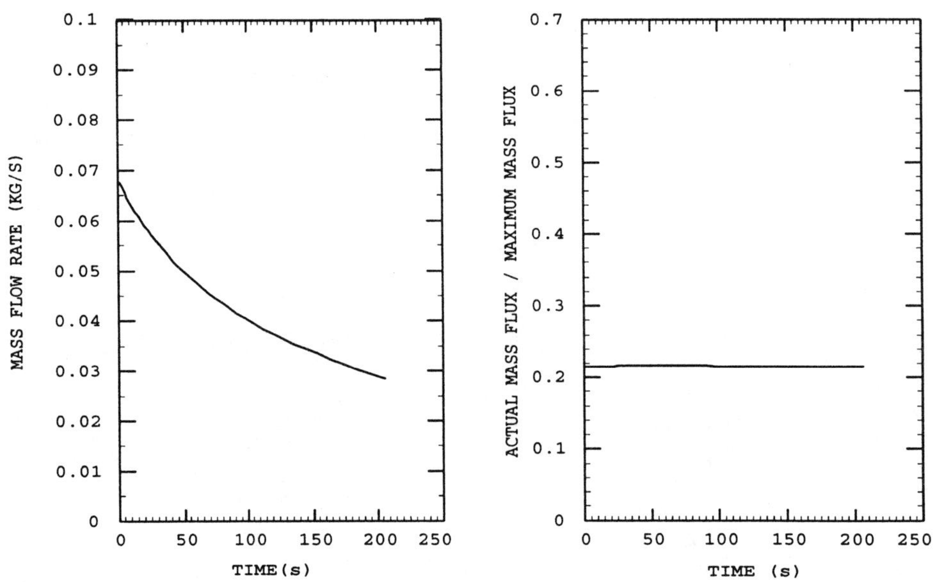

Table 7: Summary of pipe data for Nielsen's water flow experiments

Pipe Number	Pipe Diameter (mm)	Pipe Length (mm)	Pipe L/D	Pipe Roughness mm
1	79.9	1840	23	0.024
2	32.8	2000	61	0.013
3	10.6	1870	176.4	0.013

Table 8: Comparison between calculated subcooled/saturated water flows by SuperChems and experimental data

Flow (kg/s)	Pipe 1	Pipe 2	Pipe 3
Experimental	33.1	5.23	0.44
Homogeneous	27.6	4.20	0.330
Moody slip	45.0	6.45	0.43

Table 9: Calculated maximum flow capacity of inclined water pipe

Pipe Length (m)	Pipe Diameter (mm)	Pipe Angle (degrees)	Flow Rate (kg/s)
5	50.8	-45	17.76
48.9	50.8	-45	13.09
487.8	50.8	-45	12.064
5	50.8	-90	18.6
48.9	50.8	-90	15.11
487.8	50.8	-90	14.336
5	203.2	-90	410
48.9	203.2	-90	492
487.8	203.2	-90	≥ 528

Figure 25: Calculated pressure profiles for inclined water flow pipe

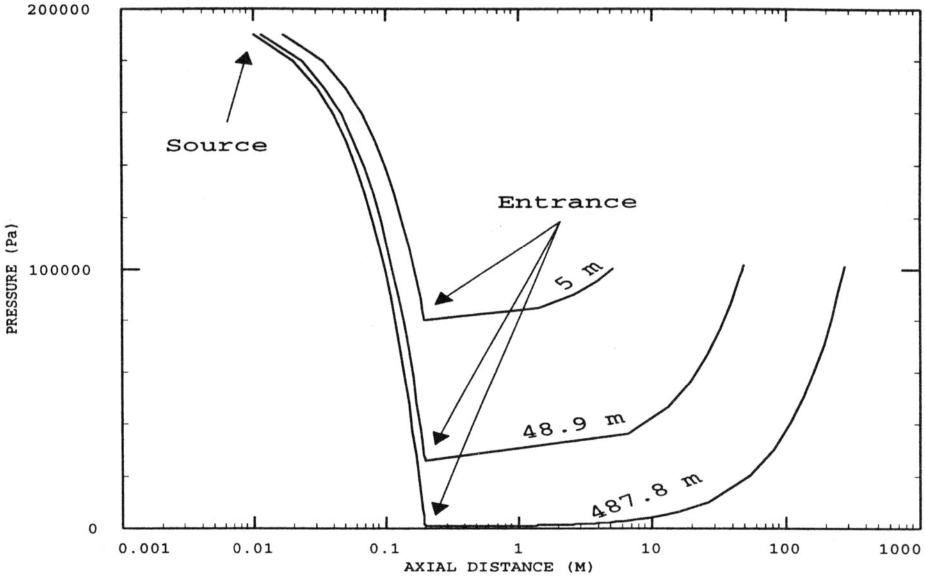

Table 10: Propane storage tank characteristics

Volume	1.893 m^3
Length	3.023 m
Outside diameter	0.950 m
Shell thickness	0.0063 m
Head thickness	0.00533 m
Maximum allowable working pressure at 611 K	1.724 MPa
Hydrostatic test pressure	2.58 MPa
Relief valve setting	1.724 Mpa
Relief valve diameter	0.0254 m
Relief valve area	0.000425 m^2

Figure 26: Sensitivity of pressure to external fire loading for non-reactive systems

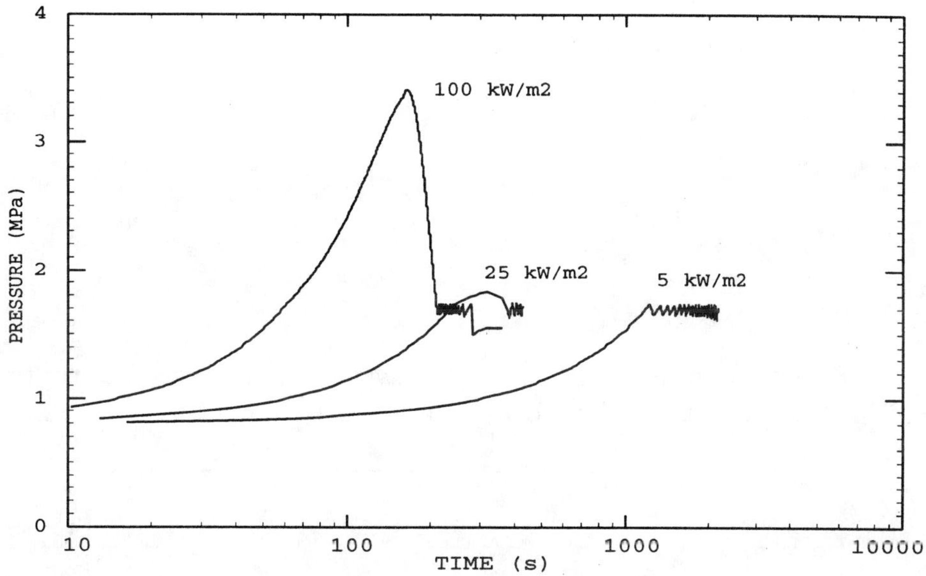

Figure 27: Sensitivity of pressure to external fire loading for reactive systems

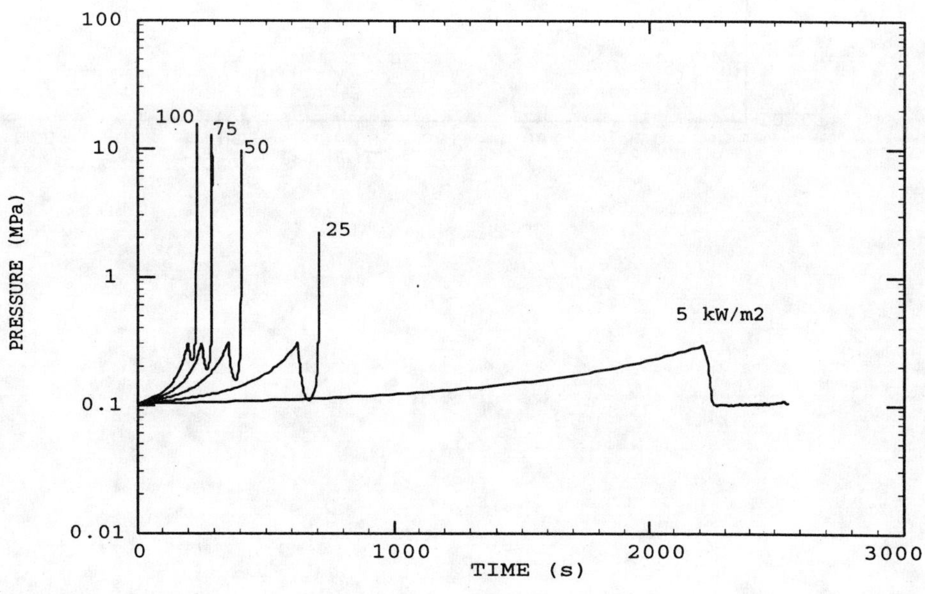

Figure 28: Sensitivity of maximum pressure level reached in vessel to rupture disk size at an external heat flux of 100 kW/m2 for the peroxide system

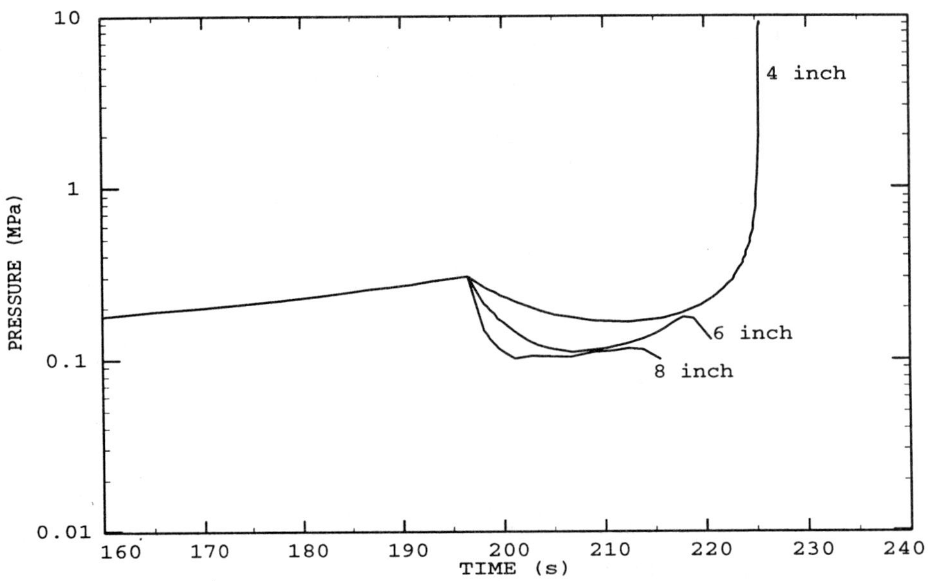

References

[1] G. A. Melhem. Advanced ers design using computer simulation. In *International Symposium on Runaway Reactions and Pressure Relief Design*, pages 502–566. AIChE, 1995.

[2] G. A. Melhem, H. G. Fisher, and D.A. Shaw. An advanced method for the estimation of reaction kinetics, scale-up and pressure relief design. *Process Safety Progress*, 14(1), 1995.

[3] O. A. Hougen and K. M. Watson. *Chemical process principles. Part I, material and energy balances.* Wiley, 1946.

[4] S. S. Grossel. *Prevention and control of accidental releases of hazardous gases, edited by V. M. Fthenakis.* Van Nostrand Reinhold, 1993.

[5] P. W. Barker. Personal communication, 1995.

[6] D. S. Nielsen. Validation of two-phase outflow. *Journal of Loss Prevention in the Process Industries*, 4:236–241, 1991.

[7] G. A. Melhem, P. A. Croce, and H. Abraham. Data summary and analysis of NFPA's BLEVE tests. *Process Safety Progress*, 12(2):76–82, 1993.

Emergency Relief Adequacy For Acrylic Polymerization Processes

A. S. Balchan, D. A. Paquet Jr., and J. A. Klein
DuPont, Wilmington, DE

ABSTRACT

Acrylic polymerizations are highly exothermic processes which must be carefully evaluated to ensure safe manufacture. In particular, failure to remove the reaction exotherm can lead to runaway polymerizations, characterized by sudden and rapid increases in reaction rate, temperature, and pressure. If not adequately vented, runaway polymerizations can quickly lead to reactor overpressure, possible vessel rupture, and release of flammable or toxic materials.

Because of complex polymerization kinetics and the large number of industrial acrylic polymer processes, practical application of DIERS two-phase flow methodology for sizing reactor emergency vents required development of a computer model to simulate specific commercial polymerizations. This model is currently used to determine the emergency vent adequacy of reactors prior to manufacture of new resins in existing equipment, as well as to assist in design of new reactor systems. The purpose of this paper is to provide guidance on practical application of DIERS technology through use of computer simulation for ensuring safe acrylic resin manufacture.

INTRODUCTION

Emergency relief systems are installed on process vessels to protect against the overpressure caused by runaway reactions and other upset events. The technology of the emergency relief of such vessels has evolved over the last decade with the development of an understanding of the capacity of relief devices to handle not only the required vapor flow, but the coincident flow of liquid as well (two-phase flow). The advances made by the original Design Institute for Emergency Relief Systems (DIERS), and the subsequent DIERS Users Group, has led to the development of recognized methods for the prediction of multi-phase flow and the methodology for sizing of relief devices under such flow conditions (Fisher et al 1992). Most reported systems are relatively simple first-order reactions or decompositions. In addition, with few exceptions, the practical applications have been limited to simple batch reactors, where overall composition is fixed throughout the batch. This paper illustrates some of the difficulties in application of standard DIERS methodology to practical cases of commercial acrylic polymerization reactions and illustrates the value of computer simulation of the reacting system in proper application of the DIERS technology to emergency relief design (Fisher 1991, Turner and Richter 1995).

The overpressure leading to emergency relief is caused either by the increasing vapor pressure of volatile species in the reaction mass during the course of an exothermic reaction ("high vapor pressure" system), or by the evolution of non-condensible gases through chemical decomposition or reaction ("gassy" system), or a combination of these two. The first of these types of systems is characterized by boiling of one or more ingredients upon relief of system pressure, which serves to cool the remaining material and limits further increases in temperature or pressure (so-called reaction tempering). In the case of high vapor pressure systems, the specific volumetric flow rate required of the emergency relief system (and hence the required relief area) is proportional to the average heat release rate per unit mass during the venting process. For low vapor pressure systems, the critical parameter in determining the required vent area is the rate of non-condensible gas generation at relief conditions.

The DIERS research led to the development of calorimeters for direct determination of both the reaction type and the values of the self-heat and/or gas generation rates (Fauske and Leung 1985, Creed and Fauske 1990) as a function of temperature for a fixed composition. Such devices are miniature batch reactors, allowing the programmed heat-up of a sample of mixed reactants under a superimposed inert gas pad. For high vapor pressure systems, the heat release rate during the venting process is taken as the average of the self-heat rates at relief activation and at temperature turnaround. These values can be obtained from tests run at a high pad pressure to suppress boiling and at the relief pressure. For low pressure systems, the rate of gas evolution at relief is determined from the rate of pressure increase in the containment vessel. While calorimetric tests are important analytical tools, they must be used with a clear understanding of their limitations.

REACTOR SAFETY

Given the potential hazards of industrial polymerizations, a comprehensive resin Reactor Safety program has been implemented worldwide at all DuPont Automotive resin manufacturing plants with the goal of achieving completely safe resin manufacturing (Balchan et al 1995). One of the key activities of the Reactor Safety program is the mandatory, routine safety screening of all acrylic resin manufacturing formulas, prior to releasing these formulas for production in plant equipment (Klein and Balchan 1996). Safety screening of acrylic resins includes checking the formula for inherent safety considerations, computer simulation to evaluate reaction exotherm and runaway reaction emergency vent sizing, and reaction calorimetry, as required. In all cases, no products may be produced in resin manufacturing equipment unless Reactor Safety has conducted the required safety screening and issued written approval for manufacture.

ACRYLIC POLYMERIZATION SYSTEMS

The sizing of emergency vents for acrylic polymerizations and the determination of the adequacy of existing vent configurations to satisfactorily relieve a runaway reaction pose a significant challenge for the safety professional. Essentially all acrylic polymerizations utilize small amounts of one or more polymerization initiators to form primary free radicals through thermal decomposition. Polymer chains are then formed from monomer addition to these free radicals. When each primary radical adds a single monomer unit, it becomes a growing polymer chain. As the reaction proceeds, almost all of the polymerization involves addition to growing polymer chains, and only very little (1%) involves addition to primary radicals. Overall, polymerization kinetics are very complex, with a large number of side reactions and very fast reaction rates (Odian 1991).

Most acrylic reactions are carried out in a solution of the volatile monomers in one or more solvents. The system vapor pressure is thus the sum of the individual solvent and monomer contributions. A number of factors makes these systems particularly challenging. First, the monomer component of the reacting mixture not only contributes significantly to the system vapor pressure, but is reacting and is thus being depleted as the polymerization proceeds. Secondly, the boiling point of the solution varies not only with system pressure, but with changing composition as the percent of polymer increases. Lastly, the system viscosity increases significantly with time as the reaction proceeds.

Few commercial acrylic polymerizations are carried out in the batch fashion of the calorimeter, as discussed below. Not only must the specific process details be taken into account, but the use of two-phase DIERS methodology demands an accurate knowledge of the kinetics of the reaction system. It can be very misleading to assume that the maximum exotherm (peak relieving rate) occurs exactly at the relief pressure. A dynamic model is the only method we have found by which the DIERS two-phase flow technology can be realistically applied to commercial acrylic polymerization reactor systems.

Polymerization Processes

Batch Process. In the true batch process, monomers, initiators and solvents are charged to the reactor at ambient temperature and heated. External heating is stopped once the exotherm begins and the heat of reaction is removed through evaporative (reflux) cooling, cooling coils, cooling jacket, or similar means. This type of process is least preferred because there is little control over finished product properties and the risk of a severe runaway is high. It is considered the most hazardous form of acrylic process.

With the calorimeters used to obtain data for emergency vent sizing, a fixed composition sample is heated adiabatically at a predetermined rate of temperature increase, and the sample temperature and enclosure pressure are measured throughout the exotherm. This particular profile of composition, temperature and reaction closely approximates the standard batch process just described with loss of cooling at the start of the exotherm.

A variation of the batch process is the addition of initiator as a very rapid ("shot") feed to a solution of monomer and solvent held at its boiling point. This type of reaction is somewhat more controlled than the first, but still has very high runaway potential. Because the initiator solution is only a small percentage of the reaction mass, the temperature remains at the boiling point during the shot addition. Since no reactants have yet been converted up to that point, the reaction rate after the initiator addition is higher than if all reactants were heated together and had partially reacted at lower temperatures.

This modification of the standard batch process is difficult to reproduce calorimetrically. While it is possible to inject a tiny amount of initiator to a preheated monomer solution in a calorimeter, in practice it has been found difficult to reproduce full-scale reactors in this way.

Semi-batch Process. A preferred process for acrylic polymerization is carried out in a batch reactor with the reactants fed in over relatively long periods of time to allow for a controlled exotherm and adequate heat removal. In these processes, either the initiator is fed into a monomer solution at its boiling point or both initiator and monomer solution are fed to the reactor as separate streams at fixed schedules while maintaining a boiling (reflux) condition in the reactor. The use of long feed schedules and adequate process control of feed rates and boil-up minimizes the potential for buildup of large quantities of unreacted reagent. The major runaway potential in these semi-batch processes is loss of external cooling with continued ingredient feed. Since the reactant composition (and reaction rate) varies continuously during this type of process, it cannot be approximated by a calorimetric test.

Continuous Stirred-Tank Process (CSTR). In this variant of the semi-batch process, initiator and monomer solution streams are fed continuously to an agitated reactor maintained at reflux temperature while finished or semi-finished product is withdrawn at the same overall rate. Again, the continuous nature of the process serves to mitigate against buildup of unreacted monomer.

Potential Upset Conditions

Total and sustained loss of cooling, without any mitigating action on the part of operations, is considered to be the most likely, and most hazardous cause of a runaway for all of these processes. Other paths to a runaway condition include:

Operational error is always possible, particularly when certain steps in a batch involve manual weighing and addition of ingredients. Errors of this type can include such things as incorrect amounts of one or more ingredients, incorrect type of ingredient, or omission of one or more ingredients.

Control error, primarily in the semi-batch and CSTR processes, can lead to a runaway condition. An incorrectly programmed rate of ingredient feed can lead to a pooling of unreacted ingredients and a drop in reactor temperature - subsequent heating can result in a runaway condition. Poor temperature control during heat-up of a batch process can also present a hazard.

In some cases, combinations of these errors are considered as part of hazard evaluation methods. Other upset conditions leading to a runaway reaction are possible, but are not typically considered as important in this type of processing.

KINETIC MODEL

A dynamic tank model for free radical solution polymerization is used to simulate runaway behavior. The model combines material and enthalpy balances with multi-component free radical polymerization kinetics (initiation, propagation, chain transfer, diffusion-controlled termination) to predict the evolving polymer architecture and reactor performance. Arbitrary feed schedules and various temperature control options permit the simulation of a wide range of operating conditions; lab scale calorimeter and manufacturing processes are readily simulated. Component thermodynamic/kinetic databases and a plant reactor equipment database facilitate the use of the model for routine safety evaluation of new formulations and/or operating conditions.

Several simulations are performed to ensure the safe manufacture of a resin prior to actual production. Simulation results of the proposed procedure determine if existing plant cooling capabilities can safely remove the peak reaction exotherm. To mimic the sudden loss of cooling, runaway simulations in which the reactor becomes adiabatic at various times identify the worst case scenario. In these simulations the predicted average heat release rate over the venting process is used either to size the emergency relief for new reactor systems or to determine the relief system adequacy for new products in existing reactor configurations, using established two-phase flow methods (Leung 1986). Homogeneous vessel venting and a 10% overpressure allowance are assumed. To determine the start of the venting process, reactor pressure is calculated assuming an ideal solution for the non-polymeric species with a Flory-Huggins based correction factor to account for the presence of polymer. A two-parameter Antoine equation is used for the pure component vapor pressures.

The model assumes that all polymerizations are the "high vapor pressure" type, that is, excess pressure for relief derives from increasing system temperature caused by the polymerization reaction. While the decomposition of the initiators used in this type of process generate either carbon dioxide or nitrogen, it is assumed that the "gassy" contribution to overpressure is negligible in comparison. Indeed calorimetric testing of each of the polymer reactions described in this paper showed that the fundamental vent rate for gas generation was, at worst, slightly more than 3% of the vapor generation rate. Calorimetric tests on many acrylic systems have shown the effect of reaction tempering to be relatively small at the venting pressures used in practice. This may be due to the relatively high viscosity of the solution at venting pressure. We normally ignore any tempering benefit in calculation of vent diameter.

We further assume that the "worst case" system failure is always total loss of external cooling (reaction goes adiabatic) and reactor venting with continued ingredient feeds at the particular time in the batch where the rate of temperature increase at relief pressure is highest (largest vent area). The model calculates the time from the start of the batch at which this condition occurs. For batch processes this occurs after all ingredients are loaded and heated (time zero). For semi-batch, or feed processes, it can occur at time zero, or at some later time, depending on specific ingredients, feed schedules and processing conditions. In the following examples, the time indicated for loss of cooling represents the maximum exotherm at relief condition in each case. We do not normally consider other major upset conditions (e.g.. ingredient mischarge or incorrect feed rate) occurring simultaneously with loss of cooling to be credible cases.

Figures 1 and 2 illustrate the temperature and pressure profiles due to a loss of cooling at 0, 50 and 100 minutes into a typical acrylic feed process, where initiator and monomer solutions are being fed simultaneously and uniformly into refluxing solvent over a 4 hour period. During normal operation, the reflux temperature decreases slowly with time as the percentage of polymer in the reactor increases. Loss of external cooling at the start of feeds (0 minutes) causes an immediate rapid rise in temperature. Loss of cooling at later times gives milder exotherms, reflecting the increasing quantity of reacted material in the reactor. As illustrated in Figure 2, the time that loss of cooling occurs can have a profound effect on the pressure profile during the runaway. Loss of cooling at the start of the feeds causes the pressure to rise sharply, exceeding the assumed MAWP of the vessel. Loss of cooling at 50 minutes will cause the relief device to open, but the final pressure will not exceed the MAWP. The final pressure from loss of cooling at 100 minutes will be below the relief pressure, and the runaway will be contained. Clearly, in this example, the worst case is loss of cooling at the start of the batch. The emergency relief vent must be adequately sized for this case in order to safely vent this pressure buildup, maintaining the reactor pressure below the MAWP.

FIGURE 1 - Computer simulation showing runaway reaction occurring at three different times.

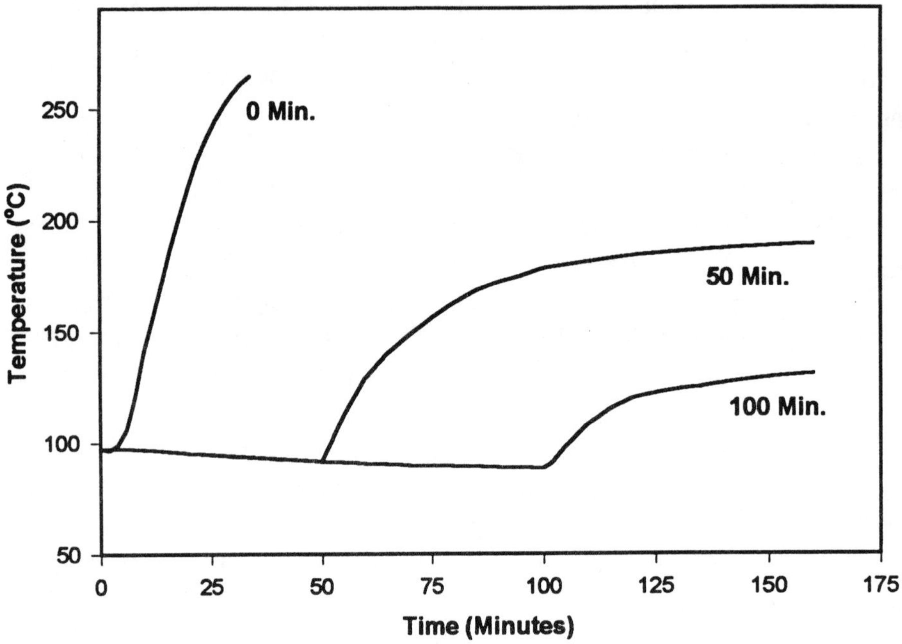

FIGURE 2 - Pressure profiles for runaway simulations shown in Figure 1.

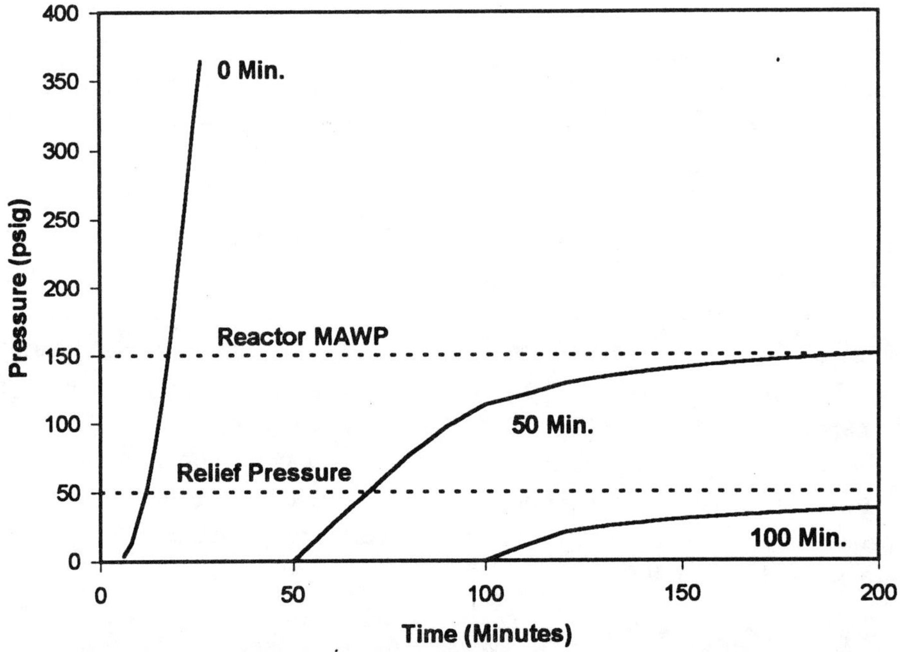

APPLICATIONS

In this section, the kinetic model is applied to several different actual cases of batch and semi-batch acrylic solution polymerization. In each case we assume that the reactor is protected with a rupture disc opening at 50 psig. Since calorimetric tests can only be run by heating a sample of fixed composition at a specific temperature ramp rate, they cannot be used to duplicate actual processing conditions for semi-batch or feed processes. For these, calorimetric tests are run with a composition representing the total combined feeds, and the results are compared with the model run at the same composition and heat-up rate. The differences between calorimetric results obtained in this way and the predicted results for actual semi-batch processes can be significant, requiring caution in the sizing of emergency relief vents.

Batch Process

Polymer A. This polymer consists of approximately 61% monomer dissolved in isopropyl alcohol and toluene. The atmospheric boiling point is about 102°C.

Figure 3 compares the measured self-heat rate at 300 psig for this polymer with that calculated by the model, assuming that all ingredients are mixed at room temperature and heated adiabatically (standard batch process), in the same fashion as the calorimeter. These results show good agreement between the measured and predicted maximum self-heat rate, which is approximately 14.5°C/minute. At lower temperatures (<100°C) the reaction rate is slowed by the effect of polymerization inhibitors present in the calorimeter sample. Although the model represents this inhibition well, the effect is relatively unimportant, since all reactions of interest occurred at temperatures at which the inhibitor had been completely reacted. The inhibitor effect was found to some degree in all polymers studied.

In the actual manufacturing process, the initiator solution is added as a shot to a monomer/solvent mixture at its boiling point, and thus the reaction rates are higher than those realized when the initiator, monomers and solvents are heated together, as shown by the model results in Figure 3. The difference between the two, as explained earlier, is due to the partial reaction of the ingredients at lower temperatures during the heat-up process.

Figure 4 illustrates the temperature and pressure rise, as calculated by the model, for the "worst case" loss of cooling immediately after the initiator shot. The rate of temperature change at relief is 20.5°C/minute vs. a maximum of 29.7°C/minute just prior to the relief point. Pressure and temperature show no further rise after depletion of reactants, with a maximum pressure of 59.7 psig. Use of a higher relief pressure of 70-80 psig, if not in excess of the MAWP of the vessel, would allow the runaway to be fully contained.

In this simple bulk case, the measured peak self-heat rate of 14.5°C/minute is fortuitously close to the calculated value of 20.5°C/minute at the assumed vent pressure, and a reasonable value of emergency vent diameter can be obtained from calorimetric data.

Semi-batch Process

Polymer B. Polymer B is 82.3% monomer in amyl acetate solvent. The atmospheric boiling point is 144°C. For this process the monomer and initiator solutions are fed simultaneously into refluxing solvent over a period of 180 minutes.

FIGURE 3 - Self-heat rate for Polymer A.

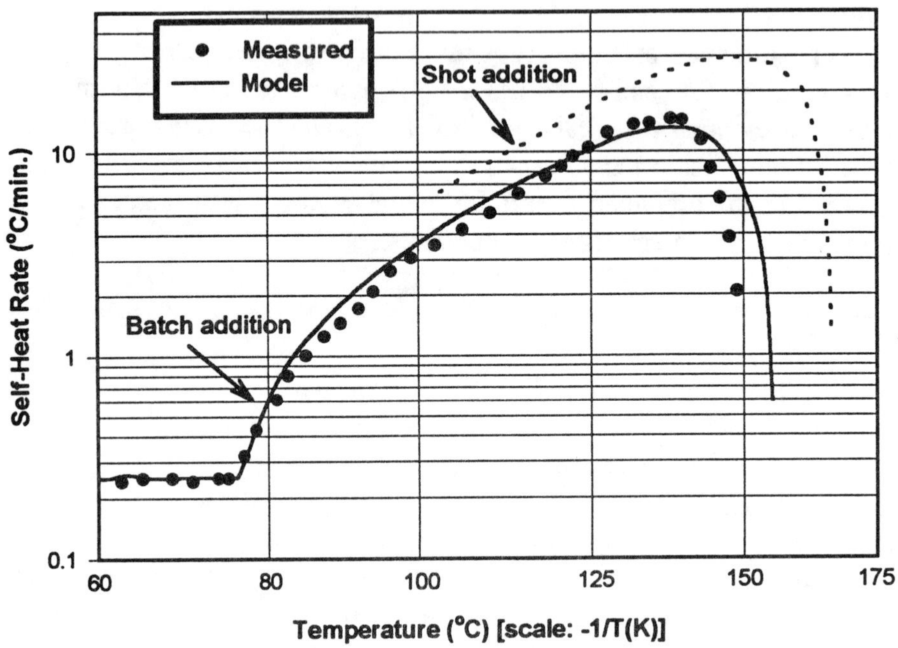

FIGURE 4 - Worst case simulation for Polymer A batch process.

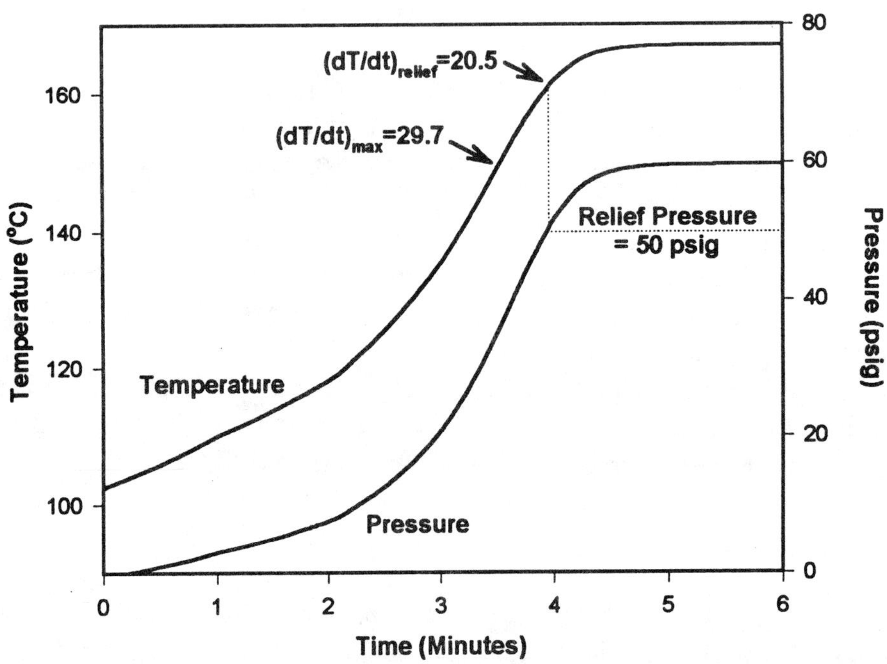

Figure 5 illustrates the temperature and pressure profiles for the "worst case" loss of cooling with the standard feed schedule. The "worst case" (highest exotherm at relief pressure) was found to be from loss of cooling 20 minutes after the start of the batch. After loss of cooling, the reaction rate initially accelerates, reflecting an increased rate of conversion of unreacted monomer already fed to the reactor. The maximum self-heat rate of 3.8°C/minute occurs toward the end of this time period, after which the temperature profile represents a combination of reaction of new monomer being fed to the process, plus an increase in the overall monomer conversion in the reactor. As seen in Figure 3, the final pressure for this case barely reaches the relief pressure of 50 psig, at which point the reactants are virtually exhausted. The self-heat rate at the relief pressure is only 0.12°C/minute.

Figure 6 compares the self-heat rate predicted by the model for the combined feeds heated adiabatically from room temperature with calorimetric results for the same mixture. The maximum temperature rise rate for the mixed feeds is in excess of 1,000°C/minute. Again, fairly good agreement between the measured and predicted result is observed. The maximum self-heat rate of >1000°C measured for the combined feeds, though, is significantly higher than the value of 3.8°C/minute predicted for the feed process. This is due to the higher monomer and initiator concentrations present in the batch model, illustrating the difficulty of directly using batch calorimetric results in sizing emergency relief vents for semi-batch processes.

This is shown more clearly in Figure 7, where the batch and semi-batch self-heat rates are compared. At the relief pressure of 50 psig (211°C), the batch process self-heat rate is over 1,000 times greater than that for the "worst case" semi-batch process, where the reaction takes place uniformly over a period of three hours. If, therefore, an emergency relief system for this process were designed solely from calorimetric data at the relief pressure, the relief area would be overestimated be about that same percentage.

Polymer C. Polymer C is 62.6% monomer in a mixed solvent of toluene and methyl ethyl ketone. In this case, a portion of the monomer/solvent charge is brought to reflux and the remaining monomer and most of the initiator is fed uniformly over 90 minutes. In the worst case shown in Figure 8, with loss of cooling at 10 minutes, the relief pressure of 50 psig is reached at about 40 minutes into the batch. At this point the self-heat rate is 1.6°C/minute, versus the maximum rate of 2.5°C/minute. The final stages of the process include a small solvent rinse at 90 minutes, followed by a final initiator feed to convert any remaining unreacted monomer. This final feed causes the temperature to reach a final peak of 200°C at about 110 minutes into the batch.

In order to best match the conditions during the first 90 minutes of the process, the overall composition corresponding to the initial monomer charge and the first monomer and initiator feeds was run in the calorimeter and compared with the model, as shown in Figure 9. Again, fairly good agreement between the calorimetric and predicted results are obtained, with a maximum self-heat rate of approximately 10°C/minute. This value is again significantly higher than the self-heat rate of 1.6°C/minute obtained from the semi-batch simulation. In addition, the final temperature reached in this case is about 140°C vs. a relief temperature of 160°C for the actual feed process. This difference is because the feed process is much more efficient in converting monomer and the second initiator feed, which cannot be duplicated calorimetrically. For this process, therefore, while the calorimetric data is in reasonable agreement with the model run under the same conditions, it is of little value in predicting the relief exotherm or determining an adequate vent diameter for the conditions of the actual process.

FIGURE 5 - Worst case simulation for Polymer B semi-batch process.

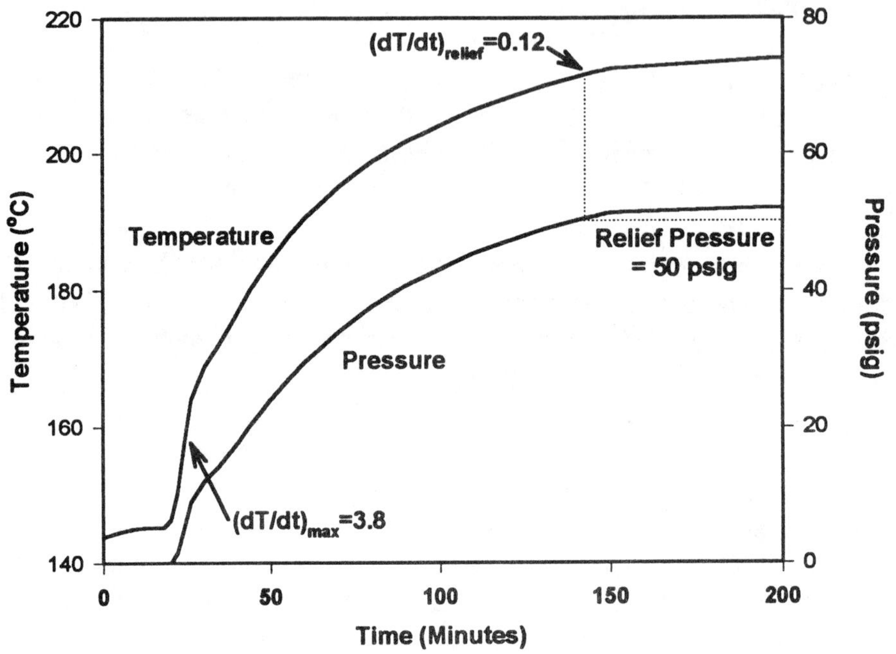

FIGURE 6 - Self-heat rate for Polymer B.

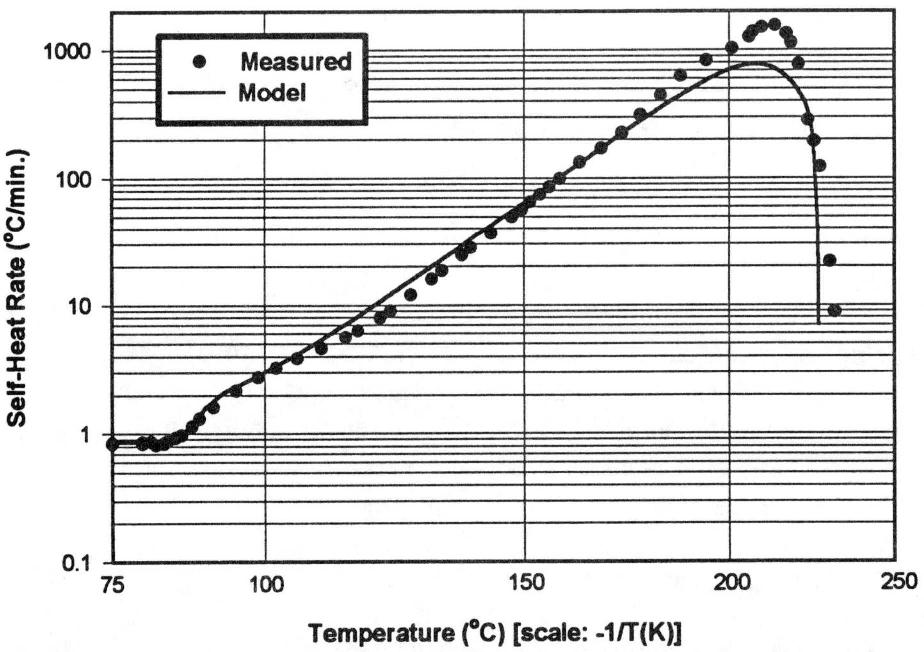

FIGURE 7 - Self-heat rate for Polymer B.

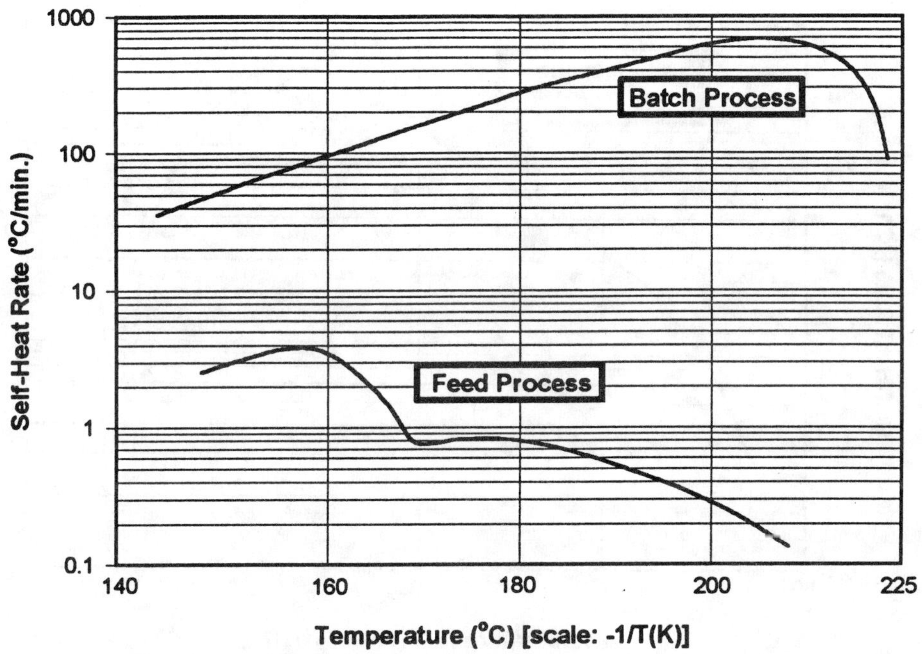

FIGURE 8 - Worst case simulation for Polymer C semi-batch process.

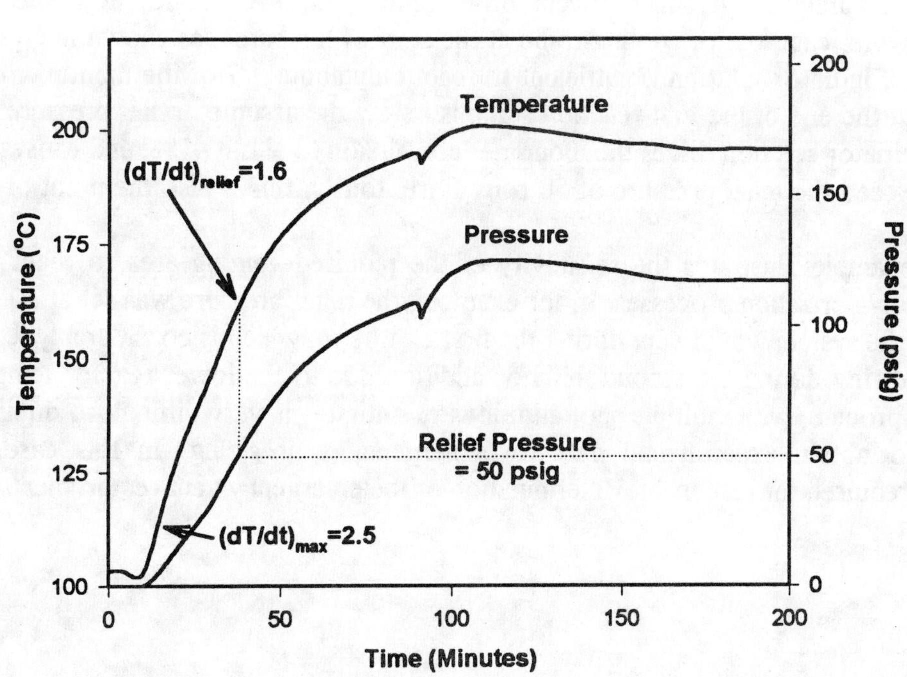

FIGURE 9 - Self-heat rate for Polymer C.

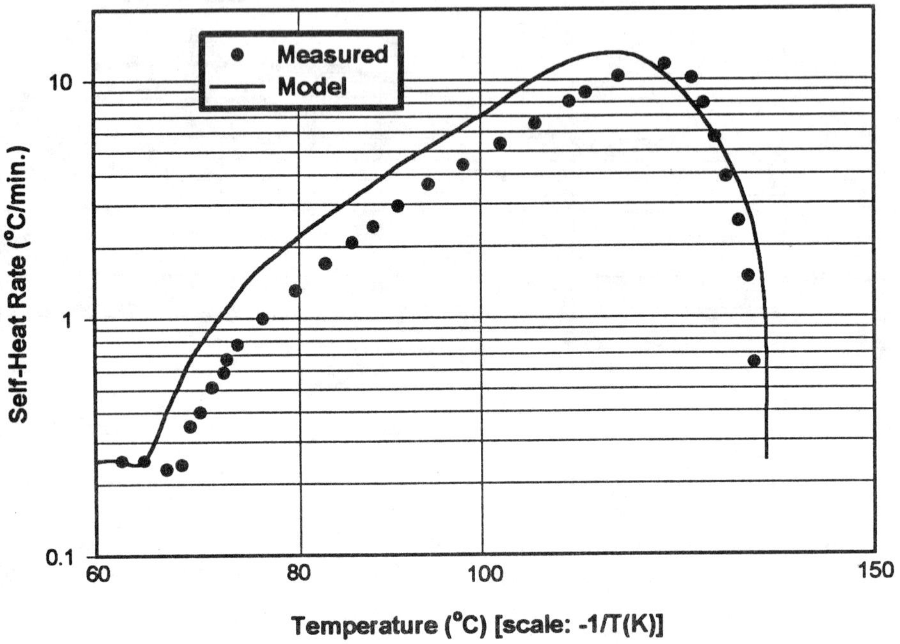

Polymer D. Polymer D is 50% monomer in a mixed solvent of isopropanol, toluene and ethyl acetate. The initiator solution is added in three stages - 25% of the total initiator is added as a shot to the monomer solution in the reactor, with the remainder added in two increments at 45 minutes and 120 minutes. A small solvent rinse follows the last initiator addition. The model indicates the worst case loss of cooling to be at the start of the batch, as shown in Figure 10. The initial charge of initiator solution is sufficient to react only about 1/3 of the monomer charge, and the pressure at the end of the first reaction stage is below the assumed relief pressure of 50 psig. The second initiator solution drives the monomer conversion to about 47%, and causes the system pressure to exceed the relief pressure of 50 psig. Fortuitously, this is also the point of the highest self-heat rate.

This example illustrates the sensitivity of the required venting area to relief pressure in commercial polymerization processes. If, for example, the relief pressure was set at 30 psig, rather than 50 psig, the system would vent during the first exotherm, which is considerably less energetic than that occurring during the second initiator addition due to the lower reactor temperature. In addition, this process, with multiple shot additions of initiator, is very difficult to duplicate with a calorimeter to obtain experimental data for emergency vent sizing. In this case, computer simulation is required for reasonable determination of the emergency vent requirements.

FIGURE 10 - Worst case simulation for Polymer D semi-batch process.

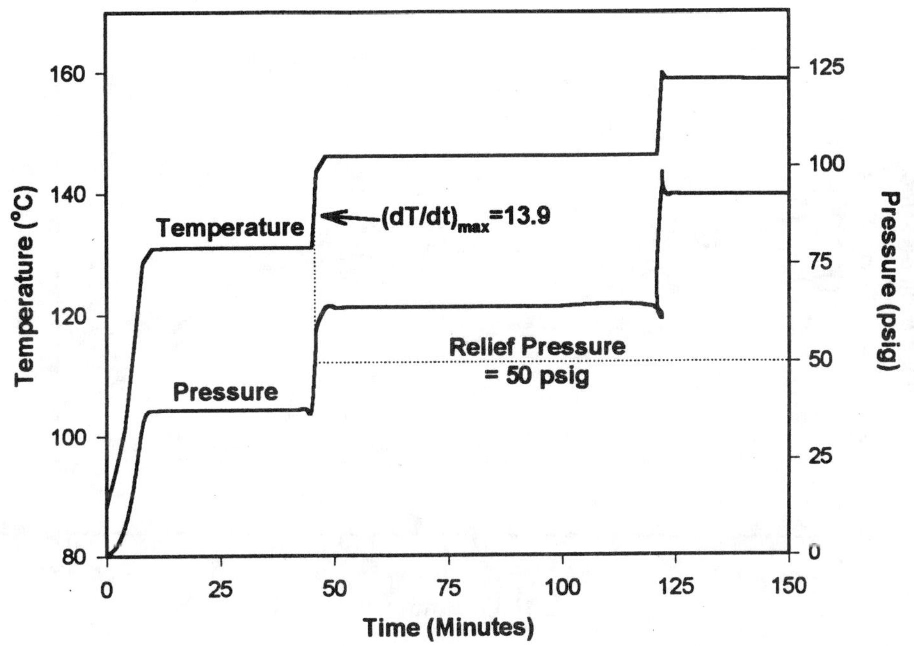

SAFETY SCREENING

As discussed in the previous section, calorimetric data for maximum self-heat rates must be used with caution when sizing emergency vents for many typical commercial acrylic polymerization processes. Good agreement between calorimetric and predicted self-heat rates for batch processes is generally observed, however, validating in many respects the complex kinetics incorporated into the computer model. For these reasons, routine safety screening of new acrylic resin processes prior to manufacture relies primarily on computer simulation. Calorimetric measurements are used to supplement the computer model, as needed, such as when good kinetic data area not available for new monomers or when stability of reactant mixtures must be determined.

Typical computer simulation results for determining emergency vent requirements for a runaway reaction of a semi-batch acrylic resin formula are shown in Figure 11. In this case, the required diameter for the emergency vent is shown for different relief pressures, as a function of the time loss of cooling occurs during the process, assuming a 10% overpressure. The emergency vent diameter requirement for this resin decreases when the runaway occurs later in the batch, due to higher monomer conversions and subsequent lower exotherms. In addition, the required emergency vent diameter also decreases as the relief pressure is increased, due to higher exotherms at the lower relief pressures. In both cases, these results are often seen for many semi-batch resins, but always depend on the specifics of the resin formula. In general, the DIERS vent areas are 2 to 4 times higher than for all-vapor venting.

FIGURE 11 - Typical emergency vent requirements for runaway reaction simulation.

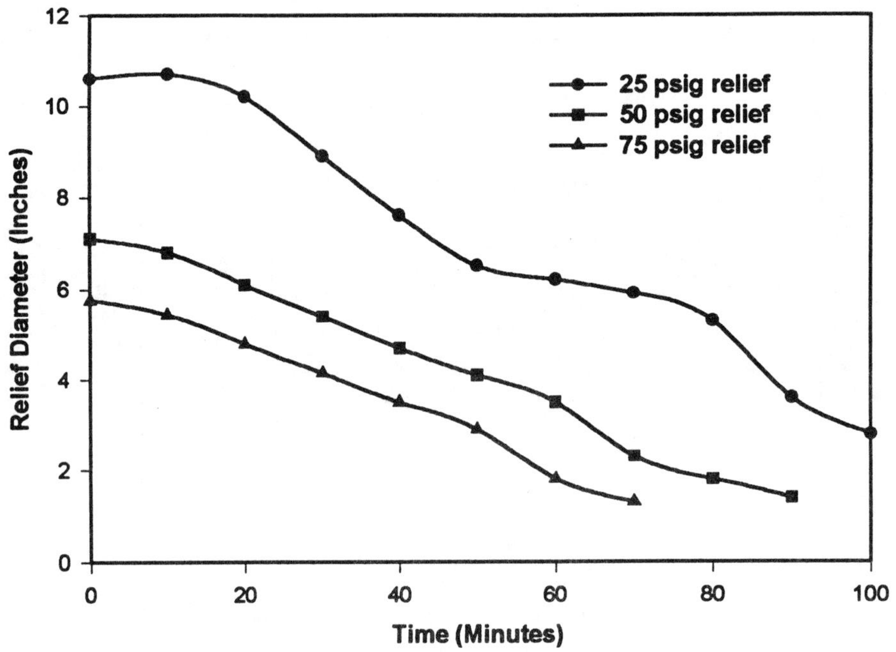

SUMMARY

A wide range of commercial acrylic resin manufacturing processes have been reviewed. In each case, the use of an experimentally-validated computer simulation model of the complex reaction kinetics is vital in determining the reaction conditions throughout the process. With the model, it is possible to determine whether reactor pressure under various assumed upset conditions will exceed the relief pressure, the exotherm at emergency relief, and the required emergency vent area for two-phase flow. In practice, the computer model helps to focus attention on those specific processes having the greatest risk potential, allowing the appropriate precautions to be made to ensure safe manufacture. The quantitative use of calorimetric data, other than to validate the model, is limited to those processes that exactly duplicate the batch conditions of the calorimetric test. In general, the use of calorimetric data for a semi-batch or continuous feed processes will lead to significant oversizing of the relief system.

REFERENCES

Balchan, A.S., Klein, J.A., and Klein, F.G. (1995). *Process Safety of Polymer Resin Manufacturing: A 20-Year Perspective*, in Loss Prevention and Safety Promotion in the Process Industries, J.J. Mewis, H.J. Pasman, and E.E. De Rademaeker, (editors), Vol. 1, p. 3, Elsevier

Creed, M.J. and Fauske, H.K. (1990). *An Easy, Inexpensive Approach to the DIERS Procedure*, Chemical Engineering Progress, March, p. 45

Fauske, H.K. and Leung, J.C. (1985). *New Experimental Technique for Characterizing Runaway Chemical Reactions*, Chemical Engineering Progress, August, p. 10

Fisher, H.G. (1991). *An Overview of Emergency Relief System Design Practice*, Plant/Operations Progress, Vol. 10, No. 1, p. 1

Fisher, H.G. et al (1992). *Emergency Relief System Design Using DIERS Technology*, DIERS Project Manual, The Design Institute for Emergency Relief Systems, American Institute of Chemical Engineers

Klein, J.A. and Balchan, A.S. (1996). *Safe Formulation and Manufacture of Acrylic Resins*, in International Conference and Workshop on Process Safety Management and Inherently Safer Processes, Center for Chemical Process Safety, p. 329

Leung, J.C. (1986). Simplified Vent Sizing Equations for Emergency Relief Requirements in Reactors and Storage Vessels, AICHE Journal, Vol. 32, No. 10, p. 1622

Leung, J.C. and Fauske, H.K. (1987). *Runaway System Characterization and Vent Sizing Based on DIERS Methodology*, Plant/Operations Progress, Vol. 6, No. 2, p.77

Odian, G. (1991). Principles of Polymerization, 3rd Edition, John Wiley.

Turner, F. and Richter, S.H. (1995). *Accurately Size Batch Reactor Relief Systems*, Chemical Engineering Progress, March, p. 33

QRA of Chemical Reaction Systems — The State of the Practice

John Farquharson, Project Manager
Scott McNutt, Technical Editor
Henrique Paula, Process Safety Engineer
Michael Roberts, Process Safety Engineer
Al Waller, Process Safety Consultant

JBF Associates, Inc.
1000 Technology Drive
Knoxville, TN 37932-3353

ARCO Chemical Company
3801 West Chester Pike
Newtown Square, PA 19073-2387

Prepared for Presentation
at the 31st Annual Loss Prevention Symposium

AIChE 1997 Spring National Meeting

(S81)\papers\93-14.laf

QRA of Chemical Reaction Systems — The State of the Practice

John Farquharson, Project Manager

Scott McNutt, Technical Editor

Henrique Paula, Process Safety Engineer

Michael Roberts, Process Safety Engineer

Al Waller, Process Safety Consultant

JBF Associates, Inc.
1000 Technology Drive
Knoxville, TN 37932-3353

ARCO Chemical Company
3801 West Chester Pike
Newtown Square, PA 19073-2387

Abstract: *This paper discusses a quantitative risk assessment (QRA) of an exothermic chemical process. Fault tree/event tree techniques were used to estimate the frequency of runaway reactions that could result in overpressure and vessel rupture. The paper focuses on the current methodology used in performing this type of study, emphasizing analytical techniques that help avoid potential time-consuming pitfalls. The challenging problem of predicting how rare sequences of events (e.g., involving failure of several engineered control and safety systems) will develop is systematically modeled using fault trees/event trees, which makes the QRA process a unique and powerful analytical tool. The results of this study provided the guidance for selecting the most appropriate action(s) to implement from a set of various design options for improving the safety of this system. Use of QRA* *has enabled plant management to have confidence that they are selecting the most cost-effective safety improvement option.*

Introduction

As regulations become more stringent and public scrutiny becomes more exacting, the chemical process industry (CPI) is searching for methods that better analyze the frequency and consequences of accident scenarios related to engineered processes. A better understanding of such accidents will help analysts identify practical ways to minimize catastrophic occurrences (Arendt, 1991) (AIChE, 1989). One such method is Quantitative Risk Assessment (QRA). When used by qualified experts teamed with personnel familiar with the process, QRAs can reveal unique insights into the relative importance of individual plant systems and components. QRAs have the advantage of providing both qualitative and quantitative results for use in making informed decisions on how to better manage the risk associated with engineered systems. Nevertheless, QRAs also have drawbacks, such as potential lack of completeness, difficulties with model validity, data uncertainty, reproductibility, and potential misunderstanding of results; some of these drawbacks are inherent to any risk analysis technique (e.g., no one can ever guarantee that *all* possible accident scenarios have been considered), and some are unique to QRA. However, these shortcomings can be anticipated and overcome through careful planning, managing expectations for the study results, selecting the appropriate team, using the right analysis methodology and software, and clearly documenting results. This article considers these difficulties and the means to address them.

Setting Up a QRA

In this paper, we use an actual QRA that we performed on a semi-batch exothermic chemical process to illustrate the QRA process, its potential benefits, and practical ways of avoiding potential pitfalls. The event of interest in this example is an exothermic decomposition reaction leading to an unrelieved pressure rise and potentially resulting in a vessel rupture. We go through the actual QRA, from customer contact to final report, to demonstrate the current method of performing CPI QRAs.

Objectives of the study

The customer contacted us, interested in having a QRA performed on a system because the company wanted to improve its understanding of potential decomposition events that could lead to an unrelieved pressure rise and eventual vessel rupture (the *accident of interest* for this study). The company also hoped that through this QRA it could gain enough training to perform similar studies at other sites. To keep the analysis focused, we met with the customer to establish specific objectives for the QRA. Together, we developed the following three-point agenda:

- Estimate the frequency of the accident of interest.

- Estimate the range of potential consequences associated with these events.

- Prepare the results in a way that helps the customer develop recommendations for reducing the frequency or mitigating the consequences associated with the accident of interest.

By defining specific goals for the QRA, we avoided overworking the problem and revising the project scope later on, both of which are wasteful and costly.

Physical scope of the study

Maintaining clear study goals is a paramount concern in QRA. In the scope for this analysis, we included the reactor system, applicable portions of the feed system (i.e., isolation capability), the reactor relief system, and the instrumentation and control systems. Utility and auxiliary systems were considered only to the extent that they affected systems included in the analysis. We excluded external events (e.g., floods, earthquakes, sabotage).

QRA project team details

After the objectives of the analysis had been set and its physical scope defined, we worked with the customer to put together an analysis team. Our personnel were chosen primarily because of their experience in the development and application of risk assessment techniques. Customer personnel from corporate process technology, plant process engineering, plant operations, plant instrumentation and control, and plant maintenance were included on the project team because of their collective experience and unique insights into the process.

Performing a QRA

Having gathered the necessary personnel and informed them of the analysis objectives and scope, we pursued the analysis objectives through these steps:

1. Developing an in-depth understanding of the process and preparing a process description

2. Developing models, specifically:

    - Identifying the initiating events that could, if not mitigated by the operators or by engineered safety features, result in the accident of interest (unrelieved pressure rise from rapid decomposition and resulting vessel rupture)

    - Developing an event tree for each initiating event that shows how the initiating event and subsequent system failures could lead to each accident sequence of interest. The combination of all unique sequences represents the total frequency for the accident of interest

    - Developing fault tree models that show how the system failures can occur

- Developing component failure, test interval, and repair data as well as human error frequencies and probabilities

3. Analyzing the frequency of occurrence for the accident of interest, specifically:

   - Estimating the frequency of occurrence for each accident sequence of interest

   - Estimating the importance values for each system

   - Estimating the importance values for each component

4. Estimating the ranges of consequences for the accident of interest

1. Process understanding and process description

Our first task was to develop an in-depth understanding of the process, then express that understanding in a written and graphic depiction (i.e., process description). The interaction of customer participants and our personnel was the most important factor in this development. A QRA team must thoroughly understand the process design and operation. The analysis team conducted this task by (1) touring the site, (2) reviewing piping and instrumentation drawings (P&IDs) and logic diagrams, and (3) interviewing process, operations, and engineering personnel.

The analyzed process is a polymerization reactor that produces batches of crude polymer product by reacting an intermediate catalyzed polymer with a monomer according to specific recipes. The reaction is exothermic; the reaction rate (and therefore the heat generation) is dependent upon reactor temperature, catalyst concentration, and free monomer concentration. It is a semi-batch process controlled by a distributed control system (DCS). For most of the batch cycle, the free monomer concentration is sufficient to potentially cause a runaway exothermic reaction and exceed the onset temperature of decomposition. Decomposition of the polymer results in rapid generation of volatiles and rapid pressure rise, which might exceed the reactor relief capacity and rupture the reactor. Figure 1 provides a simplified schematic of the polymerization reactor.

The reactor is equipped with an agitator and a tempered water loop to control the temperature within the reactor. The polymerization reactor is initially charged with an intermediate catalyzed polymer. Monomer is fed to the reactor to initiate the reaction and continued through the batch. The timing sequence and flow rates of the monomer are recipe-dependent and are controlled by the DCS. Free monomer concentration greater than a specified percentage could exceed both the heat removal capacity and the relief system capacity and result in a high reactor temperature that could cause decomposition. Both the DCS and a programmable logic controller (PLC) monitor system parameters and will isolate monomer feed upon detection of system conditions that could lead to a runaway exothermic reaction.

The reactor relief system, consisting of two pilot-operated relief valves and a lower capacity bellows relief valve, is not demanded during a routine batch. If an upset occurs in heat removal during the batch, the reactor relief system is designed to vent sufficient accumulated free monomer, which limits the heat from the exothermic reaction and provides evaporative cooling. Process modeling indicates a rapid temperature rise would occur upon a cooling upset. Each pilot-operated

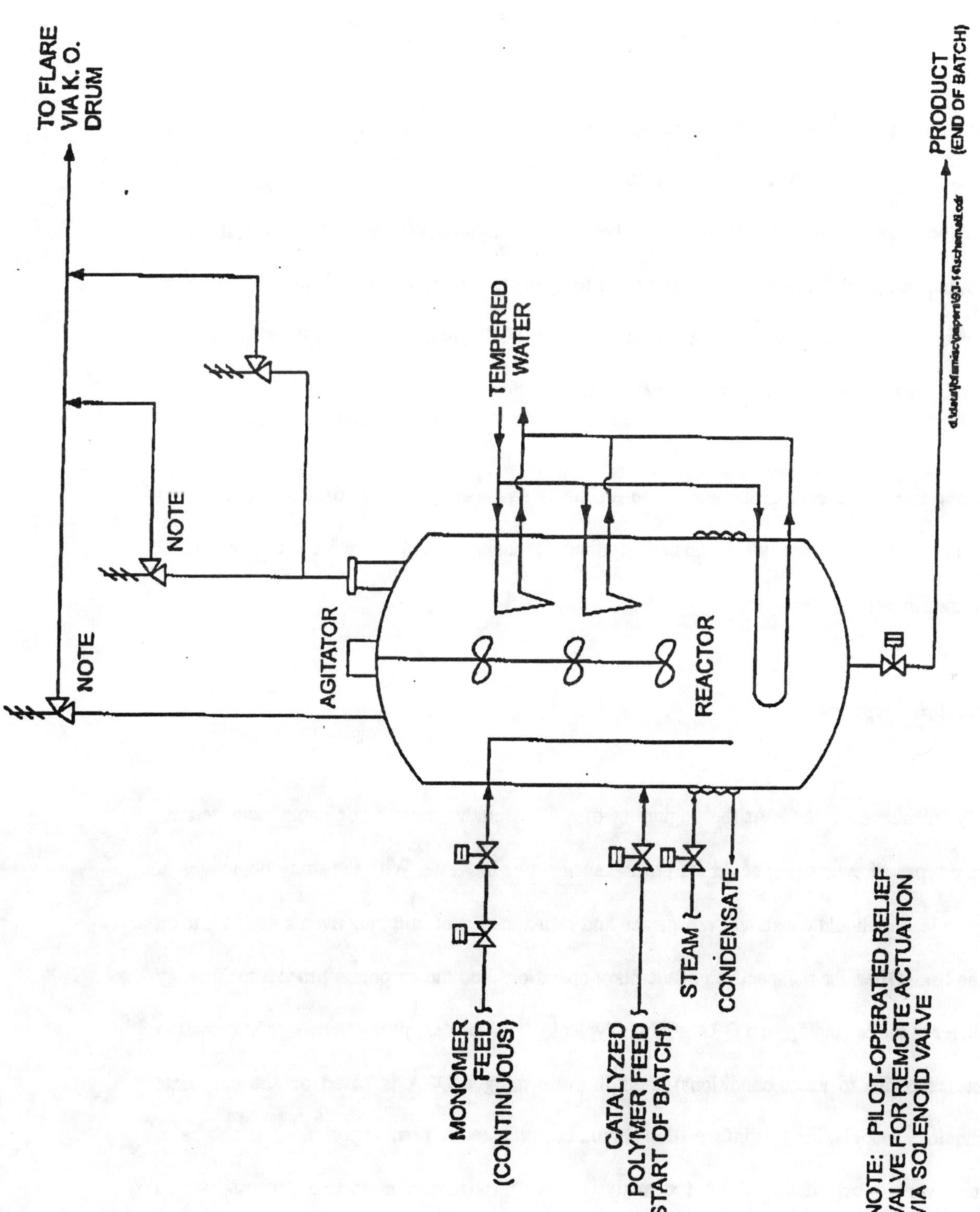

Figure 1 Semi-batch Polymerization Reactor

relief valve can be opened either mechanically through a pressure response or remotely via an electrical signal to a solenoid valve. The remote venting signal can be actuated either by the PLC shutdown system (through a specified temperature deviation or pressure signal) or by an operator's action. Successful operation of one pilot-operated relief valve will limit the temperature rise to prevent decomposition provided that the free monomer concentration is within design and that monomer feed is isolated. The lower capacity bellows relief valve provides limited venting redundancy on direct pressure should both pilot-operated relief valves fail to open; however, it is not designed to prevent the system from reaching the decomposition temperature under all conditions. The relief system is not designed to relieve the high pressure caused by a decomposition.

In addition to relief system redundancy, the reactor is equipped with redundant tempered water circulation pumps, redundant catalyst weigh cells, redundant monomer feed flowmeters, and redundant temperature transmitters.

2. Model development

Identify initiating events. At the beginning of a QRA study, most participants have general ideas as to the types of events that could lead to the accident of interest. With the study objectives in mind, the team leader should constantly reorganize and refine the list of initiating events until the team agrees on the list. Often the perspectives of the plant operations and the corporate process technology personnel differ at the beginning of a QRA study. Typically, the accident phenomenology information (e.g., system response to upset conditions) used in performing a QRA is based on the corporate process technology models, but this information should be compared to plant experience. In this case, the initiating events were identified by the analysis team through an interactive process, which includes:

- Review of the P&IDs, process control logic diagrams, and other system documentation
- Field review of the system
- Discussion with plant personnel and process technology personnel
- Brainstorming (e.g., what-if or hazard and operability [HAZOP] reviews)

Through these efforts, we eliminated several events that were at first thought to be initiating events. Table 1 shows the final list of 14 initiating events for the polymerization reactor and associated systems. As shown in Table 1, each initiating event (IE) is labeled by a unique identifier (IE1 through IE14). The initiating events are divided into two categories: **cooling upsets** and **feed upsets**. The cooling upsets are subdivided into four distinct types based on the expected different responses of the safety systems to each type. In this particular batch process, the effectiveness of the relief valves varies throughout the monomer feed phase; there are different, redundant, and/or diverse methods for the safety systems to react to a loss-of-cooling upset during different parts of the batch process. To take this into account, the types of initiating events for cooling upsets are further divided into monomer feed time-regions.

If a cooling upset were to occur during Time-Region I of the monomer feed phase of a batch, at least one of the two pilot-operated relief valves must open upon receiving a remote actuation signal to avoid a runaway to decomposition temperature. During this time region of the batch, a pressure response would not be adequate to avoid a runaway reaction because the system pressure peaks below relief setpoint pressure while the temperature continues to increase. During Time-Regions II and III, a pressure response would be adequate to avoid a runaway reaction; one of the pilot-operated relief

Table 1 Analyzed Initiating Events

Initiating Event Type	Monomer Feed Time-Region		
	I[a]	II[b]	III[c]
Cooling Upsets			
Loss of cooling	IE1	IE2	IE3
Loss of agitation	IE4	IE5	IE6
High steam flow to the jacket	IE7	IE8	IE9
Loss of site AC power	IE10	IE11	IE12
Feed Upsets			
High monomer concentration that exceeds the relief capacity due to low catalyst or low temperature	IE13		
High monomer feed rate	IE14		

[a]Safety system: Pilot operated relief valve (PORV) actuation due to remote signal.
[b]Safety system: PORV actuation due to remote signal or direct pressure.
[c]Safety system: Opening of bellows valve or PORV actuation due to remote signal or direct pressure.

valves must open in Time-Region II. In Time-Region III, operation of the smaller capacity bellows relief is adequate.

Develop event sequence models. We reviewed each initiating event to identify the system responses and/or operator actions that could help prevent the accident of interest. The accident phenomenology information provides the basis for developing the event trees. The review of all aspects of an in-progress QRA study (especially the event trees) can serve as a forum for discussion of accident phenomenology information. This type of internal review is very important and necessary to ensure proper understanding and acceptance of the QRA results.

An event tree (see Figure 2) is a graphical method of displaying the various accident scenarios that can result from a type of initiating event. The initiating event is shown at the left side of the event tree. Reading the event tree from left to right, listed along the top of the tree are system responses and/or operator actions expected to occur in response to the initiating event in the approximate order they would be demanded. Underneath the system responses/operator actions for each event tree "branch point" are alphanumeric identifiers (for this event tree: TOP11, TOP12, and TOP13) that correspond to fault trees that outline the specific combinations of equipment failures and/or operator errors that could lead to failures of the systems. To define the specific accident sequences, the event tree considers the initiating event and the subsequent system responses and/or operator actions expected to occur during the event. The branch points represent the success or failure of the system responses and/or operator actions. A successful response is represented by an upward path (success) at a branch point, and an unsuccessful response is represented by a downward path (failure) at a branch point. There are two columns on the far right of the event tree: the sequence column contains an alphanumeric identifier for each unique accident sequence; the consequence column indicates whether an accident sequence leads to a significant consequence.

Develop fault tree models. We developed fault trees for each initiating event and for each branch point in the event trees. (Figure 3 is the fault tree developed for the second branch point [TOP12] of the event tree displayed in Figure 2.) A total of 25 fault trees were developed for this QRA. A fault tree explicitly models the combinations of equipment failures and operator errors that can cause the specified failure event of interest (called the TOP event because it is at the top of the fault tree). The fault tree consists of AND and OR logic gates that model the system's failures as specific combinations of basic component failures (circles) and human errors (diamonds). Conventionally, mechanical system failures that are not developed in detail (e.g., instrument air failure) are also shown as diamonds since failures are grouped rather than further broken down into

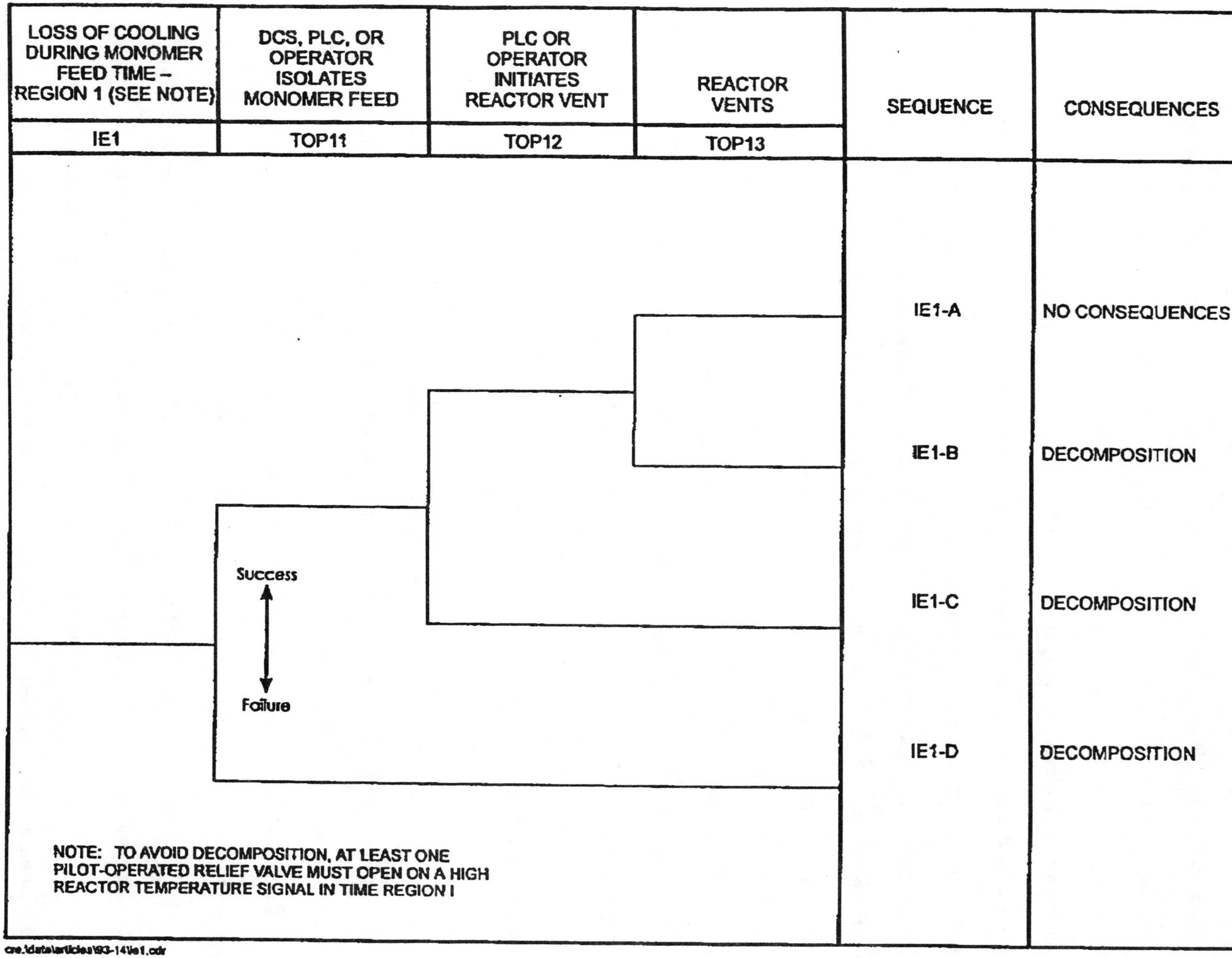

Figure 2 Event Tree for Loss of Cooling during Monomer Feed Time-Region I

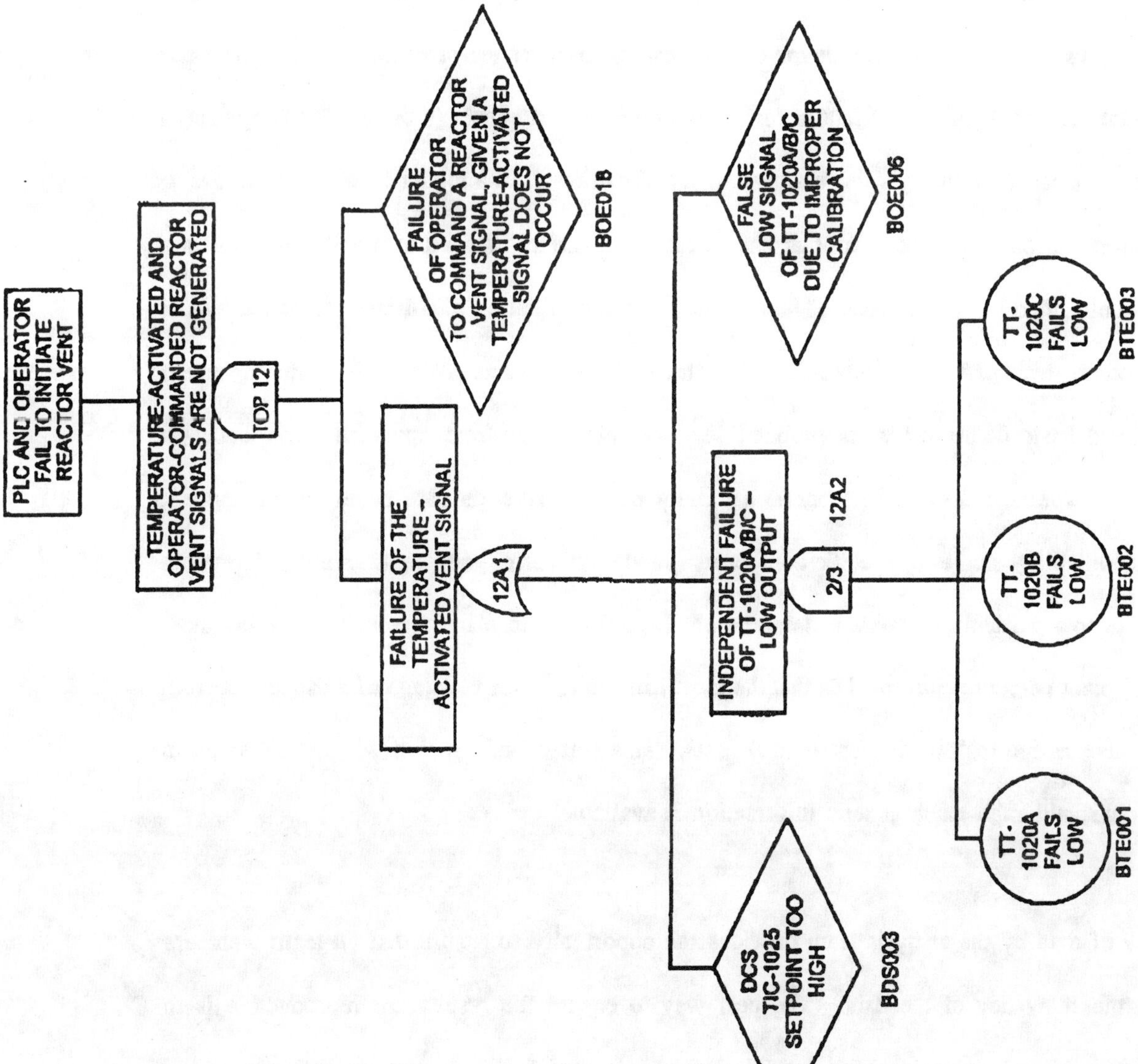

C:\ARTICLES\93-14\TOP12.CDR

Figure 3 Fault Tree for PLC and Operator Fail to Initiate Reactor Vent

components. The gate labeled 12A2 in Figure 3 is a two-out-of-three gate (any combination of two of the three input event failures will result in occurrence of the gate).

Collect component failure data, test interval and repair data, and human error frequencies and probabilities. This analysis step involves establishing frequencies of occurrence for the basic initiating events and probabilities of failure for the basic demanded events (a basic *initiating* event is a basic event that has a frequency value assigned to it and can result in a higher level initiating event; a basic *demanded* event is a basic event that has a probability value assigned to it and can result in a higher level demanded event). A total of more than 350 basic initiating plus basic demanded events were required in this QRA. The input data that the analysis team used for basic initiating event frequencies and basic demanded event probabilities were developed from operating experience (as provided from experienced plant personnel), a variety of industry-wide data bases for component failure rates, reliability studies of specific components, and engineering judgment. Some information will be unique to a particular analysis and can be obtained only from plant personnel. For instance, in this study, plant personnel informed us that the most likely reason for blockage of an air outlet from a solenoid valve is due to "dirt dauber" (wasp) nests. This sort of information will not be found in reference books, although much generic information is available.

Review of data by the analysis team provides the opportunity to ensure that all team members agree upon the relevancy of the data. The best way to ensure the results are reproducible is to ascertain that the data are as relevant as possible and also to record all assumptions, references, and methodologies that the team uses (Arendt et al., 1989). But for some component/human errors, very little information is available for developing a probability of failure value. In these instances, the analysis team made every effort to provide accurate but conservative values. It is vital to recognize that the data accuracy is not essential for success of the QRA. Although there may be uncertainty in the final results, the greatest benefits of the QRA is the relative importance of the different contributors to risk. These importance measures are often robust with respect to data accuracy. Although sometimes performed, data sensitivity analyses were not requested for this study.

3. Frequency analysis

There are two steps in performing the frequency analysis: (1) developing minimal cut sets (MCSs) and (2) calculating the results.

Develop minimal cut sets. An MCS is a combination of failure events in which, if all the failure events exist, the TOP event exists, and if any of the failure events do not exist, the TOP event does not exist because of this MCS. The collection of MCSs for each fault tree represents all possible unique combinations of failure events that the analysis team identified as potential causes of the TOP event. By definition, each MCS includes only independent failures and does not include any single component failure more than once. The analysis team used the IRRAS® computer program to build and solve the fault trees (generate the MCSs) for this analysis (Russell et al., 1992). The qualitative results (the MCSs) of the fault trees are useful because they describe the different ways that the TOP event (branch point failures) can occur. The team reviewed the fault tree MCSs to ensure that the MCSs properly represent the TOP events of interest.

An event sequence analysis is performed by first building an event sequence logic model. The event sequence logic model consists of an AND gate whose inputs are the fault tree TOP events (initiating event and branch point failures) that must occur for the event sequence to occur. IRRAS® determines the MCSs for the event sequence logic model by using the MCSs for each input fault tree.

The definition of an event sequence MCS is the same as the definition of an MCS for a TOP event. A total of more than 10,000 MCSs were generated in this QRA.

Certain event sequence MCSs must be modified to eliminate (1) mutually exclusive events and/or (2) repeated events. Mutually exclusive events are events that cannot physically exist at the same time (e.g., a temperature element "failing low" and the same temperature element "failing high"). MCSs that contain mutually exclusive events can never occur and, therefore, should be eliminated from consideration. An example of a repeated event is a component failure represented by a basic initiating event within the initiating event fault tree that is also represented as a basic demanded event within a branch point fault tree. In our analysis, the tempered water system temperature transmitter, whose failure in the low output state is modeled as a basic initiating event in the initiating event fault tree IE1 and as a basic demanded event in the engineered feature/system fault tree TOP11, is such a repeated event. We modified event sequence MCSs that contain repeated events so that the repeated basic demanded event occurs with a probability of 1.0, given that the corresponding basic initiating event has occurred. Use of a consistent naming convention is helpful in this modification step.

Calculate the results. For this study, the desired results were the expected (fractional) number of occurrences over a 1-year time period for each identified accident sequence and the associated component importance values. The modified MCSs for each event sequence were quantified to obtain the frequency of occurrence (estimated number of events per year) for each event sequence. The analysis team used BRAVO® to quantify the results (Guthrie et al., 1993). We obtained the system importance values by evaluating the contribution of the branch points (e.g., what percentage contribution to the frequency of decomposition events can be attributed to a failure of the relief valve system).

4. Consequence Analysis

Given the occurrence of a decomposition event leading to reactor vessel rupture, three types of consequence outcomes were identified for subsequent analysis: a physical explosion (resulting from rapid expansion of decomposition products to atmospheric pressure following vessel failure), a fireball, and a vapor cloud explosion. The physical explosion occurs before the vapor cloud explosion and the fireball (i.e., it is not mutually exclusive of the vapor cloud explosion and the fireball). The vapor cloud explosion and the fireball are mutually exclusive events. Because of large uncertainties in (1) the reaction history leading to a decomposition event, (2) the makeup of decomposition products, and (3) the behavior of the decomposition products following vessel failure, the consequences of the event of interest were estimated using simplified modeling approaches for thermal radiation, overpressure, and fragment projectile range.

Presenting QRA Results

For this study, we focused the vast majority of the effort on developing the frequency models rather than dwelling on the consequences of what is already known to be an unacceptable accident. Thus, the primary results of interest for this study involve the frequency of occurrence of decomposition events. The total frequency of vessel rupture (resulting from decomposition events) for the existing reactor system is estimated at the equivalent of less than 10% probability during the lifetime of the reactor. Table 2 summarizes the normalized frequency results for our analysis, specifically the percent contribution to the frequency of decomposition events. The feed upset initiating event, IE13, "High Monomer Concentration That Exceeds the Relief Capacity Due to Low Catalyst or Low Temperature," accounts for 77% of the total frequency of decomposition events and is the dominant initiating event. This result emphasizes the importance of maintaining correct batch

Table 2 Percent Contribution of Analyzed Initiating Events to the Frequency of Decomposition Events

Initiating Event Type	Monomer Feed Time-Region		
	I[a]	II[b]	III[c]
Cooling Upsets	(Percent Contribution)		
Loss of cooling	IE1 (8%)	IE2 (<0.1%)	IE3 (<0.1%)
Loss of agitation	IE4 (14%)	IE5 (<0.1%)	IE6 (<0.1%)
High steam flow to the jacket	IE7 (<0.1%)	IE8 (<0.1%)	IE9 (<0.1%)
Loss of site AC power	IE10 (1%)	IE11 (<0.1%)	IE12 (<0.1%)
Feed Upsets			
High monomer concentration that exceeds the relief capacity due to low catalyst or low temperature	IE13 (77%)		
High monomer feed rate	IE14 (e)[d]		

[a]Safety system: PORV actuation due to remote signal.
[b]Safety system: PORV actuation due to remote signal or direct pressure.
[c]Safety system: Opening of bellows valve or PORV actuation due to remote signal or direct pressure.
[d]Not quantified due to physical limitations of the feed system.

conditions. Two cooling upset initiating events that occur in Monomer Feed Time-Region I (loss of cooling and loss of agitation) account for most of the remainder of the frequency of decomposition events (the contributions of IE1 and IE4 combine for 22% of the total frequency). This result emphasizes the significance of operating in a time region that is dependent only upon remote actuation of the pilot-operated relief valves for system protection. These results also show the importance of QRA in general, because previous *qualitative* studies did not identify the magnitude of the "baseline" risk from operating outside correct batch conditions in comparison to the perceived risk of operating in Monomer Feed Time-Region I.

Table 3 provides a ranking of the engineered safety features or systems by their contribution to the frequency of decomposition events. These values are the percentage of the total frequency of the sequences that involve a failure of the specific engineered feature/system. For example, the low temperature trip is the engineered feature/system that provides protection for all IE13 sequences ("High Monomer Concentration That Exceeds the Relief Capacity Due to Low Catalyst or Low Temperature"). The primary contribution to the values displayed in Table 3 for the relief actuation initiation signal and the relief valves comes from the cooling upset initiating events that occur in Monomer Feed Time-Region I. The engineered feature/system "monomer feed isolation" has an insignificant contribution to the overall frequency. This is due to the redundancy and diversity of signals that are designed to isolate the monomer feed upon cooling upsets.

Table 3 Engineered Features/Systems Ranked by Contribution to the Frequency of Decomposition Events

Feature/System	Percent Contribution
Low temperature trip	77%
Relief valves (actuation components included)	12%
Relief valve actuation initiation signal	11%
Monomer feed isolation	<0.1%

Other quantitative frequency results that were presented in this study are a ranking of the event sequence MCSs by their percent contribution to the overall frequency and a ranking of the basic initiating/basic demanded events by their relative importance values. These rankings are valuable in identifying potential system improvements that could reduce risk. When evaluating the overall frequency and rankings, all results are considered order-of-magnitude due to the estimated nature of the input data and other uncertainties.

Conclusions: The Value of QRAs

QRA is a useful interactive process for performing an in-depth analysis of engineered chemical process systems. Used correctly, QRA can provide insight to the relative importance of individual systems and components. But to ensure that the QRA yields results of value, QRA teams must aim for balance in their work. That is, at the beginning of a QRA, one goal that should be clearly delineated is to derive just enough information for making informed decisions about the process. Too much information results in wasted effort, and too little information defeats the purpose of the QRA. Like Goldilocks, QRA teams need to search for a happy medium that is "just right."

Agreeing on objectives and limits on the study can help teams avoid the pursuit of extraneous information. Selecting the appropriate personnel to participate in the study is essential. With the right personnel, the team will begin the study with accurate system information. Having an experienced analyst (someone who can recognize and avoid potential QRA pitfalls) lead or participate in the analysis is also essential. This person can ensure correct application of QRA methods and software as well as keep the team focused on relevant material. And such a leader can be invaluable when you communicate to management what QRAs can and cannot do.

To keep the analysis on the correct agenda, a review of an in-progress QRA can serve as a forum for discussion of accident phenomenology information. This type of internal review is crucial to ensure proper understanding and acceptance of the QRA results. Once the event tree/fault tree models are completed, quantitative analysis can be performed using detailed automated modeling techniques.

The challenging problem of predicting how rare sequences of events (e.g., involving failure of several engineered control and safety systems) will develop is systematically modeled using fault trees/event trees, making the QRA process a unique and powerful analytical tool. The results of this study allowed the customer to focus on the important risk contributors, thus providing the guidance for selecting the most appropriate action(s) to implement from a set of various design options for improving the safety of this system. Use of QRA has enabled management to have confidence that they are selecting the most cost-effective safety improvement option.

REFERENCES

Arendt, J.S., "Implementing Process Safety Management Programs at Industrial Facilities", *Loss Prevention Symposium Proceedings*, AIChE 1991.

Arendt, J.S. et al., *Evaluating Process Safety in the Chemical Industry, A Manager's Guide to Quantitative Risk Assessment*, Chemical Manufacturers Association, Washington, DC, December 1989.

Guidelines for Chemical Process Quantitative Risk Analysis, Center for Chemical Process Safety, AIChE, New York, 1989.

Guthrie, V.H. et al., *BRAVO 2.0®: A PC Program for Quantitative Accident Frequency Assessment*, JBF Associates, Inc., Knoxville, TN, January 1993.

Russell, K.D. et al., *Integrated Reliability and Risk Analysis System (IRRAS®) Version 4.0*, NUREG/CR-5813, EG&G Idaho, Inc., Idaho Falls, ID, January 1992.

31st LOSS PREVENTION SYMPOSIUM

Session 42. **Use of Consequence Modeling in Risk Reduction**

Wed., March 12, 1997
8:30 - 11:45 PM
Grand Ballroom C

Ephraim Scheier, Chair
Oxy-Chem
15th Floor
5005 LBJ Freeway
Dallas, TX 75244

James J. Rooney, Vice-Chair
JBF Associates, Inc.
1000 Technology Dr.
Knoxville, TN 37932-3353

Session Introduction - E. Scheier

42a. **The Lessons Learned from the Development of a Scenario Based Approach for the Optimization of Passive Fire and Blast Division Design on a New Offshore Platform,** B. Poblete, Lloyds Register, Houston, TX; P. Caldwell, Caldwell Consulting Ltd., Insch, Scotland

42b. **Use of Consequence Analysis in Process Hazards Analysis,** R. Gardner, S. Dharmavaram, DuPont Engineering, Wilmington, DE; J. Klein, DuPont Automotive, Philadelphia, PA

42c. **Toxic Hazard Reduction with Passive Mitigation Systems,** P. Myers; R. Morgan, S. Flamberg, Four Elements, Inc., Columbus, OH

BREAK

42d. **Protection of Process Buildings and Their Occupants From Toxic and Flammable Releases,** C.E. Fryman, BP Oil Company, Cleveland, OH

42e. **The Explosion Blast Prediction Model in the Revised "Yellow Book",** W. Mercx; A. van den Berg, TNO Prins Maurits Laboratory, Rijswijk, Netherlands

42f. **Recent Developments in the Baker Strehlow VCE Analysis Methodology,** Q. Baker, C. Doolittle, G. Fitzgerald, M. Jun Tang, Wilfred Baker Engineering, Inc., San Antonio, TX

Concluding Remarks by J. J. Rooney

The Lessons Learned from the Development of a Scenario Based Approach for the Optimisation of Passive Fire and Blast Division Design on a New Offshore Platform

Benjamin R. Poblete, P.Eng.
Lloyd's Register

Paul B. Caldwell, C.Eng.
Caldwell Consulting Ltd.

Prepared for presentation at the 31st Annual Loss Prevention Symposium
Houston, TX, March 9-13
Use of Consequence Modelling in Risk Reduction

December 5, 1996

THE LESSONS LEARNED FROM THE DEVELOPMENT OF A SCENARIO BASED APPROACH FOR THE OPTIMISATION OF PASSIVE FIRE AND BLAST DIVISION DESIGN ON A NEW OFFSHORE PLATFORM

Benjamin R. Poblete
Lloyd's Register, Industry Division
15810 Park Ten Place, Suite 291
Houston, TX 77084

Paul B. Caldwell
Caldwell Consulting Limited
Wardes Grange, Insch,
Aberdeenshire, UK AB52 6LX

ABSTRACT

The objective of this paper is to outline some of the valuable lessons learned during the development of a scenario based approach for the optimization of passive fire protection and structural design, on the fire and blast divisions of a large cold water offshore platform.

This platform was the first fixed offshore exploration and producing platform off the east coast of Newfoundland. During the early detailed engineering design of this large cold water offshore platform it was highlighted that to comply with the regulations, with regards to the fire and blast protection from the potentially critical accidental events identified, would result in a significant weight and cost penalty for the installation. At that stage of the project it must be emphasised that there were no set industry guidelines on how to design for explosion overpressure or an industry standard jet fire test for an offshore installation. Therefore a systematic method of evaluating the fire and explosion risks was initiated with the Project's Formal Safety Assessment program and the results of these consequence modelling and risk analysis were the basis for the optimization of the passive fire protection and structural integrity on the primary structural bulkheads and beams. The development of this systematic thinking was based on sound loss prevention engineering judgement and the incorporation of developing analysis techniques and equipments into the installation design. It must be emphasized that the risk on the platform safety functions was successfully demonstrated to be tolerable and cost effectively reduced to as low as reasonably practicable with the application of this systematic method of evaluation.

Examples from the application of the approach on this project will be presented, along with conclusions on the lessons learned.

1 INTRODUCTION

The production of hydrocarbons on offshore installations has the potential for major fire and explosion hazards. This was demonstrated in 1988 by the Piper Alpha incident in the UK sector of the North Sea (Cullen, 1990). The subsequent design of new installations necessitated the need to ensure that the risk of any potential hazards are minimised, including explosion overpressure and jet fire hazard.

This oilfield is located off the east coast of Canada, approximately 315 kilometres southeast of St. John's, Newfoundland. The project consists of a concrete gravity base structure (GBS), a topsides with five super modules and two drilling rigs, crude oil loading facilities, purpose-built shuttle tankers, and support vessels. The five super modules consist of a process module, wellhead module, mud module, utility module and an accommodation module (Figures 1) .

The super modules will be large, approximately 60m long by 20m wide by 30m high (although the modules will be subdivided by intermediate decks about 8m high) (Figure 2) and semi-enclosed / open in the major hydrocarbon areas and enclosed in most unclassified and accommodation areas. An immediate ignition of a hydrocarbon gas or fluid would result in a potentially damaging jet or pool fire to the primary structural steel of the installation. If there were only limited explosion venting available in the modules, damaging explosion overpressure could be generated if a large, stoichiometric hydrocarbon gas and air mixture were to completely fill the volume and ignited.

The objective of the design team was to prepare an inherently safe installation. However, there were several constraints on the design process including topsides weight and project schedule. In addition since this platform was to be the first major offshore production installation in Canadian waters, there were no established precedents to follow with regards to the engineering design for explosion overpressure and fires. It must be emphasised that the approach for explosions and fires were developed uniquely and separately. The objective of this document is to highlight some of the interactions and differences between the two approaches and lessons learned throughout the design, construction and commissioning of the facility.

2 DEVELOPMENT OF APPROACH

The project team developed safety design philosophy documents and a Formal Safety Assessment (FSA) Program. The FSA Program required a systematic identification and assessment of potential hazards and wherever reasonable and practical, the hazards should be eliminated or the risks to persons, the environment and the installation should be minimised. A risk acceptance criteria were developed for assisting decision making and for determining whether the risks were As Low As Reasonably Practicable (ALARP).

M10 - PROCESS MODULE
M20 - WELLHEAD MODULE
M30 - MUD MODULE
M40 - UTILITY MODULE
M50 - SERVICE/QUARTERS MODULE

M71/72 - DRILLING DERRICKS
M73 - DRILLERS OFFICES/PIPERACK
M81 - MAIN LIFEBOAT STATION
M82 - AUX. LIFEBOAT STATION
M83 - FLARE BOOM
M84 - HELIDECK

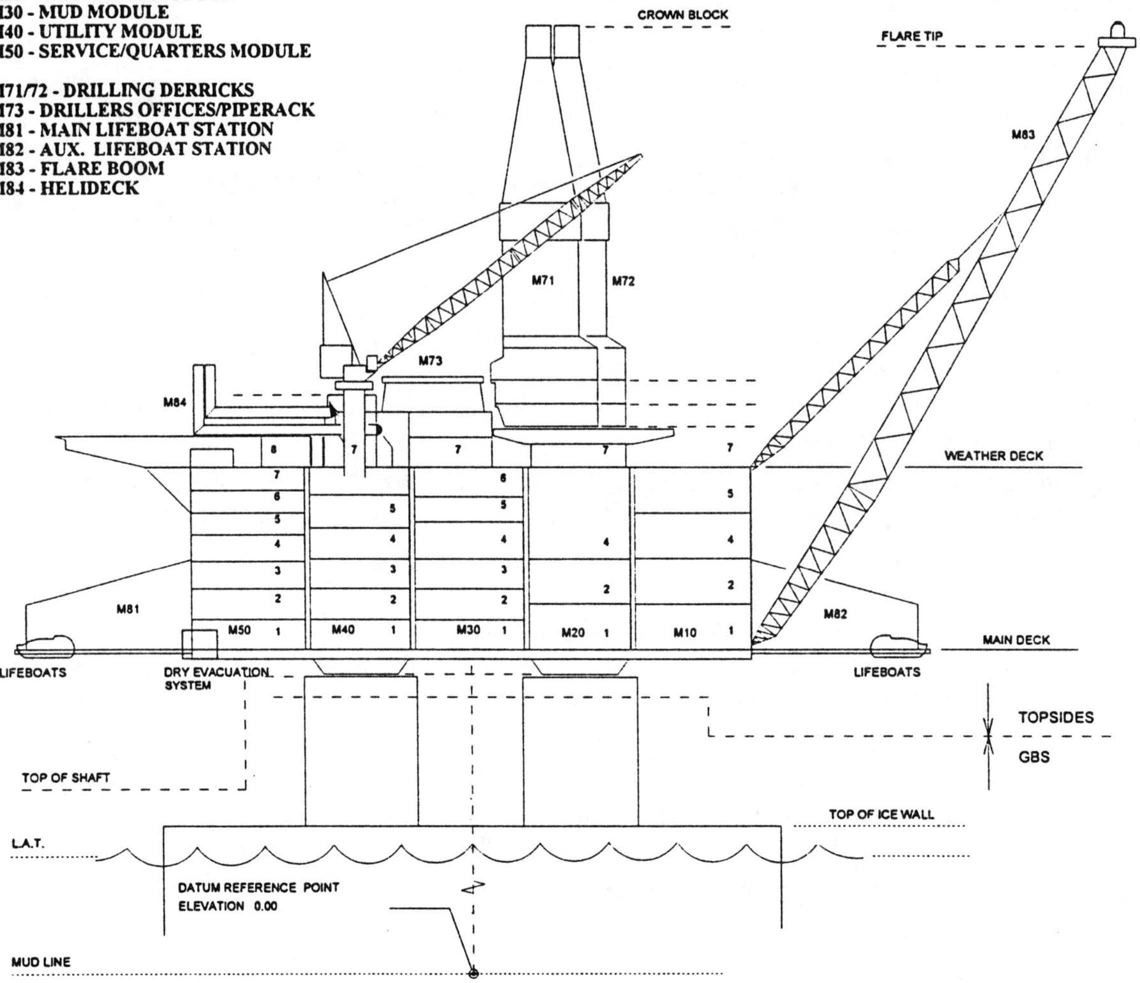

Figure 1: Layout of topsides modules.

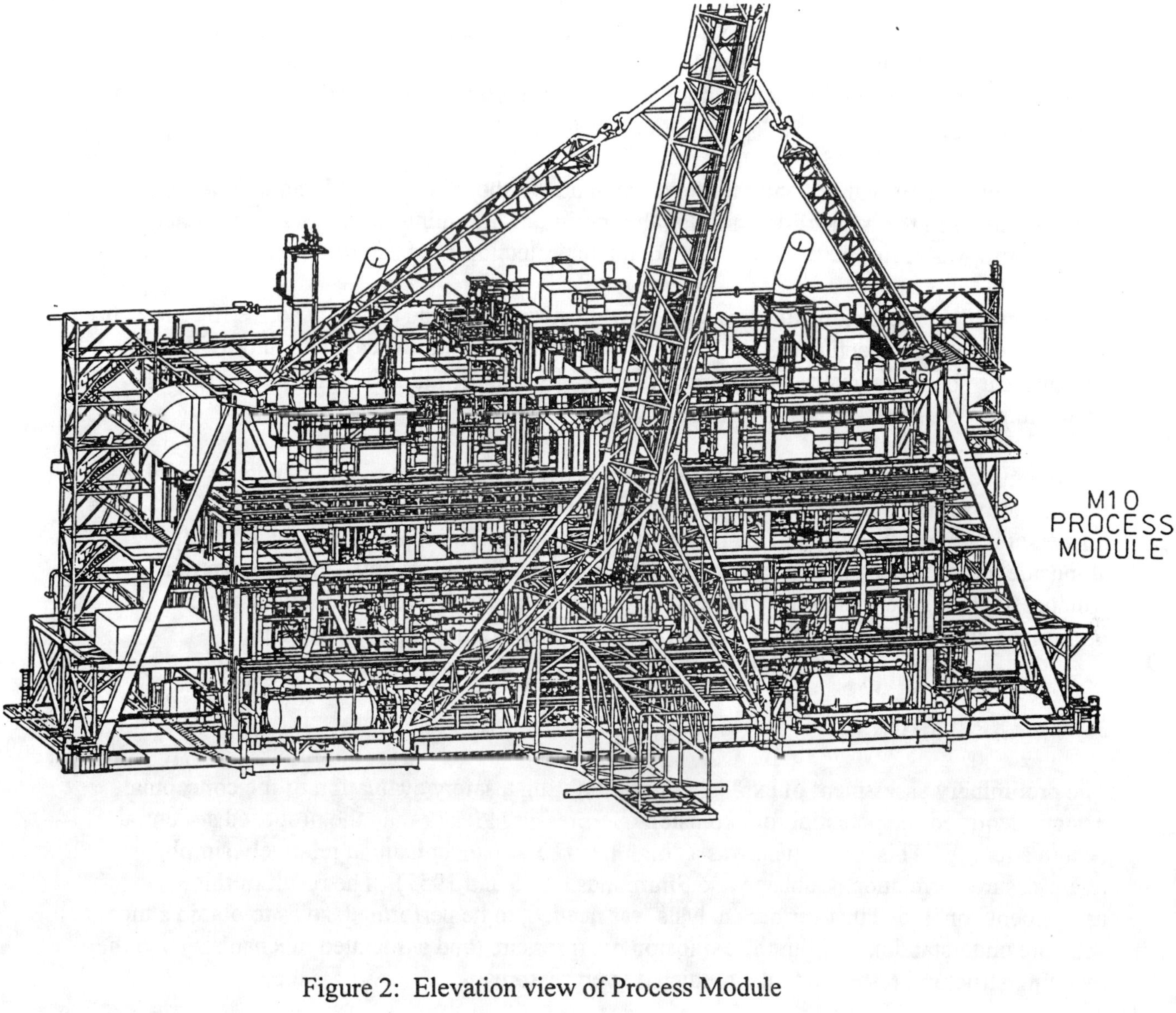

Figure 2: Elevation view of Process Module

The main emphasis of the safety design philosophy and FSA program was on a passive fire and explosion containment principle. This mean that high risk areas are separated from adjacent areas by partitions designed to control the spread of a fire or an explosion.

Particular emphasis was placed on the segregation of hydrocarbon risk areas from each other, from the accommodation and from other utility areas. The accommodation spaces, control stations, sources of emergency power, firewater pumps and escape and muster points have been given special consideration in determining the level of separation required from each other and other platform areas.

The requirements of SOLAS and the Newfoundland Offshore Petroleum Installations Regulations form the basis of determining the minimum fire rating for the platform. Each area of the platform was reviewed to ensure that adequate protection has been provided.

The Passive Fire Protection (PFP) was intended to supplement the platform active fire protection system (i.e. automatic deluge, sprinkler, etc.) where installed, but be capable on providing an approved fire barrier and structural stability in the absence (or failure) of any active fire protection.

The overall design intent of the PFP system was that it will offer up to two (2) hours stability to the platform primary and load bearing structure during a hydrocarbon fire. This will allow time for personnel to safely evacuate the platform, and for fire fighting support to be brought alongside to assist in controlling any fire. By adopting this philosophy, serious structural collapse of the platform and the subsequent further release of hydrocarbon inventory can be minimised, allowing for fire fighting operations to proceed in a controlled manner.

3 EXPLOSIONS

The preliminary assessment of explosion hazard during a safety evaluation of the conceptual design identified the potential for explosion overpressure greater than the proposed structural design strength. This assessment was a "high level" assessment using a relatively simple overpressure correlation (Cubbage and Simmonds, 1955 and 1957). The result of this assessment concluded that further analysis was needed, to be performed, so as to obtain a more accurate understanding of both the explosion overpressure (and associated mechanism) and the resulting structural response to the predicted overpressures.

The overall methodology for the assessment of explosion overpressures is shown in Figure 3. This indicates that if the predicted overpressure was less than the design strength, no further analysis of overpressure has to be performed. However, if the design strength is exceeded, then further design optimisation is required. This optimisation work can comprise of; improved venting, equipment layout modification, reduction of potential leak and ignition sources, etc.. In

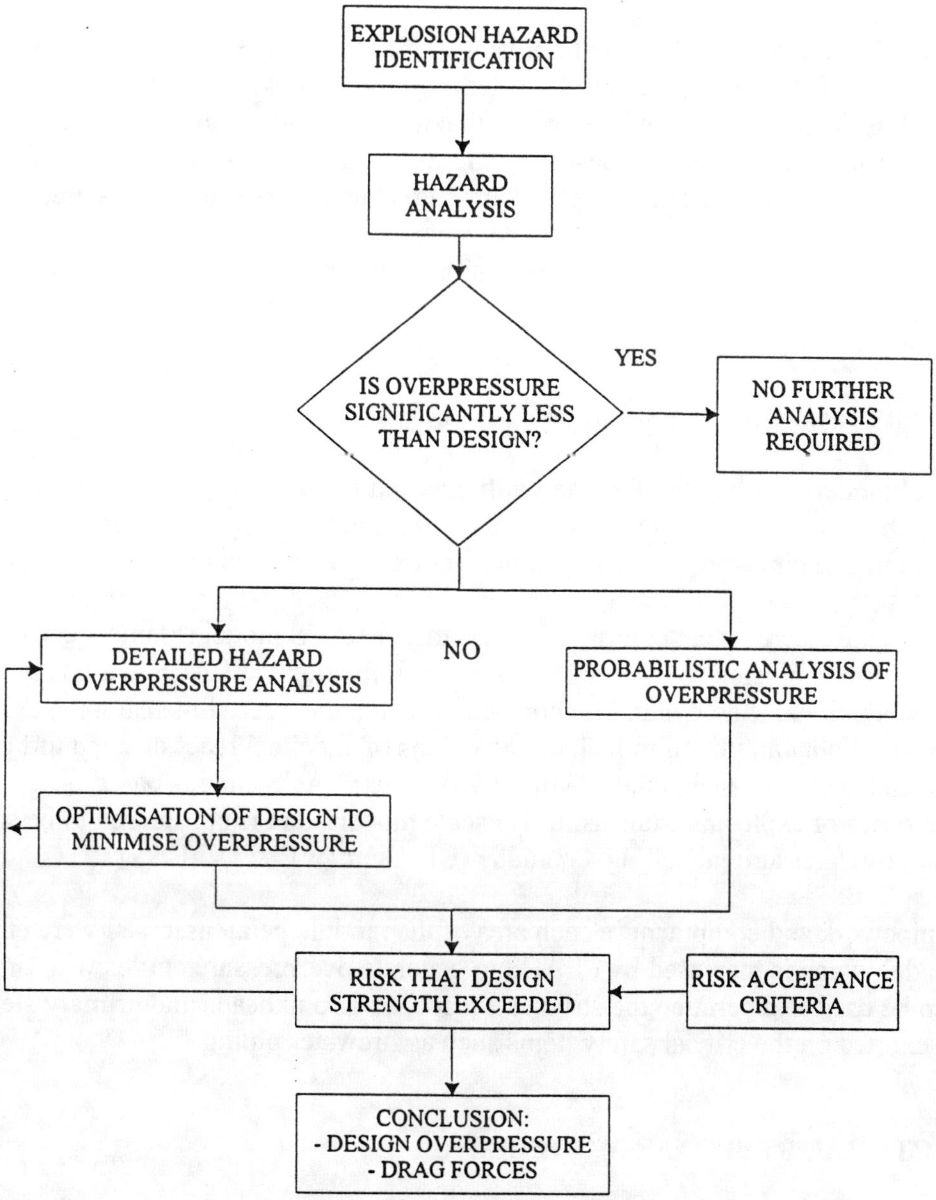

Figure 3: Overall methodology for assessment of explosion overpressures

addition to the above work, the platform structure could also be strengthened (at the penalty of weight and cost). It must be emphasised that the structural design could allow for significant permanent (plastic) deformation of the structure as long as overall structural collapse is avoided. Local buckling of secondary steel is acceptable. The plastic deformation was also acceptable with regards to susceptibility of supports of critical piping systems (hydrocarbon, firewater and vent/flare lines) in certain areas of the platform. It is acceptable to have plastic deformation with hydrocarbon fluid lines as long as there will not be a significant release to escalate an incident. The pipe diameters will also determine if the acceleration forces are due to explosion overpressure or drag forces. It may be determined that after the optimisation of the design, the explosion overpressure still exceeds the design strength. At this stage probabilistic analysis can be used to determine how often damage might occur and what the associated risks are.

The conclusion of the overall process was an explosion overpressure design basis that met the project safety criteria.

3.1 EXPLOSION OVERPRESSURE MODELLING

Simple empirical models can be used for small volumes and for relatively "open" modules with few obstacles in the flame path (UK Dept of Energy, 1989), but for large volumes with a complex equipment and pipework geometry, more sophisticated models are required.

Christian Michelsen Research were contracted to use their FLACS model (Flame Acceleration Simulator) to perform explosion overpressure computer simulations. FLACS is a three-dimensional computational fluid dynamics code which solves the three-dimensional Navier-Stokes equations, suitably amplified to include the effects of chemical reactions and turbulence, by employing a finite-volume technique (CMR, 1993). The FLACS code is based on a comprehensive series of explosion trials using 1:5 scale models, and is considered to be state-of-the-art explosion overpressure modelling technique (UK Dept. of Energy, 1989).

Details of the pipework and equipment in each area of the module being assessed were entered into a CAD model which is then used by FLACS to generate overpressure curves with time. The curves can then be used to determine the structural response of bulkheads and primary steel, and the drag loads exerted on the critical safety items such as firewater piping.

3.2 STRUCTURAL RESPONSE MODELLING

Experience of explosions in offshore modules showed that bulkheads not specifically designed to resist overpressure may fail at low overpressures (typically 0.1 bar) (Cullen, 1990). An initial assessment of Hibernia's stressed-skin bulkhead design showed that because the bulkheads carry significant loads during the transportation and operational phases, they would withstand approximately 0.4 bar overpressure.

To determine the actual failure pressures for the stress-skin bulkhead design, a detailed structural analysis program BLAST, developed by Mobil Research and Development Corporation was used (MRDC BLAST Version 1.1.0). The BLAST program evaluates the response of stiffeners or unstiffened panels subject to a time-varying load simulating an explosion. The static capacity of a panel as determined by the force-deformation relationship for an "equivalent" uniform plate is developed by the program and accounts for excursions in the inelastic range. The model has been used to simulate the response of a number of offshore structures in the North Sea. The main structural frames and floors of the modules were also analysed for explosions. This was done using linear and non-linear analysis tools.

3.3 ASSESSMENT OF EXPLOSION OVERPRESSURE

Using the two models described above, the modules where explosion overpressure presented a hazard were examined in detail. In two of the modules, it was found that the "worst case" overpressure generated, by an optimum mixture of gas in air filling the entire volume of the module, could cause an overpressure that would exceed the design strength of the bulkheads. The primary structure would not be significantly damaged, but failure of the bulkheads might allow a subsequent fire to spread, compromising the project's safety criteria.

The design was optimised by combining the following prevention and mitigation measures:

- providing a sound engineering design to minimise the potential for leaks;
- providing process control and emergency shutdown systems and gas detection and blowdown systems to minimise the inventory released;
- using louvres and explosion panels to provide maximum explosion venting where the module could not be "opened up";
- orienting large equipment items towards the vents to minimise the blockage and to allow unrestricted venting;
- strengthening structures to withstand the predicted overpressure.

In some cases it was not practical to strengthen the structures for all explosion overpressure events because of weight (and schedule) constraints. For these events, the risk of impairing the structure (which is one of the installation's main safety functions because it supports the escape routes, temporary safe refuge and evacuations systems) was predicted. The risk assessment took into account the safety systems incorporated into the design, including the blowdown and deluge system.

The risk of impairment of the structure was compared with the project's risk acceptance criteria and found to be low. An assessment of the penalty of increasing the strength of the structure (the penalty being weight and cost) was made and compared with the benefit (the reduction in frequency of events that cause impairment) in a cost benefit analysis. This demonstrated that the risk of impairment of the structural integrity had been reduced to as low as reasonably practicable (ALARP).

4 JET AND POOL FIRES

In parallel to the above explosion optimisation work, the project team attempted to save weight, on the platform, by optimising the quantity of PFP on the platform using a scenario based approach.

The Newfoundland Offshore Petroleum Installations Regulations (18 October 1993) provided operators with two alternative approaches to PFP for offshore platforms:

i) Based on the specific requirements to the fire classes for different steel structures, decks and bulkheads or

ii) Based on an analytical approach that defines the design fire loads for the different platform areas based on the site-specific, credible, fire scenarios. This approach must be supported with documentation that demonstrates an "equivalent level of safety" to the specific requirements of the regulations.

To utilise the analytical approach, the project team had to adhere to the following major guidelines during the design of this optimised PFP:

- utilise the project Concept Safety Evaluation (CSE) and Fire Risk Analysis (FRA) studies as the basis for design accidental fire scenarios;
- ensure structural redundancy and maximum allowable structural core temperature (calculated and documented by the project structural engineering group);
- select PFP materials that will withstand the actual accidental scenarios;
- utilize the computer program SUPER-TEMPCALC to determine the necessary PFP thickness based on the calculated core temperatures.

SUPER-TEMPCALC (version 3.1) is a fire-dedicated two-dimensional finite element program developed by FSD (Fire Safety Design, Lund, Sweden). The program is designed to solve two-dimensional, non-linear, transient, heat transfer differential equations incorporating thermal properties which vary with temperature. The program utilises rectangular or triangular finite elements in either rectangular or cylindrical co-ordinates. Convective and radiated heat transfer is modelled as a function of time. The program could analyse structures composing of several materials and is capable taking into consideration the heat absorbed by any existing voids that may exist in the structure. The main program consists of:

- a materials property database;
- a pre-processor which automatically creates and calculates the finite element division from the retrieved data in the property database;
- a post-processor which represents the results graphically in the form of time-temperature curves, isotherms, and temperature gradients.

The use of the scenario based approach in the optimisation of PFP relied on two critical factors:

i) the duration and heat loads provided for the program are data obtained from a Fire Risk Analysis (FRA) performed on the platform. It is assumed that the FRA considered the probability of occurrence of the various credible fire scenarios, the burning fluid properties, the distance between the fire and the exposed structural item and the effects of the drain and blowdown system on the fire incident;

ii) the actual maximum core temperature will be calculated as a function of the structural load re-distribution capabilities, the load conditions, the convection characteristics and the whether the structural item, in question, is a primary or secondary steel member. Traditionally, a maximum steel temperature of 400 °C has been used as the basis for PFP design.

4.1 OVERALL APPROACH

The overall approach for optimisation work consists of three stages (Figure 4):

i) Identify the fire scenarios;
ii) Perform structural thermal analysis;
iii) Specification of type and the extent of passive fire protection.

The PFP optimisation work went through a similar iterative process as the explosion overpressure modelling work. A description of the some of the critical decisions made during each stage will be described below. It must be emphasised that the decisions made during the PFP optimisation work had to also consider the contribution generated by the explosion modelling work.

4.2 IDENTIFICATION OF FIRE SCENARIOS

The first stage (Figure 5) of the PFP optimisation process is to clearly identify the fire scenarios that will be utilised in the study. Two separate and parallel activities were performed during this stage of the work. The first set of activities were to determine the credible fire scenarios to be utilised in the optimisation work. This information were obtained after a review of the platform CSE and FRA studies. The platform boundaries, for each of the scenarios were identified, as well as the duration and type of fire (jet or pool) that would exist within the exposed area. The positive contribution to the reduction of heat load and duration from the active fire protection system and the platform blowdown system were not included in the data for this stage of the work due to the project schedule constraint. The conservatism, in the above mentioned assumption, was therefore carried on to the next stage of the approach.

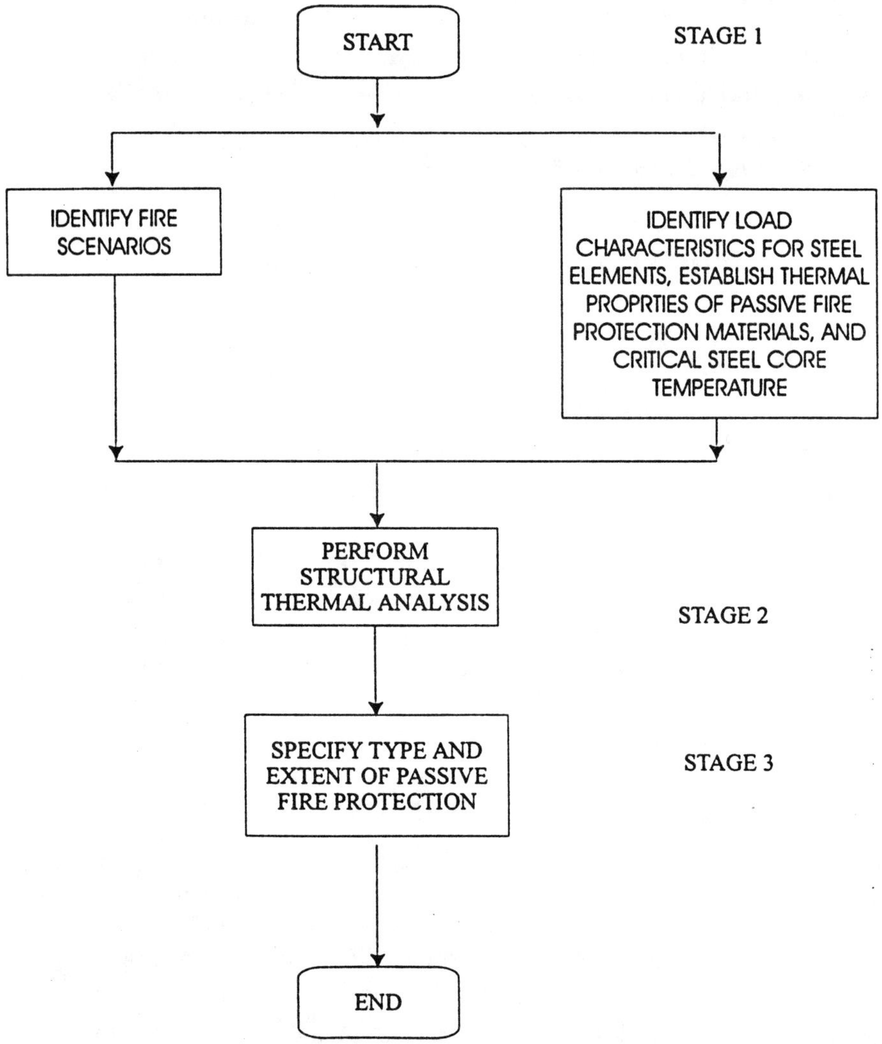

Figure 4: Outline of the overall approach to the optimisation of structural passive fire protection

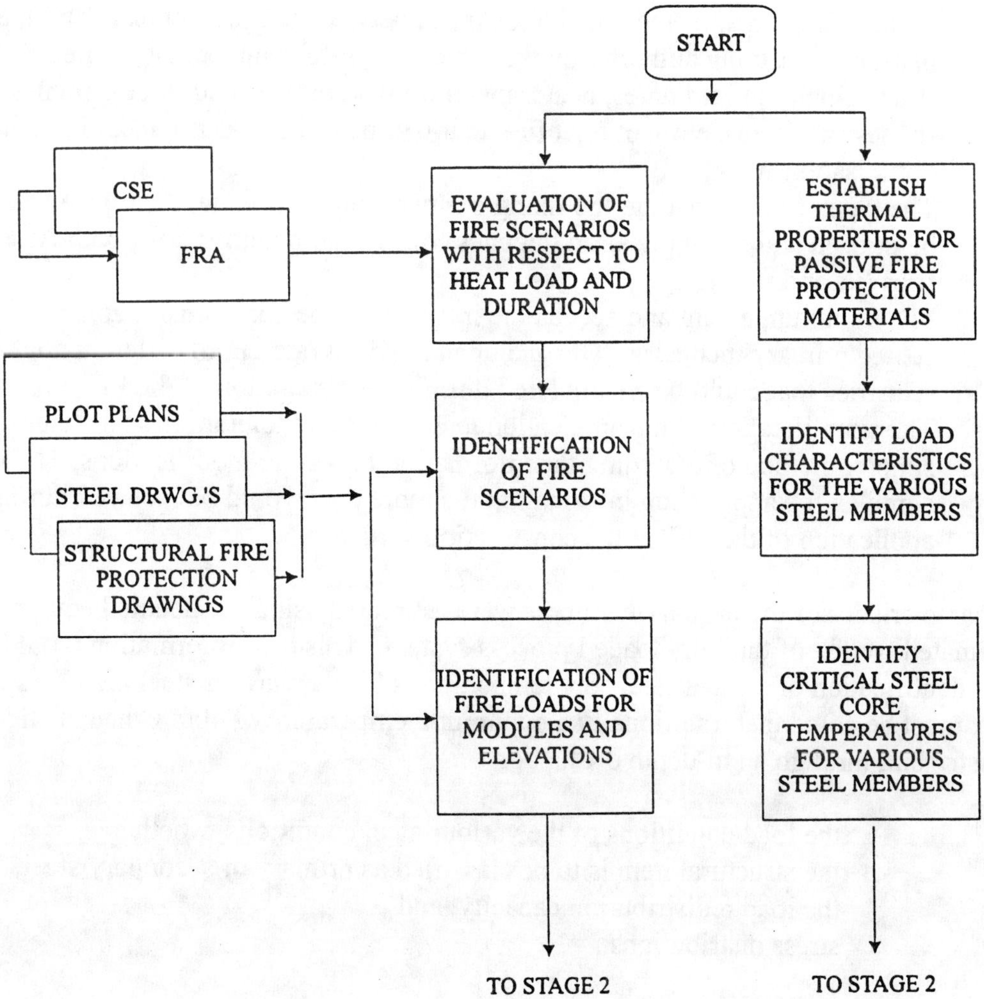

Figure 5: Stage 1: Identification of fire scenarios

The second set of activities involved the establishment of the thermal properties of the PFP materials. More information were needed to complete the data in the materials property database. The data were obtained by performing more simple and complex jet fire tests on the two types of PFP materials utilised on the platform. Some of the difficulties encountered during the data gathering phase of PFP properties included:

- Since there were no industry standard jet fire test during this stage of the optimisation an agreement had to be made between the project team and the platform certifying authority on the typical physical characteristics (i.e. flow/velocity of jet flames, heat transfer on flat surfaces and three dimensional shapes and forms, etc.), of a jet fire, that will be utilised as a standard for the optimisation work;
- The type of undercoating, the fixing or anchoring technique, the type of mesh to be utilised and the application thickness of PFP to be utilised for both type of fires (jet and pool);
- The minimum extent and acceptable integrity of the heat transfer zones from PFP coated primary members. This included the development of different application scenarios that could be encountered during the installation of the PFP;
- The simplification of the application techniques and tolerances so as to ensure a simpler method of recording the information for future modifications, The more complex the application information the more cost would be incurred during the application of the PFP at the construction yards.

The load characteristics of the primary members were estimated using the detailed equipment list and the estimated weight of the fully loaded process system. This load information would also include the identification of critical steel core temperatures for the various steel members on the platform. It must be noted that traditionally a maximum temperature of 400° C had been adopted by the industry but, through an in-depth evaluation of:

- the load conditions of the various steel conditions, whether
- the structural item is to be classified as primary or secondary steel,
- the load redistribution capacity, and
- stress distributions,

critical steel core temperatures can be determined. Therefore, the load characteristics identification had to be performed prior to the determination of the maximum allowable core temperatures.

4.3 STRUCTURAL THERMAL ANALYSIS

The second stage (Figure 6) consists of a series of iterative computer simulations which compares the calculated temperature against the critical temperature for the coated structural

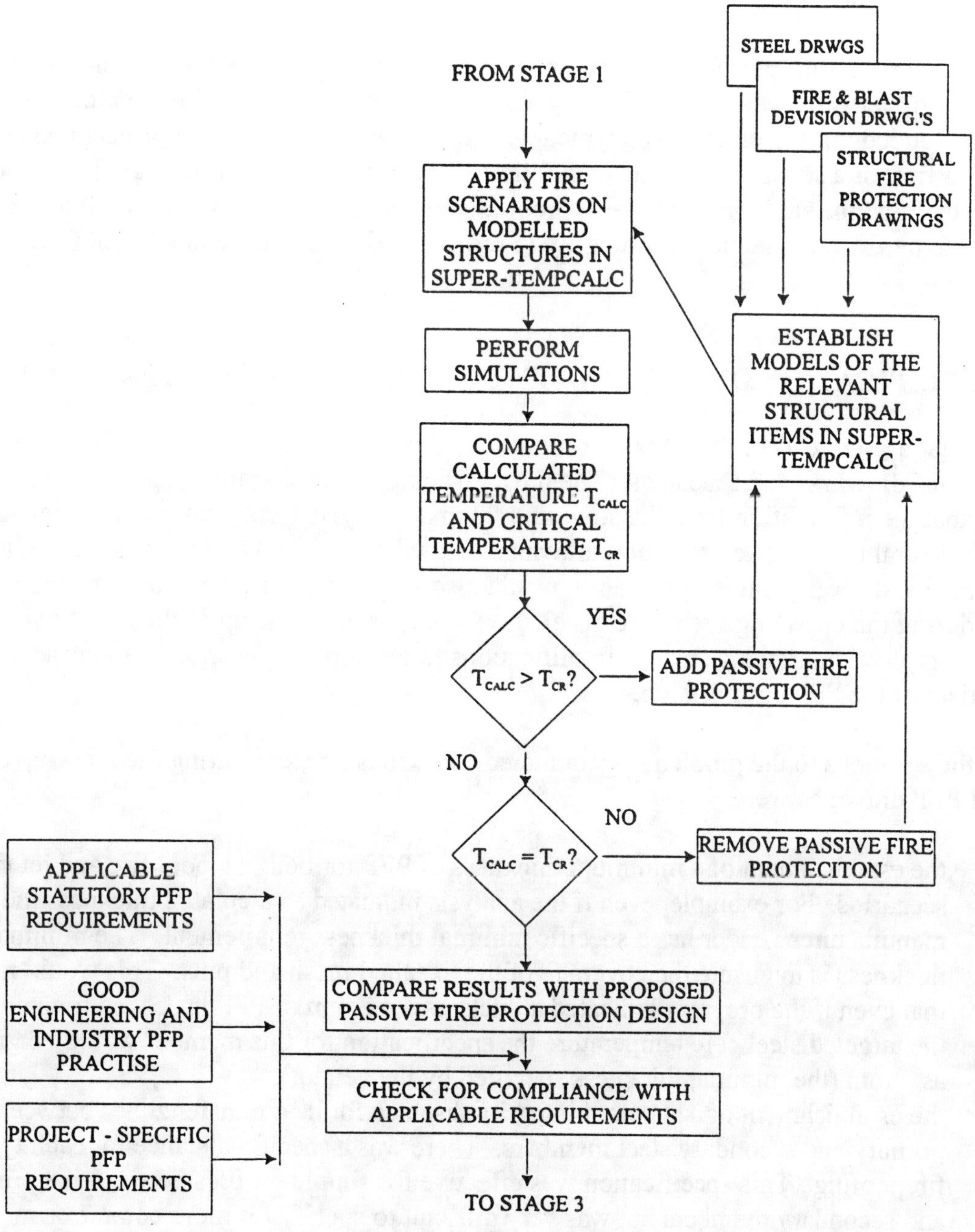

Figure 6: Stage 2: Structural thermal analysis

members. If the calculated temperature exceeded the critical temperature then PFP is added onto the structure. If the calculated temperature is lower than the critical temperature then PFP is removed. When the calculated and the critical temperature are approximately equal then the required amount (thickness) of passive fire protection has been found.

Jet fire loads were given special considerations and attention due to the jet fire erosion effects on the PFP. Jet fire erosion was dealt with using a double carbon fibre mesh double pinned onto the structure for added reinforcement. The typical jet fire scenario chosen by the project consisted of an initial jet fire for a set period of time then followed by a specified pool fire duration. Thus, it was important that the structure have the appropriate thickness of PFP to withstand the erosion effects of the jet fire and maintain its integrity to protect against the subsequent pool fire exposure.

4.4 SPECIFICATION OF TYPE AND EXTENT OF PASSIVE FIRE PROTECTION

This third stage (Figure 7) of the work incorporated the simulation results into engineering specifications, drawings and schedules. The most difficult task during this stage of the work was to ensure that the information from the structural thermal analysis were incorporated accurately into the structural passive fire protection drawings that will be utilised by the construction teams. The accuracy and access to this information would also be critical during any modifications, of the PFP, during the operating life of the installation. During the mark-up of the structural passive fire protection drawings many simplifications to the application specifications added to conservatism of the PFP optimisation.

Some of the solutions to the problems encountered by the design team during the mark-up of the structural PFP drawings were:

- the establishment of a minimum thickness of PFP for both the pool fire and jet fire scenarios. For example, even if the analysis indicated a calculated thickness the PFP manufacturer/vendor has a specific minimal thickness requirement. The minimal thickness is to ensure the coverage of the installed mesh and pins. This would mean that even if the program indicated a thickness of 2 mm of PFP is needed to maintain the targeted steel core temperature the specification for this member would be given as 4 mm (the minimal thickness specified by the vendor);
- the establishment of a standard integrity distance for heat transfer zones between primary and secondary steel members. There was a specific distance for each type of firepoofing. This specification was effective for simple profiles between the primary and secondary members but was very difficult to specify for more complex configurations and connections. The project team did not take advantage of the potential savings with more detailed analysis of each of these complex configurations due to project schedule constraint;
- the specification of different types of mesh (fiberglass and high temperature carbon fibre mesh) to support the PFP, according to the area fire scenarios. The mix in

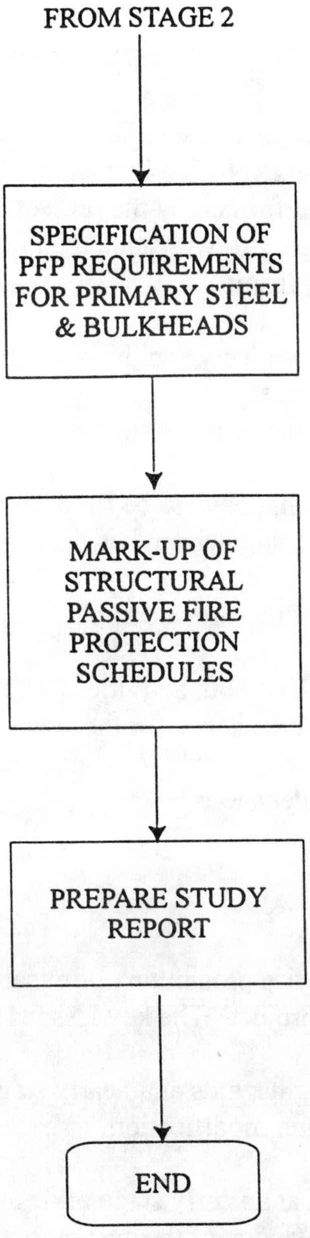

Figure 7: Stage 3: Specification of type and extent of passive fire protection

specification, according to the fire scenarios, caused some confusion in the construction yards. This was because they had no prior knowledge of the fire scenarios involved in each of the areas of the installation.

5 INTERACTIONS BETWEEN FIRE AND EXPLOSION OPTIMISATION WORK

One of the most important interaction between the explosion overpressure and PFP optimisation work was the integrity of the PFP after an explosion incident inside a module. A significant amount of inter-disciplinary work was performed by the project to ensure that critical structural steel and bulkheads will retain PFP integrity to withstand a jet or pool fire incident after and explosion incident. Discussions ranging from:

- the "pull strength" of the coating between the steel and PFP;
- to the reinforcement pin distances;
- to the degree of bending that the primary supports will provide as it responds to an explosion overpressure;
- the quantity of pipe supports that have to be fireproofed (due to its connection to primary members) after the piping system has been designed for explosion drag forces;
- to the extent and credibility of the fire and explosion scenarios;

were encountered during the parallel path of both activities. The structural criticality of the supports were the basis for all the project decisions for these work. The complex multi - dimensional interactions between criticality, weight, cost and schedule created some very interesting and challenging engineering decisions.

6 LESSONS LEARNED

The explosion overpressure and passive fire protection optimisation assessment approach was developed over the design phase of the project. The key lessons learned during that time were:

- The importance of identifying hazards at an early stage in the design when there is scope for relatively easy design modification.

- The need to establish criteria at an early stage of design. The criteria can be refined with time and improved knowledge of the design, but an initial "benchmark" for decision making is necessary.

- The need to establish and finalise early the structural design. Significant design modifications become increasingly difficult later on in the design phase because of schedule constraints. Other aspects (eg. equipment layout) can be modified later. This would mean that early on in the conceptual phase of the design a team consisting of Loss Prevention/Structural/"FLACS"/PFP specialists should be established to make decisions on the design issues relating to blast overpressure and passive fire protection design.

- Keep full documentation of early decisions and assumptions - you will almost certainly need to review them later.

- Empirical correlation for overpressure prediction should be used with caution. Comparison with state-of-the-art codes did not show any consistent over or under prediction. However, they have a role to play given the time (and cost) of running more complex models.

- Determine and agree the amount of structural redundancy that will be used and whether plastic deformation is acceptable.

- Ensure critical safety systems are able to withstand the effects of the initial explosion and a potential subsequent fire. For example, passive fire coatings on structures must remain "bonded" to the structure, deluge pipework and supports must remain intact and large hydrocarbon vessels must not be damaged.

- Consider the "worst case" events to determine potential adverse impact and to assist in the optimisation of the design.

- Be prepared to work in an iterative manner when optimising the design. Combinations of changes may have different effects, on explosion overpressure and PFP design, than single changes. However, bear in mind the overall cost and time available for optimisation and explosion overpressure assessment modelling.

- Use risk assessment to determine the acceptability of the "worst case" events if these cannot be accommodated by the design. Ensure that due account is taken of active protection systems.

- Consider using deluge on gas detection as experiments have shown this to significantly reduce explosion overpressures under certain circumstances.

- Be prepared to obtain early involvement of construction specialists during the specification of the optimised PFP. This would help pinpoint early any construction difficulties that could occur with the application of the PFP system.

7 CONCLUSIONS

The optimisation of a complex offshore design for explosion overpressure and fire can be a time consuming and lengthy process. Some significant weight and cost savings were incurred with the optimisation work while maintaining an equivalent level of safety to the early, prescriptive platform design. However, it is still very fundamental to minimise the potential for fire and overpressure from occurring and causing harm to persons or damage to the installation.

Structural redesign due to explosion overpressure and PFP optimisation needs to be done at an early stage in a project because of schedule constraints.

Empirical models for explosion overpressure can be used with caution, but not for determining the structural design basis.

"Worst case" events should be considered and risk assessment used for determining the acceptability of impairment of the structure after optimisation of the design to withstand explosion overpressure and fire.

Critical safety systems should also be designed to withstand the effects of fire and explosion and to ensure their functionality after the incidents.

8 REFERENCES

Cullen, Lord D.W. (1990). *The Public Inquiry into the Piper Alpha Disaster.* London: HMSO, Department of Energy, Cm 1310.

Cubbage, P.A. and Simmonds, W.A. (1955). An Investigation of Explosion Reliefs for Industrial Drying Ovens - I, Top Reliefs in Box Ovens. *Trans. Gas Eng.*, 105, 430.

Cubbage, P.A. and Simmonds, W.A. (1957). An Investigation of Explosion Reliefs for Industrial Drying Ovens - II, Back Reliefs in Box Ovens, Reliefs in Conveyor Ovens. *Trans. Gas Eng.*, 107.

UK Dept. of Energy (1989). Review of the Applicability of Predictive Methods to Gas Explosions in Offshore Modules. Prepared by British Gas for Dept. of Energy, *Offshore Technology Report OTH 89 312.* London, U.K.: HMSO, Department of Energy.

Christen Michelsen Research (1993). *Gas Explosion Handbook, Version 1.1, Ref No. CMR-93-A25003.* Bergen, Norway: CMR.

Mobil Research and Development Corporation. *BLAST Software for Explosion Effects on Plate Structures, Version 1.1.0, Doc. No: EES-MANU-1120.02/8-ELST.* Dallas, Texas: MRDC

Canada - Newfoundland Offshore Petroleum Board, *Installation Regulations*, 18 October 1993.

KEYWORDS:

explosion overpressure
blast overpressure
fire risk analysis
passive fire protection optimisation
risk assessment
risk criteria
explosion risks
scenario based approach
jet and pool fire risks

Use of Consequence Analysis in Process Hazards Analysis

R. J. Gardner, S. Dharmavaram, and J. A. Klein*

DuPont Engineering, Wilmington, DE
* DuPont Automotive, Philadelphia, PA

Abstract

Consequence analysis is used by PHA Teams to better understand the possible effects of hazardous events resulting from loss of engineering and administrative controls in the process being studied. Consequence analysis consists of identifying potentially hazardous events, estimating release amounts and conditions, evaluating consequences and affected areas, and determining the resulting safety and health effects on exposed persons. When properly applied, consequence analysis is a valuable tool in helping the PHA Team to make good process risk reduction decisions. This paper presents consequence analysis approaches that emphasize cost effective risk decisions in PHA's, including:

- When to use specialized models versus more approximate estimates.
- Simplified approaches for interpreting consequence analysis footprints to understand the number and severity of potential injuries.
- How to balance the level of consequence analysis effort with cost and safety factors in making PHA decisions regarding risk.

Toxic Hazard Reduction with Passive Mitigation Systems

by

Philip M. Myers, Richard S. Morgan, and Stephanie A. Flamberg

Four Elements, Inc.
355 East Campus View Blvd., Suite 250
Columbus, Ohio 43235

Prepared for Presentation at the 31st Annual Loss Prevention Symposium,
Use of Consequence Models in Risk Reduction

March, 1997

Abstract

A major component of the United States Environmental Protection Agency's 112(r) Risk Management Program Rule is the evaluation and control of potential public exposures to toxic chemicals. The rule requires identification and evaluation of a worst case accidental toxic release scenario, but provides for the inclusion of passive mitigation systems in modeling the dispersion hazard zones that are to be communicated to the public. These passive mitigation systems can be taken into account in the consequence modeling if they will withstand the initiating event that causes the accidental release and function as intended.

This paper discusses several passive mitigation systems that can be designed as an integral part of storage or processing of highly toxic chemicals. These include, but are not limited to, optimizing storage conditions, design of secondary containment, toxic vapor generation control, and enclosure design. The effectiveness of various passive mitigation systems is evaluated using consequence modeling for case studies involving highly toxic chemicals such as phosgene, chlorine, and ethylene oxide.

Key Words: Risk Reduction, Risk Mitigation, Passive Mitigation, Consequence Modeling

Introduction

Risk management is an integral part of designing and operating a processing or storage facility. Through use of a structured and systematic risk management system, senior executives can provide for protection of the public and the onsite work force, the environment, and the business. With passage of the Environmental Protection Agency's Risk Management Program Rule [EPA, Final Rule], an additional challenge has now been thrust upon senior company management—communication of the worst case scenarios to the public. Through advance planning, screening, and evaluation, the task of senior management and the public relations staff can be made much easier by minimizing the extent of the worst case hazard zones at each plant site.

Worst Case Scenarios

The EPA, in its final rule, defines the worst case as "release of the largest quantity of a regulated substance from a vessel or process line failure that results in the greatest distance to an endpoint...." At least one worst case release must be considered for both toxics and flammables for covered processes. In evaluating toxics, the dispersion endpoint that must be used, in order of preference is (1) the Emergency Response Planning Guideline 2 (ERPG-2) concentration developed by the American Industrial Hygiene Association , or (2) the Level of Concern (LOC) for Extremely Hazardous Substances (EHSs). Stable atmospheric conditions (F stability) and low wind speed (1.5 m/s) must also be used. In evaluation of the worst case release hazard zones, the EPA requires that the entire inventory be released over a period of ten minutes <u>unless passive mitigation systems are in place</u>.

The EPA defines passive mitigation as "equipment, devices, or technologies that function without human, mechanical, or other energy input" at the time of the accident. Credit can be taken for these passive mitigation systems provided they can withstand the initiating event and function as intended. Examples of passive mitigation systems that can be included in evaluation of the worst case scenarios are use of refrigeration, dikes, vapor generation control, and enclosures.

Several case study examples will be used to illustrate how passive mitigation options can be used to significantly reduce the worst case dispersion hazard distance. What initially appears to be a truly catastrophic event can be reduced to a manageable event less likely to be of concern to the surrounding community when presented in the required Risk Management Plan (RMP). Also of key importance is that real risk mitigation be achieved--and the greatest risk reduction possible for the investment.

Definition of Case Studies

Three case studies are reviewed in this work. The covered processes include the following chemicals:

- Phosgene
- Chlorine
- Ethylene Oxide

The parameters from these evaluations are "genericized" for the purposes of this presentation, while still illustrating the key points. The study options and passive mitigation alternatives are:

Option 1/Base Case—over 27,000 kg (30 tons) of material are stored at the site, under pressurized conditions.

Option 2—the storage is refrigerated and utilizes a secondary containment dike able to hold 110% of the tank capacity. The dike wall height is 0.6m (2 feet).

Option 3—the storage is refrigerated and utilizes a secondary containment dike able to hold 110% of the tank capacity. The dike wall height is increased to 1.8m (6 feet) to delay spill over of the heavy gas vapors from the dike and to minimize pool surface area.

Option 4—the storage is refrigerated and utilizes a secondary containment dike able to hold 110% of the tank capacity. The dike wall height is 0.6m (2 feet) and the dike floor and walls are insulated with a suitable low conductivity material.

Option 5—the storage is refrigerated and utilizes a secondary containment dike able to hold 110% of the tank capacity. The dike wall height is 1.8m (6 feet) and the dike floor and walls are insulated with a suitable low conductivity material.

Option 6—the storage is refrigerated and utilizes a secondary containment dike able to hold 110% of the tank capacity. The dike wall height is 0.6m (2 feet) and the dike floor and walls are insulated with a suitable low conductivity material. The storage area is also enclosed in a building of approximately 1800 m^3 with an air turnover rate of 0.5 air changes per hour.

Option 7—the storage is refrigerated and utilizes a secondary containment dike able to hold 110% of the tank capacity. The dike wall height is 1.8m (6 feet) and the dike floor and walls are insulated with a suitable low conductivity material. The storage area is also enclosed in a building of approximately 1800 m^3 with an air turnover rate of 0.5 air changes per hour.

Option 8—consideration is given to storage of the necessary inventory in one-ton pressurized containers.

Certainly, other mitigation options are possible and may be highly effective. Table 1 summarizes the study cases and options reviewed here.

Source Term Modeling

In the case of pressurized storage, the source term modeling is simple as the material is released as a gas over a period of ten minutes. For analysis of the refrigerated spills, a pool model is necessary. A pool model is chosen that evaluates the rate of vapor generation due to heat transfer effects and accounts for the delay in heavy gas vapor overflow from the secondary containment [Mudan]. This model is adjusted to also include the vapor generation due to convective heat transfer and diffusion, thereby accurately accounting for the total vapor generation rate from the pool.

Dispersion Modeling

The EPA has provided the RMP Offsite Consequence Analysis Guidance [EPA] for use by covered stationary sources. This guidance document provides simple tables and equations for use in estimating the worst case and alternative scenario hazard distances. Covered facilities using this approach will be considered in compliance. While companies may wish to begin screening analyses for some facilities using this approach, the results will generally be conservative. In layman terms this means senior management will have to communicate a larger hazard distance for the worst case scenario than actually exists. In many cases this conservatism can be significant and may artificially increase the level of community concern. Therefore, some additional modeling may be desirable to provide more realistic estimates of the hazards. There are a number of good public domain dispersion models available (several jointly sponsored/funded by the EPA) which can be used to realistically estimate the hazards. Use of such public domain models also eliminates the need to educate Local Emergency Planning Committees (LEPCs) on the finer points of a particular proprietary dispersion model. Of course, proprietary models can be used, but must be explained and justified to the LEPC and other agencies. In the case studies presented here, all dispersion modeling is carried out with the EPA sponsored DEGADIS model.

Table 1. Study Cases and Alternatives Evaluated

Alternative	Study Case - Phosgene	Study Case - Chlorine	Study Case - Ethylene Oxide
• Pressurized	P1	C1	E1
• Refrigerated • Low Dike Wall (0.6m) • Concrete Dike	P2	C2	E2
• Refrigerated • High Dike Wall (1.8m) • Concrete Dike	P3	C3	E3
• Refrigerated • Low Dike Wall (0.6m) • Insulated Dike	P4	C4	E4
• Refrigerated • High Dike Wall (1.8m) • Insulated Dike	P5	C5	E5
• Refrigerated • Low Dike Wall (0.6m) • Insulated Dike • Enclosure	P6	C6	E6
• Refrigerated • High Dike Wall (1.8m) • Insulated Dike • Enclosure	P7	C7	E7
• Pressurized One Ton Cylinders	P8	C8	E8

Case Study Results

Figure 1 depicts the change in rate of phosgene release to air in going from pressurized to refrigerated storage. In the case of pressurized storage the entire inventory is released as a gas over a ten minute period. With refrigeration, the phosgene forms a liquid pool which then evaporates at a much lower rate. In fact, the rate of phosgene release to air is reduced to 0.7% of the pressurized case. Further reductions in the emission rate are gained through the various passive mitigation options outlined previously—the effects are shown in Figure 2.

All of these alternatives become feasible and more effective once refrigeration is implemented. Option 3 (shown as P3 in Figure 2) provides an additional 53% reduction in the rate of phosgene released. Increasing the dike wall height from 0.6m to 1.8m has two effects: (1) with higher dike walls, the exposed surface area of the pool is minimized, and (2) it takes longer for the heavier than air phosgene vapors to overflow the dike. Passive mitigation Option 4 includes use a low wall dike design (0.6m), but incorporates use of low conductivity dike insulation. This option provides an additional 47% reduction in the rate of phosgene release to air as compared to the base refrigerated case. In Option 5, a high wall dike and dike insulation are used in combination to achieve an 83% reduction in the emission rate as compared to refrigeration alone. Option 6 and Option 7, which include an enclosure for the storage facilities then provide 98.7% and 99.3% reduction from the base refrigerated case. In combination, these passive mitigation measures reduce the rate of phosgene release to air from 46 to 0.002 kg/s, or by a factor of over 20,000.

Figure 1. Rate of Phosgene Release to Air - Pressurized vs. Refrigerated Case

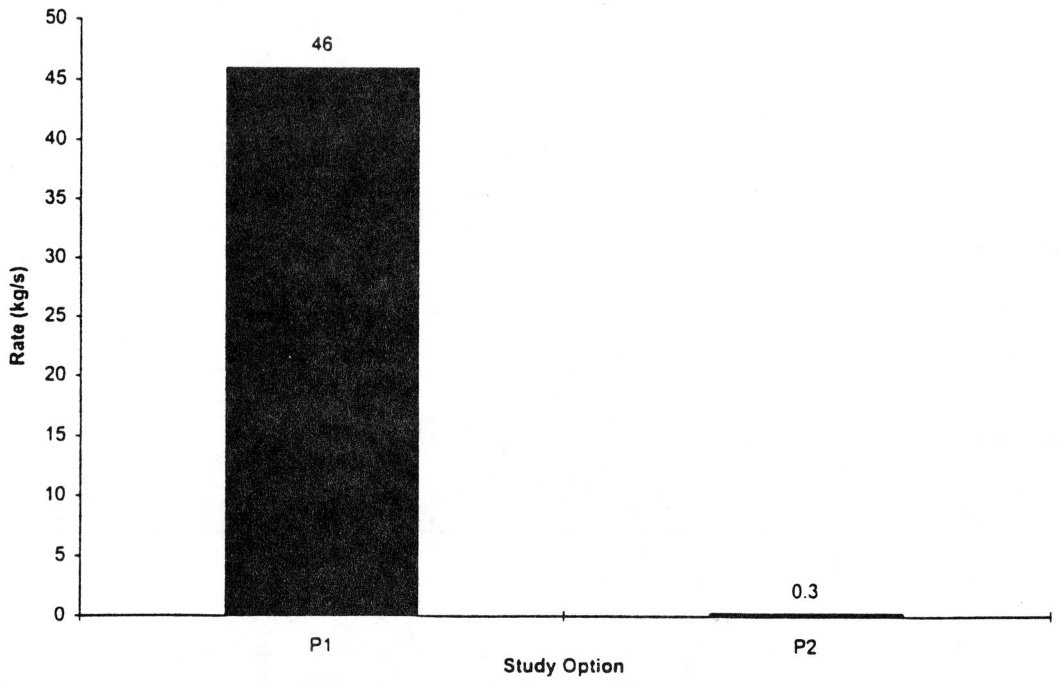

Figure 2. Rate of Phosgene Release to Air - Refrigerated Options

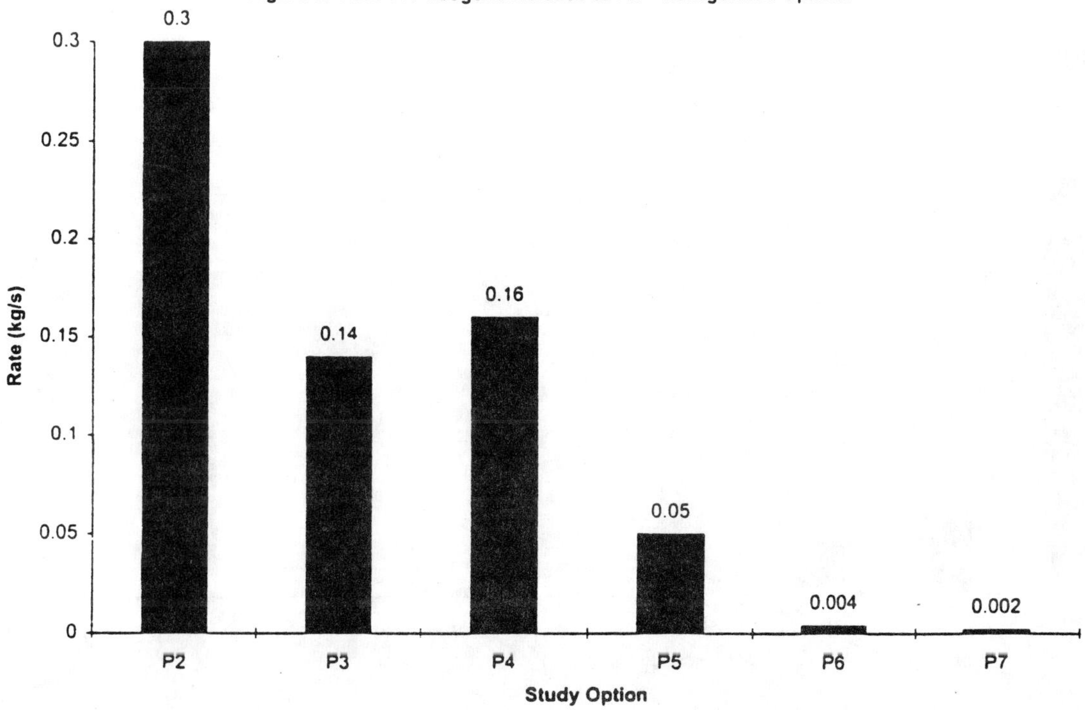

Figure 3 depicts the predicted dispersion distances based on these phosgene source rates. All dispersion modeling is conducted using the DEGADIS model and the EPA required parameters for rural terrain (3 cm surface roughness length) and endpoint concentration (ERPG-2 for phosgene is 0.2 ppm). The dispersion model is exercised for F stability and a wind speed of 1.5 m/s. In the pressurized case, the release is predicted to travel about 100km (over 60 miles) which is well outside the range of validity for dispersion models. As a point of interest, the EPA RMP Offsite Consequence Analysis Guidance tables list ">25 miles" (40km) as the hazard distance for a release rate (to air) less than 10% of the value used in this case study. As can be seen in Figure 3, by refrigeration alone, the worst case hazard distance can be reduced by 85% to a distance of 14.7km. Substantial additional mitigation of the consequence zone can be achieved through implementation of the other passive mitigation options discussed. The dispersion distance for Option 7 which includes use of refrigeration, an insulated high walled dike, and an enclosure, is reduced to 1% of the pressurized case to 1km.

This case study points out the substantial potential benefits of implementation of passive mitigation systems. Through implementation of such systems the site manager can communicate to the surrounding public that the worst case dispersion extends to a distance of only 1 km from the point of release rather than 100km. This is clearly more acceptable to the public, and places senior management in a much better position for communication and discussion of the hazards posed by the site. In addition, it is easier

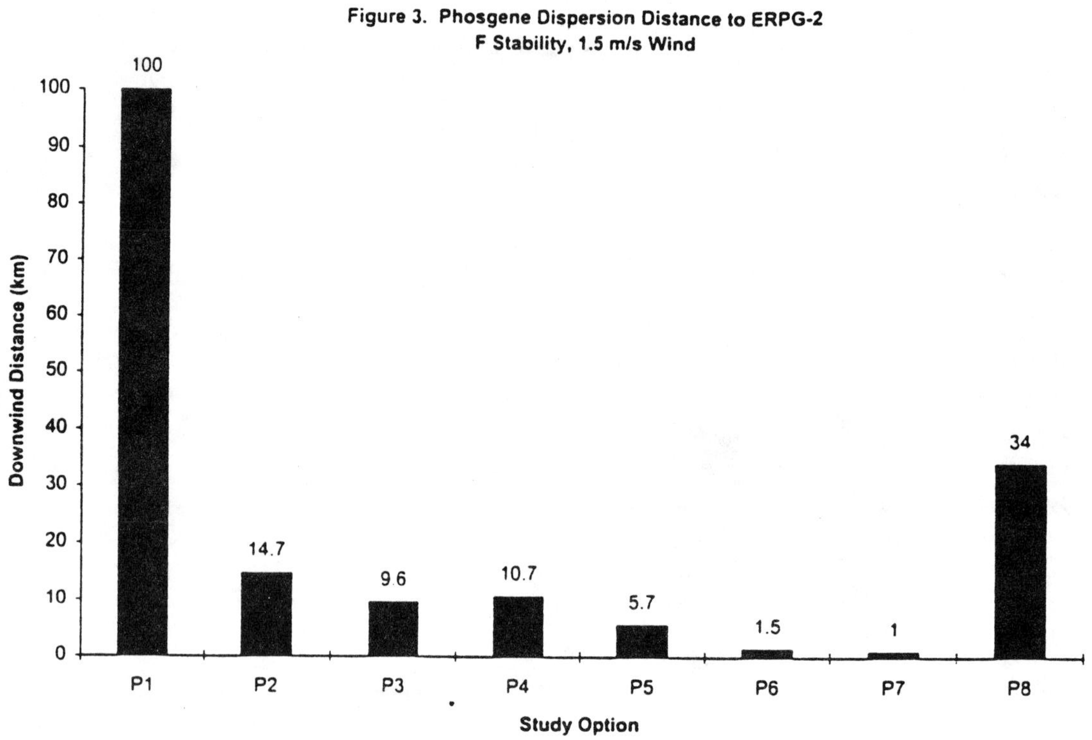

for the site to devise, implement, and test an emergency response plan for the potentially affected area. In contrast, it may be much harder to convince the LEPC that your site has an effective response plan for a chemical that has toxic impacts to a distance of 40 to 100km.

As a number of sites do store phosgene in one-ton containers, this scenario is evaluated as well, and presented as Option 8 in Figure 3. The dispersion distance is about 34 km (over 20 miles) for a ten minute release of the phosgene. Clearly, the other passive mitigation options for storage are preferred. While the maximum inventory that can be released in a worst case event is reduced, more effective consequence mitigation measures are available. In addition, while the EPA Risk Management Program Rule does not consider event likelihood or frequency, by storing the necessary inventory in one-ton containers the frequency of release is tremendously increased—increasing the overall level of risk to the onsite workforce and the public as compared to other reasonable alternatives.

Similar results are achieved for chlorine storage facilities. Figure 4 presents the chlorine emission rates for the pressurized and refrigerated storage scenarios. The volatilization rate is reduced by 99% in the refrigerated case. Figure 5 shows the further gains that can be achieved in reducing the rate of chlorine release to air for the worst case scenario. Through a combination of passive mitigation measures, the volatilization rate can be

Figure 4. Rate of Chlorine Release to Air - Pressurized vs. Refrigerated Case

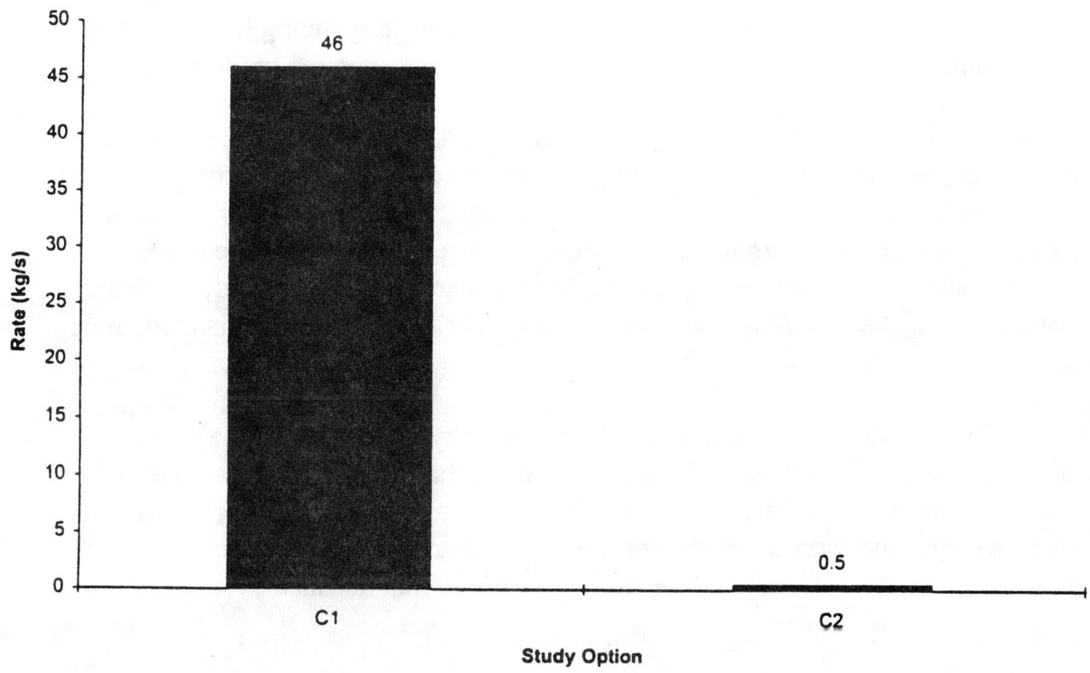

Figure 5. Rate of Chlorine Release to Air - Refrigerated Options

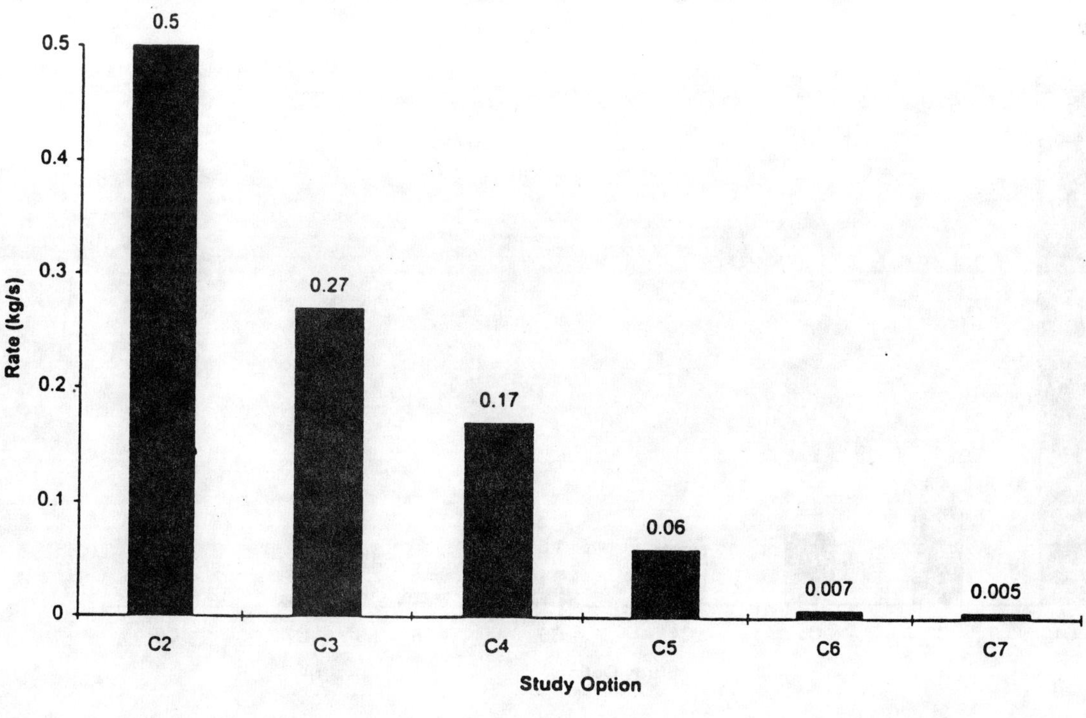

reduced from 46 kg/s in the pressurized storage case to 0.005 kg/s—a factor of nearly 10,000.

Figure 6 illustrates the benefits of passive mitigation systems in reducing the worst case dispersion distance for each of the eight options. Again, the dispersion modeling is carried out using DEGADIS with the required EPA parameters, and assumes a rural setting. The endpoint concentration for chlorine, the ERPG-2, is 3 ppm. As can be seen, the worst case dispersion distance is reduced from 24 km (15 miles) to 0.5 km. For a number of facilities, the hazard may therefore remain onsite. For still others, especially in rural areas, there may be no significant populations or receptors within the greatly reduced worst case hazard distance. As the facility risks are significantly reduced, the task of senior executives in communicating the worst case to the public can is made much easier.

Figure 7 and Figure 8 show the reduction in the rate of ethylene oxide release to air for each of the study options considered. It is demonstrated that a reduction by a factor of nearly 50,000 in the evolution rate is possible. This translates to significant reductions in the worst case dispersion hazards which are shown in Figure 9. In fact, use of refrigeration alone (Option 2) reduces the worst case dispersion distance by over 90% to

Figure 6. Chlorine Dispersion Distance to ERPG-2
F Stability, 1.5 m/s Wind

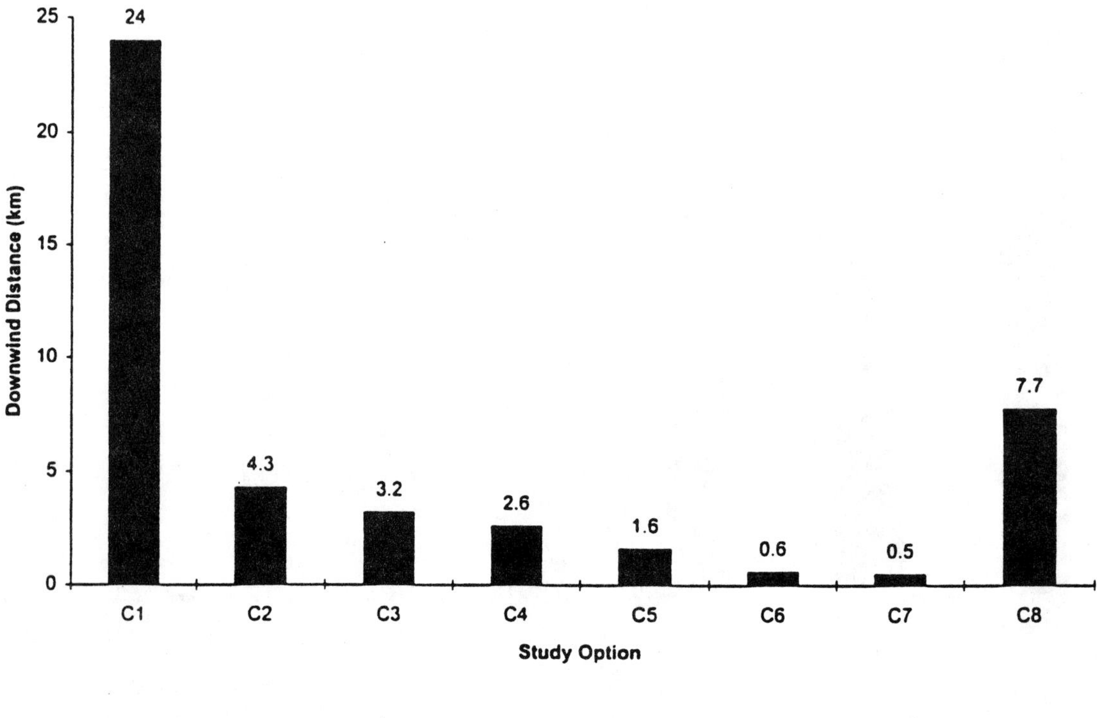

Figure 7. Rate of Ethylene Oxide Release to Air - Pressurized vs. Refrigerated Case

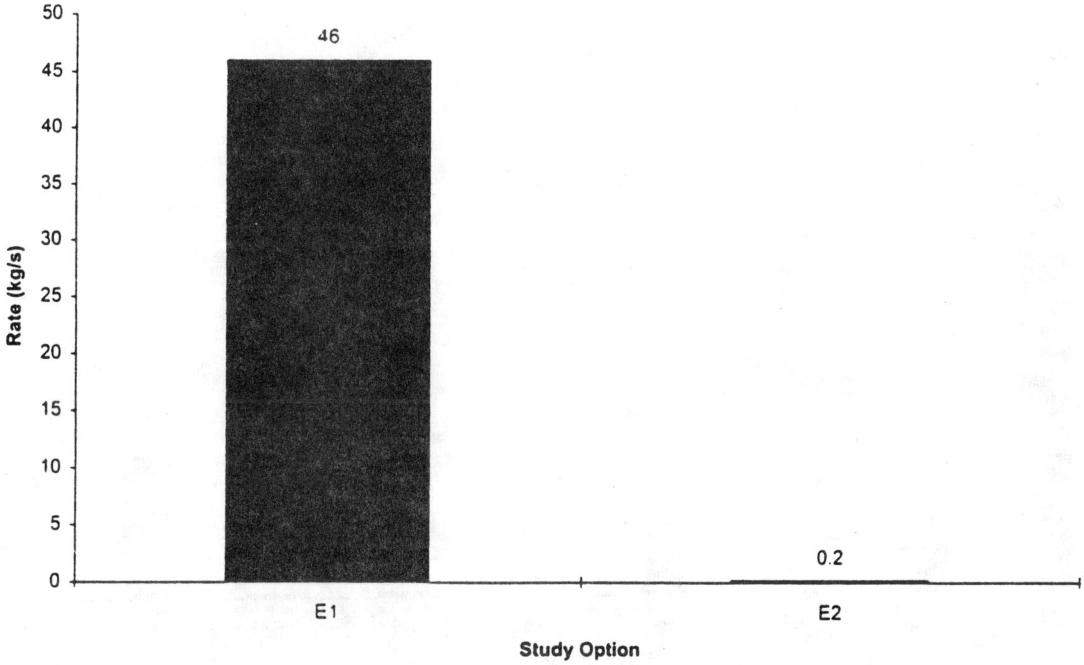

Figure 8. Rate of Ethylene Oxide Release to Air - Refrigerated Options

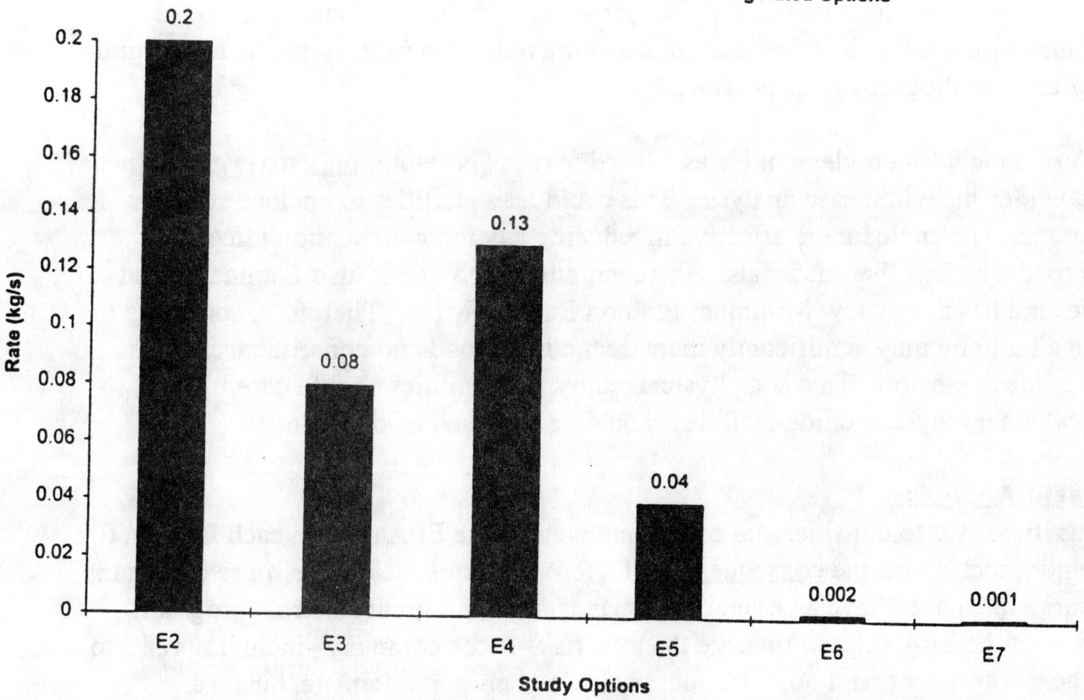

Figure 9. Ethylene Oxide Dispersion Distance to ERPG-2
F Stability, 1.5 m/s Wind

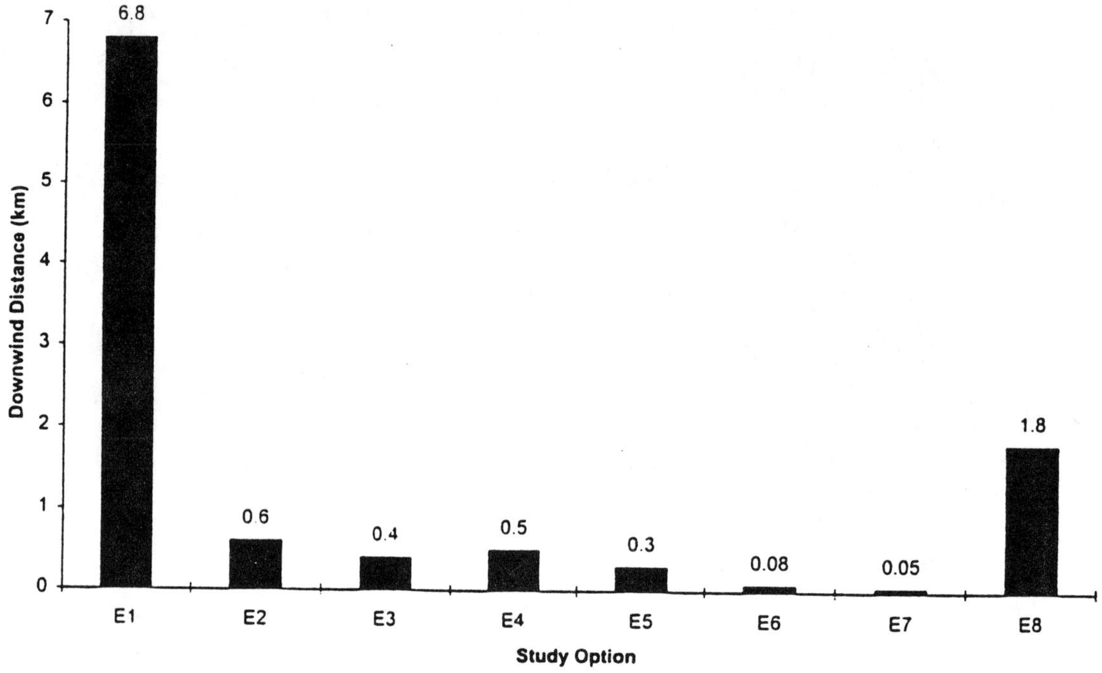

0.6km. In the case of ethylene oxide, through use of refrigeration and a high walled, insulated dike, the worst case dispersion distance can be reduced to 0.3km as compared to 6.8km for the pressurized storage case. For some facilities, this reduction in hazard distance may also reduce the RMP requirements if a reduction in Program level (Program 3 to Program 2 to Program 1) is possible.

The EPA rule includes ethylene oxide as a listed toxic. Therefore, <u>only</u> toxic effects need be evaluated for the worst case analysis. This could lead facilities to enclose ethylene oxide storage. The enclosure is effective in reducing any toxic dispersion effects for ethylene oxide, as for other materials. However, ethylene oxide is also flammable and explosive, and has a very low Minimum Ignition Energy (MIE). Therefore, confining the release in a building may significantly increase the likelihood and consequences of an ethylene oxide explosion. This is a physical reality and facilities should carefully consider whether ethylene oxide facilities should be enclosed in any manner.

<u>Cost Benefit Analysis</u>
While it is important to consider and address impacts of the EPA rule on each facility, the EPA program focuses on the <u>consequences</u> of a few scenarios. It is even more important for each organization to have an overall risk management program (internal program, irrespective of the EPA rule) to manage the total risks to the company--including risks to people, the environment, and other business risks (e.g. property damage, business

interruption). It is also essential that all investments in risk mitigation provide cost-effective means to tangible risk reduction, including both the likelihood and consequences of all possible events. Therefore, a beneficial extension of the previous analyses is to utilize a cost benefit approach when evaluating various risk mitigation options, passive or active. Investments made should be scrutinized and evaluated in the same manner as many other business decisions. A structured Cost-Benefit Analysis (CBA) approach can provide key information to decision makers for approving or rejecting alternatives considered in a systematic, documented, and objective manner. Companies often do use Cost-Benefit Analysis to:

- target resources for the highest "rate of return"
- provide a rational, justifiable basis for not implementing certain recommendations
- provide a systematic way to evaluate and control expenditures
- identify alternative cost-effective solutions
- yield a tangible reduction in risk

EPA Risk Management Program - Steps That Should be Taken Now

Many companies are reviewing and starting the EPA Risk Management Program process now as a part of other ongoing risk management activities. Although there are many operating and business pressures to delay compliance expenditures, there are potential economic benefits to beginning the process now. Those companies involved in advance planning will have many more opportunities and choices than those waiting until the last three months before RMPs must be registered. Clearly, advance planning and evaluations may be needed to make changes involving production schedules, procurement, construction, etc. and to take full advantage in implementation of passive and other mitigation systems.

Actions companies should begin now include:

- start planning the overall process and strategy, and determine which facilities and processes are currently covered by the EPA rule
- evaluate opportunities for inventory reduction to "opt out" of the regulatory RMP requirements
- identify means to reduce the RMP regulatory requirements by lowering the Program level (Program 3 to Program 2 to Program 1)
- begin developing your communication plan
- develop or work to continue good ongoing relationships with the LEPC and local community and involve the LEPC in the RMP process
- begin screening analyses for facility worst case scenarios
- ensure that initial Process Hazards Analyses (PHAs) include scenarios resulting in offsite impacts (where possible)
- in revalidation PHAs, ensure that scenarios potentially impacting offsite are included

- in all PHAs, consider and document "safeguards" including monitoring and detection specifically
- as a part of all ongoing PHAs for covered processes document possible worst case and alternative scenarios, as well as the relevant inventories, process conditions, and other necessary details for subsequent evaluation
- identify and evaluate opportunities for reducing the worst case hazard distances through passive mitigation systems

Summary

The study cases presented in this work demonstrate that significant reductions in the worst case hazard distance can be achieved through a variety of passive risk mitigation options. Even in the case of some of the most toxic chemicals, these worst case dispersion distances can be tremendously reduced. At some facilities this may result in an onsite hazard only, and reduce RMP requirements. In all cases, the task of communicating the worst case scenarios to the local community can be made much easier by minimizing the extent of the worst case. As the hazard areas are reduced, development, implementation, and testing of associated emergency response plans also becomes a much more manageable task. Through advance planning, screening, and evaluation, each company can prepare and make any needed modifications in a cost-effective manner, and while there is still ample time for implementation of passive and other mitigation systems as part of the overall company risk management efforts.

Acknowledgments

The authors wish to thank Dr. Krishna Mudan of Four Elements for his valuable contributions to this work.

References

[1] USEPA, 40 CFR Part 68, Risk Management Program Final Rule, Published June 20, 1996.

[2] Mudan, Krishna S. et al., Dike Design Alternatives to Reduce Toxic Vapor Dispersion Hazards, November, 1987.

[3] USEPA, RMP Offsite Consequence Analysis Guidance, May 24, 1996.

42D

Protection of Process Buildings and Their Occupants from Toxic and Flammable Releases

Chuck Fryman

Prepared for presentation at the 1997 Spring National Meeting
Houston, TX, March 9-13

31st Annual Loss Prevention Symposium

UNPUBLISHED

PROTECTION OF PROCESS BUILDINGS AND THEIR OCCUPANTS FROM TOXIC AND FLAMMABLE RELEASES.

C. E. Fryman,
BP Oil Company,
200 Public Square, Cleveland, OH 44114.

ABSTRACT

Petroleum refinery process buildings, their occupants and equipment are exposed to hazards from toxics and flammable releases from the various process operations at the site. Common toxic hazards include releases of hydrogen sulfide and hydrogen fluoride gases which can enter the process building through the ventilation (HVAC) system and openings, such as doors and windows. Flammable hazards include the threats of flash fires, pool fires, jet fires, BLEVEs and vapor cloud explosions, which can damage the process building, internal equipment and injure the occupants. The history of the petroleum refining and petrochemical industries shows that toxic and flammable releases occur at a relatively low frequency, but with potentially catastrophic consequences including personnel deaths and injuries, property damage, and business interruption losses. Faced with making the decision of what action should be taken to retrofit existing process buildings, BP Oil Company developed a comprehensive and systematic methodology to evaluate each petroleum refinery process building to establish the level of risk and determine both the costs and effectiveness for a range of mitigation measures appropriate to each building type. Risk mitigation measures included improvements to the process building structures, HVAC system gas detection and shutdown logic, source controls, process area gas detectors and alarm systems, window treatments, relocation of buildings and occupants, and training of personnel. The level of risk and the risk reduction for each mitigation measure were expressed in dollars per annum and included personnel, property damage and business interruption losses. The risk reduction benefit may be compared with the cost of implementation which allows the calculation of cost-benefit ratios for various mitigation measures. Addressing the highest risks first, BP Oil Company has identified cost-effective risk mitigation measures and implemented them at petroleum refinery sites.

Protection of Process Buildings and Their Occupants from Toxic and Flammable Releases

CHUCK FRYMAN

BP Oil Company
200 Public Square 7-P, Cleveland, Ohio 44114-2375

INTRODUCTION

Petroleum refinery process buildings, their occupants and equipment are exposed to hazards from toxics and flammable releases from the various process operations at the site. Common toxic hazards include releases of hydrogen sulfide and hydrogen fluoride gases which can enter the process building through the ventilation (HVAC) system and openings, such as doors and windows. Flammable hazards include the threats of flash fires, pool fires, jet fires, BLEVEs and vapor cloud explosions, which can damage the process building, internal equipment and injure the occupants.

The history of the petroleum refining and petrochemical industries shows that toxic and flammable releases occur at a relatively low frequency, but with potentially catastrophic consequences -- including personnel deaths and injuries, property damage, and business interruption losses.

Faced with making the decision of what action should be taken to retrofit existing process buildings, BP Oil Company developed a comprehensive and systematic methodology to evaluate each petroleum refinery process building to establish the level of risk and determine both the costs and effectiveness for a range of mitigation measures appropriate to each building type. Risk mitigation measures included improvements to the process building structures, HVAC system gas detection and shutdown logic, source controls, process area gas detectors and alarm systems, window treatments, relocation of buildings and occupants, and training of personnel.

The level of risk and the risk reduction for each mitigation measure were expressed in dollars per annum and included personnel, property damage and business interruption losses. The risk reduction benefit may be compared with the cost of implementation which allows the calculation of cost-benefit ratios for various mitigation measures. Addressing the highest risks first, BP Oil Company has identified cost-effective risk mitigation measures and implemented them at petroleum refinery sites.

THE ISSUES AND CONCERNS

The siting and design of these process buildings has long been a concern, but that concern was sharpened with the occurrence of several major explosions and toxic chemical releases over the past decade. These incidents have injured or killed a significant number of people and resulted in billions of dollars in property damage and business interruption losses. Also, OSHA's process safety management (PSM) rule requires that process building siting be considered in conducting process hazard analysis.

BP Oil strives to manage all types of risk, including risks of injury to people and economic losses due to fines and penalties, third party liability, property damage, and business interruption. In considering the risks associated with process building siting and design, BP Oil considered all buildings on the petroleum refinery site and not just those which were occupied. These buildings included motor control centers, electrical sub-stations, warehouses and storage locations. We wanted to include anything that could produce a loss.

We also wanted to evaluate the risks of more than just vapor cloud explosions (VCEs). Although VCEs do represent a significant risk in petroleum refineries, they by no means are the only risks involved in operating a refinery. We were also concerned about other flammable scenarios, like pool fires, flash fires, and jet fires. And, we were concerned about releases of toxic substances, like hydrogen sulfide, ammonia, chlorine, and hydrogen fluoride.

Accordingly, BP Oil developed a comprehensive risk assessment methodology to evaluate all the above-mentioned risks throughout the petroleum refinery. After assessing the risks, we evaluated the feasibility and cost-effectiveness of appropriate remedial measures, and set about implementing the remedial measures within the context of our risk management model, which is discussed later.

A STAGED APPROACH

A screening evaluation of process buildings was begun in early 1993 throughout the BP Oil worldwide network consisting of 18 petroleum refineries. This screening evaluation focused on personnel risks associated with VCEs. A relatively simple workbook was developed to assist site technical personnel in assessing the frequency and severity of VCEs, and the resulting injuries to personnel. This work, now referred to as Stage 1, resulted in a rank ordering of refineries according to their potential risks.

Because it was necessary to consider other flammable and toxic risks to personnel, and because BP Oil wanted to consider economic risks due to property damage, business interruption, liability, fines and penalties, another more comprehensive and more accurate risk assessment methodology was developed. This methodology, which we now refer to as Stage 2, was developed and piloted at one of our petroleum refineries and is now being implemented at other refineries throughout the network.

The final stage, i.e., Stage 3, of work involves implementing mitigating measures to reduce the risk to acceptable levels. These mitigating measures include items that are relatively inexpensive and easily implemented and measures for which feasibility studies need first to be done to determine the appropriate alternative. All of the mitigating measures are subjected to a risk-benefit evaluation in which the cost-effectiveness of the measure is compared to the risk reduction achieved. More details are provided later on the risk-benefit evaluation.

RISK EVALUATION MODEL

The BP Oil risk evaluation model is contained in a user-friendly Microsoft Excel workbook. The model takes into account the following main inputs.

Building construction -- the materials of construction are considered, including the foundation, framing and roof structures. We also consider the ease of egress of airborne toxic and flammable materials into occupied buildings.

Occupancy level -- the person with the maximum exposure (i.e., the person at maximum individual risk) and the exposures of other personnel (i.e., the societal or population risks). We do not set occupancy screening criteria.

Property damage -- these losses are estimated using guidance provided by refinery project engineering teams and are based on the severity of the event.

Business interruption -- these losses are estimated by the refinery commercial team based on the length of time the unit is out of operation, margin losses, and alternate operating scenarios.

Likelihood of the event -- the likelihood of a major release of flammable or toxic materials is based on the extensive history of similar events maintained in BP Oil's process safety database extending back in time to 1970.

Consequences of the event -- the consequential losses are estimated using BP's Cirrus compendium of consequence models. These model consider VCEs, various fire scenarios (pool, flash and jet), and toxic gas dispersion (heavy gas, momentum and buoyant).

Costs of mitigation measures -- a preliminary screen of the measures based on its cost-effectiveness is completed by comparing the mitigation costs against the maximum spend assuming zero residual risk.

RISK MANAGEMENT MODEL

BP Oil's risk management model starts with the identification of potential risks. These potential risks include risks to people (including workers, contractors and members of the general public), offsite impacts (including environmental cleanup and our best estimate of any outrage), property damage, business interruption, and liabilities (including punitive and civil damages).

We try to maintain an awareness of our risks through a careful analysis of the incidents we experience throughout our company, including BP's worldwide operations in oil and gas exploration and chemical manufacturing. We are also interested in learning from incidents that have happened to others in our industry, and have in place a system to collect information on incidents occurring outside of BP and share that information throughout BP. Finally, we also review the reports of studies and reviews that have been completed within BP and outside our company, including those completed by regulators, activists, special interest groups, and our competitors. All of this information is used to help us maintain what is hopefully an evergreen view of the risks facing our operations.

The next step in our risk management process is a qualitative screening of the potential risks. Risk is the product of consequence and frequency -- it answers the questions of "How big?" and "How often?". This screening identifies our best estimate of the likelihood of the event occurring anywhere in our operations and the severity of the event, allowing us to decide whether further analysis is required.

If further analysis of the risk is deemed necessary, then we will conduct a more detailed analysis of the severity of the losses and likelihood of the event occurring. This analysis might be done via a detailed quantified risk assessment (QRA) study or some other risk assessment methodology. Because of our significant experience in having conducted QRA studies of virtually every refinery process unit somewhere in our worldwide operations, we often conduct what we call SQRAs -- Semi-Quantified Risk Assessments -- which are designed to extend existing QRA studies from similar process units to the current process unit under study. Other methodologies include HazOp studies, what-if reviews, and project safety reviews.

The objectives of the risk analysis will drive the scope of the risk assessment and often the methodology used. Objectives for these studies include compliance with BP Guidelines, regulatory requirements, business requirements, and license-to-operate issues. Our BP risk acceptability guidelines set maximum individual risks for our workers and contractors, and for the offsite public. Regulatory requirements, like the OSHA PSM and the forthcoming EPA RMP regulations, require an identification of the consequences due to process operating conditions and accidental releases; although neither of these regulations require it, we would perform a risk evaluation to determine where and when to spend our money. Business requirements include our concerns for operating safely and without damaging the environment; our biggest business driven study to date dealt with the use of hydrofluoric acid (HF) in our alkylation units. Finally, license-to-operate issues are usually driven by the local community or workforce.

After the risks have been assessed, a decision needs to be taken on the actions. A decision must be made on which risk mitigation measure to implement as well as the implementation timing. Sometimes additional risk assessment study or other input is needed in order to decide the actions. We attempt to capture lessons learned coming from the decision making process (e.g., use elevated water monitors at all HF alkylation units) in our engineering and operating practices, and in our lessons learned database.

There are two inputs for deciding the actions -- the Action Classification Scheme, which incorporates our risk-benefit approach and stakeholder input. The Action Classification Scheme or Risk Benefit Evaluation (RBE) is used to aid in decision making and does not alone determine what actions are taken. RBE provides a description of the risks (severity and likelihood) from the initiating event, ranks those risks with those from other initiating events, describes the risk reduction associated with the mitigation options, and the cost-effectiveness ranking of the mitigation measures. This allows the decision to consider the relative magnitude of the risk and the amount of risk reduction for the monetary spend. RBE will be further discussed in the next section of this paper.

The other important input for deciding actions is from the stakeholders. The stakeholders include the employees, contractors, local community, company management, regulatory agencies, special interest groups, and anyone else who has an interest in the outcome of the decision. Stakeholder input is really a continuous process and is one of the most important elements of the entire risk management process. The final decision must also consider normal industry practices and our concerns about liability.

Mitigation of the risks will usually be done based on a combination of outcomes from the risk-benefit evaluations and stakeholder negotiations. We use the risk-benefit evaluation as the starting point for stakeholder negotiations. The risk-benefit evaluation presents the company's view of what is safe and appropriate -- recognizing that not everything can be done at once because of limited resources, and that everything that could be done is not cost-effective or appropriate in every situation -- so something must be done to prioritize spends. However, at the end of the day, the outcome of stakeholder negotiations determines which risk mitigations get implemented for those few issues holding strong interest to the stakeholders. For those issues, it is of course important to maintain a continuing dialog with the stakeholders to see that their interests are addressed.

RISK-BENEFIT APPROACH

BP Oil Company has been using the Risk-Benefit Approach since about 1990, beginning with our risk assessment work for our HF Alkylation units. During the 1990-92 time frame, we developed the risk management framework, implemented it at our HF Alkylation refineries, and discussed it at length with the US EPA, who was in the process of conducting a risk assessment study of HF Alkylation required by the 1990 Clean Air Act Amendments. RBE was well received by US EPA and helped to fend off additional regulation of the HF Alkylation process.

The RBE provides a defensible position for not completing all actions, i.e., deferring spending on some risk mitigations. Under the RBE philosophy, the highest risks are addressed first. And, RBE directs spending so that the most risk reduction is achieved for the monetary spend, i.e., the most risk reduction for the buck.

Use of RBE within BP Oil Company has steadily increased over the past few years. We have used it for new projects to evaluate process siting options, fire protection requirements, use of gas detectors, emergency isolation requirements, and pressure relief requirements. We have also used it for evaluating the siting and design of our process buildings and routinely use it to evaluate the risk-benefit associated with staff relocations at site. As mentioned earlier, we also plan to use RBE to evaluate the off-site impacts and follow-up actions required by the new EPA RMP regulations.

During a meeting with senior US OSHA representatives on another subject, we informally had the opportunity to discuss RBE. They were very receptive to RBE and supportive of the approach; furthermore, they indicated that we were one of the few companies to take risk-based, proactive steps to sort out the hundreds of recommendations resulting from PSM compliance activities.

The RBE begins with a risk prioritization step. Here risks are ranked according to the severity of the outcome and likelihood of occurrence of the initiating event. The severity rating scale ranges from catastrophic to minor. You should note that all the aforementioned consequential losses -- property damage, business interruption, cleanup, liability, outrage, etc. -- are included. In choosing a particular severity rating, the overriding consequences would be used to determine the rating; for example, estimated environmental cleanup costs of $100M would override an OSHA lost time accident valued at $30M.

The likelihood rating scale is also divided into four categories, ranging from highly likely to extremely unlikely. It is important to note that this scale has proven very practicable because it is easily used in the field. Most people have little problem recalling events that occur once per year, once every 5 years, or once during the 30-year life of the operation. To address the no likelihood rating, we use our lessons learned system which contains a record of the major process safety incidents occurring throughout the world back through 1971.

We next position the initiating events on a four-by-four matrix of likelihoods and consequences. The boxes of the matrix are labeled with the letters "A" through "D" which correspond to relative priorities and action requirements. High likelihood and high consequence events receive an "A" rating, whereas low likelihood and low consequence events receive a "D" or "C" rating. "B" ratings cover the interim range.

It is intended that "A" rated events receive immediate attention, including notification of senior management and a comparison to BP Guidelines. "B" items require prompt attention, normally within 3 months, "C" items are addressed after "A" and "B" items, and "D" have no requirement for further risk reduction. Mitigations for "A", "B" and "C" rated items are implemented if the costs are within the cost-benefit justification guidelines; otherwise, the mitigations are considered to be not cost-effective and are ranked low on the totem pole.

The cost-benefit justification guidelines were developed to reflect the probabilized losses that the company would suffer should an unintentional event occur. For fatalities we use a range of $1 million to $10 million per statistical fatality averted, based on reductions in societal or population risks. Lost time accidents are costed at $30,000 each and OSHA recordable accidents are costed at $3,000 each. Oil spill cleanup costs are estimated at $10,000 per barrel spilled. Business interruption and property damage losses are estimated based on the specific event, as are liability and outrage costs. The use of these cost justification guidelines allow us to place a value on the level of effort we are willing to undertake to reduce either the likelihood of the event occurring or the consequential losses. Although, please keep in mind that this is just one input to a sometimes rather complex decision making process.

The BP Risk Guidelines set acceptability criteria for the statistical chances of fatality to our workers and members of the general public. Our risk acceptability model is based on a three-tier framework of individual risk criteria. Risks in Zone A need mitigation, or at least additional risk assessment; whereas risks in Zone C require no further risk reduction efforts. Risks in Zone B require the consideration of further risk reduction if the mitigation measure is cost effective in reducing the risks. The cost-benefit justification guidelines mentioned above are the initial basis for determining cost effectiveness; optimization is then achieved by evaluating the reduction of both societal risks and mitigation costs.

This next slide shows the individual risk criteria that define the boundaries between Zones A, B and C for both our workers and members of the public. You will note there is an order of magnitude difference between the criteria for workers and members of the public; this difference reflects the fact that workers are able to exercise more control over their risks and derive more direct benefits compared with those members of the public who merely live adjacent to a facility. We have found that most all of our onsite risks to workers fall within Zone B, and thus require us to consider cost-effective risk reduction measures. Of course, this underscores the importance of our use of the RBE and the cost justification guidelines.

In order to place the consideration of spends for risk mitigations on par with spends for commercial projects, we use the same economic evaluation criteria. These economic criteria vary over time depending on the lending rates and corporate targets for financial returns. This slide shows an example of how these are used in our risk matrix. Let us consider a 15% IRR (internal rate of return) as the appropriate hurdle rate and 10-year depreciation period for the capital project. This would result in the maximum spends shown for the various cells of the matrix. For a Severity-3, Likelihood-3 rated event, we would pay no more than $3,000 for the mitigation to eliminate the risk. In actuality, we would pay less than $3,000 because there will likely be some residual risk, but $3,000 is a good screening amount.

APPLICATION TO PROCESS BUILDING STUDIES

As of the time of the writing of this paper, the Stage 2 risk evaluation has been completed at one of our petroleum refineries and is underway in two other BP Oil refineries. The risk-benefit evaluation (RBE) approach has been well received by refinery management teams as an objective input to aid their decision making.

Several personnel risk mitigation measures are being implemented at each of our refineries. These measures include application of anti-shatter film to windows to reduce hazards from flying glass, securing of portable buildings to their foundations, securing of ceiling fixtures and material/equipment stored or mounted against the outside walls of buildings, and ensuring that every person understands their role in an emergency.

Feasibility studies are under way to determine acceptable options for improving the resistance of process buildings to VCEs, pool fires and jet fires. We are also evaluating flammable and toxic gas detection and alarm systems for placement in the operating unit area and in the fresh air intake ducts of process buildings. HVAC shutdown and recirculation logic is also being evaluated for the air intake detectors.

When the technically feasible mitigation options are identified, they will be evaluated by the RBE approach to determine their cost-effectiveness in reducing risks.

SUM UP AND CLOSING

We in BP Oil Company have found the application of the risk-benefit evaluation (RBE) approach to aid our decision making, especially in deciding priorities for allocating relatively scarce monetary and personnel resources. This RBE approach can be used in both regulatory and business loss prevention situations, but its use in regulatory cases is limited to performance based regulations. Perhaps the outcome of the current risk and cost-benefit arguments in the US Congress will one day encourage the wider application of the risk-benefit approach. In the interim, BP Oil Company will continue to exploit the use of this powerful decision aiding tool.

[presents/aiche-02.doc]

The Explosion Blast Prediction Model in the Revised 'Yellow Book'

W.P.M. Mercx and A.C. van den Berg
TNO Prins Maurits Laboratory
PO Box 45, 2280 AA Rijswijk
The Netherlands

Prepared for presentation at the 31st Annual Loss Prevention Meeting;
Houston, TX, March 9-13, 1997

November, 1996

The Explosion Blast Prediction Model in the Revised 'Yellow Book'

W.P.M. Mercx and A.C. van den Berg
TNO Prins Maurits Laboratory
P.O. Box 45, 2280 AA Rijswijk
The Netherlands

1 Introduction

More than fifteen years ago, the Dutch Ministry of Social Affairs published the first edition of a report called: 'Methods for the calculation of the physical effects of the escape of dangerous material (liquids and gases)' (CPR 14E, 1979). The publication contained a number of practical models to quantify the various effects following an accidental release of a dangerous liquid or gas. Models were given for various modes of release, for evaporation, heat radiation, dispersion, vapor cloud explosion (VCE) and rupture of vessels. The publication became wellknown by its nickname 'Yellow Book'. Several revisions were published during the years after its first publication.
It was decided in 1994 to make a complete new version of the Yellow Book. The subcommittee 'Risk Evaluation' of the Committee for the Prevention of Disasters was tasked with the job and TNO was assigned to evaluate and propose physical models and to prepare the draft text for the new publication.
The completely revised Yellow Book will be published in 1997.

One of the models that was completely revised in the new version is the vapor cloud explosion blast model. The old version contained two models: the so-called 'TNO model' (Wiekema, 1980) and a correlation model based on TNT equivalency. The new Yellow Book will contain the Multi-Energy model.

The Multi-Energy model has existed already for quite a long time (van den Berg, 1985). It is generally accepted as a better model for blast predictions for vapor cloud explosion than the TNT equivalence model. However, general application of the model was hampered due to a lack of guidance on a number of issues. Because of that, the model is more a concept than a straightforward method.
Still, the lack of guidance exists. Nevertheless, the committee has recognized the potential of the model and has accepted it for inclusion in the book. A procedure was included to apply the method step by step. Due to the present lack of guidance, this procedure applies the model in a conservative way.

The next chapters discuss the differences between the old 'TNO model' and the Multi-Energy model. The procedure to apply the Multi-Energy model is described. The motivation for the various steps in the procedure are discussed. This discussion reveals the items on which proper guidance for application is missing. Existing guidance will be evaluated. The discussion clearly identifies future research topics.

These topics have to be studied before full benefit of the potential of the Multi-Energy model will be obtained.

2 Yellow Book vapor cloud explosion blast models

In order to be able to discuss the merits of the vapor cloud explosion blast models, the models will be briefly described. As the correlation model in the old Yellow Book is in fact a TNT equivalence model, this model will be neglected.

The 'TNO model'

The 'TNO model' starts with the calculation of the combustion energy of the explosive mass content of a flammable cloud. The explosive mass is the part of the cloud of which the concentration of the fuel in air is in between the explosion (flammability) limits.

A hemispherical cloud with radius r_o (m) is postulated having a total combustion energy similar to the calculated combustion energy of the explosive mass and a combustion energy density E_c of 3.5 MJ/m^3. This combustion energy is an average value for the combustion energy density of stoichiometric mixtures of most hydrocarbons in air.

Then a 'characteristic explosion length' L_o (m) is defined according to:

$$L_o = (2/3 \times \pi \times r_0{}^3 \times E_c / p_0)^{1/3} \tag{1}$$

in which:

L_o: the characteristic explosion length (m);
r_o: radius of hemispherical cloud (m);
E_c: combustion energy density (MJ/m^3);
p_o: the ambient pressure (Pa).

In order to determine the absolute peak pressure p (Pa) or dimensionless peak overpressure PS and the duration of the positive phase of the shock wave t_+ (s) at a certain distance r (m) from the center of the hemisphere, a blast chart is given. This chart is reproduced in Figure 1. The positive phase duration is made dimensionless by L_o and c_o, the ambient velocity of sound (m/s).

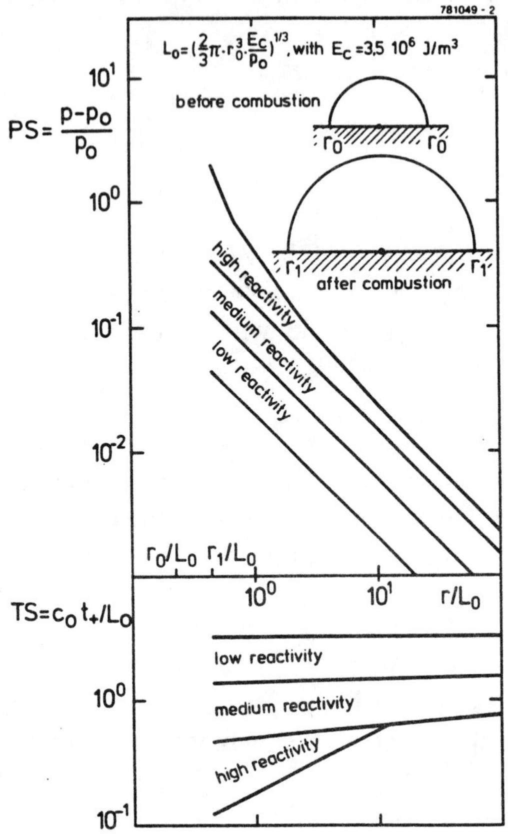

Figure 1: *Dimensionless peak overpressure PS and dimensionless positive phase duration TS for a vapor cloud explosion.*

The blast chart accounts for the effect of fuel reactivity on the blast characteristics. Three regions of reactivity are distinguished. A table is presented subdividing the most common gases into the three reactivity classes. Table 1 shows the classification of some of these gases.

Table 1: *Reactivity class of some fuels.*

Low reactivity:	Ammonia, Methane, Ethylchloride
Average reactivity:	Propane, Ethane, Ethene, n-Butane
High reactivity:	Ethylene oxide, Acetylene, Benzene

The blast chart account for the effect of flame accelerations by obstacles in congested regions by stating that the upper or lower boundary of a reactivity region should be chosen depending on whether the obstacle density is high or low.

The Multi-Energy model (ME model)
In order to start the application of the Multi-Energy model, the size and location of the flammable vapor cloud should be known as for the TNO model.
The principle difference between the Multi-Energy model and the TNO model however is that, while the TNO model assumes that all combustion energy present in the flammable cloud contributes to the explosion, the Multi-Energy model assumes that only part of the combustion energy present in the flammable cloud contributes to pressure generation in the explosion.
The part of the available combustion energy to be taken into account is the combustion energy of those parts of the cloud that are located in regions with a high obstacle density. In this way, the Multi-Energy models takes account of the basic mechanism of a gas explosion.

The basic mechanism is that flame acceleration occurs in regions with a high turbulence level prior to the arrival of the flame. The expansion flow ahead of the flame during the combustion process will create turbulence due to the interaction with obstacles present in the flow path. The flame will be accelerated to high speeds and accompanying overpressures. As a consequence, the absence of obstacles will reduce the flame speed to levels with no significant generation of overpressure.
More detailed descriptions of this basic mechanism can be found in many publications, for instance CCPS (1994). Experimental evidence is given, for instance, in Mercx et al. (1993).

If the volume of the flammable cloud entrapped in an obstructed area is known, the combustion energy E (J) contributing to the explosion can be calculated.
Blast charts are provided to calculate the blast characteristics as a function of distance from the explosion center. The charts are valid for a combustion energy density of 3.5 MJ/m^3. The charts for the determination of peak overpressure P_s (Pa) and the duration of the positive phase t_p (s) are given.
Figures 2a and 2b show the blast charts for overpressures and duration, respectively. Both the axis of each chart are made dimensionless according to:

$$P_s' = P_s / p_0, \qquad t_p' = \frac{t_p \times c_0}{(E/p_0)^{1/3}}, \qquad r' = \frac{r}{(E/p_0)^{1/3}} \tag{2}$$

in which:

P_s: peak side-on overpressure (Pa);
P_s': scaled peak side-on overpressure (-);
p_o: ambient pressure (Pa);
t_p: positive phase duration (s);
t_p': scaled positive phase duration (-);
c_o: ambient velocity of sound (m/s);
r: distance to explosion center (m);
r': scaled distance to explosion center (-);

E: available combustion energy (J).

Furthermore, the charts give an impression of the shape of the blast wave. This may be of interest when evaluating possible damage.

As in the 'TNO model', a number of lines are given in each graph. Each line starts at a certain overpressure at a distance r_o' which is the non-dimensionalized radius of a hemispherical flammable cloud with combustion energy E and a combustion energy density of 3.5 MJ/m^3.
Each line belongs to a certain 'source strength', which is an overpressure level expected to occur in the obstructed region. The source strength is influenced by the number, type and orientation of the obstacles present in obstructed region as well as of the reactivity of the fuel.

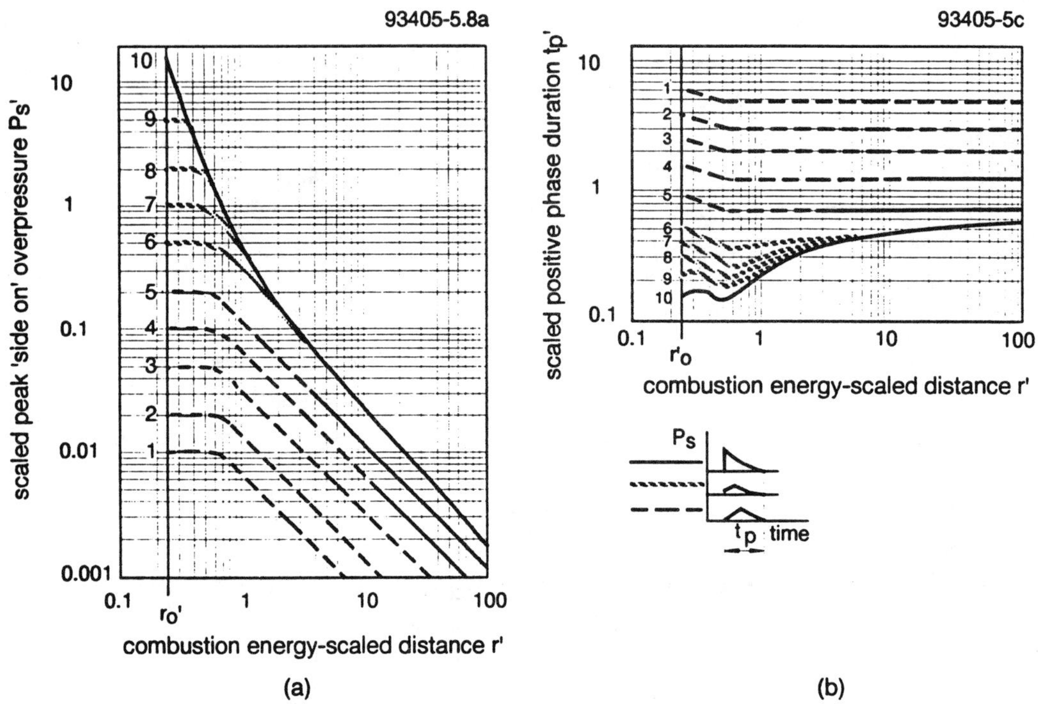

Figure 2: Blast charts for Multi-Energy model for a: peak overpressure and b: positive phase duration.

Comparing the two models

The principle difference between the TNO model and the Multi-Energy model has already been pointed out. The first model assumes that the combustion energy contributing to the explosion is the total combustion energy available in the flammable part of the cloud, while the latter assumes that only the combustion energy of the flammable cloud entrapped in congested regions contributes to the explosion.

Despite the fact that the assumption made in the Multi-Energy model is proven to be in line with gas explosion physics, it introduces a number of questions which would not arise when applying the TNO model.

- *How does one define an obstructed region ?*
- *How is the presence of several obstructed regions in the cloud dealt with? When does one have to consider them as independent relatively small explosion sources and when as a single relatively large explosion source?*
- *In case of several independent explosion sources, how does one with multiple blast waves?*

The blast charts in the TNO model clearly separate the influence of the reactivity of the fuel and the influence of obstacle obstruction on the explosion source strength. Most weight is given to the influence of the reactivity. As a result of this separation, the maximum and minimum source overpressure is limited. For instance, a methane explosion will never lead to a pressure at the edge of the cloud higher than 15 kPa (upper boundary of low reactivity region).
This may be very helpful in reducing consequences but clearly it is not true at all. It has been demonstrated in various large-scale experimental programmes (Mercx, 1994 a, b, 1996) that a methane explosion in a highly obstructed region may produce higher overpressures than, for instance, an ethylene explosion in a less obstructed region.

The consequence of the TNO model is also that when a very large flammable cloud can be produced in a non-obstructed region, considerable overpressures may still be obtained at very long distances from the explosion center. An event where this can lead to an incredible situation is the collision of an LNG tanker in the estuary of a river just before entering a harbor. An enormous flammable cloud may then develop. Situated on the sea it will hardly produce overpressure upon ignition: a flash fire will be the result. However, applying the TNO model will result in blast waves of very long duration which load hazardous industrial installations situated near the shore. (The case of a collision between a tanker loaded with a hazardous substance with another one is a compulsory case in the risk analysis of hazardous process installations in the estuary of the river Westerschelde in the Netherlands.)

The influence of reactivity and obstructed region on the source overpressure is not separated in the blast charts of the Multi-Energy model. The model recognizes the fact that a low reactivity gas may produce higher overpressures than a high reactivity gas, depending on the nature of the obstructed region.
The difficulty now becomes the choice for the source strength, i.e. the overpressure in the explosion, or, which class to pick from the available ten classes. Due to the fact that the Multi-Energy model gives more weight to the influence of the obstructed region than to reactivity compared to the TNO model, the choice for the source strength becomes more subjective than the choice for the initial starting overpressure in the TNO model.
The questions most often asked are then as follows.

- *How can I decide on which source strength to choose?*
- *Is it possible to separate the influence of reactivity from the influence of the obstructed region and how can each influence be quantified?*

Before describing how the Multi-Energy model was implemented in the new Yellow Book and how the questions relating to the application of it were avoided, by-passed, or solved, currently available guidance and ongoing research to develop guidance for the application of the Multi-Energy model will be described and discussed first.

3 Available guidance for the application of the Multi-Energy model

The most frequently posed question is which explosion class to choose for entering the blast charts of the ME model. Besides the guidance TNO presented on this aspect (CCPS, 1994), other investigators developed their own guidance for the ME model (Kinsella, 1993) or guidance for their own model based on a similar philosophy as the ME model (Cates, 1991 and Baker et al., 1994). The guidance is presented briefly first,. then some comparisons are made with experimental data.

CCPS, 1994

The most conservative choice for the explosion source strength for an obstructed region is to assume class 10. However, the choice of class 7 seems to better represent actual experience.

The unobstructed parts of the cloud should be modeled with class 1. If initial turbulence is present or turbulence is created due to a jet release, class 3 could be taken.

A better motivated choice may be obtained by comparison of the actual case with the vast amount of experimental data which is available in literature.

An alternative is to use a CFD code capable of simulating the explosion process, such as FLACS (van Wingerden et al., 1993), AutoReaGas (van den Berg et al., 1995) or EXSIM (Hjertager, 1991), to come up with a value for the overpressure in the explosion.

Kinsella, 1993

Kinsella argues that, for a stoichiometric flammable cloud, there are three factors influencing the severity of a vapor cloud explosion:

- degree of congestion (obstruction) by obstacles present in the cloud;
- ignition energy;
- degree of confinement.

Based on an investigation of experimental data, a table has been composed where these three factors are combined into 12 categories. A small region of ME model classes to choose is given for each category (Table 2). The range of source overpressure is given too.

Table 2: Initial blast strength index according to Kinsella (1993).

Blast strength category	Low energy ignition	High energy ignition	High conges-tion	Low conges-tion	No conges-tion	Parallel plane confinement	Uncon-fined	ME-model class	Source overpres-sure (kPa)
1		X	X			X		7-10	100->1000
2		X	X				X	7-10	100->1000
3			X			X		5-7	20-100
4	X	X		X		X		4-6	10-50
5		X		X			X	4-6	10-50
6		X			X	X		4-5	10-20
7	X		X				X	4-5	10-20
8		X			X		X	4-5	10-20
9	X			X		X		3-5	5-20
10	X			X			X	2-3	2-5
11	X				X	X		1-2	1-2
12	X				X		X	1	1

Baker et al. 1994

A method is presented based on the same principles as the ME model. Similar type blast charts are given, however now based on the maximum flame speed to be expected in the exploding obstructed parts of the flammable cloud.

A matrix is presented to come to the maximum flame speed in a specific case taking account of:

- the divergence of flame expansion;
- the obstacle density;
- the fuel reactivity.

The classification of fuels into reactivity categories is taken equal to the classification in the TNO model. The blast charts need the maximum flame speed expressed in a Mach number to choose the starting pressure for the blast determination as a function of distance. Table 3 presents the matrix. The derived starting pressures are given in brackets.

Table 3: Flame speed in Mach numbers and overpressures in kPa (in brackets) for soft ignition sources.

1D flame expansion		Obstacle density		
		high	medium	low
	high	5.2 (>1000)	5.2 (>1000)	5.2 (>1000)
reactivity	medium	2.265 (>800)	1.765 (400)	1.029 (200)
	low	2.265 (>800)	1.029 (200)	0.294 (40)
2D flame expansion		**Obstacle density**		
		high	medium	low
	high	1.765 (400)	1.029 (200)	0.588 (100)
reactivity	medium	1.235 (200)	0.662 (100)	0.118 (8)
	low	0.662 (100)	0.471 (80)	0.079 (4)
3D flame expansion		**Obstacle density**		
		high	medium	low
	high	0.588 (100)	0.153 (15)	0.071 (4)
reactivity	medium	0.206 (20)	0.100 (8)	0.037 (1)
	low	0.147 (15)	0.100 (8)	0.037 (1)

Baker et al. (1994) state that, although the values given in the matrix are questionable because of lack of sufficient appropriate data, the matrix itself is a good framework to incorporate improved values when these become available from future research.

Comparison of guidance with experimental data

Most experimental research on the influence of obstacles on the explosion overpressure were rather designed in an attempt to understand the governing physical processes than to quantify the effects systematically. One of the objectives of the MERGE and EMERGE projects (Mercx, 1994a and 1996) was to obtain a comprehensive and consistent database on the influence of obstacles.
Mercx et al. (1994 b) reports on these experiments.

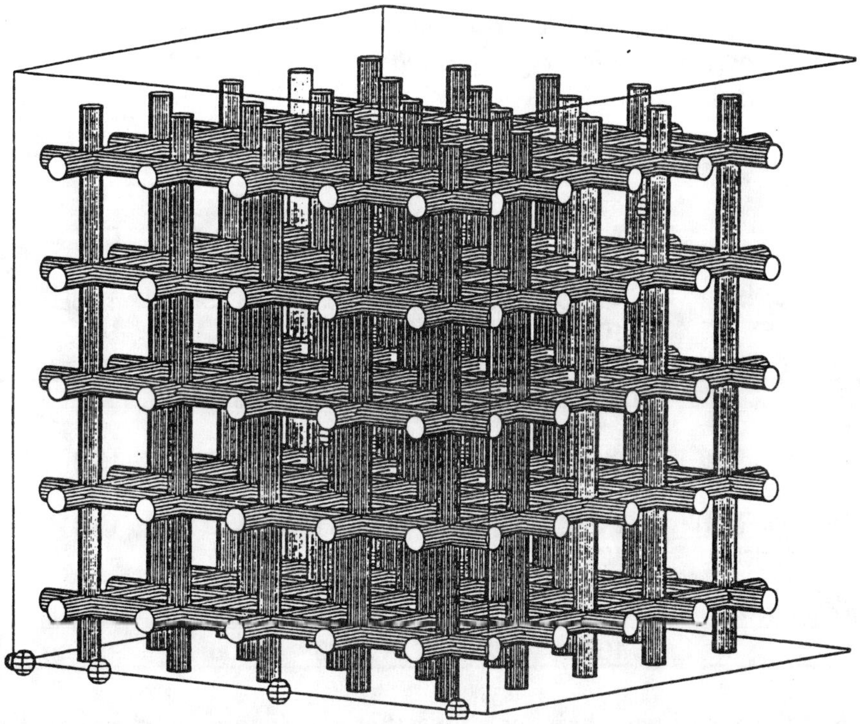

Figure 3: Typical obstacle configuration used in the MERGE and EMERGE project.

The obstacle array consisted of cylindrical tubes orientated in three opposite directions (Figure 3). Diameter of the pipes (D), heart-to-heart distance between the pipes (Pitch) and number of pipes in a plane (n) were varied. Volume blockage (VB), the ratio between total obstacle volume and total array volume was 10 and 20%. The ratio of the width and depth of the obstacle array was always one. The height was half the width. Various stoichiometric mixtures of fuel in air were used. Ignition was in the center at ground level.

Table 4 gives the average maximum overpressures measured in the so-called medium- and large-scale tests. The oxygen enrichment was to simulate a large-scale test at medium-scale such that the medium-scale test would produce similar overpressures.

Table 4 *Medium- and large-scale test results of some MERGE experiments, approximate overall array dimensions: 4 x 4 x 2 m^3 (medium-scale; type A,B,C and D) 8 x 8 x 4 m^3 (large-scale; type E, C*). Oxygen concentration in oxygen-enriched test was 22.5% prior to fuel addition.*

	Obstacle configuration					
Type	A	B	C	D	E	C*
D (mm)	43	41	86	82	168	82
Pitch (mm)	4.65D	3.25D	4.65D	3.25D	4.65D	4.65D
VB (%)	10	20	10	20	10	10
n (-)	20	30	10	15	10	20
	Average maximum overpressures (kPa)					
Fuel:						
methane	66.2	160	10.5	70.2	13.8	86.3
methane/O_2 enriched	111	-	20.8	134	-	-
methane/propane (3:1)	91.2	408	14.1	85.2	16.9	104
meth/prop O_2 enriched	123	-	18.6	196	-	-
propane	128	288	19.2	144	22.7	240
propane/O_2 enriched	236	-	23.9	260	-	-
ethene	1075	1330	37.5	306	-	-

Evaluation of the experimental results shows that a wide variety of overpressures was obtained, although the configurations appeared to be quite similar at first glance. The reactivity plays a role as well as does the obstacle configuration. The obstacle configuration cannot be characterised by volume blockage only. The number of obstacles the flame has to pass (= 1/2 x n) as well as the length of the flame path (= 1/2 x n x Pitch) should be incorporated.

Applying the available guidance to predict a maximum source overpressure will lead to an underprediction in most cases. ME model class seven (100 kPa), like Kinsella's blast strength category 3 (low energy ignition, high congestion, unconfined, 20-100 kPa), predict too low overpressures in most cases. Baker's prediction (assuming VB of 10% is a high obstacle density) is 15 kPa for methane, 20 kPa for propane and 20 kPa for ethene. These values are acceptable for the type C and E obstacle configuration, but are too low for all others.

Probably, the volume blockage ratios in most realistic industrial situations will be lower than 10%. This would give more credit to the predictions. On the other hand, the geometrical scale in realistic situations might be considerably larger. Nevertheless, none of the guidance is capable of predicting the wide variety of overpressures obtained in the experiments.

The presently available guidance addresses only one of the various questions posed in Chapter 2. Obviously the ME-model should be applied in a conservative way. Special attention should be given if the model is to be implemented in a new version of the Yellow Book as then a widespread use of the model can be expected.

4 The Multi-Energy model in the Yellow Book

The user of the Yellow Book is taken through a step-by-step procedure to apply the ME model. Table 5 gives an overview of the consecutive steps to be taken.

Table 5: Step-by-step procedure for ME model application according to Yellow Book.

Step 1:	Application constraints
Step 2:	Determine cloud size
Step 3:	Recognize potential blast sources
Step 4:	Define obstructed region(s) by estimating its boundaries or by using procedure.
	Procedure: step 4a: break-down structures into basic geometrical shapes
	step 4b: assume an ignition location
	step 4c: determine obstacle orientation
	step 4d: build-up of obstructed region
	step 4e: define a box containing obstructed region
	step 4f: subdivision into multiple boxes
	step 4g: define additional obstructed regions if appropriate
Step 5:	Estimate the source strength or class number for each region
Step 6:	Combination of obstructed regions
Step 7:	Location of unobstructed part of vapor cloud
Step 8:	Calculate radius
Step 9:	Calculate blast parameters
Step 10:	Handling of multiple obstructed regions
Step 11:	Construct blast history at specific location

It is beyond the scope of this publication to discuss each step in the procedure in detail; therefore, only some important steps are addressed here.

<u>Step 1</u>

Before starting to use the ME model one should realize that:

- the model is a simple model for blast predictions from unconfined vapor cloud explosions. Vented internal explosions should be dealt with in another manner;
- like all practical and therefore simple models, the ME model assumes point symmetry. Directional effects which will affect the blast close to the source are **neglected**. Siting of control rooms and other buildings close to the hazardous area should not be based on simple modeling;
- detonations are assumed not to occur. This means that detonation of the **whole** cloud is excluded. Detonative blast strength may be reached in highly obstructed regions.

Step 4
The obstructed region should be determined by either estimating its boundaries or by applying a detailed procedure. The detailed separate procedure is provided to determine the boundaries of an obstructed region. The procedure is intended to be used for a totally inexperienced applier of the ME model in order to avoid that he might define, for instance, the regular pattern of lampposts along a street as a obstructed region. After having applied the procedure a couple of times, one finds out that estimating the boundaries of an obstructed region is the preferable option.

Step 5
If a lower class number cannot be motivated (neither by computer codes nor by experimental evidence), a class number of 10 is recommended.
The discussion in the Committee on this item was quite elaborate and extensive. The decision to opt for a high number was based on the following.
The source strength or class number depends on both the influence of obstacle configuration and on reactivity. Recent research shows that the influence of reactivity can be quantified (Mercx et al., 1994b). Based on the fractal scaling theory developed by Shell (Taylor and Hirst, 1988), the ratios of overpressures P_2/P_1 for two different reactive mixtures (mixture 1 and 2) in a similar obstacle configuration can be derived to be:

$$P_2 / P_1 = \left(\frac{(\alpha . u_l)_2}{(\alpha . u_l)_1} \right)^{2.7} \qquad (3)$$

in which:
P overpressure in mixture 1 or 2 (Pa);
α expansion ratio of mixture 1 or 2 (-);
u_l laminar burning velocity of mixture 1 or 2 (m/s).

The equation could be used for the ME model if a reference value for a maximum overpressure of a specific gas is chosen. For instance, if it could be agreed that the maximum overpressure in **any** unconfined vapor cloud explosion of methane cannot exceed a certain limit, then all other fuels could be related to that overpressure with the equation (3). In this way, the influence of the obstacle configuration is assumed to be maximised in any case. The discussion is then how to decide on the limit. A British Gas publication (Harris and Wickens, 1989) states that overpressures of methane explosions up to four bars were found in their experiments. Using a 4 bar limit for methane implies an ME model class of almost 9. With a class 9 for the lowest reactivity, additional scaling for the reactivity makes no sense. This was the reason for the Commission to accept a class number 10.

Step 6
It may be possible that more than one obstructed region can be defined within a flammable cloud. There is presently no guidance at all on separation distances, i.e. the distance between two obstructed regions at which the explosion is separated into two separate smaller explosions instead of a single large one. Therefore, the Yellow Book neglects this option of the ME model to reduce blast. A small reduction however is allowed: if the distance between two obstructed regions is 25 m or more, than the combustion energy of this unobstructed region can be neglected for the determination of the explosion energy.
The value of 25m is very arbitrary. A possible way to arrive at a separation distance could be as follows. Due to expansion of the exploding obstructed part of the cloud, the radius of this burnt part of the cloud is twice the original one. If it can be assumed that the flame speed has been reduced to the laminar speed upon arrival at a distance of twice the original radius, then the radius of the obstructed region where the explosion started can be used as a first assumption for a value for the separation distance. Experimental research should be carried out to find out whether or not this supposition is meaningful.

Step 11
This step is intended to combine the blast of the obstructed part of the cloud with the unobstructed part. The unobstructed part will produce a much lower overpressure than the obstructed part. However, if the volume of the unobstructed part is relatively large, its positive phase duration may be very long; then it may influence the damage consequences. Therefore it is suggested to calculate the blast characteristics of both the obstructed and unobstructed region and to simply superpose both pressure histories.
Procedures given in the so-called 'Green Book' (CPR16E, 1989) to calculate the consequences of a blast wave, are capable of taking the resulting pressure history into account.

5 Ongoing research and concluding remarks

Being not satisfied with the present amount of guidance on the choice for an ME model class number, twelve European companies decided to sponsor a project called GAME, an acronym for 'Guidance for the Application of the Multi-Energy model'. In this project, a combination of parameters was devised by which the conditions, which largely determine the development of a gas explosions in obstructed environments, can be characterized (van den Berg and Eggen, 1996). The parameter combination, which should be applicable to realistic situations, was correlated to explosion overpressures available from experimental data.

The general approximate relation can be expressed as:

$$P_{max} \propto \left(\frac{VB \times L_f}{D}\right)^b \times \quad u_l^{\,2.7} \times S_c^{\,0.7} \qquad (4)$$

where:

P_{max}: maximum explosion overpressure (Pa);
VB: volume blockage ratio (-);
D: average obstacle size (m);
L_f: maximum flame path length (m);
S_c: scale factor (m);

u_l: laminar burning velocity (m/s);
b: constant which determines best fit correlations.

Figure 4 shows the correlation compared with overpressure results from several experimental programs.
If it appears possible to apply the correlation to a realistic situation, then the choice for an ME model class number is based on sound evidence. The main objective of the ongoing follow-up project GAMES (GAME Second phase) is exactly to investigate the practical applicability of the correlations derived.

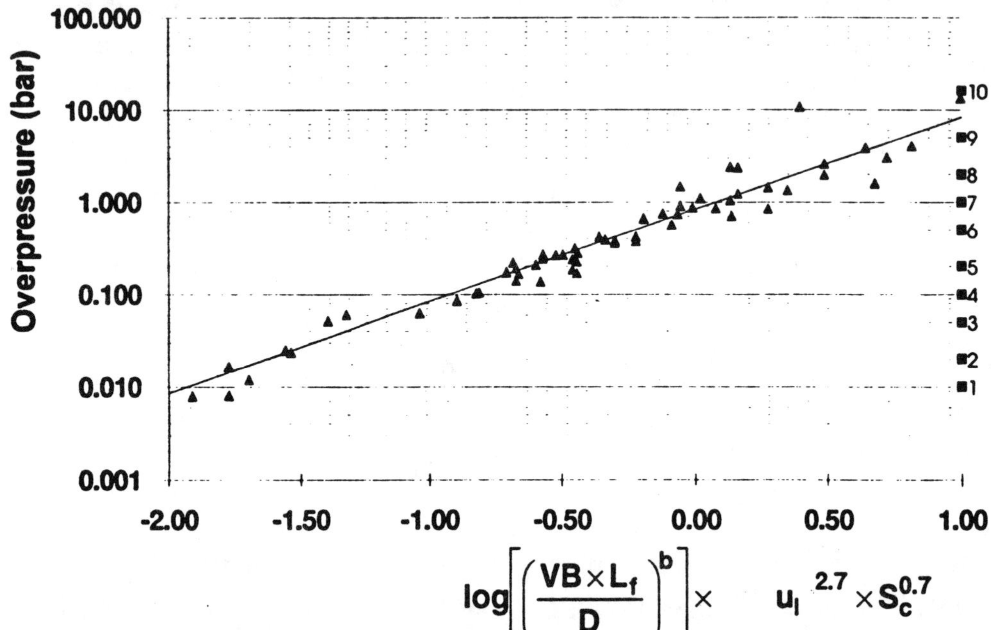

Figure 4: *Overpressures measured in MERGE tests (Mercx, 1994 a) versus correlation according to equation 4.*

It will be clear that more research is required to provide answers to the questions related to the application of the Multi-Energy and similar models. The most important problem to solve is on the determination of the separation distances. Then, it will be possible to use the full potential of the Multi-Energy model. It will then be possible to distinguish multiple explosion sources which, if they can be considered as separate, may lead to a reduction in the blast overpressures values predicted by simple modeling.

The Committee responsible for the new Yellow Book recognized the potential of the ME model to improve simple practical explosion blast predictions and accepted the model for inclusion despite the present lack of guidance for application. Application of the procedure adopted in the Yellow Book will reveal whether or not the adopted approach is too conservative. Hopefully, it will also reveal the potential of the model to the users, thereby generating interest to participate in research required to fully exploit the possibilities of the model.

References

Baker, Q.A., Tang, M.J. and Scheier, E. (1994), 'Vapor Cloud Explosion Analysis', Procs 28th Loss Prevention Symposium AIChE, Atlanta, Georgia, USA April 1994.

Berg, A.C. (1985),'The Multi-Energy method - a framework for vapour cloud explosion blast prediction', Journal of Hazardous Materials, Vol. 12, 1-10.

Berg, A.C. van den and Eggen, J.B.M.M. (1996),'GAME: Guidance for the Application of the Multi-Energy method', 2nd Int. Specialists Meeting on Fuel-Air Explosions, Bergen, Norway, June, 1996.

Berg, A.C. van den, Mercx, W.P.M., Absil, L.H.J., Mouilleau, Y. and Hayhurst, C.J. (1995),'Simulation of Vapour Cloud Explosion and Subsequent Blast Propagation in Build-up Areas', Proceedings of the International Gas Research Conference, November 1995, Cannes, France.

Cates, A.T. (1991),'Fuel Gas Explosion Guidelines', Proc. Conf. on Fire and Explosion Hazards, Moreton-in-Marsh, UK, April 1991.

CCPS (1994), 'Guidelines for Evaluating the Characteristics of Vapor Cloud Explosions, Flash Fires and BLEVE's', Centre for Chemical Process Safety, AIChE, New York, USA.

CPR 14E (1979), 'Methods for the calculation of the physical effects of the escape of dangerous materials', Commission for the Prevention of Disasters, Dutch Ministry of Social Affairs, Directorate-General of Labour, Voorburg, The Netherlands, edition 1979.

CPR 16E (1989), 'Methods for the determination of possible damage to people and objects resulting from releases of hazardous materials', Commission for the Prevention of Disasters, Dutch Ministry of Social Affairs and Labour, Voorburg, The Netherlands 1989.

Harris, R.J. and Wickens M.J. (1989),'Understanding vapour cloud explosions, an experimental study', 55th Autumn meeting of the Institution of Gas Engineers, Kensington, UK.

Hjertager, B.H. (1991),'Explosions in Offshore Modules', IChemE Symposium Series No. 124, 19-35.

Kinsella, K.G. (1993),'A rapid assessment methodology for the prediction of vapour cloud explosion overpressure', Proc. Int. Conf. on Safety, Health and Loss Prevention in the Oil, Chemical and Process Industries, Singapore, February 1993.

Mercx, W.P.M., Wees, R.M.M. van and Opschoor, G. (1993) 'Current Research at TNO on Vapour Cloud Explosion Modelling', proceedings of the 27th Annual AIChE Loss Prevention Symposium, March 29-April 1, 1993, Houston, TX, USA.

Mercx, W.P.M. (ed.), (1994a),'Modelling and Experimental Research into Gas Explosions', Overall Final Report of the MERGE project, EUcontract: STEP-CT-0111(SSMA), EU-DG-XII, Brussels, December 1994.

Mercx, W.P.M., Johnson, D.M. and Puttock, J. (1994b), 'Validation of scaling techniques for experimental vapour cloud explosion investigations', Proc. AIChE Loss Prevention Symposium, Atlanta, Georgia, USA, April 1994.

Mercx, W.P.M. (ed), (1996),' Extended Modelling and Experimental Research into Gas Explosion', Summary Report for the project EMERGE, EUcontract: EV5V-CT93-0274, EU-DG-XII, Brussels, in preparation (1996).

Taylor, P.H. and Hirst, W.J.S. (1988), 'The scaling of vapour cloud explosions: a fractal model for size and fuel type', proc of 22nd Int. Symp. on Combustion, Combustion Institute.

Wiekema, B.J. (1980),'Vapour cloud explosion model', Journal of Hazardous Materials, Vol.3, 221-232.

Wingerden, van C., Arntzen, B.J., Storvik, I.E. (1993), 'FLACS-93, a New Explosion Simulator', Proceedings of the ERA conference on Offshore Structural Design Against Extreme Loads', London, UK, November 1993.

Recent Developments in the Baker-Strehlow VCE Analysis Methodology

by

Quentin A. Baker
Craig M. Doolittle
Gary A. Fitzgerald
Ming Jun Tang

Wilfred Baker Engineering, Inc.
8700 Crownhill, Suite 310
San Antonio, TX 78209

Paper No. 42f

31st Loss Prevention Symposium
American Institute of Chemical Engineers
Houston, TX

March 9-13, 1997

Abstract

The Baker-Strehlow methodology was developed to provide an objective approach to prediction of blast pressures from vapor cloud explosions. The complete methodology was first published in 1994. Since then, it has evolved through ongoing research and use in VCE hazard analyses, facility siting studies and accident investigations. This paper presents recent developments in the Baker-Strehlow methodology. The use of the flame speed correlation is reviewed and revisions to the methodology are presented. Methods for handling multiple confined/congested zones within a vapor cloud and the ramifications of multiple blast waves on structures are discussed. The effects of varying degrees of confinement and congestion on VCE blast predictions are also addressed. Lastly, a procedure to determine fuel reactivity for mixtures is presented.

Introduction

The purpose of this paper is to present an update to the Baker-Strehlow method of predicting blast effects from vapor cloud explosions (VCEs). The complete method was first published at the 28th Loss Prevention Symposium in 1994[1] shortly after the development of a correlation for determining maximum flame speed in a VCE. Since that time, the Baker-Strehlow method has been used extensively in VCE hazard assessments in refineries and chemical plants. As expected, many practical lessons have been learned during the course of the hazard assessments, and the Baker-Strehlow method has evolved as a result. The changes have been evolutionary, not revolutionary. In keeping with the goals of the original study in which the methodology was developed, all changes have been incorporated with the intent of achieving an objective methodology to provide consistent prediction of VCE blast effects.

A brief review of the Baker-Strehlow method is in order to set the stage for subsequent discussions about revisions to the methodology. The approach adopted was a combination of the Strehlow and TNO methods. Strehlow's spherical model was chosen because blast curves are selected based on flame speed, which affords the opportunity to use empirical data in the selection The pressure and impulse curves developed by Strehlow[2] are shown in Figures 1 and 2.

The procedures from the Multi-Energy method[3] were adopted for determination of the energy term. Specifically, confinement and/or congestion is the basis of the determination for the size of the flammable vapor cloud that contributes to the generation of blast overpressure, and multiple blast sources can emanate from a single release. Fuel reactivity, confinement and obstacle density influence the reaction rate.

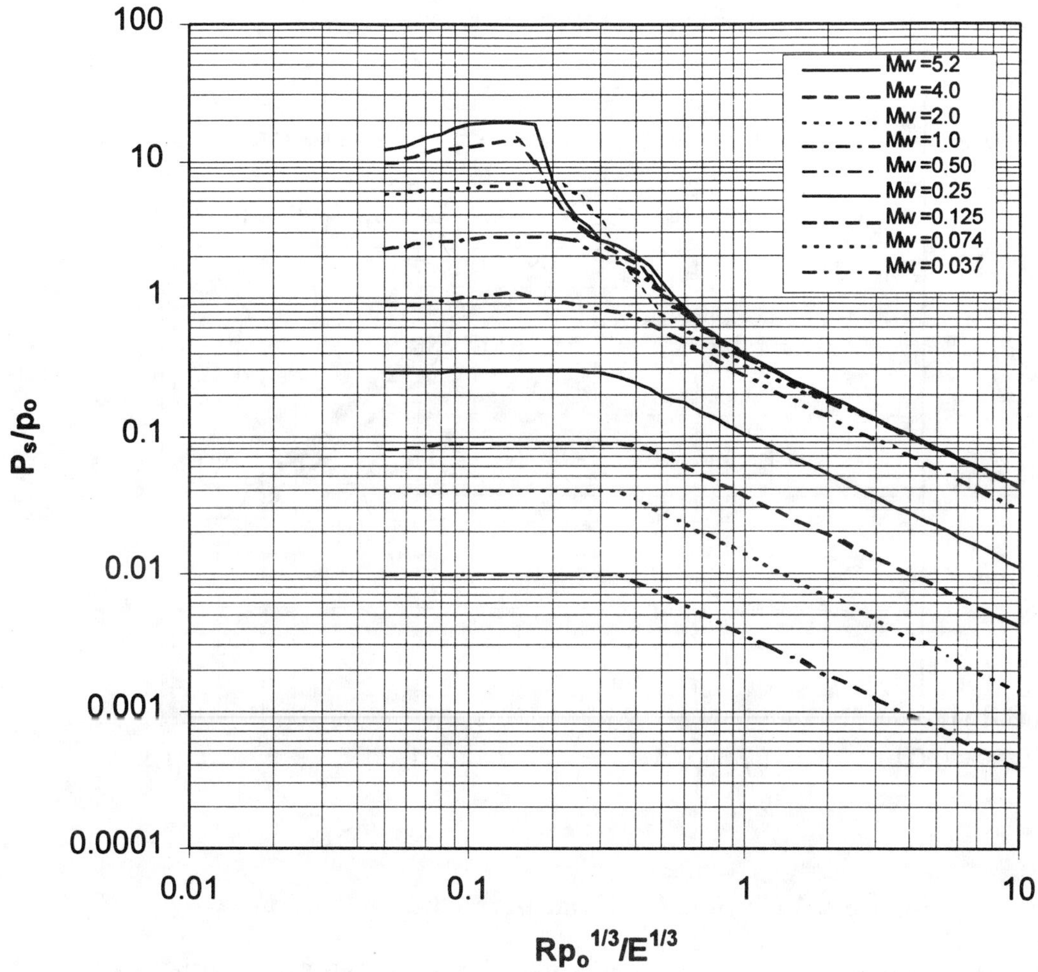

Figure 1. VCE Side-On Hemispherical Overpressure Curves

Literature was reviewed to compile experimental data as the basis for an objective determination of flame speed to be used in selecting a blast curve from Figures 1 and 2. It was found that the combined effects of fuel reactivity, obstacle density, and confinement could be correlated to the flame speed. Table 1 relates flame speed to 27 combinations of fuel reactivity, confinement (flame expansion), and obstacle density. Shown in each cell of Table 1 is flame speed Mach number based on the maximum flame speed found in literature.

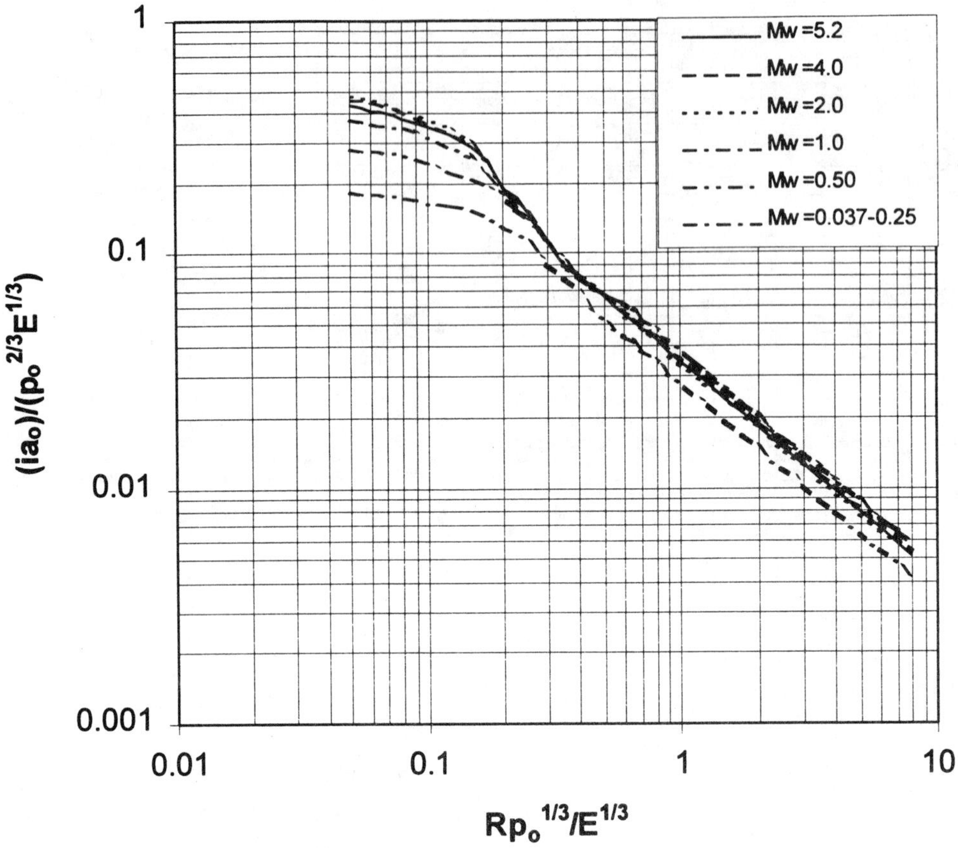

Figure 2. VCE Side-On Hemisperhical Impulse Curves

Table 1. Flame Speed in Mach Numbers (M_w) for Soft Ignition Sources

FLAME EXPANSION	FUEL REACTIVITY	OBSTACLE DENSITY		
		High	**Medium**	**Low**
1D	High	5.2	5.2	5.2
	Medium	2.27	1.77	1.03
	Low	2.27	1.03	0.294
2D	High	1.77	1.03	0.588
	Medium	1.24	0.662	0.118
	Low	0.662	0.471	0.079
3D	High	0.588	0.153	0.071
	Medium	0.206	0.100	0.037
	Low	0.147	0.100	0.037

The revisions to the Baker-Strehlow method resulting from experience gained during plant walk-downs and hazard assessments include:

- Systematic identification of areas of confinement and congestion in which a VCE could occur. These areas are termed "potential explosion sites" or "PESs."
- Selection of the level of confinement for a PES when zones of both 2D and 3D flame expansion are present within the PES.
- Deciding between 2D and 3D flame expansion when a source of 2D confinement is well elevated above ground level.
- Selecting the reactivity for a fuels that is a mixture of fuels with differing reactivities.
- Predicting blast loads on buildings when there are multiple PESs and multiple ignition sources within one vapor cloud.

The following sections address each of the above topics to provide additional guidance for the use of the Baker-Strehlow method.

Potential Explosion Sites (PESs)

A vapor cloud explosion involves a flame which moves rapidly through a fuel-air mixture. Locations of congestion and/or confinement in combination with flammable fuel vapors are conducive to developing high flame speeds which result in explosions capable of developing damaging blast overpressures. The vast majority of vapor cloud explosions producing damaging pressure waves are deflagrations, for which the flame speed does not exceed the speed of sound in the unreacted fuel/air mixture ahead of the flame front. Detonations (supersonic flame speeds) are also possible but much less likely. Research to date has shown that a flame will transition from a deflagration to a detonation only in cases where the fuel/air mixture is contained in a long pipe or enclosed culvert, in extremely large scale events with near homogeneous, stoichiometric mixtures, or in large scale, extreme congestion with a stoichiometric mixture[4]. These circumstances are typically not found in the petro-chemical industry, so only deflagrations are usually considered. The remainder of this discussion will refer to deflagrations unless otherwise stated.

A PES is defined as an area of congestion and/or confinement into which a flammable vapor cloud is introduced and ignited. Each PES is categorized based on the type of confinement and the amount of congestion present according to the categories specified by Baker et al.[1] to determine the explosion size (energy) and severity (maximum flame speed). A PES may contain several zones with different congestion and confinement categories. In this case, the zones are averaged into a single area of congestion and confinement. The maximum flame speed is then selected using the method developed by Baker et al.[1] based on the fuel reactivity, degree of confinement, and degree of congestion.

The confinement at each zone is determined by the presence of physical surfaces which significantly limit expansion of the flame front. The confinement is determined by the ability of the flame front to expand in 1, 2, or 3 dimensions. A zone is considered 3-

dimensional if the flame is free to expand in all directions, 2-dimensional if the flame is free to expand in two dimensions but restricted in the third dimension (i.e. an area with open sides and a solid deck above), and 1-dimensional if the flame front is allowed to expand only in one dimension (i.e. a pipeline or long culvert). The congestion is determined by obstacle density created by pipes, process equipment, structural members, and other physical obstacles. Congestion is evaluated as low, medium, or high. Fuel reactivity is also evaluated as low, medium, or high. Typically, the only high reactivity fuels encountered at petro-chemical sites are hydrogen, acetylene, ethylene, ethylene oxide, propylene oxide, and mixtures containing these materials. Methane, ammonia, and carbon monoxide are examples of low reactivity fuels found in industry. The majority of other fuels are medium reactivity. A detailed method of determining the fuel reactivity for single fuels and mixtures is presented in a later section.

It is important to note that during a deflagration, flames accelerate while traveling through confined and/or congested zones and decelerate after exiting such a zone. Damaging blast pressures are generated in the confined/congested areas, whereas a fireball with negligible pressure is produced in an unconfined/uncongested area. As a result, only the energy that is released by fuel within confined/congested zones is taken as the explosion energy.

Some zones for explosion scenarios are either adjoining another or very close to another (within about 5 m [16 feet]) such that the zones for an individual scenario are evaluated as one composite PES. This can be performed by determining an average confinement/ congestion category for the entire area. This average can either be based on engineering judgment or based on area volumes. Other zones are separated sufficiently (greater than 5 m [16 feet]) such that they can be considered as separate PESs. In this case each area is evaluated separately, and usually the highest pressure among the PESs is taken as the controlling pressure and the impulses are summed. An explanation of this and other methods is provided later in this paper.

Blast loads are a strong function of both explosion energy and flame speed. The energy term for each PES is calculated based on the volume of flammable fuel/air mixture trapped within the confined/congested zones of that PES as determined by a dispersion model. Figure 3 is an example of a lower flammability limit (LFL) isopleth from a dispersion model overlayed on a unit plot plan to determine the trapped volume. The mass of fuel (used in the energy calculation) within the PES is based on a stoichiometric mixture of fuel and air. The heat of combustion for the fuel is then used to calculate the resultant energy released. The resultant energy and flame speed is then used to calculate the VCE free-field overpressures and impulses using the free-field blast curves developed by Professor R. A. Strehlow.

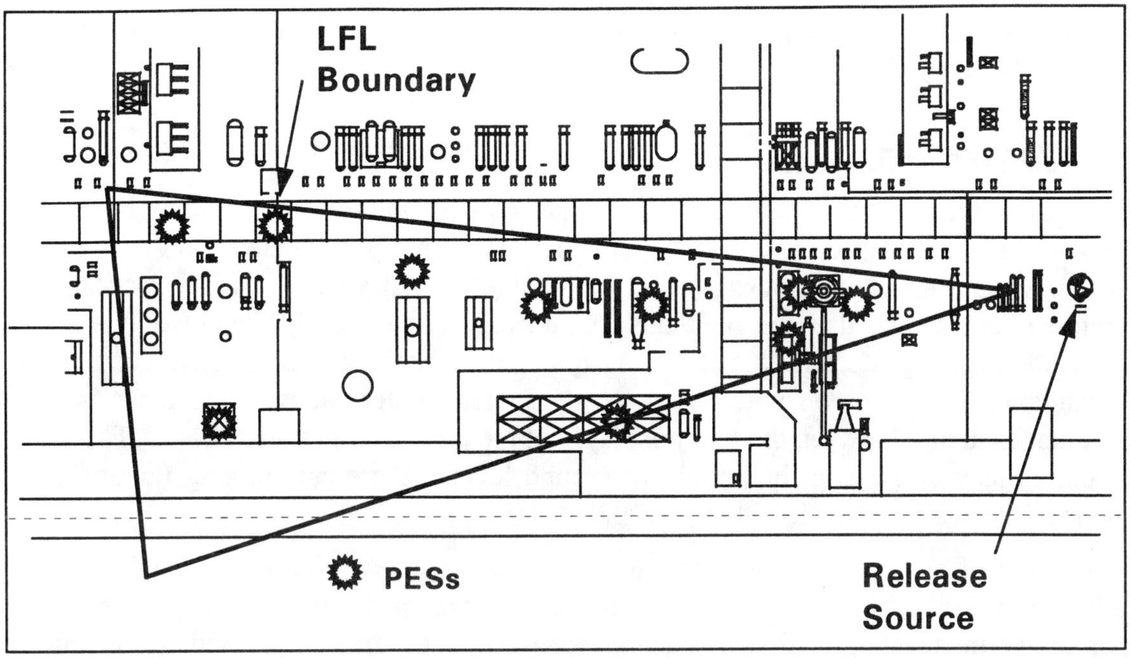

Figure 3 - Vapor Cloud Lower Flamability Limit Footprint

Mixed Zones of 2D and 3D Flame Expansion

Many process units have varying levels of confinement and congestion within the unit. Ultimately, a single confinement and congestion must be selected for each PES that best represents the entire unit. This selection proved to be particularly troublesome during hazard assessments when the selection was between 2D and 3D flame expansion. As can be seen in Table 1, there is a large difference in the predicted maximum flame speed of 2D and 3D flame expansion for similar levels of congestion and fuel reactivity. For example, if a unit is generally medium obstacle density and the fuel is medium reactivity, the difference is flame speed between 2D and 3D expansion will be Mach 0.662 versus 0.100 (See Table 1). The difference in predicted overpressures is considerable, with the maximum overpressure (inside the burning cloud) being 1.7 atm versus 0.06 for the 2D and 3D cases, respectively. The user faces the dilemma in making a selection that is either overly conservative or unconservative, with no middle ground.

The solution to this problem was to interpolate between the 2D and 3D expansion cases in Table 1, resulting in an intermediate flame speed. This intermediate level of confinement is referred to as 2½D to distinguish it as a hybrid of 2D and 3D confinement. For the above example, the flame speed for the 2½D case is the average of Mach 0.662 and 0.100, or Mach 0.381. This approach can be justified based on experiments in which flame acceleration through groups of obstacles and deceleration between rows of obstacles or entire congested zones has been documented[5,6]. It is reasonable to conclude the flame speed for confined/congested zone that is half 2D and half 3D flame expansion will be between the values that would be achieved if this same zone were entirely 2D or 3D. Experimental data is lacking to predict actual flame speeds for various configurations of 2D and 3D flame expansion (i.e., flame starting in a 3D area and propagating into a 2D area, or vice-versa), so a simple average between 2D and 3D values in Table 1 is recommended at this time.

Elevated 2D Confinement

Sources of confinement that create 2D flame expansion include solid decks on multi-level process units, roofs on process buildings, an extremely densely populated level of a pipe rack (i.e., greater than 80% blockage), and fin fan coolers. Experience with denser-than-air releases has shown that the flammable cloud may not reach the height of the 2D confinement, leaving room for the cloud to expand 3-dimensionally for a period of time. For example, it is common practice to locate fin fan coolers above pipe racks with significant clearance between the top-most layer of the pipe rack and the coolers. The fin fan coolers may be 10 m or higher above ground level. A ground-hugging flammable cloud may reach only a fraction of this height.

Investigation of elevated 2D confinement revealed two issues: flame acceleration and shock reflections. The issue concerning flame acceleration is whether the expanding vapor cloud in a deflagration can expand sufficiently to drive unburned flammable mixture up to the height of 2D confinement. If so, the 2D confinement will contribute to flame acceleration. Considering common hydrocarbons, the expansion ratio of the combustion zone is a factor of 7-8 compared to the initial volume of the unburned fuel-air mixture. An expansion ratio of 8 represents in increase in radius of either a spherical or hemispherical cloud of 2. Therefore, a hemispherical vapor cloud will expand to twice its height at completion of combustion. Clearly, a source of 2D confinement that is below 2 times the height of a vapor cloud will influence the combustion event and it is recommend that this scenario be treated as 2D when selecting a flame speed from Table 1. At this time it is not known whether sources of confinement above 2 times cloud height will influence flame acceleration appreciably, so the 2-times rule should not be considered an upper bound.

The issue of shock reflections was considered next. A case study was performed using WBE's 2-dimensional hydrocode VCE2D to investigate the effect of elevated sources of 2D confinement on blast reflections. A hemispherical cloud of stoichiometric ethylene-air mixture with a radius of 6.3 m (volume = 510 m^3) was chosen as the explosion source. A 10 m diameter solid plate was centered over the vapor cloud to provide confinement. Plate heights of 10 and 20 m were investigated. The baseline case was the same cloud without 2D confinement.

The case study revealed that the plate 20 m above the ground had no effect on either pressure or impulse at locations on the ground in either the near-field or far-field. However, the plate 10 m above ground introduced a reflection that increased pressure at all standoff distances and impulses in the near-field. The 20 m height represents 3.2 cloud radii.

Additional study is needed to gain a better understanding of confinement on both flame acceleration and shock reflections. Until such study is completed, it is recommended that sources of 2D confinement whose height are less than 3.2 times the flammable cloud height be treated as creating 2D flame expansion. Confinement above 3.2 times flammable cloud height can be treated as inconsequential to the VCE.

Fuel Reactivity

The fuel reactivity is a term used to describe the propensity of a flame to accelerate in a VCE for a given fuel. The reactivity ratings of low, medium, and high developed by TNO[7] were adopted for the Baker-Strehlow method. Factors determining fuel reactivity are laminar (fundamental) burning velocity and fuel expansion[8]. Generally, medium reactivity single component fuels have burning velocities between 45 and 75 cm/s. Low and high reactivity fuels are usually less than 45 and greater than 75 cm/s, respectively.

At the time the Baker-Strehlow method was published in 1994, it was recognized that mixtures had a flame speed proportional to the composition. Mixtures containing 3% of a high reactivity fuel were conservatively to be treated as high reactivity. Puttock[9] has now shown that 40% hydrogen in a hydrogen-propane mixture will have a burning velocity of 80 cm/s, and thus be high reactivity. This paper proposes that the burning velocities of mixtures involving fuels of different reactivity categories can be calculated using Le Chatelier's principle. The following equation will calculate the resulting burning velocity using Le Chatelier's principle.

$$V_B = \frac{100}{\frac{x_1}{V_{B1}} + \frac{x_2}{V_{B2}} + \frac{x_3}{V_{B3}} + \ldots}$$

Where:

V_B = Burning velocity of the mixture (cm/s)

x_1, x_2, x_3 = Mole % of each component in the mixture

V_{B1}, V_{B2}, V_{B3} = Burning velocity of each component in the mixture (cm/s)

For a mixture involving hydrogen (high reactivity) with propane (medium reactivity), the above equation can be solved for the minimum mole percent hydrogen to result in a burning velocity of 80 cm/s. The equation variables are the following: $V_B = 80$, $V_{B1} = 312$ (hydrogen), $x_2 = 100 - x_1$, $V_{B2} = 46$ (propane). The resulting hydrogen concentration (x_1) is 50 mole percent hydrogen. This does not exactly coorelate to Puttocks experimental results; however, it can provide a reasonable method of estimating mixture reactivities if the mixture high reactivity threshold is taken at 75 cm/s instead of 80 cm/s. Recalculating the previous problem substituting $V_B = 75$ cm/s results in a hydrogen concentration (x_1) of 45 mole percent hydrogen. Therefore, this paper concludes that a burning velocity threshold between medium and high reactivities should be 75 cm/s when using Le Chatelier's principle to coorelate with experimental data.

Puttock also presented data showing that 10% hydrogen in methane (burning velocity = 40 cm/s, low reactivity) will behave like propane (burning velocity = 46 cm/s, medium reactivity). Calculating the resultant burning velocity of this mixture using the above equation results in a $V_B = 44$ cm/s. This is close to the propane burning velocity and would qualify as a medium reactivity according to the guidelines.

As previously mentioned, there are factors other than burning velocity to determine fuel reactivity. This is why there is an area of overlap between some low and medium reactivity fuels' burning velocities in the region of 45-50 cm/s. For instance, carbon monoxide is a low reactivity fuel with a burning velocity of 46 cm/s, while n-butane is a medium reactiviy fuel with a burning velocity of 45 cm/s. The conservative approach is to take the mixure as low reactivity if the burning velocity is less than 40 cm/s or if all components in the mixure are low reactivity (regardless of burning velocity), high reactivity for burning velocities greater than 75 cm/s, and medium for all other cases.

Blast Loads from Multiple PESs and Ignition Sources

A common and difficult scenario to handle is one in which the flammable portion of a vapor cloud covers multiple PESs. Each PES within the flammable cloud will produce a blast wave, and each blast wave will act upon nearby receptors, such as control rooms. The magnitude and timing of multiple blast waves effects the response of the receptors as will be discussed below. The analyst is forced to decide how to handle multiple blast waves on a receptor to make accurate damage predictions. The assumptions concerning multiple PESs and ignition sources as well as structural response implications are discussed in the next two subsections.

Effects of Ignition Location

Large hydrocarbon releases often form vapor clouds which encompass several individual PESs. As discussed above, the Multi-Energy method predicts that damaging overpressures originate from individual PESs, and are not produced in open areas. Thus, a large vapor cloud explosion may consist of several individual confined/congested sites, each producing a finite overpressure. Because of the variability involved such as ignition location, wind direction, etc., several possible pressure-time history combinations can be produced. These combinations can be examined by considering a vapor cloud which encloses two separate PESs as shown in Figure 4. An observer is located as shown. Three possible ignition points are labeled as A, B, and C.

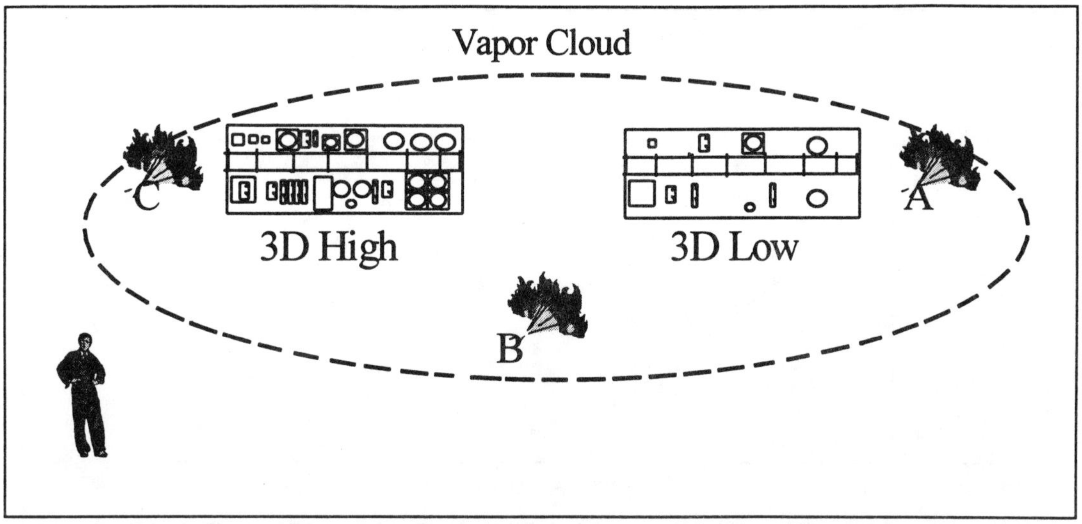

Figure 4 - Ignition Location In A Vapor Cloud

If ignition occurs at point A, the flame will propagate through the 3-D Low PES and generate an associated explosion overpressure. The flame will then decelerate to a low velocity as it moves out of the 3-D Low volume. When the flame front reaches the 3-D High PES, it will again accelerate and generate an explosion overpressure. If the volumes of the two PESs are similar, the observer would experience an idealized blast loading as shown in Figure 5.

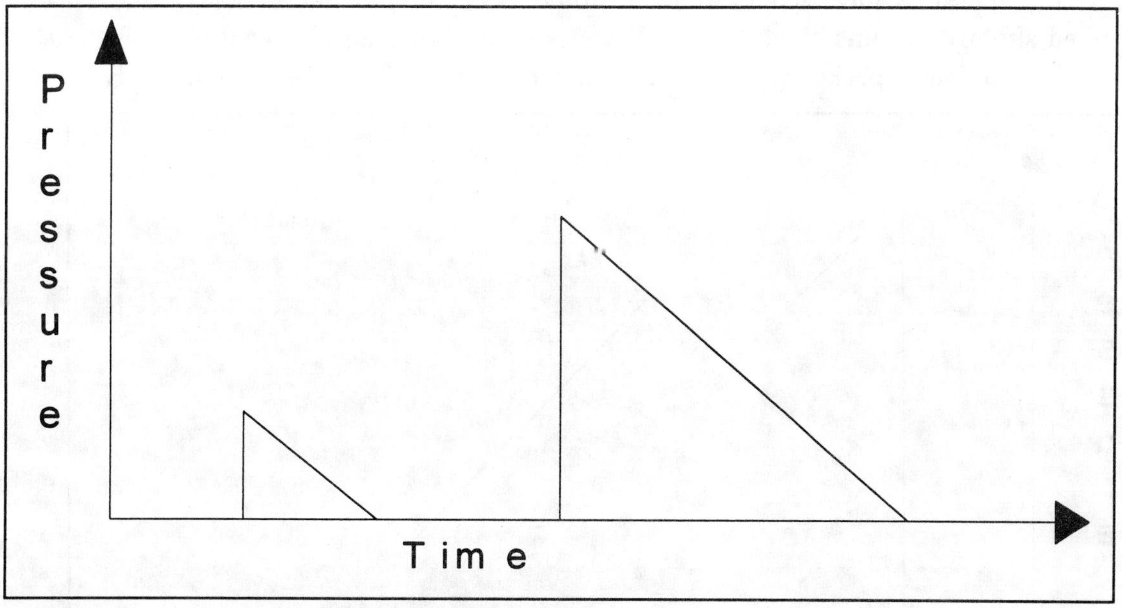

Figure 5 - Pressure Curves for Ignition at Point A

If ignition occurs a point C, the process will be similar to an ignition at point A, but the order of the PESs will be reversed. Thus, the observer would experience an idealized blast loading as shown in Figure 6.

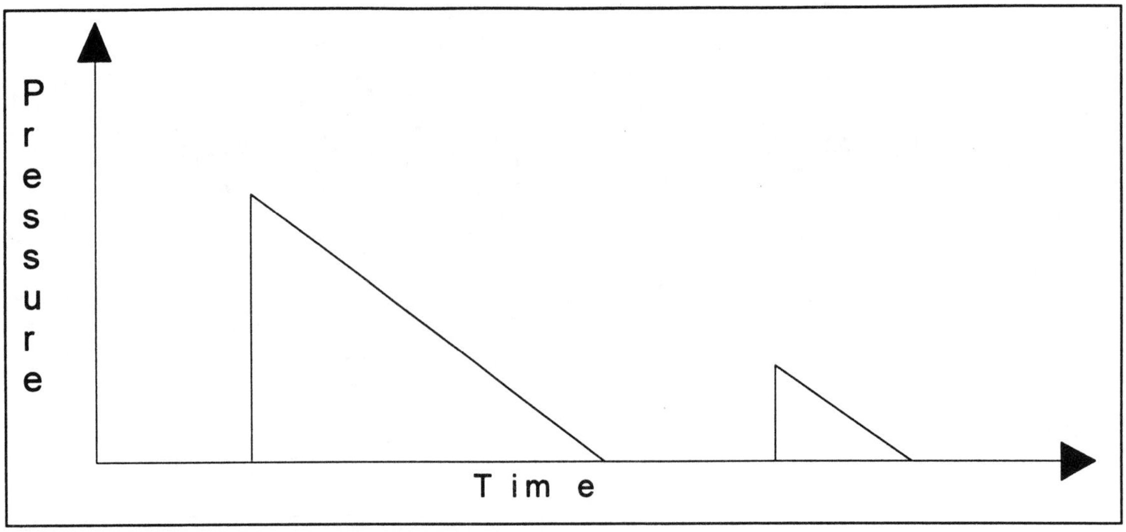

Figure 6 - Pressure Curves for Ignition at Point C

In general, two idealized blast loadings are possible for an ignition at point B. The first possible idealized blast loading would consist of a single pulse. This would be generated if the ignition point B were located such that the flame front reached both PESs at essentially the same time. If the shock wave emanating from one of the PESs was slightly lagging, it may "catch up" or coalesce with the initial shock wave due to its propagating medium having been pre-conditioned by the initial shock wave[10]. The observer would experience an idealized blast loading consisting of a single shock wave. The second possible idealized blast load would occur if the initial delay time was increased slightly. In this case, the shock waves would form an idealized blast loading consisting of a double peaked shock wave as shown in Figure 7, which we will call B'.

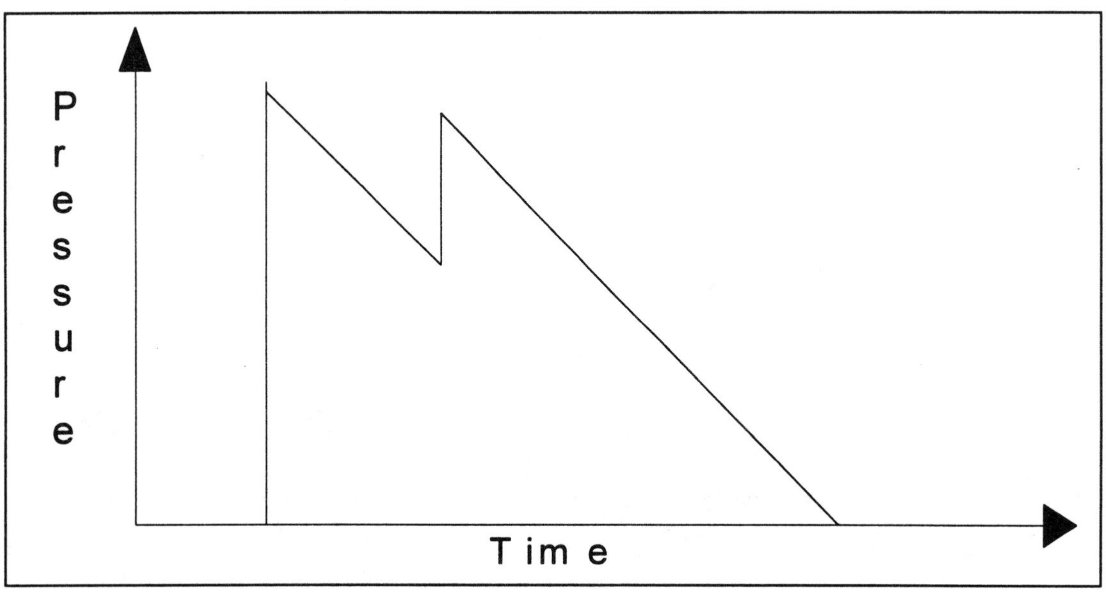

Figure 7 - Pressure Curve for Ignition at Point B'

Design Pulse and Implications

For ignition points A and C, the time delay between the blast overpressures from the two individual PESs will be a function of the distance between the PESs. For distances on the order of 5 m or more, this delay could be greater than one second. A building located at the observer's location would undoubtedly complete its response to the first applied loading before the second blast overpressure arrives. Thus, the second blast load will be applied to a pre-damaged building. Analysis of a damaged building for the second load should take into account the permanent deflection caused by the first loading. This is accomplished by adjusting limit deflection criteria accordingly.

For the ignition point B, two types of loadings may occur, the coalesced single pulse, and the closely spaced double pulse (B'). The worst case building loading would occur for a closely spaced double pulse if the second pulse would arrive when the building was in its maximum deflected state. However, for a typical metal clad steel framed building, the analysis would require several iterations because of the various natural frequencies of the different building components. Again, given the uncertainties in the loads predictions, in this case the exact shape of the pulse, this effort is typically not justified. Thus, the idealized blast load shape for an ignition at point B is often considered to be a single coalesced pulse.

Two methods for combining blast pulses from multiple sources into an equivalent triangular pulse is shown in Figure 8. The first method is to use the greater of the two peak pressures, sum their impulse, and compute duration of the equivalent triangular load. This is shown in the top of Figure 8 (Method 1). The second method is to use the duration of the most significant pulse (usually either the pulse with the highest peak pressure or the pulse with the greatest impulse, based on engineering judgment), sum their impulses, and solve for an effective peak pressure for the equivalent triangle. This is shown in the bottom of Figure 8 (Method 2).

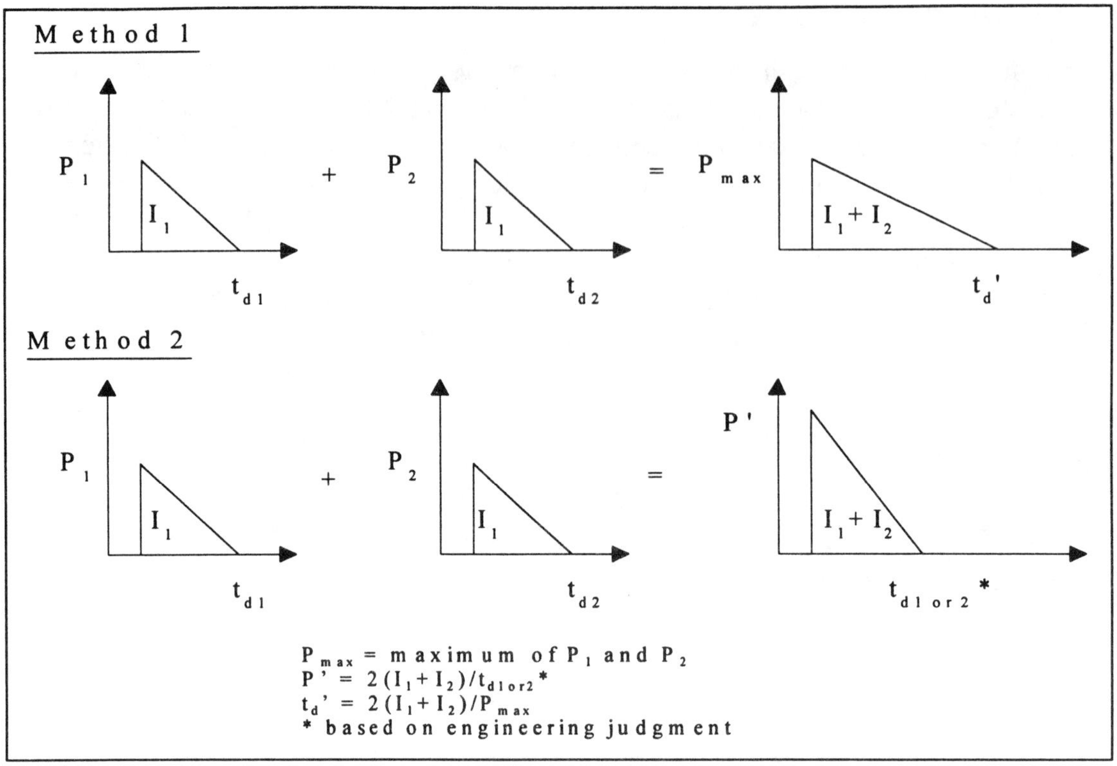

Figure 8 - Pressure Curve Representation of Methods to Combine Blast Loads from Multiple Sources

Method 1 is generally conservative; however, it is unconservative for either elastic or slightly plastic systems and when the blast duration (td, not td') is approximately equal to or less than the natural frequency. In cases where the blast duration is greater than the natural frequency, which is typical of larger VCEs, or in any case with significant plastic deformation, Method 1 tends to be conservative. Considering that the accepted practice in both Government and industry is to allow significant plastic deformation of ductile building components subjected to accidental explosions, Method 1 provides conservative results yet avoids the overly conservative predictions associated with Method 2. If an existing building is predicted to have an elastic or slightly plastic response using Method 1, a slightly greater actual response during an accidental explosion is normally considered acceptable for ductile systems.

It can be shown[11] that Method 2 is always conservative, and is very conservative in the pressure sensitive regime (load durations much greater than the natural frequency of the building or component). This is especially true for systems that are elastic or only slightly plastic (maximum deflection 1.5 times greater than elastic limit deflection or less).

The recommendation of this paper is to use Method 1 for siting analyses (maximum of individual predicted peak pressures, enhanced duration by conserving impulse) and Method 2 considered on a case-by-case basis during detailed siting analysis or the design of building upgrades. Consideration should also be given to analyzing the building for a multi-pulse loading (Figures 2-4 above) if sufficient information is available to allow accurate prediction of the load histories.

Conclusions and Recommendations

The Baker-Strehlow method of predicting VCE blast pressure and impulse has been revised as a result of lessons learned during application to hazard assessments in refineries and chemical plants. The revisions to the methodology are as follows:

1. Identification of Potential Explosion Sites (PES) during site inspections and dispersion analysis has provides a systematic approach to analyzing potential locations of VCEs and defining the confinement and congestion for each VCE. PESs also provide a means to rank hazards and explore mitigation option.
2. A PES frequently contains vary degrees of confinement and congestion. When both 2D and 3D flame expansion are possible in one PES, very large differences in blast pressure predictions result from the selection of 2D versus 3D flame expansion. An intermediate selection, termed 2½D has been added to allow interpolation of the flame speed between 2D and 3D values.
3. Elevated sources of confinement may not influence the airblast resulting from a low-lying vapor cloud. Confinement above 3.2 cloud heights can be disregarded in a VCE analysis.
4. Le Chatelier's rule has been adopted to provide an objective means determining fuel reactivity for mixtures. The following guidelines are given as to reactivity determination when using Le Chatelier's principle:
    - Low Reactivity - Mixtures with laminar burning velocities less than 45 cm/s or all components in mixture are low reactivity.
    - Medium Reactivity - Mixtures with laminar burning velocities between 45 cm/s and 75 cm/s and do not qualify as low reactivity.
    - High Reactivity - Mixtures with laminar burning velocities greater than 75 cm/s
5. Multiple PESs and ignition sources within one flammable cloud present the possibility of varying combinations of separate or coalesced shock waves. Structural response of a building is affected by the magnitude and timing of multiple shock waves. Given the uncertainty of the events that could actually occur during a VCE, combining multiple blast waves into a single pulse by preserving the highest of the peak pressures and summing the impulses is recommended for siting studies to give an estimate of structural response. Analysis of structural response is to multiple blast waves is recommended for detailed siting or building design if sufficient information is available to make accurate predictions of load histories.

References

1. Baker, Q.A., Tang, M.J., Scheier, E.A., and Silva, G.J., "Vapor Cloud Explosion Analysis," 28th Annual AIChE Loss Prevention Symposium, Atlanta, GA., April 19, 1994.
2. Strehlow, R.A., Luckritz, R.T., Adamczyk, A.A., and Shimpi, S.A., "The Blast Wave Generated by Spherical Flames," Combustion and Flame, Volume 35, 1979.
3. Vanden Berg, A.C., Opschoor, G., and Zeguwen, J.P., "Blast Effects from Vapour Cloud Explosions," International Conference on Vapour Cloud Modeling, 1987.
4. Mercx, W. P. M., "Objectives and Results of Two CEC-Sponsored Projects on Modeling and Experimental Rsearch into Vapor Cloud Explosions," International Conference and Workshop on Modeling and Mitigating the Consequences of Accidenatl Releases of Hazardous materials, CCPS, September 26-29, 1995.
5. Taylor, P.H., "Recent Research on Vapour Cloud Explosions," The 12th Annual Leeds University Short Course on Fire and Explosions, Part 2, 3-4 December 1987.
6. Van Wingerden, C.J.M., "Experimental Investigation into the Strength of Blast Waves Generated by Vapour Cloud Explosions in Congested Areas," 6th International Symposium on Loss Prevention and Safety Promotion in the Process Industries, Volume I, CCPS, European Federation of Chemical Engineers, June 1989.
7. Zeeuwen, J.P. and Wiekema, B.J., "The Measurement of Relative Reactivities of Combustible Gases," Conference on Mechanisms of Explosions in Dispersed Energetic Materials, 1978.
8. Taylor, P.H., Hirst, W.J.S., "The Scaling of Vapour Cloud Explosions: a Fractal Model for Size and Fuel Type," 22nd International Symposium on Combusion, 1988.
9. Puttock, J. S., "Fuel Gas Explosion Guidelines - the Congestion Assessment Method," 2nd European Conference on Major Hazards On and Off-shore, Manchester, UK, October 24, 1995.
10. Zaker, T.A., "Blast Pressures from Sequential Explosions," IIT Research Institute, Chicago, IL Phase Report II, Project J6166, March 25, 1969.
11. Whitney, M.G., Barker, D.B., Waclawczyk, J.H., "Structural Response to Multiple Pulse Blast Loading," 6th International Symposium on Interaction of Nonnuclear Munitions with Structures, Panama City Beach, FL, May 3, 1993.

31st LOSS PREVENTION SYMPOSIUM

Session 43. **Storage, Processing, & Transportation of Hazardous Chemicals**

Wed., March 12, 1997
1:30 - 4:35 PM
Grand Ballroom C

Stanley S. Grossel, Chair
Process Safety and Design, Inc.
41 Sussex Rd.
Clifton, NJ 07012

Joseph F. Louvar, Vice-Chair
BASF Corporation
1419 Biddle St.
Wyandotte, MI 48192-3836

Session Introduction - Stanley S. Grossel

43a. **A Detailed Reaction Study of Phosphorus Trichloride and Water,** G. Melhem; D. Reid, Arthur D. Little, Cambridge, MA

43b. **A Fire Risk Analysis Model for Assessing Options for Flammable and Combustible Liquid Products in Storage and Retail Occupancies,** J. Hall, Jr., National Fire Protection Association, Quincy, MA

43c. **A Review of Chemical Compatibility Issues,** D. Frurip; T. Hofelich, The Dow Chemical Co., Midland, MI; D. Leggett, Hazard Evaluation Laboratories, Monmouth Junction, NJ; J. Kurland, Union Carbide Corp., South Charleston, WV; J. Neimeier, Eli Lily and CO, Indianapolis, IN

BREAK

43d. **Decomposition of Thermally Unstable Chemicals in Solvents,** S. Chervina, S. Evon, G. Bodman, Eastman Kodak CO, Rochester, NY

43e. **Developing Chemical Compatibility Charts Using CHEMPAT,** D. Leggett, Hazard Evaluation Laboratories, Monmouth Junction, NJ

43f. **Improving the Effect of Atmospheric Stability Class On Hazard Zone Predictions For Accidental Hazmat Releases,** John L. Woodward, DNV Technica, Inc.

Concluding Remarks - J.F. Louvar

A Detailed Reaction Study of Phosphorus Trichloride and Water

G. A. MELHEM
D. REID
Arthur D. Little, Inc.
20 Acorn Park
Cambridge, Massachusetts 02140

ABSTRACT

This paper reports on a comprehensive literature search and small scale experimental work on the reaction characteristics of phosphorus trichloride and water. More than 30 tests were conducted, including both closed and open test cells. The water to phosphorus trichloride molar ratio was varied from 1 to 25. When in contact, water and phosphorus trichloride will form two liquid layers with a reaction starting at the interface. The impact of variables on reaction rates including the interface surface area, layer depth, and stirring were investigated experimentally. A reaction rate model that fits all the measured data is presented.

Case studies illustrating the use of this data for emergency relief systems and vent containment design are presented in reference [1].

BACKGROUND

The study reported in this paper was conducted in order to address safety and effluent handling issues with two reator vessels where PCl_3 is used. Each reactor has a volume of 5,000 gallons and operates on a batch cycle. Each reactor is protected against overpressure by a rupture disc and a vent line which leads to an effluent handling system. A scenario was identified where it is possible for a heel of phosphorus trichloride to remain in either one of the reactors undetected (below detection level) at 40 C, and for an operator to attempt to flush the vessel with water. This can lead to the generation of gaseous hydrogen chloride and excessive system pressure.

Existing experimental information for the PCl_3 and water reaction was not adequate for the development of a good working reaction model which would allow us to determine the effects of this potential problem. The objective of this study was to establish a working reaction model for the PCl_3 and water reaction and to evaluate the potential for reactant pileup and possible system overpressure. The scope of the analysis included:

- Conducting a literature review of existing PCl_3 / water data;

Figure 1: Competing Vent Containment Configurations

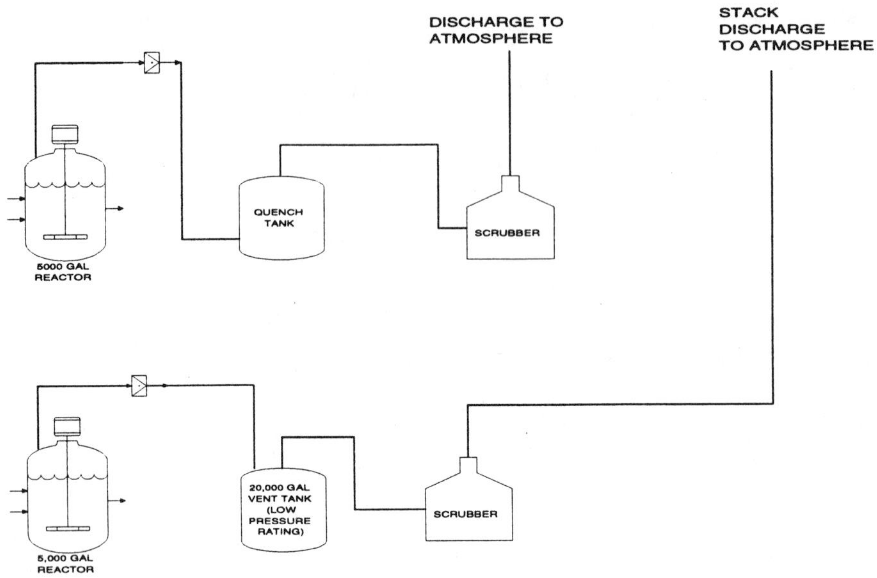

- Conducting small scale experiments to assess the impact of surface area on the reaction velocity at 40 C;
- Analyzing all data sets and establishing a reaction rate model that is sensitive to surface area, temperature and composition;
- Validating the reaction model with experimental data;

Scenario Definition

It has been recognized that the level detection equipment on the reactor vessel will not detect low levels and there is the possibility of a heel of PCl_3 remaining in the vessel after a batch run. This heel may be up to 500 gallons and there is concern that when the vessel is flushed there is the potential for water to be added. We wish to confirm that the reactors rupture discs are suitable for this duty and to determine the worst possible overpressure that may arise in associated downstream equipment in the event of water being inadvertantly added to an undetected PCl_3 heel.

In addition, two possible effluent handling configurations were considered in order to reduce to toxicity impact of the reaction products which include large amounts of HCl. Figure 1 shows a process flow diagram for one reactor and the downstream emergency venting equipment under evaluation.

LITERATURE REVIEW

Historical Perspective

Much of the basic research into the behavior of PCl_3 and water was carried out towards the end of the previous century. While some of this work is outdated, it still provides us with useful background information for this study. Following is a brief description of some publications that consider the PCl_3 and water reaction.

Mellor [2] describes the chemical properties of PCl_3 as determined by early experimental work conducted between 1810 and 1928, including the reaction with water. Davy [3] noted a rise in temperature and the gradual formation of hydrochloric acid and phosphorous acid when water reacted with PCl_3 and suggested that three moles of water was required per mole of PCl_3 as follows:

$$PCl_3 + 3H_2O \longrightarrow H_3PO_3 + 3HCl \qquad (1)$$

For excess PCl_3 cases where there is insufficient water present for the complete reaction Kraut [4] observed the formation of some red phosphorus and phosphoric acid. He suggested that phosphorous acid first formed by the above equation then reacted with the PCl_3:

$$PCl_3 + 4H_3PO_3 \longrightarrow 3H_3PO_4 + 3HCl + 2P \qquad (2)$$

so that overall, twelve moles of water reacted with every five moles of PCl_3 as follows:

$$5PCl_3 + 12H_2O \longrightarrow 3H_3PO_4 + 15HCl + 2P \qquad (3)$$

Geuther [5] said that phosphorus separation only occurred with impure PCl_3 contaminated with arsenic chloride. He proposed further intermediate reactions for the excess PCl_3 case involving the formation of phosphine (PH_3):

$$\begin{aligned} 4H_3PO_3 &\longrightarrow 3H_3PO_4 + PH_3 && (4) \\ PCl_3 + PH_3 &\longrightarrow 2P + 3HCl && (5) \end{aligned}$$

Note that the overall reaction is still represented by Equation 3. Besson [6] found that some phosphoryl monochloride formed in excess PCl_3 cases.

Carrara and Zoppelari [7] conducted a series of experiments to determine the reaction velocity of some sulphur and phosphorus compounds with water. They assumed water and PCl_3 reacted as per Equation 1 above and presented the following rate equation based on their experimental findings:

$$\left[\frac{1}{t}\right] \log \left[\frac{A}{A-x}\right] = ACO = 3 \times 10^{-4} min^{-1} \qquad (6)$$

where t = time (minutes), O = wet surface area (cm^2), x = quantity of reagent reacted, A = total quantity of decomposible liquid and C = constant. They do not state the value of A used in their PCl_3 and water experiments but analysis of their data suggests it was 10. The experiments were carried out at 5 C and were observed to be very energetic. This is not consistent with the data and rate equation presented which suggests only small quantities PCl_3 react in 70 minutes. This rate equation was found to be of little use because of the low temperature considered in the experiments and the uncertainty in the data used to derive it.

Sidgwick [8] discussed the nature of the water-trichloride reaction and suggested that the oxygen atom in a water molecule would attach itself to the lone pair of electrons on the phosphorus atom and a hydrogen and chlorine atom would then react to form HCl. This would be repeated until all chlorine atoms were displaced from the phosphorus.

Mitchell [9] observed that when PCl_3 and water were reacted together the solution first produced had much stronger reducing properties than after several hours and suggested the intermediate formation of an oxychloride, such as $P(OH)_2Cl$. He also reported a slight smell of phosphine which became more pronounced at much higher temperatures.

Vapor phase hydrolysis was investigated by Hudson [10]. He detected no reaction between water vapor and PCl_3 and suggested that the pyramidal structure of PCl_3 and the lone pair of electrons prevented co-ordination occuring with water molecules. In solution, however, the intimate contact between the molecules allows reaction.

General texts on the chemistry of phosphorus support the reactions described by the equations presented above and provide useful information about the possible reaction products. Phosphorous acid can assume either of two tautomeric forms $HP(O)(OH)_2$ and $P(OH)_3$. A literature review suggests that it will predominantly be in the $HP(O)(OH)_2$ form. Phosphorous acid decomposes when heated to give phosphoric acid and phosphine in the absence of water and phosphoric acid and hydrogen, if water is present. The equations for thermal decomposition are:

$$4H_3PO_3 \longrightarrow 3H_3PO_4 + PH_3 \qquad (7)$$

$$H_3PO_3 + H_2O \longrightarrow H_3PO_4 + H_2 \qquad (8)$$

These decomposition reactions are reported to be fairly rapid at temperatures about 250 - 275 C with Equation 7 beginning to occur at temperatures greater than 180 C. The decomposition reaction 7 may account for the observation of Mitchell [9] that some phosphine was present and supports the reaction mechanism proposed by Geuther [5] above in Equations 1, 4 and 5.

More Recent Literature

There have been relatively few recent publications that consider the chemistry of the PCl_3 and water reaction. Schnockel and Schunck [11, 12] reported the partial hydrolysis of gaseous PCl_3 to $POCl$ at high temperature and also produced molecular phosphoric acid (PHO_2), again at very high temperatures [13]. Although the kinetics of the phosphorus atom in various states has been comprehensively studied [14, 15, 16] there is little information which is useful in the context of this study. Rozhdestvenskii et al. [17] determined the coefficients of the thermodynamic potential for a large number of substances in the gas phase, including PCl_3.

General information about PCl_3, including its production, handling, uses, physical properties and toxicology, may be found in the Ullmanns Encyclopedia [18] and in the Encyclopedia of Chemical Technology [19].

Experimental Work

Several sets of experiments were carried out using ARC and RSST to investigate the PCl_3 and water reaction. These experiments are summarized in Table 1. Two types of reactions were considered, (a) reactions between PCl_3 and water and reactions between PCl_3 and hydrochloric acid (11 wt % HCl, 87 % water, 2 % other). Both stirred and non-stirred cases were conducted. In the ARC experiments, PCl_3 and water were frozen in liquid nitrogen as they were added in sequence. The reaction data was measured as water reached its melting point. This is shown later in Figures 7 through 10.

In addition, a series of four VSP experiments were conducted. These experiments are also shown in Table 1. The VSP experiments involved the gradual addition of approximately 45 gm of de-ionized water to approximately 60 gm of PCl_3 and were all stirred. The duration of the water feed varied between 5 and 10 minutes.

Expected Reaction Between PCl_3 and Water

The reviewed literature suggests that under stoichiometric or excess water conditions the reaction between PCl_3 and water will follow Equation 1 above with three moles of HCl forming for every mole of PCl_3 consumed.

The heat of reaction is calculated to be -64.6 kJ/gmole PCl_3 if HCl forms as a gas and -289.2 kJ/gmole PCl_3 if aqueous HCl forms. Note the latter case is considerably more exothermic due to the high HCl heat of solution. The heat of reaction for reaction 3 is calculated to be -47.5 kJ/gmole PCl_3 and -272.1 kJ/gmole PCl_3 for the gaseous and aqueous HCl cases respectively.

The possible formation of phospine is an abvious concern given its flammability and high toxicity. Thermochemically, phosphine has a positive heat of formation and Gibbs free energy of formation and would not therefore be expected as a reaction product. Experimentally, however, phosphine has been detected at higher temperatures. The thermal decomposition of phosphorous acid to phosphoric acid and phosphine which occurs in the absence water at temperatures of 180 C and above would seem to account for this.

Table 1: PCl_3 / Water Experiments Summary

Type	Mass Ratio H_2O/PCl_3	Mole Ratio H_2O/PCl_3	Water Type	Comments
ARC	3.208	24.456	Tap water	PCl3 shot into water†
ARC	0.395	3.012	water	Frozen, not stirred
ARC	0.394	3.000	acid	Frozen, not stirred
ARC	0.787	5.999	acid	Frozen, not stirred
ARC	0.237	1.809	acid	Frozen, not stirred
RSST	0.772	5.881	acid	Approx. 150 psig back pressure
RSST	1.192	9.084	acid	Ambient pressure (open system)
VSP	1.292	9.853	acid	Stirred, water fed over time,
VSP	1.292	9.853	DI water	Stirred, water fed over 5 mins.
VSP	1.292	9.853	DI water	Stirred, water fed over 10 mins.
VSP	1.292	9.853	DI water	Stirred, water fed over 6 mins.
VSP	1.292	9.853	DI water	Stirred, water fed over 5 mins.

Note: Acid is approximately 11 wt % HCl, 87 wt % water, 2 % miscellaenous
†: Stirred at 500 rpm

Physical Properties of Reaction Materials

Phosphorus trichloride is a colorless fuming liquid which is very toxic and hazardous. It severely burns skin, eyes and mucous membranes and may cause death if vapors are inhaled [19]. Hydrogen chloride is a clear colorless gas which is highly soluble in water. Phosphorous acid is a crystalline solid which melts at 73.6 C and is soluble in water. Phosphoric acid does not possess any specific toxic effects and is considered irritating in concentrations of 10-25 % and corrosive in concentrations greater than 25 %. Phosphine is a highly poisonous colorless gas which is reported to have a garliclike odor. Air - phosphine mixtures are explosive over a wide range, with a LFL of 1.6 % and an UFL of close to 100 %.

A more complete set of physical property data is presented for phosphorus trichloride, hydrogen chloride, phosphorous acid, phosphine and water in Table 2.

Additional Experiments

The information obtained from the literature and small scale ARC, RSST and VSP data was not sufficient to either fully understand the mechanism of the reaction between water and PCl_3 or develop a kinetic model that could be applied to the full scale working process. Further experimental work was necessary and was focused on the addition of water at ambient temperature to PCl_3 which was heated to 40 C, so as to mimic the likely plant conditions.

The objectives of these additional experiments were as follows:

Table 2: Selected Physical Properties

	Hydrogen Chloride	Phosphorus Trichloride	Phosphorous Acid	Phosphine	Water
Chemical Formula	PCl_3	HCl	H_3PO_3	PH_3	H_2O
Molecular Weight	137.35	36.47	82	34	18
Boiling Point (K)	349	188	473	185	373
Melting Point (K)	181	162 †	347	139	273
Decomp. Temp. (K)				648	
Specific Gravity	1.57	-	1.65		1.0
Water Solubility (g/g solution)	reacts	42.3 /100 (@291K)	307 /100 (@273K)		
Heat of Formation (kJ/mole) -	(gas) -287.0 (liq) -319.7	(gas) -92.3 (aq) -167.2	(aq.) 964.8		(liq) -285.8
LFL (%)				1.6	
UFL (%)				100	
Toxicity:MAK(mg/m3)				0.15	
TLV-TWA(mg/m3)				0.42	

†: Crystallization point

1. to develop a better understanding of the reaction(s) between water and PCl_3 by getting a visual confirmation of the reaction behaviour;

2. to determine the effects of key parameters such as reactant composition, reactant mole ratio, surface area and stirring on the rate of gas evolution;

3. to gather sufficient data to enable the reaction velocity to be calculated and a working kinetic model to be established.

Experimental Setup

The additional experiments were carried out using the apparatus shown in Figure 2. The apparatus was contained within a fume cupboard and all produced gas was vented from the reactor vessel directly to the atmosphere. Four different reactor vessels were used during the experiments - a dewar flask, a 50 ml beaker, a 100 ml graduated cyclinder and a 25 ml graduated cylinder. The dimensions of each vessel are shown in Table 3. An electrically driven glass rod with a right angle bend was used as a stirrer in four of the cases to investigate the effect of mechanical mixing on the reaction. Only one stirring speed, estimated to be around 300 rpm, was used.

The desired volume of PCl_3 was measured, heated to typically about 50 C and then added to the reactor vessel. The reactor was then placed on an electronic mass balance and a

Table 3: Experimental Vessel Dimensions

Vessel	Inside Diameter (mm)	Cross Sectional Area (mm^2)	Vessel Bottom	Bottom Depth (mm)	Bottom Volume (cm^3)
Dewar Flask	60	2827	elliptical	20	37.7
50 ml Beaker	38	1134	cylindrical		
25 ml Grad. Cylinder	16	201	cylindrical		
100 ml Grad. Cylinder	26	531	cylindrical		

Note, when 30 ml added to dewar, calculated top interface area is 2770 mm^2 (98 % of full area) Allowing for measurement error, assume 30 ml top interface area is 2827 mm^2

Figure 2: Experimental Apparatus

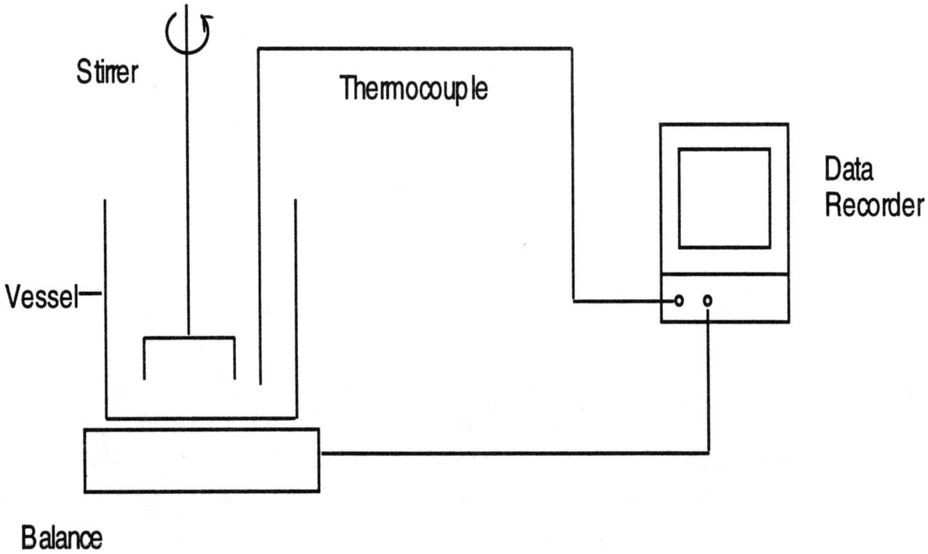

thermocouple was inserted so that it penetrated the PCl_3 layer prior to water addition. The stirrer was then started if required and a measured volume of distilled water or HCl (aqueous) at ambient temperature was added. The temperature and mass of the reactor was recorded throughout the water addition and reaction stages using a computer data logging system.

When appropriate, general observations were also recorded, although in experiments involving the dewar flask this was difficult because it was not possible to see the reactants. The dewar flask was however used for most cases as it provided approximate adiabatic conditions, thus simulating a large reactor. The heat loss rate from the other vessels, whilst greater than from the Dewar, was not considered to be a significant factor when the reaction proceeded rapidly.

Experimental Program

The following variables were considered in the experimental program:

- PCl_3 to water mole ratio - balanced, excess water and excess PCl_3 cases were considered, based on reaction given by Equation 1;
- reactor cross-sectional area - the cross-sectional area was varied between 201 mm^2 and 2827 mm^2;
- layer depth for the balanced case only;
- water only compared to water in a 20 wt % HCl solution;
- no mechanical mixing compared to mechanical stirring;
- instantaneous compared to gradual water addition.

The reactant quantities and reactor conditions for the experiments conducted are presented in Tables 4 and 5.

Experimental Results

A summary of the results obtained from all experiments is presented in Table 6. The initial mass of the reactants is recorded along side the total mass of gas lost from the reactor during the course of the reaction. The mass ratios of gas evolved / water added and gas evolved / PCl_3 are calculated for comparison. The maximum gas evolution rate recorded has been calculated per unit reactor cross-sectional area (the maximum specific mass loss rate), again to allow comparison between all the cases. The time between the first addition of water and the evolution of gas has been estimated from the mass loss against time graphs. These graphs are displayed in Figures 3 through 6.

Table 4: Experimental Reactant Quantities

Test No.	PCl_3 +	PCl_3 Moles (gmol)	H_2O Moles (gmol)	HCl Mass (g)	H_2O/PCl_3 Mole Ratio
1	H_2O	0.34	0.52	0.0	1.51
2	H_2O	0.34	1.01	0.0	2.95
3	H_2O	0.34	1.03	0.0	3.02
4	H_2O	0.34	1.01	0.0	2.94
5	H_2O	0.69	2.04	0.0	2.98
6	H_2O	0.34	2.02	0.0	5.89
7	H_2O	0.34	2.02	0.0	5.89
8	H_2O	0.34	3.03	0.0	8.84
9	H_2O	0.17	0.52	0.0	3.02
10	H_2O	0.12	0.41	0.0	3.36
11	H_2O	0.31	1.01	0.0	3.30
12	H_2O	0.34	1.03	0.0	3.00
13	H_2O	0.34	1.03	0.0	3.00
14	H_2O	0.34	1.03	0.0	3.00
15	20 % HCl	0.34	0.50	2.3	1.47
16	20 % HCl	0.34	0.52	2.3	1.51
17	20 % HCl	0.34	1.05	4.7	3.06
18	20 % HCl	0.34	0.81	3.6	2.36
19	20 % HCl	0.34	2.02	9.1	5.90
20	20 % HCl	0.17	0.50	2.3	2.93
21	20 % HCl	0.34	0.93	4.2	2.72

Note, HCl or H_2O are added at 22 C.

Table 5: Experimental Reactor Conditions

Test No	PCl_3 +	H_2O/HCl Addition	Vessel Type	Interface Area (mm^2)	Stirring (Y/N)	PCl_3 Layer Depth (mm)	H_2O Layer Depth (mm)
1	H_2O	instant	dewar	2827	no	17.3	3.3
2	H_2O	instant	dewar	2827	no	17.3	6.4
3	H_2O	instant	dewar	2827	no	17.3	6.6
4	H_2O	instant	dewar	mixed	yes	mixed	mixed
5	H_2O	instant	dewar	2827	no	27.9	13
6	H_2O	instant	dewar	2827	no	17.3	12.8
7	H_2O	instant	dewar	mixed	yes	mixed	mixed
8	H_2O	instant	dewar	2827	no	17.3	19.3
9	H_2O	instant	50 ml bk	1134	no	13.2	8.2
10	H_2O	instant	25 ml gr	201	no	49.0	36.8
11	H_2O	instant	100 ml gr	531	no	48.0	34.3
12	H_2O	gradual	dewar	2827	no	17.3	6.5
13	H_2O	gradual	dewar	mixed	yes	mixed	mixed
14	H_2O	gradual	dewar	mixed	yes	mixed	mixed
15	20 % HCl	instant	dewar	2827	no	17.3	4.0
16	20 % HCl	instant	dewar	mixed	yes	mixed	mixed
17	20 % HCl	instant	dewar	2827	no	17.3	8.3
18	20 % HCl	instant	dewar	mixed	yes	mixed	mixed
19	20 % HCl	instant	dewar	2827	no	17.3	16.1
20	20 % HCl	instant	50 ml bk	1134	no	13.2	10.0
21	20 % HCl	gradual	dewar	mixed	yes	mixed	mixed

Note, depth of PCl_3 is measured from bottom of vessel.

Table 6: Experimental Results

Test No.	PCl_3 +	Initial Reactant Mass (g)	Total Mass Loss (g)	Vapor to H_2O Mass Ratio	Vapor to PCl_3 Mass Ratio	Maximum Specific Mass Loss Rate (g/mm^2/s) $\times 10^3$	Estimated Start Time (s)	Initial PCl_3 Temp. (C)	Maximum Temp. Reached (C)	Estimated Specific Addition Rate (g/mm^2/s) $\times 10^3$
1	H_2O	56.9	16	1.7	0.34	0.156	18.3	39.6	85.2	
2	H_2O	66.3	35.2	1.9	0.75	0.675	19.2	30.6	90.2	
3	H_2O	66.7	34.9	1.9	0.74	1.470	14.9	38.6	92.5	
4	H_2O	66.2	37.0	2.0	0.79	1.540	14.3	36.0	88.5	
5	H_2O	133.0	73.4	2.0	0.78	3.230	15.5	41.0	86.0	
6	H_2O	85.4	32.0	0.9	0.68	2.140	19.7	38.0	96.3	
7	H_2O	85.4	32.0	0.9	0.68	2.500	10.7	38.0	98.0	
8	H_2O	104.6	31.0	0.6	0.66	2.360	26.7	38.4	94.7	
9	H_2O	33.4	14.0	1.5	0.59	1.450	18.7	37.3	91.8	
10	H_2O	24.6	14.5	2.0	0.86	2.710	55.2	37.7		
11	H_2O	61.3	34.8	1.9	0.83	2.330	62.0	36.0		
12	H_2O	66.6	21.0	1.1	0.45	0.057		37.0	79.0	0.0198
13	H_2O	66.6	21.0	1.1	0.45	0.070	62.0	37.0	71.0	0.0163
14	H_2O	66.6	22.0	1.2	0.47	0.054	73.0	42.0	61.0	0.0216
15	20 % HCl	56.6	17.0	1.9	0.36	0.049	17.0	38.0	68.0	
16	20 % HCl	56.9	23.0	2.5	0.49	0.0132	13.0	36.0	59.0	
17	20 % HCl	67.0	35.4	1.9	0.75	0.175	51.4	37.2	74.1	
18	20 % HCl	62.5	45.5	3.1	0.97	0.196	0.00	36.0	67.0	
19	20 % HCl	85.5	39.0	1.1	0.83	0.464	19.7	37.0	76.0	
20	20 % HCl	33.1	15.9	1.8	0.68	0.174	68.0	34.1	73.4	
21	20 % HCl	64.8	20.0	1.2	0.42	0.037	67.0	40.0	48.0	0.017

Figure 3 shows the results from the Dewar flask experiments that involved water and PCl_3. The maximum temperatures attained varied between 85 and 99 C. The delay between water addition and gas evolution varied between 14 and 27 seconds. The maximum specific gas evolution rate was similar, given the level of accuracy, in the stirred, balanced and excess water cases with an average of 2.2×10^{-3} g/mm^2/s. This was an order of magnitude greater than the excess PCl_3 case which showed a maximum rate of 1.6×10^{-4} g/mm^2/s.

Note, that temperature readings recorded for experiment No. 5 are not comparable with the other cases due to the thermocouple initially being well below the PCl_3 and water interface. In the other cases the thermocouple was placed very close to the interface.

Figure 4 shows the results from all experiments that involved HCl and PCl_3. The maximum temperatures attained varied between 59 and 75 C. The delay between water addition and gas evolution varied between 0 and 68 seconds. The maximum specific gas evolution rate was similar, given the level of accuracy, in the stirred, balanced, excess HCl and smaller reactor cases with an average of 2.3×10^{-4} g/mm^2/s. This was higher than the excess PCl_3 case with a maximum rate of 5×10^{-5} g/mm^2/s.

Figure 5 shows the results from various comparable experiments that had different reactor cross-section areas or layer depths and were not stirred. Note that temperature readings recorded for experiments No. 10 and 11 are not comparable with the other cases due to the thermocouple initially being well below the PCl_3 and water interface.

Figure 6 shows the results from comparable stirred and non-stirred cases. When water was added, the maximum specific gas evolution rate was marginally higher in the stirred cases and the delay before gas evolution and the maximum temperature achieved were slightly reduced. With HCl addition, a similar trend was evident, except that there was an increase in the maximum specific gas evolution rate by a factor of 2.5 in one case and a large reduction in the gas evolution delay in the other.

Four experiments were carried out with gradual addition of water to PCl_3. Due to limitations in measuring the very low addition rate it was difficult to draw any conclusions from this data.

Experimental Observations

When water was added quickly to PCl_3 it formed a distinct layer on top. Although there was no evolution of gas it was possible to see some schlieren effects in the water layer suggesting heat generation. After a delay period, which varied from case to case, the mixture bubbled vigorously as gas evolved. Comparing the temperature - time graphs with the gas evolved - time graphs shows that the temperature began to increase as soon as the water was added. In the course of the experiments it was evident that PCl_3 vapor was also escaping from the system in spite of the heavier PCl_3 layer being blanketed by an overlying water layer. No odor of phosphine was detected during the experiments.

Some qualitative side experiments were also conducted using a larger beaker and a large excess of water in an attempt to understand how PCl_3 vapor could escape without reacting. When the PCl_3 was added it first formed a layer at the bottom and as the reaction proceeded this layer separated into several globules. Bubbles formed at the top of these globules which often grew to several millimetres in diameter.

When the bubbles reached a critical size they rose to the top of the water layer, carrying a small droplet of PCl_3 with them. This levitation caused some mixing of the PCl_3 in the bulk of the water because a PCl_3 trail was released as the bubbles rose. Despite this some of the levitated PCl_3 reached the surface of the water where it remained as a small droplet. When enough droplets aglomerated to overcome the water surface tension the PCl_3 fell back to the bottom of the vessel beaker. This circulating effect which results in a small quantity of PCl_3 lying on top of the water layer would allow for some PCl_3 to enter the vapor phase.

Discussion of the Results

The rate of gas evolution appears to be less for the excess PCl_3 cases, for both water and HCl addition scenarios. In the balanced and excess water cases the gas evolution rate is similar.

The cross-sectional area does not appear to have had any effect on the maximum specific gas evolution rate, however in the cases where the layer depths were greater a longer time delay was observed and the maximum specific gas evolution rate was higher.

Figure 3: Experimental Results Summary. $PCl_3 + H_2O$ (dewar only).

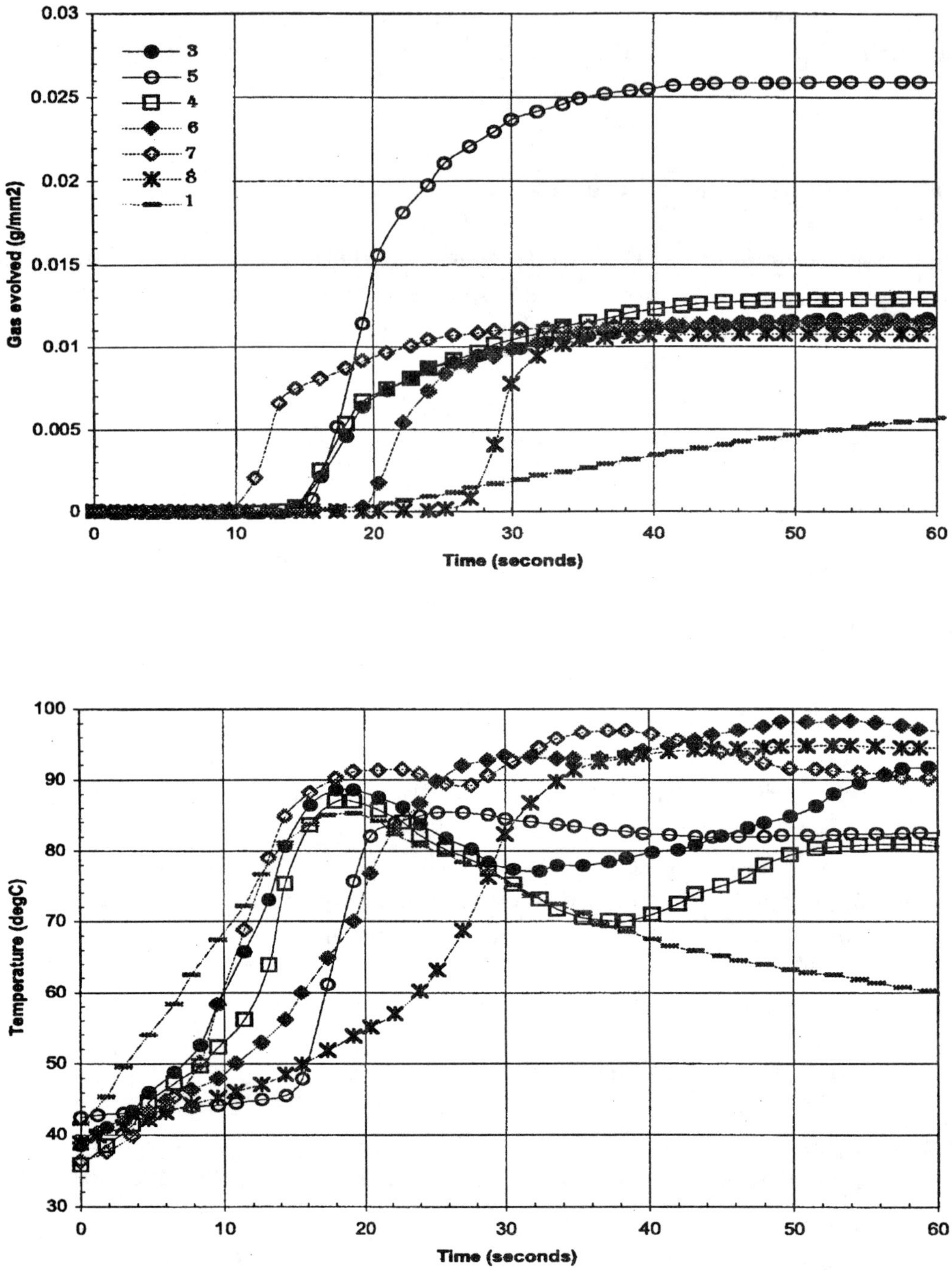

Figure 4: Experimental Results Summary. PCl_3 + 20 % HCl (dewar only).

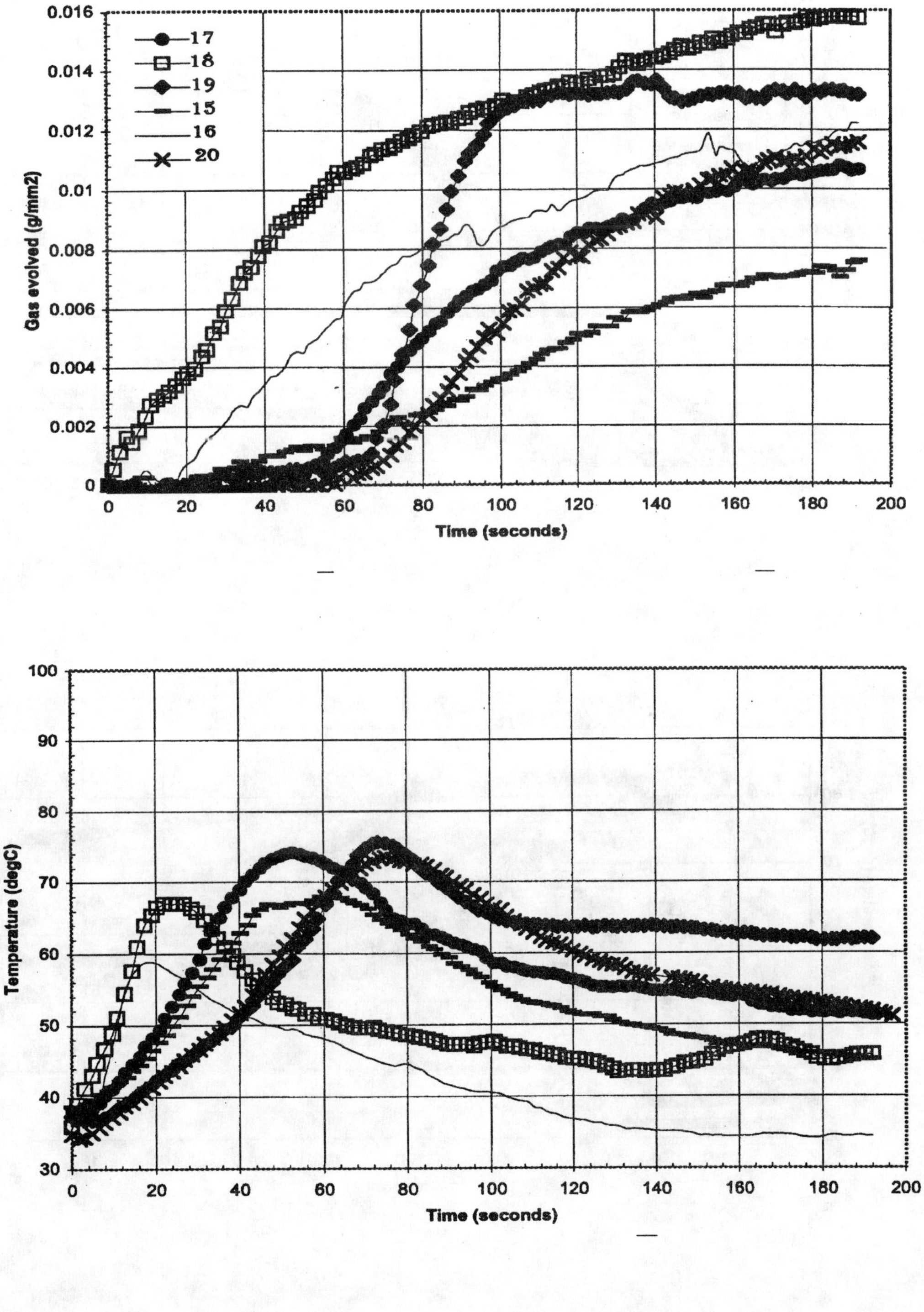

Figure 5: Experimental Results Summary. Surface Area Effects.

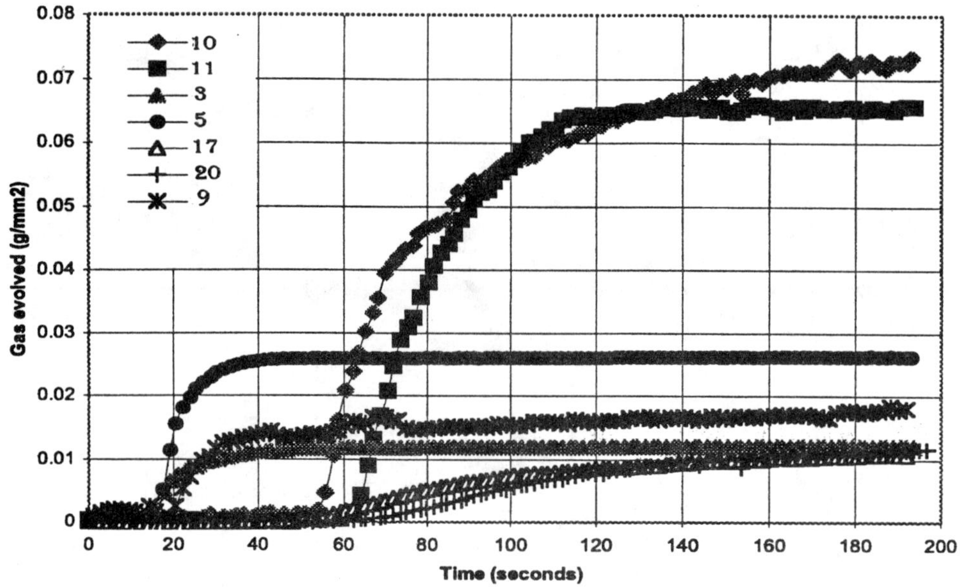

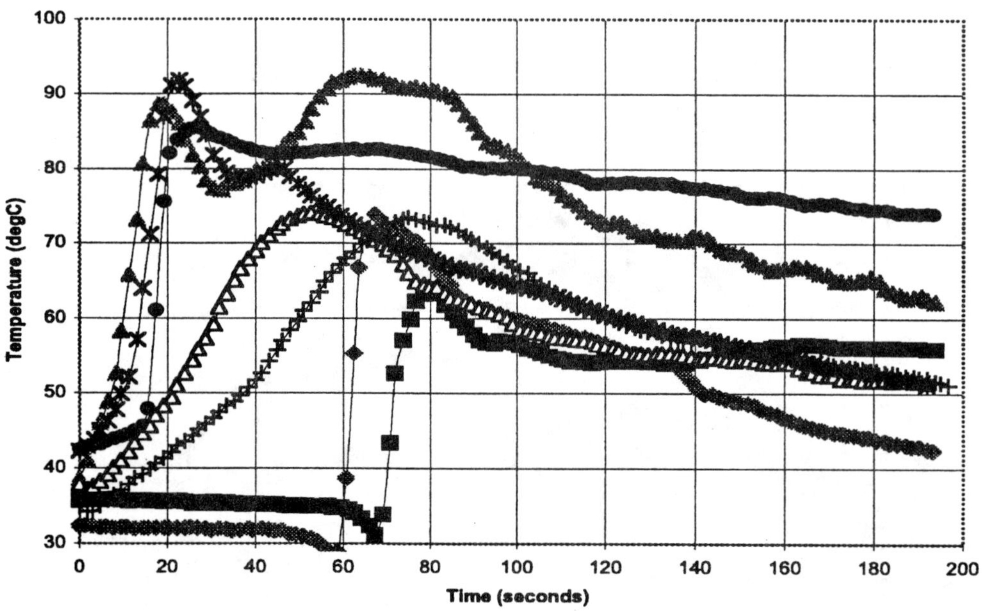

Figure 6: Experimental Results Summary. Stirring Effects.

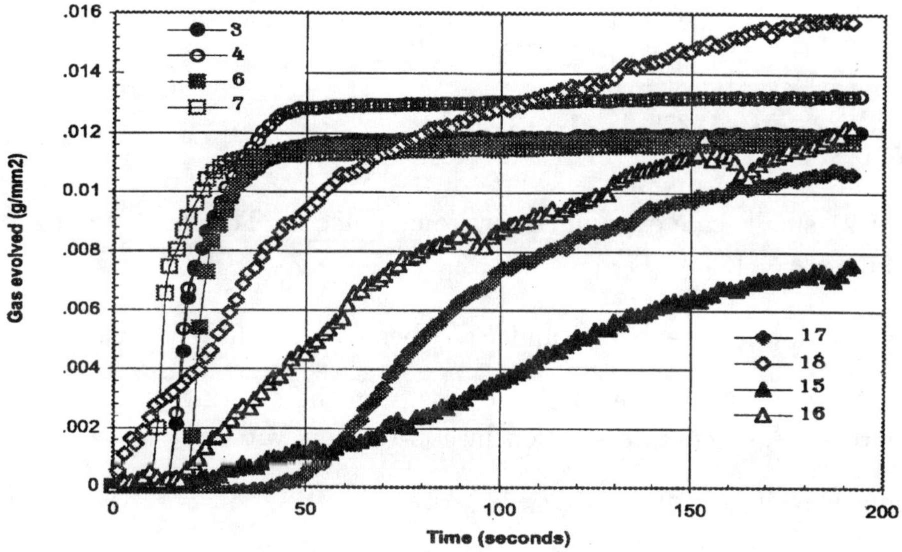

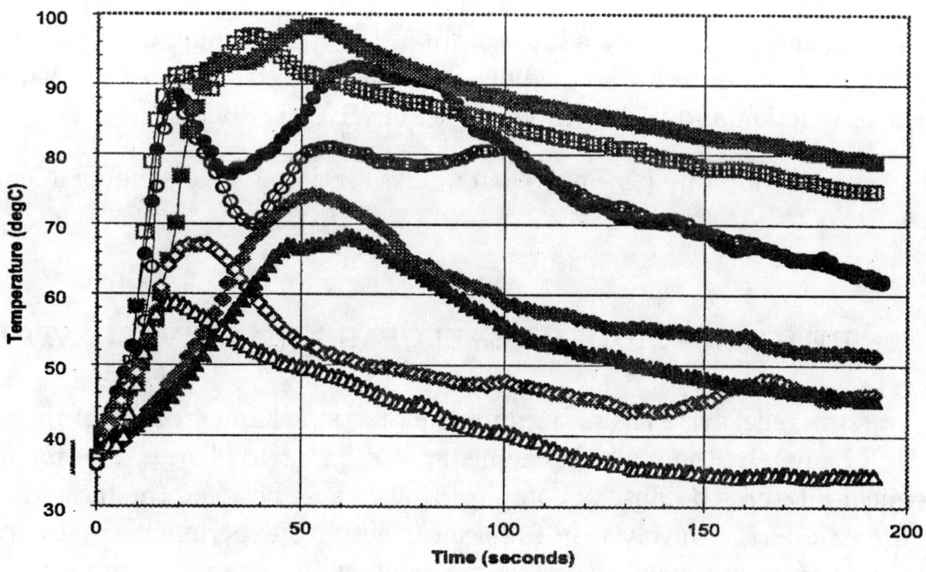

The results show that the rate of gas evolution is significantly reduced when 20 % HCl is added instead of water. Also, a higher peak temperature was achieved with the water addition cases as compared to the HCl cases.

In the water addition cases, stirring has little effect on the rate of gas evolution or temperature rise which suggests the reaction is fairly rapid.

In the gradual addition cases, the rate of gas evolution suggests that the water reacts immediately and there is no opportunity for accumulation to occur.

Experimental Conclusions

A total of 21 small scale experiments were conducted. The main findings from the experimental data are as follows:

- the reaction between water and PCl_3 begins immediately, with no delay period;
- there is a delay in the evolution of gas from the reacting mixture as HCl produced by the reaction enters into solution in non-reacted water;
- the gas evolved includes PCl_3;
- when HCl is added rather than water, the maximum temperature achieved is less and the reaction appears to proceed more slowly;
- from the stoichiometric or balanced case three stages are apparent following the addition of water. Firstly a rapid temperature rise and little loss of vapor as the reaction proceeds and all HCl enters into solution; then the temperature falls because of tempering from water and PCl_3 vapor loss; finally the temperature rises again due to the heat of reaction and little tempering as only HCl is liberated;
- the gas evolution rate per unit reactor cross-sectional area is independent of the reactor size.

REACTION MODEL DEVELOPMENT AND VALIDATION

Ideally, pressure relief systems are sized for the reaction characteristics of the chemicals in the vessel. Unfortunately, the reaction chemistry is difficult to characterise for many chemicals and chemical mixtures because of complex interactions between the individual components. The high cost and risk involved in full-scale reactivity experiments make test data scarce. Chemical engineers use small-scale tests and modelling computer codes to overcome these difficulties. A comprehensive approach which provides a reliable basis for design of reactive systems is described by Melhem et al. [20].

This advanced approach has been adopted to model the PCl_3 and water reaction as previous attempts to model this reaction by other methods had proved unsatisfactory due to nonideal solution effects and the presence of two immiscible reacting layers. The methodology used is described briefly in the following sections. Note that a greater emphasis has been placed on the reaction stoichiometry, nonideal solution behaviour and kinetic modelling as these are areas where this approach is particularly strong.

Reaction Stoichiometry

The reaction stoichiometry and the chemical identity of the products must be known before adiabatic constant volume kinetic data can be used for designing emergency relief systems. Both the stoichiometry and the chemical identity of the products are implicit in the measured temperature / pressure data at constant volume.

The pressure / temperature relation is a function of thermal inertia, liquid fill level, composition and chemical identity (vapor-liquid equilibria, liquid-vapor density, heats of formation, etc). For a specified relief device set pressure there is a unique corresponding system temperature and for reactive systems this temperature corresponds to a reaction rate. Small errors in estimating this temperature can lead to inadequate sizing of the relief device. Consequently, the pressure / volume / temperature relation of a reactive system is a fundamental and unique relation that must be accurately represented during direct scale-up or computer simulation.

A unique solution can be guaranteed by requiring that the reaction stoichiometries and chemical identities of the products are thermochemically favorable. This can be done for any proposed stoichiometry and product list by simultaneously calculating the physical and chemical equilibrium conditions at a constant volume by direct minimization of the Gibbs free energy. To simplify the equilibrium calculation the following constraints can be imposed:

- calculated pressure, volume and adiabatic temperature rises match experimental values;
- calculated reaction products are consistent with experimentally observed ratio of noncondensables / volatiles at final cooldown conditions;
- calculated reaction products agree with experimental light gas analysis findings.

This analysis allows the global stoichiometry to be identified from constant volume small-scale tests, such as the ARC.

The Expert Edition of SuperChems, the Arthur D. Little thermal hazards assessment and pressure relief design software, is used to carry out the above calculation. It has an extensive chemical database, containing thermodynamic and physical properties for over 1,200 materials, and can determine the simultaneous physical / chemical equilibria using the above method.

For this study the stoichiometry was determined to be as follows:

$$PCl_3 + 3H_2O \longrightarrow H_3PO_3 + 3HCl \qquad (9)$$

This is consistent with the experimental observations discussed earlier. Note that the maximum temperatures achieved during the experimental work were around 120 C and consequently the amount of phosphine generated by thermal decomposition of phosphorous acid, which occurs at 180 C and above, is expected to be negligible.

Reaction Model

For systems that have complicated reaction mechanisms, kinetic data can be obtained by carrying out experimental analysis in a closed test cell, such as the ARC. Once the number of reactions, stoichiometry and reaction products are determined, as described in the previous section, an iterative nonlinear regression analysis is used to determine the activation energies, reaction orders and pre-exponential factors for all reactions involved. These parameters are then adjusted to minimize the difference between the predicted and actual pressure and temperature histories.

The Expert Edition of SuperChems is used to carry out this iterative calculation. The procedure is as follows:

- estimate the reaction orders, pre-exponential factors and activation energies;
- simulate the test cell, including the heat-up period;
- compare simulation with actual test cell results and calculate the sum of the squares of the difference between the temperature / pressure, temperature / time and pressure / time histories;
- repeat until the sum of the squares is minimized.

The working model can then be used to predict the transient temperature, pressure and venting flowrate behavior of full-scale equipment by specifying the size, configuration, operating conditions etc. of the actual plant.

Using the ARC data reported earlier, the reaction was determined to be second order with respect to the concentration of water and one-third order with respect to the concentration of PCl_3. The pre- exponential factor was determined to be 2 and the activation energy 3,500 kJ/mole. The rate expression is therefore:

$$\frac{dN}{dt} = 8D^2\,[H_2O]^2\,[PCl_3]^{1/3}\exp(-3500/T) \tag{10}$$

where D is the diameter of the PCl_3/H_2O interface surface area in inches, T is the system temperature in Kelvins, and [] is the overall concentration in kmol/m^3.

Model Validation

The kinetic model was extracted from ARC data as described above. Figures 7, 8, 9 and 10 compare the actual ARC data against the model output for the following four ARC runs:

- water and PCl_3 at the balanced (3:1) ratio;
- 11 % HCl and PCl_3 at the balanced water:PCl_3 ratio (3:1);
- 11 % HCl and PC13 with excess water (6:1);
- 11 % HCl and PC13 with excess PC13 (1.8:1).

The pressure / temperature, pressure / time, rate of pressure change / time and rate of temperature change / time graphs are all considered. In all cases the model output shows good agreement.

Altough not reported in detail in this paper, the ARC model was also validated against the VSP experiments, which involved the addition of water to PCl_3 over a 5 minute period with stirring. There is reasonable agreement with the model slightly underpredicting the temperature and overpredicting the pressure. This suggests that the model is suitable for open and vent containment systems.

CONCLUSIONS

From the findings of the literature review and the additional experimental work, carried out as part of this study, a working model has been established for the PCl_3 and water reaction.

The kinetic information for the model was extracted from ARC experimental data. The model has been validated against several ARC data sets and VSP experiments and a good agreement obtained. This working model is based on global reaction stoichiometry and the experimentally determined reaction kinetics and can predict the system pressure-volume-temperature behavior, taking into account non-ideal solution behavior. It can therefore be used to predict the transient temperature, pressure and venting flowrate behavior of the full-scale equipment.

The experimental work, involving open systems, provided a visual confirmation of the reaction behavior and showed that:

1. the reaction between water and PCl_3 begins immediately, with no delay period;
2. there is a delay in the evolution of gas from the reacting mixture as HCl produced by the reaction enters into solution in non-reacted water;
3. the gas evolved includes PCl_3;
4. when HCl is added rather than water, the maximum temperature achieved is less (maximum approximately 75 C compared to 100 C) and the reaction appears to proceed more slowly;

Figure 7: Model prediction for ARC data. Water to PCl3 molar ratio = 3.012

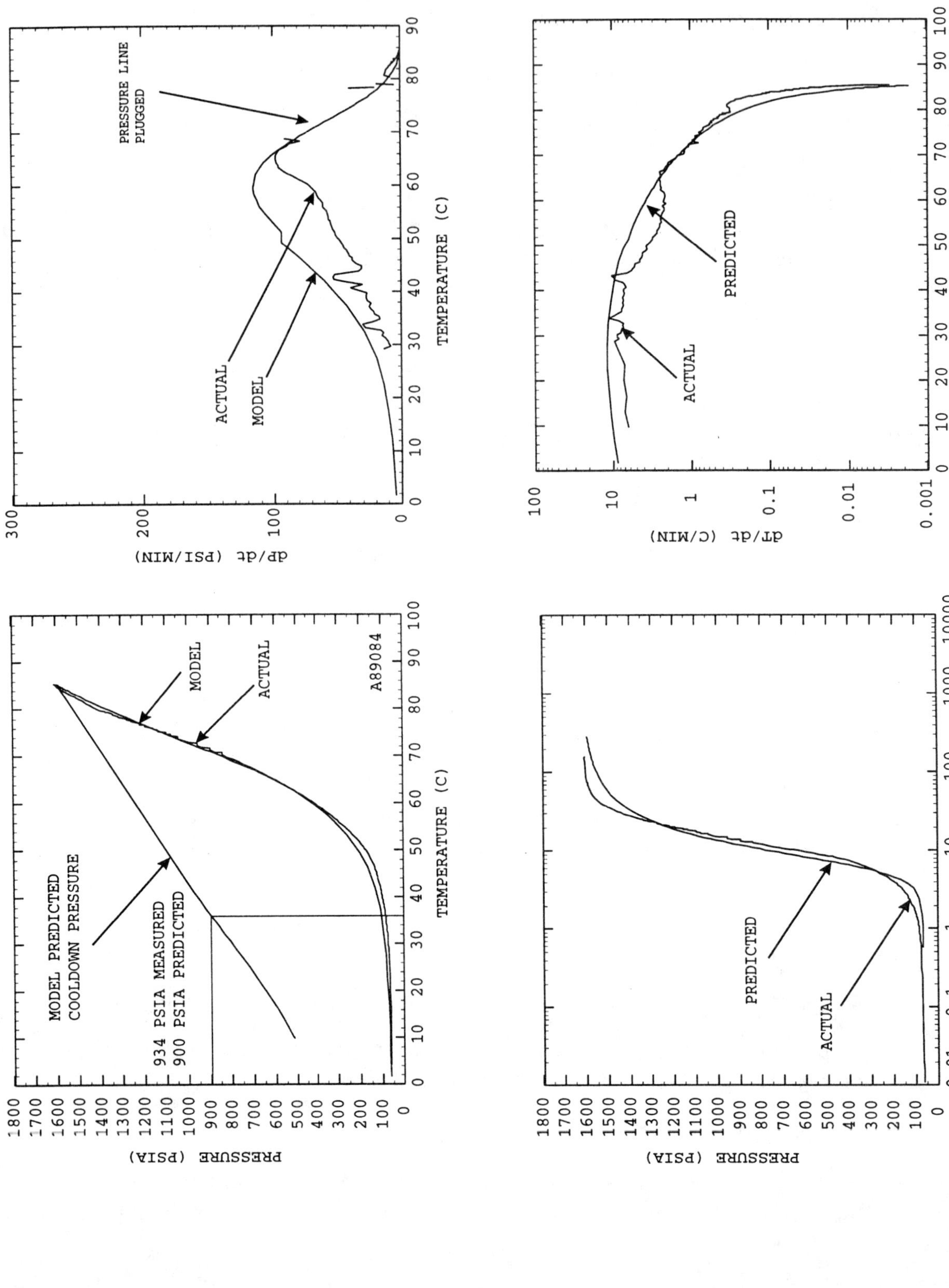

Figure 8: Model prediction for ARC data. Water to PCl3 molar ratio = 3

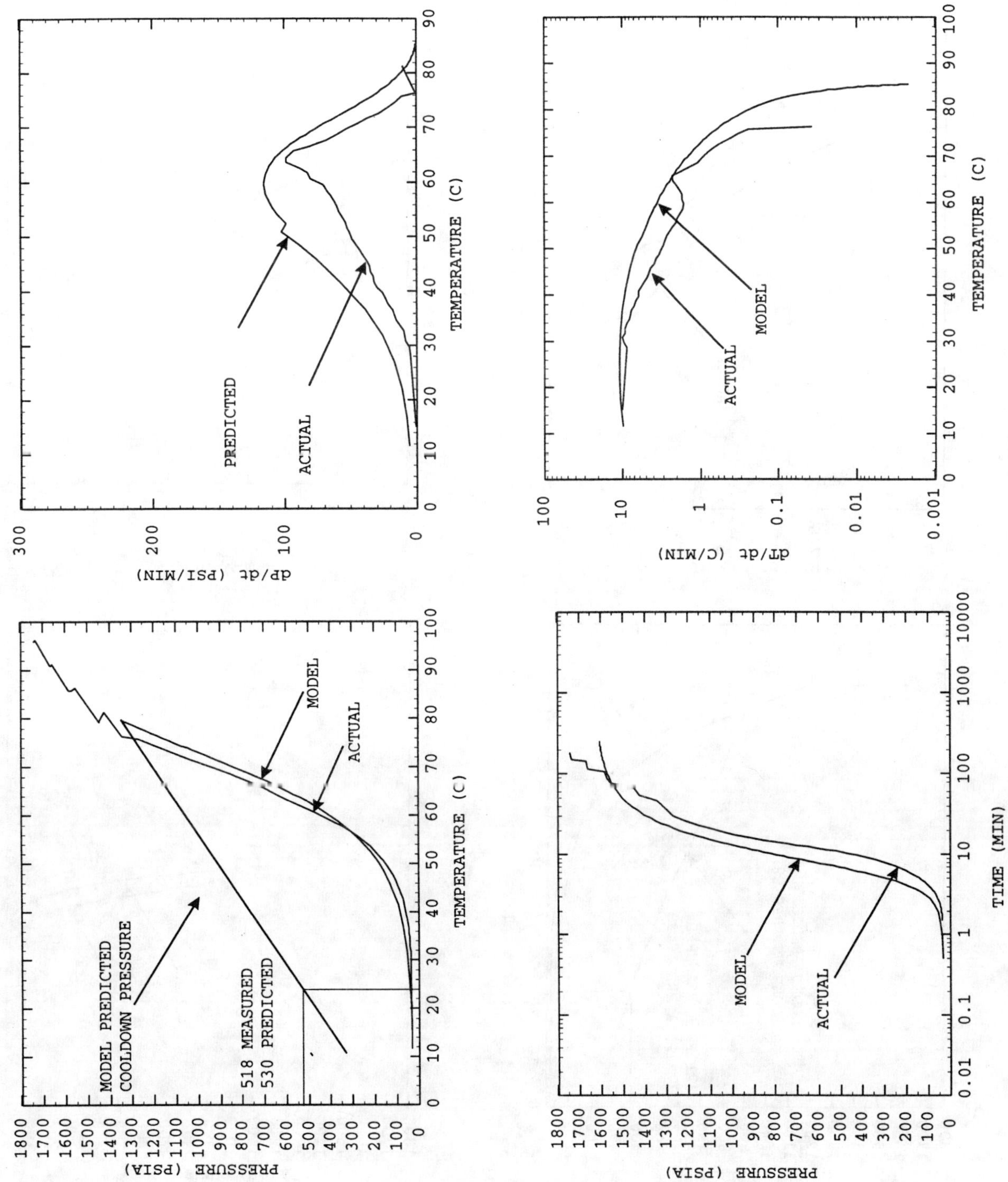

Figure 9: Model prediction for ARC data. Water to PCl3 molar ratio = 6

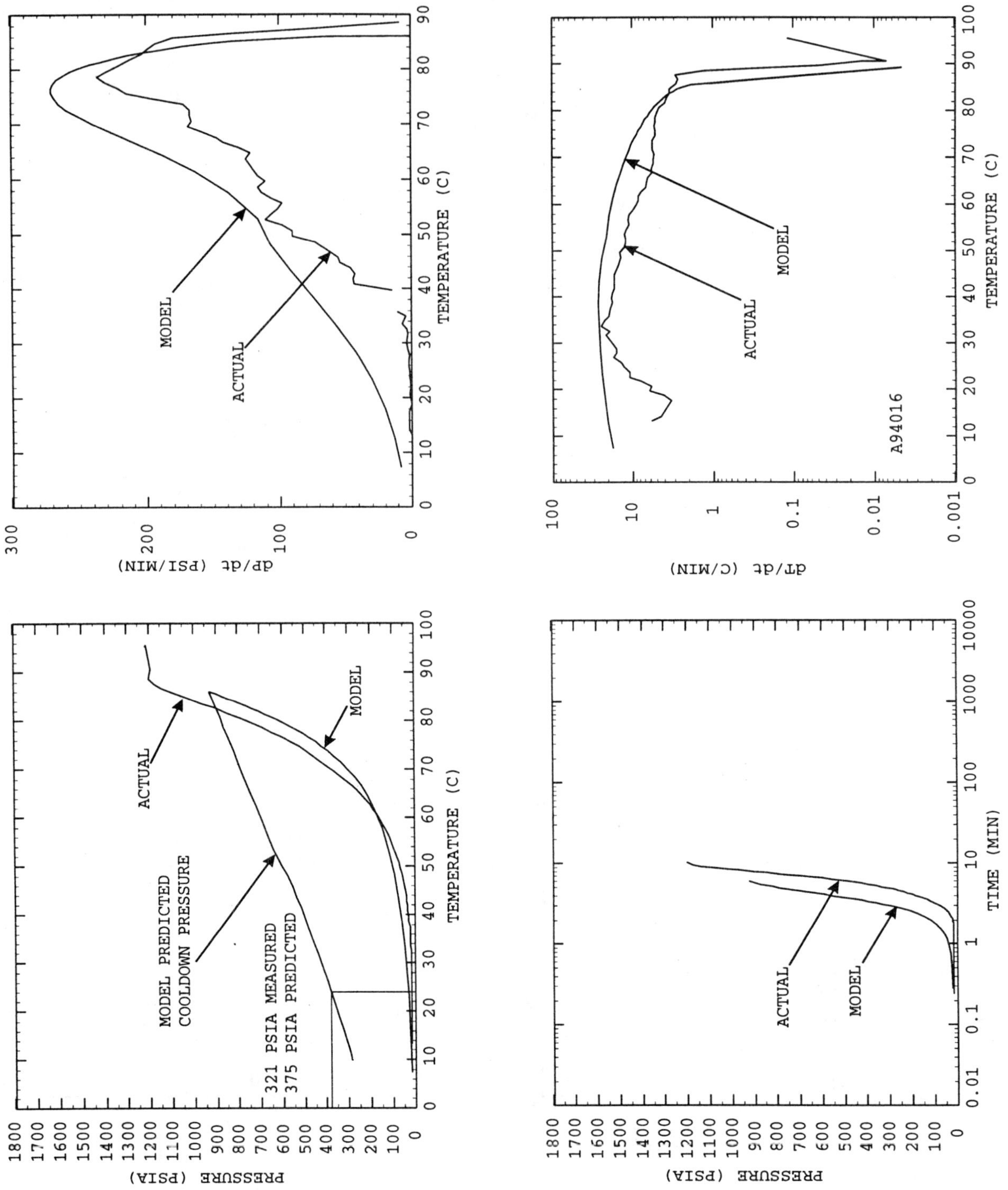

Figure 10: Model prediction for ARC data. Water to PCl3 molar ratio = 1.809

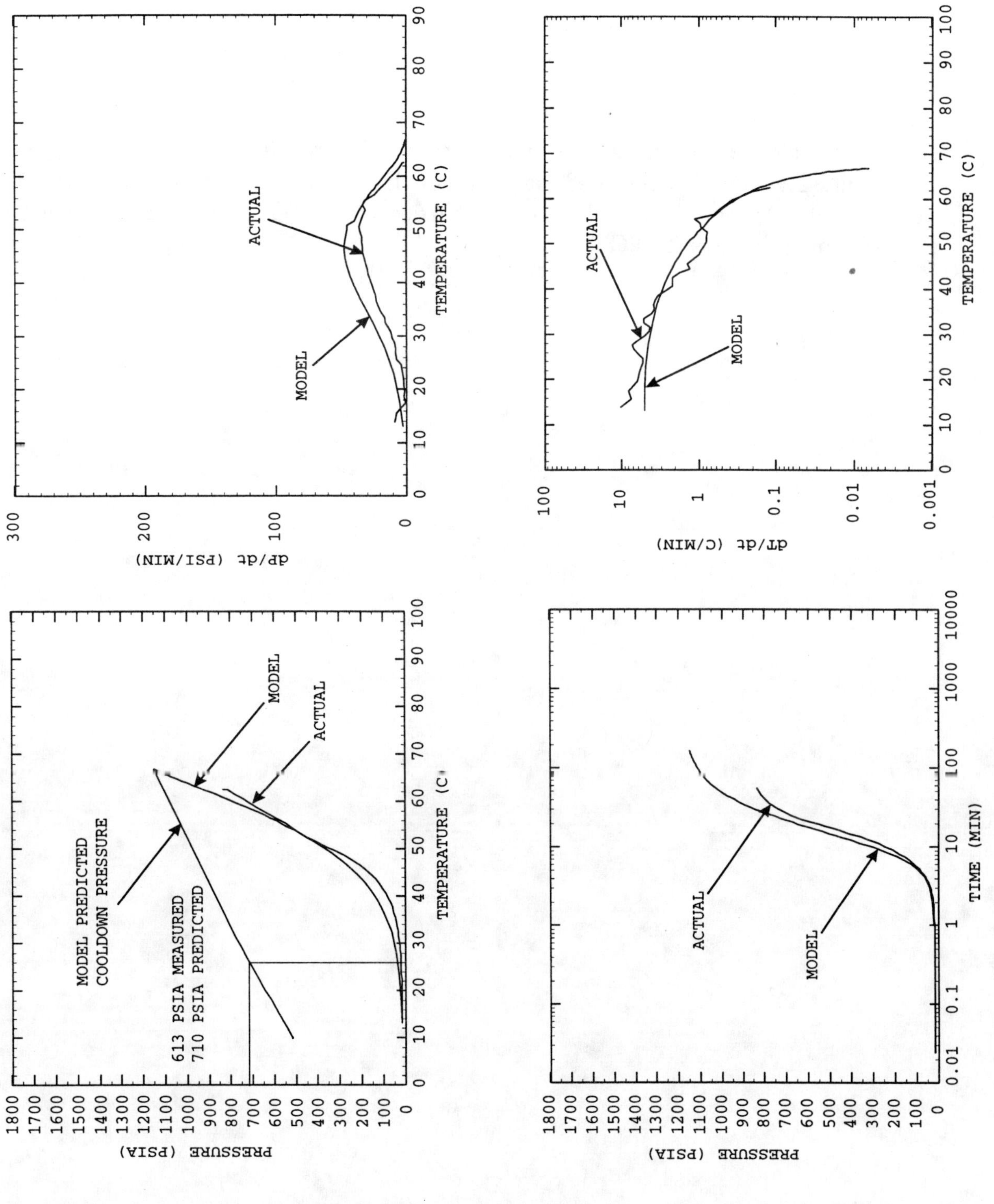

5. from the stoichiometric or balanced case three stages are apparent following the addition of water. Firstly a rapid temperature rise and little loss of vapor as the reaction proceeds and all HCl enters into solution; then the temperature falls because of tempering from water and PCl_3 vapor loss; finally the temperature rises again due to the heat of reaction and little tempering as only HCl is liberated;

6. the gas evolution rate per unit reactor cross-sectional area is independent of the reactor size.

Although not detected during these experiments, other research reported a slight odor of phosphine. The literature reviewed suggests that this is because of phosphorous acid decomposing to phosphoric acid and phosphine. This decomposition is reported to be fairly rapid at higher temperatures (250 C).

References

[1] G. A. Melhem and H. G. Fisher. An overview of superchems for diers for emergency relief systems and vent containment design. In *Loss Prevention Symposium, Houston, AIChE*, 1997.

[2] J. W. Mellor. *A Comprehensive Treatise on Inorganic and Theoretical Chemistry Vol. VIII.* 1928.

[3] H. Davy. *Phil. Trans*, 100/101, 1810/1811.

[4] K. Kraut. *Journ. Chem. Soc.*, (158):332, 1871.

[5] A. Geuther. *Ber. Liebigś Ann.*, 5.925/240.208, 1872/1887.

[6] A. Besson. *Compt. Rend.*, 111.972, 122.814, 125.771, 1032, 132.1556, 1890,1896,1897,1901.

[7] G. Carrara and I. Zoppelari. Reaction velocity of non-homogeneous systems - reaction of some sulphur and phosphorus compounds with water. *Gazz. Chim. Ital.*, 26(i):493, 1896.

[8] N. V. Sidgwick. The behaviour of the simple halides with water. *Journ. Chem. Soc*, 125:2672, 1924.

[9] A. D. Mitchell. The hydrolytic decomposition of phosphorus trichloride. *Journ. Chem. Soc.*, 127:336, 1925.

[10] R. F. Hudson. The vapour phase hydrolysis of non-metallic chlorides. In *International Congress of Pure and Applied Chemistry (London), Vol. 11*, pages 297–305, 1947.

[11] H. Schnockel and S. Schunck. *Z. Anorg. Allg. Chem.*, 548:161, 1987.

[12] H. Schnockel and H. Schunck. High temperature hydrolysis of pcl3 and pbr3. detection of molecular metaphosphoric acid. *Phosphorus and Sulfur*, 39:89–90, 1988.

[13] A. U. Acuna, D. Husain, and J. R. Wiesenfeld. Kinetic study of electronically excited phosphorus atoms, p(3 2dj, 3 2pj), by atomic absorption spectroscopy. *J. Chem. Phys.*, 58(2), January 1973.

[14] A. U. Acuna, D. Husain, and J. R. Wiesenfeld. Kinetic study of electronically excited phosphorus atoms, p(3 2dj, 3 2pj), by atomic absorption spectroscopy ii*. *J. Chem. Phys.*, 58(12):–, June 1973.

[15] D. Husain and N. K. H. Slater. Time-resolved resonance of ground state phosphorus atoms, p[3p3(4s1/2)]. *J. C. S. Faraday II*, 74:1627, 1978.

[16] M. A. A. Clyne and Y. Ono. Kinetic studies of ground-state phosphorus atoms. *J. Chem. Soc. Faraday Trans. 2*, 78:1149–1164, 1982.

[17] I. B. Rozhdestvenskii, V. N. Gutov, and N. A. Zhigulskaya. Approximation coefficients of the thermodynamic potential for substances formed by aluminium, boron, carbon, calcium, chlorine, copper, fluorine, hydrogen, potassium, lithium, magnesium, nitrogen, sodium, oxygen, phosphorus, sulphur, silicon and titanium atoms in a temperature range up to 6000 k. *Glavniiproekt Energ. Inst.*, 7:88–121, 1973.

[18] *Ullmanns Encyclopedia of Industrial Chemistry, Fifth Edition, Vol. 19.*

[19] *Encyclopedia of Chemical Technology, 3rd Edition, Vol. 17.*

[20] G. A. Melhem, H. G. Fisher, and D. A. Shaw. An advanced method for the estimation of reaction kinetics, scaleup, and pressure relief design. *Process Safety Progress*, 14(1), 1995.

A Fire Risk Analysis Model for Assessing Options for Flammable and Combustible Liquid Products in Storage and Retail Occupancies

Dr. John R. Hall, Jr.
National Fire Protection Association

Introduction

A couple of very large recent fires have raised awareness of the dangers potentially posed by large stores of flammable and combustible liquid products in the increasingly popular large warehouse-style retail occupancies. These same fires called into question the ability of sprinkler systems designed to existing codes and standards to handle the hazard.

The National Fire Protection Research Foundation (NFPRF) became involved in the discussions on what to do about this problem, and in the course of their involvement, they sought and sponsored a fire risk analysis of the problem and the options for dealing with it. The author worked with Dr. Pravin Gandhi and David Sheppard of Underwriters Laboratories, Inc., (UL) on this project. UL's work focused particularly on the development of a rate of heat release fire growth model that could be used to generalize from a body of fire test results. This modeling of the situations specific sprinkler designs could handle was then embedded in a larger decision-tree probabilistic framework, which was the focus of the author's work and will be the focus of this article.

Why Fire Risk Analysis?

Fire risk analysis has rarely been the analysis method of choice in examining proposed changes to codes and standards. Fire protection engineers tend to think in terms of fire hazard analysis, in which a design scenario is selected, a successful outcome is defined, and some combination

Printed with permission from the Fourth Quarter issue (Volume 31, Number 4) of *Fire Technology*.

of modeling, fire incident data analysis, and particularly fire testing is used to determine whether the proposed fire protection design will produce a successful outcome if presented with a fire of the design scenario type. Success tends to be sharply defined in an either/or manner. The hazard is protected or it is not. A major fire is prevented or not.

Fire risk analysis takes a different perspective. Risk can be reduced or increased, but it can never be eliminated. Realistic worst-case fire scenarios are far worse than the design scenarios used in fire hazard analysis and are beyond the capacity of any fire protection design. Success is better defined in terms of improvement, not some sort of unattainable absolute solution. And fire risk analysis can help indicate which fire protection designs produce large improvements, which ones produce small improvements, and which ones will or might make things worse or at least no better.

Fire risk analysis does not require that risks be compared to costs, let alone that reductions in risk exceed the cost of such reductions. Fire risk analysis does not require that fire protection be limited to those arrangements that will produce the largest improvements. Such decisions depend upon the value judgments made by those who are responsible for setting society's levels of acceptable risk. Fire risk analysis can indicate whether a certain level is achieveable, and how, but not whether that level is appropriate.

It should be clear that this project was not an end in itself but was part of a journey toward improved fire safety. The alternatives analyzed were all fire safety strategy choices under active consideration by code-writing bodies. This project lays the groundwork for focused attention on certain fire safety directions and implies specifications for future fire testing to confirm the adequacy of fire safety proposals for the handling of these products in these properties.

What is Fire Risk Analysis and How Does It Work?

"Fire risk" refers to any well-defined summary measure characterizing unwanted fires that may occur in terms of severity and probability. Fire risk is typically analyzed by the use of a tree-diagram method, in which the branches of the tree show how the unplanned, uncontrollable elements of a fire scenario combine with the design choices to produce or avoid a fire outcome. Every complete branch of the tree, from the first event identified to the final outcome, has an associated severity and probability that the tree is used to calculate. For every fire scenario, you can be safer by reducing its probability or its severity or both.

In terms of executing the analysis of fire risk, fire statistics can be used to help estimate probabilities and severities, but fire statistics often lack the level of detail needed for decision-making and they cannot address new products, which by definition have never been in use and so have never had the chance to be in a real fire, i.e., not a test. Fire tests can be seen as one way of estimating severity in much more detail for a particular fire scenario, that is, for the types of fires that will tend to behave like the fires in the tests. Modeling can be used to generalize from the test results, i.e., to estimate how the fire test results would have changed if the test scenario had been different. Therefore, modeling plus testing may give the best possible estimates of severity with the required level of detail for all fire scenarios.

Of all the data sources available, only fire incident data can provide real-world based estimates of the probability side. The only other alternative is estimation by (hopefully expert) panels or individuals. Therefore, a fire risk analysis framework will need to be tailored to the structure of the best available fire incident data bases and will reflect the limitations of that structure.

Fire risk analysis, then, can consist of a series of fire hazard analyses whose results are weighted by probabilities to produce an end-measure of expected severity. Fire risk analysis can also consist of a series of fire growth modeling runs in which the probabilities are summed for all scenarios that lead to unacceptable severity, and that is the approach used here. This is tantamount to adding up the probabilities of all the fire scenarios the design choice (e.g., container, packaging, suppression system) flunked, whether it flunked an actual test or was projected (by modeling) to have flunked a test that had never been conducted.

Overview of Fire Risk Analysis Structure and Calculation

Figure 1 shows the structure of the general model of the fire risk analysis calculation. As the formula at the bottom of Figure 1 indicates, the identified fire safety objective is to prevent an "uncontrollable fire of unacceptable size". Qualitatively, this objective is aimed at a runaway fire that endangers much of the building and offers the potential for a large loss. To permit analysis with available data, this is restated quantitatively in terms of fire spread beyond the room of fire origin, which is considered firm evidence of either the failure or the absence of the fire protection systems and features in place.

In this model, the tree begins with a specification of decision or choice variables and then asks five questions in sequence:

(a) Is there a fire (probability of ignition)? The probability of ignition is estimated as the ratio of the estimated annual number of fires in the property class of interest to the estimated number of facilities of that type. The numerator comes from 1980-1991 national estimates of fire statistics, computed using the usual methods. [1] The denominator is an estimate based on the number of facilities in Phoenix and Portland (OR), the two communities for which data was readily available to the project, that data then being projected to the country, based on population. Specifically, the average US population during 1980-1991 was approximately 239 million, or 165 times the 1993 population in Phoenix and Portland of roughly 1.45 million.

(b) If there is a fire, what type of fire is it (probability of scenario i)? Four scenarios were defined, as shown in Table 1. The first two represent the two likely high-challenge initiating fires in the room where large stores of product are located. The well-known fires involving these products and properties have all been arson or product spill fires. Table 1 explains how the coding structures of the available fire incident data bases were used to estimate which fires involved these scenarios. The third scenario captures all other fires in the product rooms, and the fourth scenario captures all other fires in the property.

(c) Does the fire (of scenario type i) take place in the presence of "miscellaneous favorable factors," not explicitly documented, that tend to keep fires small? These factors are an accommodation to the limited detail available in the best national fire incident data bases. The probability of such factors being present is estimated as the frequency with which fires stayed small (i.e., no flame beyond the room of origin) in the absence of a sprinkler system to stop them.

The factors may be the kind you could not include in a fire protection design (e.g., random good luck in the location of the fire or the movement of people to discover it) or the kind you could (e.g., the use of roving patrols to achieve very early discovery of any possible fires). Either way, they are, by definition, factors that cannot be explicitly identified, except that they must be applicable to the indicated fire scenario. This one fact allows many candidate factors to be eliminated from consideration as major contributors, as will be seen.

(d) <u>If there is a fire, is there a complete, operational sprinkler present?</u> The probability of having any kind of sprinkler is estimated from data on reported fires, while the probability that a sprinkler will be complete and operational is inferred as the frequency with which a sprinkler operates and fire is kept from spreading beyond the room of origin, given that the sprinkler has been tested (and so would have been expected to activate if operational) by a fire that spread beyond the immediate area of origin.

The patterns in the statistics assure that the calculation is dominated by fires involving ordinary combustibles, and the model assumes that any sprinkler design will control a fire in ordinary combustibles wherever it begins. Hence, any sprinkler design that does not control such a fire is assumed to be partial or non-operational or both.

It is assumed that in the fires reported from the field, all automatic suppression equipment consists of conventional ceiling-mounted, water-based sprinklers, because the usage of in-rack and foam-water sprinklers is still so recent and infrequent. The exact type and extent of coverage are not recorded in the reported data and so are assumed to be the most frequently used system in available tests, i.e., a system with 165°F sprinklers. The probability of operationality for foam-water systems is estimated to be 0.9345 times the probability of operationality for the corresponding water-based systems, based on analysis done at Factory Mutual Research Corporation. [2] The probability of operationality for in-rack systems is estimated to be the same as the operationality probability for ceiling-mounted systems. [3]

(e) Given (a) through (d), does an uncontrollable fire of unacceptable size result? This is the phrase used to define failure. In fire incident data, it means a fire that spread beyond its room of origin. (Flame beyond the room of origin is usually taken to mean flashover in the room of origin.) In interpreting fire tests, it means a fire that had developed to a point where it either had overwhelmed the sprinklers or would do so if allowed to continue. The two definitions are assumed to be equivalent.

The modeling assumes that the sprinklers are overpowered because the rate of heat release exceeds the capacity of the sprinklers. Other modes of failure, such as a shielded fire developing when released product flows under a storage array, are assumed to occur only under conditions where an excessive rate of heat release also occurs.

The model assumes that the miscellaneous favorable factors, if present, will always produce a small fire. The model also assumes that any fire not having either those favorable factors or a complete, operational sprinkler system to stop it will reach unacceptable size. For fires without miscellaneous favorable factors confronted with a complete, operational sprinkler, test data and modeling are used to estimate which will happen.

Table 2 summarizes the six dimensions of choice that were considered, the alternatives considered on each dimension, and the 504 combinations that were therefore analyzed. Table 3 provides a detailed example of the model's calculations by walking through the actual values calculated under (a) through (d) for some of the 36 combinations involving rack storage in general warehouses.

Calculating Probability of Uncontrolled Fire of Unacceptable Size

The approach used required that specific physical initiating fire conditions be identified to represent each fire scenario and that modeling be used to generalize from the fire tests to the predicted performance of designated fire suppression systems in the presence of that initiating fire and the various product and product storage and handling choices to be analyzed. In so doing, it was necessary to work around the fact that the reported fire data do not indicate specific physical conditions.

A non-representative NFPA fire incident data base called FIDO (Fire Incident Data Organization) does include incidents with such information, which sufficed to support the view that fires do occur that are more severe from the onset than any fire protection system can hope to control. For example, one identified fire was an incendiary fire started on the exterior walls of a building, after 34 55-gallon drums of Class I liquid, stored outside, had been opened and emptied around the building, to start and spread the fire. Total building involvement in fire was described as instantaneous.

A general estimating method, based on modeling of the rate of heat release, was developed by UL to permit estimation of sprinkler performance for more cases than the test data would allow. However, the method could not be used for all the scenarios under consideration because the data for the required delivered density (RDD) for the flammable and combustible liquids has not yet been developed. In the analysis, these cases were coded as possible successes or failures. This produced a range in the severity estimate for those cases and therefore a range, rather than a point estimate, for the estimated risk for many alternatives.

Tests already performed were characterized by UL in terms of a rate of heat release curve relative to time. Test results also indicate the effectiveness of each type of suppression system for a specific scenario. The DETACT-QS fire model, using generalized heat release rate parameters, was used for some cases to determine the effectiveness of suppression systems different from those used in the tests (e.g., 16 foot ceilings with 286°F sprinklers).

Most tests for which data was available used initiating fires involving Class I B flammable liquid in either an 8-ounce or a 5-gallon quantity. These two tests represent an 80-fold difference in the size of the initial fire (two 8-ounce quantities per pint, 8 pints per gallon, 5 gallons, or 2x8x5 = 80).

Starter fire characteristics do not have to be estimated for Scenario 4 fires (fire starting in a room without stored product), because such fires involve only ordinary combustibles. By assumption, the performance of the suppression system does not need to be modeled; if present and operational, it always works and produces success.

Scenario 1 (accelerant-driven arson fire) is assumed to have started near the product and with the use of significant flammable liquids to start it, so it is matched to the larger 5-gallon initiating-fire tests. Scenario 3 (not starting with either a product spill or accelerant-driven arson) is assumed to have started with a much more modest initiating fire, so it is matched to the 8-ounce initiating-fire tests.

Scenario 2 (product spill) does not have a standard initiating fire size, because the size of the spill may depend upon the type of container and other decision variables. The typical fire of Scenario 2 is assumed to start with a lift truck or other method being used to remove a pallet or rack of containerized product from the top of a stack. (This was the chain of events that led to one of the best-known major fires involving these products.) In the process, all of the containers in the group being moved are dislodged, leading to potential product spills through impact breaches after a fall.

Based on fire incident data, requirements of current codes and standards, and current superstore design, the model assumes that the height of the fall is 10 feet for palletized storage and 20 feet for rack storage. Drop tests are used to estimate how many containers will break under these conditions, and a calibration is made to the best known such fire so that a pallet of 1-gallon metal round (paint can style) containers dropped from 15 feet would be expected to release product from 8-10 containers.

Tests were done with sealed containers, filled with one gallon of water, in corrugated cartons with bursting strength of 200 lb/in^2. See Table 4 for a summary of results.

Work by UL on the rate of heat release modeling indicated that the results were not highly sensitive to the exact size of the initiating fire for fires greater than 2 gallons. This made it possible to match the various product spills estimated from the drop tests to the two initiating fire sizes represented in the test data. Thus, the drop tests could be taken to mean that the metal F-style container dropped 10 feet would resemble the 8-ounce test fires, the plastic F-style container dropped 10 feet would not rupture and would produce no spill fire, and the other four cases studied would resemble the 5-gallon test fires.

Estimated Fire Risk

Table 5 illustrates the results of the fire risk estimation for the case of general warehouses using rack storage. The comments that follow also apply to all the cases studied.

Nearly all of the overall fire risk comes in fires where the sprinkler system failed to operate and there were no miscellaneous favorable factors to produce a small fire in the sprinkler's absence, as opposed to fires that overwhelmed an operational sprinkler system. In order to focus on the impact of product-related choices, then, it is necessary to calculate a second set of fire risk values that would apply, assuming a complete operational sprinkler system is guaranteed to be present. These are shown in Table 6.

Some of the values shown in Table 6 are zero. This does not mean that zero risk is achieved in those cases, only that the non-zero risk occurs in the fraction of very severe scenarios that are even more severe than the most severe scenarios used in the analysis. That risk should be well below the 1/10,000 (E-4) to 1/1,000,000 (E-6) range that includes the values shown in Table 6.

The FIDO analysis cited earlier indicated that the fire risk values calculated for this project involve fires whose typical size is much smaller than the handful of well-publicized fires that made this issue a major concern in the fire community. If you take any of the property classes analyzed, the nation can expect to experience some fires spreading beyond the room of origin every year. Most of these fires will not receive attention beyond the local area where they occur, because the facilities involved are small, even marginal facilities, for which even a large proportional area of damage does not translate into a strikingly large dollar loss. Unfortunately, a better, more restrictive definition of failure cannot be easily substituted because of the limitations of available data. This concern should not apply to the risk values in Table 6, which reflect only fires that begin in a room with product and overwhelm a complete, operational sprinkler system.

Sensitivity Analysis

1. There is considerable uncertainty in the absolute level of risk estimated here. The estimates of the number of facilities in the country of each property type are very rough. And as noted, for overall risk, many fires deemed unacceptably large really are not. This uncertainty does not affect the use of the analysis to (a) identify options providing zero risk or minimum achieveable risk or (b) identify fire protection choices providing greater or lesser leverage on the magnitude of risk. It does complicate the application of any other notion of acceptable risk to the results.

2. Spill fires are overshadowed by other types of fires in the risk calculation to such a degree that factors related to spill fire probability (e.g., relative ignitability of different classes of liquids) have a negligible impact on the overall fire risk. This generalization is true even in view of the uncertainty associated with several parameters and assumptions in modeling spill fires. [4]

3. An uncontrollable fire of unacceptable size will nearly always be a fire for which there is neither a complete, operational, adequately designed suppression system to stop it nor the kinds of good luck and design factors classified under miscellaneous favorable factors. These fires will far outnumber the fires that overpower a complete, operational, adequately designed suppression system.

Most fires start in a room that does not have stored product, so it is the effectiveness of the suppression system and the presence and effectiveness of the miscellaneous favorable factors in those other rooms that dominate the overall risk calculation. This means the only major miscellaneous favorable factors must be those that apply throughout a property, which excludes any factors specific to the storage, handling or protection of the product alone.

Conclusions

1. Fires that spread beyond the room of origin -- the threshold used to identify fires large enough to be called unacceptable -- do occur and occur with some frequency. However, the typical such fire is much smaller and less consequential than the few major fires that tend to be thought of when the fire risk potential of stored flammable and combustible liquid products is being discussed. Therefore, it is important not to see the risk calculated here as being probabilities of occurrence for fires as large as those few well-publicized fires.

2. The dominant scenario in the fire risk calculation, in both the base case and most of the sensitivity analyses, is a fire that (a) starts in a room without large stores of product and (b) grows to unacceptable size by spreading beyond the first room involved in fire, because (c) there were no miscellaneous favorable factors to stop it in the room of origin, and (d) there was no sprinkler system, or at least none that was operational, in the room where the fire began.

3. Some of the miscellaneous favorable factors may reflect fire safety provisions, but many will not. These factors are called "miscellaneous" precisely because there is no data available to identify specific factors, let alone assess their relative importance. However, some insight can be gained by keeping in mind that the key miscellaneous favorable factors are

those that can prevent a large fire from developing even where there is no sprinkler system and in a room without large stores of product. Hence, no fire protection factors focused on the products or the room with the products could qualify. This supports the notion that most of the factors that matter are the random ones that cannot be designed for.

4. If the presence of a complete, adequately designed sprinkler system could be assured (e.g., by unusually effective enforcement of an appropriate code), then the fire risk associated with fires where the sprinkler failed to operate would be on the same order as the fire risk associated with fires that started small (i.e., 8 ounce fire) and overpowered the sprinklers. If in addition, the probability of operationality could be raised to 98% or better (which is higher than most field studies have found), then the fire risk associated with fires where the sprinkler failed to operate would be on the same order as the fire risk associated with fires that started large (i.e., 5 gallon fire) and overpowered the sprinklers. Fires that start large are more likely to overpower the sprinklers but are much less likely to occur, which is why the risk associated with them often is much less than the risk associated with fires that start small.

5. A suppression system like foam-water that is more effective than conventional sprinklers when it works but is more likely to fail to work will sometimes show up worse than conventional sprinklers in the overall fire risk calculation, because the probability of being operational when fire occurs is so much more important to the calculation than relative effectiveness when operating.

6. If one assumes a complete operational sprinkler system, then differences between types of containers do appear. Metal F-style containers always had zero risk for this limited definition of risk. Plastic F-style

containers had the potential for zero risk -- i.e., for some combinations, all their scenarios showed as successes or as possible successes/possible failures -- but only for certain combinations: (a) Class II liquids in fire-resistive packaging in rack storage with water-based sprinklers; (b) all liquids in fire-resistive packaging in pallet storage with water-based sprinklers; and (c) all combinations with foam-water sprinklers except for Class I liquids in regular packaging in rack storage. Metal round containers also had the potential for zero risk but had it for all combinations.

These results clearly show the potential value of additional testing to confirm or disconfirm that the scenarios that were rated as possible successes are successes.

7. The conventional fire protection engineering approach of defining the hazard and then protecting it is equivalent to demanding success in all fire scenarios but only considering scenarios in which sprinklers are operational. Perfect sprinkler operationality is not achieveable. Even with operationality rates better than most field studies have shown to occur, the risk associated with the fires that occur when sprinklers don't operate will tend to dwarf the risk preventable by changes in container and packaging choices. It is important to understand that the hazard protection approach involves choices like this.

8. The analysis indicated that for both plastic and metal containers, there appear to be packaging and suppression system options that could produce success in all scenarios studied, assuming a complete operational sprinkler system. Conversely, the results also left open the possibility that some scenarios could result in failure for plastic and some metal

containers. Modeling was insufficient to resolve these cases; additional fire testing will be needed to do so.

References

1. See, for example, John R. Hall, Jr. and Beatrice Harwood, "The National Estimates Approach to U.S. Fire Statistics," *Fire Technology*, Volume 25, Number 2 (May 1989), pp. 99-113.

2. FMRC developed probabilities of operationality, for comparative purposes only, in Table VI in B.G. Vincent, H.C. Kung and P. Stavrianidis, *Technical Report -- Fire Protection for U.S. Army Helicopter Hangars*, FMRC J.I. 0X0N1.RA, Norwood, MA: Factory Mutual Research Corporation, September 1993. The values were 0.931 for water-based systems and 0.87 for foam-water systems, giving 0.87/0.931 = 0.9345 as the scaling ratio.

3. In a conversation with the author, Larry Krasner of FMRC suggested that in-rack water-based systems be estimated to have the same probability of operationality as ceiling-mounted water-based systems, even though in-rack sprinklers are more vulnerable to damage from ordinary traffic, because such damage is likely to be spotted when it occurs, leading to swift repair and no significant change in the probability of operationality when fire occurs.

4. A sensitivity analysis was done for plastic F containers with water sprinklers. It showed that increasing the spill fire probability by a factor of 10, as might be true for some class of liquid, increased the overall fire risk by 28% for paint and varnish warehouses and 5% for general warehouses. (The other property classes were not affected at all.) This implies that even very large differences in spill probabilities by class of liquid, if they had been captured in the model, would have had only a modest impact on the calculated fire risk.

Figure 1. Overview of Fire Risk Analysis Structure and Calculation

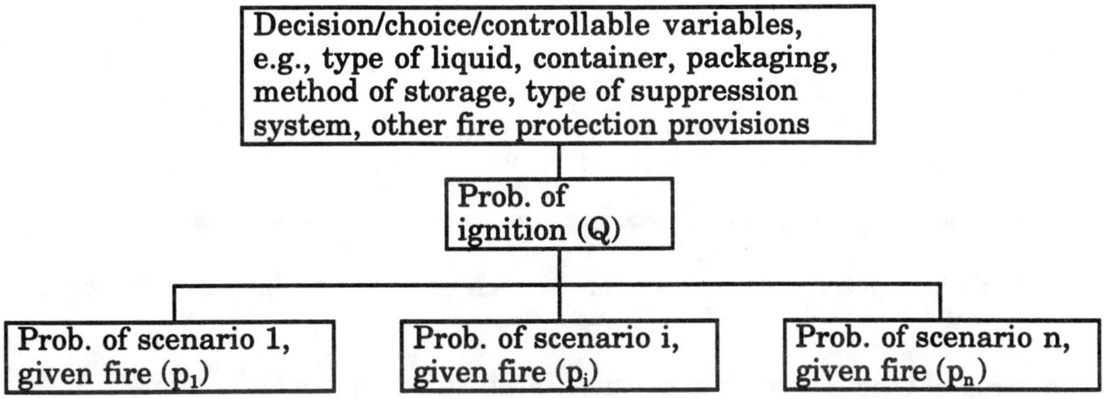

For each scenario i, the tree continues as follows:

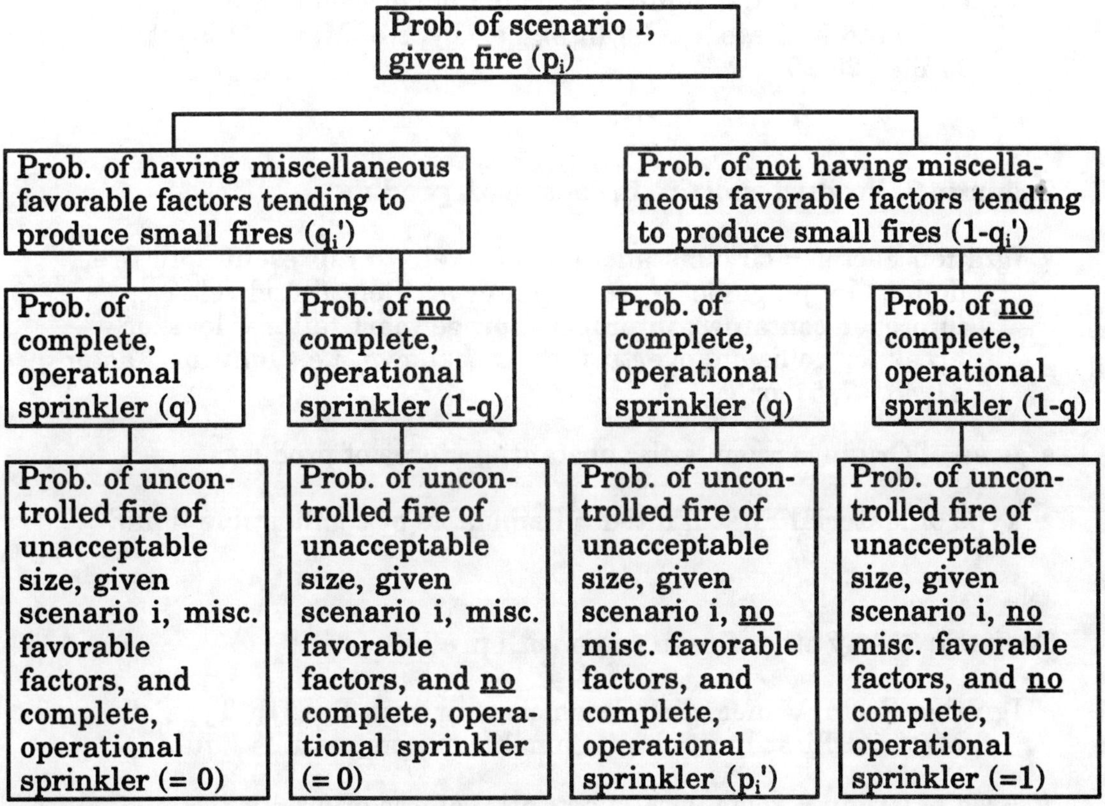

Hence, the overall probability of uncontrollable fire of unacceptable size =

$$Q\,[\,q\,(p_1\,(1-q_1')\,p_1' + \ldots + p_i\,(1-q_i')p_i' + \ldots + p_n\,(1-q_n')\,p_n') +$$

$$(1-q)\,(p_1\,(1-q_1') + \ldots + p_i\,(1-q_i') + \ldots + p_n\,(1-q_n'))]$$

Table 1. Initiating Fire Scenarios Used in Fire Risk Analysis Model

The four initiating fire scenarios used were defined in terms of categories defined by NFPA 901, *Uniform Coding for Fire Protection*, 1976 edition, which was the basis of coding in the U.S. Fire Administration's National Fire Incident Reporting System (NFIRS). This made it possible to use national fire experience data to estimate probabilities of each scenario, given that fire occurred, for each property use class.

Scenario 1. Arson fire with accelerants in room with product

- Ignition Factor = incendiary or suspicious fire, i.e., Ignition Factor 10-29

- Area of Origin = room with substantial stores of product, defined by judgment to be product storage room ; receiving, loading or shipping area ; sales or showroom area; or process or manufacturing area (Area of Origin 41, 45, 15 or 38).

- Type of Material First Ignited = flammable or combustible liquid (taken to indicate accelerants), i.e., Type of Material First Ignited 20-29

Scenario 2. Product spill fire in room with product

- Ignition Factor = circumstances most likely to represent spill fire, defined by judgment as fuel spill or accidental fuel release; improper container; improper storage; part failure, leak or break; or collision, overturn or knockdown, i.e., Ignition Factor 41, 45, 47, 51, or 71

- Area of Origin = room with substantial stores of product

- Type of Material First Ignited = flammable or combustible liquid

Scenario 3. Any other fire in room with product

- Ignition Factor ≠ incendiary/suspicious or spill fire <u>OR</u> Type of Material First Ignited ≠ flammable or combustible liquid

- Area of Origin = room with substantial stores of product

Scenario 4. Any fire not starting in room with product

- Area of Origin ≠ room with substantial stores of product

Table 2. Choice Alternatives Analyzed

1. **Occupancy/Specific Property Use (7 alternatives).** The three retail occupancies were all assumed to have 16' ceilings; the four storage occupancies were all assumed to have 30' ceilings. This assumption affects which test/modeling results are applied to each occupancy.

   - Supermarket
   - Hardware store
   - Department store
   - Packaged petroleum products storage
   - Alcohol beverage storage
   - Paint or varnish warehouse
   - General warehouse

2. **Method of Storage (2 alternatives).** Rack and pallet storage were analyzed. It was assumed that pallet storage is protected by ceiling-mounted sprinklers and rack storage is protected by both in-rack and ceiling-mounted sprinklers. This was done because of the judgment of the UL team that in-rack sprinklers are typically more effective for this storage method.

3. **Type of Suppression System (2 alternatives).** Water and foam-water systems were analyzed.

4. **Class of Liquid (3 alternatives).** Liquids were divided into Class I, II, and III, as defined in NFPA 30. Further divisions into Class I A vs. I B vs. I C or III A vs. III B were not possible with available data.

5. **Type of Container (3 alternatives).** The number of container types analyzed was limited to the number for which drop tests were performed: (a) metal F-style containers, (b) metal round containers (paint can type), and (c) plastic F-style containers. To match these with the test results, it was further specified that the metal F-style containers were for non-vented containers. All were one-gallon containers.

6. **Type of Packaging (2 alternatives).** Analyses were done for regular (cardboard) cartons and for fire resistive cartons. While there are many types of fire resistive cartons, the available test results were all or mostly of one type, and that was the basis for the analysis.

Total number of alternatives analyzed = 7x2x2x3x3x2 = 504.

Table 3. Sample Estimates of the Key Probability Parameters, for General Warehouses

A. Reported 1980-91 structure fires per year in the USA *	2,809.8
B. Estimated average number of U.S.A. facilities in 1980-91 = 165 x 1,728 general warehouse properties reported in 1993 in Phoenix and Portland	285,120
C. Q (overall probability of fire, by property use) = A / B	0.0099
D. Probability of scenario 1, given that fire occurs **	0.0099
E. Probability of miscellaneous favorable factors, given that a fire of scenario 1 type occurs = (number of reported general warehouse fires of scenario 1 with extent of flame damage confined to room of origin) /(number of reported general warehouse fires of scenario 1 with extent of flame damage known) *	0.3695
F. Probability that a sprinkler system is present = (number of reported general warehouse fires where sprinklers were reported present) / (number of reported general warehouse fires where sprinkler presence or absence was known) *	0.1635
G. Probability, if sprinklers are present, that they are complete and operational = (number of reported general warehouse fires where sprinklers were reported to have activated) /(number of reported general warehouse fires where sprinklers were reported to have been present and to have had a chance to operate, i.e., excluding fires reported or judged too small to activate detectors) **	0.7042
H. Probability that a complete, operational sprinkler system is present = F x G	0.1152

* Based on national estimates of 1980-1991 structure fires, using data from the National Fire Incident Reporting System (NFIRS) and NFPA survey

** Based on national estimates of 1980-1991 structure fires, with proportional allocation of fires with any defining dimension of Scenario 1 (ignition factor, area of origin, and type of material first ignited) unknown or unreported; the formula for three-way proportional allocation is too long to show in a table cell but the approach is standard practice in fire statistics estimation [1].

*** Based on national estimates of 1980-1991 structure fires, after adjustment of data based on plausibility rules linking fire size to reported sprinkler performance; see project report for details

Table 4. Summary of UL Drop Tests of Containers in Cartons

Container Type	Drop Height (feet)	Number of Containers with Complete Failure	Proportion of Containers with Complete Failure (Probability)	Size of Spill Fire (Analogous to)
Metal F-style	20'	19/40	0.475	Large (like Scenario 1)
Metal F-style	10'	2/40	0.050	Small (like Scenario 3)
Metal round	20'	32/40	0.800	Large (like Scenario 1)
Metal round	10'	18/36	0.500	Large (like Scenario 1)
Plastic F-style	20'	29/60	0.483	Large (like Scenario 1)
Plastic F-style	10'	0/60	0.000	No spill/ no fire

Table 5. Estimated Overall Fire Risk – Probability of Uncontrolled Fire of Unacceptable Size Rack Storage in General Warehouses

Class I liquid

Container type	Packaging type	Type of sprinkler system: Water-based	Foam-water
Plastic F-style	Regular	0.0041	0.0041-0.0042
Plastic F-style	Fire-resistive	0.0041	0.0041-0.0042
Metal F-style	Regular	0.0041	0.0041
Metal F-style	Fire-resistive	0.0041	0.0041
Metal round	Regular	0.0041	0.0041
Metal round	Fire-resistive	0.0041	0.0041

Class II liquid

Container type	Packaging type	Type of sprinkler system: Water-based	Foam-water
Plastic F-style	Regular	0.0041	0.0041-0.0042
Plastic F-style	Fire-resistive	0.0041	0.0041-0.0042
Metal F-style	Regular	0.0041	0.0041
Metal F-style	Fire-resistive	0.0041	0.0041
Metal round	Regular	0.0041	0.0041
Metal round	Fire-resistive	0.0041	0.0041

Class III liquid

Container type	Packaging type	Type of sprinkler system: Water-based	Foam-water
Plastic F-style	Regular	0.0041	0.0041-0.0042
Plastic F-style	Fire-resistive	0.0041	0.0041-0.0042
Metal F-style	Regular	0.0041	0.0041
Metal F-style	Fire-resistive	0.0041	0.0041
Metal round	Regular	0.0041	0.0041
Metal round	Fire-resistive	0.0041	0.0041

Table 6. Estimated Fire Risk
Assuming Complete, Operational Sprinkler--
Probability of Uncontrolled Fire of Unacceptable Size
Rack Storage in General Warehouses

Class I liquid

Container type	Packaging type	Type of sprinkler system: Water-based	Type of sprinkler system: Foam-water
Plastic F-style	Regular	9.3E-5	6.6E-6 - 8.7E-5
Plastic F-style	Fire-resistive	7.1E-6 - 9.3E-5	0.0 - 8.7E-5
Metal F-style	Regular	0.0	0.0
Metal F-style	Fire-resistive	0.0	0.0
Metal round	Regular	0.0 - 9.2E-6	0.0 - 8.6E-6
Metal round	Fire-resistive	0.0 - 9.2E-6	0.0 - 8.6E-6

Class II liquid

Container type	Packaging type	Type of sprinkler system: Water-based	Type of sprinkler system: Foam-water
Plastic F-style	Regular	9.3E-5	0.0 - 8.7E-5
Plastic F-style	Fire-resistive	0.0 - 9.3E-5	0.0 - 8.7E-5
Metal F-style	Regular	0.0	0.0
Metal F-style	Fire-resistive	0.0	0.0
Metal round	Regular	0.0 - 9.2E-6	0.0 - 8.6E-6
Metal round	Fire-resistive	0.0 - 9.2E-6	0.0 - 8.6E-6

Class III liquid

Container type	Packaging type	Type of sprinkler system: Water-based	Type of sprinkler system: Foam-water
Plastic F-style	Regular	9.3E-5	0.0 - 8.7E-5
Plastic F-style	Fire-resistive	8.6E-5 - 9.3E-5	0.0 - 8.7E-5
Metal F-style	Regular	0.0	0.0
Metal F-style	Fire-resistive	0.0	0.0
Metal round	Regular	0.0 - 9.2E-6	0.0 - 8.6E-6
Metal round	Fire-resistive	0.0 - 9.2E-6	0.0 - 8.6E-6

A Review of Chemical Compatibility Issues

D. J. Frurip, T. C. Hofelich, The Dow Chemical Co.,
D. J. Leggett, Hazard Evaluation Laboratories Inc.,
J. J. Kurland, Union Carbide Corp., and
J. K. Niemeier, Eli Lilly and Company

Prepared for presentation at the March, 1997 AIChE Loss Prevention Symposium,
Houston Texas

December 31, 1996

UNPUBLISHED

A Review of Chemical Compatibility Issues

D. J. Frurip, T. C. Hofelich, The Dow Chemical Co.,
D. J. Leggett, Hazard Evaluation Laboratories Inc.,
J. J. Kurland, Union Carbide Corp.,
and
J. K. Niemeier, Eli Lilly and Company

ABSTRACT

The accurate assessment of binary chemical compatibility is an important part of the safe handling, transport, and processing of industrial chemicals. This paper reviews a number of critical issues with regard to chemical compatibility which have been studied by members of ASTM Committee E27 (Hazard Potential of Chemicals). This paper focuses on compatibility charts and, in particular: tools for their preparation; the scope of their intended use; the definition of categories (entries); sources of data; their accuracy; their ease-of-use and how revisions are managed. The preparation of a compatibility chart forces the chemist or engineer to think about the hazards associated with possible binary combinations. Compatibility with adsorbents, materials of construction, water, heat, cleaning agents, etc. should not be ignored.

Key to defining compatibility is proper definition of the scenario. Test methods should be capable of providing reasonably accurate predictions of energy and gas released (and rates, where possible) as a result of inadvertent contact within the context of the defined scenario.

In general, any listing of compatibilities tends to be less accurate the more it is simple and inclusive. It is difficult to balance scientific accuracy with the desire of users for simple guidelines. Thus, public compatibility charts should be conservative. The non-expert should be encouraged to consult with experts or perform further testing if the conservative predictions becomes too restrictive in practice.

I. Introduction

Knowledge of potentially hazardous chemical incompatibilities is essential to safe and successful industrial process design in the chemical industry. Unfortunately, a significant number of reactive chemicals incidents in industry (and elsewhere) result from incompatibility issues, including waste streams, waste storage and transportation. Therefore it is imperative to consider the "reactive" compatibility of materials at an early stage in process scaleup. Some examples of where these data might be critical are:

(1) avoiding obvious hazards of inadvertent contact;

(2) complying with government regulations such as DOT HM-183 for tank trucks or Coast Guard restrictions on adjacent cargoes;

(3) recognizing reactive scenarios which may occur at elevated temperature (e.g. fire scenario).

ASTM Committee E27 on Hazard Potential of Chemicals, subcommittee E27.02 on Thermal Stability and Condensed Phases, became involved with chemical compatibility issues when reviewing ASTM D-34 Proposal P168, a "Proposed Guide for Estimating the Incompatibility of Selected Hazardous Wastes Based on Binary Chemical Reactions" (86D34). This guide was based on published work prepared for the EPA which includes a Hazardous Waste Compatibility Chart. The proposed standard had much merit, but ultimately failed to be adopted because of some errors. One of the main criticisms was that several binary combinations which are generally known to be hazardous, and published in other charts, were not indicated as such. For example phenols plus aliphatic amines, caustics and alcohols, and at least ten other potentially hazardous combinations were not flagged as hazardous. Since the ASTM and other charts are widely available, we have decided to draw upon our collective experience within the committee to point out the benefits and pitfalls of generalized charts and also to propose criteria for scenario-based judgment of chemical compatibility.

In this paper we review a number of critical issues with regard to chemical compatibility which have been studied by the authors. We will focus on compatibility charts and a number of factors relating to them. These include:

- their use in industry;
- tools for their preparation;
- the definition of categories (entries);
- sources of data and the use of published charts;
- their accuracy;
- their ease-of-use;
- experimental techniques for the determination of compatibility.

Also, in this paper, we evaluate several published charts, protocols, etc., which cover pure materials, process streams, as well as chemical wastes. Several detailed examples are presented of typical industrial cases involving the determination of chemical compatibilities.

II. Definition of Compatibility: the Scenario Dependence

All would agree to the need for accurate compatibility information and, at first, this might seem straightforward to obtain. However, there are some fundamental concepts that must be well thought out prior to determination of compatibility or to the preparation of a compatibility chart. Many of the following concepts may be found in reference 94HOF.

A. Defining the Scenario

The first concept that must be defined is the scenario for the chemical mixing. By scenario we mean a detailed physical description of the process whereby a potential inadvertent combination of materials may occur. Details such as specific amounts of materials, temperatures, confinement (closed or open system) and the times the materials will be in contact contribute to the definition of compatibility. It is certainly true that there is a great difference in the acts of combining relatively reactive materials (such as strong oxidizers with strong reducing agents) on a small scale (e.g. 0.05 ml) under controlled laboratory conditions as compared to pumping a tank-truck full of the oxidizer into a large storage tank of reducer. However, we must also recognize the potential time and temperature dependency of chemical reactions. The Arrhenius nature of reactions shows that most reactions are faster at higher temperatures. For a typical reaction with an activation energy of 20 kcal/mol, this translates into over a 6000 fold increase in rate for a 100°C increase in temperature. For a reaction with an activation energy of 35 kcal/mol (not uncommon in peroxide decomposition), the rate increase is over 4 million for the same temperature increase. These rates translate into changing half-lives from months and days at the lower temperatures to minutes and seconds at higher temperatures. For the same reasons, one cannot define compatibility in terms of a specific temperature at which two materials react.

Response and cleanup time impact the consideration of compatibility. If a mixture has a half-life of weeks should one consider it incompatible if it will not sit more than one day?

Size of the mixture affects both the total amount of energy and the rates at which that energy can be given off to the environment. A small flask may be able to lose heat fast enough to its environment relative to the amount of heat it is giving off while a rail car of the same mixture will lose heat proportionately at a much slower rate and possibly self-heat to thermal explosion.

Thus, it is imperative, when making a decision about the compatibility of two chemicals, that a scenario be defined. The scenario should contain the essential information of temperature, quantities, heat transfer characteristics, and also the length of time the materials will remain mixed.

Examples of mixing scenarios are given below:

1. Ambient temperature in summer, northern climate (~25°C); 5000 gallon scale; insulated storage tank; storage time seven days maximum, nitrogen padded headspace (chemical transport scenario).

2. Ambient temperature in a hotter, subtropical climate (~40°C), drum (55 gal) storage of mixed waste for 3 months maximum.

B. Defining Incompatibility Within the Scenario Framework

Besides defining the scenario when developing a compatibility chart, one also needs to make sure it is clearly understood what it means to state that two materials are "incompatible". In a general sense, we all know what is meant in the chemical industry when it is said that two materials are incompatible. What is implied is that there may be "undesirable" consequences of mixing these materials at a macroscopic (industrial) scale. These consequences might be, at the worst case, a fast chemical reaction or an explosion, or less severe, an undesirable temperature rise which might take the mixture above its flash point or cause a pressure increase in the system. If, however, the tank where the mixing will occur is inerted with nitrogen, and the material has an acceptably low vapor pressure increase, then even this temperature rise might not pose a practical problem.

In essence then, a working definition of compatibility needs to be defined before compatibility judgments can be effectively and accurately made. In the example scenario in the preceding paragraph, one might define two materials as being incompatible if the mixture results in a temperature rise of greater than 10° C or a gas release (or pressure increase) that would jeopardize the physical integrity of the container.

III. Preparing a Compatibility Chart

Probably the most common and convenient way to represent binary chemical incompatibility is through the use of a simple two dimensional chart or matrix. Ideally, all components of interest (including common cleaning materials, water, heat, materials of construction, etc.) are listed on both the x and y axes of the grid and the intersection of the cells in the matrix represent the consequences of each mixed pair. Presentation of the data in chart form allows for quick use, especially in a process upset (i.e. emergency) condition. An example of a simple compatibility chart is given in Fig. 1 below.

A. Use of Charts

Binary compatibility charts are an extremely useful teaching tool for new and even veteran employees. They can be especially useful in general education in the chemical plant or lab, or for areas and operations where commonly performed tasks might lead to a chemical mix such as might occur during co-shipment in compartmentalized containers, storage in common diked areas, or common waste containers in the lab. Compatibility information is essential during a Process Hazard Review (PHR, e.g.

Figure 1. A Hypothetical Compatibility Chart (from CHEMPAT).

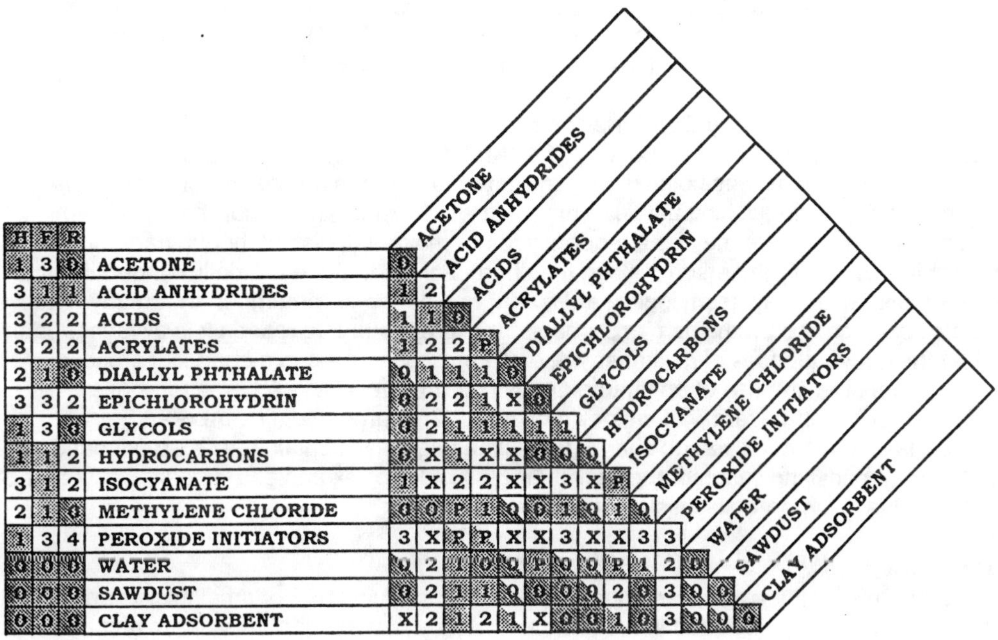

H - HEALTH F - FLAMMABILITY R - REACTIVITY 0 - MINIMAL

1 - CAUTION 2 - DANGER 3 - SEVERE DANGER 4 - EXTREME DANGER

P - POLYMERIZATION or PRECIPITATION X - UNLIKELY MIXTURE

"what if", or HAZOP). Also, the very act of preparation of a compatibility chart forces the chemist or engineer to think about the hazards associated with possible binary combinations. Thus, severe hazards might be identified which might not otherwise be determined.

B. General Guidelines for Chart Preparation

Some general guidelines for preparation of a compatibility chart are given below. Depending on the particular situation and audience for the chart, the preparer needs to consider each option carefully. Additional, complimentary guidelines for chart preparation may be found in Section III-C-2 below.

1. State the Scenario

In the preparation of a compatibility chart, one should consider stating both the scenario and the scenario based definition of incompatibility explicitly on the chart. This avoids confusion when comparing charts and also lets the chart user know explicitly what the authors mean by incompatibility.

2. Decide on a Hazard Rating Scheme

The scale of the individual binary hazards needs to be well thought out prior to chart construction. In certain instances, it may be desirable to have a simple "yes/no" scale. For regulatory compliance (HM-

183* [*] for example (94HOF)), this may be the preferred manner to construct a chart. In other instances, typical plant charts for example, more "information rich" ratings may be advantageous. For example, a numerical score of 1,2, and 3 might be appropriate with "1" indicative of a compatible mixture, a "2" might indicate a moderate hazard (say, a temperature increase), and a "3" might indicate a severe hazard. Numerous schemes have been developed for individual charts in the public literature and these are discussed later (see also the CHEMPAT section below, specifically, Table 1). There is always a danger, of course, in making the chart so complicated that its practical usefulness is lost.

3. Consider the Hazards for all Binary Combinations
The potential hazard for each binary mixture needs to be carefully considered . Avoid using blanks (empty cells) in compatibility charts since blanks may indicate that there is no hazard, or, simply that the hazard is unknown. Clearly distinguishing between an "unknown hazard" and a "non-hazard" is an important consideration. Also, there may be a tendency to not make a decision about binary mixtures which are "improbable", i.e., that the person making the chart decides that the is little likelihood that the materials could ever mix. This may seem logical at first but one needs to remember that many reactive chemicals incidents have occurred with such "improbable" mixtures. For this reason it is always prudent to at least consider the consequences of mixing any and all species involved in a chemical process. At a minimum, one should consider the use of an unambiguous hazard level rating scheme for these situations. As an example: non-hazard = 0; unknown hazard = ?; improbable mixture = X. For an improbable mixture that is quite reactive, the chart intersection might still contain the "X" designation but also contain another numerical indication of the severe hazard (X,3) for example. See also Section III C-2 (and in particular, Table1) for more discussion of hazard ratings.

4. Define the Categories
The definition of categories for the chart is an important part of construction. For small plants and operations, each chemical may be included in the chart and still be of manageable size. In fact, many times these charts may include additional useful compatibility information such as compatibility of chemicals with materials of construction, water (from process streams or from rain in diked areas), cleaning agents, sealants, and compatibility of materials with adsorbents. As an example of the latter, styrene is incompatible with many clay-based adsorbents (86HAN) and this important information can be flagged on the grid. Sometimes, "heat" might be considered as an entry to flag particularly heat sensitive materials such as monomers.

For more "general" compatibility charts, say, for a large manufacturing site, the best manner to construct a chart is to group chemicals into natural groupings based on their chemical structure. Examples of these groupings are mineral acids, aliphatic amines, monomers, water-based formulations, halogenated hydrocarbons, etc.. This is the way in which the published compatibility charts (discussed in detail below) are constructed. One problem with this manner of construction is the fact that for a number of classes, certain binary combinations might be legitimately known to be compatible (again we emphasize that this must be within the constraints of the defined scenario) whereas other combinations within these two groups may not be. It may be best to provide the worst case compatibility rating with a list of compatible exceptions.

5. Document How the Decisions are Made
Backup and supporting data should be easily accessible for chart users and for easy chart updates. If testing was performed to make a decision about a particular binary combination in a chart, then reference this test in the chart. As mentioned earlier, exceptions to the chart should also be noted. The CHEMPAT software (discussed below) provides easy access to these data.

C. Tools for Preparation of Compatibility Charts

Standard and commonly available software can be quite useful in the preparation of a compatibility chart. Two examples are discussed below.

---

[*] This legislation forbids the shipment of incompatible cargoes in adjacent sections of a compartmentalized tank truck.

1. Spreadsheet Software
Commonly available spreadsheet formats are amenable to the preparation of compatibility charts. Microsoft EXCEL™, for example, may be used to set up such a chart quite conveniently. The chart may then be made accessible via a network server to all those involved in a common operation.

2. CHEMPAT
Recently, a Dow Chemical developed software package called CHEMPAT has been made available to the public* by the AIChE (95CHE). This package offers a convenient way to manage compatibility information.

The main incentive for the development of CHEMPAT came from the fact that in the past, chemical compatibility charts, or reactivity matrices, had been prepared by many groups within a chemical manufacturing facility, in paper chart format. These charts were produced manually, based on data and experience of plant operating personnel. However, paper charts do not allow for the inclusion of substantiating data for the hazard assessments of each binary pair of chemicals. Also, the modification of the chart, due to a change in the process and/or the addition or removal of chemical(s), was difficult, time consuming, and often not done.

CHEMPAT provides a straightforward and consistent way of developing and documenting this required information. It is a DOS based program that has been designed to organize and display information regarding the consequences of inadvertent mixing of process materials. The program produces a compatibility, or inter-reactivity, chart and manages the user-supplied information that is required by CHEMPAT. Because this process is computer-based, data integrity is assured and maintaining up-to-date charts is straightforward.

Producing a CHEMPAT chart requires that every combination of process materials be reviewed, with regard to the hazards of inadvertent mixing. A hazard level is then assigned which, in turn, provides an action plan for all personnel in case of such an event. Therefore, the process of developing the chart acts as an effective part of hazard evaluation of plant operations. The use of the chart, in daily operations, provides workers with a reliable source of pertinent information regarding the process chemical interactions in situations that are not part of normal operating procedures.

CHEMPAT recommends seven steps to be used when developing compatibility charts. Some of the information below is redundant to the information presented in Section IIIB above but it is presented here for completeness since these steps more clearly define the mechanism by which a chart is prepared in CHEMPAT.

Steps to Prepare a Compatibility Chart via CHEMPAT

Step 1. Decide who will build the chart: The compatibility chart may be developed by a hazard evaluation group, by the client, or preferably as a collaborative effort.

Step 2. Select the process expert: This individual must be familiar with all aspects of the plant's processes, and will be responsible for outlining the physical or chemical concerns which may become factors in the event of an inadvertent chemical mixture.

Step 3. Define the mixing scenario: The process expert, in cooperation with the hazard evaluation group, will examine the process, including related activities such as water treatment, waste handling, etc., in order to develop a list of components used by the plant and the related mixing scenarios.

*Dow donated the program to the chemical industry and AIChE volunteered to administer the distribution.

Step 4. Gather the data: Complete and credible information on the mixing hazards for the compounds should be obtained from the open literature, process experts, or by explicit testing, as discussed in Section IV.

Step 5. Assign hazard levels: After all mixture data have been accumulated, define the hazard levels. CHEMPAT utilizes a rating system of 0-4 ("0" representing a low hazard, "4" representing an extreme hazard). Table 1 shows the suggested CHEMPAT hazard ratings. The hazard ratings should not be under or over cautious. The chart should be predominantly a combination level zero and level one with *some,* typically in the range of 0 - 10%, level two hazards. If a *significant* number of level three and level four hazards have been assigned, the overall assessment and assignment scheme needs to be reviewed. (See also the discussion above, Section IIIB, about assignment of hazard ratings.)

Once the hazard levels have been determined, the plant will define appropriate emergency response plans. Therefore, the response of the emergency team may, at the time of the inadvertent mixing, be in direct relation to the level of the assessed hazard. A hypothetical set of actions, for each hazard level, are presented in Table 1.

Step 6. Generate Compatibility Database and Chart:
The information on each of the mixtures is entered into CHEMPAT. A special, location-specific designation may be used to highlight a unique inter-reactivity hazard (such as "no hazard if dry, dangerous reaction if wet"). There is no program limit on the amount of information that may be stored.

The printed chart (illustrated in Figure 1, gray-scale rendition) illustrates the potential hazards of the defined mixtures for a typical production facility. The triangular region holds the compatibility ratings for the binary mixtures. All ratings are represented with color coding to indicate the appropriate hazard level when using a color printer. The chart also indicates the compound's NFPA Health, Flammability and Reactivity ratings.

Step 7. Distributing the information: Distribution of the CHEMPAT chart and database occurs in two steps. First the process expert receives a complete copy of the software package and database. That person is able to modify the database as the process changes and will serve as a continuing focal point of the compatibility project. Second, user versions, with restrictions on the ability to modify the information, are installed on centrally located personal computers in key areas of the process plant. The personnel who will use the compatibility chart and software are trained on the User Version of the program.

D. Sources of Compatibility Information

When making a decision about compatibility, within the context of the defined scenario, a number of options are available to aid in the process. These include the scientific literature (including published compatibility charts) and local "experts" (including synthetic chemists and those with experience in the process). Several publicly available compatibility charts are reviewed in detail in Section V below.

1. Using the Public Literature

The public scientific literature is a good place to start but sometimes these data are relatively obscure. Many sources are available but by far the best source of such data is Bretherick's "Reactive Chemical Hazards" (96BRE). This compilation has been in print for at least 20 years and is still published in book form. This work has recently been made available for PC searches and now is also available in Windows format. The Windows version allows for logical searches (example: chlorine AND amine). The author,

Table 1. Hazard Levels and Typical Associated Emergency Response Actions for Use in CHEMPAT.

Hazard Rating	Hazard Level	Suggested Emergency Response
0	**Minimal**	Report inadvertent mixing event to supervision; no further action necessary.
1	**Caution**	Report event to supervision; devise and implement plan(s) to manage the situation; no emergency procedures to be initiated
2	**Danger**	Report event to supervision; prepare to initiate unit emergency plan if needed; notify personnel in immediate area; consider halting normal activities until extent of situation is fully assessed.
3	**Severe Danger**	Report event to supervision; initiate unit emergency plan; notify all plant personnel; cease normal activities until extent of situation is fully assessed; consider need to evacuate the plant; report event to plant industrial security and other ex-plant Emergency Response groups.
4	**Extreme Danger**	Initiate unit emergency plan; notify all plant personnel to evacuate the area; cease normal activities, if possible, before evacuating; report event to plant industrial security and other ex-plant Emergency Response groups once evacuation is underway or complete.

Les Bretherick, has scoured the available literature, much of it obscure, to summarize the data for individual compounds, mixtures, and certain "classes" of materials (such as nitro aromatics, etc.). There is also an extensive discussion of the general field of Reactive Chemicals hazards including the basics of hazard evaluation, kinetic factors, adiabatic systems, reactivity vs. composition and structure, reaction mixtures, and protective measures. This work covers 4,800 elements or compounds with an additional 5,000 secondary entries involving two or more materials, with 30,000 cross references.

If the material is a pure substance (i.e. as opposed to a chemically complex process stream), a good source for compatibility information in the U.S. is the Material Safety Data Sheet. These data sheets are compiled by the manufacturers and may present detailed, known incompatibility problems which might be otherwise difficult to find.

As an example, Fig. 2 shows a section of an MSD Sheet for 35% hydrogen peroxide. As can be seen, a number of significant incompatibilities are listed.

Other good sources of chemical reactivity and compatibility information include: Merck Index (83MER); "*Encyclopedia of Explosives and Related Items*" (60FED; this multi-volume set is unfortunately out of print); "*Dangerous Properties of Industrial Materials*" (84SAX); "*Fire Protection Guide to Hazardous*

Figure 2: Excerpt from a Material Safety Data (MSD) Sheet for 35% H_2O_2

35% HYDROGEN PEROXIDE, H_2O_2

SECTION 7: HANDLING AND STORAGE

Incompatible with many transition metals including: chromium, manganese, silver, platinum, palladium. May form potentially explosive compounds with ketones, ethers, carboxylic acid anhydrides and alcohols. May react explosively upon mixing with hydrazine, aniline and glycerine. Sodium borate, urea, sodium carbonate, triethylamine, sodium fluoride and sodium pyrophosphate form addition compounds analogous to hydrates with hydrogen peroxide which can crystallize and may react violently by grinding or warming with oxidizable materials.

Materials" (91NFP); "*Guidelines for Chemical Reactivity Evaluation and Application to Process Design*" (95CCP); and finally one should not overlook the myriad of college text books on organic and inorganic chemistry which provide valuable information about chemical synthesis ("what reacts with what").

2. Synthetic Organic and Inorganic Chemists
These individuals are trained to know "what reacts with what" and under what conditions. For those unfamiliar with synthetic chemistry, these experts should be consulted frequently.

3. Knowledgeable Process Personnel
There is no substitute for experience in a complicated chemical process. Those who have "been around" for years will be a good source of historical compatibility information. Capturing this information in a compatibility chart will help ensure this information is not lost.

4. CHEMMIX
Another tool for use in making compatibility decisions is the software package CHEMMIX (89CHE). This relatively inexpensive DOS or Macintosh based program is based on the ASTM P-168 chart (discussed in more detail in Section V) and provides a systematic method for determining binary compatibility. Basically, the user enters the chemical name or class of the materials of interest (there is a database of 1700 chemicals which have been pre-sorted into classes). The program then "analyzes" the materials simultaneously. In essence, the program considers every possibly binary combination and predicts the consequences of mixing those binaries and assigns the appropriate hazard codes (heat generation, fire, explosion, etc.). Whereas this software is really no more predictive than a paper copy P-168 chart, it does provide a useful and efficient way to look at many different materials simultaneously. There is also available from the same company which distributes CHEMMIX a handy sliding chart Chemical Reaction Guide which is a convenient way to present the same data from the P-168 chart.

IV. Experimental Tools for the Determination of Compatibility

In certain cases, an experimental determination of the compatibility of two (or more) materials is the most prudent approach. This may be necessary if the compounds or molecular types cannot be found in one of the published charts, if the data are not provided by the manufacturer, or if the scale and scenario of mixing is such as to warrant an actual verification of the hazards involved. In other words, there is always a risk associated with the prediction/estimation of the consequences of mixing and the key question becomes whether or not the owner of the process is willing to accept that risk.

One key consideration in the safe design of an experimental compatibility test is scale. Some of the test schemes proposed by or actually used by the published charts involve large quantities of materials (up to 300 ml) which may become a hazard to the tester if the mixture were to be highly reactive. Some of the test schemes do require a smaller scale pretesting but these are typically nonquantitative. Thus, we believe, it is prudent to start the testing at small enough scale so as to minimize the potential dangers to the test operator. Also, in order to apply the results of compatibility tests to various scenarios, quantitative data suitable for scaleup are needed. One such quantitative testing scheme using 2-200 mg quantities is described below.

There are a number of methods one may use to experimentally assess the consequences of mixing two or more materials. One technique used at a number of chemical companies is described below. In essence, this experimental technique involves Differential Scanning Calorimetry (DSC) testing of the individual components and the mixture and also Mixing Calorimetry. The typical method and philosophy of this approach is as follows:

1. Determine some measure of the inherent thermal stability of the individual components using DSC.

2. Determine the instantaneous energy release (heat and gas evolution) upon mixing of the two components using heat-of-mixing calorimetry.

3. Determine the relative thermal stability of the mixture (as compared and contrasted to the individual components) by DSC.

This experimental formalism allows an evaluation of materials which might react very quickly upon mixing and also materials which require more elevated temperature or an induction period to react. Neither test alone, i.e. mixing calorimetry or DSC, is adequate to fully evaluate the reactivity hazards associated with a mixture. For example, a DSC on a mixture of HCl(aq) and NaOH(aq) would show *no* exothermic activity (to ca. 400°C) since the acid-base neutralization takes place virtually instantaneously and the resultant solution is simply NaCl(aq). A mixing calorimetry test could quantify the heat of neutralization and thus more accurately define the hazards associated with this mixture. If a reaction is autocatalytic, the mixing calorimetry might not detect any heat due to the characteristic induction times for this type of process. The DSC test would probably be able to detect the reaction since it subjects the sample to a continuous temperature ramp.

DSC is a common technique for thermal characterization of materials and is widely used in industry for the initial screening of samples in so-called reactive chemicals hazard evaluation and details have been published (95CCP). Basically, a DSC experiment (run in the typical temperature scanning mode) measures heat release (or uptake) from a small sample as a function of temperature as the sample is subjected to a constant temperature ramp. Typical conditions include a temperature ramp from 30-450°C at 10°C/min. using ca. 1 mg samples.

While there are numerous choices available for DSC sample encapsulation, many have found the best to be the sealed glass capillary tube described by Whiting et al. (88WHI). It is generally best to choose an encapsulation technique which minimizes volatilization of the sample. The purpose of the low headspace capillary tubes is to suppress vaporization endotherms from volatile samples which may mask or affect other chemical processes in the sample. Also, the avoidance of metal containers minimizes the potential reactivity problems between sample and container.

Some special considerations should be given to planning the DSC experiment on the mixture. As discussed below, the heat-of-mixing experiment only detects heat changes which are kinetically fast enough to be detected during the course of the experiment (typically several minutes), at the temperature of the mixing experiment (typically 10-60°C). The DSC of the mixture can detect autocatalytic processes or processes which occur relatively rapidly at slightly elevated temperatures.

Hence it is very important that the mixture be made immediately prior to the DSC experiment. Where possible, it is best to use the same mixture from the heat-of-mixing experiment. In general, it is advisable to initially mix at ca. 50/50 by volume but overall it may be prudent to test other ratios as well (e.g. 20/80 and 80/20.; see Example 3 in Section VII).

A mixing calorimetry device has been designed, constructed and developed within Dow (Midland) laboratories and referred to as the "2-Drop" calorimeter. A basic schematic is shown in Figure 1 of reference 94HOF. Of course, other suitably constructed mixing calorimetry devices may be employed for test purposes and some are commercially available.

Perhaps the easiest part of the above experimental procedure is the actual testing. The hardest part sometimes is a straightforward and unambiguous interpretation of the results. Clearly, if there is no substantial mixing heat (or gas generation) and the DSC trace of the mixture is ostensibly a superposition of the DSC traces of the individual components, no incompatibilites exist regardless of the scenario. On the other hand, if substantial incompatibility exists between the substances, then a large mixing heat may be observed and/or the DSC trace may show a large exotherm at relatively low temperature (with respect to the individual component DSC traces). Unfortunately both these behaviors are rare in practice and, more often than not, the results fall in between the two extreme cases just described. The analyst then may have a more difficult time is assessing the results. It may be desirable to estimate the thermokinetics of the incompatibility reaction and from these data, predict the conditions (temperature, volume, etc.) whereby a mixture might become a hazard (on a large scale) using heat gain - heat loss models. This technique has been described in reference 94HOF.

Finally, we again caution the reader not to perform testing for compatibility on a scale which might endanger the analyst. Direct mixing of several hundred ml of two materials into a Dewar flask might answer the question but if the samples are highly reactive, the consequences of such an experiment might be hazardous. The techniques described above use microliter quantities, minimizing the hazards to the operator.

V. Public Compatibility Charts

In the ideal world one could reference an easy-to-use one page public document and quickly determine without a doubt whether mixing two materials could create a hazard. Such documents have been attempted in the form of 'global' compatibility charts. These charts attempt to predict the outcome of mixing many possible pairs of chemicals. However, the real world of chemical compatibility is often too complex for such simplification. Nonetheless, global compatibility charts, even with their limitations, are valuable tools and are good starting points in a search for compatibility information.

For this publication, several published charts were compared (See Table 3). The two most comprehensive charts are the U.S. Coast Guard (93COA), a Norwegian chart NVC 4-75 (75STE), and an ASTM product, the P-168 chart (86D34). The Norwegian chart also covers a large number of materials, but it was superseded by the USCG chart which used the Norwegian information as well as additional information. The Finnish chart (92RAH) is noteworthy in that it attempts to address the hazards of mixing household chemicals.

The charts reviewed have some similarities in their approach (see Table 2). All attempt to categorize materials into reactivity groups. This categorization is dependent primarily on functional groups (e.g., alcohols, organic acids, esters) of the compounds. All are designed only for near room temperature and pressure application and only consider binary mixtures. They attempt to be conservative by considering the consequences of the worst-case combination between reactivity groups.

The charts show many differences in selection of reactivity groups. The USCG chart is more narrow in its definition of groups than is P-168. For example, 'Caprolactam solution', 'Epichlorohydrin', 'Sulfuric Acid', 'Nitric Acid', 'Ammonia', and 'Vinyl Acetate' are all individual categories in the USCG chart, whereas these materials are grouped with other compounds in the P-168 chart. However, P-168 has several reactivity groups not found in the USCG chart (e.g., metal groups, organophosphates, inorganic sulfides, metal nitrides). Also, the USCG approach is different in that it considers some categories to be relatively non-reactive and thus considers binary mixtures between these categories to be non-hazardous. Their matrix doesn't include cells for these mixtures.

Table 2: Similarities Between Published Compatibility Charts.

- **Reactivity groups are used**
- **Charts are displayed as a matrix showing consequences of binary mixtures.**
- **Charts consider only binary mixtures**
- **Scenarios are limited to near room temperature (<55 °C) and ambient pressure conditions**
- **Charts attempt to be conservative: the most reactive materials are considered for each binary mixture**
- **Charts contain warnings regarding application**

Because the number of compounds of interest for compatibility purposes is almost infinite, quite a bit of generalization is necessary to place all compounds in less than 50 groups as these charts have. For example, the P-168 chart places both fluoroacetic acid and valeric acid, $CH_3(CH_2)_3COOH$, in the same reactivity group. Fluoroacetic acid is a much stronger acid than valeric acid and there are cases in which it will present different hazards (usually more severe) than valeric acid. Because of the necessary generalization, conservatism was necessary and the worst-case was assumed when developing the charts. For each binary combination an attempt was made to consider the individual compounds within the reactivity groups which would result in the most severe consequence. The hazard published in the charts is usually based on this worst-case combination. However the USCG chart indicates that some binary combinations are non-hazardous except for specific combinations of individuals within the two groups.

The proper use of the charts require careful selection of the reactivity group. For example, if one is using the P-168 chart to answer compatibility questions regarding a carbamate (RNH_2COOH) with ammonia, the carbamate could be classified either as having amine and ester functionalities or as a carbamate. Neither the amines or esters are said to be incompatible with ammonia. However, mixing carbamates with ammonia is said to result in heat and gas. Each chart is accompanied by an index of compounds and instructions on their placement into reactivity groups. It is important to use these indices to ensure that materials are placed in proper groups. Since the indices don't include every conceivable compound it is sometimes necessary to use one's judgment in selecting a group or groups for a compound. Testing kits are described for detecting the major functional groups in a mixture that has unknown composition (84WOL). Detection limits are said to be 5-20%.

The charts all include a caveat that concentration, synergistic, and antagonistic effects have been assumed not to influence the reactions. In other words, consequences could differ due to inclusion of more components which may tend to dilute, catalyze, or inhibit the reaction.

The chart recommendations for various binary pairs differ. As an example, for compatibility with amines the USCG and P-168 charts give different recommendations in 6 cases (alcohols/glycols, esters, ketones, nitriles, phenols/cresols, and epoxides).

The P-168 chart is based largely on general literature references, case histories, and inferences rather than actual testing data. This tends to make the chart over-conservative when attempting to predict what will happen at room temperature and pressure. For example, the combination of aldehydes and unsaturated hydrocarbons is said to result in heat generation. Associated documentation gives a reference and states "at elevated temperatures, a Diels-Alder type reaction can take place between acrolein and 1,3-butadiene and may be exothermic." The referenced book states that the reaction takes place at 100°C. The rate of reaction is perhaps very slow at temperatures less than 55°C (the temperature often used as a maximum for material handling and transportation purposes) and in many cases heat transfer would be sufficient to prevent a temperature excursion. For the reasons just mentioned the P-168 chart tends to be more conservative than the USCG chart.

Even when charts utilize solely test data there is still the possibility of error. This is due to the fact that the testing schemes used in chart preparation may or may not be aligned with the actual scenario at hand. For example, in the USCG test, mixtures are observed for a minimum of 30 minutes following mixing. However, in a more adiabatic (no heat loss and low thermal inertia, phi) situation the reaction may cause a runaway if given enough time. This is especially true in the case of autocatalytic reactions in which reaction products catalyze the reaction. The key question that must always be answered is does heat removal capabilities exceed heat generation capabilities in the scenario being reviewed. Table 4 gives information on testing schemes recommended by the charts.

The P-168 chart attempts to give specific information on hazards resulting from mixing materials, rather than simply stating that materials are incompatible. P-168 uses the following consequences: heat generation, fire, gas generation, gas/toxic, gas/flammable, explosion, violent polymerization, and solubilization of toxic substances. The USCG chart simply indicates known incompatibilities.

The charts could do a better job documenting the rationale for concluding a pair is compatible or incompatible. For instance, the USCG chart indicates that aliphatic amines and alcohols are incompatible but it does not give any rationale for this. Amines and alcohols are generally considered compatible at room temperature except for perhaps a hydrogen bonding interaction which might liberate heat. Likewise, P-168 does not indicate what conditions or compounds will cause amines and esters to generate gases. Again, amines and esters are generally compatible at room temperature. Carey and Sundberg (90CAR) indicates that amines and esters can react but only under more severe conditions.

The charts should not be considered the final word on compatibility, but simply guides. They are easy to use and, along with consulting with local experts, are a good place to start. The charts should be used to flag potential problems, but a blank cell in a chart should never be taken to mean that there is no hazard with the combination. A blank may simply indicate that the chart preparers had no information available. This is especially true in the ASTM P-168 chart where there are numerous cases (at least 12) where no reactivity is indicated (blank cells) in the chart for binary mixtures but the mixtures may be hazardous. It is important that all available charts be referenced since the charts sometimes differ. Other sources of information (see Section III above) should be referenced as well. When in doubt actual testing should be performed as recommended in Section IV. In addition, the chart recommendations must be used with consideration to the scenario under which the two materials are to be mixed. Section VI below gives examples of how the charts can be used misused. While published charts can be very useful in hazards assessment, their data must be applied carefully.

Table 3: Characteristics of Some Published Compatibility Charts. Refer to Table 2 for Common Traits Among Charts. References Can be Found at the End of the Paper.

Title	Organization/ Preparer(s)	First Published/ Last Revision	Scenario	Definition of Hazard	Number of Reactivity Groups	Source of Data
Norwegian Maritime Research Chart	Norwegian Maritime Research - Steensland, Vossgard, Askheim	1974/1974 (74STE)	Mixing of adjacent freighter cargoes.	T rise > 50°C, gas evolution, exotherm at < 70°C, or anticipation that toxic gas could be produced.	38 (25 major categories and 13 subcategories)	Norwegian testing, Dow testing from early 70's (77FLY), USCG NVC 5-70 (predecessor of NVC 4-75), generalizations based on reactivity information.
NVC 4-75	U.S. Coast Guard	1975/ 1993 (93COA)	Mixing of adjacent freighter cargoes	T rise > 25°C or gas evolution	43 total groups. 36 used on chart.	Mostly from Dow testing performed in early 70s (77FLY). Judgment was also applied when data was lacking. Some data (about 25%) was taken from Norwegian testing
P-168	U.S. Environmental Protection Agency (EPA) - Hatayama	1980/ 1980 (86D34)	Mixing of hazardous wastes.	Heat or gas generation, polymerization, solubilization of toxic substances. Required magnitude of each not defined.	41	Case histories and generalizations based on basic chemical reactivity information.
Finnish chart	Paavo Ristola, Ltd. - Rahkonen, Malmen, Ettala.	1992 (92RAH)	Mixing of household hazardous wastes.	Not defined.	10	Case histories, P-168, NFPA 491M (75NFP), Bretherick's Handbook (earlier editions to 96BRE).

Table 4. Compatibility Testing and Evaluation Schemes.

Method	Testing and Evaluation Scheme
Norwegian Maritime	Dewar flask; 1:1 molar ratio; final volume 10 ml; sealed system; gas generation determined by final pressure; observation for at least 10 minutes; if result is negative at room temperature, the mixture is heated to 70 °C at 4 °C/min on a Differential Thermal Analysis (DTA) instrument; mixture is considered hazardous if an exotherm is detected.
Dow - early 70's	300 ml Dewar flask; 1:1 molar ratio; final volume 10 ml; if result is negative at room temperature, the mixture is heated to 200°C at 20°C/min on a Differential Thermal Analysis (DTA) instrument. If DTA showed an exotherm at less than 46°C, further Dewar testing is performed.
US Coast Guard (USCG)	Dewar flask with magnetic stir bar; 20 ml total sample, at binary ratios of 2:18, 10:10, and 18:2; Observation for a maximum of 30 minutes; If temperature rise is <25 °C and no gas is evolved a test is performed at 50 °C on either 10 ml of most reactive ratio at room temperature or if no reaction observed a 1:1 ratio is used.
Dow - early 90's	Three types of tests are used for screening: 1) determination of inherent thermal stability of individual components using DSC; 2) determination of the instantaneous energy release (heat and gas evolution) using a heat of mixing apparatus; 3) determination of the relative thermal stability of the mixture (as compared and contrasted to the individual components) by DSC. Additional testing is sometimes recommended. The thermokinetic data from these tests are used in a model that compares heat generation with heat removal for a given scenario. Criteria for a hazard is defined as any temperature rise above 10 °C during the lifetime of the scenario or any generation of gas. See reference 94HOF and discussion in Section IV of this paper for details.

VI. The Special Case of Wastes

Of special concern is the compatibility of wastes and this is where our ASTM work began with the reevaluation of ASTM Proposal P168. Wastes are often poorly (and/or improperly) defined whereas most other process streams in industry are often carefully characterized to meet specifications. The common compatibility classification by consideration of functional groups (as discussed in detail above) is often untenable for wastes. Characterization of the waste, therefore, is often a necessary first step. Unfortunately, the variety of wastes is virtually limitless, given the economic advantages of compositing wastes for disposal. Compounding the problem is the fact that in many chemical processes there are usually more wastes than other process streams. The atomic composition in the waste may be known by material balance but there is little incentive to determine structures, which could be quite complex and numerous. The specifications on waste tend to be restricted to those items covered by RCRA (materials that are ignitable, toxic, reactive and corrosive) plus viscosity, value as fuel and solids content which affect their use in boilers. In principle, compatibility comes into play under RCRA when wastes are mixed (by the generator).

Of course, an unambiguous classification could be based on experimental testing (heats of mixing and thermal stability tests of the mixture). Direct experimental testing has obvious time saving advantages to a full characterization of the chemical composition of each and every waste, especially if the waste stream varies in composition over time as the process changes (optimizes).

VII. Examples

The following several examples serve to illustrate that compatibility issues in industry are sometimes not as straightforward to resolve as might appear at first. In every case below, keep in mind that experimental means to determine the compatibility should be considered (the larger the scale, the larger the risks associated with a predicted determination of compatibility).

Example 1: Because of the addition of a new chemical process to an existing plant, you are contemplating new chemical storage in an existing diked area. A key reactive chemicals question in these situations is the mutual compatibility of the materials stored in the common area. Incidental leaks or spills may lead to unwanted hazardous reactions which might jeopardize the integrity of all the inventory. The three new chemicals being added are styrene, 3% aqueous sodium chloride, and triethylamine. Currently, the dike contains storage for 10% aqueous hydrogen peroxide, methylene chloride, and 2% aqueous sodium carbonate. The volumes of storage involved are on the order of 5000 liters each.

Step one would be to consult those with a good knowledge of chemistry Step two would be to look at the publicly available compatibility charts mentioned above. For example, the Coast Guard chart shows two hazards associated with storage of these materials. First, polymerizable materials such as styrene can easily be initiated by peroxides and thus in this case may be considered highly incompatible. Secondly, aliphatic amines and chlorinated aliphatics can be quite reactive. Right away, we see that it probably would not be desirable to store some of these materials in an area with a common dike. We also note that reference to several other sources would confirm the hazards associated with the chemicals mentioned. These would include Bretherick (96BRE), and Material Safety Data Sheets (or equivalent) provided by the manufacturers.

No apparent incompatibilities exist between the other materials and sodium chloride solution. Your recommendation in this case might be to allow the storage of the brine in the existing dike but to consider isolated storage for the other new materials. Due to the fact that styrene's potential energy is so large (polymerization heat is nearly -9000 joule/g), it probably is prudent to isolate this material.

Example 2: Technical Services at your company wishes to make a change to a process. Currently room temperature methyl chloroformate (MeO-CO-Cl) is added via carboys through the manway to a -50 °C reaction mixture containing predominately dimethylformamide (Me_2NCHO). The change calls for pumping room temperature MCF from drums into the -50 °C DMF solution and subsequently rinsing the drums into the reaction vessel using DMF. Should the change be made?

The question here is whether or not MCF and DMF are compatible even at room temperature. If the published compatibility charts are consulted, one comes to an incorrect conclusion. The P-168 guide classifies MCF as an ester and a halogenated organic, DMF is classified as an amide. The chart does not indicate a compatibility problem with this combination. All indications are that mixing the two materials will not create a hazard. *However, this very process change resulted in an accident.* When DMF was used to rinse the MCF drum and the drum was closed pressure began to build. The drum ruptured releasing noxious methyl chloride. Methyl chloride is also flammable, but a fire did not occur. Bretherick's Handbook (96BRE) does indicate a problem with binary combinations of MCF and DMF. Under the entry "Acyl Halides" (of which MCF is a member) Bretherick's says "This group tends to react violently with protic organic solvents, water, and the aprotic solvents, dimethylformamide and dimethyl sulfoxide....".

<u>Example 3</u>: Production wishes to combine 20 gallons of waste which is predominately formic acid with the contents of a nearly full 55 gallon waste drum containing predominately methylene chloride.

All known references (P-168 and US Coast Guard charts; NFPA 491M; MSDS) show no indication that mixing formic acid and methylene chloride would create a hazard. Consultation with an organic chemist leads one to suspect that a complex may be formed in the mixture with a release of the heat of complexation. In addition, not all constituents in the wastes are known and therefore a quick lab trial is performed at a 1:1 (by volume) ratio. The heat of mixing test shows a 10°C adiabatic temperature rise. This is below what is considered a hazard in the US Coast Guard test. However, if the methylene chloride and formic acid are mixed in a different ratio, say 95:5 instead of 1:1, and the limiting reagent is found in the methylene chloride the resulting temperature rise might be 25°C rather than 10°C. This is due to increasing the moles of limiting reactant and also to the large difference in heat capacity between the two materials. Formic acid has a heat capacity of about 2.17 J/g/°C and methylene chloride has a heat capacity of about 1.19 J/g/°C (both heat capacities at room temperature). If the materials are mixed at 20°C under this scenario, the boiling point of the mixture (about 41°C) might be reached. This example illustrates the point that the criteria for safety used for shipping cargo (i.e., the USCG criteria) may not be appropriate for other circumstances.

<u>Example 4</u>: A clearly marked railcar containing a glycol ether is accidentally pumped into a 2MM lb. storage tank of styrene. The error is noticed only after the styrene polymerization process is effectively quenched and the entire process is stopped. You are consulted as to the magnitude of the hazard associated with this large vessel containing an inadvertent mix.

None of the published charts indicate any incompatibilities between styrene (a polymerizable material) and glycol ethers (classified as an ether, and an alcohol/glycol). Certainly the evidence at hand (the quenching of the polymerization process) suggests no acceleration of styrene's polymerization tendency when mixed with the glycol ether. However, due to the large scale involved with a highly energetic monomer like styrene (and thus a significant risk to the plant), and the fact that glycol ethers can form peroxides, you recommend experimental verification of your initial assessment. An ARC (Accelerating Rate Calorimetry) test verifies the effective dilution effect of the glycol ether and the lack of significant peroxide initiators. The results of additional microcalorimetry testing of the contaminated material under ambient conditions allow you to make the final determination that the mixture may be safely transported for disposal.

VIII. Conclusions

We hope that this paper will assist those in the chemical industry and elsewhere who are frequently confronted with the tasks of compiling a compatibility chart or making a decision about the compatibility of two chemicals. Whereas most of the procedural aspects regarding compatibility issues are straightforward, there are numerous pitfalls to be aware of. The most important message regarding the use of published charts is that they are general compilations, not complete (a blank may indicate no hazard or an unknown hazard!), and consequently they should always be used as a guide only. When the risk of a prediction is unacceptably high, experimental testing should always be considered.

In summary then, we hope that the reader is cognizant of the following important points regarding compatibility issues:

1. Compatibility is scenario dependent and the definition of a scenario is critical to accurate compatibility assessment.

2. A number of useful tools are available for the mechanics of chart preparation (including CHEMPAT).

3. Several standard published sources of compatibility information are available.

4. Published charts are helpful but their predictions must be carefully considered.

VIII. Acknowledgments

The authors would like to thank the members of ASTM Committee E27 for their helpful discussions and insight relating to the topic of chemical compatibility which greatly helped in the preparation of this manuscript.

IX. References

60FED B.T. Fedderoff et al., "Encyclopedia of Explosives and Related Items", Picatinny Arsenal, Dover, N.J., 1960. (out of print)

75NFP NFPA 491M Manual of Hazardous Chemical Reactions, 1991 edition, A Compilation of Chemical Reactions Reported to be Potentially Hazardous., National Fire Protection Association, Quincy, MA, 1991. Arranged alphabetically by chemical name with over 3550 mixtures of two or more chemicals reported to be potentially dangerous at ordinary or moderately elevated temperatures. Citations to the original reports are given. Last revised in 1975.

74STE Steensland O.; Vossgard, R.; Askheim, N. E., "Reactivity Between Chemical Cargoes", Norwegian Maritime Research 1974, 2 (2), 21-34.

77FLY Flynn, J. P.; Morrissette, M. D., "Development of a Compatibility Guide for the Water Transport of Bulk Chemical Cargoes", Journal of Hazardous Materials 1977, 1 (4) 331-342.

83MER "The Merck Index, An Encyclopedia of Chemicals, Drugs, and Biologicals", Tenth Ed., M. Windholz, ed., Merck & Co., 1983.

84SAX N.I. Sax, "Dangerous Properties of Industrial Materials", Sixth Ed., Van Nostrand Reinhold Co., 1984.

84WOL "Design and Development of a Hazardous Waste Reactivity Testing Protocol", Wolbach, C.D.; Whitney, R.R.; Spannagel, U.B., Acurex Corp., Energy and Environ. Div., Mountain View, CA, USA, (1984) 155 pp. NTIS, Springfield, Va (USA). PB84-158807.

86D34 ASTM (American Society of Testing and Materials) Committee D-34, Proposal-168,"Proposed Guide for Estimating the Incompatibility of Selected Hazardous Wastes Based on Binary Chemical Reactions", and included Minority Report, March 1986. See also: "A Method for Determining the Compatibility of Hazardous Wastes", Hatayama, H.K.; Chen, J.J.; De Vera, E.R.; Stephens, R.D.; Storm, D.L., CA Dept. Health Serv., Berkeley, (1980) 165 pp. NTIS, SPRINGFIELD VA.. PB80-221005. Also published as Report No: EPA-600/2-80-076, April 1980, 165 pages.

86HAN S. Hankin, J. Hazard. Matl., vol. 13, p. 217 (1986); see also the discussion in Frurip. et al., Proc. Safety Prog., vol. 12(2), p. 79 (1995).

88WHI L.F. Whiting, M. LaBean, and S. Eadie, Thermochim. Acta, vol. 136, p231-245 (1988).

89CHE CHEMMIX software available from Environmental Software & Systems Inc., P.O,. Box 1182, Bowling Green OH 43402-1182; phone (and FAX) 419 353-8540.

90CAR F. Carey and R.J. Sundberg, "Advanced Organic Chemistry", 3rd Ed., Plenum, N.Y., 1990.

91NFP "Fire Protection Guide on Hazardous Materials", 10th Ed., National Fire Protection Association, 1 Batterymarch Park, P.O. Box 9101, Quincy MA, 02269-9101, 1991.

92RAH "Work Safety In Collection And Transport of Hazardous Wastes", Rahkonen, P.; Malmen, Y.; Ettala, M., Paavo Ristola Ltd., Terveystie 2, SF-15870 Hollola, Finland, WASTE MANAGE. RES., (1992) vol. 10, no. 2, pp. 217-226.

93COA "Compatibility of Cargoes", published in 46 CFR Ch. 1 (10-1-93 Edition) - Part 150 with changes that appeared in the Federal Register dated April 11, 1994 as a Final Rule by Coast Guard (CGD 92-100) for Bulk Hazardous Materials. See also: "Compatibility Guide for Adjacent Loading of Bulk Liquid Cargoes (Final Report)", Report No. USCG-D-156-75.

94HOF T.C. Hofelich, D.J. Frurip, and J.B. Powers, "The Determination of Compatibility via Thermal Analysis and Mathematical Modeling", Proc. Safety Prog., vol.13 (4), pp. 227-233 (1994).

95CCP "Guidelines for Chemical Reactivity Evaluation and Application to Process Design", CCPS, AIChE, New York, 1995.

95CHE CHEMPAT, Copyright 1995, The Dow Chemical Co., Distributed in cooperation with the American Institute of Chemical Engineers. See also: Gay D. M.; Leggett, D. J., "Enhancing Thermal Hazard Awareness with Compatibility Charts", Journal Test. Eval. 1993, 21 (6), 477-480.

96BRE L. Bretherick, "Bretherick's Handbook of Reactive Chemical Hazards", Fifth Ed., Butterworths, 1996.

Decomposition of Thermally Unstable Chemicals in Solvents

Sima Chervina, Skip E. Evon, and Glenn T. Bodman

Health, Safety and Environment, Eastman Kodak Company

Rochester, NY 14652-6270

Abstract

Knowledge of the decomposition reactions of chemicals when tested neat versus in mixtures with various solvents is essential for process safety studies. The decomposition of several mercaptotetrazole derivatives was studied by differential scanning calorimetry, both neat and in seven different solvents. Most mixtures exhibited destabilization of the chemical in the solvent compared with the neat material, resulting in a lower decomposition onset temperature in the mixture. It was demonstrated that the decomposition mechanism and kinetics changed significantly when these chemicals were mixed with solvents. The highly variable effects on thermal stability of mixing thermally unstable chemicals with solvents demonstrates the value of thermal stability testing of reaction mixtures in performing process hazard identification.

1. Introduction

An important part of assessing hazards of any chemical handling operation such as synthesis, distillation, purification, filtration, drying, storage, shipment, etc., is characterization of the thermal stability of the chemicals involved. The relative thermal stability of chemicals is usually assessed by studying chemical decomposition reactions using small-scale specialized calorimeters such as the differential scanning calorimeter (DSC) and accelerating rate calorimeter (ARC), and comparing the observed decomposition phenomenon such as the onset temperature, the amount of energy released, and the rate of energy release.

Identification of the decomposition characteristics of a neat chemical is the first step in identifying the potential hazards of its use in manufacturing [1-3]. This information will generally be sufficient for assessing thermal hazards of operations where only neat chemicals are involved (grinding, storage, shipment). Historically, it was also considered sufficient, or even conservative, for reaction mixture hazard evaluation, because diluting a chemical with solvent was assumed to slow decomposition. More recently, emphasis has been placed on the necessity of also testing chemical/solvent mixtures or reaction mixtures that will occur during processing. This approach recognizes the possibility that the decomposition characteristics may be different under these conditions.

The results of our previous studies [4-6] demonstrated that decomposition of nitrobenzene derivatives changes significantly in solvents and that even a small amount of solvent (< 10%) can decrease the thermal stability of a chemical significantly. The decrease of the decomposition onset temperature of the ten nitrobenzene derivatives tested varied between 40°C and 180°C when mixed with sulfuric acid [4]. Mixing with solvents such as aqueous acid or base solutions, propyl acetate, acetone, toluene, and water resulted in onset temperature decreases between 10°C an 50°C for several chemicals tested. An increase in the maximum rate of heat release during decomposition was observed for several chemicals in mixtures with sulfuric and methanesulfonic acids, alkaline water solutions, acetone, and water [4, 5]. The results demonstrated the value of studying the behavior of additional classes of thermally unstable chemicals in mixtures with solvents.

The results of studying the decomposition of three mercaptotetrazoles in seven common solvents are presented in this paper. The emphasis was placed on comparing the thermal hazards in the mixtures with the thermal hazards of neat materials and on assessing whether a significant change in the decomposition reaction mechanism and kinetics occurred in the mixture.

2. Experimental

DSC experiments were conducted on each of three mercaptotetrazole derivatives in the pure state and in binary mixtures with seven commonly used solvents. The mercaptotetrazole/solvent combinations tested are summarized in Table 2. The mixtures were prepared by inserting 0.5 ± 0.1 mg of a chemical and 0.5 ± 0.1 mg of a solvent in a glass capillary [7], after which the capillary was centrifuged and sealed under nitrogen. The actual weight percent of mercaptotetrazole in the mixtures varied from 44% through 54%. For neat chemical testing, a sample size of approximately 1 mg was used. The DSC experiments were conducted at 10°C/min heating rate using the TA Instruments Model 2910 Differential Scanning Calorimeter. The instrument was calibrated using a dedicated 1 mg indium standard in a glass capillary at 10°C/min heating rate. The cell constant, onset slope and temperature correction factors are automatically updated by the TA2200 DSC computer.

The following decomposition characteristics were determined:

T_o - decomposition onset temperature (the temperature at which a 0.025 W/g deflection from the established baseline is first observed), °C

R - maximum rate of heat evolution (determined from the peak height divided by the sample mass), W/g of mixture

T_m - peak temperature–temperature corresponding to the maximum rate of heat evolution, °C

The relative hazards of chemical/solvent mixtures were characterized by comparing the observed decomposition onset temperature, the peak temperature, and the maximum rate of heat evolution of the neat chemical (T_{on}, T_{mn}, R_n) with those of the chemical in the solvent (T_{os}, T_{ms}, R_s), where the subscripts "n" and "s" refer to "neat" and "solvent", respectively.

3. Results and Discussion

Two generic types of behavior can be assumed when comparing decomposition in solution to that of the neat chemical:

- Dilution effect--the decomposition mechanism and kinetics do not change in the mixture, so the change in decomposition characteristics is associated entirely with the change in the concentration of the decomposing species (in this case the mixture will represent less of a thermal hazard than the neat material).
- Destabilization effect--substantially different decomposition characteristics than those predicted by a simple dilution effect (in this case the mixture will represent more of a thermal hazard than the neat material). These mixtures should be identified and studied to characterize and understand potential process hazards.

Attempting to discriminate between 'dilution' effects and 'destabilization' effects and identifying whether a change in decomposition mechanism and kinetics occurred in the mixture can be done using the following approach:

1. Derive kinetic equations of decomposition for the neat chemicals,
2. Using the derived kinetic equations, calculate the decomposition parameters in mixture assuming a dilution effect,
3. Compare the calculated decomposition parameters with the actual test results.

The application of this approach and the results obtained are presented below.

1. The decomposition kinetics of three neat chemicals were studied using the isothermal DSC method described in [8, 9]. All three chemicals demonstrated an autocatalytic decomposition mechanism which can be described by the following kinetic model:

$$r_1 = A_1 \exp(-E_1 / RT) C_A$$
$$r_2 = A_2 \exp(-E_2 / RT) C_A^n C_B^m$$

The parameters of the model were determined and are summarized in Table 1 [8].

Table 1

Decomposition Mechanism and Kinetics of Mercaptotetrazoles

Chem	Mol Wt	Autocatalytic Reaction				1-st Order Rxn	
#		n	m	E_2, J/mol	A_2, s^{-1}	E_1, J/mol	A_1, s^{-1}
1	178	2	1.5	139000	4.27E+14	184000	1.47E+19
2	235	1.4	1.6	217000	2.62E+22	114000	1.14E+9
3	313	0.7	0.7	196000	3.60E+20	160000	4.20E+14

2. The derived kinetic equations (Table 1) were used to predict the decomposition of the neat chemicals and their 50% solutions (assuming a simple dilution effect) in a screening DSC experiment at a 10°C/min heating rate. The predicted DSC curves, presented in Figs. 1, 3, and 5, demonstrated that in the case of a dilution effect, the DSC curve of the mixture will have:

- a higher decomposition onset temperature,
- a higher decomposition peak temperature,
- a lower rate of decomposition.

The specific differences in the decomposition onset and peak temperatures between the 50% mixtures and the neat chemicals ($\Delta T_o = T_{os} - T_{on}$, $\Delta T_m = T_{ms} - T_{mn}$) and the ratios of the maximum rates of heat evolution by the neat chemicals to those of the mixture (R_n/R_s) for all three chemicals were calculated from the predicted curves and can be found in Table 2.

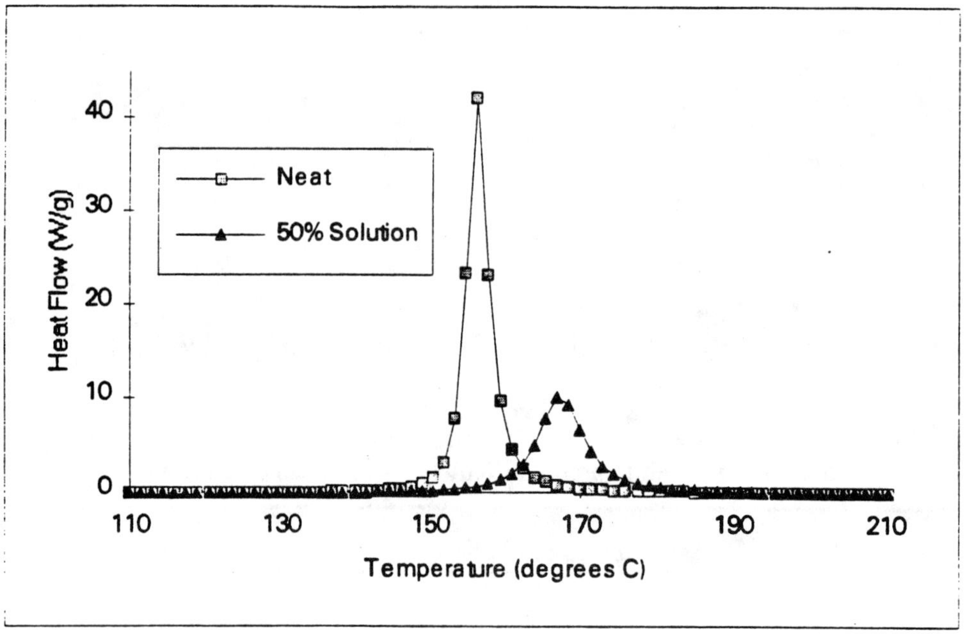

Fig. 1. Simulation of Chemical 1 decomposition (neat and in solution) in screening DSC runs

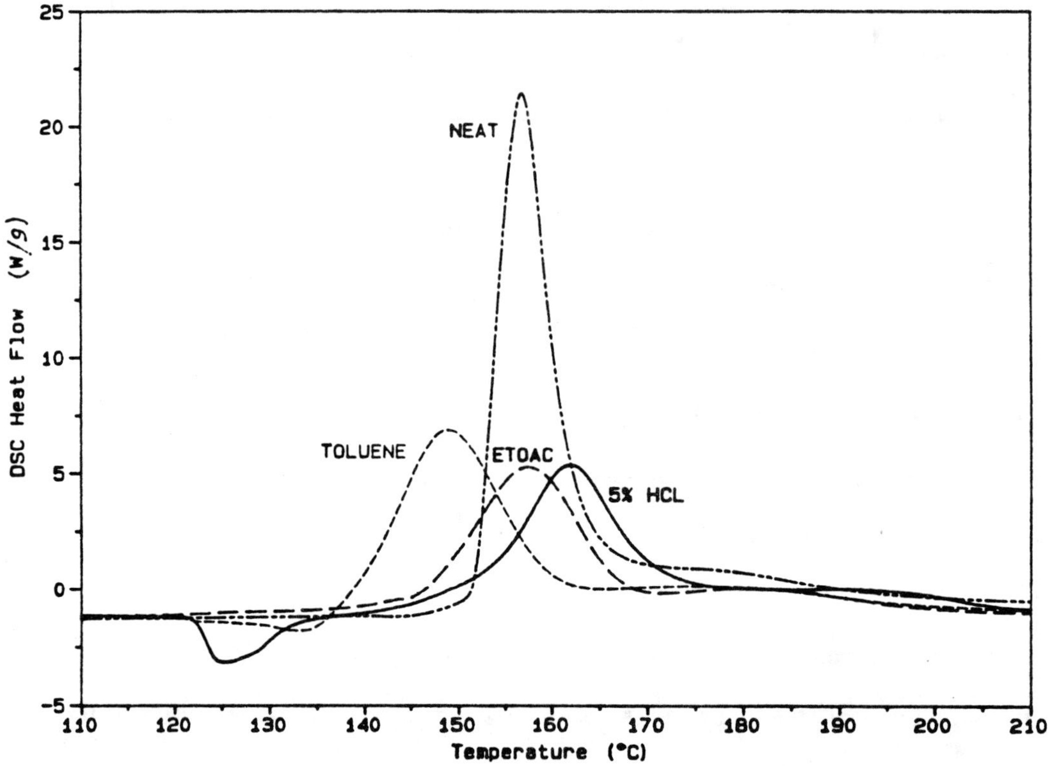

Fig. 2. Screening DSC runs of Chemical 1, neat and in binary mixtures with three solvents

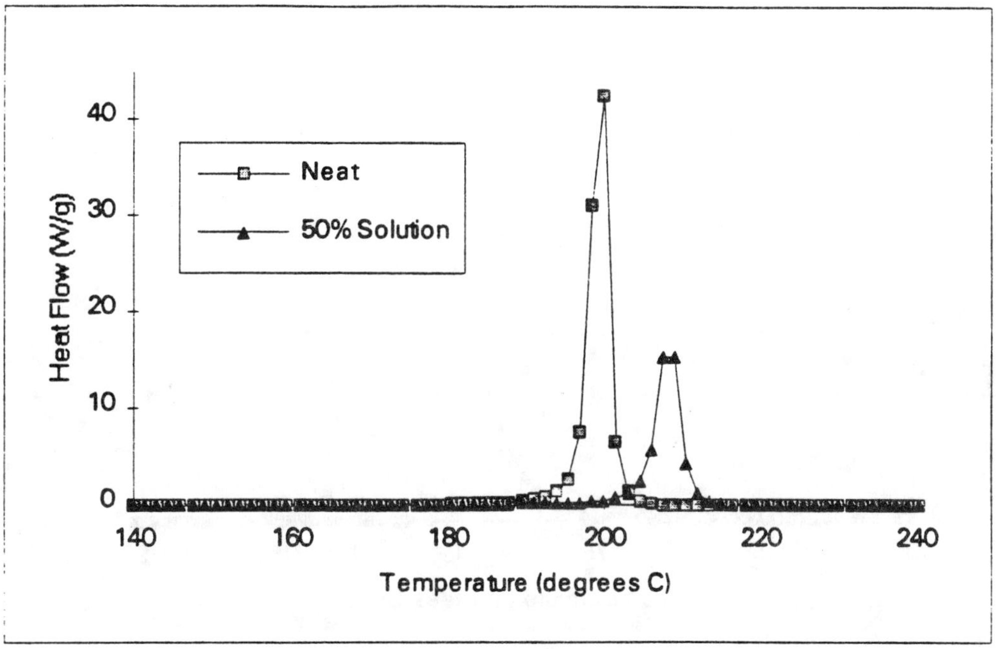

Fig. 3. Simulation of Chemical 2 decomposition (neat and in solution) in screening DSC runs

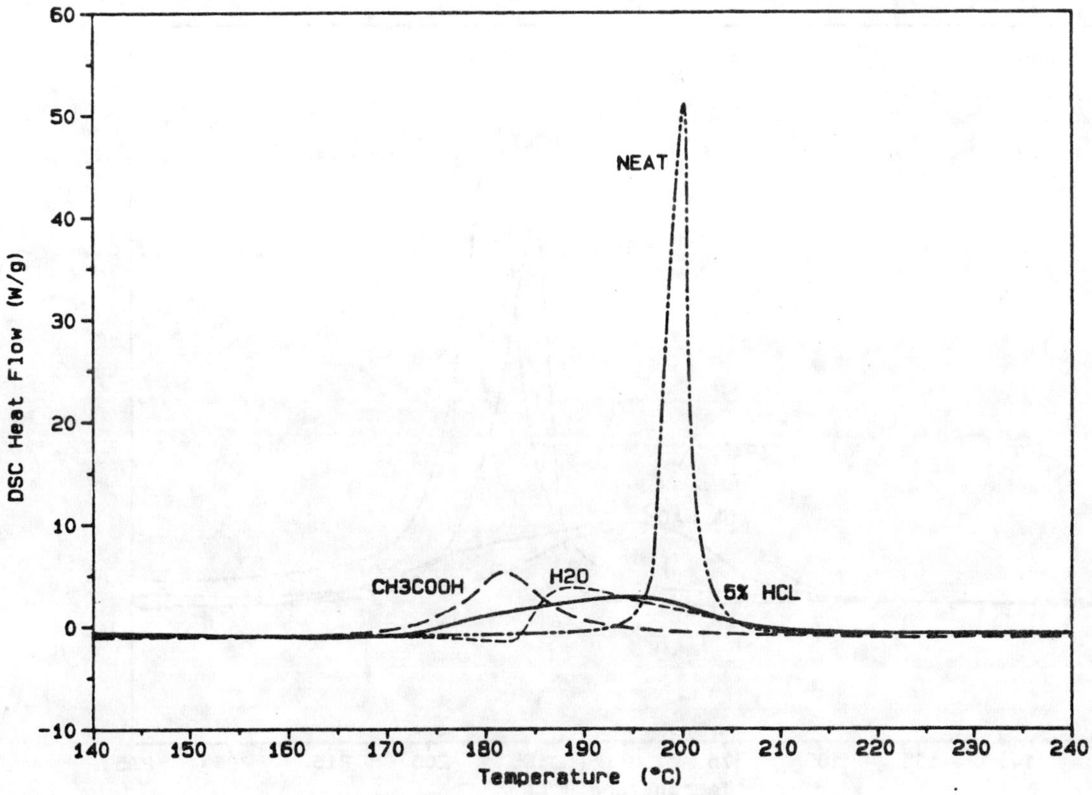

Fig. 4. Screening DSC runs of Chemical 2, neat and in binary mixtures with three solvents

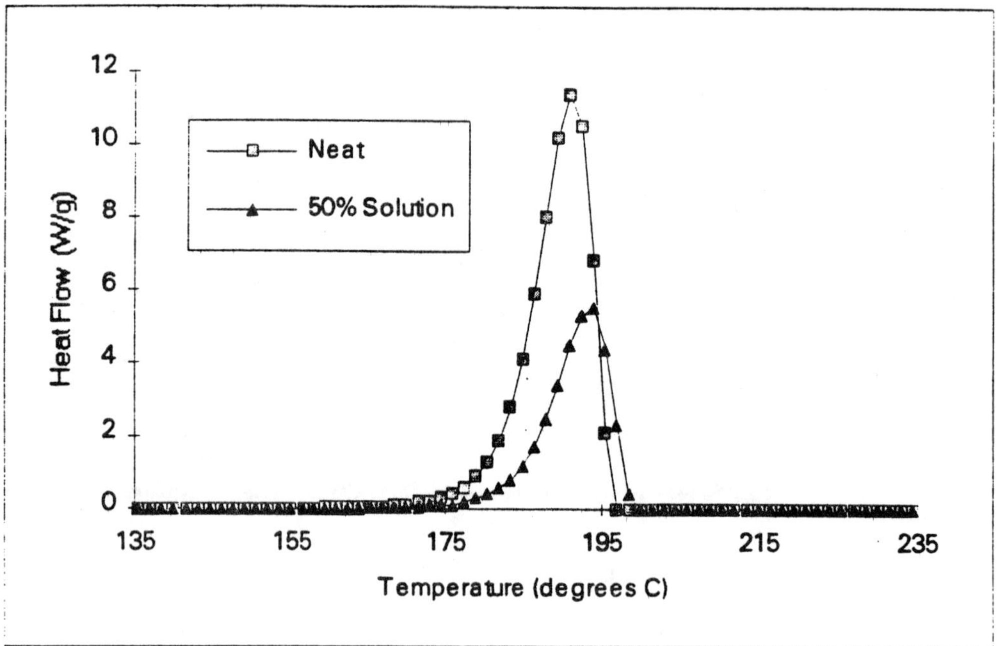

Fig. 5. Simulation of Chemical 3 decomposition (neat and in solution) in screening DSC runs

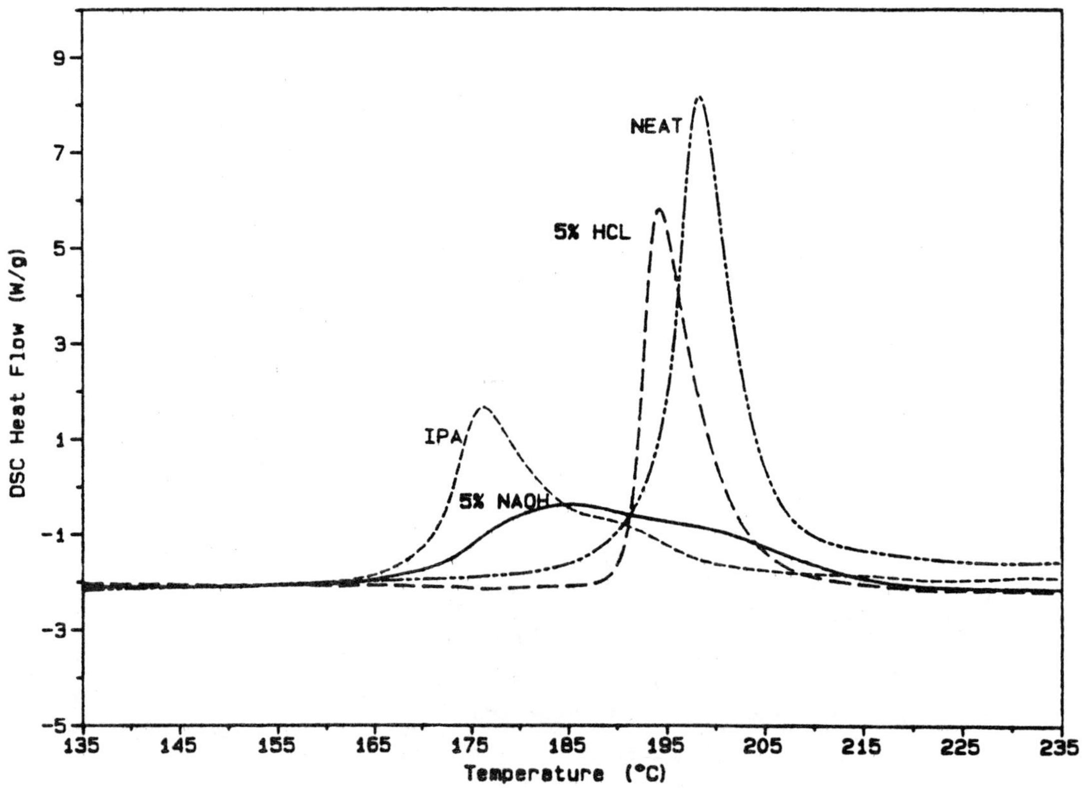

Fig. 6. Screening DSC runs of Chemical 3, neat and in binary mixtures with three solvents

3. The experimental DSC overlay curves of the neat chemicals and their binary mixtures with different individual solvents, presented in Figs. 2, 4, and 6, clearly show a destabilization phenomenon: the decomposition of most mixtures takes place at temperatures lower than decomposition of the neat chemicals.

The effects observed in the mixtures are compared to each other and to those of the predicted simple "dilution effect" in Table 2.

A destabilization effect (decrease in the thermal stability of the mixture compared with the neat chemical) is indicated by one or more of the following:

- the experimental ΔT_o is smaller than the calculated one,
- the experimental ΔT_m is smaller than the calculated one,
- the experimental R_n/R_s is smaller than the calculated one.

Table 2

Change in Decomposition Characteristics in Mixtures versus Neat Chemicals

	Chemical 1			Chemical 2			Chemical 3		
Solvent	ΔT_o	ΔT_m	R_n/R_s	ΔT_o	ΔT_m	R_n/R_s	ΔT_o	ΔT_m	R_n/R_s
Calculated	**6**	**11**	**4.16**	**11**	**8**	**2.81**	**6**	**3**	**2.07**
5% NaOH	-23	8	4.66	-27	-19	29.71	-22	-12	5.73
5% HCl	-18	5	3.51	-14	-8	14.81	5	-3	1.25
Acetic Acid	-26	5	3.71	-27	-19	8.23	-20	-18	1.95
Isopropanol	-22	6	3.69	-28	-18	7.03	-23	-22	2.67
Ethyl Acetate	-31	-1	3.5	-17	-11	6.42	-25	-18	2.01
Toluene	-10	-9	2.75	-15	-5	4.81	-12	-9	2.25
Water	-24	5	3.4	6	-11	11.04	19	-5	1.53

Analyzing the data summarized in Table 2, it can be seen that all mixtures tested have at least one decomposition characteristic that indicates destabilization, most of the mixtures have two indicators, and several have all three (e.g., all mixtures of Chemical 1 except NaOH solution). The average decrease of the decomposition onset temperature in mixtures was about

25°C, and the maximum rate of heat evolution was lower in all mixtures compared to the neat chemicals. This demonstrates that the decrease in thermal stability of mercaptotetrazoles in solvents is not as dramatic as that seen for nitrobenzene derivatives [5], but still has the potential to lead to a hazardous situation if not identified.

This clearly demonstrates the importance and necessity of chemical/solvent mixture testing when assessing thermal hazards of such processes including drying, distillation, crystallization, and of reaction mixture testing when assessing the thermal hazards of chemical syntheses.

In the case of destabilization phenomenon, it is evident that the reaction observed in the mixture may be different than that observed for the neat chemical. It is not always clear whether the added solvent catalyzes the decomposition reaction, forms an intermediate with the chemical that was less thermally stable, changes the reaction media (e.g., dipole moment of the solvent) which destabilizes the chemical, or whether a chemical reaction with the added solvent occurs. There was no attempt in this study to explain the chemistry leading to the phenomenon observed. However, considering the reaction mixture behavior from the thermal runaway perspective, the critical information is not what type of reaction occurred, but rather the rate and temperature range over which heat is evolved.

4. Conclusions

In recent years, emphasis in textbooks and in the literature has increased regarding the necessity of evaluating decomposition hazards of reaction mixtures in addition to the hazards identified through the testing of individual chemicals. However, literature data are limited to a description of the behavior of nitrobenzene derivatives in various solvents. This paper provides further information on the behavior of a different chemical class of thermally unstable materials in various solvents. Even though the observed destabilization of mercaptotetrazoles in solvent was not as dramatic as that observed for nitrobenzene derivatives, it provides further evidence of the need for such reaction mixture testing. While by no means definitive, the results lay a foundation for predicting the effects of chemical/solvent interactions, and demonstrate the need for such testing as part of process hazard evaluation.

Acknowledgments

The authors would like to thank Paul R. Kingsley for carrying out the experimental work necessary to complete this project.

References

[1] T. Grewer, Thermochim. Acta, 187 (1991) 133.

[2] T. Ando, Y. Fujimoto and S. Morisai, J. Hazard. Mater., 28 (1991) 251.

[3] W. J. Fenlon, Plant/Oper. Prog., 3 (1983) 197.

[4] S. Chervina, S. E. Evon, G. T. Bodman, Thermochim. Acta, 1997 (to be published).

[5] S. Chervina, S. E. Evon, G. T. Bodman, J. Therm. Anal., 1997 (to be published).

[6] S. Chervina, S. E. Evon, G. T. Bodman, The Program Book of the 11th Congress on Thermal Analysis and Calorimetry, Philadelphia, PA, August 1996.

[7] L. F. Whiting, M. S. Labean and S. S. Eadie, Thermochim. Acta, 136 (1988) 231.

[8] S. Chervina, G. T. Bodman, The Proceedings of the 30th Annual Loss Prevention Symposium, New Orleans, LA, February 26-28, 1996.

[9] S. Chervina, G. T. Bodman, The Proceedings of the 24th North American Thermal Analysis Society Conference, San Francisco, CA, September 10-13, 1995.

Developing Chemical Compatibility Charts Using CHEMPAT

David J Leggett,

Hazard Evaluation Laboratories, Inc.,
7 Deer Park Drive
Monmouth Junction
NJ 08852-1921

Presented at:

AIChE 31st Annual Loss Symposium Symposium
March 9 - 13, 1997
Houston Texas

Developing Chemical Compatibility Charts Using CHEMPAT

David J Leggett, Hazard Evaluation Laboratories, Inc.

I. INTRODUCTION

OSHA's Process Safety Management Standard, 29 CFR 1910.119 (September 1992) requires that process hazard evaluations be performed on covered processes. The regulations contain 14 areas for consideration, six of which are related to the need to provide comprehensive information of all process materials, including the consequences of inadvertent mixing of process materials. This information can be made available to those involved in the manufacturing process through the development of an Interaction Matrix. A brief overview of the subject is given in a publication by the Center for Chemical Process Safety (CCPS)[1]. A previous paper in this session has provided a general discussion of the significant issues involved when considering chemical compatibility.

In 1996 AIChE announced the availability of the computer package CHEMPAT that provides a straightforward method of generating and documenting the interaction matrix, or Compatibility Chart, and accompanying database. CHEMPAT was developed, and used internally, by The Dow Chemical Company since 1987. It was donated by Dow to AIChE in 1995 for use by the chemical process industries.

This paper outlines an approach used to develop chemical compatibility information through the use of CHEMPAT.

II. BACKGROUND

A chemical compatibility chart provides, in a compact, easy-to-read format, hazard information for the inadvertent mixing of chemicals. The goals of sound, technical management in developing compatibility charts may be summarized as:

1. Provide a reference guide for actions to be taken, by plant personnel, in the event of an inadvertent mixing of process materials.
2. Provide a consistent source of reactive chemicals hazard data.
3. Assist in the Reactive Chemicals (RC) training of plant personnel.
4. Help identify where data gaps may exist in chemical reactivity information.
5. Provide the tools to perform a hazard analysis with regard to inadvertent mixing.

Developing the database and accompanying chart for a chemical manufacturing facility is a structured hazard evaluation. The thought process required demands a thorough and exhaustive examination of the ways the plant handles its process materials. The chart may also be used as a reference guide for selecting the right spill absorbent, or to indicate appropriate materials of construction with regard to chemical compatibility.

In the past, chemical compatibility charts, or reactivity matrices, have been prepared by many groups within a chemical manufacturing facility, in paper chart format. These paper charts were produced manually, based on data and experience of plant operating personnel. However, paper charts do not allow for the inclusion of substantiating data for the hazard assessments of each binary pair of chemicals. Also, the modification of the chart, due to a change in the process and/or the addition or removal of chemical(s), was difficult, time consuming, and often not done.

CHEMPAT provides a straightforward and consistent way of developing and documenting this required information (Figure 1 shows a completed chart.) It is a PC-DOS based program that has been designed to organize and display information regarding the consequences of inadvertent mixing of process materials. The program produces a compatibility, or inter-reactivity, chart and manages the user-supplied information that is required by CHEMPAT.

The package would be used by technical personnel in a production environment, batch or continuous, pilot through world-scale plant, who are responsible for developing the Process Hazard Evaluation. Typically, this individual would be the safety officer, a safety and loss prevention specialist, a plant engineer or a senior plant foreman. The actual production of the chart and database may be assigned to junior personnel once the data have been acquired and the potential hazard evaluated.

III. STEPS IN DEVELOPING THE COMPATIBILITY CHART

The technique used to develop compatibility charts comprises seven steps which are outlined in the following section. Normally, the compatibility chart is prepared by two people with experience in hazard evaluation and assessment. Once prepared, the compatibility information is presented to operations personnel during regular training sessions. The user version of the chart is also installed on local PCs so that the information used in the chart is readily accessible.

III.1. Decide Who Will Build The Chart

After plant personnel have determined the need to develop a compatibility chart the company's hazard evaluation group is contacted to initiate the project. Alternatively, help may be obtained from a consulting company specializing in chemical process hazard evaluation. Various levels of support may be available from complete compatibility chart development by the hazard evaluation consulting group to complete development by the client.

III.2. Select The Process Expert

The process expert is the contact through whom the chart will be developed. This individual must be intimately familiar with all aspects of the plant's processes. This person will also be responsible for outlining the physical or chemical concerns which may become factors in the event of an inadvertent mixing.

III.3. Define The Mixing Scenario

This is the most important step. The process expert, in cooperation with the hazard evaluation group, will examine the process including related activities such as water treatment, waste handling, etc., in order to develop a list of components used by the plant and the related mixing scenarios. This list should be as comprehensive as possible. However, since a plant may comprise several different unrelated areas, separate lists, and scenarios, could be composed for each area. For example, the plant QC lab is usually separated from the process materials. Plants may develop, because of their layout and size, a separate chart for the water treatment processes, including only treatment chemicals, materials of construction, and spill absorbents.

In considering the components do not limit the list to only process chemicals. It should include raw materials, products, process chemicals, absorbents, insulating materials, materials of construction, waste streams, laboratory chemicals, air and water, and heat.

III.4. Gather The Data

Once the compound list has been made, information on both the pure compounds and the potential mixtures is obtained. Pure compounds information is often extracted directly from the Material Safety Data Sheets. Complete and credible information on the mixing hazards for the compounds is obtained usually from the open literature or by explicit testing, as described below.

Literature sources include: Existing company databases including incident reports, Material Safety Data Sheets, the NFPA's *Fire Protection Guide to Hazardous Materials*,[2] Sax's *Dangerous Properties of Industrial Materials*,[3] Bretherick's *Handbook of Reactive Chemical Hazards*,[4] and the U.S. Coast Guard's *Chemical Hazard Response Information System* (CHRIS).[5] Another valuable information source is from experts familiar with the reactivity of the mixture. The credentials of this "expert opinion" should be documented along with the opinion given.

Experimentally, DSC and heat of mixing calorimetry, and/or equivalent techniques, provide comparable data at minimum cost. In some instances it will be necessary to obtain data from adiabatic calorimetry, especially if mixing scenarios involve prolonged storage. This approach will resolve most questions regarding potential incompatibilities. Questions that remain after screening tests are complete may be resolved using adiabatic, Dewar, heat flow, or reaction calorimetric techniques. For process owners without local testing facilities, AIChE's Center for Chemical Process Safety [6] maintains a list of consulting groups, some of which have the required testing expertise.

Once the data have been obtained the following information should be recorded in the CHEMPAT database: test description and method, date of evaluation, testing location, and thermochemical data. Any additional information of interest for the particular location/situation should be included.

III.5. Assign Hazard Levels

After all mixture data have been accumulated the hazard levels are defined. The role of the hazard evaluation group is to act as technical consultant(s). However, the plant's process expert decides on the final rating. In this way, the process, plant, or laboratory has the responsibility of determining hazard levels. The hazard evaluation group will often counsel the plant on being under- or overly-conservative in its ratings.

III.5.a. Hazard potential and hazard levels: A chemical compatibility chart, displayed in prominent places throughout the plant, provides guidance for all personnel if an unplanned mixing of chemicals occurs. It is, therefore, important that plant personnel are trained on the significance of the various hazard levels and appropriate actions to be taken. A hypothetical set of actions, for each hazard level, are presented in Table 1.

In practice, each plant will establish a set of actions, based on the associated hazard levels, for a variety of mixing scenarios. Once the actions to be taken have been linked to the mixture scenarios, the complete compatibility database and chart may be easily developed. Factors to consider when establishing plausible mixing scenarios include temperature, likely time-line for the scenario, quantities involved, pressure rating of the container and vent size, and materials of construction.[7] These factors, taken in the context of the results of the mixture screening work, provide the factual basis for setting the response to the particular mixing scenario.

When assigning the response to the hazard level for any mixture ask the question, "What prudent and conservative actions need to be taken to maintain the safety of personnel, the integrity of the plant and the health of the business?" The data that have been gathered regarding the chemical and physical results of the mixing, together with the plant's in-place lines of defense and emergency response procedures for that particular event, are used to assess the potential hazard. Other factors, such as proximity of the mixing event to storage areas, tank farms, sources of ignition, etc., are also included. Figure 2 illustrates this process, as a flow chart.

III.5.b. Hazards For Pure Compounds: Although we are focused on mixing reactivity hazards the inclusion of health and flammability ratings provides useful additional safety information.

The logic and rating system used by CHEMPAT is similar to that described in NFPA reports.[8] Three hazard ratings are also assigned to pure materials. These data may be obtained from standard ratings, MSD sheets or other company information.

III.5.c. Hazards Due To Mixing: Only the *additional* hazards of mixing need to be considered. Materials which are themselves highly reactive, for example, reactive monomers, may generate a mixture with little or no *additional* reactivity. For companies making the same product at different locations mixing hazards will likely be plant-specific. This is because, for a given reactant, concentration or mass of reactants used, temperatures, pressures, plant equipment, etc., will vary among locations, and from process to process. The relative degree of hazard, a specific process location, will therefore vary. Ratings are assigned on a scale of 0 to 4, as shown in Table 1. Mixing hazard ratings should be based on Reactive Chemicals concerns, and not, in the first instance, product quality concerns.

Once the hazard levels have been determined, the plant will implement appropriate emergency response plans to handle the potential situations. As noted earlier, the hazard levels on the chart now become the action keys for personnel to follow in the event of inadvertent mixing.

On completion of the chart, it is reviewed jointly by plant supervision and the hazard evaluation and/or safety and loss prevention groups. The chart should be predominantly a combination level zero and level one with *some* level two hazards. "Some" is obviously arbitrary, but is typically in the range of 0 - 10%. If a *significant* number of level three and level four hazards (>2% of total combinations) have been assigned, the overall assessment and assignment scheme needs to be reviewed.

It is possible that the hazard assessment has been too conservative resulting in too many "3"s and "4"s. However, it is also possible that the number of "3"s and "4"s is an accurate reflection of the potential hazards that may exist in the plant. In the latter case, *the plant should undertake an orderly program to reduce the potential for hazardous interactions of materials.*

The process expert, who has the responsibility to assemble the compatibility database and chart, will use all features of the program. Once the database is complete, the User Version is distributed to the locations within the process area. This version has no editing capabilities. Thus the database information is secure and can only be viewed.

III.6. Generate Compatibility Database and Chart

In order for CHEMPAT to be an effective information tool, it is necessary that the steps described above be followed as closely as possible. Assuming this to be the case, then the task remaining is data entry.

The information on each of the mixtures is entered into CHEMPAT. A special, location-specific designation may be used to highlight a unique inter-reactivity hazard (such as "no hazard if dry, dangerous reaction if wet"). There is no program limit on the amount of information that may be stored. This enhances the chart's use as a reference and training tool.

The printed chart (illustrated in Figure 1, gray-scale rendition) illustrates the potential hazards of the defined mixtures for a typical production facility. The triangular region displays the compatibility ratings for the binary mixtures. All ratings may be represented with color coding to indicate the appropriate hazard level when using a plotter. Plant information is included in the chart's header, with definitions placed below the chart drawing. Additionally, the program can generate a copy of the textual information used to produce the conclusions represented in the chart (Figures 3 and 4). This chart also indicates the Health, Flammability and Reactivity of the compound to the left of the compound names.

III.7. Distributing The Information

The final step in producing the compatibility chart is getting the information out to the personnel who need it. With the computer-based software program, this distribution has two parts. First, the process expert, who has served as a focal point, receives a complete copy of the software package and database. That person is able to modify the database as the process changes and will serve as a continuing focal point of the compatibility project. Second, the User Version is distributed to key areas of the process plant or laboratory. The program is installed on centrally located personal computers. In this installation, restrictions are set on the capability to modify the information. In this way, the risk of unauthorized changes will be minimized.

Finally, the personnel who will see and use the compatibility chart and software program are familiarized and trained in the use of the chart and, as is appropriate, the User Version of the program.

IV. OVERVIEW OF THE CHEMPAT PROGRAM

Producing a compatibility chart requires that every combination of process materials be reviewed, with regard to the potential for inadvertent mixing. A hazard level is then assigned which, in turn, provides an action plan for all personnel in case of such an event. The computer package CHEMPAT has been designed to simplify the management and storage of chemical compatibility information.

Use of the program involves first entering the information about the components. This sets up the structure of the database. Once this step is complete the compatibility information for each binary combination of components is then entered. Any pair of components may be selected so that the information may be entered as it is gathered. The compatibility chart may be printed or plotted at any time in the development of the chart. The program is available from AIChE and comes with a 60 page manual that covers all aspects of program use.

V. STRATEGY FOR DEVELOPING A COMPATIBILITY CHART

Figure 1 shows the compatibility chart that was developed for a production facility involved in the manufacture of a group of related resins. The original list of components comprised 47 specific chemicals. This meant that the chart and database would require considering 1,128 combinations. The large number of combinations proved to be daunting to the production group, as well as the RC group. Therefore, a strategy was adopted of grouping the individual chemicals into families of related chemicals. This resulted in fourteen components, six of which were material groups. The properties of the eight specific chemicals were sufficiently different, or reactive, to warrant separate attention. The spill adsorbents were kept separate to emphasize the differences in reactivity between sawdust and the clay-based adsorbent.

With the project reduced to a more tractable size, 105 combinations, reactivity data was gathered from a variety of sources, including expert opinion. Data for seventeen combinations was missing or found to be lacking in reliability. These combinations were tested using heat of mixing and DSC. The results of these tests were adequate to provide a reliable hazard assessment and no further testing was conducted. However, it is entirely possible that further testing, such as adiabatic calorimetry or heat flow calorimetry, would be necessary to adequately characterize the system.

The complete set of incompatibility data was entered into CHEMPAT and the hazard ratings reviewed with plant personnel. It is important not to overlook this step as this is the point in the process where buy-in with plant folks is obtained. This is also when the plant decides how to relate the hazard ratings, generated by the testing group with the actions to be taken by personnel should inadvertent mixing occur. The work involved with this particular compatibility chart and database was quite typical taking about 15 man-days to complete.

VI. SUMMARY

OSHA's Process Safety Management Standard, 29 CFR 1910.119 (September 1992) requires that process hazard evaluations be performed on covered processes. The regulations contain 14 areas for consideration, six of which are related to the need to provide comprehensive information of all process materials, including the consequences of inadvertent mixing of process materials. CHEMPAT provides, in a compact, easy-to-read format, this type of hazard information. One of the major goals of the chart is to provide a reference guide for actions to be taken, by plant personnel, in the event of inadvertent mixing of process materials. The program also has the provision to display, health, flammability, and reactivity ratings, often available from National Fire Protection Association (NFPA) publications

VII. REFERENCES

[1] *Guidelines for Hazard Evaluation Procedures,* 2nd Edition, Center for Chemical Process Safety, American Institute of Chemical Engineers, 345 East 47th St., New York, NY 10017.

[2] *Fire Protection Guide on Hazardous Materials,* 10th Edition, National Fire Protection Association, Quincy, MA, 1991.

[3] Lewis, R.J., *Dangerous Properties of Industrial Materials*, 8th Edition, Van Nostrand Reinhold, New York, 1992.

[4] Bretherick, L., *Handbook of Reactive Chemical Hazards*, 4th Edition, Butterworth-Heinemann Ltd., Oxford, 1992.

[5] Commandant Instruction M.16465.12A, *Chemical Hazard Response Information System (CHRIS)*, U.S. Department of Transportation, Washington, D.C., 1974.

[6] Center for Chemical Process Safety, American Institute of Chemical Engineers, 345 East 47th St., New York, NY 10017.

[7] Hofelich, T.C., D.J. Frurip, and J.B Powers, "The Determination of Compatibility via Thermal Analysis and Mathematical Modeling", *Process Safety Progress*, Vol **13,** No. 4, October **1994,** 225-233.

[8] NFPA 325, *Fire Hazard Properties of Flammable Liquids, Gases , and Volatile Solids,* 1994 Edition, National Fire Protection Association, Quincy, MA 02269, 1990.

D.J. Leggett, Ph.D., Grad.RSC, CChem is technical director for Hazard Evaluation Laboratories, Inc., Monmouth Junction, NJ (908/329-3090; Fax: 908/329-3023; e-mail: 104657.22125 @compuserve. com). Prior to joining HEL Inc he worked for Dow Chemical as manager of the Hazard Evaluation labs at Freeport, Texas, providing support for more than 85 plants and processes. Leggett has many years experience in all areas of process hazard assessment and analysis. He has assisted clients by providing all necessary Reactive Chemicals data and interpretation in order to comply with OSHA PSM standard. He is a member of AIChE, ACS, and The Royal Society of Chemistry. He has published over 40 papers, and one book. He has a Ph.D. in analytical chemistry from University of Waterloo, Ontario, Canada.

Table 1. Hazard Level and Associated Actions (Typical).

Hazard Level	Action Classification
0	Report inadvertent mixing event to supervision; no further action necessary.
1	Report event to supervision; devise and implement plan(s) to manage the situation; no emergency procedures to be initiated
2	Report event to supervision; prepare to initiate unit emergency plan if needed; notify personnel in immediate area; consider halting normal activities until extent of situation is fully assessed.
3	Report event to supervision; initiate unit emergency plan; notify all plant personnel; cease normal activities until extent of situation is fully assessed; consider need to evacuate the plant; report event to plant industrial security and other ex-plant ER groups.
4	Initiate unit emergency plan; notify all plant personnel to evacuate the area; cease normal activities, if possible, before evacuating; report event to plant industrial security and other ex-plant ER groups once evacuation is underway or complete.

Table 2. Definition of Hazard Levels, Description and Examples

Hazard Level	Definition	Description of Typical Event(s)	Example
0	Minimal (Note 1)	No observable reaction will occur between materials that are mixed at process temperatures (and somewhat above)	75% sulfuric acid + 95% sulfuric acid
1	Caution (Note 2)	An observable reaction may occur on mixing of the materials at (or somewhat above) process temperatures	50% sodium hydroxide + water
2	Danger	A reaction will occur on mixing of the materials, with accompanying energy release, and may result in venting of materials.	ethylenediamine and methylene chloride
3	Severe Danger	A violent reaction will occur on mixing and is likely to result in a sizable loss, possibly including process equipment integrity.	chlorine and acetylene
4	Extreme Danger	A violent reaction will occur on mixing likely including a deflagration or detonation	liquid oxygen and grease
P	Precipitate (Note 3)	It is suggested that this special class of hazard be used for materials that may form a precipitate (or polymer), either on standing or on mixing	uninhibited styrene
S	Special	In addition to the above "standard" hazard levels, each database is allowed to include one special hazard classification	
X	Excluded (Note 4)	This hazard rating is for mixtures that are highly unlikely to occur	
{blank}	Not Known	All entries that are not otherwise marked are classified as NOT KNOWN, indicating that no information is available for that mixture	

Note 1: A small heat of mixing may be allowed, provided that it does not elevate temperatures to the point of causing other reactions, unacceptable pressure rises, or process upsets. A 10° to 20°C temperature rise may be acceptable for materials handled in metal process equipment, but may be unacceptable for materials normally handled in FRP containers.

Note 2: This may only be a heat of mixing or heat of adsorption. However, if the reaction is not expected, it may lead to a Reactive Chemicals incident .

Note 3: The hazard may not come directly from the precipitation or polymerization reaction, but may result from the failure of process lines, drains, pumps, valves, etc., which may become clogged.

Note 4: These mixtures are marked with an "X" on the finished chart. This category may be difficult, in an absolute sense, to reliably assign. By design intent the compatibility chart and database deal with events that are statistically unlikely to occur. Therefore, careful consideration must be given when assigning the classification "unlikely to occur" since that assignment might overlook a potential hazard. The category is often assigned to pipeline chemicals or to chemicals contained in widely separated tank farms. However, the category should not be used for chemicals that are delivered and/or stored in different warehouses

Production Facility Compatibility Chart

CHEMPAT - CHEMICAL COMPATIBILITY CHART

SITE:Development Facility. UPDATED:Sept. 17,1996 at 8:55. 100% FILLED.
SITE CONTACT:Plant Expert.

H	F	R		ACETONE	ACID ANHYDRIDES	ACIDS	ACRYLATES	DIALLYL PHTHALATE	EPICHLOROHYDRIN	GLYCOLS	HYDROCARBONS	ISOCYANATE	METHYLENE CHLORIDE	PEROXIDE INITIATORS	WATER	SAWDUST	CLAY ADSORBENT
1	3	0	ACETONE	0													
3	1	1	ACID ANHYDRIDES	1	2												
3	2	2	ACIDS	1	1	0											
3	2	2	ACRYLATES	1	2	2	P										
2	1	0	DIALLYL PHTHALATE	0	1	1	1	0									
3	3	2	EPICHLOROHYDRIN	0	2	2	1	X	0								
1	3	0	GLYCOLS	0	2	1	1	1	1	1							
1	1	2	HYDROCARBONS	0	X	1	X	X	0	0	0						
3	1	2	ISOCYANATE	1	X	2	2	X	X	3	X	P					
2	1	0	METHYLENE CHLORIDE	0	0	P	1	0	0	1	0	1	0				
1	3	4	PEROXIDE INITIATORS	3	X	P	P	X	X	3	X	X	3	3			
0	0	0	WATER	0	2	1	0	0	P	0	0	P	1	2	0		
0	0	0	SAWDUST	0	2	1	1	0	0	0	0	2	0	3	0	0	
0	0	0	CLAY ADSORBENT	X	2	1	2	1	X	0	0	1	0	3	0	0	0

H - HEALTH F - FLAMMABILITY R - REACTIVITY 0 - MINIMAL

1 - CAUTION 2 - DANGER 3 - SEVERE DANGER 4 - EXTREME DANGER

P - POLYMERIZATION or PRECIPITATION X - UNLIKELY MIXTURE

Above data are valid for this site only. Do not use elsewhere.

Triangular region shows mixing hazards. Partial shading indicates not tested.

Figure 1. Chemical Compatibility Chart For a Typical Production Facility

Figure 2. FlowChart Overview of Hazard Evaluation and Assignment Process

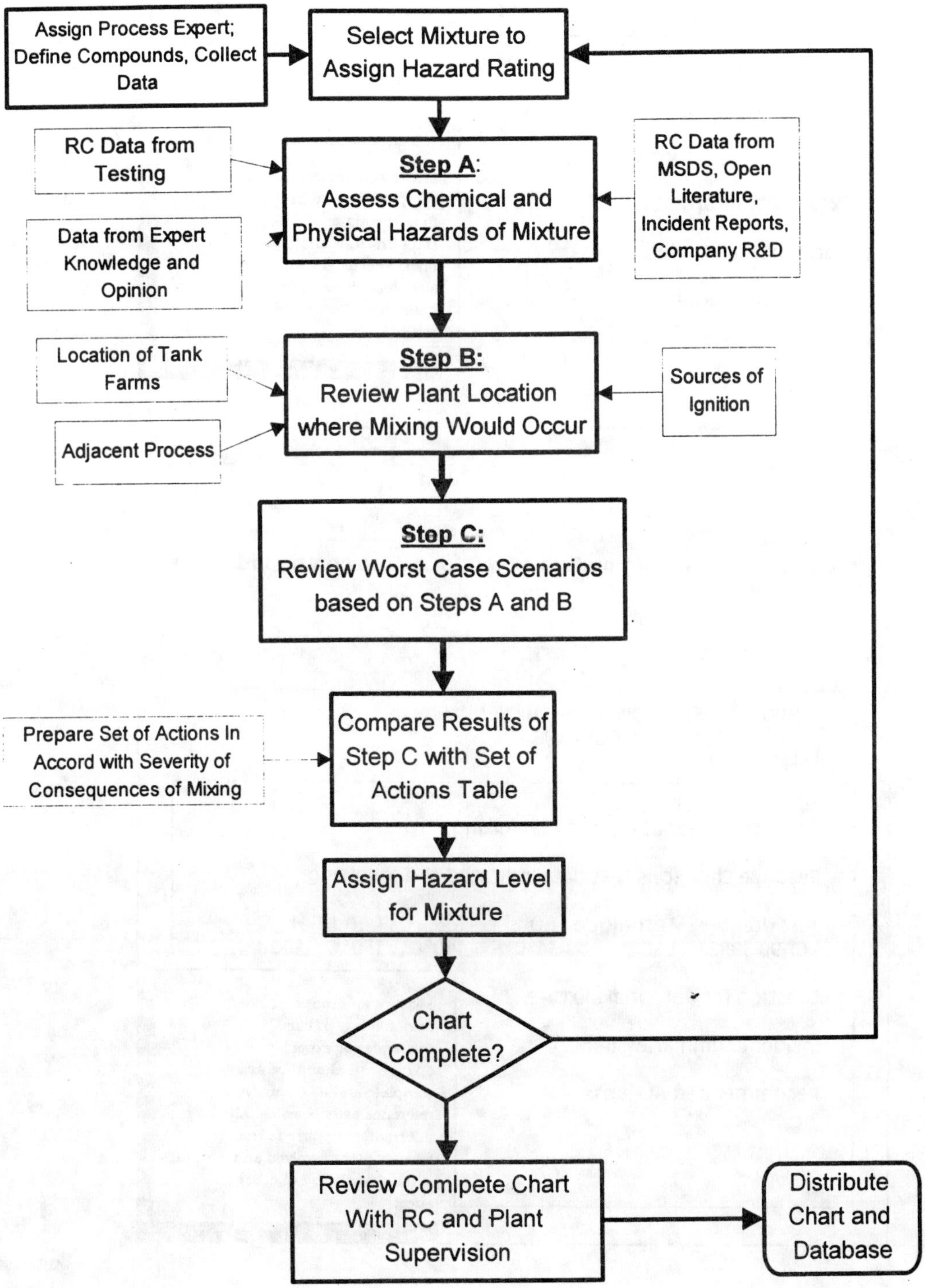

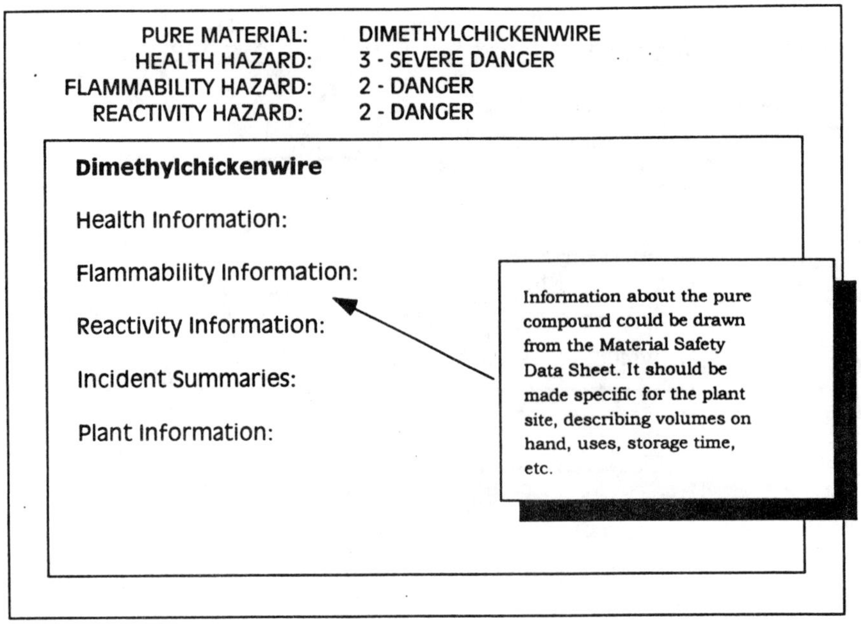

Figure 3. Example of information stored for a pure compound

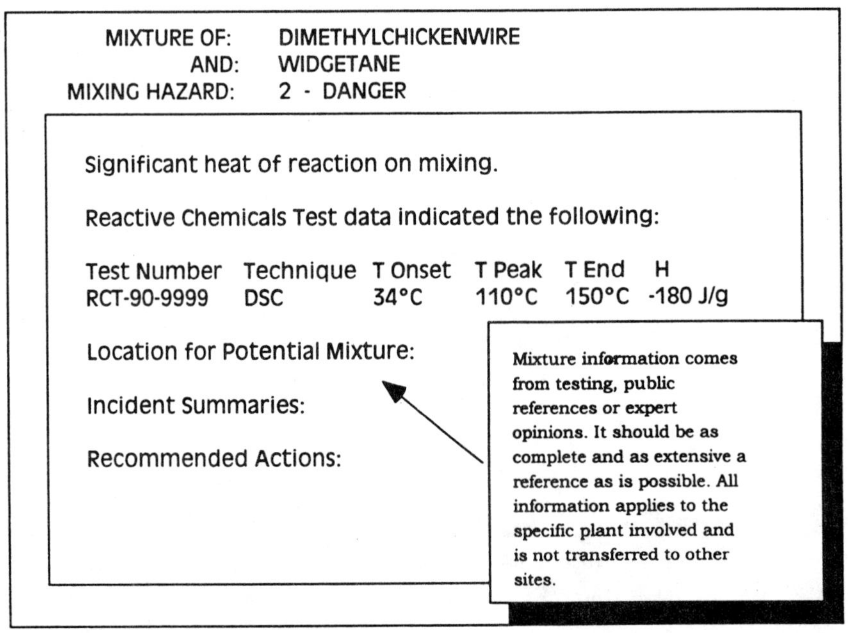

Figure 4. Example of information stored for a mixture

IMPROVING THE EFFECT OF ATMOSPHERIC STABILITY CLASS ON HAZARD ZONE PREDICTIONS FOR ACCIDENTAL HAZMAT RELEASES

John L. Woodward
DNV Technica, Inc.
40195 County Center Drive, Suite 200
Temecula, CA 92591
(909) 694-5790

For AIChE Loss Prevention Symposium- March 9-13, 1997
Houston, Texas

1.0 INTRODUCTION

Predicting hazard zones for accidental releases of hazardous materials has long been important for planning emergency responses and for risk analysis. Applying such predictions to transportation of hazardous materials requires improved techniques and understanding of the fundamentals of atmospheric dispersion. This is because typical transportation routes pass through a wide variety of temperatures, wind patterns, and atmospheric stabilities. The focus of this paper is on recent improvements in understanding the effect of atmospheric stability on dispersion.

Atmospheric stability effects have become the focus of interest in recent developments. For example, the EPA has published guidelines for complying with provisions of the Risk Management Programs for Chemical Accident Release Prevention, known as the RMP regulations. The RMP regulations require that off-site consequence analysis must include treatment of at least one worst-case (Program 1 requirement) or a worst-case and a more credible case alternative (Programs 2 and 3 requirement). Worst-case has been defined as the release of the largest inventory (i e contents of largest isolatable vessel) of toxic or flammable material over a ten minute period under conditions of a 1.5 m/s wind speed and atmospheric stability F. For these conditions the EPA provides tabulated dispersion modeling estimates in a general guideline. To encourage improved technology, the RMP regulations allow industry to develop improved models with more accurate capability.

The experimental data base supporting heavy gas modeling is quite sparse, because of the expense of full-scale field tests. In particular, the number of experiments conducted at F stability has been woefully inadequate. This is mainly because such tests must be conducted at night or in the early morning under low wind speed conditions. The percentage of the time a region experiences F stability ranges typically from 3% to as much as 60% at a pump station on the Alyeska pipeline.

Fortunately, a large body of experimental data on heavy cloud dispersion has recently been released by the German government [Heinrich, et. al. 1988,1989] in which atmospheric stability is addressed, for the first time, in a reasonably substantial manner. We discuss these findings here, along with a discussion of model improvements generated therefrom and compare our model with two others used to generate the EPA tabulated predictions.

2.0 WHAT IS F STABILITY AND WHY IS IT SIGNIFICANT?

Each of us is familiar with the atmospheric conditions evident in mountain valleys in the early morning following a cloudless night with calm winds. An invisible ceiling seems to have been lowered during the night, compressing visible air pollutants into a stagnant layer. This condition is defined as Pasquill-Gifford F Stability. **Figure 1** is a typical photograph of such conditions, in this case showing a cloud layer that has settled into the valley as seen from the road leading down from Arrowhead Lake near San Bernadino, California.

Technically, F stability is defined as a stable atmospheric condition in which vertical turbulent mixing tends to be suppressed by the vertical temperature profile which is sub-adiabatic. This is explained in Appendix A along with a standard correlation for assigning stability class.

With F stability, we obtain significantly larger hazard distance predictions than with neutral D stability, and even more than with unstable A stability, as **Figure 2** illustrates. This is for two example releases: (1) a release of 1000 lb./min. of a two-phase aerosol of anhydrous ammonia from a typical compressor at 180 psig dispersing to a toxic end point of 200 vppm; (2) a release of 100 lb./min. of a two-phase aerosol of chlorine dispersing to a toxic end point of 3 vppm. Interestingly, the two curves virtually overlap.

3.0 INFERRING THE EFFECT OF STABILITY CLASS ON DISPERSION

Essentially all dispersion models must treat four modes of dispersion: high momentum jet mixing, dilution across evaporating pool, heavy gas dispersion, and passive dispersion. Atmospheric stability affects the latter three modes.

3.1 Stability Class Effect Well-Established for Passive Dispersion

Passive dispersion applies with dilute vapor clouds where the presence of the contaminant does not influence the normal turbulence in the atmosphere. Passive dispersion models use the analytic solution for a point source release, a gaussian form, giving the familiar bell-shaped concentration profiles illustrated in **Figure 3**. Extensive experiments have been conducted with tracer gases in low concentration for air pollution modeling which resulted in the commonly-used plots of the cross wind standard deviation, σ_y, and the vertical standard deviation, σ_z. An example plot of σ_y against of the plume increases proportional to σ_y times σ_z, independent of the rate of the release, depending only on the extent of normal atmospheric fluctuations.

Figure 4 reveals a sensitivity between A and F stability represented by the spacing between the curves. Taking points from these curves at several distances downwind and plotting the ratio of each value thus found to that at D stability gives the result shown in Figure 5. **Figure 5** represents a multiplier for stability class effects. Thus, passive dispersion is predicted to have just half the dispersion rate with F stability as with D stability. On the other hand, dispersion at stability class A is predicted to have 3 to 4 times the air entrainment rate as dispersion at stability class D.

Since the curves for A and B stability in Figure 6 tend asymptotically toward a nearly constant value at large distances, we take these large distance values to plot against stability class in **Figure 6**. In Figure 6 the passive dispersion curve is compared with a heavy gas dispersion curve. Before obtaining the test data of Heinrich et al we assumed that the heavy gas curve was identical with the passive dispersion curve. The new test data justify for heavy gas dispersion, an effect of atmospheric stability that is less sensitive than for passive dispersion. This is consistent since heavy gas dispersion is known to suppress atmospheric dispersion.

For completeness, the effect of stability class on heavy gas dispersion occurs in models which use standard atmospheric vertical wind profile theory, including ours, in two ways. Vertical wind profiles are calculated using friction velocity, u*, which is a parameter related to surface roughness. We take advantage of a considerable body of literature on normal atmospheric turbulence by making our top surface entrainment velocity proportional to u*. As **Figure 7** illustrates, this introduces a factor ranging from about 0.9 for E and F stability to between 1.1 and 1.3 for A stability. The factors in Figures 6 and 7 both multiply to accentuate or reduce entrainment rates. Of these factors the Figure 6 curve is adjustable to describe dispersion.

3.2 Recent Tests Signficantly Expand Data Base for F Stability

Table 1 gives a summary of the major experiments supporting dispersion models, illustrating how few experiments were made at F stability before the German government-sponsored tests by TUV reported by Heinrich et al. Table 1 uses tests selected as of good quality by Hanna et al (1991a,b) and included in their "modeler's data base". The German government sponsored these propane release tests specifically to improve our understanding of atmospheric stability on dispersion. The TUV tests were conducted over a two year period, near the town of Meppen, close to the Dutch border. The tests were over absolutely flat open farmland with few bushes as limited by the farmer's permission, and many were made in the early morning. Of the 60 tests, 25 were at F stability and 7 at E stability.

The TUV test data are summarized in **Table 2**, along with our model predictions. Ambient temperatures are taken from Heinrich's original report, and vary appreciably. Some tests discharged the two-phase propane through a cyclone (details not clearly disclosed) to slow the release velocity, since otherwise jet momentum tends to dominate the dispersion. An undetermined, but substantial amount of rainout (droplet fallout) occurred when using the cyclone. Since the expansion zone velocity was not measured and is not readily calculated, we set it at a low value, 10 m/s, for all runs which used a cyclone. The expansion zone is that part of a plume just past the vena contracta (from an orifice release). For high pressure releases, the velocity is normally sonic at the vena contracta and increases to supersonic speeds as the pressure decays from the choke pressure to atmospheric. When the plume impacts surfaces, as in a cyclone, a large fraction of the discharge momentum dissipates. Rainout fractions and vapor rates are calculated using our latest model the Unified Dispersion Model (UDM) as detailed by Woodward and Papadourakis (1995).

There is always considerable uncertainty in determining stability class. For example, **Figure 8** plots the experimentally-derived Richardson number calculated from measured vertical gradients using Equation A.1 against stability class. There is a correlation evident in this plot, but also considerable variability illustrating both the large scatter in the data and the difficulty of assigning stability class.

3.3 Experimental Data Indicate a Stability Class Effect on Dispersion

The TUV data reveal that the effect of stability class on dispersion is measurable and discernable, albeit with a large uncertainty level. **Figure 9** plots, for tests with the same discharge rate (6 kg/s), the extrapolated distance to the 2.1% LFL concentration for propane, termed the lower flammable distance (LFD). The LFD tends to increase with Richardson Number (stability class) as expected. The next question is how well do models predict this effect?

4.0 MODEL PREDICTIONS DESCRIBE TUV EXPERIMENTS

4.1 Predictions Made with Unified Dispersion Model (UDM) Fit Available Test Data

Our latest-generation consequence-evaluation model, the UDM is documented by Woodward (1990), Woodward and Cook (1995), Woodward and Papadourakis (1995), and Woodward, Cook, and Papadourakis (1995). This model is incorporated into our consequence and risk programs, PHAST and SAFETI. In an extension of previous capabilities, it is able to treat a heavy aerosol cloud which becomes buoyant and lifts off. **Figure 10** illustrates that this model fits well the ammonia aerosol test data from the Desert Tortoise experiments at Frenchmen's Flats, Nevada. **Figure 11** illustrates the model fit for some Goldfish HF tests.

4.2 Uncertainties in TUV Data Produced by Different Analyzer Results and Extrapolation of Data

The TUV data are not entirely consistent, partly because the two different types of analyzers used give different results, as **Figures 12** and **13** illustrate. These plot TUV Tests 42 and 43, which were run at nearly identical conditions. The discharge rates are equal, and the wind speed nearly so. Consequently, the model predictions are nearly the same for these cases. Yet, the catalytic sensor data are considerably higher in Test 42 than in Test 43. The preferred sensors are the eleven infrared (IR) spectrometers, whose values are plotted as the solid triangles in Figures 12 and 13. These data as originally reported were erroneous because of interference by ice crystals in the plume but have been corrected. There were forty catalytic sensors, whose values are plotted as the open triangles. The model agrees with the catalytic sensors in Figure 12 and with the IR sensors in Figure 13.

Figure 12 also illustrates that extrapolation was used to obtain "observed" LFD values. Heinrich et al draw a straight line through their data (on a log-log plot) and extrapolate or interpolate to obtain their observed LFD values tabulated in Table 3. These uncertainty bands are, on the whole, quite broad, more typical of Figure 13 than of Figure 12. Without the guidance of a model, it is difficult to choose the slope of these extrapolation lines. We improved the original estimates for Table 3, but prefer to plot predictions through the measured data as an indicator of model performance.

4.3 Modeled Stability Class Effect Illustrated

Model predictions are reasonably sensitive to the stability class effect, as **Figure 14** illustrates for TUV Run 52. The predictions for F stability are to the right of those for D stability. Over a larger distance, the differences become substantial.

4.4 Predictions by UDM of TUV Data are Unbiased

Predictions are compared with observations for all TUV cases in **Figures 15 and 16** at each atmospheric stability class. For Figure 15 an unbiased prediction has a value of 1.0, plotting the ratio of predicted to observed LFD. For Figure 16 an unbiased prediction has a value of 2.1%, the LFL concentration, since we plot the predicted concentration at the observed LFD. By either measure, the model predictions are very nearly unbiased.

5.0 COMPARISON WITH EPA GUIDANCE DOCUMENTS

The general EPA Guidance document [EPA (a)] provides a table of distance to the toxic end point for a generic material. Using the specified toxic end point for ammonia of 0.14 mg/liter (200 ppm), the generic tables give the curve in **Table 5** and **Figure 17** for 10 minute releases with rural conditions under worst-case meteorology (1.5 m/s, F stability). An ammonia-specific Guidance Document [EPA (b)] gives predictions by the SACRUNCH model developed by Science Applications, Inc., also given in Table 5 and Figure 17. For comparison, we calculated dispersion distances using typical conditions for an ammonia compressor at both the high-pressure and the low pressure receiver as cited in the ammonia-spccific Guidancc Document. These conditions, along with the other model inputs assumed and calculated are listed in **Table 4**. For a given receiver pressure and temperature, the discharge velocity and mass flux are constant, so the mean drop size remains constant as well. To obtain the release rates listed in the EPA Guidance document, vary the effective puncture diameter.

Figure 17 and **Table 5** show the UDM predictions for centerline concentration nearly coincide with those of the EPA Guidance document. Both disagree with the predictions of the SACRUNCH model by about a factor of two.

The UDM predicts that the ammonia cloud remains negatively buoyant beyond the 200 ppm concentration for all stability classes. For low releases, the UDM predicts the ammonia cloud will warm and become positively buoyant, or lift off after the droplets evaporate for unstability and neutral atmospheric conditions as shown in Figure 18. The lift-off occurs closer with A stability, as would be expected since the largest vertical mixing occurs with A stability.

With small discharge rates rainout is predicted to not occur. The predicted percent rainout increases with increasing discharge rate, as shown in Table 5.

6.0 CONCLUSIONS

Atmospheric stability class is known to be a sensitive parameter for dispersion models, for both passive and heavy-gas dispersion. An extensive, and under-exploited set of experiments were sponsored by the German government specifically directed toward understanding the effect of stability class on dispersion. These tests, conducted by the research company TUV, are evaluated here using DNV Technica's latest dispersion model, the Unified Dispersion Model (UDM). There is considerable scatter in the TUV data, particularly between two different types of propane analyzers, even after correcting the original infra-red analyzer data for interference by ice crystals. We were able to obtain essentially unbiased modeling predictions by carefully tuning the UDM to all available experimental data and then reducing the originally-postulated sensitivity to stability class. Example comparisons of UDM predictions with the Desert Tortoise ammonia and the Maplin Sands propane test data show excellent agreement.

The UDM predicts that ammonia clouds released at low elevations from temperature and pressures typical of ammonia compressor receivers will usually rain out a small fraction of the mass and will spread as a heavy aerosol cloud until well past the 200 ppm toxic endpoint. UDM predictions agree well with the distances to this toxic endpoint specified by the general EPA Guidance document for Risk Management Programs. Both predictions disagree with those of the SACRUNCH model listed in the ammonia-specific EPA Guidance Document. These discrepancies should be resolved by considering all available experimental data with various materials.

Table 1

Summary of Major Experiments Supporting Dispersion Models

A. Dense Gas Experiments

			Number	of	Tests		
Experiments	Material	A	B	C	D	E	F
Desert Tortoise	Ammonia				3	1	
Goldfish	HF				3		
Maplin Sands	Propane				(8)		
Maplin Sands	LNG				(4)		
Burro	LNG			5	2	1	
Coyote	LNG			2	1		
Thorney Island contin.	Freon 12					1	1
Thorney Island instant.	Freon 12				6	2	1
Lyme Bay	Chlorine		2	1	1		
Dow	Chlorine				(3)		
Subtotals		0	2	8	31	5	2
TUV (Heinrich)	Propane	3	6	7	11	8	25
TOTALS		3	8	15	42	13	27

B. Representative Passive Gas Experiments

			Number	of	Tests		
Experiment	Material	A	B	C	D	E	F
Prairy Grass	SO_2	3	2	7	19	3	6
Hanford continuous	Kr-85			3		1	1
Hanford instantaneous	Kr-85			3	1	1	1
... numerous others							
TOTALS		3	2	13	20	5	8

Table 2
Summary of TUV Propane Experiments

Test No.	Cyclone or not	Wind Speed, m/s	Stability Class	Ri. Number	Discharge Rate (kg/s)	Duration sec	Rainout mass %	Vapor Rate kg/s
11	no	0.2	A	-3.26	2.8	300	0	2.8
12	cyc	1.6	A	-2.11	3.9	300	10.7	3.6
14	no	2.9	B	-0.08	29.5	200	0.3	29.5
15	cyc	3.2	C	-0.035	27.5	270	24	21.3
16	cyc	3.5	D	0.146	31.0	20		36.1
18	no	3.7	B	-0.32	19.0	200	0.15	18.9
19	cyc	3.7	B	-0.25	21.0	180	21.5	16.8
20	no	2.2	A	-0.25	31.0	185	0.6	30.9
21	cyc	2.5	B	-0.14	29.5	210	25.6	22.1
22	no	1.6	F	1.75	2.7	240	0.13	2.7
24	no	5.4	D	0.0103	3.0	240	0	3
26	no	4.8	D	-0.729	3	300	0	3
27	no	2	E	-0.014	3	300	0.12	3
28	cyc	1.4	E	0.069	3	360	9.5	2.8
29	no	5.1	D	-0.6	2.4	240	0	2.4
30	no	4.4	D	-0.086	30	23	1.7	29.5
31	no	3.8	D	-0.264	2.4	170	0	2.4
32	no	2.5	D	-0.113	2.4	240	0	2.4
33	cyc	2.9	D	-0.096	2.4	300	2.7	2.4
34	no	1.7	E	0.023	2.4	150	0.05	2.4
35	cyc	3.7	D	-0.14	6	210	10.5	5.7
36	no	2.4	E	0.43	6	150	0.4	6
37	cyc	1.6	E	0.331	6	267	21.5	5
42	no	4	D /D	-0.2	6	150	0	6
43	cyc	4.7	D /D	-0.3	6	150	33.3	4.4
44	no	2	F /D	0	6	150	0.002	6
45	no	2.1	F /D	0.6	30	120	0.1	29.7
46	cyc	1.7	C /E?	0.2	30	60	21	23.9
51	no	0.2	F /E	1	6	200	0	6
52	cyc	1.5	F /D	0.8	6	300	17	5.8
53	no	0.6	F	3.7	10	300	0.07	10

Table 2, continued
Summary of TUV Propane Experiments

Test No.	Cyclone or not	Wind Spd, m/s	Stability Class	Ri Number	Discharge Rate (kg/s)	Duration sec	Rainout mass %	Vapor Rate kg/s
54	cyc	0.7	F	1.2	10	210	2.4	9
55	no	0.6	F	0.7	2	205	11	1.9
56	no	0.3	E /F	-99	15	510	0.04	15
57	no	0.4	F	1.2	10	300	0.03	10
58	cyc	0.5	F	1.9	10	360	22	8.5
59	no	0.5	E /D	0.9	6	300	0.007	6
60	cyc	0.2	E /D	0.4	6	300	16	5.8
61	no	0.1	F /F	40	2.5	600	0.01	2.5
62	cyc	0.2	F /F	42	6	60	27	46.1
63	cyc	0.4	F /F	44	12	120	24	11.4
64	cyc	0.1	F /D	3	6	540	19	5.1
65	no	0.3	F /E	0.2	2.4	600	0.01	2.4
66	no	0.2	F /E	1	6	360	18	5.2
67	cyc	0.5	F /D	3.2	2.9	420	12	2.7
68	cyc	0.7	F /D	-0.1	6.2	600	17	5.4
69	no	0.4	F /F	0.3	2.4	600	0.01	2.4
70	no	1.6	F /E	9	6	600	0.01	6
71	cyc	0.8	F /E	0.2	6	600	16	5.3
72	cyc	0.9	F /D	0	6	600	16	5.3
73	no	2.7	B /C	-2.7	6	300	0	6
74	cyc	3	B /C	-1.1	6	300	8.8	5.7
75	no	0.6	F /E	0.1	36	30	0.1	35
76	no	0.3	F /E	0.7	36	60	0.1	35.1
78	cyc	0.8	F /E	2.9	36	240	26	26.7
79	no	4	C /B	-0.5	36	80	0.001	35.7
80	no	3.2	C /B	-1.2	7.5	120	0	7.5
81	cyc	4	C /B	-1.4	5.3	40	24	41
82	no	3.4	C /D	-0.1	36	120	0.004	35.8
83	cyc	3.4	C /B	-0.3	36	120	20	29.2

Table 3
Summary of TUV Experiments and UDM Modeling

Test No.	Observed Extrapolated min. LFD (m)	Observed Extrapolated max. LFD (m)	UDM m to LFL	UDM % Conc. @Column 2 distance	UDM % Conc. @Column 3 distance
11	13	56	34.2	5.5	1.5
12	60	81	54.4	1.75	1.01
14	75	210	121	3.04	1
15	150	190	138.5	1.82	1.19
16	175	195	189	2.5	2.08
18	120	200	97.9	1.66	0.761
19	80	130	113	3.82	1.64
20	165	260	77	0.29	0.1
21	158	220	128	1.43	0.785
22	32	58	38.5	2.33	1.55
24	70	95	28	0.842	0.636
26	42	66	29	1.36	0.94
27	50	76	39.3	1.8	1.17
28	50	75	70	3.57	1.86
29	25	75	27	2.4	0.77
30	88	205	91	2.3	0.39
31	36	52	27	1.58	1.2
32	36	52	29.7	1.84	1.4
33	50	64	54	2.4	1.56
34	50	64	32.9	1.6	1.27
35	75	110	86	2.61	1.39
36	80	135	65.2	1.71	0.835
37	60	90	86.5	3.84	1.95
42	50	130	39.8	1.69	0.75
43	60	82	74.8	3	1.8
44	60	160	45.9	1.8	0.692
45	115	210	138	2.33	1.44
46	87	137	145	5.23	2.32
51	82	150	52.4	1.62	0.805
52	72	140	92.2	3.18	1.01
53	70	190	84	2.44	0.5

Table 3, cont.
Summary of TUV Experiments and UDM Modeling

Test No.	Observed Extrapolated min LFD (m)	Observed Extrapolated max LFD (m)	UDM m to LFL	UDM % Conc. At Column 2 Distance (m)	UDM % Conc. At Column 3 Distance (m)
54	102	160	116	2.62	1.21
55	43	98	64	4.01	1.04
56	95	180	97	2.15	0.8
57	92	175	80	2.2	0.31
58	88	116	115.8	3.3	2.1
59	110	180	47.6	1.1	0.54
60	82	150	102	3.03	1.1
61	46	110	31.4	1.79	0.057
62	160	280	217	3.45	1.39
63	130	230	131	2.12	0.815
64	80	180	59	1.17	0.247
65	50	95	29	1.55	0.84
66	130	165	102	1.4	0.99
67	80	100	74.5	1.87	1.28
68	75	140	95.5	3.12	1.09
69	70	90	28.7	1.18	0.88
70	70	140	47.8	1.17	0.83
71	80	200	93	2.72	0.566
72	80	130	92.5	2.67	1.17
73	40	130	38.6	2.02	0.655
74	58	70	71.1	2.98	2.14
75	125	170	155	2.6	1.88
76	100	220	150	3.11	1.27
78	160	270	177	2.48	1
79	140	260	120	1.88	0.95
80	70	145	45	1.48	0.717
81	170	190	188	2.5	2.06
82	135	270	126	2	0.895
83	130	205	161	3.07	1.36

Table 4
Conditions Modeled for Ammonia Releases

Location	Temp. F	Temp. K	Pressure psig	Pressure bara	Percent Vapor Flash, kg/kg
High Pressure Receiver	95	308	180	13.1	20.2
Low Pressure Receiver	20	266.5	34	3.27	8.34

Assumed Conditions

Release Elevation	2.0 m	6.1 ft
Ambient Pressure	1.0 atm.	1.0 atm.
Ambient Temp.	298 K	77 F
Relative Humidity	70 %	70 %
Reference Height	10 m	30.5 ft
Surface Roughness Length	3 mm	0.12 inches
Solar Flux	0 W/m^2	0 BTU/ft^2

Calculated Conditions

Receiver:	High Pressure	Low Pressure
Expansion velocity	284 m/s	117 m/s
Sauter mean drop diameter	58 μm	190μm

Table 5

Hazard Distances from the EPA Guidance Documents
Compared with Predictions by the UDM

Release Rate, lb./min.	Release Rate, kg/s	Equiv. Hole Diam. Mm	EPA Distance miles (a)	UDM Distance miles	SACRUNCH Distance miles (b)
1	0.0076	0.64	0.10	0.088	
2	0.0151	0.90	0.16	0.131	
5	0.0378	1.43	0.26	0.215	
10	0.0756	2.02	0.37	0.312	
30	0.2268	3.50	0.65	0.556	
50	0.378	4.52	0.83	0.721	
100	0.756	6.40	1.17	1.03	0.587
250	1.89	10.1	1.79	1.64	0.814
1000	7.56	20.2	3.23	3.42	1.80
2500	18.9	32.0	4.74	5.43	2.84
5000	37.8	45.2	6.11	7.71	4.05
10,000	75.6	64.0	8.29	11.0	
20,000	151.2	90.5	10.77	15.8	

(a) EPA, "RMP offsite consequence analysis guidance," May 24, 1996.

(b) EPA, 'Model risk management program and plan for ammonia refrigeration", by Science Applications International Corp., May 1996, Table 2.1.

APPENDIX A

Theoretical Basis for F Stability

Figure A.1 illustrates the theoretical basis for stable atmospheric conditions. The left-most line plots the adiabatic lapse rate, ALR, as the normal, neutral buoyancy temperature profile in the atmosphere. A particle of air rising must cool by expansion by the perfect gas law, and the rate of such cooling follows the ALR at about one degree Celsius per 100 meters. During the night, particularly on a cloudless night, the ground cools by radiation, and the atmospheric temperature gradient falls below the ALR giving a stable atmosphere. Thus, a finite parcel of air displaced from any point on this cooler profile, say from point A to point B follows the slope of the ALR from A to B and leaves this parcel of air at any height surrounded by air that is warmer, and hence less dense. So, there is an energy barrier suppressing rising air pockets, hence less turbulence for vapor cloud dispersion.

Several empirical correlations have been suggested to infer stability class. The most commonly used is given in **Table A.1**, which uses wind speed, solar insolation, and cloud cover as correlators. Stability class depends directly on vertical turbulence profiles which are difficult to measure. It depends indirectly on vertical temperature and wind speed gradients, which can be measured. These gradients can be expressed through a dimensionless number, the Richardson number, defined for the Heinrich et al tests as:

$$\mathrm{Ri} = \frac{g}{T(2)} \frac{\dfrac{[T(16) - T(0.5)]}{(16 - 0.5)} - \sigma}{\left[\dfrac{u(16) - u(0.5)}{16 - 0.5}\right]^2} \qquad \text{(A.1)}$$

where:

σ = the adiabatic lapse rate (-0.01 K/m)

g = acceleration of gravity, m/s^2.

Ri = Richardson number, ratio of modified temperature gradient to squared wind speed gradient, normalized.

T(z) = atmospheric temperature at height z

u(z) = wind speed at height z

Table A.1
Correlation of Atmospheric Stability Class
According to Gifford (1976)

Wind Speed m/sec.	Day Strong	Time Solar Moderate	Insolation Slight	Night Time Thin overcast or >1/2 low clouds	Conditions < 3/8 cloudiness
<2	A	A-B	B	-	-
2-3	A-B	B	C	E	F
3-4	B	B-C	C	D	E
4-6	C	C-D	D	D	D
>6	C	D	D	D	D

VII REFERENCES

EPA, "RMP offsite consequence analysis guidance," May 24, 1996 (a).

EPA, 'Model risk management program and plan for ammonia refrigeration", by Science Applications International Corp., May 1996 (b).

Gifford, F. A., "Turbulent diffusion typing schemes: A review", Nucl. Saf. 17 (1), 1976. Cited also in Hanna, S. R., G. A. Briggs, R. P. Hosker Jr., "Handbook on atmospheric diffusion", Technical Information Center, U.S. Dept. Of Energy, DOE/TIC-11223, p. 28, 1982.

Hanna, S. R., D. G. Strimaitis, J. C. Chang, "Evaluation of 14 hazardous gas models with ammonia and hydrogen fluoride field data", J. Haz. Mat., 26 (2) 127-158, 1991a.

Hanna, S. R., D. G. Strimaitis, J. C. Chang, "Hazard response modeling uncertainty (a quantitative method)", Report to API, Sept., 1991b.

Heinrich, M. E. Gerold, and P. Wietfeldt, "Large-scale propane release experiments over land at different atmospheric stability classes", J. Haz. Mat., 20, 287-301, 1988. Corrigendum, J. Haz. Mat., 22, 407-413, 1989. For more details see report by same authors, "Praxisgerechte Bestimmung der Zundentfernungen bei der Freisetzung schwere Gases", BMFT 326-7591-01-RG 8402, TUV (Technischer Uberwachungs-Verein Norddeutschland e. V.), Hamburg, May 1988.

Woodward, J. L., "An integrated model for discharge rate, pool spread, and dispersion from punctured process vessels", J. Loss Prevention in the Process Industries, Vol. 3, No. 1, pp. 33-37, Jan. 1990)

Woodward, J.L. and J. Cook, "Tuning a complex suite of dispersion models", International Conference and Workshop on Modeling and Mitigating the Consequences of Accidental Releases of Hazardous Materials, New Orleans, LA, Sept. 26-29, 1995.

Woodward, J.L. , J. Cook and A. Papadourakis, "Modeling and validation of a dispersing aerosol jet", J. Hazardous Materials, 44, 185-207, Dec. 1995.

Woodward, J. L. and A. Papadourakis, "Reassessment and reevaluation of rainout and drop size correlation for an aerosol jet", J. Hazardous Materials, 44, 209-230, Dec. 1995.

Figure 1 Photograph Illustrating Typical F Stability Conditions in the Early Morning

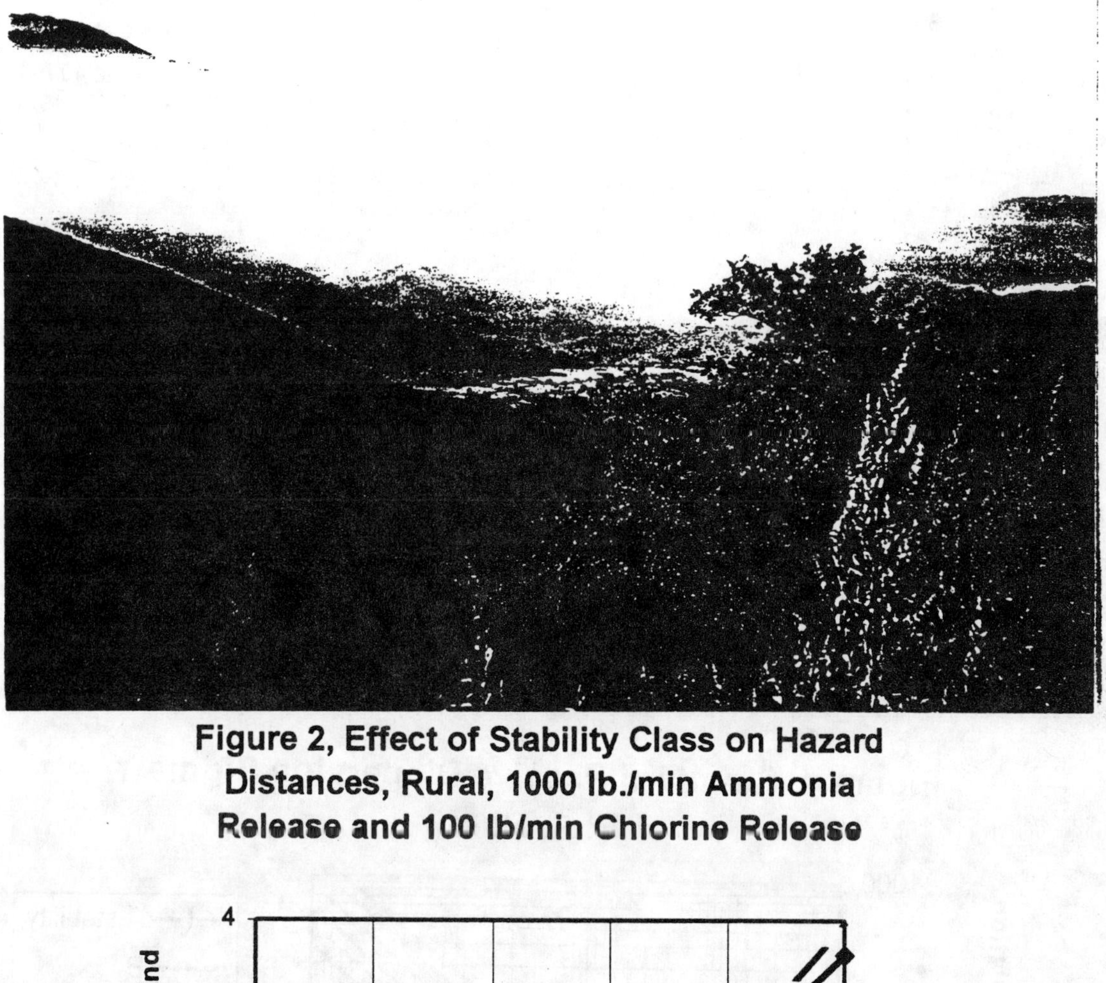

Figure 2, Effect of Stability Class on Hazard Distances, Rural, 1000 lb./min Ammonia Release and 100 lb/min Chlorine Release

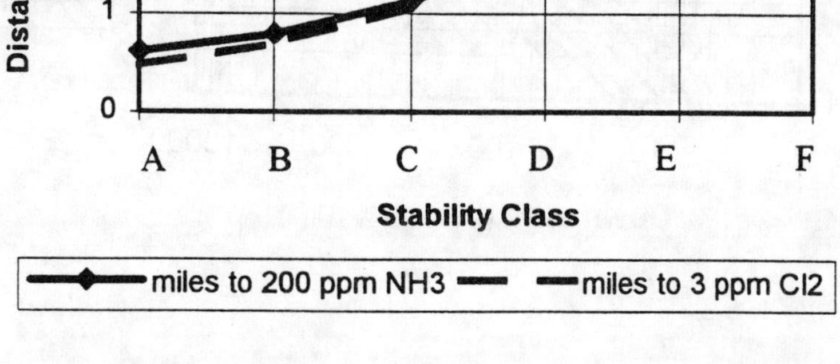

Figure 3 Example of Passive Dispersion Model Profiles

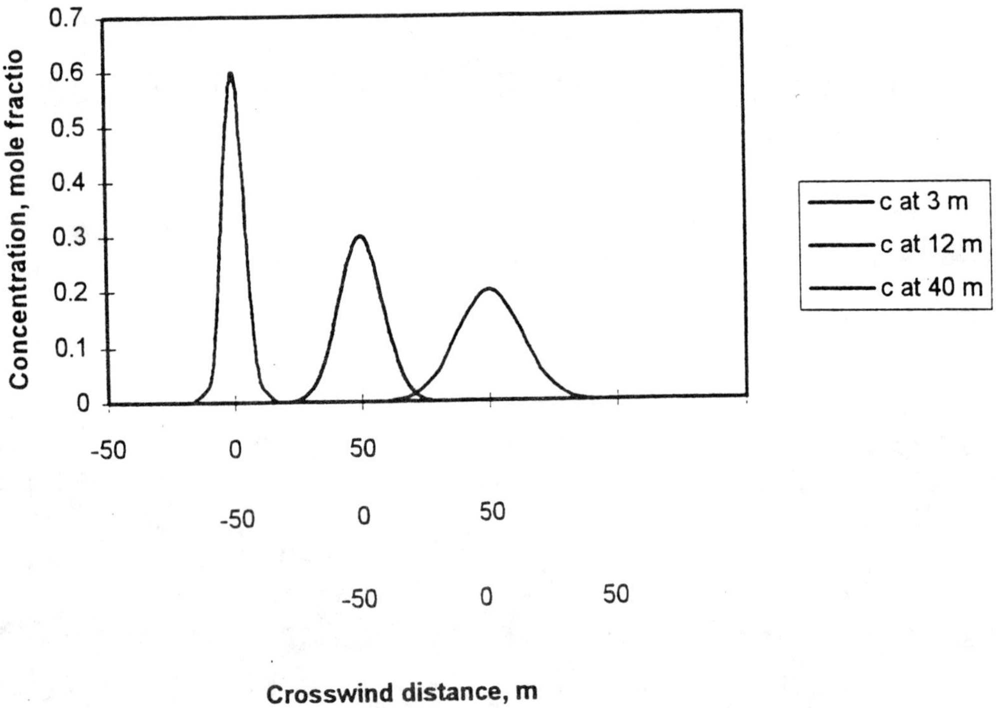

Figure 4, Standard Passive Dispersion Sigma Y Plot, Rural

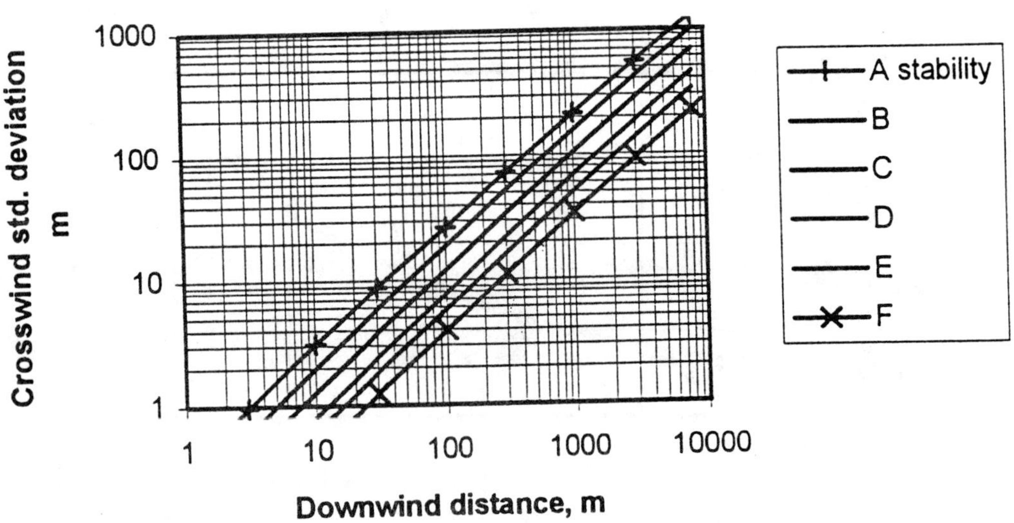

Figure 5, Inferring Stability Class Multiplier for Passive Dispersion

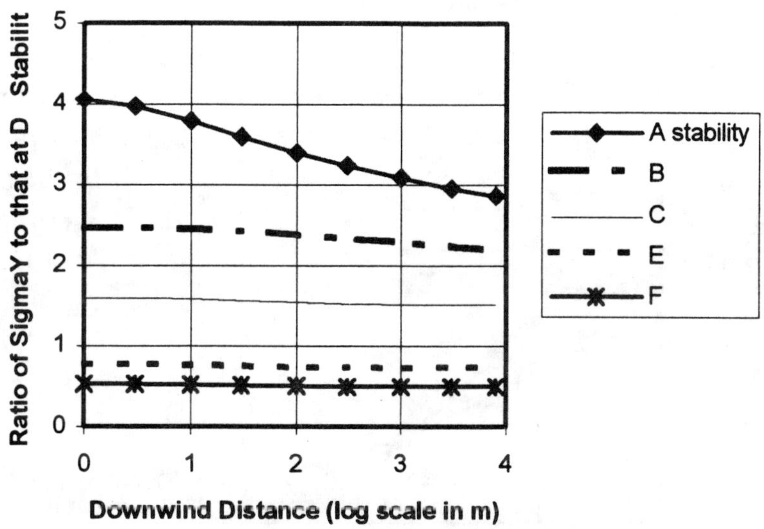

Figure 6, Dispersion Constant Multipliers for Effect of Stability Class

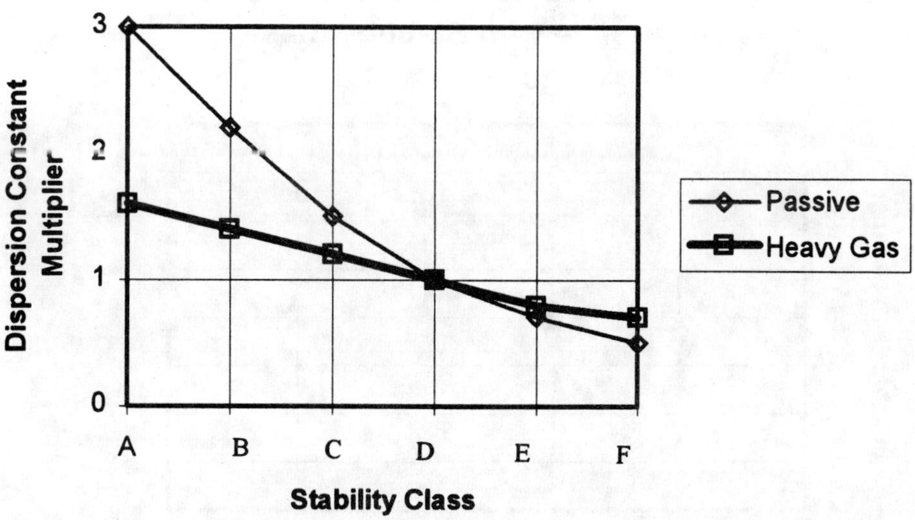

Figure 7, How Stability Class Affects Heavy Gas Dispersion Through Friction Velocity Parameter

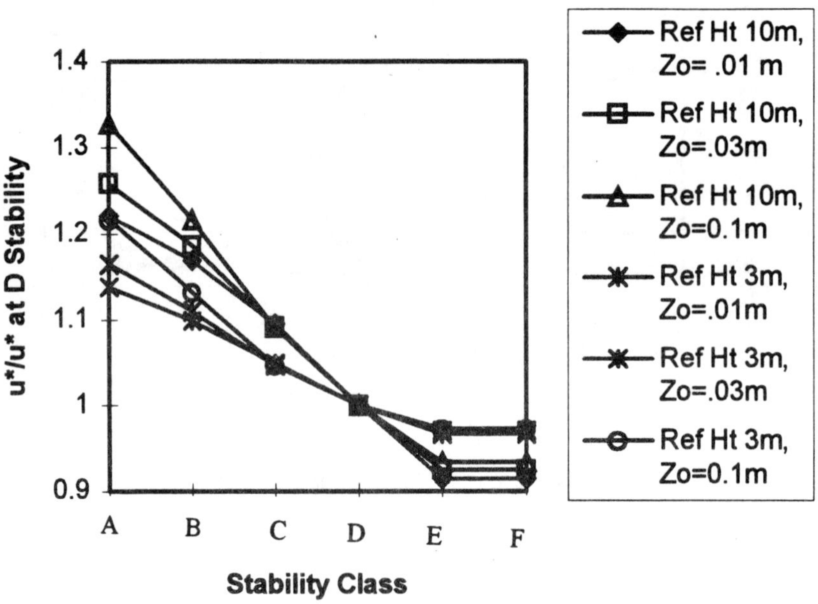

Figure 8, Measured Richardson Numbers Correlate with Stability Class

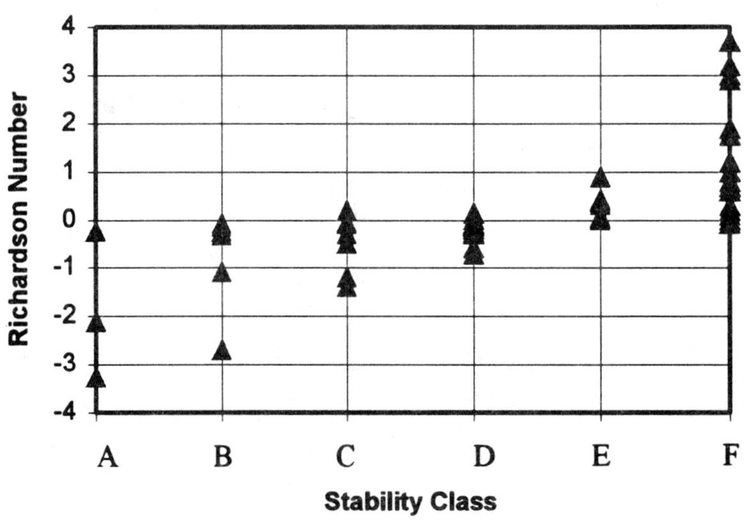

Figure 9, Extrapolated Minimum & Maximum LFD for 6 kg/s Releases vs Richardson No.

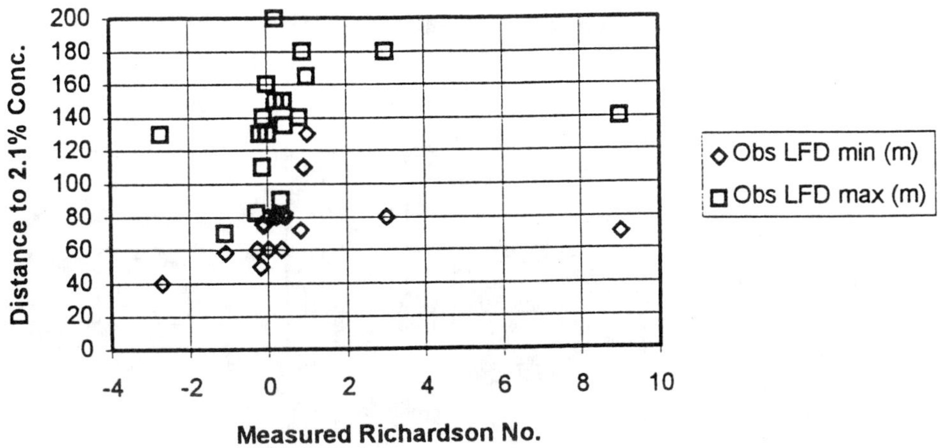

Figure 10A, Comparison of UDM with Desert Tortoise Run 1

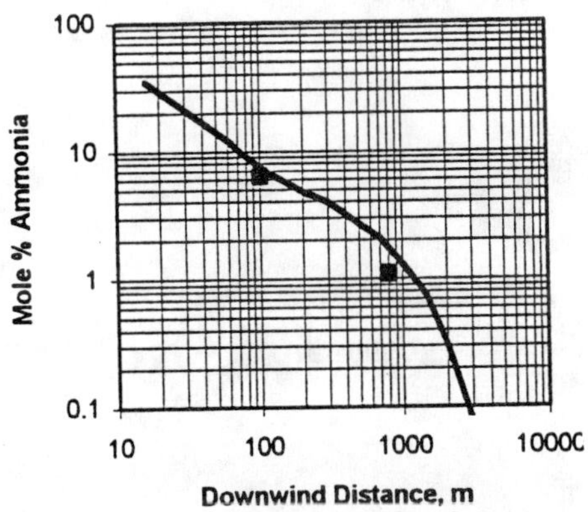

Figure 10B, Comparison of UDM with Desert Tortoise Run 2

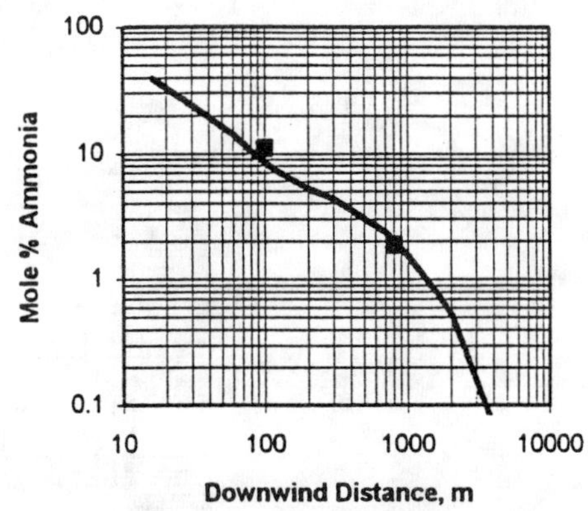

Figure 10C, Comparison of UDM with Desert Tortoise Run 3

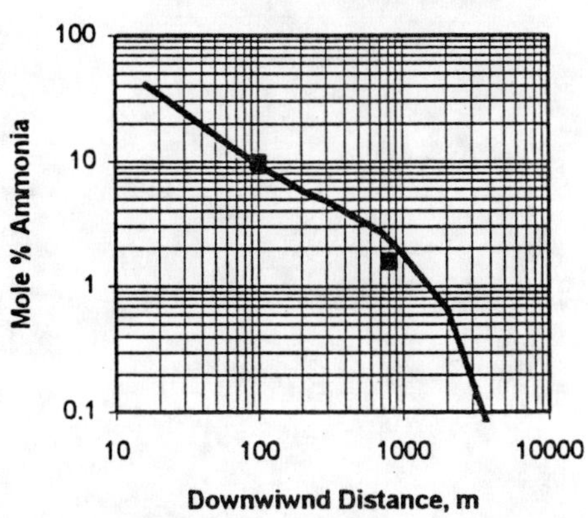

Figure 10D, Comparison of UDM Predictions with Desert Tortoise Run 4 Data

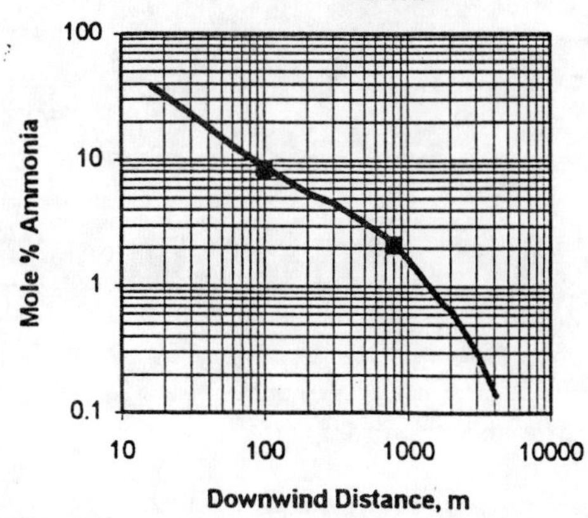

Figure 11A, Comparison of UDM with Goldfish HF Run 2

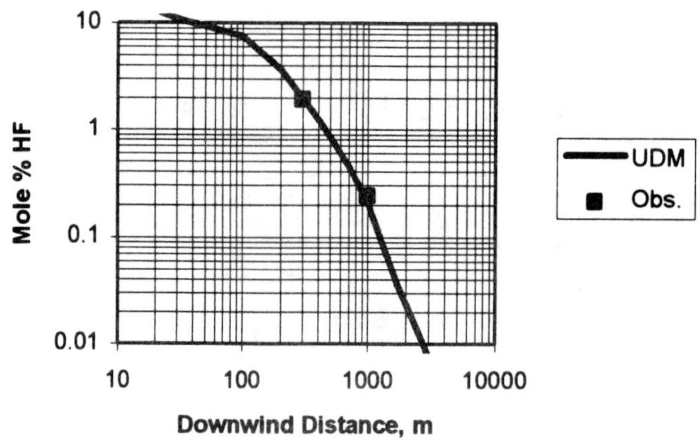

Figure 11B, Comparison of UDM with Goldfish HF Run 3

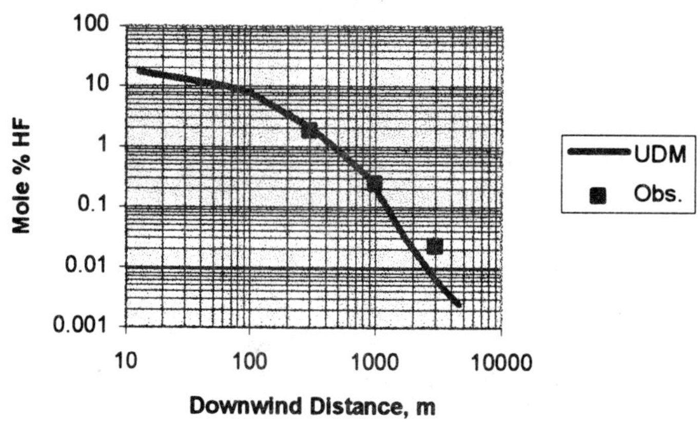

Figure 12 Predictions Pass Through Catalytic Analyzer Data, TUV Run 43

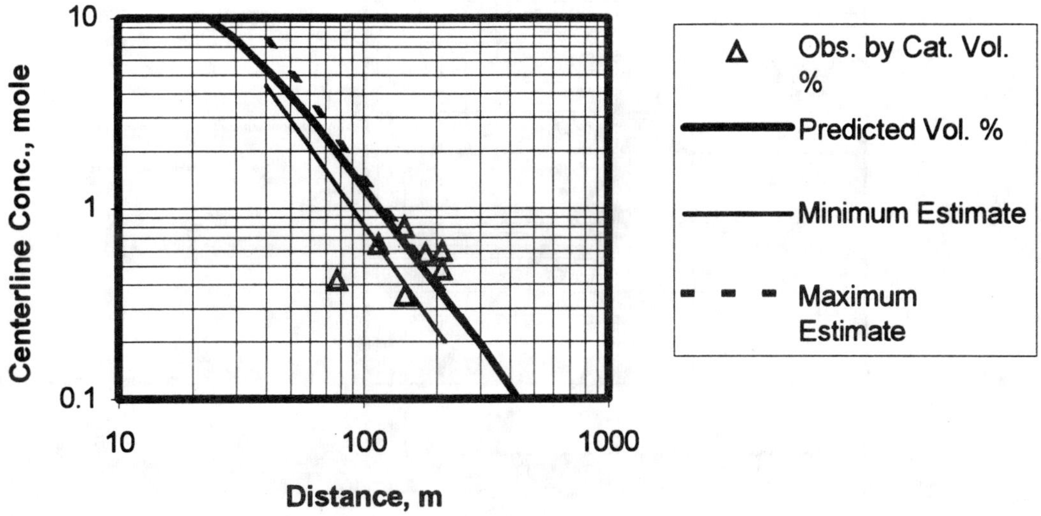

Figure 13, Observation Scatter for Different Type Analyzers, TUV Run 42

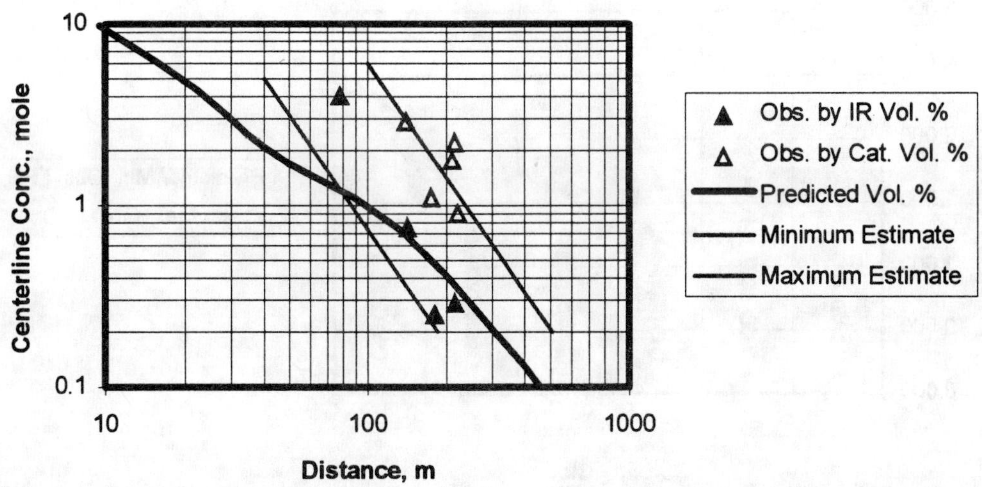

Figure 14, Sensitivity of Predictions to Stability Class, TUV Run 52

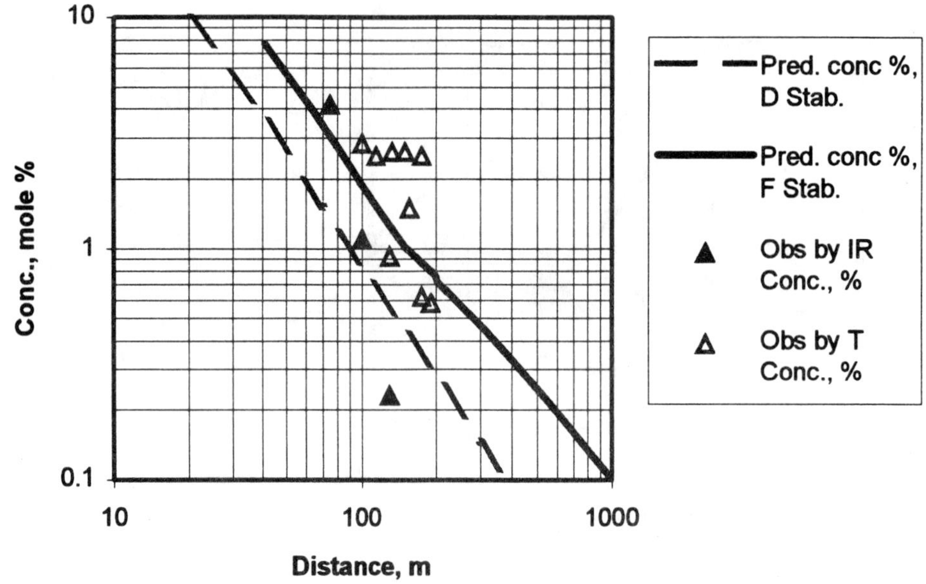

Figure 15 Predicted/Observed LFD at Each Stability Class

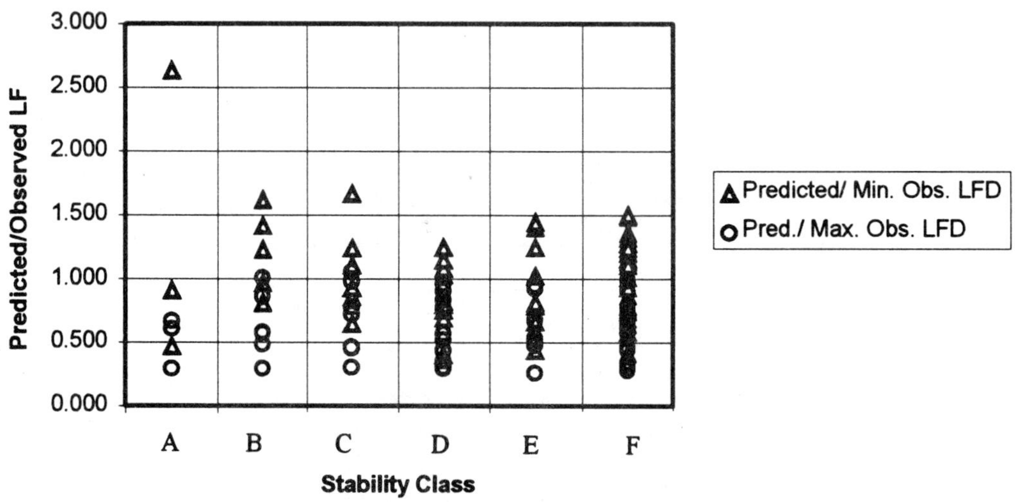

Figure 16 Predicted Concentration at Observed LFD

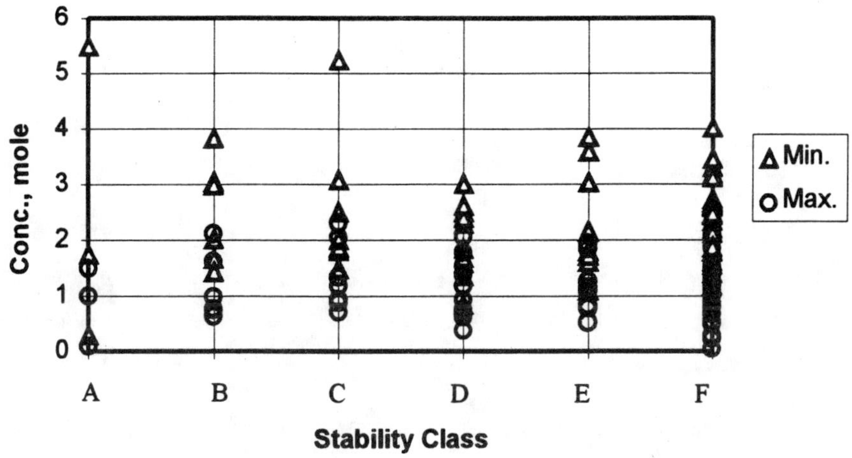

Figure 17 Comparison of Predictions of UDM and SACRUNCH Model with EPA Guidelines, Rural

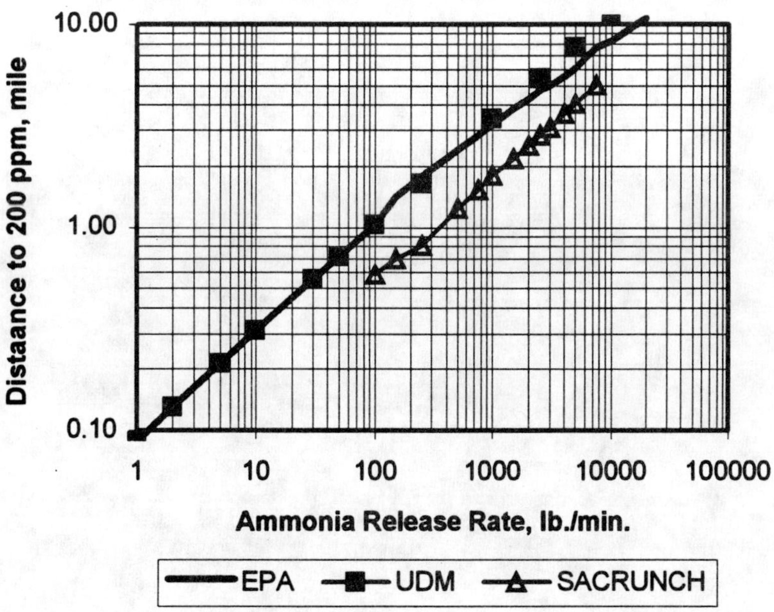

31st LOSS PREVENTION SYMPOSIUM

Session 44. **Case Histories & Miscellaneous Topics**

Thurs., March 13, 1997
8:30 - 10:30 AM
Grand Ballroom C

Robert W. Ormsby, Chair
Air Products & Chemicals, Inc.
7201 Hamilton Blvd.
Allentown, PA 18198-1501

Roy E. Sanders, Vice-Chair
PPG Industries, Inc.
P.O. Box 1000
Lake Charles, LA 70602

Session Introduction - R.W.Ormsby

44a. **Carbon Disulfide Incidents During Viscose Rayon Processing,** D. Hall, Global Environmental Solution, Radford, VA; L. Losee, Global Environmental Solutions, Magno, UT

44b. **Safety Considerations When Treating VOC Streams with Thermal Oxidizers,** F. Self; J. Hill, Bechtel, Houston, TX

44c. **No Good Deed Goes Unpunished: Case Studies of Incidents and Potential Incidents Caused by Protective Systems,** D. Hendershot, Rohm and Haas CO, Bristol, PA; A. Dowell, Rohm and Haas, Deerpark, TX

44d. **Case Histories of Two Incidents Following Process Safety Reviews,** W. Howard, Process Safety Consultant, St. Louis, MO

Concluding Remarks - R.E. Sanders

Symposium Closing - Gary Page, Symposium Vice-Chair

Carbon Disulfide Incidents During Viscose Rayon Processing

Donal S. Hall
Lyman A. Losee
Global Environmental Solutions
An Alliant Techsystems Inc. Operation
4100 South 8400 West, Annex 16
Magna, Utah 84044

Prepared for Presentation at the AIChE 31st Annual Loss Prevention Symposium,
Case Histories and Miscellaneous Topics

March 1997

UNPUBLISHED

Abstract

Two incidents occurred in a viscose rayon manufacturing process about a month apart during the same step of a semi-monthly cleaning operation in a carbon disulfide recovery box. The operation removes sulfides that accumulate on equipment surfaces during the week. In both incidents, carbon disulfide was released from the recovery box into the vapor exhaust system and the area surrounding the recovery box. Ignition occurred and the resulting vapor flash propagated throughout the vapor exhaust system and the operating bay. Both incidents caused extensive equipment and building damage and minor injuries to personnel.

Investigation and recommendations after the first incident, although appropriate, did not eliminate the root cause of the problem. Additional information obtained during the investigation of the second incident showed that a combination of several modifications and subtle changes in that particular process line, in conjunction with the colder temperatures of fall and winter, resulted in the incidents. The changing of the critical balance of temperature and pressure in the closed system and sudden volatilization of accumulated carbon disulfide in an unlikely location were triggered by startup of the cleaning operation. Once the carbon disulfide vapor pressure vented from the closed system numerous ignition sources were available.

The discussion will focus on the importance of management of change for seemingly small, logical changes, and on the importance of attention to small details in accident investigation.

Process description

Carbon disulfide (CS_2) is the primary solvent used in viscose rayon manufacturing. Production of high quality rayon requires thorough removal of all residual CS_2. Economics and prudent environmental policies require recovery and reuse of as much CS_2 as possible. The final step in CS_2 recovery takes place at the CS_2 recovery box unit where the incidents occurred. The purpose of this unit is to volatilize residual CS_2 by heating the rayon fiber process slurry, flowing from the fiber cutters, to 90 °C by direct steam injection. The recovery box operates at a slight vacuum with water and flapper seals to prevent CS_2 from escaping the unit. The CS_2 vapors are drawn through a condenser, liquefied, and returned to the storage area. The rayon process utilizes massive ventilation systems to remove and dilute any CS_2 vapors which may escape from the process.

The recovery box has an internal weir which forms the rayon fibers into a cake which floats on the transfer fluid. Transport fluid is injected into the rear of the recovery box to provide a laminar (smooth) flow stream over the weir. The fiber cake exits the unit onto a takeaway filter belt. The fluid is collected and returned to the cutting machine discharge flume. Figure 1 depicts a flow diagram of the recovery unit and recirculating system.

Due to the nature of the process, sulfur compounds accumulate on the internal surfaces of the recovery box and impede fiber cake flow. A periodic "sulfiding" operation is conducted to break up and remove the compounds from the internal surfaces. This cleaning operation is conducted every two weeks, or sooner if needed, based on operator observation of cake flow. To clean the recovery box, a cleaning solution is pumped into the box and permitted to recirculate until the box is clean. The cleaning liquid and contaminated transfer solution is removed and fresh transfer liquid is put into the recovery box recirculating system.

Incident Descriptions

Incident #1

The first incident occurred in November during startup of the weekly sulfiding operation. Operators had discontinued process flow and opened the weir supply valve which was normally closed as a result of a previous process change implemented in May. They then proceeded to prepare the recovery box for cleaning. Several seconds after the weir supply valve was opened the recovery box unexpectedly pressurized and expelled CS_2 gas into the bay where it was ignited. The CS_2 vapor was probably ignited by the heat from a low lying steam line located nearby. The resulting flash in the recovery box propagated through the vapor recovery system to the exhaust fans, where a secondary flash occurred in the exhaust plenum. A portion of the exterior building wall near the recovery box area was blown out. (The operating line involved was next to this wall).

Findings and Changes as a Result of Incident #1

The investigation of this first incident determined that the most probable cause was accumulation of CS_2 vapors due to failure of an exhaust fan. One of two fans was believed to have been shut down prior to the incident for repairs. A collection of liquid CS_2 and sudden buildup of CS_2 vapor in the recovery box was thought to be unlikely, due to the high operating temperatures (90°C) maintained inside this unit.

Incident #2

The second incident, which took place two months later in January, and also occurred at the same point in the start-up of the sulfiding operation. Fiber slurry flow had been stopped and the weir supply valve opened as in incident #1. Approximately 3 seconds after opening the weir fluid supply valve, the CS_2 recovery box pressurized and began venting steam and mist into the operating bay. Operators immediately began exiting the area. A vapor explosion occurred in the bay which blew out the temporary plywood wall erected after the previous incident. Operators observed the ventilation plenum rise and flame shoot out. The plenum and ducting connected to the ventilation system were extensively damaged.

HAZMAT/firefighting personnel extinguished what appeared to be a small pool of liquid CS_2 burning in a low spot on the floor, and another burning pool in the outlet pan of the CS_2 recovery box. Burning liquid CS_2 dripping from broken condensate system tail gas lines outside the building had to be extinguished. These lines enter the main ventilation plenum which is located outside the building.

New Information Resulting from Incident #2

- A pool of burning CS_2 was observed on the floor and at the outlet of the CS_2 recovery box by HAZMAT/firefighting personnel.
- Flame was observed exiting the cutting machine feed chutes before the vapor flash occurred in the finishing bay.

- Burning liquid CS_2 dripped onto the ground from the condensate system tail gas lines which were broken loose from the exhaust plenum when the plenum was shifted by the pressure of the internal vapor flash.

Common Factors to Both Incidents

- Both incidents occurred at the beginning of the "sulfiding" operation, prior to addition of cleaning solution.
- CS_2 vapors escaped/vented from the recovery box into the finishing bay and ventilation system.
- Vapor flash occurred in the finishing bay.
- Secondary vapor flash occurred in the ventilation system.
- Period of cold weather prior to incident.

Subtle Process Changes Contributing to the Incidents

Flow Changes Behind Weir

The original design of the CS_2 recovery box required transport fluid flow over a weir into the back of the box to even out the fiber cake exiting the distributor located at the peak of the weir. Due to increased production rates, processing problems occurred. Because of the flow split to the cutters and into the back of the recovery box, the required temperature (90°C) could not be maintained in the recovery box. An operational process review determined that product quality could be maintained without flow into the back of the recovery box. This change permitted all transport fluid to go through the cutter discharge flume and be heated by the P-Trap steam injector. It was noted that this change created a stagnant zone of transfer fluid behind the weir (about 40 gallons). The stagnant zone was not believed to be a problem because of the high operating temperature of the recovery box and regular flushing when the recovery box was "sulfided".

Water Seal Changes

The recovery box utilizes a water seal to maintain internal vacuum. Due to different lid levels, flow was drained from the upper lid at the back of the recovery box (above the weir and distributor) to the lower lids over the heating zone of the recovery box. Fibers from the rayon process continually plugged the overflow and resulted in water running down the outside of the recovery box. The outer water seal lip was raised to prevent external overflow. This permitted any overflow to go into the recovery box. No significant cooling problems were anticipated due to the low flow of cooling water (1.1-2.5 gpm) vs high flow through the recovery box and the high operating temperatures.

Weather (Temperature) Effects On The Operating Line

The operating line where the incident occurred was located next to an outside wall which was constructed of brick and sheet metal. Operating personnel indicated that the area around this operating line was colder during cool weather than the rest of the facility. The seal water was supplied from the plant water conditioning system which was not heated. The water temperature varied based on outside conditions. The water temperature was found to be between 14-16°C at the time of the incident.

Why Did These Changes Result In The Two Incidents?

The investigation determined that the stagnant zone in the recovery box was only heated by conduction through the weir from the heating zone of the recovery box. Overflow of seal water into the recovery box resulted in the formation of a liquid contact condenser and cooling of the liquid in the stagnant zone. As the seal water overflowed into the stagnant zone, it formed a sheet of water which flowed down the back panel of the recovery box. This resulted in steam and CS_2 vapors being condensed and carried into the stagnant zone. Actual temperature monitoring of a recovery box in operation and heated to 88-90°C without fiber flow was performed. The results of this test determined that the cold seal water flowing down the back of the recovery box was sufficient to lower and maintain the temperature in the stagnant zone at 22°C, which is below the 46°C evaporation/boiling point of CS_2.

When operating personnel opened the weir flow supply valve, liquid CS_2, which had accumulated in the stagnant zone, was displaced (pushed over the weir) into the heating zone by the incoming flow and by the boiling action from contact with the hot incoming transport fluid. Once the main mass of liquid CS_2 was displaced into the heating zone, it immediately began to boil, generating excessive quantities of vapor which the recovery box could not contain. Some liquid CS_2 flowed out of the recovery box, was picked up by the transport fluid recirculating system, and pumped into the cutter flume. In the flume, the CS_2 liquid/vapors were mixed with air and provided contact with an ignition source (likely a high temperature surface). The resulting vapor flash propagated through the flume vapor pickup system into the recovery box outlet vapor exhaust system, into the main vapor exhaust manifold, and produced the overpressurization and damage observed by employees and investigators.

Additional Factor Which May Have Increased The Severity Of Both Incidents

The condensate system tail gas lines discharge directly into the exhaust plenum outside the manufacturing building. These stainless steel lines safely vent any CS_2 vapors which cannot be condensed by the system. The lines are over 70 feet long and were not insulated. CS_2 tail gases condensed and accumulated in a low point in the tail gas lines due to the cold weather. The CS_2 condensate slowly flowed into the exhaust plenum and either vaporized or accumulated under water in the pit of the plenum. The investigation team believed that this additional quantity of CS_2 may have contributed to the severity of the vapor flash observed in the exhaust plenum. This accumulation is not believed to contribute to the incident which occurred inside of the manufacturing process.

Recommendations

Viscose rayon manufacture requires that viscose be made in advance so that it can be properly aged. Shutdown of the continuous viscose process requires extensive cleanup and downtime. Due to the short shelf life of aged viscose, it must either be used within several hours or discarded. The incident required the facility to fill its viscose discard storage to capacity. In order to prevent a long-term shutdown of the facility during the investigation, the investigation team worked with production personnel and developed a series of short-term recommendations for the safe restart of several production lines. This permitted the aged viscose to be spun into rayon fibers and prevented a long-term shutdown of the viscose manufacturing process.

Short-term Recommendations for Restart

1) Perform a preoperational safety startup review of each line to be restarted.

2) Re-establish hot flow into the area behind the weir to prevent accumulation of liquid CS_2.

3) Vent CS_2 vapors from the recovery box until incident investigation is complete.

4) Minimize number of personnel in the operating area.

5) Set up one operating line to perform tests for incident scenario confirmation.

Upon completion of tests and confirmation of the investigation team findings, a series of long-term recommendations were issued to prevent the reoccurrence of a similar incident.

Long-term Recommendations

1) Redesign the CS_2 recovery trough to prevent accumulation of liquid CS_2 behind the weir.

2) Modify CS_2 seal water system to provide heated seal water.

3) Review recovery box and ventilation system design to assure that flammable vapors are adequately controlled.

4) Heat trace condensate system tail gas lines to prevent condensed liquid CS_2 from being vented to the exhaust plenum.

5) Review Management Of Change program and make modifications to assure that changes are thoroughly reviewed.

6) Perform/update hazards analyses of CS_2 recovery box and condensate systems to assure that all changes previously implemented do not adversely affect process safety.

Lessons Learned

Importance of Management of Change

In all chemical and manufacturing facilities, change is a way of life. We are always striving to make a process work more efficiently with less impact on the environment. These changes must be methodically reviewed to assure that they do not affect process safety and product quality. The method used to review changes must examine all of the factors which affect a facility. In the case of this Major Rayon Manufacturer, the changes to the CS_2 recovery system were made early in the year during warm weather. No adverse effects were anticipated, and none were observed until the weather got colder. It is critical that all changes be reviewed for anticipated conditions which a plant or process may experience. These include variations of internal process conditions, facility conditions, and external conditions, both normal and abnormal.

It is very important that new changes be reviewed in two ways. First, the new change must be reviewed against the results of all previous changes. This review should examine and determine the impact of the new change on the existing "as is" system components and process conditions upstream and downstream, no matter how insignificant. Second, the new change must be reviewed against the potential effects introduced by other new changes which are planned/proposed. This second review is crucial to the identification of undesirable events due to the simultaneous implementation and interaction of multiple changes. Usually, no single change results in a problem. It is the combination of multiple changes, each one safe and acceptable in itself that results in the development of undesirable conditions.

Review of new changes, against the actual process and process conditions, for undesirable cumulative or synergistic effects may only be done through the use of an effective **Management of Change** system which documents and maintains a historical record of all changes and their effect upon the process.

Importance of Observing Small Details in Accident Investigations

In the case of this incident, which had occurred twice, no one fact resulted in an immediate answer of why the event occurred and what actually happened. It was the overwhelming evidence supplied by observing small details which helped to finalize the scenario of what caused the incident to occur. The following observations resulted in resolution of this incident and confirmation of the primary scenario.

Incident Investigation Observations

- Burning around the cutter flume sample hole. (Supported ignition trail and burning inside cutter flume)
- Globules of condensed CS_2 in water seal troughs. (Preserved by the cold water)
- Flowing of seal water down inside of CS_2 recovery boxes.
- No flow to backside of weir.
- Damage to weir supply line gaskets. (Believed to be caused by contact with liquid CS_2)
- Connection between the CS_2 recovery box and damaged exhaust ducting not connected to the CS_2 recovery box via a wastewater drain that ran under the CS_2 recovery box. (Air flow was countercurrent to water flow and permitted vapors to be drawn into the unconnected pickup on the far side of the finishing bay.)

Operator interviews were also critical to the resolution of this incident. All operators, who had observed or been involved in the incident in any way, were interviewed. Many facts were correlated between operator statements, and their observations helped to develop the time lines which supported the primary scenario. It is critical that interviews with personnel involved in the incident be conducted in an open but objective manner. It is also important not to feed facts to the personnel being interviewed as they may tend to tell you what you want to hear. The following operator observations were critical to this investigation:

Operator Observations (interviews)

- Cutter operators saw fire in cutters then observed flame and noise from a vapor flash exiting the finishing bay into the cutting bay.
- CS_2 recovery box operators indicated that the box began rumbling and crackling like the steam had just been turned on to heat the box up for a run. Then the box pressurized and blew water out of the water seals.
- Operators indicated that the line where the incident occurred smelled worse than any other line. (Evidence of CS_2 escaping ventilation system.)

- HAZMAT/firefighters observed and extinguished a pool of burning CS_2 at the outlet of the recovery box and on the floor beside the recovery box.
- Observation and timing of vapor flash in finishing bay and exhaust plenum.

Observations of pressure and temperature were logged by the process control system. These provided some insight and support into the incident. Data from the previous incident was available and supported the findings of the second investigation.

Control System Observations

- Time at which weir flow valve was opened.
- Time at which pressurization of CS_2 recovery box occurred.
- Sudden rise in CS_2 concentration outside of recovery box.
- Data from previous incident.

Summary

The purpose of this paper was to reinforce the need for a thorough Management of Change program by presenting an example of how innocuous process changes can combine and result in an incident. A Management of Change system is the only way to consistently identify potential hazards introduced by process and design changes. However, Management of Change systems must be continually maintained and improved to assure that the system does not stagnate and become a cursory function.

The second investigation was able to obtain detailed information necessary to develop the scenario which resulted in closure of both incidents. In some cases, the evidence is not sufficient to point clearly to any developed scenario. However, it must be stressed that all incident details, no matter how small, must be included in the main body of evidence and objectively evaluated against all developed scenarios. Only in this way can we determine what has happened and why.

A Note About The Major Rayon Manufacturer

The authors would like to thank the investigating team, operating personnel and the Major Rayon Manufacturer who permitted this paper to be prepared and presented.

The Major Rayon Manufacturer took the steps necessary to prevent a repeat accident and presented the incident investigation findings to all operating personnel to enhance their safety knowledge and demonstrate management commitment to safe operation of their facilities and processes.

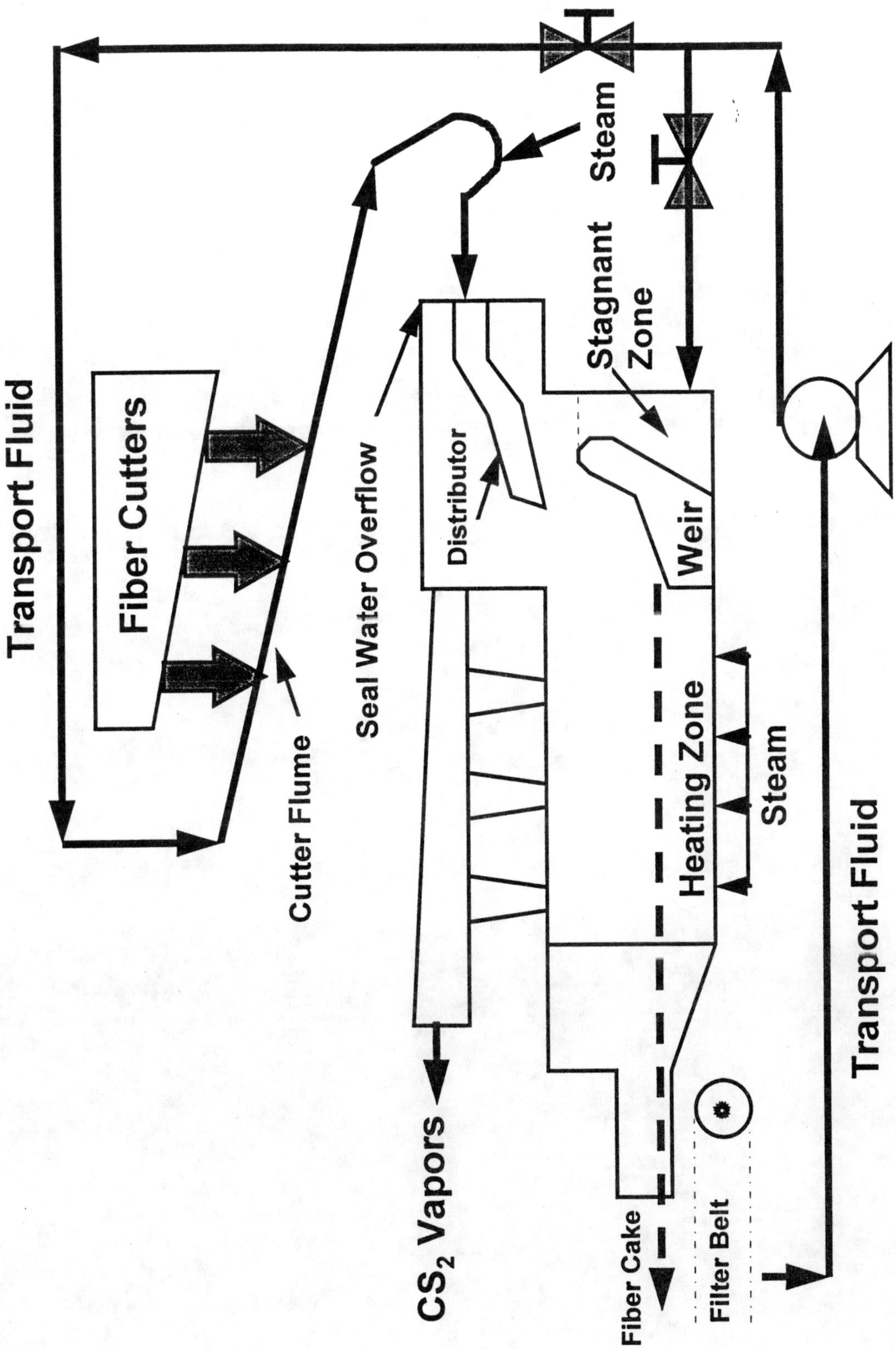

Figure 1. Flow Diagram of CS_2 Recovery Box

SAFETY CONSIDERATIONS WHEN TREATING VOC STREAMS WITH THERMAL OXIDIZERS

Freeman E. Self and Jerry D. Hill
Bechtel Corporation, Houston, Texas

Prepared for presentation at the
AIChE 31st Annual Loss Prevention Symposium - AIChE 1997 Spring National Meeting
Case Histories and Miscellaneous Topics

Houston, Texas, March 1997

ABSTRACT

Using several case histories, this paper will identify and discuss the technical and safety issues associated with designing an oxidation system for volatile organic compounds (VOCs). Descriptions of various incidents involving thermal oxidizers and vapor collection systems will reveal where problems have occurred in the past. The paper will also characterize what would generally be considered good engineering practice for such equipment.

In order to meet the VOC destruction efficiency mandated by many of the current air quality regulations, thermal oxidation is becoming increasingly prevalent at many facilities. It is a proven technology that accommodates the characteristics of almost any hydrocarbon vapor stream and operates over a wide range of applications.

The case histories presented are from both the popular chemical engineering literature and company experience. Short illustrations and citations are provided.

METHODOLOGY

Thermal oxidation is a proven technology for treating VOC streams. However evidence from many published case histories shows that this technology can create a potential explosion hazard. One possible explanation for past incidents involving oxidizers is the lack of sufficient design experience to supplement theory and prevent inappropriate application. Another is that the typical process hazard analysis may not uncover or recognize the importance of certain circumstances.

Therefore, we are approaching the safety considerations from a case history perspective. The case history method has unique advantage in that it dramatically illustrates the limits of design. Utilizing case histories, we will present a risk profile followed by suggested controls and safeguards.

The ideas in the paper should be generally applicable to all gas phase oxidizer systems, whether catalytic, flame-based thermal, or flameless thermal. In this paper we are primarily concerned with flame-based thermal oxidizers.

COMPONENTS OF THE SYSTEM

During the discussion, reference is made to certain components of a typical vapor collection and oxidation system and the following terminology will be used. The *source* of the vapor to be collected and processed is usually a storage vessel, process vent, tank car, or marine vessel. If the source is a marine vessel, then the U.S. Coast Guard regulations promulgated in Title 33 Part 154 must be followed. The source ties into the *vapor collection system* that transports the vapors to a *vapor destruction* or *vapor recovery* unit. The tie-in point is termed the *facility vapor connection* point. For marine service, it is located at the berth and is a distinct tie-in connection. For permanent land systems, the tie-in point is not specifically identified, but in general means a location in the piping close to the source. The group of components consisting of the tie-in point (vapor connection point), the vapor collection system, and the vapor destruction / vapor recovery unit is termed the *facility*.

THERMAL AND CATALYTIC OXIDIZERS FOR GAS STREAMS

There are four major types of oxidizers utilized for VOC reduction in gas streams.

Thermal Oxidizer Utilizing Flames for Flammable Concentrations Above the LEL
This has traditionally been the workhorse of the petrochemical industry and historically the classification included open flares, enclosed flares (sometimes called combustors), incinerators, etc. They are designed to burn gas in the explosive range (gases below the LEL are enriched) using an open flame at about 1,800 °F to 2,000 °F. Destruction efficiencies of 99%+ are standard for well designed and operated oxidizers. Heat exchangers using recuperative (plate or shell and tube) cross exchangers are often utilized to reduce the temperature of the stack gas and preheat the fuel. Regenerative heat

exchangers, utilizing a bed of high density packing to transfer heat, are used for large flows, high heat recovery requirements (up to 95% thermal recovery efficiency) and high temperature (2,000° F).

Thermal Oxidizer Utilizing Flames for Flammable Concentrations Below the LEL

As environmental control assumed more importance and streams often had small concentrations of VOCs, thermal oxidizers were developed to burn these streams. They were developed in the late 1960s as inexpensive solutions for destruction of solvent vapors primarily in the coatings, printing, metal finishing industries, but have now been adopted by the CPI.

A primary burner operating on fuel gas is used to combust small concentrations of VOCs in the feed gas at 1,500° F. The feed gas is typically air or nitrogen with the concentration of VOCs below 25% of the LEL. If the heating value of the feed stream is high, the primary burners will shutdown and the oxidizer will burn flameless. The oxidizer will shutdown for very high heating values to avoid damage to the equipment. Destruction efficiencies of 99% are typically specified. The oxidizers usually are provided with integral heat recovery. Regenerative types are most prevalent because of the cost and high thermal heat recovery efficiency (up to 95%). They utilize a bed of high density packing to absorb heat and alternatively switch the feed and stack gas through the packing for heat recovery. Recuperative cross exchange is also utilized but it has lower thermal efficiency of 80%.

Thermal Oxidizers Which are Flameless

Flameless thermal oxidizers represent relatively new technology developed to burn low concentrations of VOC (typical 1% volume) with low NOx and carbon monoxide emissions. They are capable of destroying a wide variety of organic compounds. These oxidizers have a bed of porous inert material that serves as the substrate for the oxidation process. The bed is pre-heated prior to introducing the fumes. The design of the matrix forces the fume stream to pass through a reaction zone maintained at 1600-1850°F. This temperature is lower than a flame temperature, resulting in low NOx formation. Heat released by the exothermic oxidation reaction is absorbed on the matrix, providing the thermal momentum needed to maintain the process. Flameless thermal oxidizers typically can provide high VOC destruction efficiency of up to 99.9%. The oxidizers are sometimes equipped with integral heat recovery for energy conservation. Costs are comparable to other types of thermal oxidizers.

Catalytic Oxidizers

Catalytic oxidizers were developed in the early 1960s to combust streams with VOCs below the lower explosive limit. As expected with catalytic processes, the technology and catalysts are continuously being improved. The stream is preheated, traditionally using recuperative cross exchange, and fed to the reactor, which is filled with the catalyst embedded on an inert substrate. When the temperature of the stream reaches the "kickoff off" temperature, the VOCs are oxidized. The kickoff temperature is a function of the type of material being oxidized, but is approximately 600° F to 900° F. Combustion efficiencies of 95-99% are typical. Because the operating temperature is so low, high thermal efficiencies of heat recovery are not important and small recuperators are used for heat recovery. Hybrids utilizing both catalytic and thermal oxidation are common. Catalytic oxidizers are cost effective and easy to maintain, with startup the primary operating concern. There are many manufacturers of catalysts and each has their special operating characteristics. For each particular type of compound to be combusted, a different catalyst may be selected. Additionally, care must be used to avoid compounds in the stream that will poison the catalyst.

RISK PROFILE

The incidents reviewed in preparing this paper are all different; however, they share many similar characteristics. A summary of some of the common elements has been organized into the Risk Profile. Often, the causes of the incidents were hard to identify and anticipate, suggesting that important issues of design are subtle and may be overlooked. Some of the conclusions found were:

- Thermal oxidation can be a potential explosion hazard. However, due to low pressures and small amounts of fuel, when incidents occurred the extent of physical damage and injury has generally been small, although some were catastrophic.
- Many of the systems were complicated, collecting waste streams from various sources, containing varying amounts of hydrocarbons, oxygen, and inerts.
- The controls and interlocks are numerous and complicated.
- The flows and compositions are variable and transient. Expected flows and concentrations did not always present a true picture of the range, especially excursion conditions during start-up or unexpected operating circumstances.
- In some cases, a "slug" or "puff" of hydrocarbons was generated within the explosive range, transported downstream, and ignited. A potential for a "slug" of hydrocarbons within the explosive range is enhanced with variable and unpredictable compositions and flows.
- Blowers are often necessary to boost the feed pressure, contributing to complicated control schemes and possible sources of ignitions.
- Ignition came from a variety of sources. These included the incinerator, the blowers, induced electrical charge from lightning or welding, static electricity, heating of pyrophoric iron sulfide deposits, heat of adsorption from hydrocarbons being adsorbed on activated carbon granules, liquid phase reactions, or a combustion device used to produce inert gas.
- Field equipment and instrumentation did not always respond exactly as anticipated.
- Due to operating conditions and calibration, instrumentation did not operate as expected.
- Detonation flame arrestors and seal pots may be used to isolate flashbacks, but may underperform due to previous damage or poor design.
- Materials of construction, insulation, etc. sometimes contributed to incidents.

GOOD ENGINEERING PRACTICE

While the problems described above are highly varied, at this time there are no consistent, established industrial standards for the design of vapor collection and destruction systems.

Probably the most technically detailed requirements for these systems are for marine operations and come from the U.S. Coast Guard. They have been promulgated in the Code of Federal Regulations as Title 33 Part 154 *Regulations for Facilities Transferring Oil or Hazardous Materials in Bulk.* These regulations will be referenced as 33CFR154.

The use of marine vapor control systems introduces potentially significant new hazards to the loading and ballasting operations of tank vessels. The primary hazards are cargo tank-over or underpressurization, overfill and spillage, fire, explosion, and detonation. During loading, the displaced vapors change over time from an oxygen-rich mixture at the beginning of filling, to a hydrocarbon-rich mixture near the end. The Coast Guard regulations address these hazards by development of safety standards for using vapor control systems.

Some of the key design features included in the Coast Guard Regulations are the use of gas analyzers, detonation arrestors, flame arrestors, and either an enriching, diluting, or inerting system. Before these rules were proposed, regulations did not address the hazards in a comprehensive manner. While there were regulations for piping and electrical equipment, there were no regulations on training and qualifications of persons operating vapor control systems, or no safety regulations applicable to facility vapor control system installations. The purpose of the regulations is to provide standards for the safe design and operation of vapor control systems and provide qualifications for personnel operating vapor control systems.

The effort to develop the regulations took three years. In 1986, the Coast Guard funded a National Research Council Commission on Engineering and Technical Systems-Marine Board to study the technical, safety, and economic aspects of vapor control systems. In response to the NRC recommendations, the Chemical Transportation Advisory Committee Subcommittee on Vapor Control was formed to develop standards for the design and operation of vapor control systems. Following the issue of the draft safety provisions, the Coast Guard funded a failure modes and effects analysis and a worst case scenario analysis. The API has also funded a qualitative risk analysis.

CONTROLS AND SAFEGUARDS

Based on an evaluation of the various case histories that were reviewed and the Coast Guard's recommend design, several suggested design features have been developed.

Segregating Streams

Segregation of complicated streams helps in the following ways:

- The system is simplified to reduce transient and uncertain compositions and flows.
- Operation is safer since there is less chance of an excursion into the flammable range when the variability of streams is reduced.
- The consequence of an incident is lower and potential damage is reduced since segregation reduces the volume of flammable material in each system.
- Understanding of the process is enhanced, reducing training requirements, and decreasing the chance of operating error.
- Operation of the system may continue even if one of the sections is out of service.
- Segregation often results in the installation of additional blowers. However, there are several advantages. Blower operation is more stable since the variation in flow is reduced, making the operation more reliable. Total operating costs may be reduced if the additionally blowers are operated intermittently.
- When the blower is included as the field equipment in the emergency shutdown logic, the shutdown reliability may be enhanced since blower shutdown is considered more reliable than control valve closing.
- Cross contamination may be reduced for multiple product lines.

Minimizing Flammable Mixtures In The Vapor Collection System

Sources of explosive mixture include unexpected production of flammable vapors, lack of controls, failure of instrumentation, poor operating practice, etc. Materials in the flammable range may be minimized with the following methods:

Provide for the Range of Flows and Compositions. Review operation and provide a design suitable for transient and uncertain compositions and flows.

Inerting, Enriching, or Diluting. The primary methods to keep the concentration of vapor mixture out of the flammable region are to (1) inert with nitrogen or other inert gas to keep oxygen below flammable range, (2) dilute with air below the lower flammable limit, or (3) enrich with flammable gas above the upper flammable limit. For point of reference, the limits of flammability for methane are 5 volume % of methane in air for LEL and 15 volume % for UEL. The range of flammability increases in a pure oxygen atmosphere because the nitrogen diluent is no longer present. Flammability limits are typically quoted at atmospheric pressure and there is little change for the slight increased pressures experienced with vapor collection systems. However, the range may widen considerably with temperature and the flammability limits should be adjusted in high temperature service. Some of the important design elements of these features are:

<u>Operation and Location of Injection Point for Inerting, Enriching, or Diluting Gas.</u> If inerting, enriching, or diluting gas is used, the gas injection point should be as close to the vapor source as possible and provide for complete mixing of the gas and vapor within a short distance of the injection point. Static mixers are sometimes used but with proper mixing nozzles they are not necessary.

<u>Inerting System Design.</u> Nitrogen or other inert gas may be utilized to keep oxygen below the flammable envelope. Control may be accomplished by either (1) oxygen analyzers (primary plus redundant) or (2) a flow orifice that controls the amount of inert gas and an oxygen analyzer for alarm and shutdown. If the system deviates from an acceptable range, an alarm should sound, and if the deviation expands, the system should shutdown. For marine systems, an alarm is activated when the oxygen concentration exceeds 8% (volume).

Oxygen ingress to the system should be avoided in order to stay below the flammable envelope and prevent the heating of pyrophoric iron sulfide which may be present. Avoid operating the collection system at a vacuum, especially if sources of air are present, such as vacuum relief valves or sleeve type pipe couplings. If it is not practical to avoid air ingress, use a dedicated analyzer to measure the presence of a flammable atmosphere.

Enriching System Design. A compatible enriching gas may be utilized to keep vapor above the upper flammable limit. Control may be accomplished by either (1) oxygen analyzer (primary plus redundant), (2) hydrocarbon analyzer (primary plus redundant) or (3) a flow orifice that controls the amount of enriching gas and an analyzer (oxygen or hydrocarbon) for alarm and shutdown. If the system deviates from an acceptable range, an alarm should sound, and the system should shutdown if the deviation expands. For marine systems, an alarm is activated when the hydrocarbon concentration falls below 170% (volume) of the upper flammable limit.

Backflow to the source from the collection system may cause a mixture at the source (which is initially below the lower explosive limit) to become flammable. Backflow prevention, such as check valves, and manual isolation should be used to prevent backflowing any high pressure enriching gas to the source. Additionally, oxygen ingress to the system should be prevented to stay above the upper flammable limit. Control is similar to that for inerting systems.

Air Diluting System Design. Air may be utilized to keep the vapor below the lower flammable limit. Control may be accomplished by either (1) hydrocarbon analyzer (primary plus redundant) or (2) a flow orifice that controls the amount of dilution air and a hydrocarbon analyzer for alarm and shutdown. If the system deviates from an acceptable range, an alarm should sound, and the system shutdown if the deviation expands. For marine systems, an alarm is activated when the hydrocarbon concentration exceeds 30% (volume) of the lower flammable limit.

Cargo Vapor Shut off Valve for Marine Service. For marine service, the facility needs to be able to isolate the vapor collection system from the marine vessel in case of system malfunction. A cargo vapor shutoff valve must be located in the facility vapor collection line upstream of the detonation arrestor or where the inerting, enriching, or diluting gas stream is introduced. According to 33CFR154.810, this cargo vapor shutoff valve must conform to certain requirements as to closing time, alarms, automatic activation, remote operation, valve seats, etc. The valve must be able to be operated or activated manually in case of failure of the automatic shutdown system.

Flammable Mixtures at Source. At times, the vapors from the source will be in the flammable region. For example, in loading operations, the displaced vapors change over time from an oxygen-rich mixture at the beginning of filling, to a hydrocarbon-rich mixture near the end.

<u>Detonation Flame Arrestor</u>. If the situation occurs where a vapor crosses into the flammable region, then the possibility exists that these vapors may be ignited from an ignition source at the facility. A detonation arrestor should be provided at the facility vapor connection tie-in point.

<u>Misting</u>. Misting in tanks, due to agitation, may also cause the fluid in the vapor space to be in the flammable region, even if the liquid hydrocarbon is stored below its flash point. Liquid agitation should be avoided to reduce misting of hydrocarbon in the vapor space.

Deposition of Flammable Chemicals in the Collection System. Deposition of flammable materials should be avoided. A discussion of the hazards of pyrophoric iron sulfide deposits is presented in the next section on ignition sources. Another example is provided in the case histories of phthalic anhydride condensation and deposition on duct work and internal insulation.

Eliminating or Preventing Sources of Ignition

Eliminating or preventing sources of ignition helps avert an explosion if a flammable mixture was to be introduced in the vapor collection system. The following discussion outlines some of these potential ignition sources and various methods of mitigating them.

Static Electricity and Lightning. Static electricity and lightning can be a source of ignition.

Vents and Detonation Flame Arrestors. Vents to the atmosphere from the vapor collection system, excluding the incinerator combustion stack or relief vents through pressure relief devices, should have detonation arrestors.

Static Electricity. The vapor collection piping should be grounded and the piping should be electrically continuous to avoid ignition by static electricity, welding, lightning, or mechanical sparks (refer to section on blowers). During loading operations, the initial liquid loading flow rate should be limited to reduce the possibility of generating static electricity. Case histories on static electricity are cited in Thomas H. Pratt, and John G. Altharton, *Some Electrostatic Considerations in the Transportation of Flammable Liquids*, Process Safety Progress, Vol 15(3), Fall 1996, pp 173-177. Two guidelines are API RP 2003 *Protection Against Ignitions Arising Out of Static, Lightning, and Stray Currents*, 5th Edition, March 1991, and NFPA 77 *Recommended Practice on Static Electricity*, 1993 Edition, National Fire Protection Association. API has also published a special report *Static Electricity Research Project* April 29, 1994.

Blowers and Compressors. Blowers or compressors add energy to the stream and, as the case histories show, can be a source of ignition. Design features for reciprocating compressors, screw compressors, liquid ring compressors, centrifugal compressors, centrifugal fans, and lobe blowers are given in 33CFR154. These features include alarms and materials of construction.

Alarms and Shutdowns. Depending on the type of blower, alarms for lube oil pressure and level, shaft bearing temperatures, and discharge gas temperature should be provided to warn of impending failure which may produce a spark, or warn of higher than design temperatures in the system.

Detonation Flame Arrestors. Detonation flame arrestors are recommended before the tank source or loading blower, and the inlet and outlet of booster blowers, particularly if there exits a possibility that the system can be in the flammable range. Although all of the above blower types are utilized, liquid ring blowers are becoming very popular because considered on the safer types.

Thermal Oxidizers. Thermal oxidizers can be a source of ignition. These recommendations to reduce the ignition potential are applicable to either a thermal or catalytic oxidizer where an open flame is present.

Seal Drums. Provide a seal drum at the inlet to the vapor destruction system. Seal drums require careful design to provide an adequate, stable, and continuous seal. One major problem is an inadequate distributor. The purpose of the distributor is two fold. First, it is utilized to dampened the pulsations due to variations in flow to keep the level of the seal constant. Secondly, it must be designed to disperse the gas into bubbles which contain the flammables and cool them quickly.

Detonation Flame Arrestors. Provide a detonation arrestor in the vapor collection line. However, plugging is a typical concern for detonation flame arrestors. Plugging may be from pipe rust or polymer type compounds. Since the plugging is a function of the material, it is difficult to estimate a cleaning frequency, but 2 months to 2 years appears representative, depending on the plugging material. Strainers upstream of detonation flame arrestors, having low pressure drop across the strainer, are a typical solution.

The case histories emphasize that maintenance, and replacement of damaged detonation flame arrestors, is very important for proper operation. Any detonation flame arrestor that is suspected of having a flame burning on the flame arrestor, or the detonation flame arrestor has been otherwise damaged, should undergo an inspection. The detonation flame arrestor should be provided with controls, such as thermocouples, to detect the presence of a flame on the arrestor and shutdown the system. Information on detonation flame arrestors and flame arrestors may be found in 33CFR154, *Appendix A to Part 154-Guidelines for*

Detonation Flame Arrestors and *Appendix B to Part 154-Standard Specification for Tank Vent Flame Arrestors.* Three publications from the American Petroleum Institute are API Publication 2028 *Flame Arrestors in Piping Systems*, 2nd Edition, December 1991, API Recommended Practice 12N *Recommended Practice for Operations, Maintenance and Testing of Firebox Flame Arrestors*, Second Edition, November 1994, and a special report *Special Report on Detonation Arrestor Safety: "Mitigation of Explosion Hazards of Marine Vapor Recovery Systems,"* R.E. White an C.J. Oswald, Southwest Research Institute, October 1992.

Quick Closing Isolation Valves. To isolate the thermal oxidizer in the event of a shutdown or flameout, two redundant isolation valves should be provided at the inlet to the thermal oxidizer.

Anti-flashback Burners. Burners which offer flashback protection may be available on some thermal oxidizers. Consult the vendor for this option.

Vapor Recovery Units. Vapor recovery units include absorption, adsorption, condensation, etc. These units in general are not sources of ignition, so that detonation flame arrestors are not required at the inlet to the vapor recovery unit.

Activated Carbon Adsorption. Activated carbon adsorption is often used for vapor recovery but can be a potential ignition source. As hydrocarbons are adsorbed on the activated carbon, the heat of adsorption is released. Although not as selective as hydrocarbons, oxygen is also adsorbed. The adsorbed hydrocarbons will slowly react with the oxygen, possibly creating a local "hot spot." If the circumstances are right, the local temperature may reach a point that the oxidation of the hydrocarbon is self sustaining which may lead to combustion of the carbon. Oxidation is enhanced with compounds such as ketones, aldehydes, and organic solvents that react freely with oxygen. Contributing factors are inadequate deactivation of the activated carbon before introducing hydrocarbons, and operations with long periods of no flow through the carbon bed. Flow tends to carry off excessive reaction heat and keep hydrocarbon levels below the flammability limit.

Safeguards are dependent on the type of system and operation, but typically include temperature and hydrocarbon indicators, shutdown interlocks, flame arrestors at the inlet to be carbon bed, and snuffing connections for water, steam, or nitrogen. For more information, refer to the information provided by Industrial Risk Insurers in *IR Information* IM.9.6.2.2 *Solvent Recovery with Activated Carbon* February 1990; *CEP Technical Manual* Loss Prevention Volume 12 March 1978 published by the American Institute of Chemical Engineers; and the following case studies.

Vapor Balancing Unit. Vapor balancing units transfer the vapor from the vessel being loaded to the vessel being unloaded via the vapor collection system. Detonation flame arrestors should be positioned at the source or facility connection point, and on the vapor return lines at the facility storage vessels. Overfill protection should be provided at the facility storage tanks to prevent liquid from entering the vapor return lines.

Combustion Devices Used To Produce Inert Gas. Combustion devices are occasionally used to produce a gas for inerting. The device may be an open flame combustor or a gas engine where the exhaust is used for the inert gas. These devices should be isolated by a hydraulic seal and a non-return (check valves) valve.

Pyrophoric Iron Sulfide Deposits. In systems containing sulfur, monitoring for iron sulfide scale or deposits in tanks and vapor collection line should be performed and any suspect deposits should be removed. Particularly for inerted systems, iron sulfide deposits may build-up, and if air is introduced into the system, heating of the pyrophoric iron sulfide may occur. Inerted closed systems, such as inerted marine vessels, should have positive pressure to avoid introduction of air from the collection system or atmosphere.

Other. Other sources of ignition include liquid phase reactions, smoking, and liquid overfill and subsequent ignition. Vapor collection piping should be held to a maximum of 350°F during normal operation. Oxygen analyzers which operate at high temperatures, such as zirconia oxide or thermagnetic oxidizers should not be used. Avoid these sources of heat that are potential ignition sources.

Avoiding Liquid Carryover or Condensation in Lines

Equipment should be protected from liquid carried over from the source or condensed in the system. Options for eliminating condensation are heat tracing, introduction of hot stream, and introduction of dry stream. Heat tracing is economical but should be avoided for substances that may polymerize. The introduction of another stream may not be feasible under certain operations and cold conditions. If condensation occurs, it may be handled by sloping the lines to a low point drain or knockout drum. If a liquid knockout vessel is employed, high level is a concern. Controls for this arrangement should include level indication, high level sensor and alarm, and redundant high-high level sensor and controller to shutdown any blowers before they can be damaged by liquid carryover. If in marine service, this high-high level sensor must also close the cargo vapor shutoff valve. The design of liquid knockout vessels plus an extensive bibliography are discussed in API RP 521 *Guide for Pressure Relieving and Depressuring Systems*, 3rd Edition, November 1990.

Instrumentation And Controls

Analyzers used for control of inerting/enriching/diluting gas should provide a location and sample speed sufficient to allow for shutdown devices to operate before an ignition source is reached. They should be located as close to the injection point as possible, but still where the vapor mix is homogeneous, and they should have a response time of 30 seconds from the time the vapor is sampled.

Reliability of the instrumentation is crucial. For inerting, enriching, diluting systems, redundant controls are recommended as explained above. Additional redundancy may be provided by low pressure or flow sensors on the inerting, enriching, or diluting gas streams. The analyzers should sample continuously to provide constant monitoring of the operation. Instrumentation reliability should be reviewed carefully and the risks if the instrumentation should fail should be assessed.

Case histories indicate that reliability may be reduced due to operating conditions. A connection for injecting calibration gas and provisions for ensuring a good sample, such as drains or tracing, should be available.

The selection of whether to use oxygen or hydrocarbon analyzers depends on the type of products, whether the source(s) are inerted or not, and whether inerting, enriching, or diluting is employed. In handling multiple products, a single hydrocarbon analyzer may not be sufficient to cover the range and multiple analyzers (with redundancy) may be required. If an oxygen analyzer is employed for marine vessels which may or may not be inerted, then a procedure for ensuring the correct set point is necessary.

The operating ranges for inerting, enriching, and diluting systems are graphically presented in Figure 1. The hydrocarbon-oxygen-nitrogen flammability diagram has been adopted from *Fundamentals of Fire and Explosion*, by Daniel R. Stull, AIChE Monograph Series Volume 73(10), 1977. The dotted line indicates the oxygen-nitrogen concentration of air.

These diagrams show the operating range and set point for alarms depending on the type of protection (inerting, enriching, diluting), whether the source is inerted or not, and the type of analyzer used for control and alarm.

The examples utilize methane as the hydrocarbon and the set points are taken from 33CFR154. If the VOC of concern if different, then the flammability diagram and set points should be reviewed.

Information on process instrumentation is provided in American Petroleum Institute API RP 551 *Process Measurement Instrumentation* 1st Edition May 1993, API RP 552 *Transmission Systems* 1st Edition October 1994, API RP 554 *Process Instrument and Control* 1st Edition September 1995, and API 555 *Process Analyzers* 1st Edition October 1995. These publications replace API RP 550 *Manual on Installation of Refinery Instruments and Control Systems,* which is out of print.

Figure 1. Operating Ranges for Inerting, Enriching, and Diluting Systems

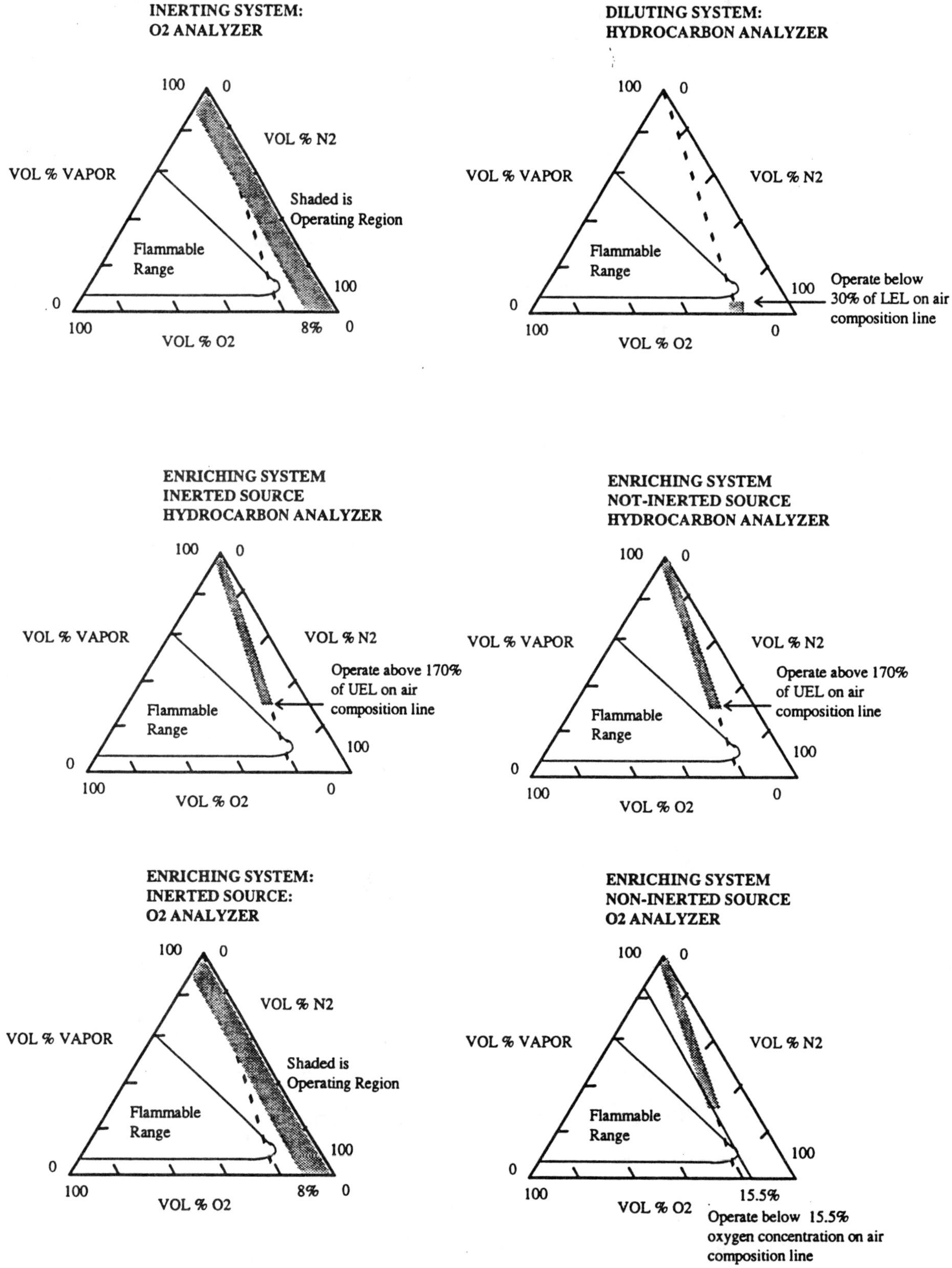

APPENDIX
CASE HISTORIES

Following are short summaries of case histories. Only enough detail has been provided to highlight the important points. Further information may be found in the cited literature. It should be pointed out that many of these cases are complicated and the complete causes, especially the ignition sources, were not always positively identified. This demonstrates one of our original premises that even with well-designed systems, care must taken to include subtle issues of safety. We welcome more case histories. Anyone wishing to share cases please contact the authors.

Case 1: High Production of Flammables and Ignition in Thermal Oxidizer

A thermal oxidizer system combusted the fumes from a pilot plant, solvent stills, and numerous reactors at a chemical manufacturing facility in Texas. The sources and thermal oxidizer were connected by an uninterrupted duct. The inlet to the incinerator was equipped with LEL monitoring equipment and alarms. An automatic system was installed to provide dilution air in the event of VOC concentrations greater than 50% of the lower explosive limit.

The concentration of toluene to the collection system from one of the reactors was well into the flammable range relative to the dilution air flow rate. The toluene flowed to the thermal oxidizer where it ignited producing an explosion in the duct. The LEL monitoring equipment failed to alarm the situation. The instrument was discovered to be located close to the thermal oxidizer, limiting time for detection and response. Further, the monitoring analyzer was found to read only 50% of the correct value for a tested toluene sample, although it was accurately calibrated for the specified calibration gas.

Reference: "Avoiding Duct Explosions-System Changes can Lead to Disaster". Harri K. Kytomaa, Chemical Processing, July 1996, page 78-81.

Case 2: High Production of Flammables and Ignition in Thermal Oxidizer

A regenerative thermal oxidizer at a Texas refinery combusted off gas streams from specific tanks in a waste water treatment unit, including an API separator, a flocculation tank, and a dissolved nitrogen flotation unit. A valve on a distillation unit draining to the waste water tank was left open too long and butanes and butenes moved to the tank. The vapor stream from the waste water tank, due to high liquid hydrocarbon flows to the tank, contained a high percentage of hydrocarbons (25% higher than originally predicted), which migrated to the incinerator and ignited. The flashback was contained by the seal drum and no damage occurred ahead of it, even though the unit periodically experienced very turbulent conditions in the seal drum which could have blown the seal before the flash back occurred, rendering the seal ineffective. The rupture disk on the thermal oxidizer plenum blew, reducing the damage to the thermal oxidizer.

The unit suffered from other problems. The hydrocarbon analyzers provided poor response which was traced to mist in the analyzer lines. Additionally, the seal drum experienced very turbulent conditions under high gas flow rates. One of the treatment units, which provided dilution air, could be manually turned down by the operators. Another dilution air control valve had to be sized to provide for sufficient makeup dilution air when this occurred.

Reference: Private communication.

Case 3: Fuel Rich Slug of Flammables in Collection System and Ignition in Incinerator

A waste gas incinerator at a chemical facility in New Jersey combusted waste streams collected by a central system. The incinerator combusted waste gases from the "slurp system" in the primary zone and liquid wastes plus another large air/hydrocarbon stream in the secondary zone. The slurp system drew air contaminated with flammables from sumps, vessels, etc. The incinerator had been down briefly, but long enough for flammables to accumulate in the stagnant slurp system. During restart, a fuel rich "slug" of combustible vapor/air mixture passed to the combustion air blower, ignited, and destroyed the blower and piping. The source of ignition may have been the blower or incinerator.

Reference: "Incinerator Problems and How to Prevent Them", D.F.Ready, and R.F. Schwab, AIChE 14th Loss Prevention Symposium-Volume 14-A CEP Technical Manual, 1981, AIChE. Paper was also given at AIChE 14th Loss Prevention Symposium, held in conjunction with the 88th AIChE National Meeting, June 1980, Philadelphia.

Case 4: Fuel Rich Slug of Flammables in Collection System and Ignition in Incinerator

A waste gas incinerator, at chemical facility in Texas combusted waste vents collected by a central system. The incinerator was designed to use natural gas as a primary fuel to combust waste gases from a reactor and minor concentrations of VOCs from tank working losses and distillation system vents. This latter system handled hydrocarbons at 10% of the lower explosive limit and provided the primary source of combustion air. The system was equipped with a booster fan, analyzers, interlocks, and a flame arrestor. Just before the incident, the reactor feeding waste gas to the incinerator tripped off line. Through a series of valve switching, organics built up in the system and a fuel rich "puff" stream was rapidly introduced into the incinerator and ignited. A secondary explosion also occurred; the source was not determined but was possibly a hot spot in the system or sparks generated by the fan wheel contacting the housing. Other possible contributing causes were also identified. Extensive damage was done to the flame arrestor, fan, valves, and incinerator piping. There were no injuries.

Reference: "Flashback from Waste Gas Incinerator into Air Supply Piping", S.E.Anderson, A.M. Dowell, J.B. Mynaugh, Plant/Operation Progress, Vol 11(2), April 1992, p 85.

Case 5: Fuel Rich Slug of Flammables in Collection System and Ignition in Thermal Oxidizer

A thermal oxidizer combusted VOC fumes at a chemical resin plant in New England, including those from a continuous reactor. The exhaust from the sources, including the continuous reactor, was boosted via fans in the source exhausts. A 600 foot long, 60 inch diameter duct was used to collect the fumes. To this system was added the exhaust of four batch polyester reactors via a new steam-traced duct. No fan was installed in the line because it was believed that ample dilution was provided by the fan of the continuous reactor. There were no interlocks between the batch reactors and the continuous reactors.

One of the batch reactors, used for an acetone dilution process, exhausted a significant volume of acetone vapor. But because the continuous reactor and its fan were down, there was no dilution stream and almost pure acetone vapor filled the new duct and spilled into other stagnant source branches. Subsequently, the fan in one of the stagnant branches started up and blew heated air which mixed with the high concentration acetone stream and traveled down the main duct to the thermal oxidizer. A minute passed before a large explosion occurred at the thermal oxidizer; the one minute delay coincided with the time it took for the stream to travel down the main duct to the thermal oxidizer. The damage patterns showed that ignition probably occurred at the thermal oxidizer fan. Blowout panels in the main duct ruptured, the thermal oxidizer incurred minor deformation, and two expansion joints suffered damage. The reactors were down for one month following the explosion.

Reference: "Avoiding Duct Explosions-System Changes Can Lead to Disaster". Harri K. Kytomaa, Chemical Processing, July 1996, page 78-81.

Case 6: Phthalic Anhydride Plant Explosion

A tail gas incinerator at a chemical facility in New Jersey combusted phthalic anhydride tail gases from a condenser. The incinerator was designed to use natural gas as a primary fuel to combust tail gases containing phthalic anhydride. The tail gases were the primary source of combustion air. The tail gases were preheated to 1100° F through a series of two cross exchangers. The tail gas duct work between the two exchangers was internally insulated with a rock wool insulation. Excessive phthalic anhydride condensation in the condenser carried over and was absorbed in the rock wool internal insulation. A solid piece of phthalic anhydride carried over to the incinerator, ignited, and exploded. No pressure protection was provided and the explosion tore open some of the ductwork.

Reference: "Incinerator Problems and How to Prevent Them", D.F.Ready, and R.F. Schwab, AIChE 14th Loss Prevention Symposium-Volume 14-A CEP Technical Manual, 1981, AIChE. Paper was also given at AIChE 14th Loss Prevention Symposium, held in conjunction with the 88th AIChE National Meeting, June 1980, Philadelphia.

Case 7: Phthalic Anhydride Plant Explosion

A tail gas recuperative incinerator at a chemical facility in New Jersey combusted phthalic anhydride tail gases. The incinerator was down and the tail gases were passed through a scrubber and sent to the atmosphere. The butterfly valve cross connecting the scrubber with the incinerator leaked and phthalic anhydride crystals were deposited at the inlet in the combustion air line at the incinerator. When the tail gases were switched from the scrubber to the incinerator, the phthalic anhydride and tail gases exploded, blowing out the expansion joint between the recuperator and incinerator. Solid phthalic anhydride had caked the rupture disk on the inlet to the recuperator and it did not operate.

Reference: "Incinerator Problems and How to Prevent Them", D.F.Ready, and R.F. Schwab, AIChE 14th Loss Prevention Symposium-Volume 14-A CEP Technical Manual, 1981, AIChE. Paper was also given at AIChE 14th Loss Prevention Symposium, held in conjunction with the 88th AIChE National Meeting, June 1980, Philadelphia.

Case 8: Waste Water Tank Explosion

A 900,000 gallon waste water tank at a Texas refinery received material from many sources in the plant's propylene oxide-styrene monomer unit, including off-spec product. The tank contained a lower layer of peroxides and upper layer approximately eleven feet deep of hydrocarbon liquids. The vapor space in the tank vented to a vapor compressor. The tank was blanketed with nitrogen, had a pressure/ vacuum vent, and an oxygen analyzer in the vapor space. The oxygen analyzer was to indicate if the peroxides had decomposed to organics and oxygen, which is a slow decomposition reaction that occurs at all but very cold temperatures.

Six months before the incident, the vapor recovery system had been added to the tank; previously the tank vented to atmosphere. Several weeks before the incident, the vapor compressor failed and was removed for repair. The nitrogen sweep continued but now vented hydrocarbons and nitrogen to the atmosphere. Three days prior, the strip chart on the oxygen analyzer "dropped to zero" but was not repaired. Two days prior, the neighbors complained of odors, since the vapor recovery unit was down and the tank was now venting to atmosphere.

Thirty-four hours before the explosion, the operators turned off the nitrogen blanket reasoning that this increased atmospheric dissemination of the odorous material and that the piping and instrumentation had to be modified. Oxygen from the peroxide decomposition bubbled up through the hydrocarbon layer, creating a flammable mixture within the explosive range in the vapor space.

Just before midnight, a crew started the vapor recovery unit. An ignition source touched off an explosion, blowing the roof off the tank and flattening the tank walls. A burning wall of hydrocarbon liquid spilled out of the crushed tank, killing seventeen people, leveling an area the size of a city block, and shutting down fifteen percent of US styrene monomer capacity. The blower was most probably the source of ignition, although another possible source was a vacuum truck parked next to the tank.

There are many references for this incident. A readily available reference is "ARCO Spells Out Causes of Channelview Blast", Oil & Gas Journal, 14 January 1991, p 22.

Case 9: Ignition by Heat of Adsorption from Carbon Canister

At a large refinery in California, a cone roof storage tank 20 ft. in diameter by 29 ft. high received waste effluents containing amounts of hydrocarbons from a biosolids thickener, DAF unit, and an API separator. A four-inch line connected the normal vent to a 55 gallon activated carbon adsorber. Due to adsorption of hydrocarbons, the heat of adsorption ignited carbon granules which subsequently ignited hydrocarbon vapors in the vapor space of the tank. An explosion ensued and the roof of the tank separated and landed 200 feet away. There were no injuries.

Reference: "Carbon-Initiated Effluent Tank Overpressure Incident", R.E. Sherman, K.C. Crawford, T.M. Cusick, and L.V. Csengery, Process Safety Progress, Vol 15(3) Fall 1996, p 148-149. Also presented as paper (6a) at the AIChE Summer National Meeting, Boston, MA, August 2, 1995.

Case 10: Large Temperature Rise in Catalytic Oxidizer Reactor

A catalytic oxidation reactor was used to combust the ethylene from a carbon dioxide stream at 1000° F. Suddenly, the air stream to the reactor was cut off, but the ethylene/carbon dioxide stream continued to flow for a short period of time. The ethylene cracked and laid down carbon on the catalyst. On the subsequent startup, there was a large temperature rise as the carbon burned off. No damage was done to the catalyst or vessel since the carbon laydown period was short.

Reference: Private communications.

Case 11: Explosions in Oil Storage Tanks

The generic incidents are based on ten different explosion investigations by Westech Industrial Ltd, Calgary, Canada. Westech estimates that there are 25-30 oil storage tank explosions in Canada each year.

Air may be drawn into the tanks from inadequate startup procedures with incomplete purging. Alternatively, air may be drawn in through production, when the makeup gas rate is less than the production rate, and outside air infiltrates through the vacuum relief device, missing thief hatches, or leaking seals. The most common source of ignition is the flare. Other sources of ignition are static discharge from improperly grounded tanks or the use of non-metal piping. Oxidation of ferric sulfide that has coated the inside of a tank or pipe wall is not a very common ignition source in tank explosion incidents. Failure of the flame arrestor between the flare and tank has been implicated in most of the tank explosions investigated. These failures are due to improper design and construction, external damage and lack of maintenance, use outside the tested range or lack of certification.

Reference: "Safeguards Cut Tank Explosion Risk During Gas Flaring". Ken Lapp and Nick Roussakis, Oil & Gas Journal, Vol 87(33), 14 August 1989, p 41-44. For further guidelines, see "Explosion Hazards in Crude Distillate Storage Tanks Assessed", A. Tamimi, Oil & Gas Journal, Vol 88(47), 19 November 1990, p 53-55.

No Good Deed Goes Unpunished:

Case Studies of Incidents and Potential Incidents Caused by Protective Systems

A. M. Dowell, III, P.E.
Rohm and Haas Texas Incorporated
Deer Park, TX

D. C. Hendershot
Rohm and Haas Company
Bristol, PA

Prepared for Presentation at the
American Institute of Chemical Engineers
31st Annual Loss Prevention Symposium
Houston, TX
Paper 44c

March 13, 1997

November 1996

UNPUBLISHED

No Good Deed Goes Unpunished: Case Studies of Incidents and Potential Incidents Caused by Protective Systems

A. M. Dowell, III, P.E.
Rohm and Haas Texas Incorporated
Deer Park, TX

D. C. Hendershot
Rohm and Haas Company
Bristol, PA

"The chief cause of problems is solutions."

- Eric Sevareid

ABSTRACT

In the course of chemical process and plant design, engineers will identify potential hazardous incidents. These potential incidents may be identified through special hazard analysis reviews and procedures, or by the design team in the course of design activities. To manage and control those hazards, the team will modify the initial design, often by adding on additional protective devices and systems — alarms, interlocks or active protective systems. However, any change in a system, even a change intended to prevent or mitigate a potential hazardous incident, also has the potential to introduce new hazards, or new mechanisms by which existing hazards can result in an incident. A number of case studies illustrating this point will be reviewed. The examples illustrate the importance of a management of change program, which must consider all changes including the addition of safety devices and systems, and which must thoroughly consider all potential effects on the system.

INTRODUCTION

When we add a new safety device onto an existing system, or to the design of a new system, the desired result is increased safety. That is certainly the intent of the designer when he adds the safety device. But, any change, even the addition of a new safety feature, has the potential to introduce new failure modes and scenarios. The designer, who is focusing on a particular hazard or failure mode when he specifies the new safety device, may not recognize other potential new failure modes. In some cases, the new failure scenarios introduced by the safety device may be more serious than the failure scenarios the device is intended to prevent. In these cases the system with the safety device may actually have a higher risk than the original design. A designer must always remember that "no good deed goes unpunished" (Powers, 1989). A good deed of adding a new safety device has the potential to punish by introducing new failure scenarios. A modified system must be thoroughly reviewed to ensure that all failure modes have been considered in the design. In this paper we will present a number of case studies which show how incidents or potential incidents may actually be caused by safety devices. Some of these case studies are based on actual incidents. Others were identified in the course of process hazard analysis studies before any incident actually occurred — the preferred way to identify potential incidents!

STORAGE OF STONE COLUMNS IN THE MIDDLE AGES

Focusing on a single aspect of a problem, and resolving that issue without reviewing the impact of the solution on the entire system has resulted in failures of engineered systems for many years. Our first example was reported by Galileo (1638), as described by Petroski (1992, 1994). In Galileo's time, stone columns for building projects needed to be stored for some period of time prior to use. If the column was stored on the ground, the parts of the column in contact with the ground would tend to stain and discolor from contact with the soil. The discoloration was difficult to remove, resulting in an unsightly appearance of the building in which the column was finally used. To prevent discoloration, it was common practice to store the columns off the ground, supported by piles of timbers or stones at each end [Figure 1(a)]. Sometimes a column would break in the middle under its own weight [Figure 1(b)]. A worker, seeing the failed columns, suggested that an additional protective feature be added to the column storage system — a third support in the middle, as shown in Figure 1(c). This would prevent the failure of the column in the middle by providing additional support. Everybody thought this was a great idea, and so a third, center support was provided under the columns.

What actually happened to a number of the columns after they had been stored for several months using the additional support? It was found that many of them still broke in the middle, but in many cases the failure mode was different. The supports were made from timbers or stones, and they would settle from the weight of the columns as they sat outside in the weather. When the columns were first set on the supports, each of the three supports was in load-bearing contact with the column, and the system worked as intended. However, as time went by and the supports and ground under the column settled, it was extremely unlikely that the three supports would all settle at exactly the same rate. After some time, the actual support of most of the columns was as shown in Figure 1(d) — held up by only two supports, either the supports on each end as in the original design, or by one end support and one middle support. In fact, some columns were actually balanced on the center support, not in contact with either end support. The center support was an "add on" device which was newer, and did not deteriorate and settle as quickly as the older end supports. The columns held up by one end support and (or) the middle support broke by the failure mode shown in Figure 1(e).

In this example, an additional safety device, the third column support, was added with the expectation that it would reduce the frequency of column failures. In fact it did not — the columns could still fail by the original failure mode, and a new failure mode was introduced by the presence of the "safety feature". It is even possible that the columns would be more prone to failure by the new failure mode [Figure 1(e)].

This example illustrates the importance of a full understanding of how a system works in designing safety features. The addition of a new safety feature changes the system. While the new safety feature provides a protection against a particular failure mechanism, any change in a system, even the addition of a safety device, introduces new failure modes and mechanisms. It is essential the entire system be reviewed from the standpoint of its complete functional requirements after any change is made, even the addition of a safety device. If the designers had understood the failure mechanisms which actually occurred, they could have identified other potential solutions, or additional safety precautions to protect against all failure modes.

TOO MANY FLOW METERS!

A process required that Raw Material A, a highly reactive, corrosive, and toxic chemical, be fed at a controlled rate to a reactor. The flow rate was extremely critical — if the flow rate exceeded a specified value there would be a potential for a rapid runaway reaction. The original design included a metering pump for Raw Material A followed by two flow meters to monitor the Raw Material A flow rate, as shown in Figure 2. A high flow rate on either flow meter (FAH 1 or FAH 2) would close a shutoff valve and stop the metering pump. The focus of the designer's attention when specifying this design was the hazard of a runaway reaction in the reactor.

A quantitative risk analysis (QRA) of this system was done, in part because of the potential runaway reaction hazard. The QRA evaluated the system very broadly, including all identified hazards:

- runaway reaction hazard
- fire hazard in case of leak or spill
- hazard of exposure to toxic vapors
- hazard from exposure to corrosive materials

All phases of plant operation were also considered:

- startup
- normal operation
- normal shutdown
- emergency (interlock) shutdown

The QRA results indicated that the second flow meter reduced the risk of a runaway reaction by less than 0.5%. The QRA also indicated that the startup and shutdown phases were major contributors to the risk, and that spurious trips caused by failure (false high indication) of the second flow meter might actually increase the overall risk of runaway reaction (although the actual number of potential spurious trips was not estimated). Furthermore, the flow meter required maintenance and calibration, with a potential exposure of operators and mechanics to Raw Material A, a very toxic and corrosive material, which could cause serious injury. Overall the conclusion was that the addition of the second safety device — flow meter FAH 2 — would in fact result in an increase in the risk of the system when all potential hazards and plant operating modes were considered. The second flow meter was eliminated from the final design.

USE OF RUPTURE DISKS AND PRESSURE RELIEF VALVES IN SERIES

The following example is not new — the issues are well known and appropriate solutions are well documented in industry standards such as API 520 (API, 1990) and the ASME vessel code (ASME, 1995). However, the phenomenon is not clear to many engineers, and we have explained it many times in the course of process hazard analysis studies. Therefore it is useful to document the concern and appropriate solution as another example of a protective feature introducing a new hazard to a system.

A rupture disk is frequently installed in series with a pressure relief valve, as shown in Figure 3, for a number of reasons. These might include:

- The material in the vessel may be extremely corrosive, and a relief valve resistant to the material may not be available, or may be extremely expensive. The rupture disk acts as a protective barrier between the corrosive material and the relief valve.

- The material in the vessel may contain solids that could foul the working parts of the relief valve, resulting in the relief valve failing to open on demand.

- The material in the vessel may be prone to form polymer or other tars due to chemical reaction of vapors or of liquid condensed from the vapors.

- Environmental regulations may require fugitive emission monitoring at the outlet of a relief valve. A rupture disk under the relief valve may eliminate the fugitive emission monitoring requirement.

In the above situations, the vessel could be protected from overpressurization by using a rupture disk alone. However, the rupture disk-relief valve combination offers the potential to minimize the discharge of material in the event of a vessel overpressure. The relief valve can close, stopping the discharge, when the vessel pressure returns to normal. Once the rupture disk bursts, the flow will continue until the vessel pressure reaches ambient pressure.

These are "good deeds" — when specifying the design of Figure 3, the designer is focused on ensuring that the relief valve will work if there is a demand on it, and also on minimizing the discharge of hazardous material to the downstream treatment equipment and, potentially, the outside environment.

But, the designer must also consider the possible punishment for this good deed. If not properly designed, the system in Figure 3 can result in nearly doubling the pressure at which the relief system will activate. Assume that the normal pressure in the vessel is P_1. The pressure between the rupture disk and the relief valve, P_2, is assumed to be ambient (0 psig), as is the pressure downstream of the relief valve, P_3. Assume that the rupture disk and relief valve are both designed to open somewhat above the normal operating pressure — say at P_4 ($P_4 > P_1$). Figure 4 shows the pressures at various points in the system during normal operation for an example case. This system will function as designed — if a process upset results in the vessel pressure P_1 increasing to P_4, the rupture disk will burst, and the relief valve, set to open at P_4, will also open, protecting the vessel from overpressure. The pressures at various locations in the system during an emergency overpressure situation are shown in Figure 5 and Figure 6. When the pressure in the vessel falls to less than P_4, the relief valve should close, minimizing the release of hazardous material.

What can go wrong? Consider what happens if there is a small, pinhole leak in the rupture disk. The small leak causes a pressure increase in the pipe between the relief valve and the rupture disk. P_2 will no longer be 0 psig, but will eventually increase until it is equal to the vessel pressure, P_1. The pressure will remain in the piping between the rupture disk and the relief valve because the relief valve set pressure of P_4 has not been exceeded, so the relief valve will not open. The pressures at various locations in the system in case of a pinhole leak in the rupture disk are as shown in Figure 7. Now what happens when a system upset results in an increase in the vessel pressure P_1? A rupture disk is a differential pressure device — it bursts when the pressure on the upstream side exceeds the downstream pressure by the specified bursting pressure.

Before the relief devices open, the pressure in the vessel P_1 can be approximated by

$$P_1 = dp(\text{rupture disk}) + dp(\text{relief valve})$$

$dp(\text{rupture disk}) = P_1 - P_2$, that is, the pressure in the vessel minus the pressure between the disk and the valve.

$dp(\text{relief valve}) = P_2 - P_3$, that is, the pressure between the disk and the valve minus the ambient pressure.

For the rupture disk to burst,

$$P_1 - P_2 = P_4, \text{ or } 50 \text{ psi}$$

For the relief valve to open,

$$P_2 - P_3 = P_4, \text{ or } 50 \text{ psi}$$

If there is a pinhole leak in the rupture disk, P_2 (between the disk and valve) approaches P_1, the vessel pressure. Substituting,

$$\begin{aligned} P_1 \text{ in the vessel } &= (P_1 - P_2) + (P_2 - P_3) \\ &= P_4 + P_4 \\ &= 50 + 50 \\ &= 100 \text{ psig, in the worst case.} \end{aligned}$$

Figure 8 and Figure 9 show the pressures that may occur at various locations in the system during an overpressure event if there is a pinhole leak in the rupture disk. The pressure in the vessel can rise to 90 psig or higher for this example case, sufficient to burst the expansion joint on the pipe connected to the vessel, which is rated for 60 psig.

For a pinhole leak in the rupture disk, some have suggested that a slow increase in the vessel pressure, P_1, would allow time for the relief valve to open slightly when P_2 reaches the relief valve set pressure, P_4. However, the relief valve will reseat with P_2 still at the relief valve set pressure, P_4. And, the vessel pressure P_1 can still increase to as much as twice P_4, because the rupture disk with a pinhole leak prevents P_2 from increasing as rapidly as P_1. Thus, a slow increase followed by a rapid increase in the vessel pressure may give the highest pressure in the vessel.

As stated, this concern is well known, although surprisingly unfamiliar to many engineers. Some ways of dealing with this issue include:

- Providing a small hole, vented to a safe place at atmospheric pressure, in the pipe between the rupture disk and the relief valve to prevent pressure buildup. However, fugitive emissions may have to be monitored for the small hole.

- Monitoring the pressure in the pipe between the rupture disk and relief valve, either with a pressure gauge which is periodically checked or by a pressure sensor with a high pressure alarm

Both of these alternatives are recognized as acceptable design options by API 520 (API, 1990) and the ASME vessel code (ASME, 1995). Both require a management system to ensure that the protective features are not compromised by plugging of the hole or failure of the instruments or alarms. It is also essential that personnel understand the reason for the protective systems so that they know the proper response for an alarm or observation of high pressure between the rupture disk and relief valve, and so that the systems are not defeated by a future change.

Note the trade-off between monitoring the relief valve outlet for fugitive emissions and monitoring the pressure in the pipe between the rupture disk and relief valve.

EXPLOSION CAUSED BY EXPLOSION SUPPRESSION SYSTEM

A plastics manufacturing plant included a grinder to eliminate oversize plastic particles from the final product. The plastic powder was being conveyed to and from the grinder by an air conveying system, and there was a potential for a dust explosion. Because of the location of the grinder and its associated piping, it was not practical to protect the system with explosion vents, and a chlorofluorocarbon (CFC) suppression system was designed to protect the equipment against dust explosion, following the design requirements of NFPA 69 (NFPA, 1992). Figure 10 shows the grinder and its suppression system as a general schematic. The pressure sensor was designed to detect the onset of a dust explosion and rapidly release the CFC into the grinder and its associated piping to quench the explosion before it generated enough pressure to damage the equipment. The suppression system is a safety feature designed to protect the equipment and personnel should a dust explosion occur.

This system operated without incident for many years, and the suppression system was never challenged — no dust explosion occurred. Then, after a number of years, the grinder did explode, and the cause was the suppression system itself! Process upsets elsewhere in the manufacturing facility resulted in water getting into the plastic grinding system. The water accumulated in the bottom of the piping below the grinder and eventually reached a point where the water pressure was sufficient to activate the CFC suppression system. Because the pipe and ducts were partially filled with water, and the plastic powder was wet and did not flow freely, the CFC suppressing agent was unable to flow easily through the system. The pressure of the CFC released into the grinder was sufficient to overpressurize it, resulting in failure of the grinder. Fortunately, the area was unoccupied at the time and there were no injuries.

Again, this is an example of the importance of identifying and evaluating all potential failure modes when designing a system. The explosion suppression system was designed and specified assuming that the grinder and its associated piping would be filled with dry, free-flowing powder if the system was triggered. The scenario in which water entered the system, both triggering the suppression system and simultaneously restricting the ability of the CFC suppressing agent to flow freely through the process resulting in overpressurization of the grinder, was not recognized. The result of this incident was a redesign of this particular system, and a thorough review of other similar systems throughout the company to search for similar hazards (Bernard, et. al., 1997).

ACUTE HEALTH HAZARD CAUSED BY AN ENVIRONMENTAL IMPROVEMENT

Thirty to fifty years ago in the USA, chemical plants frequently had a local vent for each tank and distillation column. In this example from Dowell (1996), a feed tank of API design was vented to a lifting lid for both normal and emergency vent (Figure 11). The distillation column of 1 atmosphere (15 psig) design had its normal vent to the product tank. Since one of the compounds in the column was toxic with a TWA[1] of 10 ppm, the column's emergency vent went to a tall stack.

As the concern for the environment increased during the 1960-1990 period, the plant installed collection headers for the normal vents in the process and sent them to a flare at the perimeter of the unit. This approach was a good deed to eliminate the emissions from normal operations of emptying and filling tanks, ambient temperature changes, etc. The feed tank emergency vent continued to be the lifting lid, since the API design of the tank would withstand only a few ounces of pressure. By 1990, the system had evolved to that shown in Figure 12.

However, the environmental good deed introduced a new hazard. The changes had interconnected the normal vents of distillation column and the feed tank. When there was a pressure excursion in the distillation column — say, from low coolant flow to the condenser — the small pipe in the long run to the flare did not have enough capacity to relieve all the pressure. The lifting lid on the feed tank opened at a few ounces of pressure, venting toxic gas in the tank farm at an elevation of 9.5 m (about 20 feet). Releases from the lifting lid affected personnel in this and neighboring units with an acute health hazard, requiring evacuations (Dowell, 1994).

To avoid such incidents, unit procedures specified a quick column shut-down for high column pressure or temperature. Additionally, one cause of high distillation column pressure had been identified as low coolant flow and a low coolant flow interlock (shown by "FL" in a diamond) had been added to trip the steam to the column. Probably an incident had been caused by low coolant flow. However, several other causes of high column pressure existed and there were no interlocks to protect against them.

An additional good deed to increase the column design pressure to 3.3 atm (50 psig) was planned in the early 1990s. The increase in column design pressure and relief valve set pressure would make the release from the feed tank worse. Process hazard analysis studies identified additional causes of high column pressure, and interlocks were installed for high coolant temperature, high column temperature, and high column pressure. These interlocks tripped the steam and are shown as diamonds within ovals in Figure 12.

In 1996, inherently safer design features were installed as shown in Figure 13 — the feed tank was replaced with a new tank designed for 3.3 atm (50 psig). The feed tank normal vent continued to go to the flare and the feed tank emergency vent was routed to the tall stack via a dedicated line.

[1] 8 hour Time Weighted Average

This system:

- eliminates environmental releases from normal operations,
- eliminates low level releases in the tank farm,
- greatly reduces releases from the tall stack by the combination of active interlocks and passive vessel design pressure.

This case study teaches:

- review the impact of any change on the whole system, including start-up, shut-down, upsets, and loss of utilities,
- a system can be no stronger than its weakest point,
- if active layers of protection are needed, provide detection and response for all the failure modes and ensure the personnel and interlocks can respond in time to prevent the undesired event.

OTHER EXAMPLES

New hazards can be introduced as an unanticipated side effect of a modification which was originally designed as a safety feature. Some other examples from various areas of technology include:

- In the 1970s the United States government required an interlock system on automobiles which would not allow the engine to be started unless the front seat belts were buckled. This system worked by using weight sensors in the front driver's and passenger's seats. If there was weight on the seat and the seat belt was not buckled, the automobile engine could not be started. Of course, this also meant that the engine could not be started if the driver put any weight — a bag of groceries, a briefcase — on the passenger seat. The unanticipated consequence of this system was that it actually made many drivers or passengers less likely to buckle their seat belts. People would leave the seat belts buckled all of the time so the interlock would not prevent starting the car. It was then easier to sit on top of the buckled seat belt rather than going to the extra trouble of unbuckling the belt, then re-buckling it.

- Another recent controversy in the automobile industry is associated with front passenger seat air bags. Air bags were added to deal with the behavior issue of many people who would not wear seat belts. Air bags are intended to protect the occupants of the automobile from injury in a high speed head-on collision and the requirements include the ability to protect a passenger who is not wearing a seat belt. This requirement means that the air bag must deploy extremely rapidly. Of course, as experience accumulated, it became painfully clear that passengers who did not wear their seat belts were actually endangered by the air bag, thus obviating one of the major drivers for the air bag initiative. As if that wasn't enough, it was later found that the rapidly inflating air bag is capable of seriously injuring or killing small children (and, the press reports, small women). There is a danger of serious injury, even in a low speed collision in which the child would most likely not be injured seriously if the air bag were not installed. The National Highway Traffic Safety Administration (NHTSA) now estimates that two children are being killed by air bags for every child who is saved, and the NHTSA is (as of November 1996) trying to decide what action should be taken (O'Donnell and Healey, 1996). Also note that seat belts, if worn, protect against injuries in side collisions or second collisions (for example, a collision with another vehicle followed by a collision with a tree), but air bags can only protect against the first severe head-on collision.

- When sizing rupture disks and relief valves, the design engineer is often focused on the hazard of overpressurization of the reactor or vessel. There is a temptation to over-design the emergency relief device (make it larger than really necessary). However, the designer must also consider the consequences of the discharge of material, particularly into the external environment. For example, a rupture disk which is too large will release material at a much higher rate and may create a toxic hazard, or, in some cases, a flammable vapor cloud hazard. A smaller disk may provide adequate protection for the vessel and significantly reduce or even eliminate toxicity or flammability hazards resulting from the relief device discharge.

- Chlorofluorocarbon (CFC) refrigerants are being phased out because of their adverse environmental effects. One unexpected consequence has been that some people have discovered that a mixture of propane and butane can be used to replace increasingly expensive CFCs in car air conditioners, creating a fire and explosion hazard in case of a leak from the air conditioning system (Riemerman, 1994).

SUMMARY

There is no substitute for a full understanding of how a system works.

Any change to a system, including adding a safety feature (a good deed), introduces new failure modes and mechanisms (punishment). First pass intuitive analysis may give an inaccurate perspective — remember the stone columns and the pinhole leak in the rupture disk. The changed system must be thoroughly reviewed to understand the new failure modes and to protect against them. It is more effective to gain this understanding before the change is made (management of change, process hazard analysis) than to figure it out during an incident investigation.

ACKNOWLEDGMENTS

To Stan Anderson, Kathy Pearson-Dafft, and Frank Worley for their insightful review.

DISCLAIMER

Although we believe the information contained in this paper is factual, no warranty or representation, expressed or implied, is made with respect to any or all of the content thereof, and no legal responsibility is assumed therefore. The examples shown are adapted from actual experience and are simply for illustration; as such they do not necessarily represent Rohm and Haas Company guidelines. The readers should use data, methodology, and guidelines that are appropriate for their situations.

REFERENCES

American Petroleum Institute (API) RP 520 (1990). *Sizing, Selection, and Installation of Pressure-Relieving Devices in Refineries.* Part I - Sizing and Selection. 5th Edition. (July). Washington, D. C.: American Petroleum Institute.

American Society of Mechanical Engineers (ASME) (1995). *Boiler & Pressure Vessel Code.* Section VIII, Division I, p. 93. New York: American Society of Mechanical Engineers.

Bernard, L., F. Brodie, D. Ludwig, A. Ness, and K. Weidner (1997). "Vent Access Restriction for Solids Handling Systems." *Proceedings of the 31st Annual Loss Prevention Symposium,* March 10-13, 1997, Houston, TX, Paper 40b. New York: American Institute of Chemical Engineers.

Dowell, A. M., III, (1994). "Low Cooling Tower Level Causes Evacuation." *Process Safety Progress,* 13, 1, (July), 114-122.

Dowell, A. M., III, (1996). "Vent Systems: Life Cycle and Inherently Safer Concepts." *International Conference and Workshop on Process Safety Management and Inherently Safer Processes,* sponsored by the Center for Chemical Process Safety, October 8-11, 1996, Orlando, FL, Workshop F: Case Studies on Inherent Safety: Cost Benefit Analysis; Life Cycle Cost.

Galileo (1638). *Dialogues Concerning Two New Sciences,* trans. H. Crew and A. de Salvio. New York: McGraw-Hill (Republished in 1963).

National Fire Protection Association (NFPA) NFPA 69 (1992). *Explosion Prevention Systems.* Quincy, MA: National Fire Protection Association.

O'Donnell, J., and J. R. Healey (1996). "Feds Poised for Emergency Air-Bag Action." *USA Today* (November 11) Sec. B, p. 1.

Petroski, H. (1992). "History and Failure." *American Scientist* 80, 6 (November-December), 523-26.

Petroski, H. (1994). *Design Paradigms: Case Histories of Error and Judgment in Engineering.* New York: Cambridge University Press.

Powers, G. J. (1989). "A Short Course on Risk and Reliability Assessment by Fault Tree Analysis," Carnegie Mellon University, Pittsburgh, PA. (April 25).

Riemerman, P. A (1994). "Keeping Cool Gets Costlier." *The Intelligencer,* Doylestown, PA (July 10) Sec. A, p. 6.

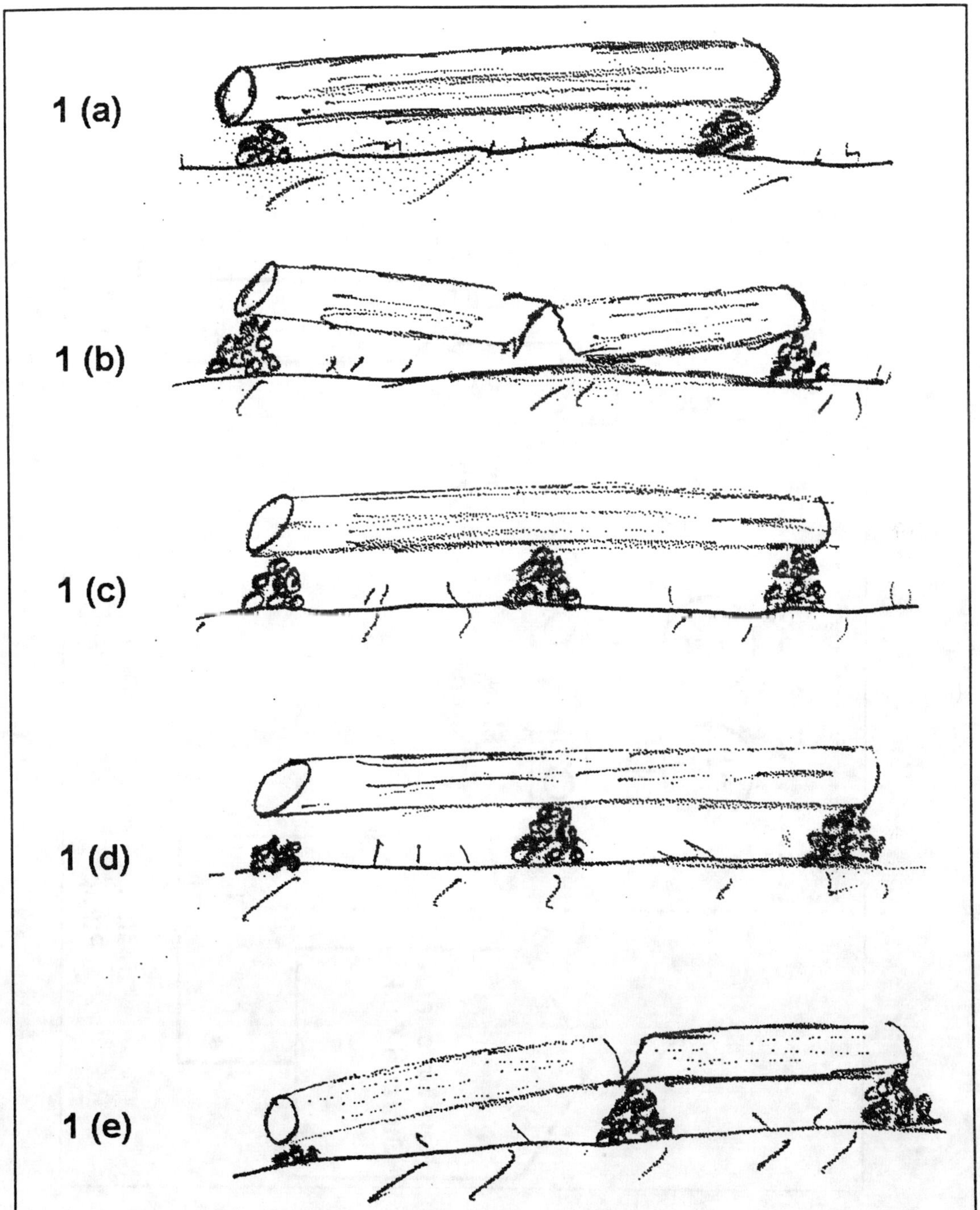

Figure 1: Failure modes of columns supported off the ground to prevent discoloration from contact with the soil

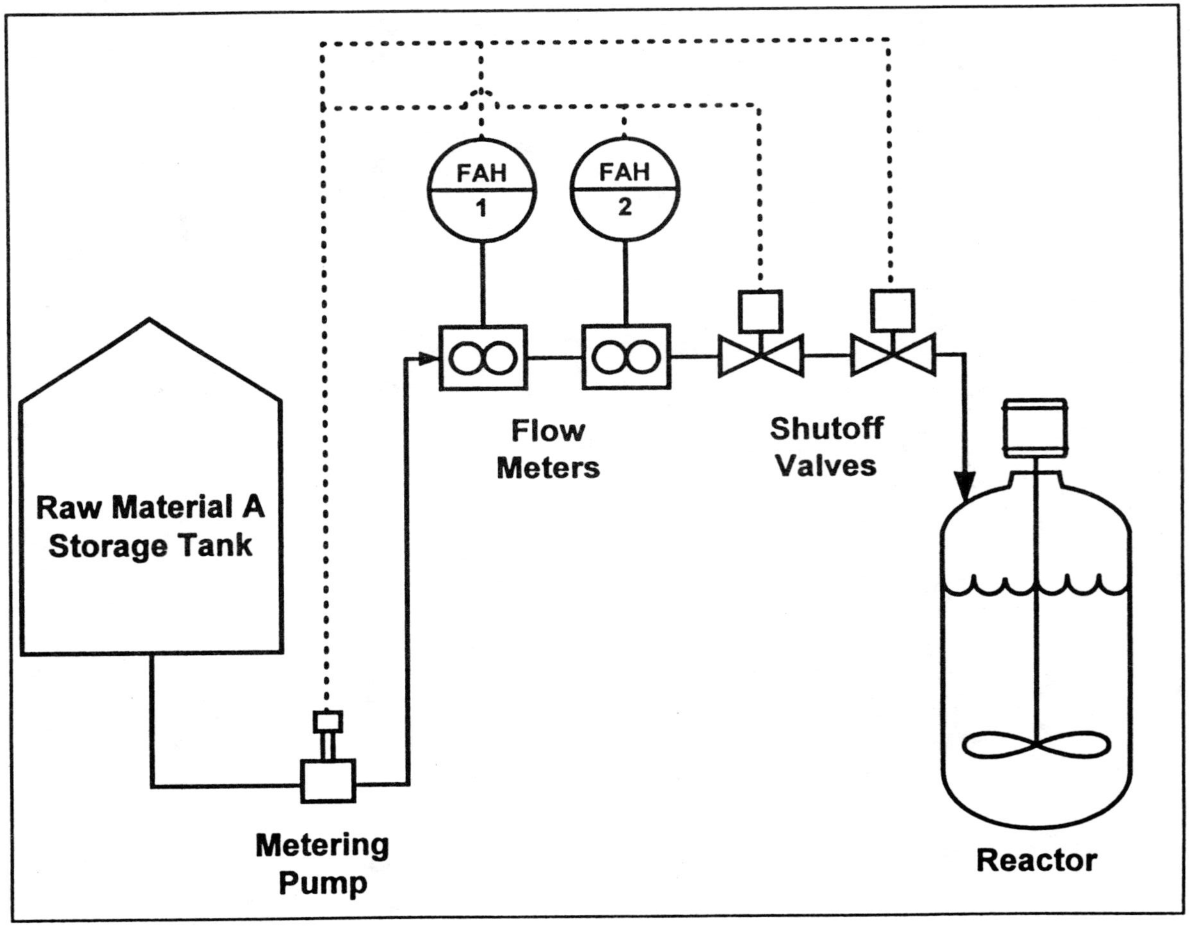

Figure 2: Proposed metering system for Raw Material A

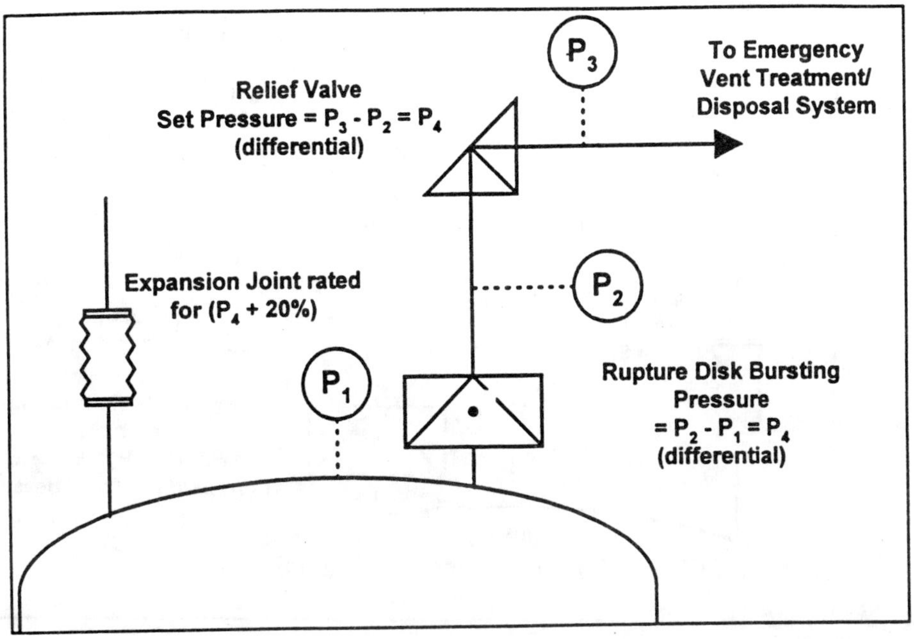

Figure 3: Emergency relief system with rupture disk and relief valve in series

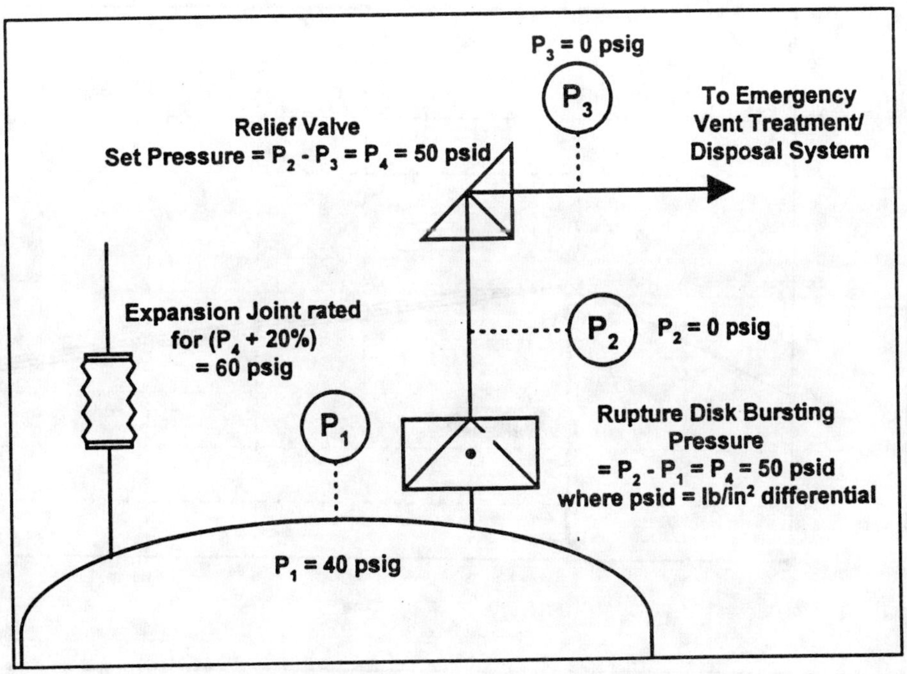

Figure 4: Pressure during normal operation with intact rupture disk

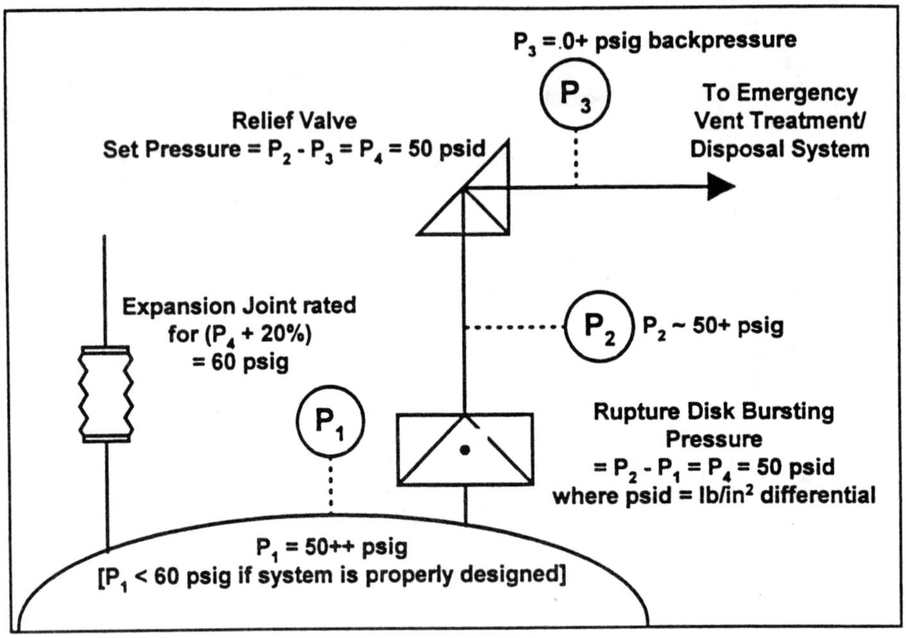

Figure 5: Pressure when emergency relief system activated due to high pressure in the vessel, rupture disk intact prior to overpressure challenge

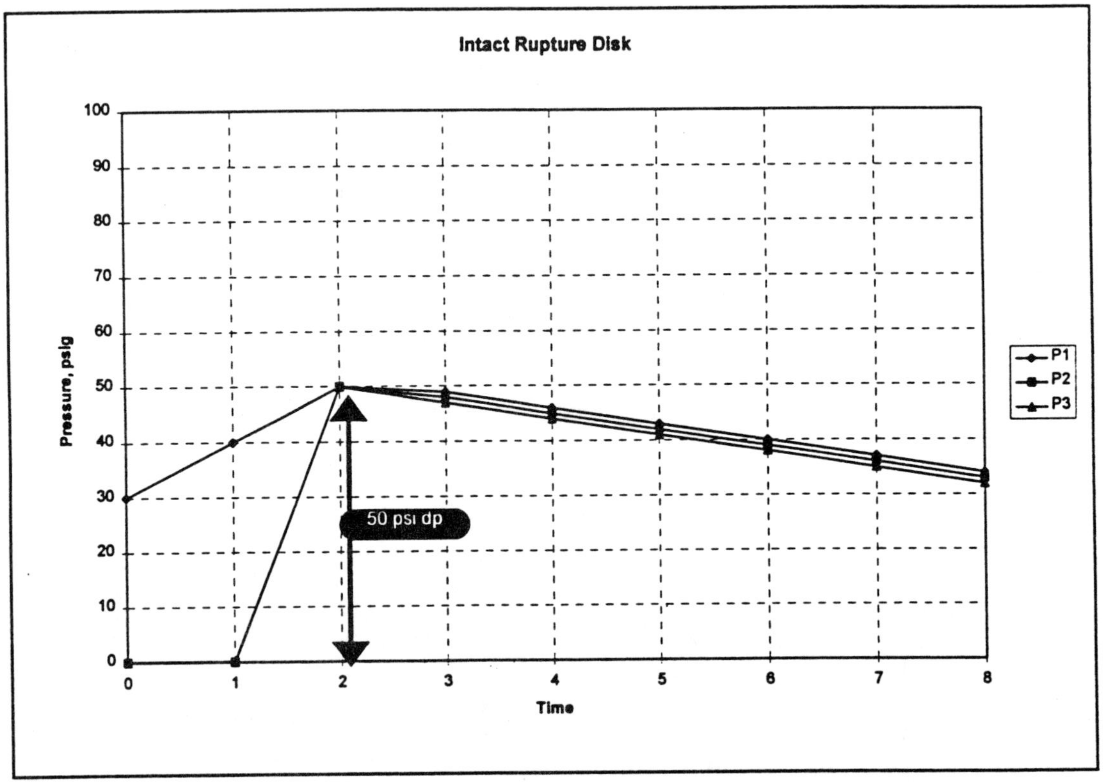

Figure 6: Pressure graph for intact rupture disk prior to overpressure

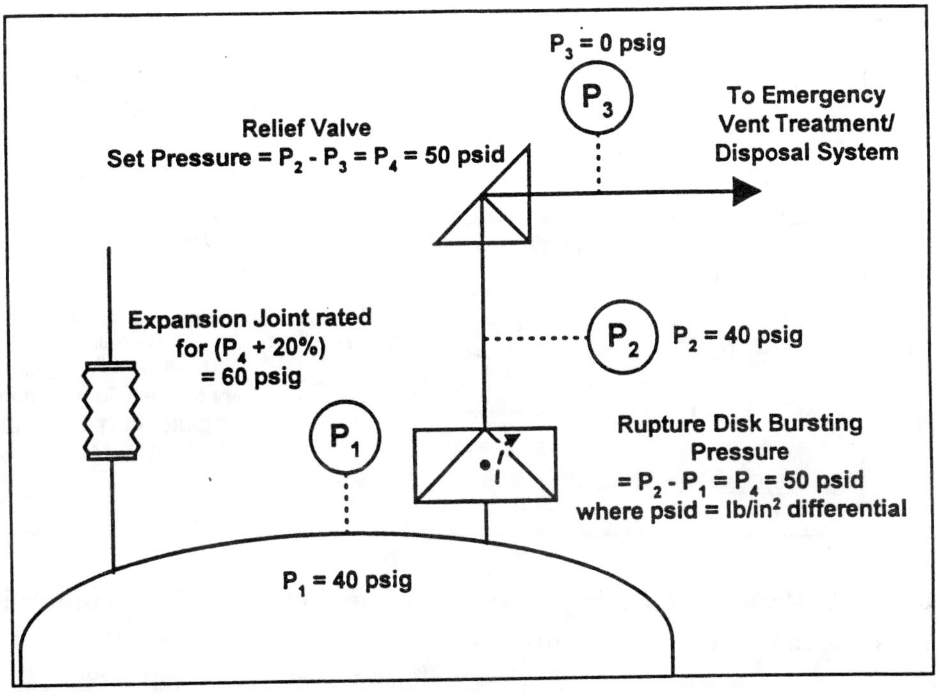

Figure 7: Pressure during normal operation with a pinhole leak in the rupture disk

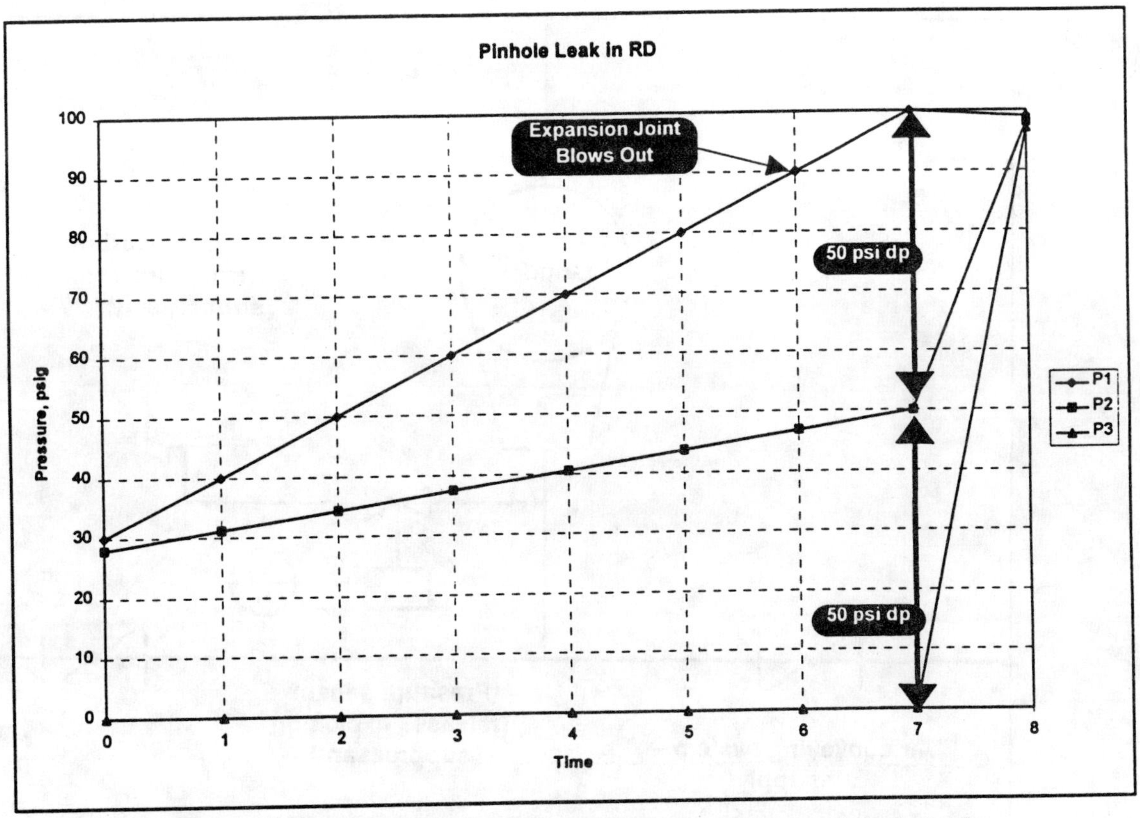

Figure 8: Pressure graph for rupture disk with a pinhole leak

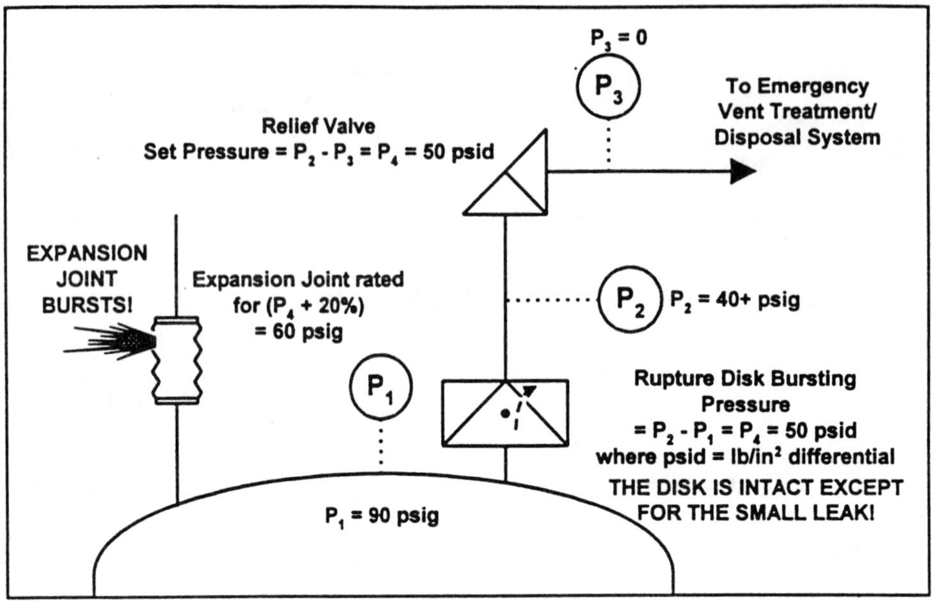

Figure 9: Pressure when high pressure occurs in vessel with a pinhole leak in the rupture disk — the rupture disk does not burst until the vessel pressure reaches ~90 psig

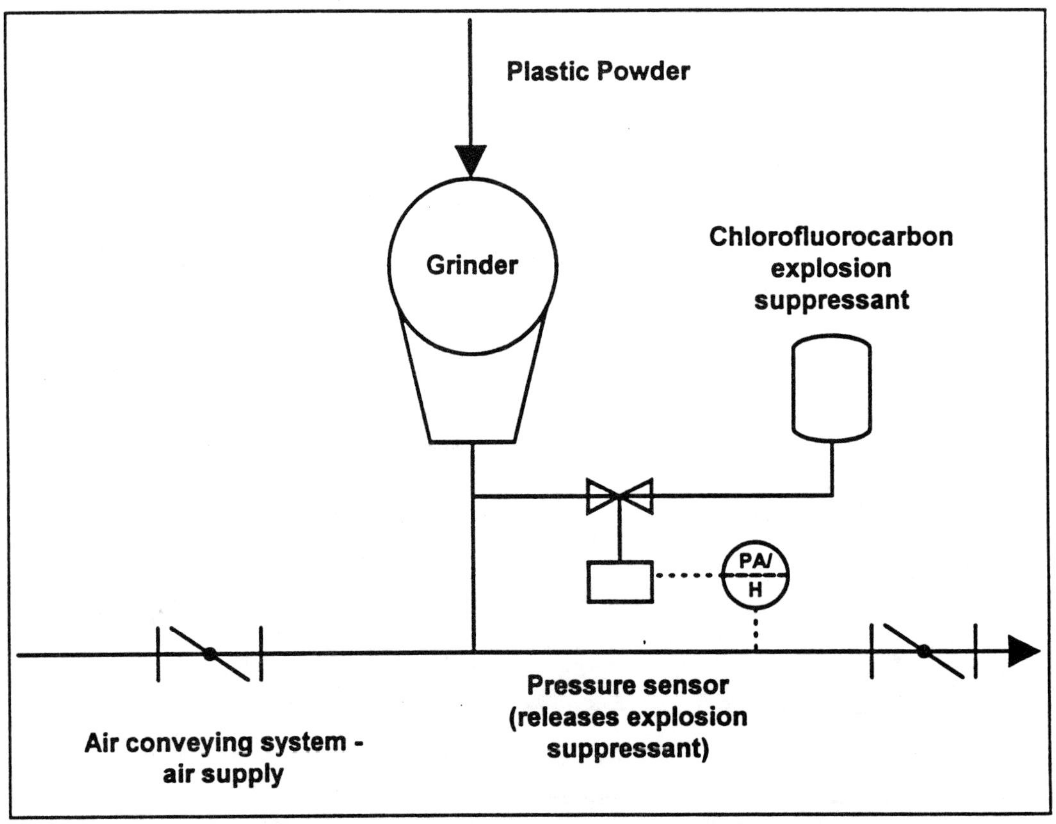

Figure 10: Plastics grinder with explosion suppression system

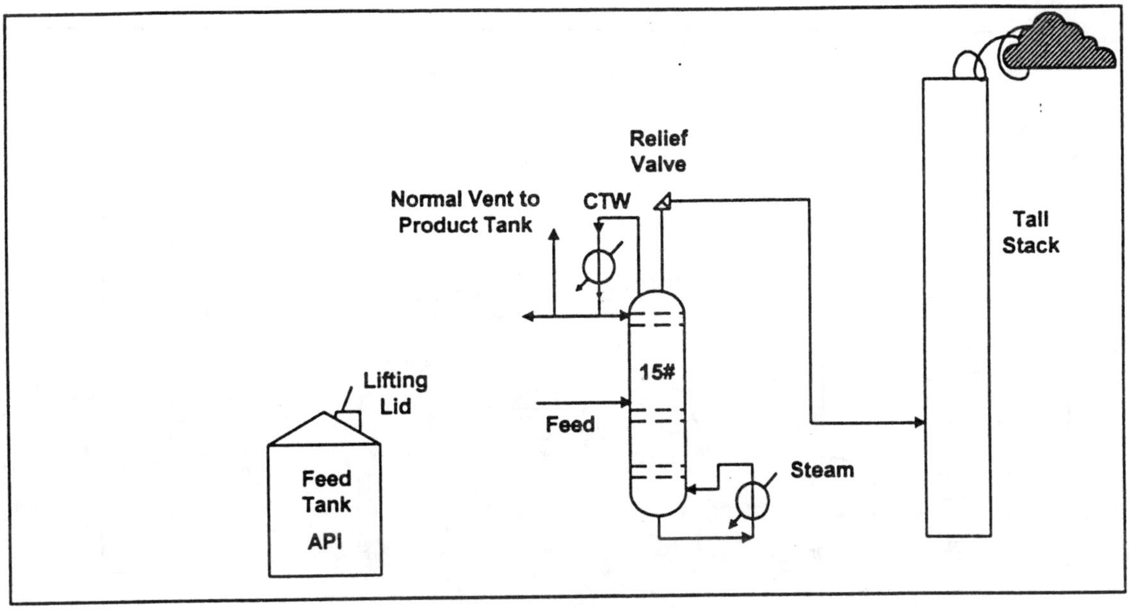

Figure 11: Original design with local normal vents

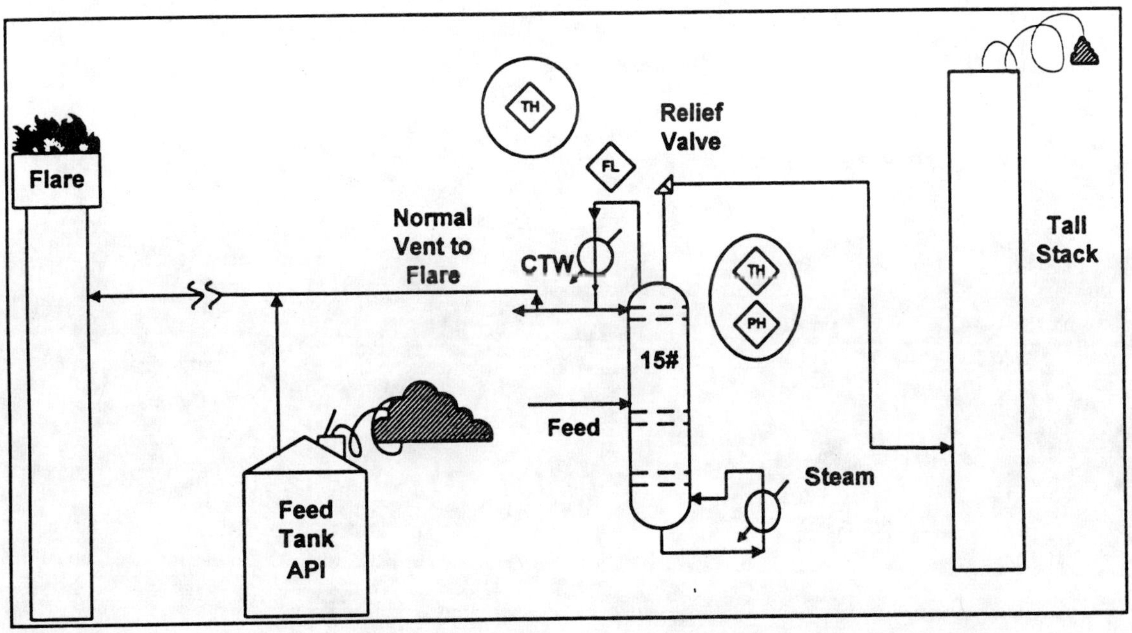

Figure 12: Design after retrofits to improve the environment. Now there is an acute health hazard near the feed tank during a distillation column upset.

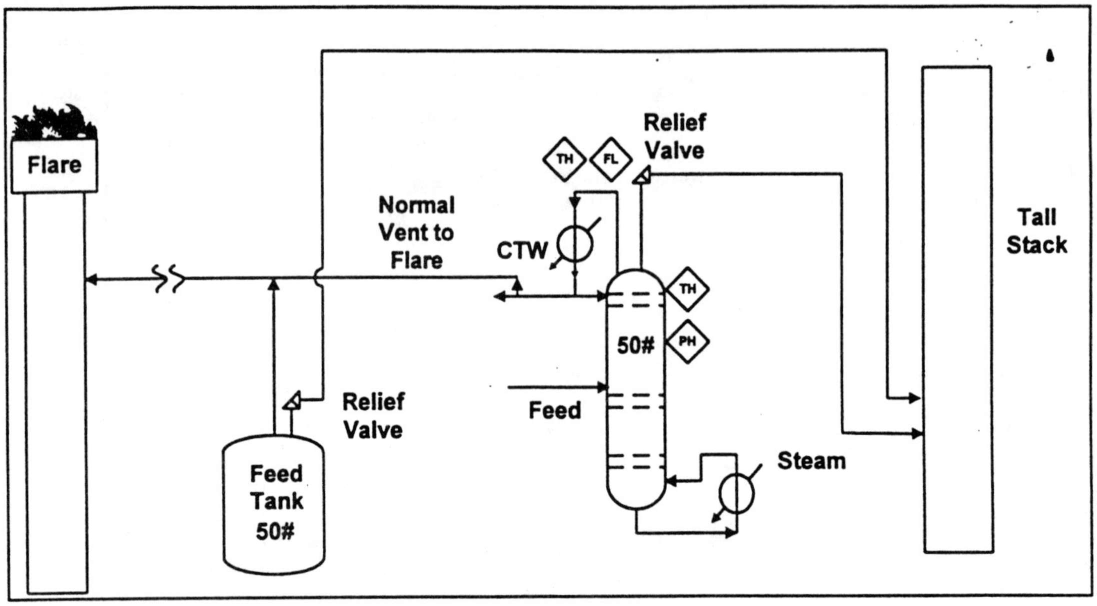

Figure 13: Design with inherently safer features to improve the environment and protect employee health.

CASE HISTORIES OF TWO INCIDENTS FOLLOWING PROCESS SAFETY REVIEWS

Walter B. Howard, Ph.D., P.E., C.S.P.

1415 Bopp Road

St. Louis, MO 63131-4135

Phone (314) 822-8247

Prepared for presentation at the AIChE 31st Annual
Loss Prevention Symposium
Case Histories and Miscellaneous Topics

March 13, 1997

UNPUBLISHED

A. INTRODUCTION

This paper discusses two incidents, both of which occurred after process safety reviews had taken place. Certain key recommendations were made for inherently safer operations, but the people in charge had decided not to implement them. In this connection it is important to note the three P's of safety: Safety of People, Property, and Profits. Safety hazards in a process operation can involve any or all of these. The first incident involved safety of property and profits, i.e., significant losses in both; the second primarily involved safety of profits. Both incidents occurred well before the days of OSHA and EPA.

B. FIRST INCIDENT

This incident occurred in another country. It involved an exothermic reaction that was carried out in two successive batch steps. The first took place in a conventional kettle-type reactor. After completion of this step the still-reacting mixture passed to a second special vessel for completion of the reaction. Figure 1 illustrates this second reactor. There were no additions to the reaction mixture at this stage. On occasional batches there was insufficient temperature control at this second stage. This resulted in a runaway reaction with energy release and boiling of vapor from the reaction mixture. (There was no thermal decomposition.) The evolved vapor came out through the vent at the top of the reactor and thence into the air. The vapor was of a moderately high-boiling combustible liquid which would then condense in the air to a dense, flammable fog. There had been no ignitions of these fogs. The reactors were located inside rooms.

This two-stage process was based on one developed by the parent company in the U.S. It was used by several of the company's plants in various countries. As a result of a process safety review some years before the incident to be described, the company had issued a non-binding yet strong recommendation to open up all of the facilities employing the process by removing the enclosing walls. The economic justification was the resulting inherently safer process, although those specific words were not in vogue at that time. The staff of the particular plant involved in the incident to be described had decided not to implement the recommendation, possibly because one wall was already partly open. The parent company had not yet developed an improved modification of the process itself.

In the particular incident to be described the above-mentioned runaway reaction did occur in the second stage reactor. The time was about midnight. A heavy aerosol fog accumulated to a depth of about two feet above the floor of the room. Suddenly the fog was ignited by an unknown sourçe. The opening in one wall was inadequate to sufficiently relieve the pressure developed by the combustion. As a result much of the building blew apart destructively. The incident resulted in severe losses due to property damage and loss of profits from lost sales.

The conclusions from the incident include the following:

B.1. The holding of process safety reviews, and the issuance of recommendations therefrom, do not ensure implementation of the recommendations, especially in countries other than the U.S. It has also been the author's repeated experience that this is sometimes true of small plants of large companies, in the same country, which may have effective loss prevention practices in large plants.

B.2. Loss prevention practices instituted and implemented by companies can be negated by decisions made consciously by people at some level of process administration, design, or operation. People at all levels of administration, design, and operation need training to include broad safety considerations as an integral part of their decision-making process.

B.3. A potentially hazardous occurrence which has happened many times without deleterious effect is never-the less dangerous and can result in a severe incident.

C. <u>SECOND INCIDENT</u>

A process to produce an organic compound involved

several steps of batch operations. Following one of the steps it became necessary to remove an impurity. A research chemist had, on laboratory scale, developed a solvent extraction process to do this. The solvent used was acetone. The appropriate engineering group had designed suitable equipment to implement the process. The appropriate process safety review was being held.

During the review it was pointed out that the proposed acetone solvent added a new flammability hazard to the process. Neither the research chemist nor the engineering design group had considered the flammability hazard. The chemist stated that acetone was the only suitable solvent, and the engineering group pointed out that the acetone would at least be inside closed equipment. Several of those in the review, including plant representatives, continued to press the issue. The chemist insisted that he was thoroughly familiar with all the chemistry involved, and that there was absolutely no other suitable solvent. It was finally agreed that the review report would point out the flammability hazard and would recommend that a search continue for a less hazardous solvent, even while acetone was utilized in the extraction process.

Soon after the new process began the plant people learned that the acetone was reacting to some extent with some of the process ingredients. The chemist had only carried out small scale tests without full identification of all the compounds in the batch after the extraction. He had overlooked the fact that acetone, although a good solvent for many purposes, is also a reactive chemical. Instead of helping, the extraction operation was greatly hindering the process and, as a result, was costing the company in lost profits.

Interestingly the research chemist quickly found another, less flammable solvent: methylene chloride. At that time there were no concerns about any potential carcinogenicity hazard of the solvent. The general health hazards of breathing the solvent vapors were of course recognized. Upon further review of the substitute solvent it was concluded that the new process was inherently safer to a satisfactory degree. It was therefore implemented.

The conclusions from the incident include the following:

C.1. Processes operating with flammable materials are hazardous even if those materials are contained in closed equipment.

C.2. Sometimes the development of inherently safer processes can come about in an unexpected manner. We must be grateful for safety improvements regardless of how they may come about!

C.3. Safety of profits is often thought of in terms of calamitous events such as explosions and fires. Unexpected and severe loss of profits can occur from process decisions made contrary to known technology.

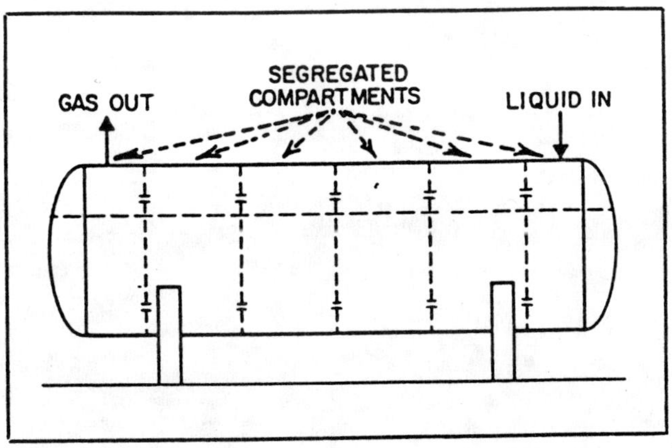

Figure 1. Special Reactor